WEST INDIAN HERPETOLOGY

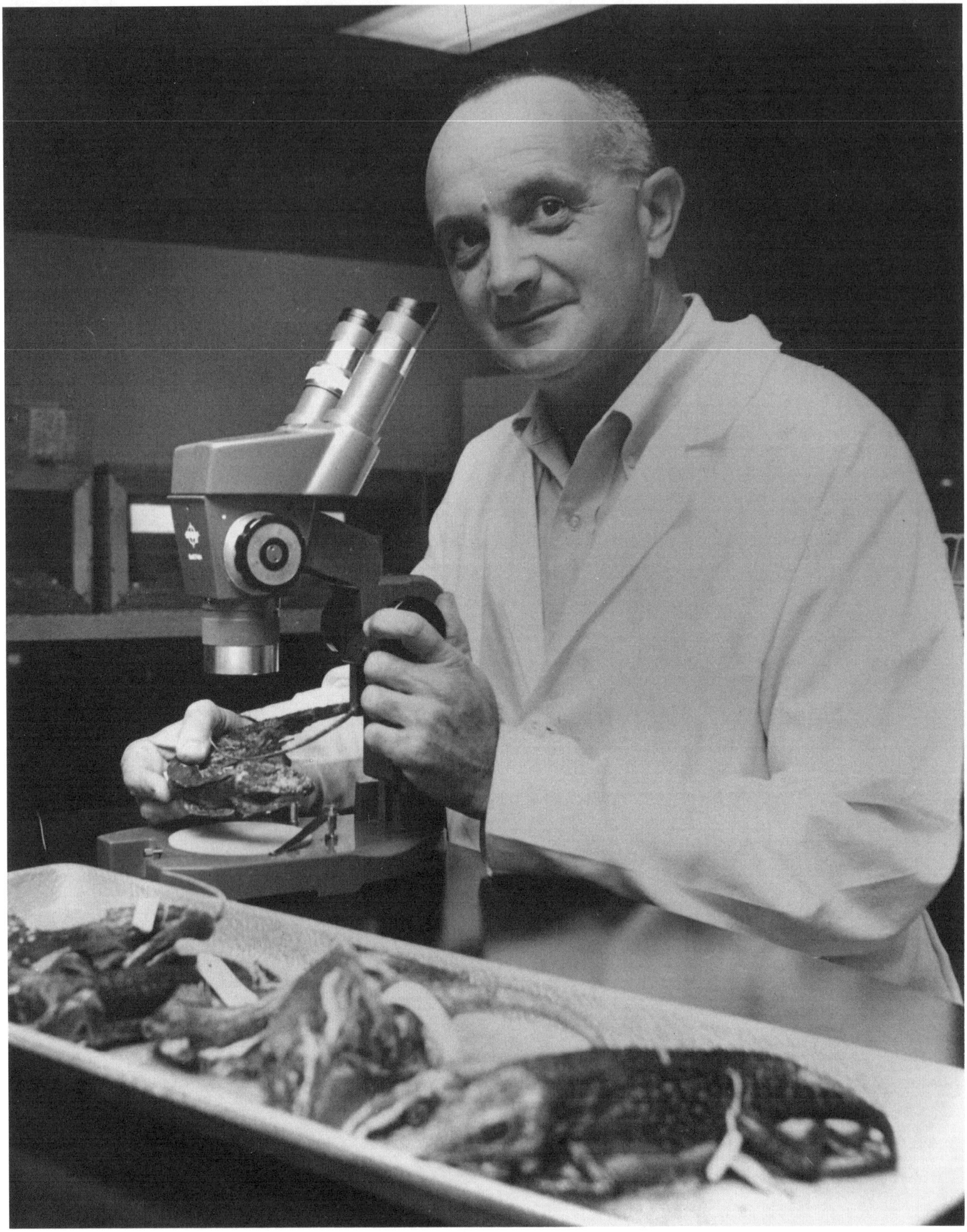

Albert Schwartz

Contributions to

West Indian Herpetology

A Tribute to Albert Schwartz

Edited by Robert Powell and Robert W. Henderson

Foreword by Thomas W. Schoener

Published by the Society for the Study of Amphibians and Reptiles • 1996

SERIES EDITOR'S NOTE

The Society is grateful to Robert Powell and Robert W. Henderson for organizing the 1994 symposium which provided the basis for this volume and for their careful editing of the manuscripts. We are also obliged to the persons and organizations whose financial contributions have made possible the inclusion of the color plates; they are listed on the patrons' page (page 8). The series editor thanks Richard Glor for his editorial assistance.

The frontispiece, a photograph of Albert Schwartz taken in the 1970s, is provided through the courtesy of Michael H. Strahm. The illustration on the title page is of *Anolis equestris* of Cuba, drawn by Jean Gabriel Prêtre (from Ramón de la Sagra's "Historia Física, Política y Natural de la Isla de Cuba," Paris, 1838–1861). The headpiece to the chapters was drafted by Christopher L. Lane; it depicts *Cyclura cornuta* from the Dominican Republic. The tailpiece to each chapter figures *Eleutherodactylus schwartzi* of the Virgin Islands; it was portrayed by Karen Bertrand.

CONTRIBUTIONS TO HERPETOLOGY, VOLUME 12

KRAIG ADLER, *Editor*
TIMOTHY D. PERRY, *Associate Editor*

Volumes in the *Contributions to Herpetology* series can be purchased from the Publications Secretary, Robert D. Aldridge, Department of Biology, St. Louis University, 3507 Laclede, St. Louis, Missouri 63103, USA (*telephone*: area code 314, 977–3910 or 977–3916; *fax*: area code 314, 977–3658; *e-mail*: ssar@sluvca.slu.edu). A list of all Society publications, including those of The Ohio Herpetological Society and the *Catalogue of American Amphibians and Reptiles*, is printed at the end of this book; additional copies of this list are available from Dr. Aldridge. *Contributions* are published irregularly and ordered by separate subscription, although Society members receive a substantial pre-publication discount. Authors who wish to have manuscripts considered for publication in the *Contributions* series should contact the Editor: Kraig Adler, Cornell University, Section of Neurobiology and Behavior, Seeley G. Mudd Hall, Ithaca, New York 14853–2702, USA.

Members of the Society receive a quarterly technical journal (*Journal of Herpetology*) and a quarterly news-journal (*Herpetological Review*). Currently, dues are US$30.00 for students, $40.00 for all others, world-wide; institutional subscriptions are $70.00. Additional $30.00 for airmail delivery outside the USA. Society members receive substantial discounts on *Herpetological Circulars, Facsimile Reprints*, and on books in the *Contributions* series. The *Catalogue* is available by separate subscription. Apply to the Society's Treasurer, Karen L. Toepfer, P.O. Box 626, Hays, Kansas 67601–0626, USA. Overseas customers should make payments in USA funds or by International Money Order. All persons may charge to MasterCard or VISA (include account number and expiration date).

Library of Congress Catalog Number: 95–71882. ISBN: 0–916984–37–0.
Production specifications are given on the last page of this book.

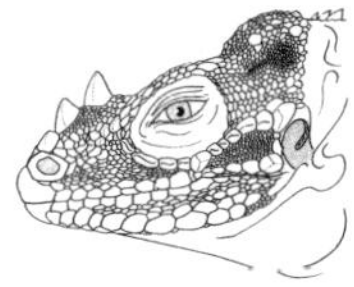

Table of Contents

Patrons of This Volume 8

Foreword .... *Thomas W. Schoener* 9

Introduction .... *Robert Powell and Robert W. Henderson* 11

Albert Schwartz

Remembrances of Albert Schwartz:

- Albert Schwartz: Dawning of a Herpetologist .... *William E. Duellman* 15
- Recollections of Albert Schwartz .... *Richard Thomas* 16
- A Singular Devotion: Albert Schwartz and the West Indies .... *Robert W. Henderson* 18

Herpetological Publications of Albert Schwartz .... *Robert W. Bourgeois* 21

West Indian Herpetology

A Brief History of West Indian Herpetology .... *Robert Powell and Robert W. Henderson* 29

An Annotated Checklist of West Indian Amphibians and Reptiles .... *Robert Powell, Robert W. Henderson, Kraig Adler, and Harold A. Dundee* 51

Color Plates of West Indian Amphibians and Reptiles .... following page 94

Evolution, Systematics, and Biogeography

The Origin of West Indian Amphibians and Reptiles .... *S. Blair Hedges* 95

Systematics and Biogeography of Eastern Caribbean *Eleutherodactylus* (Anura: Leptodactylidae): Consensus from a Multidisciplinary Approach ... *Hinrich Kaiser* 129

The Relationships of the Hispaniolan Frogs of the Subgenus *Pelorius* (*Eleutherodactylus*: Leptodactylidae) .... *John D. Lynch* 141

A Karyological Perspective on the Monophyly of the Hylid Genus *Osteopilus* .... *Karen Anderson* 157

Current Status of Biogeography of the West Indian Turtles in the Genus *Trachemys* (Emydidae) .... *Michael E. Seidel* 169

Relationships among West Indian Geckos of the Genus *Sphaerodactylus*: A Preliminary Analysis of Mitochondrial 16s Ribosomal RNA Sequences ... *Carla Ann Hass* 175

Evolution, Systematics, and Biogeography (continued)

Careening Island and the Goat Islands: Evidence for the Arid-Insular Invasion Wave Theory of Dichopatric Speciation in Jamaica .... *James Lazell* 195

Crossing Bond's Line: The Herpetofaunal Exchange between the Eastern Caribbean and Mainland South America .... *John C. Murphy* 207

Ecology and Behavior

The Role of Amphibians and Reptiles in a West Indian Rain Forest Food Web... *Douglas P. Reagan* 217

Patterns of Parental Care in Frogs of the Genus *Eleutherodactylus* .... *Daniel S. Townsend* 229

Mate Selection in the Puerto Rican Frog, *Eleutherodactylus coqui* .... *Pamela T. Lopez* 241

Biology of the Puerto Rican Cave-dwelling Frog, *Eleutherodactylus cooki*, and Some Recommendations for its Conservation .... *Rafael L. Joglar, Patricia A. Burrowes, and Neftalí Rios* 251

Larval Habitat, Ecology, and Parental Investment of *Osteopilus brunneus* (Hylidae) .... *Rebecca L. Thompson* 259

Occurrence of the Nematode *Skrjabinoptera scelopori* in the Cuban Treefrog, *Osteopilus septentrionalis*: Mainland and Island Comparisons .... *Walter E. Meshaka, Jr.* 271

The Evolution of Sexual Size Dimorphism in the Teiid Lizard *Ameiva plei*: A Test of Alternative Hypotheses .... *Ellen J. Censky* 277

Morphology, Ecology, and Behavior of the Twig Anole, *Anolis angusticeps* .... *Duncan J. Irschick and Jonathan B. Losos* 291

A Test of Assortative Mating between Sibling Lizard Species, *Anolis websteri* and *A. caudalis*, in Haiti .... *Thomas A. Jenssen* 303

Evidence of Spatial Niche Partitioning by a Hispaniolan Lizard Community in a Xeric Habitat .... *Robert Powell, John S. Parmerlee, Jr., and Donald D. Smith* 317

Diets of West Indian Racers (Colubridae: *Alsophis*): Composition and Biogeographic Implications ..... *Robert W. Henderson and Richard A. Sajdak* 327

Ecological Patterns in Greater Antillean Macrostomatan Snake Assemblages, with Comments on Body-size Evolution in *Epicrates* (Boidae).... *Javier A. Rodríguez-Robles and Harry W. Greene* 339

Conservation

Species Richness and Biogeography of the Herpetofauna in the Exuma Cays Land and Sea Park, Bahamas .... *C. Kenneth Dodd, Jr. and Richard Franz* 359

Declining Amphibian Populations in Puerto Rico .... *Rafael L. Joglar and Patricia A. Burrowes* 371

The Current Status of Sea Turtles in Haiti .... *José A. Ottenwalder* 381

CONSERVATION (continued)

CONSERVATION STRATEGY FOR THE JAMAICAN IGUANA, *CYCLURA COLLEI* .... *Peter Vogel, Richard Nelson, and Rhema Kerr* 395

CONSERVATION OF *EPICRATES MONENSIS* ON THE SATELLITE ISLANDS OF PUERTO RICO .... *Peter J. Tolson* 407

CORALLUS HORTULANUS (SERPENTES: BOIDAE) IN THE POST-COLUMBIAN WEST INDIES: NEW HABITATS, NEW PREY SPECIES, AND NEW PREDATORS .... *Robert W. Henderson, R. Allan Winstel, and Joel Friesch* 417

CONSERVATION OF THE AMERICAN CROCODILE (*CROCODYLUS ACUTUS*) IN THE DOMINICAN REPUBLIC .... *Andreas Schubert and Gloria Santana* 425

ABSTRACTS

ABSTRACTS OF ADDITIONAL PAPERS PRESENTED AT THE SYMPOSIUM IN HONOR OF ALBERT SCHWARTZ:

Pre-Columbian Human Influences on the Biogeography of Amphibians and Reptiles in the Yucatan Peninsula .... *Julian C. Lee* 435

Population Differentiation of *Iguana delicatissima* and *I. iguana* in the Lesser Antilles .... *Mark L. Day and Roger S. Thorpe* 436

Long-term Study of Activity Patterns of Puerto Rican Rain Forest Frogs .... *Margaret M. Stewart* 437

Developmental Migrations and Habitat Use of Green Turtles in the Bahamas .... *Karen A. Bjorndal and Alan B. Bolten* 440

Harmful Prey Aversion in Hatchling *Anolis lineatopus*: Learning the Good Things First.......... *Peter Vogel* 441

Anoline Diversity in Three Differentially Altered Habitats in the Sierra de Baoruco, Dominican Republic, Hispaniola .... *Lori A. Lenart and Scott P. Sowell* 442

Learning and Memory in *Anolis* Lizards .... *Sharoni Shafir and Jonathan Roughgarden* 443

The Relationship between Habitat Use and Sexual Dimorphism among Caribbean Anoles .... *Marguerite A. Butler and Jonathan B. Losos* 444

Social Control of Male Reproductive Success in *Ameiva exsul* .... *Allen R. Lewis, Grissell Tirado, Antonio Aponte, Javier Blanco, and José Sepulveda* 445

INDEX

INDEX TO SCIENTIFIC NAMES 451

PATRONS

The editors of this volume and the Society for the Study of Amphibians and Reptiles gratefully acknowledge the support of these individuals and organizations. Their help made possible the 1994 symposium and the publication of this book in honor of Albert Schwartz.

BENEFACTORS

THE CENTRAL FLORIDA HERPETOLOGICAL SOCIETY

DAVID C. LEBER

MICHAEL H. STRAHM

TOLEDO ZOOLOGICAL SOCIETY

FRIENDS

KRAIG ADLER
DUANE C. AND NANCY C. BENTON
ELLEN J. CENSKY
JOSEPH T. AND SUZANNE L. COLLINS
JULIAN C. LEE
DONALD, JEANNE, AND TONNIE WARFIELD
ROBERT F. WILKINSON, JR.

ORGANIZATIONS

AVILA COLLEGE
THE CENTER FOR NORTH AMERICAN AMPHIBIANS AND REPTILES
KANSAS CITY HERPETOLOGICAL SOCIETY
MILWAUKEE PUBLIC MUSEUM
THE NEW ENGLAND HERPETOLOGICAL SOCIETY

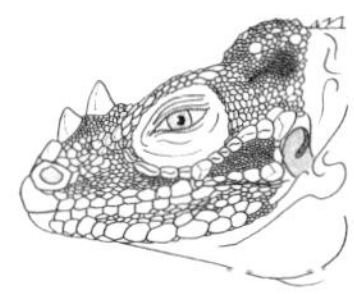

FOREWORD

THOMAS W. SCHOENER[1]

ALBERT SCHWARTZ, ONE OF THE LAST OF A DYING breed, has died. He was a pure taxonomist, not a systematist in the modern or even older sense of the word, but a describer and identifier of taxa. His research output was enormous and phylogenetically broad; not only did he write 217 herpetological papers, but he produced significant works dealing with other animals as well, including a 580-page compendium on Hispaniolan butterflies. The introduction to this volume reports that 622 species of West Indian amphibians and reptiles have been described; of Schwartz's papers on that subject, 185 were primarily taxonomic, giving an average of 3.4 species per taxonomic paper. A more meaningful figure might be the number of species he described, which is 86, nearly 14 percent of the total known fauna! From the average of 2.2 papers per described species, one might imagine that he described many subspecies as well, and indeed he did; the number is 279, so we get an average of 1.5 subspecies per taxonomic paper. Given variation about means, and that Schwartz did not deal with all 622 species, some of his papers would be expected to be very long — and some are: a 47-page paper on the *Leiocephalus carinatus* complex and an 86-page paper on the *Anolis equestris* complex, for example. Toward the end of his career, substantially or entirely nontaxonomic works became somewhat more abundant, but most of these were coauthored, including the ultimate, monumental "Amphibians and Reptiles of the West Indies" with Robert Henderson.

In contrast to Schwartz's work, of the 27 chapters published in this volume, only four deal with systematics, and none is a taxonomic paper per se. A simple reason for the latter is that taxonomic papers are published in rather specialized places, of which this volume is not conceivably an example. But what does it mean that the great preponderance of papers honoring an eminent taxonomist is collectively so "untaxonomic" in nature? Another simple reason is that this volume and the symposium from which it grew had dual purposes, one to honor Albert Schwartz, the other to give some idea of the present state of West Indian herpetology. But must this duality then be considered blatantly schizoid, or is there a substantial connection?

To try for an answer, let us list some conceptual areas to which West Indian herpetology has made major contributions: phylogeny reconstruction, functional morphology, predator-prey relationships, adaptive significance of territoriality, reproductive strategies, patterns in sexual size dimorphism, mating behavior, evolution of body size, origin of geographic distributions, resource partitioning, comparative thermoregulation, and evolutionary stasis. Find one of these topics that could have gotten off the ground without knowing the identity of the species. Taxonomy, the supplier of grist to Schwartz's mill, is fundamental to nearly everything in ecology, evolution, and behavior.

To become more personal, I first met Al Schwartz in 1966 when I was a beginning graduate student. I needed to measure specimens of *Anolis* from the extensive private collection located in his Coral Gables (Florida) home. Al was incredibly helpful and courteous, devoting nearly continuous attention to my activities. This was despite (or perhaps because of) my initial tendency to return all specimens to their jar en masse, bunched like asparagus. He had me carefully inserting them seriatim by the end. He was, however, glad to get a

[1] Section of Evolution and Ecology, Division of Biological Sciences, Storer Hall, University of California, Davis, California 95616, USA.

large series of *A. angusticeps* that I had just collected from South Bimini, coincidentally the subject of a paper he was then finishing; even more coincidentally, a paper by Irschick and Losos on Bimini *angusticeps* appears in this volume. And I was glad to get measurements for numerous species of *Anolis*, some quite rare, for my series of papers on size patterns, papers which (among many others) could not have been written without the efforts of people who describe and identify taxa.

If Schwartz's breed is a dying one, will this matter to science? Maybe taxonomists in fact have completed the bulk of their task, nearly all the species having been described (never mind identification for now). In actuality, this is clearly untrue for the West Indian herpetofauna. Hedges (this volume: figure 1) plots the number of known species by year and concludes that "the rate of species discovery has not yet declined" for West Indian amphibians and reptiles. Among certain other groups, the situation may be considerably worse. To give a firsthand example, I spent the latter part of 1992 collecting and counting spiders on 64 islands of the Great Barrier Reef. My purpose was ecological, but my greatest need was taxonomic, as apparently few of these islands had ever been visited for spiders before. Quite unlike Schwartz, I published 117 papers before my first taxonomic one, albeit coauthored. It describes a rather widespread Australian species whose genus, formerly monotypic, was known only from over 2000 miles to the north. I was more excited about this paper than many others I had recently produced, I think because it reported a new entity — not just the arrangement (or rearrangement) of already existing data, but a genuine, first-order discovery. Discovery is, of course, fundamental to science, and unlike many ancillary manipulations, it is lasting.

Appropriately, this volume ends with a section devoted to conservation. Species in the West Indies, like those nearly everywhere else, are at great risk — consider the grim news about Haiti in the introduction, for example. Somehow even worse than the loss of known species is the likelihood for many groups that species will be driven to extinction by human activities without ever becoming known. Thanks to Al Schwartz, those like him, and their discoveries, most West Indian amphibians and reptiles have been spared this same fate of anonymous extinction.

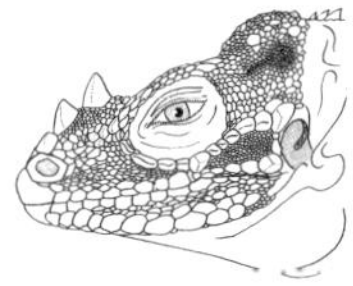

INTRODUCTION

WE CANNOT IMAGINE A MORE APPEALING REGION in which to study amphibians and reptiles than the West Indies. The biologist who chooses to work in the Antilles gets to spend sun-filled days on islands that, more often than not, are bordered by palm-lined beaches, where near-desert conditions may exist at sea level, but interior mountain ranges ascend to rugged rain-drenched forests, bougainvillaea vines splash garish colors along fences and ramshackle houses, palm fronds silhouetted against a tropical moon dance in the tradewinds, lizards occur at densities greater than anywhere else in the world (sometimes more than 2 per m^2), and seeing 10 or more snakes in a day is not unusual.

What was once described as a depauperate fauna has proved to be wonderfully diverse; the current tally is 622 species, of which 572 are endemic to the region, and spectacular discoveries continue. New species of *Eleutherodactylus* and *Sphaerodactylus* still are being described annually; the first dendrobatid frog for the region was described from Martinique in 1994; a candidatc for thc smallcst frog (9–10 mm SVL) in the world was recently collected in Cuba; and a new species of *Diploglossus* was discovered in Cuba 157 years after the last was described. Also, biochemical techniques are identifying cryptic species, producing new information regarding relationships within and between genera, and putting origins into the context of time.

Because of the diversity and population densities of the herpetofauna on West Indian islands, this region has proven to be an ideal natural laboratory for studies of behavior and ecology. Similarly, the diversity and distributions of amphibians and reptiles have provided data for controversial topics in West Indian biogeography (e.g., dispersal versus vicariance). The Antilles have been a haven for lizard ecologists for over 30 years, frog ecology has begun to receive its due attention in the past 15 years, and snake ecology in the West Indies is gathering momentum.

However, a negative aspect exists as well. West Indian environments have been subjected to the influence of non-native Americans, almost always negatively, longer than any other land masses in the New World. Habitat destruction is rampant and forests are rapidly disappearing in order to produce charcoal and to plant crops; Haiti, admittedly the most extreme example, retains less than 1% of its original forest cover. The introduction of the mongoose is correlated with the decline or disappearance of several snake taxa and populations; the entire known range of several species is now less than 0.5 km^2. The introduction of cats, dogs, and rats has had a catastrophic effect on some *Cyclura* populations. Human numbers are booming and development (largely for the tourist trade) continues unabated. Although some versatile frogs and reptiles respond favorably to habitat modification, far more species fare less well. Therefore, investigations focusing on the West Indian herpetofauna, be they to document species diversity or population densities, can have important implications for the conservation of an amazing fauna.

And what of the future? We must continue to document the biodiversity of these alluring islands, be it in the laboratory utilizing biochemical tools, or by turning rocks and logs by day and hunting at night with a headlamp in *terra incognita*. Exploration for new species should be encouraged; the "Golden Age" of discovery in the West Indies is far from over. We still know next to nothing about the natural history of 90% of the herpetofauna; its diversity, the localized distributions of some species and the multi-island distributions of others, and the spectacularly high population densities of many species will continue to provoke endless questions and provide many opportunities to address them.

And finally, we must address the critical questions of habitat degradation and destruction, introduction of alien predators and/or competitors, while simultaneously considering the needs of the human population. Defining the policy that allows for human prosperity without sacrificing a remarkable biota will be a challenge we can only hope will be met.

THE SYMPOSIUM

Considering the amount of research activity that has become focused on the West Indian herpetofauna, we have long felt the need for a gathering of interested persons in order to keep each abreast of what the others, especially those working with different organisms, were doing without the inevitable delays before their work appeared in print.

On 31 July and 1 August 1994, the Society for the Study of Amphibians and Reptiles sponsored, as part of its 37th Annual Meeting at the University of Georgia, an event entitled: "West Indian Herpetology: A Symposium in Honor of Albert Schwartz" (Figure 1). Sadly, the catalyst for this meeting was the death of Albert Schwartz on 18 October 1992. More than that of any single person (including Gundlach, Poey, Barbour, or Bond), Schwartz's name has become associated with West Indian zoology. His contributions to Antillean herpetology in particular are undeniable, and anyone studying any aspect of the biology of amphibians and reptiles in the West Indies will initially or eventually consult one or more of Schwartz's 195 papers.

The symposium had a dual purpose: (1) to honor the memory of Albert Schwartz and his many contributions to West Indian herpetology; and (2) to bring together biologists who are currently studying amphibians and reptiles in the West Indies for an exchange of ideas and information. Not everyone invited to participate in the symposium was able to do so, but we were gratified by the overwhelmingly positive response to our invitations. We were also pleased by the number of people who referred others to us and by those who approached us and asked to participate. Not everyone who participated submitted a paper for possible publication in this volume. Some had already committed their papers elsewhere, and others felt their research was still at too preliminary a stage for publication.

FIGURE 1. The logo used to promote the 1994 symposium honoring Albert Schwartz. The Volkswagen was Al's field vehicle of choice. He had unshakeable faith in this modest vehicle's traction and ground clearance, especially on the rugged terrain and unpredictable roads of Hispaniola.

Dedication

The 1994 symposium was dedicated to Albert Schwartz, the late dean of West Indian herpetologists. For biographical information see the remembrances in this volume by William E. Duellman, Richard Thomas, and Robert W. Henderson (and references cited therein), and those by Henderson, Robert Powell, and Michael Strahm in the symposium program. This book also is dedicated to Albert Schwartz, in recognition of his myriad contributions to West Indian herpetology and for the many students and colleagues he helped in uncounted ways. The following is an exerpt from the dedication in the symposium program:

Simply stated, Albert Schwartz (1923–1992) was the single most significant contributor to our knowledge of West Indian zoology. Al devoted the bulk of his life to understanding the diverse fauna of the hundreds of keys and islands in the West Indies — and succeeded beyond what could have been expected of any one person. He was an astute and accurate observer, and although first and foremost a taxonomist, his field notes and approximately 230 [herpetological] publications were filled with anecdotes regarding ecology and behavior. Other papers dealt with butterflies, birds, and mammals. But Al's contributions were not limited to published books and papers. His enthusiasm for the West Indies, carried to the end of his life, touched many persons, a number of whom have continued to explore islands, large and small, in an effort to answer the never-ending questions of when, how, and why animals, especially amphibians and reptiles, are as they are.

Organization

We have attempted to retain the organization of the symposium in structuring this volume. After an introductory section containing a historical perspective of West Indian herpetology and remembrances of Albert Schwartz, contributions are gathered into sections entitled Evolution, Systematics, and Biogeography; Ecology and Behavior; and Conservation. Some chapters logically could have been placed in more than one section; we used our best judgment and ask the authors' (and readers') forebearance with the inevitable lack of agreement. Each chapter begins with an English abstract, and for international readers (and the West Indies are truly an international arena) we include at the end of each chapter a Spanish resumen and a French résumé. After the three major sections we provide abstracts of symposium presentations for which corresponding chapters are not included in this book. An index to scientific names completes the volume.

Acknowledgments

We take this opportunity to thank the many persons involved in organizing the 1994 symposium. The local committee co-chairs were Joshua Laerm and J. Whitfield Gibbons. Martina Conrad had the unenviable responsiblity of putting the program together; she did so admirably. Amy Edwards and Elizabeth McGhee were helpful in many ways, as were other members of the organizing committee. Lynne D. Houck, then president of SSAR, went out of her way to accommodate a symposium whose many participants spilled over the limits imposed by what was scheduled to be a one-day event.

Organizations and individuals whose contributions helped to offset the costs of the symposium and the publication of this volumc arc listcd scparately in this volume (page 8). However, the generosity of the Central Florida Herpetological Society, David C. Leber, Michael H. Strahm, and the Toledo Zoological Society warrant special mention. We also want to acknowledge prominently the SSAR for its financial support.

We especially thank Kraig Adler, editor of the SSAR *Contributions to Herpetology* series, whose experience and advice were immensely valuable in bringing this volume to print. Those, however, are qualities one would expect from a series editor; for his support we cannot express enough gratitude.

John S. Parmerlee, Jr. provided a tremendous service by assuming responsibility for digitizing, revising, and occasionally regenerating the line drawings that are included in nearly every chapter. Without his help, the production process would have taken much longer, the cost would have risen substantially (or, conversely, the quality would have been noticably lower), and we would not have been nearly as proud of the final product.

When authors were unable to do so, Javier A. Rodríguez-Robles, University of California at Berkeley, and Martin Ouellet, Redpath Museum, McGill University, provided translations of abstracts into Spanish and French, respectively. We thank them for their time and effort.

While editing this volume we have called upon many colleagues to review manuscripts. The quality of the papers presented herein would have suffered considerably without the benefit of the generous assistance of Aaron M. Bauer, Ellen J. Censky, Ronald I. Crombie, Brian I. Crother, Martha L. Crump, James R. Dixon, C. Kenneth Dodd, Jr.,* William E. Duellman, Carl H. Ernst, Henry S. Fitch,* Darrel R. Frost, L. Lee Grismer, Craig Guyer, S. Blair Hedges,* W. Ronald Heyer,* John B. Iverson,* Thomas A. Jenssen, Hinrich Kaiser,* James D. Lazell, Jr., Julian C. Lee, Allen R. Lewis,* Jonathan B. Losos, John D. Lynch, Roy W. McDiarmid, Scott Monks, Randall D. Mooi, José A. Ottenwalder,* A. Stanley Rand, Douglas P. Reagan, Javier A. Rodríguez-Robles,* Rodolfo Ruibal, Richard A. Seigel, Hobart M. Smith, Judy Stamps, Margaret M. Stewart,* John B. Thorbjarnarson, Peter J. Tolson, Daniel S. Townsend, Linda Trueb, Laurie J. Vitt, Peter Vogel, Harold K. Voris, David B. Wake, Lawrence L. Woolbright, and William J. Wrenn. Those persons whose names are marked with an asterisk reviewed at least two manuscripts. Particularly deserving for assistance above and beyond the call of duty are Ken Dodd, Blair Hedges, and Hinrich Kaiser. Each reviewed at least three manuscripts, always meeting our deadlines (some with very abbreviated turnaround times), gave freely of their expertise, and actually asked if they could do more.

The individual editors of this volume also would like to recognize individuals and institutions for their assistance.

Robert Powell: Beverly J. Powell and Michael A. Powell for tolerating my herpetological inclinations and my prolonged absences in the field; Avila College for instituting a sabbatical policy just in time for me to devote adequate time to this volume without losing touch with the little sanity my friends and acquaintances still attribute to me, for providing considerable support for the many tasks associated with a project of this magnitude, and for financial assistance in attending the symposium in Athens, Georgia; John S. Parmerlee, Jr., upon whose computer wizardry I called more than once; and Robert W. Henderson, whose wry humor and hard work have made this volume possible, and whose positive response to my question in October of 1992: "What can I do to honor Al?" got us into this project in the first place.

Robert W. Henderson: Rose M. Henderson for rarely complaining about the, at times, almost daily stops at the post office on the way home from work, in order that manuscripts and reviews got to authors and reviewers expeditiously and economically; Robert W. Bourgeois for, as usual, dozens of reasons; Joanne Peterson (photography) and Susan Otto (library) at the Milwaukee Public Museum for many favors, always granted promptly; the administrative personnel at the Museum (particularly Allen M. Young) for being supportive of my time commitment to this project (but, then, I didn't tell them about all of the postage they paid); and Bob Powell for unhesitatingly assuming many of my responsibilities for the symposium when, for personal reasons, I was unable to attend, and for hardly making me feel guilty about it.

ROBERT POWELL
15 August 1995 ROBERT W. HENDERSON

ADDRESSES OF THE VOLUME EDITORS

Robert Powell, Department of Natural Sciences, Avila College, 11901 Wornall Road, Kansas City, Missouri 64145–1698, USA.

Robert W. Henderson, Section of Vertebrate Zoology, Milwaukee Public Museum, 800 West Wells Street, Milwaukee, Wisconsin 53233–1478, USA.

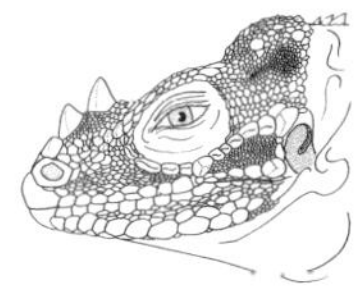

Remembrances of Albert Schwartz

Albert Schwartz: Dawning of a Herpetologist

William E. Duellman[1]

THE WEATHER WAS COLD AND RAINY ON THE SLOPES of Grandfather Mountain in North Carolina in March 1950. I had accompanied Al Schwartz and Frank Banfield, both graduate students in mammalogy at the University of Michigan, to the southern Appalachians. Al was adamant about heading south during spring break in order to escape what he considered to be the absolutely intolerable winter weather in Michigan. Frank, a Canadian, had never been so far south. I, a lowly undergraduate student, was the guide, because I had collected previously in the region. Al and Frank were trying to trap small mammals, but the weather seemed to be restricting small mammal activity. I turned my attention to searching under logs on hillsides and rocks along streams; obviously *Plethodon* and *Desmognathus* were readily captured. Disgusted with lines of empty snap traps, Al muttered something to the effect: "After driving all this way, I want to collect something, even if it doesn't have fur." Soon he was enjoying turning rocks and logs and filling bags with salamanders. Little did I realize then that I was witnessing Al's obsession with collecting.

Later that day, as rain turned to snow, we were ensconced in a run-down motel room. Al and Frank had prepared their few mammals, and I was busy drowning salamanders in warm water (Chloretone having not come into common use for anesthetizing amphibians). Al stated that because he had collected so many of the salamanders he should help prepare the specimens, but he complained about the smell of the formaldehyde. So he smoothed out a paper towel wet with water in a tray and proceeded to follow my example of laying out salamanders with their bodies and tails straight, their limbs properly positioned, and their digits spread. Commenting that preparing salamanders was much easier than mammals, Al proudly demonstrated a tray of perfectly laid out specimens. He carefully covered the specimens with another wet paper towel and poured formalin into the tray, whereupon all of the salamanders curled their bodies and tails and proceeded to crawl. That incident might well have been the end of Al's brief herpetological career!

Shortly after returning to Ann Arbor, Al started planning a trip throughout the western United States and asked if I was interested in accompanying him (all expenses paid). Our primary goal was to obtain a diversity of mammals, but I was told that I could collect amphibians and reptiles as well. In early June 1950, we left Michigan in a new station wagon for Miami, Florida. There we picked up Ray Porter and a utility trailer filled with camping paraphernalia, traps, and other assorted collecting gear. The trip encompassed west Texas, southern New Mexico and Arizona, California to Washington, and thence eastward to Idaho, Montana, Wyoming, and South Dakota before returning to Ann Arbor 13 weeks after our departure.

I learned much about mammals and the preparation of study skins, but my emphasis was on amphibians and reptiles. In west Texas and in southern New Mexico and Arizona, our camping was made not only uncomfortable, but at times hazardous, by torrential rains. These, however, brought out thousands of *Bufo* and *Scaphiopus*. Many choruses of these toads were encountered while we drove roads at night, finding many snakes, and to Al's delight, mammals — especially kangaroo rats

[1] Natural History Museum and Department of Systematics and Ecology, The University of Kansas, Lawrence, Kansas 66045, USA.

(*Dipodomys*), one of Al's favorites. By day, Al enjoyed collecting lizards, especially *Cnemidophorus* and *Callisaurus*, and became proficient with .22-caliber dust shot. By the end of that trip, Al was an enthusiastic herp collector.

In the fall of 1950, Al decided that his dissertation would be on the mammals of southern Florida, where he had collected extensively before entering The University of Michigan. Before leaving Ann Arbor in December, he acquired necessary supplies and equipment for collecting herps and informed the Division of Herpetology that he intended to increase their meager holdings from southern Florida. He was true to his word, for when he returned in September of the next year he brought an immense collection of herps, many times larger than his mammal collection. Al was excited about his herpetological finds — many previously unrepresented species, significant range extensions, and new taxa. He was especially thrilled with his large series of the black newt from the Everglades, and we collaborated on the description of *Diemictylus* (=*Notophthalmus*) *viridescens piaropicola*.

By this time, Al had become an inveterate herp collector and was spending more and more time delving into the herpetological literature. His waning interest in mammals was evidenced by the fact that he never published his dissertation on the mammals of South Florida. After Al took a position at the Charleston Museum, we began our study on the herpetofauna of South Florida and spent two summers together in the field. While we were writing the manuscript, Al undertook several systematic studies on amphibians and reptiles from the southeastern United States and made his first trip to Cuba.

Perhaps it was fortuitous that bad weather on Grandfather Mountain negatively influenced mammal collecting and that then, and during the following summer in the Southwest, I was able to entice Al away from warm, furry creatures. These events, seemingly trivial at the time, were the dawning of the herpetological Al Schwartz, who for three decades scoured the West Indies for specimens and contributed so much to our knowledge of the herpetofauna in those fascinating islands. In retrospect, I am pleased that by introducing Al to collecting amphibians and reptiles, I made a significant (albeit vicarious) contribution to the herpetology of the West Indies.

•

RECOLLECTIONS OF ALBERT SCHWARTZ

RICHARD THOMAS[1]

HARRY FREEMAN, AN ICHTHYOLOGIST AT THE UNIversity of South Carolina, introduced me to Al shortly after he had taken his first position subsequent to receiving his Ph.D. Al was the curator of vertebrates at the Charleston Museum. I was a high school student interested in herps, but without a clear idea of how I might turn that interest into an occupation. At that time, Al was very actively involved in fieldwork throughout the Southeast and was very taken with the idea of finding new information about species in the field: new distributional records, resolution of problems regarding uncertain range boundaries, and discovering new taxa. He was also an excellent general naturalist who had a good working knowledge of plants and other groups of animals, whether vertebrates or invertebrates. Since he had spent a good deal of time in southern Florida, he was particularly at home with the natural history of that area. South Florida, of course, is floristically somewhat of a transition zone to the West Indies. In retrospect, I believe it was almost inevitable that Al would work in the Antilles, although he was totally occupied with North American herps and mammals during the mid-1950s.

From the first time I met Al and throughout his life he was very intense and energetic. He had an expostulatory, almost explosive laughter, and the sense of humor to go with it. He was a tremendous story-teller and drew much of his material from the oddities that happened during field work.

One physical problem that plagued Al throughout his life was periodic bouts of tachycardia. In the field it was not unusual to find him lying in the middle of a dirt road waiting for his heartbeat to subside from 200 or so beats per minute to normal.

[1] Department of Biology, University of Puerto Rico, Río Piedras, Puerto Rico 00931, USA.

Once it had done so, he would get up and resume collecting.

Al never liked flying, but his dislike became particularly acute after one unsettling small plane landing on La Désirade off Guadeloupe. It took two tries to land because a crosswind kept blowing the plane out to sea and away from the landing strip. After that he tried to give up flying completely and get to and around the West Indies by boat. But that didn't work, and thereafter he resorted to tranquilizers before boarding a plane.

Nature has us exploratory field biologists on a variable reinforcement schedule — the most powerful inducement to continuing a behavior. You turn rock after rock, you go out night after night listening for frogs, you poke into all the odd habitats you can find. You don't know when and you don't really know what, but you do know that eventually something is going to really blow your mind. It should be evident that Al really loved fieldwork, and I am sure part of the continuing attraction, aside from the relatively abstract objective of better understanding the fauna, was the unexpected surprises that turn up again and again. For him, that sudden realization that you are looking at a species that is completely unknown scientifically was — no matter how often it happened — one of the most mind-rattling experiences.

I remember on many occasions hearing Al's growing (by the second) realization of discovery. I would usually hear a drawn out "I'll be damned" uttered low but with intensity, followed by a similarly drawn out "No kidding." These utterances were not necessarily directed at any one person, instead they would accompany the dawning realization that he had caught a specimen of an undescribed species or had just found something otherwise unusual.

I particularly recall Al finding the first specimen that we had collected of the peculiar Hispaniolan anoline *Chamaelinorops* (=*Anolis*) *barbouri*. We were night collecting, and he found one sleeping on a leaf in a coffee grove on the Sierra de Baoruco in the Dominican Republic. I wasn't on the spot for the immediate discovery, but he later recounted that he thought: "How odd, a little dried-up dead *Anolis* on a leaf!" As one doesn't often find dead lizards in the field, he went to pick it up and discovered that it was a live *Chamaelinorops*.

Al discovered early on that you could best collect most anoles at night, which meant you did not have to spend an inordinate amount of time during the day looking for them. I'm sure that Al found this out while scanning the vegetation for frogs. I recall vividly that during a Cuban trip in 1957 we were collecting interesting grass anoles in this fashion; and giant anoles, which can be especially difficult to collect by day, are often much more easily found at night. Al made a great effort to find giant anoles in Cuba, and eventually published several papers, recognizing three species where previously there had been only one. He later did similar studies of giant anoles in the Dominican Republic.

In Cuba, Al used a distinctive aluminum-bodied panel truck as a traveling "laboratorio de biología." He often paid for specimens if the local people would collect them. The truck was so recognizable that when he returned to a town, the kids would often run after him shouting "chipojo, chipojo," that being the local name in parts of Cuba for giant anoles. The other kind of "chipojo" is the chamaeleon-like *Chamaeleolis* (=*Anolis*), and they are particularly cryptic. Although they are not uncommon in some areas and can be found sleeping at night, it was a good while after Al began Cuban fieldwork that he got his first specimen. He recounted that a bunch of kids in one place were being nuisances and bringing in all kinds of lizards claiming they were "chipojos" and even pelting the truck with common anoles. In exasperation, Al was shooing them away, but one little boy was being so very insistent that Al finally looked at the lizard wrapped in a leaf. The lizard, much to Al's surprise, was his first *Chamaeleolis*.

Occasional feedback from colleagues made Al realize that some people thought he was "inventing" species or, in the case of subspecies, describing "microgeographic races" (or somesuch) that should not be described. Partly, I believe, this was due to the fallacious inference that, as island faunas are known to be smaller than those on the mainland, that many undescribed species or even subspecies simply could not exist. Also, people experienced with the broader and often not so clear-cut patterns of infraspecific variation on continents were reluctant to believe that so much localized variation could be found on relatively small islands. Al would sometimes say, half joking, half despairing, on writing another paper describing another species, "No one will believe us!"

One thing Al did think was that you should not be too timid. During our last conversation, a month or so before his death, I was lamenting that I had a new species of anole, but only one specimen. He said, "Well, remember, all you need is a holotype!" He frequently said that you should go ahead and take a chance. By this he meant that you shouldn't dither around too much if you have a few specimens but think you have something undescribed. An assumption here is that you are experienced with these animals and your insight is probably better than anyone else's on the subject.

Probably the most important thing I learned from Al was the importance of having first-hand knowledge of the animals on which you work, including the habits, colors in life, habitats, and physiography of the areas they inhabit. The insight from this knowledge is extremely important for the resolution of taxonomic problems. Of course he realized that someone has to work on the taxonomy of widespread groups in which field experience with most of the species is precluded. But that was not the sort of thing he wanted to do. For him, knowledge of the animals in the field was paramount. He would not have felt comfortable doing a systematic revision purely from museum specimens. He often bemoaned and not infrequently roared with laughter at some of the gaffes made by our predecessors in the West Indies (usually lumping several very good species into one). These were typically mistakes that could have been avoided if the person had actually seen the behavior and habitats of the animals. In fact, sometimes the error could have been avoided by looking more closely at the preserved specimens. He would often remark that some author was treating the distribution of species as if the maps in their publications actually were the territory, as if Hispaniola, for example, were really a featureless plain. What looks like a small distance between locality dots on a postage-stamp-sized map may in reality be a xeric valley separating populations in mesic uplands. This kind of knowledge enables you to understand the variations you see.

Al was an incredibly hard and consistent worker. He could teach a heavy course load and still write papers. He had a routine, and when it was time to work he would simply sit down at the microscope or typewriter and start working — no warm-up, no blocks; work, no matter what the mood. He wrote rapidly and his first drafts tended to be very clean. Even as physical infirmities began to plague him more and more, his output remained high. Another thing he did was to avoid as much as possible the distractions that take us away from research. He taught and taught well, but he did not get very much caught up in administration and he didn't go to society meetings. I think it seemed perverse to him that meetings were held during the summer when all biologists worth their salt should be in the field.

I have not touched on other aspects of Al's personality and life, such as the amazing number of people that he mentored and influenced, despite not being affiliated with a graduate institution. The lessons he passed on can be drawn from his life. They may seem like truisms, but many of us could follow Al's example and be better for it.

1. Know those animals with which you work firsthand and in the field. This knowledge is one of your greatest resources as a systematist and leads to many other kinds of questions.
2. Have a broad goal that focuses your efforts.
3. Establish a routine and work consistently.
4. Recognize distractions and arrange your life to minimize them.
5. Be willing to gamble a little. Have the courage of your convictions.

•

A SINGULAR DEVOTION: ALBERT SCHWARTZ AND THE WEST INDIES

ROBERT W. HENDERSON[1]

SINCE THE UNTIMELY DEATH OF ALBERT SCHWARTZ, I have been asked to prepare remembrances or obituaries for several journals and I have done so wil-

[1] Section of Vertebrate Zoology, Milwaukee Public Museum, Milwaukee, Wisconsin 53233, USA.

lingly. Although Al's death saddened me, it has given me pleasure to recall conversations with him and to review our collaborative affiliation. It also reminded me how fortunate I have been to have had prolonged collaborations with two remarkable zoologists: Henry S. Fitch and Al. The former is a gentle, soft-spoken man devoted to his family, who has had a long association with a major university and ample time to devote to his research. The second was a life-long bachelor, full of warmth and good humor but also capable of public tantrums and occasional rudeness, who spent most of his professional career at a junior college that provided no support for his research. Cumulatively, I have spent months in the field with Fitch; although I spent only two weeks in the field with Schwartz, I talked with him on the phone for hundreds of hours over the span of a decade. I have not encountered two people more dedicated to field work, despite advancing years and, in Schwartz's case, crippling arthritis. The two men were ideal collaborators for similar reasons: (1) enthusiasm for the subject (they did what they did not because they had to, but because it gave them pleasure); (2) professional generosity (always willing to share information and authorship; I believe this can be attributed to self-confidence and to their productivity, which allowed them to be generous with co-workers because they knew that when that paper was finished, another would come along in a few days, weeks, or months), and (3) punctuality (if data or manuscript were to be put in the mail on a certain day, I knew it would be; I also knew that the same courtesy was expected of me). Both men have had a significant influence on the careers of many students, and each has produced a body of knowledge that will stimulate research for decades to come. Although both published on organisms in addition to amphibians and reptiles (Fitch on spiders, birds, and mammals; Schwartz on butterflies, birds, and mammals), I find inspiration in their devotion to herpetology and in their productivity. Henry Fitch at 85 years of age keeps going (albeit a little slower), and I am happy that he was honored by his colleagues and students during his lifetime (Seigel et al. 1984). Al Schwartz is no longer with us, but the posthumous tribute paid him at the symposium held at the University of Georgia in Athens, and the contributions that appear in this volume, are testimony to the high esteem in which he and his contributions to West Indian herpetology are held.

Al's singular devotion was instrumental to his success. He did not dabble nor did he ever take a dilettante's attitude. Whatever he pursued he did so with tremendous energy and enthusiasm, and he appreciated others who shared that enthusiasm. His first field trip to the West Indies occurred in the winter of 1954, when he traveled to Cuba; after that, he never looked back. Every summer was spent in the West Indies; and for many years, fieldwork and writing were full-time activities. In all, he and his field associates collected more than 80,000 herpetological specimens from at least 133 islands. From 1956 to 1991, Al produced 195 publications that dealt with some aspect of West Indian herpetology (Bourgeois, this volume). From 1964 through 1985 he published at least five West Indian papers per year, and in a couple of years ten or more appeared. During that span he not only described 86 species and 279 subspecies of amphibians and reptiles, he also described nearly two dozen butterflies (Borkin 1994) and three bats, and also produced several papers on the West Indian avifauna. He published a number of papers with Orlando Garrido that dealt with several vertebrate classes on Cuba, and wrote two papers on West Indian biogeography that focused on the herpetofauna (1978, 1980). His 580-page "Butterflies of Hispaniola" appeared in 1989. When assessing Al's contributions to West Indian biology (and herpetology in particular), I believe it is important to examine the body of research produced rather than individual publications. Although he produced a number of major contributions (e.g., his checklist with Richard Thomas in 1975), the whole of his contributions is greater than the sum of the parts. Today, very few publications in West Indian herpetology lack references to his work.

Although frogs (28.7% of his West Indian publications that addressed only a single order or suborder of animals), amphisbaenians (0.6%), and snakes (14.4%) all received attention from Al, 102 of his publications (56.4%) dealt exclusively with lizards, and when pressed to choose highlights of his field work, lizards always were the featured subjects. His first *Chamaeleolis* (=*Anolis*), his first encounter with *Chamaelinorops* (=*Anolis*) *barbouri* after years of failure, the discoveries of *Anolis fowleri* and *Anolis sheplani*, his enthusiastic verbal description of male *Anolis b. bahorucoensis*, and his good-natured envy of Richard Thomas for having discovered *Sauresia* (=*Celestus*) *agasepsoides* were telltale indications of his saurophilic bent.

Within the field of herpetology during the past half century, several people have come to be identified with certain taxonomic groups as a result of their prolific and competent research. Among those who come readily to mind are Archie F. Carr (sea turtles), William E. Duellman (hylid frogs), Laurence M. Klauber (rattlesnakes), Janis A. Roze (coral snakes), David B. Wake (Neotropical salamanders), and Ernest E. Williams (anoline lizards). Far more unusual in herpetology is the association of one person with a single geographic area, without also associating that person with a particular taxonomic group; perhaps no other person is more closely associated with a single geographic area than Albert Schwartz. His devotion to the West Indian fauna was singular. From the mid-1950s to the early 1990s his name became synonymous with West Indian herpetology. And not just frogs or lizards or snakes, but with the herpetology of the area. He tried to do it all, and was amazingly successful at accomplishing what he set out to do. Here was a man that devoted himself to a geographic region: if he was to study birds, they would be West Indian birds; if he was to work on butterflies, they would be butterflies that occur in the West Indies. He even avoided working on problems that would require examination of material from outside of the region.

I was unable to attend the symposium honoring Al. It would have been exciting to listen to the presentations of research projects and to have participated in the discussions that they stimulated. And what an appropriate and worthy tribute to the memory of Al and to his many contributions to West Indian herpetology the symposium proved to be. Throughout the presentations and tributes to Al, during the touching remembrances, I would have had the same thought over and over. Even if he weren't dead, even if he were in the pink of good health, the son-of-a-gun wouldn't have bothered to attend this event. Either he would have been in the field or he would have stayed in his Miami office surrounded by his books and field notes, in a room adjacent to rooms that, until recently, overflowed with tens of thousands of butterflies, frogs, lizards, birds, and mammals from dozens of West Indian islands that he and his associates collected over a period of 35 years. Puffing on a cigarette, perhaps with a rum and coke in hand, he would have thought about the symposium being held in his honor 600 miles to the north, and he would have been absolutely delighted.

The last time I saw Al was on October 9, 1992 in a hospital in Miami. I was on my way home from Trinidad where I had been collecting tree boas, and I also had butterflies for Al. I had only recently convinced him that he should expand his geographic arena, that knowledge of the butterfly fauna on Trinidad might shed additional light on relationships in West Indian taxa. He agreed with my rationale (which had a self-serving motive, of course) and, when I saw him in the hospital, he queried me about my success. I knew he would be enthusiastic about the butterflies I had brought him, but he never saw them. Al Schwartz died quietly on October 18, 1992. Who knows in what wonderful new directions those Trinidad butterflies might have taken him.

LITERATURE CITED

Borkin, S. S. 1994. Albert Schwartz (1923–1992). Jour. Lepidop. Soc. 48: 267–271.

Schwartz, A. 1978. Some aspects of the herpetogeography of the West Indies, p. 31–51. *In* F. B. Gill (ed.), Zoogeography in the Caribbean. Acad. Nat. Sci. Philadelphia, Spec. Publ. no. 10.

—. 1980. The herpetogeography of Hispaniola, West Indies. Stud. Fauna Curaçao Carib. Isl. 189: 86–127.

—. 1989. Butterflies of Hispaniola. Univ. Florida Press, Gainesville.

— and R. Thomas. 1975. Check-list of West Indian amphibians and reptiles. Carnegie Mus. Nat. Hist. Spec. Publ. 1: 1–216.

Seigel, R. A., L. E. Hunt, J. L. Knight, L. Malaret, and N. L. Zuschlag (eds.). 1984. Vertebrate Ecology and Systematics: A Tribute to Henry S. Fitch. Univ. Kansas Mus. Nat. Hist. Spec. Publ. no. 10.

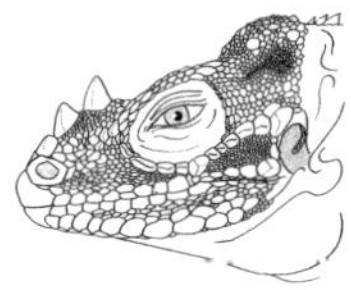

HERPETOLOGICAL PUBLICATIONS OF ALBERT SCHWARTZ

ROBERT W. BOURGEOIS[1]

WEST INDIAN CONTRIBUTIONS

1956

1. (with L. H. Ogren). A collection of reptiles and amphibians from Cuba, with the description of two new forms. Herpetologica 12: 91–110.

1957

2. A new species of *Eleutherodactylus* (Amphibia: Leptodactylidae) from Cuba. Proc. Biol. Soc. Washington 70: 209–212.
3. A new species of boa (genus *Tropidophis*) from western Cuba. Amer. Mus. Novit. 1839: 1–8.

1958

4. Four new frogs of the genus *Eleutherodactylus* (Leptodactylidae) from Cuba. Amer. Mus. Novit. 1873: 1–20.
5. A new gecko of the *Sphaerodactylus decoratus* group from Cuba. Proc. Biol. Soc. Washington 71: 27–36.
6. Another new large *Eleutherodactylus* (Amphibia: Leptodactylidae) from western Cuba. Proc. Biol. Soc. Washington 71: 37–42.
7. Another new species of *Eleutherodactylus* (Amphibia: Leptodactylidae) from western Cuba. Jour. Washington Acad. Sci. 48: 127–131.
8. A new subspecies of *Anolis equestris* from eastern Cuba. Herpetologica 14: 1–7.
9. A new frog of the *auriculatus* group of the genus *Eleutherodactylus* from western Cuba. Herpetologica 14: 69–77.
10. (with G. R. Zug). Variation in the species of *Cadea* (Amphisbaenidae), and a record of *C. blanoides* from the Isla de Pinos. Herpetologica 14: 176–179.

1959

11. A new species of frog of the *Eleutherodactylus ricordi* group from central Cuba. Amer. Mus. Novit. 1926: 1–16.
12. Variation in lizards of the *Leiocephalus cubensis* complex in Cuba and the Isla de Pinos. Bull. Florida State Mus. 4: 97–143.
13. The status of *Eleutherodactylus pinarensis* and a new species of the genus from western Cuba. Herpetologica 15: 61–69.
14. The Cuban lizards of the species *Leiocephalus carinatus* (Gray). Sci. Publ. Reading Public Mus. Art Gallery 10: 1–47.
15. A new species of toad, *Bufo cataulaciceps*, from the Isla de Pinos and western Cuba. Proc. Biol. Soc. Washington 72: 109–120.

1960

16. (with R. Thomas). Four new snakes (*Tropidophis*, *Dromicus*, *Alsophis*) from the Isla de Pinos and Cuba. Herpetologica 16: 73–90.
17. Nine new Cuban frogs of the genus *Eleutherodactylus*. Sci. Publ. Reading Public Mus. Art Gallery 11: 1–50.
18. The large toads of Cuba. Proc. Biol. Soc. Washington 73: 45–56.
19. Variation in the Cuban lizard *Leiocephalus raviceps* Cope. Proc. Biol. Soc. Washington 73: 67–82.
20. A new subspecies of *Leiocephalus stictigaster* Schwartz from central Cuba. Proc. Biol. Soc. Washington 73: 103–106.
21. (with R. J. Marsh). A review of the *pardalis-maculatus* complex of the boid genus *Tropidophis* of the West Indies. Bull. Mus. Comp. Zool., Harvard Univ. 123: 49–84.

1961

22. A review of the *Sphaerodactylus scaber* group of Cuba. Herpetologica 17: 19–26.

[1] Section of Vertebrate Zoology, Milwaukee Public Museum, Milwaukee, Wisconsin 53233, USA.

1963

23. A new subspecies of *Tropidophis greenwayi* from the Caicos Bank. Breviora 194: 1–6.

1964

24. *Anolis equestris* in Oriente Province, Cuba. Bull. Mus. Comp. Zool., Harvard Univ. 131: 405–428.
25. New subspecies of *Leiocephalus* from Cuba. Quart. Jour. Florida Acad. Sci. 27: 211–222.
26. Three new species of frogs (Leptodactylidae, *Eleutherodactylus*) from Hispaniola. Breviora 208: 1–15.
27. *Diploglossus costatus* Cope (Sauria, Anguidae) and its relatives in Hispaniola. Sci. Publ. Reading Public Mus. Art Gallery 13: 1–57.

1965

28. (with R. Thomas). Subspeciation in *Sphaerodactylus copei*. Quart. Jour. Florida Acad. Sci. 27: 316–332.*
29. Two new species of *Eleutherodactylus* from the eastern Cordillera Central of the República Dominicana. Carib. Jour. Sci. 4: 473–484.*
30. Variation and natural history of *Eleutherodactylus ruthae* on Hispaniola. Bull. Mus. Comp. Zool., Harvard Univ. 132: 481–508.
31. A review of *Sphaerodactylus vincenti* on the southern Windward Islands. Carib. Jour. Sci. 4: 391–409.
32. A new Cuban *Eleutherodactylus* of the *auriculatus* group. Herpetologica 21: 27–31.
33. Two new subspecies of the anguid lizard *Wetmorena* from Hispaniola. Proc. Biol. Sci. Washington 78: 39–48.
34. Geographic variation in two species of Hispaniolan *Eleutherodactylus*, with notes on Cuban members of the *ricordi* group. Studies Fauna Curaçao Carib. Isl. 22: 98–123.
35. (with R. Thomas). The genus *Darlingtonia* (Serpentes) in Hispaniola, including a new subspecies from the Dominican Republic. Breviora 229: 1–10.
36. A review of the colubrid snake genus *Arrhyton* with a description of a new subspecies from southern Oriente Province, Cuba. Proc. Biol. Soc. Washington 78: 99–114.
37. A new species of *Eleutherodactylus* (Amphibia: Leptodactylidae) from the Sierra de Baoruco, República Dominicana. Proc. Biol. Soc. Washington 78: 165–168.
38. (with R. Thomas). Hispaniolan snakes of the genus *Dromicus* (Colubridae). Rev. Biol. Trop. 13: 59–83.
39. A new subspecies of the gecko *Sphaerodactylus microlepis*. Herpetologica 21: 261–269.
40. The *Ameiva* (Reptilia: Teiidae) of Hispaniola. I. *Ameiva lineolata* Duméril and Bibron. Carib. Jour. Sci. 5: 45–57.

1966

41. Geographic variation in *Sphaerodactylus notatus* Baird. Rev. Biol. Trop. 13: 161–185.*
42. The relationships of four small Hispaniolan *Eleutherodactylus* (Leptodactylidae). Bull. Mus. Comp. Zool., Harvard Univ. 133: 371–399.
43. The *Leiocephalus* (Lacertilia, Iguanidae) of Hispaniola. I. *Leiocephalus melanochlorus* Cope. Jour. Ohio Herpetol. Soc. 5: 39–48.
44. (with R. F. Klinikowski). The *Ameiva* (Lacertilia, Teiidae) of Hispaniola. II. Geographic variation in *Ameiva chrysolaema* Cope. Bull. Mus. Comp. Zool., Harvard Univ. 133: 427–487.
45. (with R. Thomas). *Sphaerodactylus* (Gekkonidae) in the Greater Puerto Rico region. Bull. Florida State Mus. 10: 193–260.
46. (with R. Thomas). The *Sphaerodactylus decoratus* complex in the West Indies. Brigham Young Univ. Sci. Bull. 7: 1–26.
47. Snakes of the genus *Alsophis* in Puerto Rico and the Virgin Islands. Studies Fauna Curaçao Carib. Isl. 23: 175–227.

1967

48. A review of the genus *Dromicus* in Puerto Rico and the Virgin Islands. Stahlia 9: 1–14.
49. Frogs of the genus *Eleutherodactylus* in the Lesser Antilles. Studies Fauna Curaçao Carib. Isl. 24: 1–62.
50. (with R. Thomas). The *monticola* group of the lizard genus *Anolis* in Hispaniola. Breviora 261: 1–27.
51. The *Leiocephalus* (Lacertilia, Iguanidae) of Hispaniola. II. The *Leiocephalus personatus* complex. Tulane Studies Zool. 14: 1–53.
52. The *Ameiva* (Lacertilia, Teiidae) of Hispaniola. III. *Ameiva taeinura* Cope. Bull. Mus. Comp. Zool., Harvard Univ. 135: 345–375.
53. (with O. H. Garrido). A review of the Cuban iguanid lizard *Leiocephalus macropus* Cope. Sci. Publ. Reading Public Mus. Art Gallery 14: 1–41.
54. The *Leiocephalus* (Lacertilia, Iguanidae) of the southern Bahama Islands. Ann. Carnegie Mus. 39: 153–185.

1968

55. (with O. H. Garrido). Cuban lizards of the genus *Chamaeleolis*. Quart. Jour. Florida Acad. Sci. 30: 197–220.*
56. The geckos (*Sphaerodactylus*) of the southern Bahama Islands. Ann. Carnegie Mus. 39: 227–271.
57. Two new subspecies of *Ameiva* (Lacertilia, Teiidae) from Hispaniola. Herpetologica 24: 21–28.

* The date shown on the publication is for the previous year.

58. The *Leiocephalus* (Lacertilia, Iguanidae) of Hispaniola. III. *Leiocephalus schreibersi*, *L. semilineatus*, and *L. pratensis*. Jour. Herpetol. 1: 39–63.
59. (with O. H. Garrido). Four new subspecies of *Leiocephalus stictigaster* from Cuba. Natl. Mus. Canada Nat. Hist. Pap. 37: 1–23.
60. (with O. H. Garrido). An undescribed subspecies of *Leiocephalus raviceps* Cope (Sauria: Iguanidae) from western Cuba. Proc. Biol. Soc. Washington 81: 23–30.
61. Geographic variation in the New World gekkonid lizard *Tarentola americana* Gray. Proc. Biol. Soc. Washington 81: 123–141.
62. Geographic variation in the Hispaniolan frog *Eleutherodactylus wetmorei* Cochran. Breviora 290: 1–13.
63. (with O. H. Garrido). Anfibios, reptiles y aves de la Península de Guanahacabibes. Poeyana, ser. A. 53: 1–68.
64. The Cuban lizards of the *Anolis homolechis* complex. Tulane Stud. Zool. 14: 140–184.
65. Geographic variation in *Anolis distichus* Cope (Lacertilia, Iguanidae) in the Bahama Islands and Hispaniola. Bull. Mus. Comp. Zool., Harvard Univ. 137: 255–309.

1969

66. (with R. Thomas). A review of *Anolis angusticeps* in the West Indies. Quart. Jour. Florida Acad. Sci. 31: 51–69.*
67. A review of the Hispaniolan lizard *Anolis coelestinus* Cope. Carib. Jour. Sci. 9: 33–38.
68. Two new subspecies of *Leiocephalus* from Hispaniola. Jour. Herpetol. 3: 79–85.
69. (with D. W. Buden). Reptiles and birds of the Cay Sal Bank, Bahama Islands. Quart. Jour. Florida Acad. Sci. 31: 290–320.*
70. The status of the name *Holotropis microlophus* Cocteau. Copeia 1969: 620–621.
71. (with O. H. Garrido). Anfibios, reptiles y aves de Cayo Cantiles. Poeyana, ser. A. 67: 1–44.
72. The Antillean *Eleutherodactylus* of the *auriculatus* group. Studies Fauna Curaçao Carib. Isl. 30: 99–115.

1970

73. *Sphaerodactylus notatus*. Cat. Amer. Amphib. Rept.: 90.1–90.2.
74. A new species of large *Diploglossus* (Sauria: Anguidae) from Hispaniola. Proc. Biol. Soc. Washington 82: 777–788.
75. A new species of gecko (Gekkonidae, *Sphaerodactylus*) from Hispaniola. Jour. Herpetol. 4: 63–67.
76. A systematic review of *Ameiva auberi* Cocteau (Reptilia, Teiidae) in Cuba and the Bahamas. I. The Cuban subspecies. III. Discussion. Ann. Carnegie Mus. 41: 45–117, 152–168.
77. A systematic review of *Uromacer catesbyi* Schlegel (Serpentes, Colubridae). Tulane Studies Zool. Bot. 16: 131 149.
78. *Darlingtonia haetiana* (Serpentes, Colubridae): a new subspecies. Herpetologica 26: 324–331.
79. The subspecies of *Eleutherodactylus schmidti* Noble (Anura: Leptodactylidae). Carib. Jour. Sci. 10: 109–118.

1971

80. A systematic account of the Hispaniolan snake genus *Hypsirhynchus*. Studies Fauna Curaçao Carib. Isl. 35: 63–94.
81. A new species of bromeliad-inhabiting galliwasp (Sauria: Anguidae) from Jamaica. Breviora 371: 1–10.
82. A new subspecies of *Leiocephalus personatus* (Sauria, Iguanidae). Herpetologica 27: 176–182.
83. (with O. H. Garrido). The status of *Anolis alutaceus clivicolus* Barbour and Shreve. Carib. Jour. Sci. 11: 11–15.
84. A new species of *Eleutherodactylus* (Amphibia, Leptodactylidae) from Hispaniola. Ann. Carnegie Mus. 43: 25–31.
85. *Anolis distichus*. Cat. Amer. Amphib. Rept.: 108.1–108.4.
86. (with J. D. Lynch). Taxonomic disposition of some 19th century leptodactylid frog names. Jour. Herpetol. 5: 103–114.
87. Two new subspecies of *Diploglossus costatus* from Hispaniola (Sauria, Anguidae). Jour. Herpetol. 5: 161–165.

1972

88. (with O. H. Garrido). The lizards of the *Anolis equestris* complex in Cuba. Studies Fauna Curaçao Carib. Isl. 39: 1–86.
89. A new subspecies of *Sphaerodactylus decoratus* (Sauria, Gekkonidae) from the Bahama Islands. Herpetologica 28: 247–251.
90. The native toads (Anura, Bufonidae) of Hispaniola. Jour. Herpetol. 6: 217–231.
91. (with O. H. Garrido). The Cuban *Anolis spectrum* complex (Sauria, Iguanidae). Proc. Biol. Soc. Washington 85: 509–522.
92. (with G. Peters). Ein neuer, Bromelien bewohnender Kugelfingergecko (Gekkonidae: *Sphaerodactylus*) aus Oriente/Cuba. Mitt. Zool. Mus. Berlin 48: 393–399.

1973

93. (with D. C. Fowler). The Anura of Jamaica: a progress

report. Studies Fauna Curaçao Carib. Isl. 43: 50–142.

94. A new subspecies of *Ameiva chrysolaema* (Sauria, Teiidae) from Haiti. Herpetologica 29: 101–105.
95. A third species of the Hispaniolan *shrevei* group of *Sphaerodactylus* (Sauria, Gekkonidae). Proc. Biol. Soc. Washington 86: 35–40.
96. Six new species of *Eleutherodactylus* (Anura, Leptodactylidae) from Hispaniola. Jour. Herpetol. 7: 249–273.
97. *Sphaerodactylus*. Cat. Amer. Amphib. Rept.: 142.1–142.2.
98. A new species of montane *Anolis* (Sauria, Iguanidae) from Hispaniola. Ann. Carnegie Mus. 44: 183–195.

1974

99. A new species of primitive *Anolis* (Sauria, Iguanidae) from the Sierra de Baoruco, Hispaniola. Breviora 423: 1–19.
100. An analysis of variation in the Hispaniolan giant anole, *Anolis ricordi* Duméril and Bibron. Bull. Mus. Comp. Zool., Harvard Univ. 146: 89–146.
101. *Eleutherodactylus planirostris*. Cat. Amer. Amphib. Rept.: 154.1–154.4.
102. (with O. H. Garrido). A new Cuban species of *Sphaerodactylus* (Gekkonidae) of the *nigropunctatus* complex. Proc. Biol. Soc. Washington 87: 337–343.
103. (with B. R. Sheplan). Hispaniolan boas of the genus *Epicrates* (Serpentes, Boidae) and their Antillean relationships. Ann. Carnegie Mus. 45: 57–143.
104. (with R. Thomas). The status of *Sphaerodactylus gilvitorques* Cope and of *Sphaerodactylus nigropunctatus* Gray (Sauria: Gekkonidae). Jour. Herpetol. 8: 353–358.

1975

105. New subspecies of *Sphaerodactylus copei* Steindachner (Sauria, Gekkonidae) from Hispaniola. Herpetologica 31: 1–18.
106. (with O. H. Garrido). A reconsideration of some Cuban *Tropidophis* (Serpentes, Boidae). Proc. Biol. Soc. Washington 88: 77–90.
107. Variation in the Antillean boid snake *Tropidophis haetianus* Cope. Jour. Herpetol. 9: 303–311.
108. (with C. J. McCoy). A new name for a Cuban *Ameiva* (Reptilia: Teiidae). Herpetologica 31: 240.
109. A new subspecies of *Anolis baleatus* Cope (Sauria: Iguanidae) from the República Dominicana. Florida Sci. 38: 30–35.
110. (with R. Thomas). A check-list of West Indian amphibians and reptiles. Carnegie Mus. Nat. Hist. Spec. Publ. 1: 1–216.
111. (with R. I. Crombie). A new species of the genus *Aristelliger* (Sauria: Gekkonidae) from the Caicos Islands. Proc. Biol. Soc. Washington 88: 305–313.

1976

112. (with R. Franz). A new species of *Sphaerodactylus* (Sauria: Gekkonidae) from Hispaniola. Proc. Biol. Soc. Washington 88: 367–372.
113. Two new species of Hispaniolan *Eleutherodactylus* (Leptodactylidae). Herpetologica 32: 163–171.
114. Variation and relationships of some Hispaniolan frogs (Leptodactylidae, *Eleutherodactylus*) of the *ricordi* group. Bull. Florida State Mus. 21: 1–46.
115. (with S. J. Incháustegui). A new species of *Diploglossus* (Reptilia, Lacertilia, Anguidae) from Hispaniola. Jour. Herpetol. 10: 241–246.
116. A new species of *Sphaerodactylus* (Sauria, Gekkonidae) from the República Dominicana. Florida Sci. 39: 65–70.
117. Variations in the Hispaniolan colubrid snake *Uromacer frenatus* Günther (Reptilia, Serpentes, Colubridae). Jour. Herpetol. 10: 319–327.
118. (with D. A. Rossman). A review of the Hispaniolan colubrid snake genus *Ialtris*. Studies Fauna Curaçao Carib. Isl. 50: 76–102.

1977

119. (with R. Thomas). Two new species of *Sphaerodactylus* (Reptilia, Lacertilia, Gekkonidae) from Hispaniola. Jour. Herpetol. 11: 61–66.
120. *Eleutherodactylus abbotti*. Cat. Amer. Amphib. Rept.: 191.1–191.2.
121. *Eleutherodactylus alcoae*. Cat. Amer. Amphib. Rept.: 192.1–192.2.
122. A new subspecies of *Eleutherodactylus wetmorei* Cochran (Anura: Leptodactylidae) from northern Haiti. Herpetologica 33: 66–72.
123. (with R. Thomas). Three new species of *Sphaerodactylus* (Sauria: Gekkonidae) from Hispaniola. Ann. Carnegie Mus. 46: 33–43.
124. (with M. H. Strahm). Osteoderms in the anguid lizard subfamily Diploglossinae and their taxonomic importance. Biotropica 9: 58–72.
125. (with M. Carey). Systematics and evolution in the West Indian iguanid genus *Cyclura*. Studies Fauna Curaçao Carib. Isl. 53: 15–97.
126. The geckoes (Sauria, Gekkonidae) of the genus *Sphaerodactylus* of the Dominican Península de Barahona, Hispaniola. Proc. Biol. Soc. Washington 90: 243–254.
127. *Eleutherodactylus apostates*. Cat. Amer. Amphib. Rept.: 201.1.
128. A new species of *Sphaerodactylus* (Sauria, Gekkonidae) from Isla Monito, West Indies. Proc. Biol. Soc. Washington 90: 985–992.
129. A new subspecies of *Anolis baleatus* (Sauria: Iguanidae) from Isla Saona, República Dominicana. Florida Sci. 40: 401–405.

1978

130. Some aspects of the herpetogeography of the West Indies. Acad. Nat. Sci. Philadelphia Spec. Publ. 13: 31–51.
131. A new species of aquatic *Anolis* (Sauria, Iguanidae) from Hispaniola. Ann. Carnegie Mus. 47: 261–279.
132. The Hispaniolan *Anolis* (Reptilia, Lacertilia, Iguanidae) of the *hendersoni* complex. Jour. Herpetol. 12: 355–370.
133. *Eleutherodactylus armstrongi*. Cat. Amer. Amphib. Rept.: 208.1–208.2.
134. (with E. D. Graham, Jr.). Status of the name *Sphaerodactylus cinereus* Wagler and variation in "*Sphaerodactylus stejnegeri*" Cochran. Florida Sci. 41: 243–251.
135. (with R. Thomas and L. D. Ober). First supplement to a check-list of West Indian amphibians and reptiles. Carnegie Mus. Nat. Hist. Spec. Publ. 5: 1–35.

1979

136. A new species of cybotoid anole (Sauria, Iguanidae) from Hispaniola. Breviora 451: 1–27.
137. A new species of *Eleutherodactylus* (Amphibia, Anura, Leptodactylidae) from northwestern Haiti, Hispaniola. Jour. Herpetol. 13: 199–202.
138. (with E. D. Graham, Jr. and J. J. Duval). A new species of *Diploglossus* (Sauria: Anguidae) from Hispaniola. Proc. Biol. Soc. Washington 92: 1–9.
139. A new species of *Leiocephalus* (Reptilia: Iguanidae) from Hispaniola. Proc. Biol. Soc. Washington 92: 272–279.
140. *Eleutherodactylus audanti*. Cat. Amer. Amphib. Rept.: 224.1–224.2.
141. The status of *Uromacer frenatus* and *Uromacer dorsalis* (Reptilia: Serpentes: Colubridae). Herpetologica 35: 207–216.
142. The herpetofauna of Ile á Cabrit, Haiti, with the description of two new subspecies. Herpetologica 35: 248–255.
143. The status of Greater Antillean *Phyllodactylus* (Reptilia, Gekkonidae). Jour. Herpetol. 13: 419–426.

1980

144. (with S. J. Incháustegui). The endemic Hispaniolan lizard genus *Chamaelinorops*. Jour. Herpetol. 14: 51–56.
145. (with S. J. Cullom). Variation in the Hispaniolan giant anole *Anolis barahonae* with the description of two new subspecies. Herpetologica 36: 93–98.
146. (with E. D. Graham, Jr.). The *shrevei* group of Hispaniolan *Sphaerodactylus* (Reptilia, Gekkonidae). Tulane Studies Zool. Bot. 22: 1–15.
147. Two new species of *Eleutherodactylus* (Amphibia, Anura, Leptodactylidae) from Hispaniola. Ann. Carnegie Mus. 49: 103–112.
148. *Eleutherodactylus auriculatoides*. Cat. Amer. Amphib. Rept.: 246.1–246.2.
149. *Eleutherodactylus bakeri*. Cat. Amer. Amphib. Rept.: 247.1.
150. The herpetogeography of Hispaniola, West Indies. Studies Fauna Curaçao Carib. Isl. 61: 86–127.
151. Variation in Hispaniolan *Anolis whitemani* Williams. Jour. Herpetol. 14: 399–406.

1981

152. *Eleutherodactylus brevirostris*. Cat. Amer. Amphib. Rept.: 259.1.
153. *Eleutherodactylus counouspeus*. Cat. Amer. Amphib. Rept.: 260.1–260.2.
154. (with O. H. Garrido). A review of the Cuban members of the genus *Arrhyton* (Reptilia, Serpentes, Colubridae). Ann. Carnegie Mus. 50: 207–230.
155. (with O. H. Garrido). Las salamanquitas cubanas del género *Sphaerodactylus* (Sauria, Gekkonidae). l. El grupo *copei*. Poeyana 230: 1–27.
156. Variation in Hispaniolan *Anolis olssoni* Schmidt (Reptilia, Sauria, Iguanidae). Milwaukee Public Mus. Contrib. Biol. Geol. 47: 1–21.

1982

157. *Eleutherodactylus darlingtoni*. Cat. Amer. Amphib. Rept.: 284.1.
158. *Eleutherodactylus eunaster*. Cat. Amer. Amphib. Rept.: 285.1.
159. (with F. Gali). A new subspecies of *Leiocephalus personatus* from the República Dominicana. Jour. Herpetol. 16: 177–179.
160. (with O. H. Garrido). A new species of *Sphaerodactylus* (Reptilia, Sauria, Gekkonidae) from eastern Cuba. Proc. Biol. Soc. Washington 95: 392–397.
161. (with R. W. Henderson). *Anolis cybotes* (Reptilia, Iguanidae): the eastern Hispaniolan populations. Milwaukee Public Mus. Contrib. Biol. Geol. 49: 1–8.
162. *Eleutherodactylus flavescens*. Cat. Amer. Amphib. Rept.: 296.1–296.2.
163. *Eleutherodactylus fowleri*. Cat. Amer. Amphib. Rept.: 297.1.

1983

164. *Eleutherodactylus furcyensis*. Cat. Amer. Amphib. Rept.: 309.1.
165. *Eleutherodactylus glandulifer*. Cat. Amer. Amphib. Rept.: 310.1.
166. The *difficilis* complex of *Sphaerodactylus* (Sauria, Gekkonidae) of Hispaniola. Pt. 1. *Sphaerodactylus difficilis*, *S. clenchi*, and *S. lazelli*. Bull. Carnegie Mus. Nat. Hist. 22: 5–30.
167. (with R. Thomas). The *difficilis* complex of *Sphaerodactylus* (Sauria, Gekkonidae) of Hispaniola. Pt. 2. *Sphaerodactylus savagei*, *S. cochranae*, *S. darlingtoni*,

S. armstrongi, *S. streptophorus*, and conclusions. Bull. Carnegie Mus. Nat. Hist. 22: 31–60.

168. (with R. Thomas). Variation in Hispaniolan *Sphaerodactylus* (Sauria, Gekkonidae), p. 86–98. *In* A. G. J. Rhodin and K. Miyata (eds.), Advances in Herpetology and Evolutionary Biology. Essays in Honor of Ernest E. Williams. Mus. Comp. Zool., Harvard Univ., Cambridge, Massachusetts.
169. *Eleutherodactylus glanduliferoides*. Cat. Amer. Amphib. Rept.: 322.1.
170. *Eleutherodactylus glaphycompus*. Cat. Amer. Amphib. Rept.: 323.1.
171. *Eleutherodactylus haitianus*. Cat. Amer. Amphib. Rept.: 335.1–335.2.
172. *Eleutherodactylus heminota*. Cat. Amer. Amphib. Rept.: 336.1.

1984

173. *Eleutherodactylus hypostenor*. Cat. Amer. Amphib. Rept.: 346.1.
174. (with R. W. Henderson). *Uromacer*. Cat. Amer. Amphib. Rept.: 355.1.
175. (with R. W. Henderson). *Uromacer catesbyi*. Cat. Amer. Amphib. Rept.: 356.1–356.2.
176. (with R. W. Henderson). *Uromacer frenatus*. Cat. Amer. Amphib. Rept.: 357.1–357.2.
177. (with R. W. Henderson). *Uromacer oxyrhynchus*. Cat. Amer. Amphib. Rept.: 358.1–358.2.
178. (with R. W. Henderson). A guide to the identification of the amphibians and reptiles of Hispaniola. Spec. Publ. Milwaukee Public Mus. Biol. Geol. 4: 1–70.
179. (with R. W. Henderson and S. J. Incháustegi). Guía para la identificación de los anfibios y reptiles de la Hispaniola. Mus. Nac. Hist. Nat. Santo Domingo Ser. Monogr. 1: 1–128.

1985

180. (with D. K. Lee). Four new subspecies of *Ameiva auberi* (Sauria, Teiidae) from the Bahama Islands. Ann. Carnegie Mus. 54: 11–21.
181. (with D. R. Martinez and J. A. Lucio). Topografía interna de las culebras del género *Uromacer* (Colubridae). Caribaea 1: 48–59.
182. (with R. W. Henderson). A guide to the identification of the amphibians and reptiles of the West Indies exclusive of Hispaniola. Milwaukee Public Mus., Milwaukee, Wisconsin.
183. (with O. H. Garrido). The Cuban lizards of the genus *Sphaerodactylus* (Sauria, Gekkonidae). Milwaukee Public Mus. Contrib. Biol. Geol. 62: 1–67.
184. (with S. J. Incháustegui and R. W. Henderson). Hispaniolan giant *Diploglossus* (Sauria, Anguidae): description of a new species and notes on the ecology of *D. warreni*. Amphibia-Reptilia 6: 195–201.

1986

185. (with R. W. Henderson). The diet of the Hispaniolan colubrid snake, *Darlingtonia haetiana*. Copeia 1986: 529–531.
186. *Darlingtonia*, *D. haetiana*. Cat. Amer. Amphib. Rept.: 390.1–390.2.

1987

187. (with R. W. Henderson and T. A. Noeske-Hallin). Food habits of three colubrid tree snakes (genus *Uromacer*) on Hispaniola. Herpetologica 43: 241–248.
188. (with R. W. Henderson, T. A. Noeske-Hallin, and J. A. Ottenwalder). On the diet of the boa *Epicrates striatus* on Hispaniola with notes on *E. fordi* and *E. gracilis*. Amphibia-Reptilia 8: 251–258.
189. (with R. W. Henderson, B. I. Crother, T. A. Noeske-Hallin, and C. R. Dethloff). The diet of the Hispaniolan snake *Antillophis parvifrons* (Colubridae). Jour. Herpetol. 21: 330–334.

1988

190. (with R. W. Henderson). West Indian amphibians and reptiles: a check-list. Milwaukee Public Mus. Contrib. Biol. Geol. 74: 1–264.
191. (with R. W. Henderson, T. A. Noeske-Hallin, and B. I. Crother). The diets of Hispaniolan colubrid snakes. II. Prey species, prey size, and phylogeny. Herpetologica 44: 55–70.
192. (with F. Gali and A. Suarez). A new subspecies of *Leiocephalus personatus* (Sauria: Iguanidae) from Haiti. Proc. Biol. Soc. Washington 101: 1–3.

1989

193. (with J. F. Jacobs). A new subspecies of *Celestus costatus* (Sauria: Anguidae) from the República Dominicana. Jour. Herpetol. 23: 193–195.
194. A review of the cybotoid anoles (Reptilia: Sauria: Iguanidae) from Hispaniola. Milwaukee Public Mus. Contrib. Biol. Geol. 78: 1–32.

1991

195. (with R. W. Henderson). Amphibians and Reptiles of the West Indies: Descriptions, Distributions, and Natural History. Univ. Florida Press, Gainesville.

NON–WEST INDIAN CONTRIBUTIONS

1949

1. (with W. A. Babis). Extension of range of *Crotalus lepidus klauberi*. Copeia 1949: 74.

1950

2. Extension of range of the water snake *Natrix taxispilota*. Copeia 1950: 60.

1952

3. *Hyla septentrionalis* Dumeril and Bibron on the Florida mainland. Copeia 1952: 117–118.
4. (with W. E. Duellman). The taxonomic status of the newts, *Diemictylus viridescens*, of peninsular Florida. Bull. Chicago Acad. Sci. 9: 219–227.
5. A new race of *Pseudobranchus striatus* from southern Florida. Nat. Hist. Misc. 115: 1–9.

1953

6. A new subspecies of crowned snake (*Tantilla coronata*) from the southern Appalachian Mountains. Herpetologica 9: 153–157.

1954

7. A record of the Atlantic leatherback turtle (*Dermochelys c. coriacea*) in South Carolina. Herpetologica 10: 7.
8. (with R. Etheridge). New and additional herpetological records from the North Carolina Coastal Plain. Herpetologica 10: 167–171.
9. The salamander *Aenides aeneus* in South Carolina. Copeia 1954: 296–298.

1955

10. A clutch of eggs of *Aneides hardyi* (Taylor). Herpetologica 11: 70.
11. The chorus frog *Pseudacris brachyphona* in North Carolina. Copeia 1955: 138.
12. (with T. M. Uzzell, Jr.). The status of the turtle *Kinosternon bauri palmarum* Stejneger with notes on variation in the species. Jour. Elisha Mitchell Soc. 71: 28–35.
13. The diamondback terrapins (*Malaclemys terrapin*) of peninsular Florida. Proc. Biol. Soc. Washington 68: 157–164.

1956

14. The relationships and nomenclature of the soft-shelled turtles (genus *Trionyx*) of the southeastern United States. Charleston Mus. Leaflet 26: 1–21.
15. Geographic variation in the chicken turtle *Deirochelys reticularia* Latreille. Fieldiana, Zool. 34: 461–503.
16. (with J. R. Harrison, III). A new subspecies of gopher frog (*Rana capito* LeConte). Proc. Biol. Soc. Washington 69: 135–144.

1957

17. "Albinism" in the salamander *Amphiuma means*. Herpetologica 13: 75–76.
18. Chorus frogs (*Pseudacris nigrita* LeConte) in South Carolina. Amer. Mus. Novit. 1838: 1–12.
19. Variation and natural history of *Plethodon jordani clemsonae* Brimley. Copeia 1957: 94–107.

1958

20. (with W. E. Duellman). Amphibians and reptiles of southern Florida. Bull. Florida State Mus. 3: 181–324.

1971

21. (with G. R. Zug). *Deirochelys reticularia*. Cat. Amer. Amphib. Rept.: 107.1–107.3.

1975

22. (with D. F. Austin). Another exotic amphibian in Florida, *Eleutherodactylus coqui*. Copeia 1975: 188.

Bourgeois, R. W. 1995. Herpetological publications of Albert Schwartz, p. 21–27. *In* R. Powell and R. W. Henderson (eds.), Contributions to West Indian Herpetology: A Tribute to Albert Schwartz. Society for the Study of Amphibians and Reptiles, Ithaca (New York). Contributions to Herpetology, volume 12.

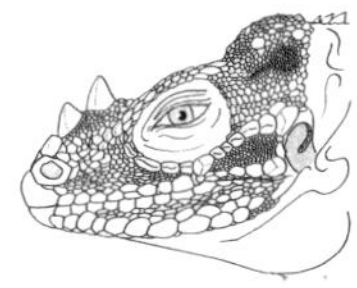

A Brief History of West Indian Herpetology

Robert Powell[1] and Robert W. Henderson[2]

Introduction

As a prelude to this volume, we wish to place the contributions contained in the following chapters into a perspective that can be gained only by considering the history of West Indian herpetology. We have not tried to be exhaustive in listing every contributor. Instead, we have attempted to provide an overview of significant persons and events during different historical periods. Inevitably, our selections are highly subjective and other students of West Indian herpetology might have chosen different persons and events to illustrate the many and varied contributions that have characterized efforts to understand this wonderfully diverse herpetofauna.

Pre-Historic Events

Reptiles played important roles in the daily lives of peoples in the pre-historic West Indies. Humans have inhabited the Greater Antilles since perhaps as early as 7000 BP, and the Lesser Antilles since about 5000 BP (Rouse 1989). The most thorough review of pre-Columbian human-reptile interactions is that by Pregill et al. (1994) for the Lesser Antilles, and Watters (1989) posed provocative pre-historic versus historic issues using reptiles as examples. Based on the frequency of their occurrence at excavations of cultural sites, marine turtles (Cheloniidae) and iguanas (*Iguana* spp.) undoubtedly were widespread food sources. In the Greater Antilles, evidence suggests that the Guanahatebey and Taino ethnic groups similarly exploited marine turtles and iguanas (presumably *Cyclura* spp.) as food (García Valdes 1948; Rouse 1948; Wing 1989). An excellent discussion of Archaic and Ceramic age Amerindian-caused extinctions and extirpations (and the difficulties involved in demonstrating those events) in the Lesser Antilles can be found in Pregill et al. (1994). Evidence from Antigua (Pregill et al. 1988) provided strong support for the pre-historic extinction and extirpation of vertebrates, and also for the introduction of species from South America to the Lesser Antilles. Censky (1988) discussed the distribution of the tortoise, *Geochelone carbonaria*, in the West Indies and suggested that this species may have been transported to the islands by Amerindians as a source of food, but Watters (1989) proposed that early European explorers introduced tortoises to the Lesser Antilles.

Columbus and the Early Colonial Period

Columbus' arrival in the islands initiated a European-dominated era of substantial changes. His letters and diaries provided descriptions of, in many instances, a West Indies that no longer exists. In 1493, on Hispaniola, he wrote (*in* Major 1870) of "trees of a thousand kinds and tall, and they seem to touch the sky," describing forests that were long ago sacrificed for ship timbers. Columbus and his crew most certainly encountered members of the West Indian herpetofauna, and his writings (*in* Major 1870; Morison 1963) contain frequent references to turtles and iguanas, surely in reference to marine turtles and *Cyclura* spp. (and possibly *Iguana* spp.).

The veracity of the anecdotal natural history observations by early European chroniclers must be viewed with a critical eye. For example, Gonzalo

[1] Department of Natural Sciences, Avila College, Kansas City, Missouri 64145, USA.

[2] Section of Vertebrate Zoology, Milwaukee Public Museum, Milwaukee, Wisconsin 53233, USA.

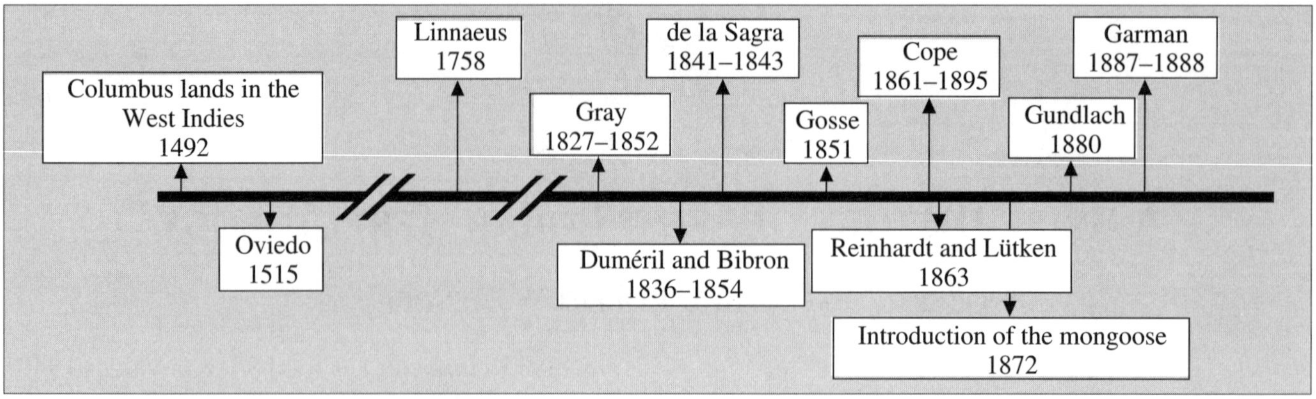

FIGURE 1. Timeline of selected events in West Indian herpetology from the arrival of Columbus through the end of the 19th century. Ranges of years indicate periods during which that person published on West Indian amphibians and reptiles.

de Oviedo, who lived in the Caribbean region from 1512–1557 and, although generally credited with attempting a more systematic approach than his contemporaries, described a snake encountered "at the foote of the Mountains called Pedernales" on Hispaniola, that he measured and found "to be more than twenty feet long, and somewhat more than a mans fist in bigness: and although she had three or foure deadly wounds with a Sword, yet dyed she not, nor stunke the same day, in so much as her blood continued warm all that time" (*in* Stoudemire 1959). Whereas the longest snake species on Hispaniola is *Epicrates striatus*, with a maximum total length of about 3 m, Oviedo seems likely to have exaggerated somewhat. In the Lesser Antilles, Captain René Laudonniere (*in* Hakluyt 1904) recounted an episode that occurred on Dominica in about 1564: ". . . it happened on a day that certaine of our men desirous to see some new things in these strange countries, walked though the woods; and following still the little rivers side, they spied two serpents [almost certainly *Boa constrictor*] of exceeding bigness, which went side by side overthwart the way. My souldiers went before them thinking to let them from going into the woods; but the serpents nothing at all astonied at these gestures glanced into the bushes with fearful hyssings; yet for all that, my men drew their swords and killed them, and found them afterward 9 great foote long, and as big as a mans leg."

Much later, Mark Catesby (1743) published a natural history which included the Bahamas. His descriptions and illustrations were remarkably accurate for the period, and the style of his work anticipated that of naturalists who were to make notable contributions a century later.

TABLE 1. Authorities who have described at least 10 currently-recognized species of endemic West Indian amphibians and reptiles. Columns are as follows: Name, with the range of years in which publications have appeared in parentheses; Anura (A); Testudines (T); Lacertilia and Amphisbaenia (L); Serpentes (S); and total number of species described, with the percentage of the total endemic herpetofauna (572 species as of August 1995: 158 amphibians and 414 reptiles) listed in parentheses.

Name (publication years)	A	T	L	S	Total	(%)
A. Schwartz (1957–1985)	44	0	37	5	86	(15.0)
E. D. Cope (1861–1895)	10	0	36	13	59	(10.3)
R. Thomas (1965–)	10	0	22	14	46	(8.0)
T. Barbour (1910–1942)	9	1	22	3	35	(6.1)
D. M. Cochran (1923–1942)	11	0	21	2	34	(5.9)
G. Bibron (1820–1854)	3	0	20	8	31	(5.4)
A. M. C. Duméril (1836–1854)	3	0	20	5	28	(4.9)
O. H. Garrido (1972–)	0	0	24	3	27	(4.7)
S. B. Hedges (1987–)	10	0	13	4	27	(4.7)
B. Shreve (1936–1968)	11	0	11	1	23	(4.0)
C. Grant (1931–1959)	3	0	16	0	19	(3.3)
S. Garman (1887–1888)	0	0	16	2	18	(3.1)
G. K. Noble (1915–1935)	8	0	10	0	18	(3.1)
E. R. Dunn (1920–1928)	15	0	1	0	16	(2.8)
J. E. Gray (1827–1852)	0	1	15	0	16	(2.8)
E. E. Williams (1959–)	3	0	13	0	16	(2.8)
L. Stejneger (1899–1916)	3	0	8	3	14	(2.4)
W. Peters (1862–1878)	2	0	6	3	11	(1.9)
K. P. Schmidt (1919–1928)	8	1	0	1	10	(1.7)

FIGURE 2. Philip H. Gosse, a 19th century naturalist in Jamaica (portrait courtesy of L. R. Croft).

CAROLUS LINNAEUS

At least some specimens collected in the Neotropics found their way into the hands of Linnaeus, who published the first formal names of amphibians and reptiles found in the West Indies. Although he described only one West Indian endemic (*Typhlops lumbricalis*), Antillean species named by Linnaeus (1758), but not necessarily from specimens collected in the islands, are *Bufo marinus*, *Caiman crocodilus*, *Ameiva ameiva*, *Cnemidophorus lemniscatus*, *Iguana iguana*, *Hemidactylus turcicus*, *Boa constrictor*, and *Corallus hortulanus*.

19TH CENTURY NATURALISTS

Through the 19th century (Figure 1), natural history explorations in the Neotropics produced a number of detailed accounts in the "A Naturalist in . . ." genre. Certainly a prominent example for the West Indies is Gosse's (1851) "A Naturalist's Sojourn in Jamaica," in which the most detailed and appealing accounts dealt with Jamaica's rich lizard fauna. Other observations provided an important historical context; for example, his pre-mongoose observations on the abundance of the probably-now-extinct racer *Alsophis ater*. Gosse (Figure 2) described the "Gray Snake," *Natrix capistrata* (= *A. ater*), and gave the maximum snout-vent length as 16 inches (= 40.6 cm); described a diet including *Anolis* and gekkonids; and observed that "Though not rare, it can scarcely be called abundant . . . I have found it scattered in distant and varying localities, in the lowlands and on the mountain-tops."

Other substantial contributions in the naturalist genre include those of Ober (e.g., 1880; "The islands, especially the shores, are teeming with lizards of every color, of every variety of marking, and of all sizes.") and, to a lesser degree, Bell (1889). Toward the end of the 19th century, Juan Gundlach (Figure 3) published his "Erpetologia Cubana" in 1880; this work included interesting and useful anecdotes regarding distribution, abundance, and behavior of some members of the Cuban herpetofauna.

The massive "Histoire Physique, Politique et Naturelle de l'Ile de Cuba" of Ramón de la Sagra bridged the gap between this genre and purely sys-

FIGURE 3. Juan Gundlach (1864), an eminent 19th century Cuban naturalist (Academy of Natural Sciences of Philadelphia, courtesy of Marsha Gross and Kraig Adler).

tematic texts. Addressing the physical and political as well as natural history, this monumental work was published in several volumes plus an atlas and in two editions, one in Spanish and the other in French. The herpetological sections dealing with turtles, crocodiles, and lizards were written by J.-T. Cocteau. Gabriel Bibron, taking over the task after Cocteau's death, wrote the sections on snakes and amphibians. Although some uncertainty exists regarding the actual dates of publication (Smith and Grant 1946; Vanzolini 1977; Adler 1989), Adler (1989) noted that "this book contains some of the most exquisite handcolored plates of reptiles ever published."

19th Century Taxonomists

Also during this period, specimens were collected for a number of the world's foremost museums where curators, many of whom never visited the West Indies, were largely responsible for describing and naming many of the species of the region (Table 1). In the introduction to their 1991 compendium, Schwartz and Henderson compared the number of species named in each of six West Indian geographic areas (each of the four Greater Antilles: Cuba, Hispaniola, Jamaica, and Puerto Rico; the Bahamas; and the Lesser Antilles) by 20-year intervals beginning in 1748. Particularly active periods occurred from 1848–1867 on Cuba, Hispaniola, and Puerto Rico and from 1868–1887 in the Lesser Antilles. Prominent among those responsible were Samuel Garman and Edward Drinker Cope. Garman (Figure 4) produced a series of papers on West Indian reptiles and included field observations as well as taxonomic descriptions in his accounts. Garman also initiated the long tradition of West Indian fieldwork by herpetologists at Harvard University's Museum of Comparative Zoology, a tradition of over a century that continues to the present day. Cope, in a series of papers extending over nearly four decades (1861–1895), amazingly accounted for over 10% of the endemic amphibian and reptilian species. John Edward Gray (British Museum) named 15 currently-recognized lizard species from 1827–1852; and Constant Duméril and Gabriel Bibron (Muséum National d'Histoire Naturelle, Paris), primarily in their "Erpétologie Générale," from 1836–1854, are attributed individually or together with naming 5.4% of the Antillean herpetofauna. Other notable contributors constitute a veritable "who's who" in herpetology during this period. These individuals represented myriad European and American institutions and included B.-G.-É. Lacepède, George Shaw, F.-M. Daudin, Georges Cuvier, Blasius Merrem, J. J. von Tschudi, Hermann Schlegel, J. T. Cocteau, Edward Hallowell, Albert Günther, Wilhelm Peters, J. T. Reinhardt, Christian F. Lütken, Johann G. Fischer, M.-F. Bocourt, Hermann Meerwarth, and George A. Boulenger.

Introduction of the Mongoose

In 1872 the small Indian mongoose (*Herpestes auropunctatus*) was introduced into the Antilles on Jamaica (Espeut 1882). This efficient predator is now known from 29 islands in the West Indies (Hoagland et al. 1989) and, based on largely cir-

FIGURE 4. Samuel Garman, herpetologist at the Museum of Comparative Zoology, Harvard University (Museum of Comparative Zoology, Harvard University, courtesy of Roxane Coombs and Kraig Adler).

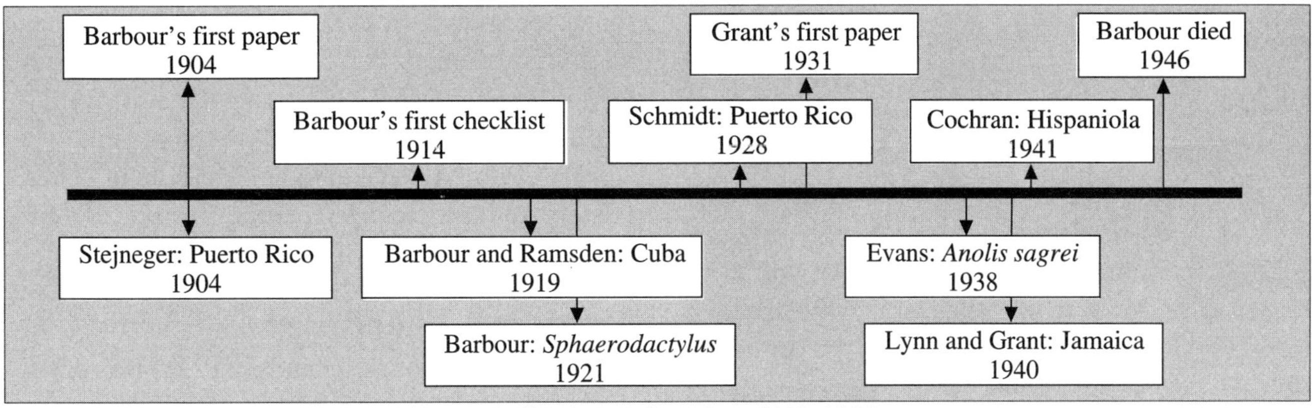

FIGURE 5. Timeline of selected events in West Indian herpetology from 1900–1946.

cumstantial evidence, is believed responsible for the extinction, extirpation, or severe decline in numbers of several species of lizards and snakes (Baskin and Williams 1966; Sajdak and Henderson 1991; Henderson 1992). For example, after 1930, specimens of the Jamaican racer *Alsophis ater* were no longer added to museum collections (Henderson 1992); by 1940, Grant reported that he had not met "anyone who had seen a live specimen in years." As of this writing (December 1995), despite intensive fieldwork, no specimen of *A. ater* has been catalogued into a collection in over 60 years.

THE PERIOD 1900 THROUGH WORLD WAR II

In Schwartz and Henderson's 1991 correlation of species named within 20-year periods, the span from 1918–1937 was notable, with only the Lesser Antillean region showing a decline. These early 20th century efforts (Figure 5) were dominated by the contributions of Thomas Barbour (Figure 6), who is attributed individually or with collaborators with naming over 6% of the frogs, lizards, and snakes. Barbour's first West Indian paper appeared in 1904. He later compiled a series of checklists (1914, 1930, 1935, 1937) and published several important synoptic works, of which "The Herpetology of Cuba" with Charles T. Ramsden in 1919 and his 1921 monograph on *Sphaerodactylus* were particularly notable. Emmett R. Dunn, in his tribute to Barbour (1946), noted ". . . one odd love of TB's, — the fascination that tiny lizards and frogs had on this man of more than average stature. The labor and expense devoted to *Sphaerodactylus* shows where his heart lay, and the last letter I received from him was about *Sminthillus* [=*Eleutherodactylus limbatus*]." Unfortunately, for observations on natural history, he often relied on secondhand accounts or made offhand judgments on the abundance of many species; he subsequently has been chastised by several authors (e.g., Curtis 1947; Lazell 1972). Nevertheless, Barbour's impact cannot be underestimated. In addition to his prolific publications dealing with amphibians and reptiles not only of the West

FIGURE 6. Thomas Barbour (1938), herpetologist at Harvard University, whose work in the West Indies dominated the first half of the 20th century (photograph by Raymond B. Cowles, courtesy of Kraig Adler).

FIGURE 7. Doris M. Cochran (right), curator at the National Museum of Natural History, Washington, D.C., and author of the book, "Herpetology of Hispaniola" (1941); here in the field in Brazil with Adolfo and Bertha Lutz in 1935 (from the Howard K. Gloyd collection, courtesy of Kraig Adler).

Indies, but of North America and beyond, he provided early and essential support of the journal *Copeia*, and guidance and often financial support to many notable herpetologists. He also continued and fostered the long and ongoing relationship between the Museum of Comparative Zoology at Harvard University and West Indian herpetology.

Prominent among those influenced by Barbour were Archie F. Carr, Emmett R. Dunn, and G. Kingsley Noble. Carr co-authored with Barbour an important monograph entitled "Antillean Terrapins" in 1940. Dunn, whose travels in the West Indies were funded by Barbour, described nearly 10% of the region's frogs. G. Kingsley Noble alone, with W. G. Hassler, and with G. C. Klingel named a total of eight anuran and ten saurian species.

Leonhard Stejneger (National Museum of Natural History, Washington) named 14 species of amphibians and reptiles from the region and, in 1904, authored "Herpetology of Porto Rico." However, the prominence of the National Museum during this time was largely due to the efforts of Doris M. Cochran (Figure 7), who not only described 11 currently recognized frogs, 18 lizards, two snakes, and an amphisbaenian, but with her book, "Herpetology of Hispaniola" (1941), wrote the most comprehensive herpetofaunal account in the region.

Karl P. Schmidt (American Museum of Natural History and the Field Museum of Natural History) is credited with naming ten West Indian amphibians and reptiles, and his "Amphibians and Land Reptiles of Porto Rico, with a List of Those Reported from the Virgin Islands" (1928) was not superseded for 50 years. W. G. Lynn named five frogs, and in 1940 wrote a guide to the amphibians of Jamaica. Benjamin Shreve described 11 anurans, 11 lizards, and a snake alone or in collaboration with Barbour or Ernest E. Williams.

Notable exceptions to the focus on taxonomy and anecdotal natural history during this period were the ecological publications of G. N. Wolcott (1923) on diets of Puerto Rican lizards, G. K. Noble and H. T. Bradley's (1933) treatise on lizard mating behavior, and Evans' (1938) field study of territoriality in *Anolis sagrei* on Cuba. Robert Mertens (Natur-Museum Senckenberg) chronicled his Hispaniolan experiences in a series of publications, several of which were reminiscent of the naturalist genre so prevalent in the previous century.

Chapman Grant (Figure 8), arguably second only to Barbour in characterizing this period in West Indian herpetology, had until recently (Smith 1986) not been accorded the respect commensurate with his contributions. This may have been due to the lack of any direct association with an academic institution, but it was odd, especially in light of his roles in founding in 1936 the journal *Herpetologica* and later, in 1946, the Herpetologists' League. Grant produced 82 publications on West Indian herpetology from 1931–1960 and covered topics as diverse as taxonomy and behavior. Hobart M. Smith, in his 1986 tribute to Grant, suggested that this proliferation of work on herpetological topics may well have been due to Grant's "irrepressible iconoclastic drive" for "shooting-down" the tempting targets provided by "then-recent monographs (such as Schmidt's 1928 Herpetology of Puerto Rico; Barbour and Ramsden's 1919 Herpetology of Cuba; and Barbour's 1930 list of the Antillean herpetofauna)."

FIGURE 8. Chapman Grant, 20th-century specialist on West Indian herpetology and founder of the Herpetologists' League, shown here in 1947 before a portrait of his grandfather, General Ulysses S. Grant. The rifle had been confiscated by Dominican authorities in Santo Domingo on a trip during which Grant had planned to collect reptiles for the national museum in Santo Domingo (with permission of the San Diego Historical Society, Union-Tribune Collection).

Notable among Grant's publications were guides to the reptiles of Jamaica (1940) and to the herpetology of the Cayman Islands (1941). Unlike most of his predecessors and many peers, Grant, a noted outdoorsman, spent significant time in the field in the West Indies, first in the context of his duties as a military officer and later at his own expense. As a result, his work was based largely on personal observations and his species descriptions benefited tremendously from his having observed and collected many of the specimens himself. Although in this respect somewhat of an anomaly, like Garman before him, Grant was to establish a pattern for the latter half of this century.

THE LAST HALF-CENTURY

Again using Schwartz and Henderson's 1991 compendium of species named during 20-year periods as a point of reference, the greatest activity on Hispaniola and Cuba occurred from 1958–1979, and lesser but still striking bursts of activity are noted for Puerto Rico and the Lesser Antilles. Because of the tremendous quantity of work and the growing number of persons involved, we made the decision to limit inclusion in this portion of the history to those persons who have published a minimum of five papers dealing with West Indian amphibians and/or reptiles. We have not attempted to be exhaustive and include everyone merely on this basis; nor have we been completely consistent, including several persons whose work, though small in quantity, we agreed had sufficient impact that their omission would have been a significant oversight. Also, for the sake of clarity and convenience, we generally have discussed the contributions of each person when they first appear in the chronology, although many have worked in the West Indies for extended periods.

The number of new species being described from the West Indies (especially from Cuba and Hispaniola) continues to increase (Hedges, this volume: Figure 1), but the emphasis of research in West Indian herpetology has shifted dramatically from taxonomic to non-taxonomic subjects, most prominently ecology and evolution. Using the bibliography and literature cited in Schwartz and Henderson (1991) for the period 1900–1990, and a survey of a small group of journals (*Caribbean Journal of Science*, *Copeia*, *Herpetologica*, *Herpetological Monographs*, *Journal of Herpetology*) for the period since 1990, we graphed (Figure 9) the relative percentages of taxonomic versus non-taxonomic papers for each decade of this century (excluding the first, as all three publications were taxonomic). Through the 1960s, taxonomic papers dominated; the 1970s was a decade of transition, with taxonomic and non-taxonomic contributions about equal in number; by the 1980s the emphasis had essentially reversed itself from the earlier period. Correspondingly, the number of published contributions has increased dramatically (in our sample) from 11 in the 1910s, 45 in the 1950s, 150 in the 1960s, and 171 in the 1970s to 264 in the 1980s. The only decade in which the number dropped from the previous decade was the 1940s, and that probably can be attributed to World War II.

Why did such a dramatic shift in research emphasis occur? We believe two reasons apply, one general to biology, the other specific to the West Indies. Ecology had come into its own in the 1960s, and conducting studies with an ecological emphasis in the West Indies is eminently logical as the region is fascinating and geographically convenient to the United States (the source of most publications). The more important reason for the new emphasis, however, were the contributions of Albert Schwartz and Ernest E. Williams. Their prodigious output of publications describing one new species after another indicated to many that the West Indies was an arena rich in potential for ecological studies. In the 1960s, students under Williams' influence traveled to the region to study what Gosse (1851) had described a century earlier as the "exceedingly interesting" anoles. Currently, ecological and behavioral studies of species of *Eleutherodactylus* are proliferating, but work on the third most speciose West Indian genus, *Sphaerodactylus*, waits for some clever ecologist with a desire to investigate these diminutive lizards.

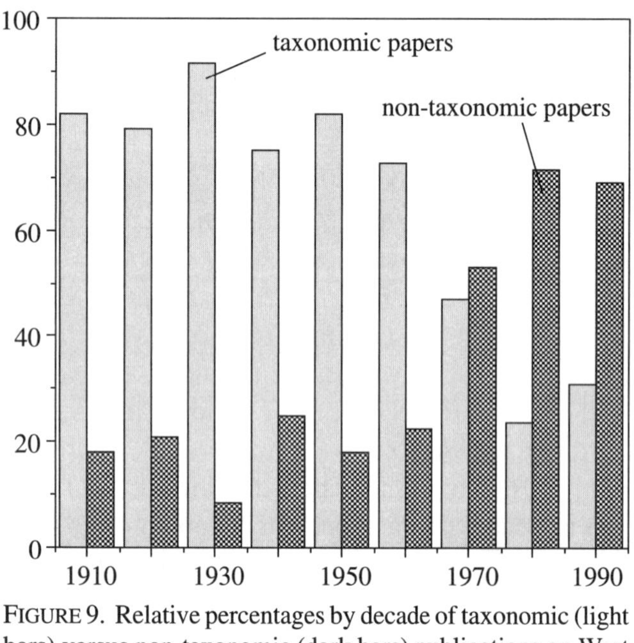

FIGURE 9. Relative percentages by decade of taxonomic (light bars) versus non-taxonomic (dark bars) publications on West Indian herpetology since 1900.

Schwartz and Williams have dominated recent herpetological work in the West Indies to an even greater degree than did Barbour and Grant in the first half of this century. They have left a legacy of publications and of students and colleagues whose work they influenced, supported, and fostered. Directly or indirectly their touch can be felt in the contributions of nearly everyone working in the region today.

Albert Schwartz

In the early 1960s Albert Schwartz (Figure 10) submitted a proposal for funding to the National Science Foundation (USA) proposing "to do a herpetological survey of the entire West Indies by making collections, describing new taxa, and addressing taxonomic problems in virtually all of the genera of amphibians and reptiles" (Duellman et al. 1993). The proposal was rejected. One reviewer claimed that it was "preposterous" for anyone to think that they could do what Schwartz had proposed. He spent the next 25-plus years proving them wrong.

Schwartz's focus was broad (understanding the West Indian herpetofauna) but his approach was to deal almost exclusively with the "nitty-gritty" and to let the broader theoretical foundation take care of itself. As a result, one cannot isolate individual efforts that define his style or his work; instead, one must accept that the body of his publications as a whole transcends any single piece.

Alone and with students and colleagues, Schwartz described 86 species, accounting for nearly 15% of the currently recognized West Indian herpetofauna, plus many subspecies. These accurate and detailed descriptions provided insights into the tremendous diversity and variation to be found in the West Indies and forced a reevaluation of the widely-accepted prior assumption that the endemic Antillean fauna, like that of many islands, was depauperate. Nowhere was this more evident than in the checklists and synoptic works he produced in 1975 (with Richard Thomas), 1988 (with Robert W. Henderson), and in 1991 (again with Henderson). The herpetofauna portrayed therein was far more

FIGURE 10. Albert Schwartz (left) and Richard Thomas examining a live *Hypsirhynchus ferox*. The photograph was taken by S. Blair Hedges in 1983 at the Hotel Cervantes in Santo Domingo, Dominican Republic.

extensive than the much shorter lists published by Barbour earlier in the century.

In the 1950s, Schwartz's work focused on the Cuban herpetofauna, an effort the Cuban revolution of 1960 only impaired, but did not stop. With Orlando Garrido he continued to publish on Cuban vertebrates as his field work shifted progressively to the Bahamas, Jamaica, and the Lesser Antilles, eventually focusing to a substantial degree on Hispaniola. The most physiographically and zoologically diverse of the Antilles, Schwartz once described Hispaniola as "an island of islands," in reference to the many types of diverse habitats dramatically separated from each other by areas of diametrically opposite habitats. In response to those who doubted that a single, relatively small island could harbor such a diverse herpetofauna, Schwartz responded that they must never have visited this remarkable place.

The legacy of Albert Schwartz is that of a prolific taxonomist who dealt not only with amphibians and reptiles, but with butterflies (a particular love later in his life), birds, and mammals — especially bats, but he was also an excellent naturalist who believed strongly that taxonomy was more than the categorization of museum specimens, that understanding relationships required seeing animals in their natural habitats, understanding their behavior, and appreciating colors and patterns that cannot be revealed in jars. This belief was the foundation for his love of the field, testament to which is given by the amount of time he spent there and by the many anecdotes told by all who accompanied him even once over the years.

Schwartz's life and work has been chronicled recently by Duellman et al. (1993), Henderson (1993*a*, 1993*b*), Borkin (1994), and elsewhere in this volume; but mainly by the fact that hardly anyone who has worked with West Indian amphibians and reptiles in the last 35 years has failed to be influenced by his work. Although never affiliated with a graduate institution, Schwartz touched many persons. Some were students who became herpetologists or professionals in other zoological disciplines. The lives of many others merely were enriched by their contact with him. Many were colleagues whose insights were enhanced or whose interests in the West Indies were ignited by Schwartz's enthusiasm and support. Regardless, all whose lives he touched benefited from his unparalleled insights into the details he alone saw so clearly.

Ernest E. Williams

Ernest Williams' (Figure 11) early work dealt primarily with chelonian anatomy and systematics, and it was some of this work on fossils from Cuba that initiated his work in the West Indies. A 1950 trip to Jamaica expanded his interest to include lizards, and he collaborated in several early works with Garth Underwood and Rodolfo Ruibal. Upon assuming the position of curator at the Museum of Comparative Zoology (MCZ) at Harvard University in 1957, Williams focused much of his attention on West Indian anoles, the MCZ collection of which was largely attributable to Barbour. The breadth of this attention is what sets Williams apart from so many others working in the Antilles. He has chosen to go beyond taxonomy to examine the ecology, behavior, and the evolutionary and biogeographic history of anoles and other groups. In addition, he promoted the application of new ideas and methods. Cytoge-

FIGURE 11. Ernest E. Williams, biologist at the Museum of Comparative Zoology, Harvard University, in 1962 (photograph by M. Graham Netting, courtesy of Kraig Adler).

netic and biochemical tools were first extensively applied to West Indian problems by Williams and his associates, and the roles of behavior and thermoregulation were first explored in the context of evolution and biogeography. And although never limited to anoles or to the West Indies, the tremendous volume of work on these animals in this region has come to characterize Williams' work through the present day.

Although difficult, if not impossible, to summarize Williams' impact by focusing on one or a few ideas or publications, one series of efforts addressing fundamental theoretical issues stands out. His classic 1969 paper on the ecology of colonization was instrumental to subsequent studies on coadaptation, exclusion, fragmentation, intra-island radiations, and species boundaries, interactions, and packing. This abundance of analytical work cannot be separated from the detailed studies of resource partitioning initiated by Oliver (1948) and expanded immensely by B. B. Collette, A. Stanley Rand, Thomas W. Schoener, and Timothy C. Moermond, thermoregulatory studies by Rodolfo Ruibal and later by Paul E. Hertz and Raymond B. Huey, and especially the "ecomorph" concept of Rand. This inseparability is a further tribute to Williams, who with his "*Anolis* Newsletters" (1974, 1977), which ema-

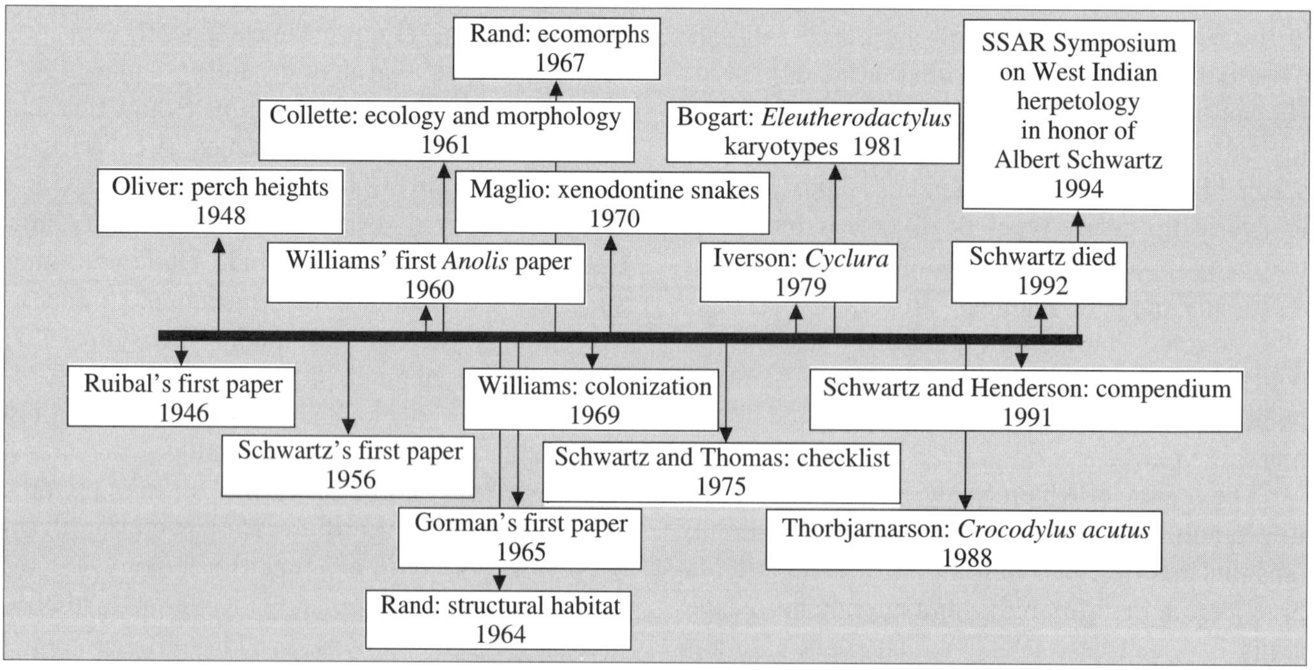

FIGURE 12. Timeline of selected events in West Indian herpetology from 1946 through the present.

FIGURE 13. Rodolfo Ruibal (1946), Cuban-born American herpetologist, shown here in the field (photograph by Charles M. Bogert, courtesy of Kraig Adler).

nated from a report to the National Science Foundation (USA), encouraged all persons working with these profoundly interesting animals to share their insights and to benefit mutually from the opportunity to engage in a forum that long pre-dated the modern exchanges increasingly prevalent in this information age.

Post–World War II and the 1950s

The modern era (Figure 12) began in 1946 with the first publication by Rodolfo Ruibal (Figure 13), a Cuban and the first native West Indian to publish extensively on the herpetofauna. Ruibal described three Antillean species, but made his most important contributions as an ecologist. Another early and important contribution was James A. Oliver's (1948) paper on Bahamian anoles, in which he introduced the ecological dimension of perch heights. This period also saw the dawning of the careers of other herpetologists whose influence in the West Indies extends beyond their publications. Most notable among these was, of course, Albert Schwartz, whose first West Indian paper (with L. H. Ogren), on a collection of amphibians and reptiles from Cuba, appeared in 1956. All of his 14 other publications through 1959 dealt with the Cuban herpetofauna as well. Ernest Williams' first West Indian paper was published in 1950.

Garth Underwood, who described three West Indian lizards (one with Williams), published his first West Indian paper in 1952, but is best known for his synoptic papers on anoles (1959) and reptiles (1962) of the Lesser Antilles and, with Williams (1959), his monograph on the anoles of Jamaica. What makes these papers so memorable are the frequent insights into the ecology and behavior of the animals that only an astute observer working in the field could have noted. P. J. Darlington in 1938 had examined the origin of the Greater Antillean fauna and this evidence weighed heavily in his classical 1957 treatise on zoogeography. Juan A. Rivero of Puerto Rico, best known as a taxonomist specializing on the genus *Eleutherodactylus* from throughout the Caribbean region (but primarily in Venezuela), published his first West Indian paper in 1959. His 1978 guide, "Los Anfibios y Reptiles de Puerto Rico," is replete with natural history.

The 1960s

The decade began auspiciously with Williams' first *Anolis* paper and with Richard Thomas' first West Indian publication (with Schwartz), both in 1960. In the course of these 10 years, the focus of

FIGURE 14. Orlando H. Garrido, Cuban zoologist. Photograph by S. Blair Hedges at Pennsylvania State University.

Schwartz's work shifted from Cuba to elsewhere in the West Indies, eventually to center on Hispaniola. Several other workers, destined to devote much of their attention to the Antillean herpetofauna, first published during this decade; and several important concepts and techniques were first applied to West Indian amphibians and reptiles.

Other than Schwartz, Richard Thomas (Figures 10 and 15) is the most prolific taxonomist of the post-war years. Alone or with co-authors (often with Schwartz and, more recently, with S. Blair Hedges), Thomas has described 10 frogs, 21 lizards, an amphisbaenian, and 14 snakes. Yet, even in his purely descriptive papers, Thomas routinely includes references to habitats and behaviors, to the extent that the reader comes to expect them along with descriptions of new taxa. Most recently (with Rafael Joglar in 1995) he has written a history of Puerto Rican herpetology.

Cuban zoologist Orlando H. Garrido (Figure 14) also has had a long history of association with Albert Schwartz; their first collaborative paper was published in 1967. Garrido has described 27 species (six with Schwartz and nine with Hedges), all from Cuba, but his inclusion of natural history data, especially in his many faunal lists of various regions of and islands off the coast of Cuba, brings his work to life. His checklist with M. L. Jaume in 1984 of the Cuban herpetofauna was a particularly important contribution. Garrido is largely responsible for seeing our knowledge of the Cuban herpetofauna grow substantially, despite the difficulties imposed by the political realities emanating from the Cuban revolution in 1960.

George C. Gorman's (Figure 16) first West Indian paper appeared in 1965; he is best known for his studies of saurian relationships (e.g., 1973) and for the use of cytological techniques. Yet he con-

FIGURE 15. Members of the Turquino Expedition at Santiago, Cuba in August 1989. Fifteen new species of amphibians and reptiles were collected during this trip and a similar effort in 1990. From left to right: Riberto Arencibia, Antonio Perez-Asso, Alberto Estrada, Emilio Alfaro, Octavio Pérez-Beato, Richard Thomas, and S. Blair Hedges (photograph by Richard Thomas).

FIGURE 16. George C. Gorman (left), who pioneered many cytological and molecular techniques while working on West Indian amphibians and reptiles, photographed with his son Ari at the University of Puerto Rico in 1972. A. Stanley Rand (right), who with Ernest E. Williams applied the ecomorph concept to West Indian lizards of the genus *Anolis*, in Panama in 1993 (photograph by Michael J. Ryan).

sistently provided glimpses of the natural history of his focal animals that transcended the laboratory tools he was pioneering. As a matter of fact, Gorman is equally renowned as an ecologist and has published several excellent behavioral studies.

Genetic data (Table 2) first were applied to problems in West Indian herpetology during the second half of this decade. Chromosome studies predominated, especially by Gorman and colleagues, who karyotyped many species of reptiles (especially *Anolis*). One of the first protein electrophoretic studies of natural populations of organisms was on West Indian *Anolis* (Gorman and Dessauer 1965). This technique was to provide valuable insights into the relationships of Antillean amphibians and reptiles during the next 25 years.

A. Stanley Rand (Figure 16) defined structural and climatic aspects of the anoline lizards of Puerto Rico (1964) and Jamaica 1967*b*), respectively — and biologists have been refining these ideas ever since. Rand determined that species could be ecologically segregated on the basis of perch height, perch diameter, and climatic habitat. Building on Collette's 1961 work relating morphology and habitat, Rand noted that species in similar structural habi-

TABLE 2. Numbers of papers reporting chromosomal and molecular data on West Indian amphibians and reptiles arranged by five-year period (since 1965) and methodology. Abbreviations: chromosome studies (C), protein electrophoresis (P), immunological investigations (I), and DNA sequencing (D).

Five-year period	C	P	I	D	Total
1965–1969	12	3	0	0	15
1970–1974	8	3	1	0	12
1975–1979	3	5	1	0	0
1980–1984	5	9	8	0	22
1985–1989	3	8	6	0	17
1990–1994	3	7	8	7	25
Total	34	35	24	7	100

tats shared morphological similarities; in so doing, he conceived the ecomorph concept, which he (with Williams 1969) and Williams (1972, 1983) refined and expanded. Ecomorphs are species of different phyletic origins but with similar morphological adaptations to similar niches. Although not intrinsically new, the concept had never previously been applied to groups below the familial level or to groups not inhabiting broadly separated geographic regions. Vanzolini (1983) noted that the concept's "... application to members of a single genus within a single zoogeographic subregion required an enormous fund of information."

Studies of structural aspects of anole habitats proliferated and, more recently, the concept has been applied to West Indian snakes by Robert W. Henderson, Peter J. Tolson, and colleagues. Rand (1967*a*) also published one of the few autecologies dealing with West Indian taxa (on *Anolis lineatopus*), and T. Preston Webster (1969) published ecological observations on *A. occultus* in Puerto Rico.

FIGURE 17. Thomas W. Schoener, population and community ecologist, in the Bahamas in the early 1980s (photograph by Ethan Temeles).

James D. Lazell, Jr., whose first paper appeared in 1961, works primarily in the Lesser Antilles and the Virgin Islands and has published a series of largely taxonomic papers (but usually replete with ecological annotations); in 1983 he wrote a biogeography of the British Virgin Islands. Harold Heatwole and collaborators began a series of papers on physiological ecology (e.g., 1965), natural history, systematics, and biogeography (e.g., 1966, with F. MacKenzie).

Beginning in 1967, Thomas W. Schoener (Figure 17) provided much of the foundation for work on lizard communities with his detailed studies of distributions, habitats, and various niche dimensions in a series of papers on Bahamian communities. A population and community ecologist, Schoener's work with West Indian lizards serves to reinforce the suitability of these organisms for a wide array of ecological studies. These lizards' high visibility and often phenomenally dense populations lend themselves ideally to testing the conceptual framework of ideas within these disciplines. One key example was the 1983 study with Amy Schoener, in which he analyzed lizard distributions on over 500 Bahamian islands, primarily from an ecological perspective.

Studies of amphibians and reptiles of the West Indies continued to play important roles in the broader discipline of biogeography and that of islands in particular. R. H. MacArthur and E. O. Wilson's (1963) important work on island biogeography and R. Levins and Harold Heatwole's (1963) paper are examples of publications in this area that benefited tremendously from the efforts of herpetologists working in the West Indies. Williams' classical 1969 study of colonization, albeit with new interpretations by Williams and others, has served as the basis for many questions posed and addressed by a number of currently active herpetologists.

The 1970s

The decade opened with the publication of V. J. Maglio's (1970) study of xenodontine snake phylogeny and zoogeography; this paper may be the

FIGURE 18. Margaret M. Stewart in the mid-1980s at El Verde Field Station in Puerto Rico (photograph by George E. Martin).

most frequently-cited of any dealing with the Antillean ophidiofauna. Schwartz continued to publish on a variety of taxa from throughout the West Indies and his first major synoptic efforts appeared: a checklist of West Indian amphibians and reptiles in 1975 (with Thomas) and his first herpetogeography in 1978.

The ecomorph concept came into its own with Williams' (1972) treatise and with more and more applications to the anoline communities of the Greater Antilles. Timothy C. Moermond (1979) carried the concept even farther by including the consideration of orientation and movements of anoles within a structural habitat matrix.

The significant role played by thermal regimes in these concepts led to varied and increasingly sophisticated studies, especially of *Anolis* thermal physiology. Building on Ruibal's 1961 paper on five species of Cuban lizards and on Rand's observations, Raymond B. Huey (since 1974) and Paul E. Hertz (since 1979) have concentrated their efforts on better understanding the thermal biology of Antillean anoles.

In 1973 Judy A. Stamps published the first in a long series of papers dealing with the behavior and socioecology of *Anolis aeneus* on Grenada, and is single-handedly responsible for it having become the ecologically best-understood of all species of *Anolis*.

Although stereotyped behavioral displays of lizards have long been documented (see for example the many contributions of Charles Carpenter), Gorman was among the first (beginning in 1968) to apply these studies to a variety of West Indian anoles and to begin the process of systematically and statistically analyzing displays to augment ecological and systematic studies. Foremost among the etho-ecologists currently working in the region, along with Stamps, is Thomas A. Jenssen. His work since 1973 often has interwoven classical ecology with increasingly sophisticated and refined analyses of behavior.

Margaret M. Stewart (Figure 18) and many co-workers (including A. S. Rand and F. H. Pough, and especially D. S. Townsend and L. L. Woolbright), in a series of papers since 1979, have produced exquisitely detailed studies of *Eleutherodactylus coqui* on Puerto Rico. As a result, this particular frog is the best-known of all West Indian amphibians. Furthermore, Stewart's efforts pioneered the now-growing field of ecological studies on West Indian frogs.

Jonathan Roughgarden and colleagues (since 1974) have addressed the ecology of Lesser Antillean anoles, and have contributed theoretical, ecological, and historical perspectives in efforts to clarify the uncertain faunal affinities with the Greater Antilles and/or South America. Roughgarden's (1995) treatment of *Anolis* lizards is a synthesis of ecological data placed in a historical context. Two other rare autecologies, by W. M. Carey (of *Cyclura pinguis* in 1975) and John B. Iverson (of *C. carinata* in 1979), appeared during this decade.

The early part of this decade saw a continuation of karyotypic studies (again largely by Gorman and associates, and most summarized by Gorman 1973). Protein electrophoretic studies, especially by T. Preston Webster (on Hispaniolan and Bahamian *Anolis*) and Gorman (on Lesser Antillean *Anolis*), began to appear more frequently (Table 2). The first application (to reptiles) of the immunological technique of microcomplement fixation of albumin was used in a study by Gorman et al. (1971) that included West Indian species.

In 1976, D. E. Rosen proposed the vicariance theory of Caribbean biogeography. Interest in obtaining phylogenetic information in order to test this theory increased dramatically, although the results of such studies did not start appearing until the next decade.

The 1980s

Although by this time the focus of the West Indian herpetological literature had shifted firmly toward ecology and evolution, not all groups have been equally well-studied. Snakes have not received the attention devoted to lizards, but since 1980 Peter J. Tolson has devoted much of his research to various aspects of the biology of West Indian species in the boid genus *Epicrates*. He has examined historical relationships within the genus (1987), worked on the reproductive behavior and physiology in most species, and more recently has devoted many of his efforts toward conservation. Also since 1980, Robert W. Henderson has been publishing papers on West Indian snake ecology, especially predator-prey relationships. He has devoted several years to aspects of the biology of West Indian populations of the arboreal boid *Corallus hortulanus*.

However, Henderson's contributions to Antillean herpetology do not end with his work on snakes; several recent papers have focused heavily on conservation, and since 1982, in collaboration with Albert Schwartz, he has produced identification guides (1984; 1984, also with S. J. Incháustegui; 1985), a new checklist (1988), and the compendium of West Indian herpetology (1991) mentioned previously. More recently, Javier Rodríguez-Robles (in collaboration with Manuel Leal and Richard Thomas) has made important contributions to our understanding of *Alsophis* behavior and ecology.

Gregory K. Pregill (since 1981) has investigated West Indian Pleistocene herpetofaunas, the impact of humans on the distribution of amphibians and reptiles, and biogeographic and systematic problems in general.

S. Blair Hedges (Figure 15) since 1982 has been an active field biologist who often ventures far from the beaten track (frequently in the company of Richard Thomas) and uses the newest biochemical tools to address questions pertaining to the West Indian herpetofauna. In addition to descriptions of 27 new species (often with Thomas and/or Garrido) and papers dealing with conservation of the herpetofauna, Hedges has published biogeographical treatises (e.g., 1989 on *Eleutherodactylus*, this volume) lending new insights into the often confounding relationships of Antillean amphibians and reptiles.

Jonathan Losos (since 1985), working primarily with the saurian fauna, has addressed a variety of topics, including ecology, behavior, and evolutionary relationships. Enlivening the exchange of ideas attempting to explain the origin and nature of the Lesser Antillean lizard fauna is the debate regarding the relative importance of size adjustment (character displacement) or size assortment (competitive exclusion) among sympatric *Anolis* species. Schoener (1969, 1988) and Williams (1972) suggested that character displacement had occurred, but size assortment was also important. Roughgarden, in a series of papers with several co-workers, beginning in 1983, proposed that a taxon-cycle might be operative, although problems with the concept were addressed by Roughgarden and Pacala (1989) and by Losos (1990). A conservative alternative has been presented by Losos in several subsequent papers (e.g., 1992).

Michael E. Seidel began his West Indian work with a 1984 paper on the status of turtles of the genus *Trachemys* on Hispaniola (with Sixto Incháustegui), and then followed a 1988 systematic revision of West Indian sliders with papers on population ecology and growth of *T. decussata* on Grand Cayman Island (1990).

Cuban zoologist Alberto R. Estrada (Figure 15) began to publish during this decade, on taxonomic and ecological topics; subsequently he has become the foremost expert on the frogs of Cuba. Octavio Pérez-Beato (Figure 15) has worked extensively with Cuban *Anolis*.

John B. Thorbjarnarson (1988) produced another rare West Indian autecology, on a Haitian population of *Crocodylus acutus*. Since 1989, Robert Powell (often with John S. Parmerlee, Jr., Donald D. Smith, and numerous students) has focused his efforts on aspects of niche dimensions of Hispaniolan lizards. Rafael Joglar has studied extensively the anuran fauna of Puerto Rico, with an additional element provided by an extensive interest in conservation matters.

Interest in chromosome studies declined during this decade (Table 2), although for the first time a large number of West Indian amphibians (21 species of *Eleutherodactylus*) were karyotyped (Bogart 1981). At the same time, the number of electrophoretic and immunological studies surged, some including West Indian species only peripherally (e.g., several papers on Neotropical xenodontine snakes by John Cadle). In Cuba, several protein electrophoretic studies on endemic reptiles (primarily *Anolis*) were published by Georgina Espinosa, Pérez-Beato, and colleagues. Other major contributors were Gorman, D. Shochat, and H. Dessauer (on *Anolis*) and S. Blair Hedges (on *Eleutherodactylus*). Much of the focus of these studies had shifted from the primarily intraspecific studies of the 1970s to interspecific investigations emphasizing historical biogeography.

The biogeographical dispersal-versus-vicariance debate became perhaps the greatest controversy in West Indian herpetology. Pregill (1981) was one of the first to counter Rosen's (1976) vicariance model, using fossil evidence to argue in favor of a dispersal origin for most groups of Antillean amphibians and reptiles. Numerous phylogenetic studies (e.g., Tolson 1987; Kluge 1989; Joglar 1989) of the West Indian herpetofauna (and other organisms) published during this decade addressed this issue directly or indirectly, but no clear consensus was reached. Much of the work on West Indian biogeography, in which amphibians and reptiles played such prominent roles, was summarized in a 1989 volume edited by Charles A. Woods.

A separate, but related controversy emerged over relationships among *Anolis* species. A striking discordance was found between the classical studies based on osteology (Williams 1976; the unpublished doctoral dissertation of Richard Etheridge) and the chromosomal and molecular data, especially that derived from serum albumin studies (Shochat and Dessauer 1981; Gorman et al. 1984). Like the vicariance-versus-dispersal controversy over the origins of West Indian amphibians and reptiles, this debate continued into the next decade.

The 1990s

This decade brought the end of an era with Schwartz's death in 1992, less than two years after publication of the massive compendium (with Henderson) of descriptions, distributions, and natural history of the Antillean herpetofauna. The Society for the Study of Amphibians and Reptiles sponsored a symposium in honor of Albert Schwartz during its joint annual meeting with the Herpetologists' League at the University of Georgia in August 1994. This volume is an extension of that symposium.

But the study of West Indian amphibians and reptiles goes on as many herpetologists continue to find that this region and its fauna serve as ideal "laboratories" for myriad systematic and ecological studies; and the legacy of Al Schwartz attracts "new blood" to the scene. Foremost among the new arrivals, Hinrich Kaiser (since 1992), working primarily with leptodactylid frogs, uses a multidisciplinary approach to systematic problems, and has found strong evidence to support a dual origin of Lesser Antillean anurans.

The emphasis on protein electrophoretic and immunological studies (many by Hedges and colleagues) continued into the early part of this decade, and largely identified dispersal as the predominate mechanism for the origin of most West Indian amphibians and reptiles (Hedges et al. 1992). One of the first mitochondrial DNA sequence studies on reptiles included the Cuban lizard *Cricosaura typica* (Hedges et al. 1991). By the middle of this decade, this latest technique is replacing protein electrophoresis as the method of choice for obtaining phylogenetic data on the Antillean herpetofauna.

The Present and Future

Midway through the 1990s, biogeographical issues are not totally resolved, the emphasis on non-taxonomic subjects continues, and papers with an ecological bent dominate; but an increasing awareness of the fragility of island faunas, including those of the West Indies, has resulted in more and more field work on the Antillean herpetofauna stressing conservation. As the dominant vertebrate classes in the region, amphibians and reptiles undoubtedly will figure prominently in any programs aimed at conservation biology in these alluring islands. Schwartz and Henderson (1991), in an "Envoi" to their compendium, wrote: "Ecological awareness that is now widespread did not exist a few years ago, and in many scientific circles the conservation of biodiversity is given top priority for policy advocacy." Nearly every herpetologist working in the West Indies since the middle of this century has published or commented publicly on one or more of the many critical concerns facing the herpetofauna as a direct result of human population growth, development, and the accompanying and seemingly inevitable alterations of habitat to which so many species apparently cannot adjust.

However, a few persons' efforts in this area stand out. For many years, C. Kenneth Dodd, Jr. and Richard Franz, independently and as collaborators, and many co-workers have been involved in a variety of studies and surveys with a focus on conservation issues. Richard Thomas, James D. Lazell, Jr., Peter J. Tolson, Robert W. Henderson (often with Richard A. Sajdak), S. Blair Hedges, and Rafael Joglar are among those mentioned previously for their other contributions, but who have been deeply involved in the conservation of the West Indian herpetofauna. Building on the foundation of work by Archie F. Carr (University of Florida), Karen A. Bjorndal heads a long list of researchers currently dealing with the problems facing marine turtles throughout the Antilles and beyond. Peter Vogel's work with the Jamaican iguana, *Cyclura collei*, thought to be extinct until rediscovered only a decade ago in the Hellshire Hills, is noteworthy.

In the Dominican Republic, Sixto J. Incháustegui has long been associated with herpetological studies, has trained several students now actively involved with Dominican conservation agencies, collaborated with Schwartz in describing two new Hispaniolan lizards, co-authored in 1984 with Henderson and Schwartz the Spanish-language guide to Hispaniolan amphibians and reptiles, but may have his greatest long-term impact as a result of his active involvement in the "Grupo Jaragua," an organization devoted to defining, shaping, and protecting Dominican national parks. Another Dominican, José A. Ottenwalder, now with the United Nations Development Programme, has been involved since the 1970s with the conservation of Hispaniolan vertebrates, including *Cyclura* spp., crocodiles, and marine and freshwater turtles. Those of us whose professional lives revolve around the diverse West Indian herpetofauna can only hope that our efforts and those of the persons listed here succeed.

Acknowledgements

We are deeply indebted to S. Blair Hedges, who not only provided insightful comments, but contributed substantially to the content by suggesting an alternative approach to our coverage of the modern era and by filling in the considerable gaps resulting from our ecological bias. Kraig Adler, Rodolfo Ruibal, and Randall D. Mooi also made a number of valuable comments on various versions of this chapter.

Literature Cited

Adler, K. 1989. Herpetologists of the past, p. 5–141. *In* K. Adler (ed.), Contributions to the History of Herpetology. Soc. Study Amphib. Rept., Oxford, Ohio. Contrib. Herpetol. no. 5.

Barbour, T. 1904. Batrachia and Reptilia from the Bahamas. Bull. Mus. Comp. Zoöl., Harvard Univ. 46: 55–61.

—. 1914. A contribution to the zoögeography of the West Indies, with especial reference to amphibians and reptiles. Mem. Mus. Comp. Zoöl., Harvard Univ. 44: 209–359.

—. 1921. *Sphaerodactylus*. Mem. Mus. Comp. Zoöl., Harvard Univ. 47: 217–278.

—. 1930. A list of Antillean reptiles and amphibians. Zoologica (N. Y.) 11: 61–116.

—. 1935. A second list of Antillean reptiles and amphibians. Zoologica (N. Y.) 19: 77–141.

—. 1937. Third list of Antillean reptiles and amphibians. Bull. Mus. Comp. Zoöl., Harvard Univ. 82: 77–166.

— and A. F. Carr. 1940. Antillean terrapins. Mem. Mus. Comp. Zoöl., Harvard Univ. 54: 381–415.

— and C. T. Ramsden. 1919. The herpetology of Cuba. Mem. Mus. Comp. Zoöl., Harvard Univ. 47: 71–213.

Baskin, J. N. and E. E. Williams. 1966. The Lesser Antillean *Ameiva*. Stud. Fauna Curaçao Carib. Isl. 89: 143–176.

Bell, H. J. 1889 (1970). Obeah; Witchcraft in the West Indies. Negro Univ. Press, Westport, Connecticut.

Bogart, J. P. 1981. Chromosome studies on *Sminthillus* from Cuba and *Eleutherodactylus* from Cuba and Puerto

Rico (Anura: Leptodactylidae). Roy. Ontario Mus., Life Sci. Contrib. 129: 1–21.

Borkin, S. S. 1994. Albert Schwartz (1923–1992). Jour. Lepidop. Soc. 48: 267–271.

Carey, W. M. 1975. The rock iguana, *Cyclura pinguis*, on Anegada, British Virgin Islands, with notes on *Cyclura ricordi* and *Cyclura cornuta* of Hispaniola. Bull. Florida St. Mus. Biol. Sci. 19: 189–233.

Catesby, M. 1743. The Natural History of Carolina, Florida and the Bahama Islands: Containing the figures of birds, beasts, fishes, serpents, insects and plants: Particularly the forest-trees, shrubs, and other plants, not hitherto described, or very incorrectly figured by authors. Together with their descriptions in English and French. To which are added, observations on the air, soil, and waters; With remarks upon agriculture, grain, pulse, roots, &c. To the whole is prefixed a new and correct map of the countries treated of. Printed at the expense of the author, 1731–1743: 2 vols. London.

Censky, E. J. 1988. *Geochelone carbonaria* (Reptilia: Testudines) in the West Indies. Florida Sci. 50: 108–114.

Cochran, D. M. 1941. The herpetology of Hispaniola. Bull. U. S. Natl. Mus. no. 177.

Collette, B. B. 1961. Correlations between ecology and morphology in anoline lizards from Havana, Cuba and southern Florida. Bull. Mus. Comp. Zool., Harvard Univ. 125: 137–162.

Curtis, A. 1947. Prevalence of snakes in Haiti. Herpetologica 3: 224.

Darlington, P. J. 1938. The origin of the fauna of the Greater Antilles, with discussion of dispersal of animals over water and through the air. Quart. Rev. Biol. 13: 274–300.

—. 1957. Zoogeography: The Geographical Distribution of Animals. John Wiley and Sons, New York.

Duellman, W. E., R. Thomas, and R. W. Henderson. 1993. Albert Schwartz 13 Sept. 1923–18 Oct. 1992. Copeia 1993: 927–932.

Duméril, A. M. C. and G. Bibron. 1834–1854. Erpétologie Générale ou Histoire Naturelle Complète des Reptiles. 10 volumes. Fain et Thunot, Paris.

Dunn, E. R. 1946. Thomas Barbour 1884–1946. Copeia 1946: 1–3.

Espeut, W. B. 1882. On the acclimatization of the Indian mongoose in Jamaica. Proc. Zool. Soc. London 1882: 712–714.

Evans, L. T. 1938. Cuban field studies on territoriality of the lizard, *Anolis sagrei*. Comp. Psychol. 25: 97–125.

García Valdes, P. 1948. The ethnography of the Ciboney, p. 503–505. *In* J. H. Steward (ed.), Handbook of South American Indians. Vol. 4. The Circum-Caribbean Tribes. Smithsonian Inst. Bur. American Ethnology, Bull. no. 143.

Garrido, O. H. and M. L. Jaume. 1984. Catálogo descriptivo de los anfibios y reptiles de Cuba. Doñana, Acta Vert. 11: 5–128.

Gorman, G. C. 1965. Interspecific karyotypic variation as a systematic character in the genus *Anolis* (Sauria: Iguanidae). Nature 208: 95–97.

—. 1968. The relationships of *Anolis* of the *roquet* species group (Sauria: Iguanidae)–III. Comparative study of display behavior. Breviora 284: 1–31.

—. 1973. The chromosomes of the Reptilia, a cytotaxonomic interpretation, p. 347–424. *In* A. B. Chiarelli and E. Capanna (eds.), Cytotaxonomy and Vertebrate Evolution. Academic Press, New York.

— and H. C. Dessauer. 1965. Hemoglobin and transferrin electrophoresis and relationships of island populations of *Anolis* lizards. Science 150: 1454–1455.

—, C. S. Lieb, and R. H. Harwood. 1984. The relationships of *Anolis gadovi*: albumin immunological evidence. Carib. Jour. Sci. 20: 145–152.

—, A. C. Wilson, and M. Nakanishi. 1971. A biochemical approach towards the study of reptilian phylogeny: evolution of serum albumin and lactate dehydrogenase. Syst. Zool. 20: 167–185.

Gosse, P. H. 1851. A Naturalist's Sojourn in Jamaica. Longmans, London.

Grant, C. 1940. The herpetology of Jamaica. II. The reptiles. Bull. Inst. Jamaica Sci. 1: 63–148.

—. 1941. The herpetology of the Cayman Islands. Bull. Inst. Jamaica Sci. 2: 1–65.

Gundlach, J. 1894. Erpetologia Cubana. Contribución a la Erpetología Cubana. G. Monteil, Habana.

Heatwole, H. 1965. Comparisons of critical thermal maxima of two species of Puerto Rican frogs of the genus *Eleutherodactylus*. Physiol. Zool. 38: 1–8.

— and F. MacKenzie. 1966. Herpetogeography of Puerto Rico. IV. Paleogeography, faunal similarity and endemism. Evolution 21: 429–438.

Hedges, S. B. 1982. Caribbean biogeography: implications of recent plate tectonic studies. Syst. Zool. 31: 518–522.

—. 1989. Evolution and biogeography of West Indian frogs of the genus *Eleutherodactylus*: slow-evolving loci and major groups, p. 305–370. *In* C. A. Woods (ed.), Biogeography of the West Indies: Past, Present, and Future. Sandhill Crane Press, Gainesville, Florida.

—, R. L. Bezy, and L. R. Maxson. 1991. Phylogenetic relationships and biogeography of xantusiid lizards inferred from mitochondrial DNA sequences. Molec. Biol. Evol. 8: 767–780.

—, C. A. Hass, and L. R. Maxson. 1992. Caribbean biogeography: molecular evidence for dispersal in West Indian terrestrial vertebrates. Proc. Natl. Acad. Sci. (USA) 89: 1909–1913.

Henderson, R. W. 1992. Consequences of predator introductions and habitat destruction on amphibians and reptiles in the post-Columbus West Indies. Carib. Jour. Sci. 28: 1–10.

—. 1993*a*. Albert Schwartz (1923–1992). Carib. Jour. Sci. 29: 1–3.

—. 1993*b*. Albert Schwartz (1923–1992): a memoriam. Herpetol. Nat. Hist. 1: 95–97.

— and M. H. Binder. 1980. The ecology and behavior of vine snakes (*Ahaetulla, Oxybelis, Thelotornis, Uromacer*): a review. Milwaukee Public Mus. Contrib. Biol. Geol. 37: 1–38.

— and A. Schwartz. 1984. A guide to the identification of the amphibians and reptiles of Hispaniola. Milwaukee Public Mus. Spec. Publ. Biol. Geol. 4: 1–70.

—, —, and S. J. Incháustegui. 1984. Guía para la identificación de los anfibios y reptiles de la Hispaniola. Mus. Nac. Hist. Nat. Ser. Monogr. 1: 1-128.

Hertz, P. E. 1979. Comparative thermal biology of sympatric grass anoles (*Anolis semilineatus* and *A. olssoni*) in lowland Hispaniola (Reptilia, Lacertilia, Iguanidae). Jour. Herpetol. 13: 329–333.

Hoagland, D. B., G. R. Horst, and C. W. Kilpatrick. 1989. Biogeography and populations biology of the mongoose in the West Indies, p. 611–633. *In* C. A. Woods (ed.), Biogeography of the West Indies: Past, Present, and Future. Sandhill Crane Press, Gainesville, Florida.

Huey, R. B. 1974. Behavioral thermoregulation in lizards: importance of associated costs. Science 184: 1001–1003.

Iverson, J. B. 1979. Behavior and ecology of the rock iguana *Cyclura carinata*. Bull. Florida State Mus. Biol. Sci. 24: 175–358.

Jenssen, T. A. 1973. Shift in the structural habitat of *Anolis opalinus* due to congeneric competition. Ecology 54: 863–869.

Joglar, R. L. 1989. Phylogenetic relationships of the West Indian frogs of the genus Eleutherodactylus: a morphological analysis, p. 371–408. *In* C. A. Woods (ed.), Biogeography of the West Indies: Past, Present, and Future. Sandhill Crane Press, Gainesville, Florida.

Kaiser, H. 1992. The trade-mediated introduction of *Eleutherodactylus martinicensis* (Anura: Leptodactylidae) on St. Barthélémy, French Antilles, and its implications for Lesser Antillean biogeography. Jour. Herpetol. 26: 264–273.

Kluge, A. G. 1989. A concern for evidence and a phylogenetic hypothesis of relationships among *Epicrates* (Boidae, Serpentes). Syst. Zool. 38: 7–25.

Laudonniere, R. 1904. The voiage of Captaine René Laudonniere to Florida in 1564. *In* R. Hakluyt (ed.), The Principal Navigations, Voyages, Traffiques, and Discoveries of the English Nation. Vol. 9. Reprint. ed., Maclehose, Glasgow.

Lazell, J. D., Jr. 1961. A new species of *Sphaerodactylus* from northern Haiti. Breviora 139: 1–5.

—. 1972. The anoles (Sauria: Iguanidae) of the Lesser Antilles. Bull. Mus. Comp. Zool., Harvard Univ. 143: 1–115.

—. 1983. Biogeography of the herpetofauna of the British Virgin Islands, with description of a new anole (Sauria: Iguanidae), p. 99–117. *In* A. G. J. Rhodin and K. Miyata (eds.), Advances in Herpetology and Evolutionary Biology: Essays in Honor of Ernest E. Williams. Mus. Comp. Zool., Harvard Univ., Cambridge, Massachusetts.

Levins, R. and H. Heatwole. 1963. On the distribution of organisms on islands. Carib. Jour. Sci. 3: 173–177.

Linnaeus, C. 1758. Systema Naturae per Regna Tria Naturae, Secundum Classes, Ordines, Genera, Species, cum Characteribus, Differentiis, Synonymis, Locis. 10th ed., Vol. 1. L. Salvii, Stockholm.

Losos, J. B. 1985. Male aggressive behavior in a pair of sympatric sibling species. Breviora 484: 1–30.

—. 1990. A phylogenetic analysis of character displacement in Caribbean *Anolis* lizards. Evolution 44: 558–569.

—. 1992. The evolution of convergent structures in Caribbean *Anolis* communities. Syst. Biol. 41: 403–420.

Lynn, W. G. 1940. The herpetology of Jamaica. I. Amphibians. Bull. Inst. Jamaica Sci. 1: 1–59.

MacArthur, R. H. and E. O. Wilson. 1963. An equilibrium theory of insular biogeography. Evolution 17: 373–386.

Maglio, V. J. 1970. West Indian xenodontine colubrid snakes: their probable origin, phylogeny, and zoogeography. Bull. Mus. Comp. Zool., Harvard Univ. 141: 1–54.

Major, R. H. (transl. and ed.). 1870. Select Letters of Columbus. Hakluyt Soc., London.

Moermond, T. C. 1979. Habitat constraints on the behavior, morphology, and community structure of *Anolis* lizards. Ecology 60: 152–164.

Morison, S. E. 1963. Journals and Other Documents on the Life and Voyages of Christopher Columbus. Heritage Press, New York.

Noble, G. K and H. T. Bradley. 1933. The mating behavior of lizards: its bearing on the theory of sexual selection. Ann. New York Acad. Sci. 35: 25–100.

Ober, F. A. 1880. Camps in the Caribbees: The Adventures of a Naturalist in the Lesser Antilles. Lee and Shepard, Boston, Massachusetts.

Oliver, J. A. 1948. The anoline lizards of Bimini, Bahamas. Amer. Mus. Novit. 1383: 1–36.

Pregill, G. K. 1981. Cranial morphology and the evolution of West Indian toads (Salientia: Bufonidae): resurrection of the genus *Peltophryne* Fitzinger. Copeia 1981: 273–285.

—, D. W. Steadman, S. L. Olson, and F. V. Grady. 1988. Late Holocene fossil vertebrates from Burma Quarry, Antigua, Lesser Antilles. Smithsonian Contr. Zool. no. 463.

—, D. W. Steadman, and D. R. Watters. 1994. Late Quaternary vertebrate faunas of the Lesser Antilles: historical components of Caribbean biogeography. Bull. Carnegie Mus. Nat. Hist. no. 30.

Rand, A. S. 1964. Ecological distribution in anoline lizards of Puerto Rico. Ecology 45: 745–752.

—. 1967*a*. Ecology and social organization of the iguanid lizard *Anolis lineatopus*. Proc. U. S. Natl. Mus. 122: 1–79.

—. 1967*b*. The ecological distribution of the anoline lizards around Kingston, Jamaica. Breviora 272: 1–18.

— and E. E. Williams. 1969. The anoles of La Palma: aspects of their ecological relationships. Breviora 327: 1–17.

Rivero, J. A. 1959. Two new species of *Eleutherodactylus* from Puerto Rico. Breviora 103: 1–6.

—. 1978. Los Anfibios y Reptiles de Puerto Rico. Univ. Puerto Rico, Mayagüez.

Rosen, D. E. 1976. A vicariance model of Caribbean biogeography. Syst. Zool. 24: 431–464.

Roughgarden, J. 1974. Niche width: biogeographic patterns among *Anolis* lizards. Amer. Nat. 108: 429–442.

—. 1995. Anolis Lizards of the Caribbean: Ecology, Evolution, and Plate Tectonics. Oxford Univ. Press, New York.

—, D. Heckel, and E. R. Fuentes. 1983. Coevolutionary theory and the biogeography and community structure of *Anolis*, p. 371–410. *In* R. B. Huey, E. R. Pianka, and T. W. Schoener (eds.), Lizard Ecology: Studies of a Model Organism. Harvard Univ. Press, Cambridge, Massachusetts.

— and S. Pacala. 1989. Taxon cycle among *Anolis* lizard populations: review of evidence, p. 403–432. *In* D. Otte and J. Endler (eds.), Speciation and its Consequences. Sinauer, Sunderland, Massachusetts.

Rouse, I. 1948. The Arawak, p. 507–546. *In* J. J. Steward (ed.), Handbook of South American Indians. Vol. 4. The Circum-Caribbean Tribes. Smithsonian Inst. Bur. Amer. Ethnol., Bull. no. 143.

—. 1989. Peopling and repeopling of the West Indies, p. 119–135. *In* C. A. Woods (ed.), Biogeography of the West Indies: Past, Present, and Future. Sandhill Crane Press, Gainesville, Florida.

Ruibal, R. 1946. A new *Sphaerodactylus* from the Dominican Republic. Amer. Mus. Novit. 1308: 1–4.

—. 1961. Thermal relations of five species of tropical lizards. Evolution 15: 98–111.

Sagra, R. de la. 1839–1843. Histoire Physique, Politique et Naturelle de l'Ile de Cuba. Arthus Bertrand, Paris.

Sajdak, R. A. and R. W. Henderson. 1991. Status of West Indian racers in the Lesser Antilles. Oryx 25: 33–38.

Schmidt, K. P. 1928. Amphibians and land reptiles of Porto Rico, with a list of those reported from the Virgin Islands. New York Acad. Sci. 10: 1–160.

Schoener, T. W. 1967. The ecological significance of sexual dimorphism in size in the lizard *Anolis conspersus*. Science 155: 474–477.

—. 1969. Size patterns in West Indian *Anolis* lizards: I. Size and species diversity. Syst. Zool. 18: 386–401.

—. 1988. Testing for nonrandomness in sizes and habitats of West Indian lizards: choice of species pool affects conclusions from null models. Evol. Ecol. 2: 1–16.

— and A. Schoener. 1983. On the voluntary departure of lizards from very small islands, p. 491–498. *In* A. G. J. Rhodin and K. Miyata (eds.), Advances in Herpetology and Evolutionary Biology: Essays in Honor of Ernest E. Williams. Mus. Comp. Zool., Harvard Univ., Cambridge, Massachusetts.

Schwartz, A. 1978. Some aspects of the herpetogeography of the West Indies, p. 31–51. *In* F. B. Gill (ed.), Zoogeography in the Caribbean. Acad. Nat. Sci. Philadelphia, Spec. Publ. no. 10.

— and O. H. Garrido. 1967. A review of the Cuban lizard *Leiocephalus macropus* Cope. Reading Public Mus. Art Gallery 14: 1–41.

— and R. W. Henderson. 1982. *Anolis cybotes* (Reptilia, Iguanidae): the eastern Hispaniolan populations. Milwaukee Public Mus. Contrib. Biol. Geol. 49: 1–8.

— and —. 1985. A guide to the identification of the amphibians and reptiles of the West Indies exclusive of Hispaniola. Milwaukee Public Mus., Milwaukee, Wisconsin.

— and —. 1988. West Indian amphibians and reptiles: a checklist. Milwaukee Public Mus. Contrib. Biol. Geol. 74: 1–264.

— and —. 1991. Amphibians and Reptiles of the West Indies: Descriptions, Distributions, and Natural History. Univ. Florida Press, Gainesville.

— and L. H. Ogren. 1956. A collection of reptiles and amphibians from Cuba, with the description of two new forms. Herpetologica 12: 91–110.

— and R. Thomas. 1960. Four new snakes (*Tropidophis*, *Dromicus*, *Alsophis*) from the Isla de Pinos and Cuba. Herpetologica 16: 73–90.

— and —. 1975. A check-list of West Indian amphibians and reptiles. Carnegie Mus. Nat. Hist. Spec. Publ. no. 1.

Seidel, M. E. 1988. Revision of the West Indian emydid turtles. Amer. Mus. Novit. 2918: 1–41.

—. 1990. Growth and population characteristics of the slider turtle, *Trachemys decussata*, on Grand Cayman Island. Jour. Herpetol. 24: 191–196.

— and S. J. Incháustegui Miranda. 1984. Status of the trachemyd turtles (Testudines: Emydidae) on Hispaniola. Jour. Herpetol. 18: 468–479.

Shochat, D. and H. C. Dessauer. 1981. Comparative immunological study of albumins of *Anolis* lizards of the Caribbean islands. Comp. Biochem. Physiol. 68A: 67–73.

Smith, H. M. 1986. Chapman Grant, *Herpetologica*, and The Herpetologists' League. Herpetologica 42: 1–32.

— and C. Grant. 1946. The proper names for some Cuban snakes; an analysis of dates of publication of Ramon de la Sagra's Historia Natural de Cuba, and of Fitzinger's Systema Reptilium. Herpetologica 14: 215–222.

Stamps, J. A. 1973. Display and social organization in female *Anolis aeneus*. Copeia 1973: 264–272.

Stejneger, L. 1904. The herpetology of Porto Rico. Rept. United States Natl. Mus. 1902: 551–724.

Stewart, M. M. 1979. The role of introduced species in a Jamaican frog community. Act. Symp. Internatl. Ecol. Trop. 1: 111–146.

Stoudemire, S. A. (transl. and ed.). 1959. Natural History of the West Indies, by Gonzalo F. de Oviedo. Univ. North Carolina Press, Chapel Hill.

Thomas, R. and R. Joglar. 1995. The herpetology of Puerto Rico, past, present, and future. Ann. New York Acad. Sci.: in press.

Thorbjarnarson, J. B. 1988. The status and ecology of the

American crocodile in Haiti. Bull. Florida State Mus. Biol. Sci. 33: 1–86.

Tolson, P. J. 1980. Comparative reproductive behavior in four species of snakes of the boid genus *Epicrates*, p. 87–97. *In* Proc. 4th Reptile Symp. Captive Propagation and Husbandry. Zool. Consortium, Inc., Thurmont, Maryland.

—. 1987. Phylogenetics of the boid snake genus *Epicrates* and Caribbean vicariance theory. Occas. Pap. Mus. Zool. Univ. Michigan 715: 1–68.

Underwood, G. 1952. Introduction to the study of Jamaican reptiles (Part X). Nat. Hist. Notes Nat. Hist. Soc. Jamaica 53: 97–105.

—. 1959. The anoles of the eastern Caribbean (Sauria, Iguanidae). Pt. III. Revisionary notes. Bull. Mus. Comp. Zool., Harvard Univ. 121: 191–226.

—. 1962. Reptiles of the eastern Caribbean. Carib. Affairs, new ser. 1: 1–192.

— and E. E. Williams. 1959. The anoline lizards of Jamaica. Bull. Inst. Jamaica Sci. 9: 1–48.

Vanzolini, P. E. 1977. An annotated bibliography of the land and fresh-water reptiles of South America (1758–1975). Vol. I (1758–1900). Mus. Zool., Univ. São Paulo, São Paulo.

—. 1983. Introduction, p. vii–x. *In* A. G. J. Rhodin and K. Miyata (eds.), Advances in Herpetology and Evolutionary Biology: Essays in Honor of Ernest E. Williams. Mus. Comp. Zool., Harvard Univ., Cambridge, Massachusetts.

Watters, D. R. 1989. Archeological implications for Lesser Antilles biogeography: the small island perspective, p. 153–165. *In* C. A. Woods (ed.), Biogeography of the West Indies: Past, Present, and Future. Sandhill Crane Press, Gainesville, Florida.

Webster, T. P. 1969. Ecological observations on *Anolis occultus* Williams and Rivero (Sauria, Iguanidae). Breviora 312: 1–5.

Williams, E. E. 1950. *Testudo cubensis* and the evolution of Western Hemisphere tortoises. Bull. Amer. Mus. Nat. Hist. 99: 541–560.

—. 1960. Notes on Hispaniolan herpetology. 1. *Anolis christophei*, new species, from the Citadel of King Christophe, Haiti. Breviora 117: 1–7.

—. 1969. The ecology of colonization as seen in the zoogeography of anoline lizards on small islands. Quart. Rev. Biol. 44: 345–389.

—. 1972. The origin of faunas. Evolution of lizard congeners in a complex island fauna: a trial analysis. Evol. Biol. 6: 47–89.

— (ed.). 1974. The Second *Anolis* Newsletter. Mus. Comp. Zool., Harvard Univ., Cambridge, Massachusetts.

—. 1976. West Indian anoles: a taxonomic and evolutionary summary. 1. Introduction and a species list. Breviora 440: 1–21.

— (ed.). 1977. The Third *Anolis* Newsletter. Mus. Comp. Zool., Harvard Univ., Cambridge, Massachusetts.

—. 1983. Ecomorphs, faunas, island size, and diverse end points in island radiations of *Anolis*, p. 326–370. *In* R. B. Huey, E. R. Pianka, and T. W. Schoener (eds.), Lizard Ecology: Studies of a Model Organism. Harvard Univ. Press, Cambridge, Massachusetts.

Wing, E. S. 1989. Human exploitation of animal resources in the Caribbean, p. 137–152. *In* C. A. Woods (ed.), Biogeography of the West Indies: Past, Present, and Future. Sandhill Crane Press, Gainesville, Florida.

Wolcott, G. N. 1923. The food of Porto Rican lizards. Jour. Dept. Agr. Porto Rico 7(4): 5–38.

Woods, C. A. (ed.). 1989. Biogeography of the West Indies: Past, Present, and Future. Sandhill Crane Press, Gainesville, Florida.

Powell, R. and R. W. Henderson. 1996. A brief history of West Indian herpetology, p. 29–50. *In* R. Powell and R. W. Henderson (eds.), Contributions to West Indian Herpetology: A Tribute to Albert Schwartz. Society for the Study of Amphibians and Reptiles, Ithaca (New York). Contributions to Herpetology, volume 12.

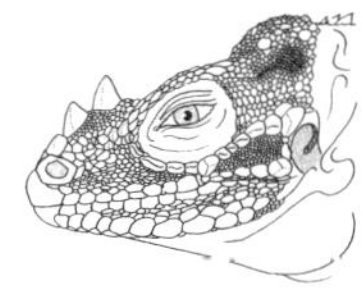

An Annotated Checklist of West Indian Amphibians and Reptiles

Robert Powell,[1] Robert W. Henderson,[2] Kraig Adler,[3] and Harold A. Dundee[4]

SINCE THE PUBLICATION OF THE MOST RECENT WEST Indian checklist (Schwartz and Henderson 1988) and the compendium on West Indian herpetology by Schwartz and Henderson (1991), the known number of species has increased considerably. Furthermore, neither of these sources nor any other single list has provided complete citations of the original descriptions.

To remedy this, we have included in this tribute to Albert Schwartz an updated list of all 622 species known to occur in the West Indies and of which descriptions have been published (Hedges, this volume, has indicated that an additional 11 species currently are being described). We have defined the West Indies to include the Bahama Islands, Turks and Caicos Islands, Cayman Islands, Greater Antilles and satellites (including the entire Puerto Rico Bank), Lesser Antilles, and islands in the western Caribbean (see below). We have excluded Trinidad and Tobago, which are geologically, floristically, and faunistically continental (for a current list of amphibians and reptiles found on these islands and their distribution patterns, see Murphy, this volume).

For each species listed we provide: (1) an abbreviated annotation regarding the distribution of that species within the West Indies (ranges outside of the West Indies are not indicated): BAH = Bahama Islands, CAY = Cayman Islands, CUB = Cuba and satellites, HISP = Hispaniola and satellites, JAM = Jamaica and satellites, LA = Lesser Antilles, N = Navassa Island, PRB = Puerto Rico Bank (including Mona Island and the Virgin Islands), TC = Turks and Caicos Islands, WC = Western Caribbean (Swan Island, San Andrés, and Providencia), WI = West Indies (widespread); (2) a notation indicating species introduced to (but not within) the region (INT); (3) information on the holotype and type-locality of species not included in previous lists; (4) the full citation of the original description (museum acronyms are as in Schwartz and Henderson 1991 unless otherwise indicated); (5) the current (since 1991) status of any recently suggested generic reallocations; (6) the basis for, or the citation pertinent to, any elevation of subspecies to species status since 1991; (7) as the best source of pertinent literature regarding any taxon, information regarding available accounts in the *Catalogue of American Amphibians and Reptiles* (CAAR) on species found in the West Indies; and (8) note the plate number in bold type of those species illustrated in this volume.

Entries are listed alphabetically in groupings corresponding to class, order (and suborder for squamate reptiles), and family. We have chosen to use the iguanian lizard families proposed by Frost and Etheridge (1989), despite concerns expressed by Hedges (this volume) and others. Although this taxonomy may be subjected to future changes, we believe that it better represents the true situation than would grouping all iguanians in the family "Iguanidae."

We thank S. Blair Hedges for his invaluable help in compiling this list.

[1] Department of Natural Sciences, Avila College, Kansas City, Missouri 64145–1698, USA.

[2] Section of Vertebrate Zoology, Milwaukee Public Museum, Milwaukee, Wisconsin 53233–1478, USA.

[3] Division of Biological Sciences, Cornell University, Ithaca, New York 14853–2702, USA.

[4] Tulane Museum of Natural History, Tulane University, Belle Chasse, Louisiana 70037–3098, USA.

Class Amphibia, Order Anura

Family Bufonidae

Bufo cataulaciceps. *Bufo cataulaciceps* Schwartz 1959: 110. • CUB • Schwartz, A. 1959. A new species of toad, *Bufo cataulaciceps*, from the Isla de Pinos and western Cuba. Proc. Biol. Soc. Washington 72: 109–119. Recently reassigned to *Bufo* from *Peltophryne* (Hedges et al. 1992*b*; also see Hedges, this volume).

Bufo empusus. *Peltaphryne empusa* Cope 1862: 344. • CUB • Cope, E. D. 1862. [A curious Cuban bufonid (*Peltaphryne empusa*)]. Proc. Acad. Nat. Sci. Philadelphia 14: 344. Recently reassigned to *Bufo* from *Peltophryne* (Hedges et al. 1992*b*; also see Hedges, this volume).

Bufo fluviaticus. *Bufo fluviaticus* Schwartz 1972: 226. • HISP • Schwartz, A. 1972. The native toads (Anura, Bufonidae) of Hispaniola. Jour. Herpetol. 6: 217–231. Recently reassigned to *Bufo* from *Peltophryne* (Hedges et al. 1992*b*; also see Hedges, this volume). CAAR: Powell, R. and G. K. Pregill. 1991. No. 507.

Bufo fractus. *Bufo güntheri fractus* Schwartz 1972: 218. • HISP • Holotype, USNM 189235; type-locality, "0.7 mi. (1.1 km) W. Higüey, La Altagracia Province, República Dominicana" • Schwartz, A. 1972. The native toads (Anura, Bufonidae) of Hispaniola. Jour. Herpetol. 6: 217–231. Recently reassigned to *Bufo* from *Peltophryne* (Hedges et al. 1992*b*; also see Hedges, this volume). Elevated herein from subspecific status based on: (1) differences in pattern, advertisement call, and ecology (*B. fractus* is mesophilic, whereas *B. guentheri* is a xerophile); (2) a hiatus of approximately 100 km within which no toads (other than *B. marinus*) are known to occur between populations of this species and those of *B. guentheri*; and (3) comments in Pregill (1981) and Powell (1993). CAAR: Powell, R. 1992. No. 535 (as *Peltophryne guentheri fracta*).

Bufo fustiger. *Bufo peltacephalus fustiger* Schwartz 1960: 47. • CUB • Schwartz, A. 1960. The large toads of Cuba. Proc. Biol. Soc. Washington 73: 45–55. Recently reassigned to *Bufo* from *Peltophryne* (Hedges et al. 1992*b*; also see Hedges, this volume).

Bufo guentheri. *Bufo güntheri* Cochran 1941: 8. • HISP • Cochran, D. M. 1941. The herpetology of Hispaniola. Bull. United States Natl. Mus. no. 177. Recently reassigned to *Bufo* from *Peltophryne* (Hedges et al. 1992*b*; also see Hedges, this volume). CAAR: Powell, R. 1992. No. 535.

Bufo gundlachi. *Bufo gundlachi* Ruibal 1959: 2. • CUB • Ruibal, R. 1959. *Bufo gundlachi*, a new species of Cuban toad. Breviora 105: 1–14. Recently reassigned to *Bufo* from *Peltophryne* (Hedges et al. 1992*b*; also see Hedges, this volume).

Bufo lemur. *Peltaphryne lemur* Cope 1868 (1869): 311. • PRB • Cope, E. D. 1868 (1869). Sixth contribution to the herpetology of tropical America. Proc. Acad. Nat. Sci. Philadelphia 20: 305–313. Recently reassigned to *Bufo* from *Peltophryne* (Hedges et al. 1992*b*; also see Hedges, this volume).

Bufo longinasus. *Bufo longinasus* Stejneger 1905: 765. • CUB • Stejneger, L. 1905. Description of a new toad from Cuba. Proc. United States Natl. Mus. 28: 765–767. Recently reassigned to *Bufo* from *Peltophryne* (Hedges et al. 1992*b*; also see Hedges, this volume).

Bufo marinus. *Rana marina* Linnaeus 1758: 211. • WI • Linnaeus, C. 1758. Systema Naturæ per Regna Tria Naturæ, Secundum Classes, Ordines, Genera, Species cum Characteribus, Differentiis, Synonymis, Locis. Tomus I. Editio decima. Reformata. L. Salvii, Holmiae. CAAR: Easteal, S. 1986. No. 395. **Plate 1B.**

Bufo peltocephalus. *Bufo peltocephalus* Tschudi 1838: 52. • CUB • Tschudi, J. J. von. 1838. Classification der Batrachier, mit Berucksichtigung der fossilien Thiere dieser Abtheilung der Reptilien. Petitpierre, Neuchâtel. Recently reassigned to *Bufo* from *Peltophryne* (Hedges et al. 1992*b*; also see Hedges, this volume). **Plate 1A.**

Bufo taladai. *Bufo taladai* Schwartz 1960: 51. • CUB • Schwartz, A. 1960. The large toads of Cuba. Proc. Biol. Soc. Washington 73: 45–55. Recently reassigned to *Bufo* from *Peltophryne* (Hedges et al. 1992*b*; also see Hedges, this volume).

Family Dendrobatidae

Colostethus chalcopis. *Colostethus chalcopis* Kaiser, Coloma, and Gray 1994: 24. • LA • Holotype, NMC (Canadian Museum of Nature) 33675; type-locality, "approximately 3 km (by road) NE Morne Rouge, Martinique, French Antilles (about 14°48' N, 61°8' W, approximate elevation 500 m)" • Kaiser, H., L. A. Coloma, and H. M. Gray. 1994. A new species of *Colostethus* (Anura: Dendrobatidae) from Martinique, French Antilles. Herpetologica 50: 23–32. **Plate 1C**.

Family Hylidae

Calyptahyla crucialis. *Hyla crucialis* Harlan 1826: 64. • JAM • Harlan, R. 1826. Descriptions of several new species of batrachian reptiles, with observations on the larvae of frogs. Amer. Jour. Sci. Arts 10: 53–65. Hedges (this volume) has indicated that this species may be reassigned to the genus *Osteopilus*, but also see Anderson (this volume). **Plate 1D.**

Hyla cinerea. *Calamitae* [sic] *cinereus* Schneider 1799: 174. • PRB • INT • Schneider, J. G. 1799. Historiae Amphibiorum Naturalis et Literariae. Fasciculus Primus: Ranas, Calamitas, Bufones, Salamandras et Hydros. F. Frommanni, Jenae.

Hyla heilprini. *Hyla heilprini* Noble 1923: 1. • HISP • Noble, G. K. 1923. Six new batrachians from the Dominican Republic. Amer. Mus. Novit. 61: 1–6. **Plate 1E.**

Hyla marianae. *Hyla marianae* Dunn 1926: 129. • JAM • Dunn, E. R. 1926. The frogs of Jamaica. Proc. Boston Soc. Nat. Hist. 38: 111–130. Hedges (this volume) has indicated that this species may be reassigned to the genus *Osteopilus*, but also see Anderson (this volume).

Hyla pulchrilineata. *Hyla pulchrilineata* Cope 1869: 163. • HISP • Cope, E. D. 1869. Seventh contribution to the herpetology of tropical America. Proc. Amer. Philos. Soc. 11: 147–169. Hedges (this volume) has indicated that this species may be reassigned to the genus *Osteopilus*, but also see Anderson (this volume).

Hyla squirella. *Hyla squirella* Bosc *in* Daudin 1800: [9], plate [5] lower figure (individual pages and plates are not numbered). • BAH • INT • Bosc d'Antic, L. A. G. *in* F.-M. Daudin. 1800. Histoire Naturelle des Quadrupèdes Ovipares. Livr. 1. Fuchs and Delalain, Paris. CAAR: Martof, B. S. 1975. No. 168.

Hyla vasta. *Hyla vasta* Cope 1871: 219. • HISP • Cope, E. D. 1871. Ninth contribution to the herpetology of tropical America. Proc. Acad. Nat. Sci. Philadelphia 23: 200–224. Hedges (this volume) has indicated that this species may be reassigned to the genus *Osteopilus*, but also see Anderson (this volume).

Hyla wilderi. *Hyla wilderi* Dunn 1925: 161. • JAM • Dunn, E. R. 1925. A new tree-toad from Jamaica. Occas. Pap. Boston Soc. Nat. Hist. 5: 161–162. Hedges (this volume) has indicated that this species may be reassigned to the genus *Osteopilus*, but also see Anderson (this volume).

Osteopilus brunneus. *Hyla brunnea* Gosse 1851: 361. • JAM • Gosse, P. H. 1851. A Naturalist's Sojourn in Jamaica. Longman, Brown, Green and Longmans, London. **Plate 1F.**

Osteopilus dominicensis. *Hypsiboas dominicensis* Tschudi 1838: 30. • HISP • Tschudi, J. J. von. 1838. Classification der Batrachier, mit Berucksichtigung der fossilien Thiere dieser Abtheilung der Reptilien. Petitpierre, Neuchâtel.

Osteopilus septentrionalis. *Hyla septentrionalis* Duméril and Bibron 1841: 538 (substitute name for

Trachycephalus marmoratus Duméril and Bibron 1841, not *Hyla marmorata* Laurenti 1768). • BAH, CAY, CUB, LA, PRB • Duméril, A. M. C. and G. Bibron. 1841. Erpétologie Générale ou Histoire Naturelle Complète des Reptiles. Vol. 8. Libr. Encylopédique Roret, Paris. CAAR: Duellman, W. E. and R. I. Crombie. 1970. No. 92. **Plate 1G**.

Pseudacris crucifer. *Hyla crucifer* Wied, 1838: 275. • CUB • INT • Wied-Neuwied, M. A. P. zu. 1838. Reise in das innere Nord-Amerika in den Jahren 1832 bis 1834. Vol. 1. J. Hoelscher, Coblenz.

Scinax rubra. *Hyla rubra* Daudin 1802: 19. • LA • INT • Daudin, F.-M. 1802. Histoire Naturelle des Rainettes, des Grenouilles et des Crapauds. Levrault, Paris.

Family Leptodactylidae

Eleutherodactylus abbotti. *Eleutherodactylus abbotti* Cochran 1923: 93. • HISP • Cochran, D. M. 1923. A new species of *Eleutherodactylus* from the Dominican Repuplic [sic]. Proc. Biol. Soc. Washington 36: 93–94. CAAR: Schwartz, A. 1977. No. 191.

Eleutherodactylus acmonis. *Eleutherodactylus acmonis* Schwartz 1960: 42. • CUB • Schwartz, A. 1960. Nine new Cuban frogs of the genus *Eleutherodactylus*. Sci. Publ. Reading Public Mus. Art Gallery 11: 1–50.

Eleutherodactylus albipes. *Eleutherodactylus albipes* Barbour and Shreve 1937: 383. • CUB • Barbour, T. and B. Shreve. 1937. Novitates cubanae. Bull. Mus. Comp. Zoöl., Harvard Univ. 80: 377–387.

Eleutherodactylus alcoae. *Eleutherodactylus alcoae* Schwartz 1971: 26. • HISP • Schwartz, A. 1971. A new species of *Eleutherodactylus* (Amphibia, Leptodactylidae) from Hispaniola. Ann. Carnegie Mus. 43: 25–31. CAAR: Schwartz, A. 1977. No. 192.

Eleutherodactylus alticola. *Eleutherodactlylus alticola* Lynn 1937: 89. • JAM • Lynn, W. G. 1937. Two new frogs from Jamaica. Herpetologica 1: 88–90.

Eleutherodactylus amadeus. *Eleutherodactylus amadeus* Hedges, Thomas, and Franz 1987: 943. • HISP • Hedges, S. B., R. Thomas, and R. Franz. 1987. A new species of *Eleutherodactylus* (Anura, Leptodactylidae) from the Massif de la Hotte, Haiti. Copeia 1987: 943–949.

Eleutherodactylus amplinympha. *Eleutherodactylus amplinympha* Kaiser, Green, and Schmid 1994 (1995): 2220. • LA • Holotype, KU 221259; type-locality, "near Freshwater Lake, Dominica, West Indies (ca. 61°20' W, 15°20' N; elevation ca. 800 m." • Kaiser, H., D. M. Green, and M. Schmid. 1994 (1995). Systematics and biogeography of eastern Caribbean frogs (*Eleutherodactylus*: Leptodactylidae), with a description of a new species from Dominica. Canadian Jour. Zool. 72: 2217–2237.

Eleutherodactylus andrewsi. *Eleutherodactylus andrewsi* Lynn 1937: 88. • JAM • Lynn, W. G. 1937. Two new frogs from Jamaica. Herpetologica 1: 88–90.

Eleutherodactylus antillensis. *Hylodes antillensis* Reinhardt and Lütken 1862: 209. • PRB • Reinhardt, J. and C. F. Lütken. 1862. Bidrag til det vestindiske Öeriges og navnligen til de dansk-vestindiske Öers Herpetologie. Vidensk. Meddel. Naturhist. Foren. Kjøbenhavn 10–18: 153–291.

Eleutherodactylus apostates. *Eleutherodactylus apostates* Schwartz 1973: 262. • HISP • Schwartz, A. 1973. Six new species of *Eleutherodactylus* (Anura, Leptodactylidae) from Hispaniola. Jour. Herpetol. 7: 249–273. CAAR: Schwartz, A. 1977. No. 201.

Eleutherodactylus armstrongi. *Eleutherodactylus armstrongi* Noble and Hassler 1933: 2. • HISP • Noble, G. K. and W. G. Hassler. 1933. Two new species of frogs, five new species and a new race of lizards from the Dominican Republic. Amer. Mus. Novit. 652: 1–17. CAAR: Schwartz, A. 1978. No. 208. **Plate 1H**.

Eleutherodactylus atkinsi. *Eleutherodactylus atkinsi* Dunn 1925: 165. • CUB • Dunn, E. R. 1925. New frogs from Cuba. Occas. Pap. Boston Soc. Nat. Hist. 5: 163–166.

Eleutherodactylus audanti. *Eleutherodactylus audanti* Cochran 1934: 164. • HISP • Cochran, D. M.

1934. Herpetological collections made in Hispaniola by the *Utowana* Expedition, 1934. Occas. Pap. Boston Soc. Nat. Hist. 8: 163–188. CAAR: Schwartz, A. 1979. No. 224.

Eleutherodactylus auriculatoides. *Eleutherodactylus auriculatoides* Noble, 1923: 3. • HISP • Noble, G. K. 1923. Six new batrachians from the Dominican Republic. Amer. Mus. Novit. 61: 1–6. CAAR: Schwartz, A. 1980. No. 246.

Eleutherodactylus auriculatus. *Hylodes auriculatus* Cope 1862: 152. • CUB • Cope, E. D. 1862. On some new and little known American Anura. Proc. Acad. Nat. Sci. Philadelphia 14:151–159.

Eleutherodactylus bakeri. *Eleutherodactylus bakeri* Cochran 1935: 369. • HISP • Cochran, D. M. 1935. New reptiles and amphibians collected in Haiti by P. J. Darlington. Proc. Boston Soc. Nat. Hist. 40: 367–375. CAAR: Schwartz, A. 1980. No. 247.

Eleutherodactylus barlagnei. *Eleutherodactylus barlagnei* Lynch 1965: 2. • LA • Lynch, J. D. 1965. A new species of *Eleutherodactylus* from Guadeloupe, West Indies. Breviora 220: 1–7.

Eleutherodactylus bartonsmithi. *Eleutherodactylus bartonsmithi* Schwartz 1960: 10. • CUB • Schwartz, A. 1960. Nine new Cuban frogs of the genus *Eleutherodactylus*. Sci. Publ. Reading Public Mus. Art Gallery 11: 1–50.

Eleutherodactylus bresslerae. *Eleutherodactylus bresslerae* Schwartz 1960: 45. • CUB • Schwartz, A. 1960. Nine new Cuban frogs of the genus *Eleutherodactylus*. Sci. Publ. Reading Public Mus. Art Gallery 11: 1–50.

Eleutherodactylus brevirostris. *Eleutherodactylus brevirostris* Shreve 1936: 95. • HISP • Shreve, B. 1936. A new *Anolis* and new Amphibia from Haiti. Proc. New England Zoöl. Club 15: 93–99. CAAR: Schwartz, A. 1981. No. 259.

Eleutherodactylus brittoni. *Eleutherodactylus brittoni* Schmidt 1920: 179. • PRB • Schmidt, K. P. 1920. Contributions to the herpetology of Porto Rico. Ann. New York Acad. Sci. 28: 167–200.

Eleutherodactylus caribe. *Eleutherodactylus caribe* Hedges and Thomas 1992: 191. • HISP • Holotype, USNM 314177; type-locality, "2.6 km SW Dame-Marie, Dépt. de la Grand'Anse, Haiti, 0 m" • Hedges, S. B. and R. Thomas. 1992. A new marsh-dwelling species of *Eleutherodactylus* from Haiti (Anura: Leptodactylidae). Jour. Herpetol. 26: 191–195.

Eleutherodactylus cavernicola. *Eleutherodactylus cavernicola* Lynn 1954: 400. • JAM • Lynn, W. G. 1954. Description of a new frog of the genus *Eleutherodactylus* from Jamaica, B.W.I. Jour. Washington Acad. Sci. 44: 400–402.

Eleutherodactylus chlorophenax. *Eleutherodactylus chlorophenax* Schwartz 1976: 168. • HISP • Schwartz, A. 1976. Two new species of Hispaniolan *Eleutherodactylus* (Leptodactylidae). Herpetologica 32: 163–171.

Eleutherodactylus cochranae. *Eleutherodactylus cochranae* Grant 1932: 325. • PRB • Grant, C. 1932. A new frog from the Virgin Islands. Jour. Dept. Agric. Puerto Rico 16: 325–326.

Eleutherodactylus cooki. *Eleutherodactylus cooki* Grant 1931 (1932): 145. • PRB • Grant, C. 1931 (1932). A new frog from Puerto Rico. Jour. Dept. Agric. Puerto Rico 16: 145–148. **Plate 1I.**

Eleutherodactylus coqui. *Eleutherodactylus coqui* Thomas 1966: 376. • PRB • Thomas, R. 1966. New species of Antillean *Eleutherodactylus*. Quart. Jour. Florida Acad. Sci. 28: 375–391.

Eleutherodactylus corona. *Eleutherodactylus corona* Hedges and Thomas 1992: 351. • HISP • Holotype, USNM 310824; type-locality, "Caye Paul (10.7 km WNW Les Platons), Dépt. du Sud, Haiti, 1120 m" • Hedges, S. B. and R. Thomas. 1992. Two new species of *Eleutherodactylus* from remnant cloud forest in Haiti (Anura, Leptodactylidae). Herpetologica 48: 351–358.

Eleutherodactylus counouspeus. *Eleutherodactylus counouspeus* Schwartz 1964: 2. • HISP • Schwartz, A. 1964. Three new species of frogs (Leptodactylidae, *Eleutherodactylus*) from Hispaniola. Breviora 208: 1–15. CAAR: Schwartz, A. 1981. No. 260.

Eleutherodactylus cubanus. *Eleutherodactylus parvus* Barbour and Shreve 1937: 386. • CUB • Barbour, T. and B. Shreve. 1937. Novitates cubanae. Bull. Mus. Comp. Zoöl., Harvard Coll. 80: 377–387.

Eleutherodactylus cundalli. *Eleutherodactylus cundalli* Dunn 1926: 121. • JAM • Dunn, E. R. 1926. The frogs of Jamaica. Proc. Boston Soc. Nat. Hist. 38: 111–130.

Eleutherodactylus cuneatus. *Hylodes cuneatus* Cope 1862: 152. • CUB • Cope, E. D. 1862. On some new and little known American Anura. Proc. Acad. Nat. Sci. Philadelphia 14: 151–159.

Eleutherodactylus darlingtoni. *Eleutherodactylus darlingtoni* Cochran 1935: 368. • HISP • Cochran, D. M. 1935. New reptiles and amphibians collected in Haiti by P. J. Darlington. Proc. Boston Soc. Nat. Hist. 40: 367–375. CAAR: Schwartz, A. 1982. No. 284.

Eleutherodactylus dimidiatus. *Hylodes dimidiatus* Cope 1862: 151. • CUB • Cope, E. D. 1862. On some new and little known American Anura. Proc. Acad. Nat. Sci. Philadelphia 14: 151–159.

Eleutherodactylus dolomedes. *Eleutherodactylus dolomedes* Hedges and Thomas 1992: 356. • HISP • Holotype, USNM 309780; type-locality, "Caye Paul (10.7 km WNW Les Platons), Dépt. du Sud, Haiti, 1120 m" • Hedges, S. B. and R. Thomas. 1992. Two new species of *Eleutherodactylus* from remnant cloud forest in Haiti (Anura, Leptodactylidae). Herpetologica 48: 351–358.

Eleutherodactylus eileenae. *Eleutherodactylus eileenae* Dunn 1926: 212. • CUB • Dunn, E. R. 1926. Additional frogs from Cuba. Occas. Pap. Boston Soc. Nat. Hist. 5: 209–215.

Eleutherodactylus emiliae. *Eleutherodactylus emiliae* Dunn 1926: 213. • CUB • Dunn, E. R. 1926. Additional frogs from Cuba. Occas. Pap. Boston Soc. Nat. Hist. 5: 209–215.

Eleutherodactylus eneidae. *Eleutherodactylus eneidae* Rivero 1959: 4. • PRB • Rivero, J. A. 1959. Two new species of *Eleutherodactylus* from Puerto Rico. Breviora 103: 1–6.

Eleutherodactylus etheridgei. *Eleutherodactylus etheridgei* Schwartz 1958: 16. • CUB • Schwartz, A. 1958. Four new frogs of the genus *Eleutherodactylus* (Leptodactylidae) from Cuba. Amer. Mus. Novit. 1873: 1–20.

Eleutherodactylus eunaster. *Eleutherodactylus eunaster* Schwartz 1973: 250. • HISP • Schwartz, A. 1973. Six new species of *Eleutherodactylus* (Anura, Leptodactylidae) from Hispaniola. Jour. Herpetol. 7: 249–273. CAAR: Schwartz, A. 1982. No. 285.

Eleutherodactylus euphronides. *Eleutherodactylus urichi euphronides* Schwartz 1967: 6. • LA • Holotype, MCZ 43229; type-locality, "Grand Etang, 1700 ft., St. Andrew Parish, Grenada" • Schwartz, A. 1967. Frogs of the genus *Eleutherodactlyus* in the Lesser Antilles. Stud. Fauna Curaçao Carib. Isl. 24: 1–62. Elevated from subspecific status by Kaiser et al. (1994 [1995]).

Eleutherodactylus flavescens. *Eleutherodactylus flavescens* Noble 1923: 2. • HISP • Noble, G. K. 1923. Six new batrachians from the Dominican Republic. Amer. Mus. Novit. 61: 1–6. CAAR: Schwartz, A. 1982. No. 296.

Eleutherodactylus fowleri. *Eleutherodactylus fowleri* Schwartz 1973: 255. • HISP • Schwartz, A. 1973. Six new species of *Eleutherodactylus* (Anura, Leptodactylidae) from Hispaniola. Jour. Herpetol. 7: 249–273. CAAR: Schwartz, A. 1982. No. 297.

Eleutherodactylus furcyensis. *Eleutherodactylus furcyensis* Shreve and Williams 1963: 329. • HISP • Shreve, B. and E. E. Williams. 1963. The herpetology of the Port-au-Prince region and Gonave Island, Haiti. Part II. The frogs. Bull. Mus. Comp. Zool., Harvard Univ. 129: 302–342. CAAR: Schwartz, A. 1983. No. 309.

Eleutherodactylus fuscus. *Eleutherodactylus fuscus* Lynn and Dent 1943: 235. • JAM • Lynn, W. G. and J. M. Dent. 1943. Notes on Jamaican amphibians. Copeia 1943: 234–242.

Eleutherodactylus glandulifer. *Eleutherodactylus glandulifer* Cochran 1935: 367. • HISP • Cochran, D. M. 1935. New reptiles and amphibians collected in Haiti by P. J. Darlington. Proc. Boston Soc. Nat. Hist. 40: 367–375. CAAR: Schwartz, A. 1983. No. 310.

Eleutherodactylus glanduliferoides. *Eleutherodactylus glanduliferoides* Shreve 1936: 96. • HISP • Shreve, B. 1936. A new *Anolis* and new Amphibia from Haiti. Proc. New England Zoöl. Club 15: 93–99. CAAR: Schwartz, A. 1983. No. 322.

Eleutherodactylus glaphycompus. *Eleutherodactylus glaphycompus* Schwartz 1973: 257. • HISP • Schwartz, A. 1973. Six new species of *Eleutherodactylus* (Anura, Leptodactylidae) from Hispaniola. Jour. Herpetol. 7: 249–273. CAAR: Schwartz, A. 1983. No. 323.

Eleutherodactylus glaucoreius. *Eleutherodactylus cundalli glaucoreius* Schwartz and Fowler 1973: 73. • JAM • Schwartz, A. and D. C. Fowler. 1973. The Anura of Jamaica: a progress report. Stud. Fauna Curaçao Carib. Isl. 43: 50–142.

Eleutherodactylus gossei. *Eleutherodactylus gossei* Dunn 1926: 118. • JAM • Dunn, E. R. 1926. The frogs of Jamaica. Proc. Boston Soc. Nat. Hist. 38: 111–130.

Eleutherodactylus grabhami. *Eleutherodactylus grabhami* Dunn 1926: 121. • JAM • Dunn, E. R. 1926. The frogs of Jamaica. Proc. Boston Soc. Nat. Hist. 38: 111–130.

Eleutherodactylus grahami. *Eleutherodactylus grahami* Schwartz 1979: 200. • HISP • Schwartz, A. 1979. A new species of *Eleutherodactylus* (Amphibia, Anura, Leptodactylidae) from northwestern Haiti, Hispaniola. Jour. Herpetol. 13: 199–202.

Eleutherodactylus greyi. *Eleutherodactylus greyi* Dunn 1926: 213. • CUB • Dunn, E. R. 1926. Additional frogs from Cuba. Occas. Pap. Boston Soc. Nat. Hist. 5: 209–215.

Eleutherodactylus griphus. *Eleutherodactylus griphus* Crombie 1986: 146. • JAM • Crombie, R. I. 1986. Another new forest-dwelling frog (Leptodactylidae: *Eleutherodactylus*) from the Cockpit Country of Jamaica. Trans. San Diego Soc. Nat. Hist. 21: 145–153.

Eleutherodactylus gryllus. *Eleutherodactylus gryllus* Schmidt 1920: 172. • PRB • Schmidt, K. P. 1920. Contributions to the herpetology of Porto Rico. Ann. New York Acad. Sci. 28: 167–200.

Eleutherodactylus guanahacabibes. *Eleutherodactylus guanahacabibes* Estrada and Novo Rodríguez. 1985: 2. • CUB • Estrada, A. R. and J. N. Rodríguez. 1985. Nueva especie de *Eleutherodactylus* del grupo *ricordi* (Anura: Leptodactylidae). Poeyana 303: 1–10.

Eleutherodactylus guantanamera. *Eleutherodactylus guantanamera* Hedges, Estrada, and Thomas 1992: 76. • CUB • Holotype, KMNHNCU 589; type-locality, "El Molino (7 km W Palenque), Guantánamo Prov., Cuba, 405 m" • Hedges, S. B., A. R. Estrada, and R. Thomas. 1992. Three new species of *Eleutherodactylus* from eastern Cuba, with notes on vocalizations of other species (Anura: Leptodactylidae). Herpetol. Monogr. 6: 68–83.

Eleutherodactylus gundlachi. *Eleutherodactylus gundlachi* Schmidt 1920: 3 (substitute name for *Eleutherodactylus plicatus* Barbour 1914, not *Hylodes plicatus* Günther 1901 [=*Eleutherodactylus rhodopis* Cope]). • CUB • Schmidt, K. P. 1920. Some new and rare amphibians and reptiles from Cuba. Proc. Linnaean Soc. New York 33: 3–8.

Eleutherodactylus haitianus. *Eleutherodactylus haitianus* Barbour 1942: 179 (substitute name for *Eleutherodactylus intermedius* Cochran 1941, not *Eleutherodactylus intermedius* Barbour and Shreve 1937). • HISP • Barbour, T. 1942. Two preoccupied names. Copeia 1942: 179. CAAR: Schwartz, A. 1983. No. 335.

Eleutherodactylus hedricki. *Eleutherodactylus hedricki* Rivero 1963: 2. • PRB • Rivero, J. A. 1963. *Eleutherodactylus hedricki*, a new species of frog from Puerto Rico (Salientia, Leptodactylidae). Breviora 185: 1–7.

Eleutherodactylus heminota. *Eleutherodactylus bakeri heminota* Shreve and Williams 1963: 325. • HISP • Shreve, B. and E. E. Williams. 1963. The herpetology of the Port-au-Prince region and Gonave Island, Haiti. Part II. The frogs. Bull. Mus. Comp. Zool., Harvard Univ. 129: 302–342. CAAR: Schwartz, A. 1983. No. 336.

Eleutherodactylus hypostenor. *Eleutherodactylus hypostenor* Schwartz 1965: 498. • HISP • Schwartz, A. 1965. Variation and natural history of *Eleutherodactylus ruthae* on Hispaniola. Bull. Mus. Comp. Zool., Harvard Univ. 132: 479–508. CAAR: Schwartz, A. 1984. No. 346.

Eleutherodactylus inoptatus. *Leptodactylus inoptatus* Barbour 1914: 252. • HISP • Barbour, T. 1914. A contribution to the zoögeography of the West Indies, with especial reference to the amphibians and reptiles. Mem. Mus. Comp. Zoöl., Harvard Univ. 44: 209–359.

Eleutherodactylus intermedius. *Eleutherodactylus intermedius* Barbour and Shreve 1937: 384. • CUB • Barbour, T. and B. Shreve. 1937. Novitates cubanae. Bull. Mus. Comp. Zoöl., Harvard Coll. 80: 377–387.

Eleutherodactylus ionthus. *Eleutherodactylus varians ionthus* Schwartz 1960: 6. • CUB • Holotype, AMNH 63414; type-locality, "6 mi. E La Maya, Santiago de Cuba Prov., Cuba" • Schwartz, A. 1960. Nine new Cuban frogs of the genus *Eleutherodactlyus*. Sci. Publ. Reading Public Mus. Art Gallery 11: 1–50. Elevated from subspecific status by Hedges et al. (1992*a*).

Eleutherodactylus jamaicensis. *Eleutherodactylus jamaicensis* Barbour 1910: 287. • JAM • Barbour, T. 1910. Notes on the herpetology of Jamaica. Bull. Mus. Comp. Zoöl., Harvard Univ. 52: 271–301.

Eleutherodactylus jasperi. *Eleutherodactylus jasperi* Drewry and Jones 1976: 161. • PRB • Drewry, G. E. and K. Jones. 1976. A new ovoviviparous frog, *Eleutherodactylus jasperi* (Amphibia, Anura, Leptodactylidae) from Puerto Rico. Jour. Herpetol. 10: 161–165.

Eleutherodactylus johnstonei. *Eleutherodactylus johnstonei* Barbour 1914: 249. • JAM, LA • Barbour, T. 1914. A contribution to the zoögeography of the West Indies, with especial reference to the amphibians and reptiles. Mem. Mus. Comp. Zoöl., Harvard Coll. 44: 209–359. CAAR: Kaiser, H. and J. D. Hardy, Jr. 1994. No. 581.

Eleutherodactylus jugans. *Eleutherodactylus jugans* Cochran 1937: 312 (substitute name for *Leptodactylus* [= *Eleutherodactylus*] *darlingtoni* Cochran 1935: 372, not *E. darlingtoni* Cochran 1935: 368). • HISP • Cochran, D. M. 1937. A necessary change in an amphibian name. Jour. Washington Acad. Sci. 27: 312.

Eleutherodactylus junori. *Eleutherodactylus junori* Dunn 1926: 120. • JAM • Dunn, E. R. 1926. The frogs of Jamaica. Proc. Boston Soc. Nat. Hist. 38: 111–130.

Eleutherodactylus karlschmidti. *Eleutherodactylus karlschmidti* Grant 1931: 55. • PRB • Grant, C. 1931. A new frog from Porto Rico. Copeia 1931: 55–56.

Eleutherodactylus klinikowskii. *Eleutherodactylus klinikowskii* Schwartz 1959: 62. • CUB • Schwartz, A. 1959. The status of *Eleutherodactylus pinarensis* and a new species of the genus from western Cuba. Herpetologica 15: 61–69.

Eleutherodactylus lamprotes. *Eleutherodactylus lamprotes* Schwartz 1973: 253. • HISP • Schwartz, A. 1973. Six new species of *Eleutherodactylus* from Hispaniola. Jour. Herpetol. 7: 249–273.

Eleutherodactylus leberi. *Eleutherodactylus leberi* Schwartz 1965: 27. • CUB • Schwartz, A. 1965. A new Cuban *Eleutherodactylus* of the *auriculatus* group. Herpetologica 21: 27–31.

Eleutherodactylus lentus. *Hylodes lentus* Cope 1862: 151. • PRB • Cope, E. D. 1862. On some new and little known American Anura. Proc. Acad. Nat. Sci. Philadelphia 14: 151–159.

Eleutherodactylus leoncei. *Eleutherodactylus leoncei* Shreve and Williams 1963: 335. • HISP • Shreve, B. and E. E. Williams. 1963. The herpetology of the Port-au-Prince region and Gonave Island, Haiti, part II. The frogs. Bull. Mus. Comp. Zool., Harvard Univ. 129: 302–342.

Eleutherodactylus limbatus. *Phyllobates limbatus* Cope 1862: 154. • CUB • Cope, E. D. 1862. On some new and little known American Anura. Proc. Acad. Nat. Sci. Philadelphia 14: 151–159.

Eleutherodactylus locustus. *Eleutherodactylus locustus* Schmidt 1920: 174. • PRB • Schmidt, K. P. 1920. Contributions to the herpetology of Porto Rico. Ann. New York Acad. Sci. 28: 167–200.

Eleutherodactylus lucioi. *Eleutherodactylus lucioi* Schwartz 1980: 105. • HISP • Schwartz, A. 1980. Two new species of *Eleutherodactylus* (Amphibia, Anura, Leptodactylidae) from Hispaniola. Ann. Carnegie Mus. 49: 103–112.

Eleutherodactylus luteolus. *Litoria luteola* Gosse 1851: 366. • JAM • Gosse, P. H. 1851. A Naturalist's Sojourn in Jamaica. Longman, Brown, Green and Longmans, London.

Eleutherodactylus mariposa. *Eleutherodactylus mariposa* Hedges, Estrada, and Thomas 1992: 78. • CUB • Holotype, MNHNCU 591; type-locality, "5.4 km WSW La Tagua, Meseta del Guaso, Cuba, 720 m" • Hedges, S. B., A. R. Estrada, and R. Thomas. 1992. Three new species of *Eleutherodactylus* from eastern Cuba, with notes on vocalizations of other species (Anura: Leptodactylidae). Herpetol. Monogr. 6: 68–83.

Eleutherodactylus martinicensis. *Hylodes martinicensis* Tschudi 1838: 77. • LA • Tschudi, J. J. 1838. Classification der Batrachier, mit Berucksichtigung der fossilen Thiere dieser Abtheilung der Reptilien. Petitpierre, Neuchâtel. CAAR: Kaiser, H. and J. D. Hardy, Jr. 1994. No. 582. **Plate 1K.**

Eleutherodactylus melacara. *Eleutherodactylus melacara* Hedges, Estrada, and Thomas 1992: 70. • CUB • Holotype, MNHNCU 252; type-locality, "Estribo Turquino, 1–2 km N Pico Turquino, Santiago de Cuba Prov., Cuba, 1770 m" • Hedges, S. B., A. R. Estrada, and R. Thomas. 1992. Three new species of *Eleutherodactylus* from eastern Cuba, with notes on vocalizations of other species (Anura: Leptodactylidae). Herpetol. Monogr. 6: 68–83.

Eleutherodactylus minutus. *Eleutherodactylus minutus* Noble 1923: 4. • HISP • Noble, G. K. 1923. Six new batrachians from the Dominican Republic. Amer. Mus. Novit. 61: 1–6.

Eleutherodactylus monensis. *Hylodes monensis* Meerwarth 1901: 39. • PRB • Meerwarth, H. 1901. Die westindischen Reptilien und Batrachier des naturhistorischen Museums in Hamburg. Mitt. Naturhist. Mus. Hamburg 18: 1–41.

Eleutherodactylus montanus. *Eleutherodactylus montanus* Schmidt 1919: 519. • HISP • Schmidt, K. P. 1919. Descriptions of new amphibians and reptiles from Santo Domingo and Navassa. Bull. Amer. Mus. Nat. Hist. 41: 519–525.

Eleutherodactylus neodreptus. *Eleutherodactylus neodreptus* Schwartz 1965: 165. • HISP • Schwartz, A. 1965. A new species of *Eleutherodactylus* (Amphibia: Leptodactylidae) from the Sierra de Baoruco, República Dominicana. Proc. Biol. Soc. Washington 78: 165–168. This species soon will be allocated to the synonymy of *E. audanti* (S. B. Hedges, manuscript).

Eleutherodactylus nortoni. *Eleutherodactylus nortoni* Schwartz 1976: 165. • HISP • Schwartz, A. 1976. Two new species of Hispaniolan *Eleutherodactylus* (Leptodactylidae). Herpetologica 32: 163–171.

Eleutherodactylus nubicola. *Eleutherodactylus nubicola* Dunn 1926: 116. • JAM • Dunn, E. R. 1926. The frogs of Jamaica. Proc. Boston Soc. Nat. Hist. 38: 111–130.

Eleutherodactylus orcutti. *Eleutherodactylus orcutti* Dunn 1928: 1. • JAM • Dunn, E. R. 1928. Two new frogs from Jamaica. Proc. United States Natl. Mus. 74, art. 22: 1–2.

Eleutherodactylus oxyrhynchus. *Hylodes oxyrhynchus* Duméril and Bibron 1841: 622. • HISP • Duméril, A. M. C. and G. Bibron. 1841. Érpétologie Générale ou Histoire Complète des Reptiles. Vol. 8. Libr. Encyclopédique Roret, Paris.

Eleutherodactylus pantoni. *Eleutherodactylus pantoni* Dunn 1926: 119. • JAM • Dunn, E. R. 1926. The frogs of Jamaica. Proc. Boston Soc. Nat. Hist. 38: 111–130.

Eleutherodactylus parabates. *Eleutherodactylus parabates* Schwartz 1964: 9. • HISP • Schwartz, A. 1964. Three new species of frogs (Leptodactylidae, *Eleutherodactylus*) from Hispaniola. Breviora 208: 1–15.

Eleutherodactylus parapelates. *Eleutherodactylus parapelates* Hedges and Thomas 1987: 269. • HISP • Hedges, S. B. and R. Thomas. 1987. A new burrowing frog from Hispaniola with comments on the *inoptatus* group of the genus *Eleutherodactylus* (Anura: Leptodactylidae). Herpetologica 43: 269–279.

Eleutherodactylus patriciae. *Eleutherodactylus patriciae* Schwartz 1964 (1965): 474. • HISP • Schwartz, A. 1964 (1965). Two new species of *Eleutherodactylus* from the eastern Cordillera Central of the Republica Dominicana. Carib. Jour. Sci. 4: 473–484.

Eleutherodactylus paulsoni. *Eleutherodactylus paulsoni* Schwartz 1964: 5. • HISP • Schwartz, A. 1964. Three new species of frogs (Leptodactylidae, *Eleutherodactylus*) from Hispaniola. Breviora 208: 1–15.

Eleutherodactylus pentasyringos. *Eleutherodactylus pantoni pentasyringos* Schwartz and Fowler 1973: 114. • JAM • Schwartz, A. and D. C. Fowler. 1973. The Anura of Jamaica: a progress report. Stud. Fauna Curaçao Carib. Isl. 43: 50–142.

Eleutherodactylus pezopetrus. *Eleutherodactylus pezopetrus* Schwartz 1960: 37. • CUB • Schwartz, A. 1960. Nine new Cuban frogs of the genus *Eleutherodactylus*. Sci. Publ. Reading Public Mus. Art Gallery 11: 1–50.

Eleutherodactylus pictissimus. *Eleutherodactylus pictissimus* Cochran 1935: 371. • HISP • Cochran, D. M. 1935. New reptiles and amphibians collected in Haiti by P. J. Darlington. Proc. Boston Soc. Nat. Hist. 40: 367–375.

Eleutherodactylus pinarensis. *Eleutherodactylus pinarensis* Dunn 1926: 213. • CUB • Dunn, E. R. 1926. Additional frogs from Cuba. Occas. Pap. Boston Soc. Nat. Hist. 5: 209–215.

Eleutherodactylus pinchoni. *Eleutherodactylus pinchoni* Schwartz 1967: 45. • LA • Schwartz, A. 1967. Frogs of the genus *Eleutherodactylus* in the Lesser Antilles. Stud. Fauna Curaçao Carib. Isl. 24: 1–62.

Eleutherodactylus pituinus. *Eleutherodactylus pituinus* Schwartz 1964 (1965): 479. • HISP • Schwartz, A. 1964 (1965). Two new species of *Eleutherodactylus* from the eastern Cordillera Central of the Republica Dominicana. Carib. Jour. Sci. 4: 473–484.

Eleutherodactylus planirostris. *Hylodes planirostris* Cope 1862: 153. • BAH, CAY, CUB • Cope, E. D. 1862. On some new and little known American Anura. Proc. Acad. Nat. Sci. Philadelphia 14: 151–159. CAAR: Schwartz, A. 1974. No. 154.

Eleutherodactylus poolei. *Eleutherodactylus poolei* Cochran 1938: 93. • HISP • Cochran, D. M. 1938. A new species of frog from Haiti. Proc. Biol. Soc. Washington 51: 93–94.

Eleutherodactylus portoricensis. *Eleutherodactylus portoricensis* Schmidt 1927: 2. • PRB • Schmidt, K. P. 1927. A new tree-frog from Porto Rico. Amer. Mus. Novit. 279: 1–3.

Eleutherodactylus probolaeus. *Eleutherodactylus pictissimus probolaeus* Schwartz 1965: 110. • HISP • Schwartz, A. 1965. Geographic variation in two species of Hispaniolan *Eleutherodactylus*, with notes on Cuban members of the *ricordi* group. Stud. Fauna Curaçao Carib. Isl. 22: 98–123.

Eleutherodactylus rhodesi. *Eleutherodactylus rhodesi* Schwartz 1980: 108. • HISP • Schwartz, A. 1980. Two new species of *Eleutherodactylus* (Amphibia, Anura, Leptodactylidae) from Hispaniola. Ann. Carnegie Mus. 49: 103–112.

Eleutherodactylus richmondi. *Eleutherodactylus richmondi* Stejneger 1904: 593. • PRB • Stejneger, L. 1904. The herpetology of Porto Rico. Rept. United States Natl. Mus. 1902: 551–724.

Eleutherodactylus ricordii. *Hylodes ricordii* Duméril and Bibron 1841: 623. • CUB • Duméril, A. M. C. and G. Bibron. 1841. Érpétologie Générale ou Histoire Complète des Reptiles. Vol. 8. Libr. Encyclopédique Roret, Paris.

Eleutherodactylus ronaldi. *Eleutherodactylus ronaldi* Schwartz 1960: 14. • CUB • Schwartz, A. 1960. Nine new Cuban frogs of the genus *Eleutherodactylus*. Sci. Publ. Reading Public Mus. Art Gallery 11: 1–50.

Eleutherodactylus rufifemoralis. *Eleutherodactylus rufifemoralis* Noble and Hassler 1933: 4. • HISP •

Noble, G. K. and W. G. Hassler. 1933. Two new species of frogs, five new species and a new race of lizards from the Dominican Republic. Amer. Mus. Novit. 652: 1–17. CAAR: Ruder, J. M., R. Powell, and J. S. Parmerlee, Jr. 1995. No. 601.

Eleutherodactylus ruthae. *Eleutherodactylus ruthae* Noble 1923: 6. • HISP • Noble, G. K. 1923. Six new batrachians from the Dominican Republic. Amer. Mus. Novit. 61: 1–6.

Eleutherodactylus schmidti. *Eleutherodactylus schmidti* Noble 1923: 5. • HISP • Noble, G. K. 1923. Six new batrachians from the Dominican Republic. Amer. Mus. Novit. 61: 1–6.

Eleutherodactylus schwartzi. *Eleutherodactylus schwartzi* Thomas 1966: 386. • PRB • Thomas, R. 1966. New species of Antillean *Eleutherodactylus*. Quart. Jour. Florida Acad. Sci. 28: 375–391. **Plate 1J.**

Eleutherodactylus sciagraphus. *Eleutherodactylus sciagraphus* Schwartz 1973: 259. • HISP • Schwartz, A. 1973. Six new species of *Eleutherodactylus* (Anura, Leptodactylidae) from Hispaniola. Jour. Herpetol. 7: 249–273.

Eleutherodactylus semipalmatus. *Eleutherodactylus semipalmatus* Shreve 1936: 94. • HISP • Shreve, B. 1936. A new *Anolis* and new Amphibia from Haiti. Proc. New England Zoöl. Club 15: 93–99.

Eleutherodactylus shrevei. *Eleutherodactylus urichi shrevei* Schwartz 1967: 13. • LA • Holotype, MCZ 43230; type-locality, "Lowrt, 1000 ft., St. Andrew Parish, St. Vincent" • Schwartz, A. 1967. Frogs of the genus *Eleutherodactylus* in the Lesser Antilles. Stud. Fauna Curaçao Carib. Isl. 24: 1–62. Elevated from subspecific status by Kaiser et al. (1994 [1995]).

Eleutherodactylus sierramaestrae. *Eleutherodactylus sierramaestrae* Schmidt 1920: 3. • CUB • Schmidt, K. P. 1920. Some new and rare amphibians and reptiles from Cuba. Proc. Linnaean Soc. New York 33: 3–8.

Eleutherodactylus sisyphodemus. *Eleutherodactylus sisyphodemus* Crombie 1977: 194. • JAM • Crombie, R. I. 1977. A new species of frog of the genus *Eleutherodactylus* (Amphibia: Leptodactylidae) from the Cockpit Country of Jamaica. Proc. Biol. Soc. Washington 90: 194–204.

Eleutherodactylus symingtoni. *Eleutherodactylus symingtoni* Schwartz 1957: 210. • CUB • Schwartz, A. 1957. A new species of *Eleutherodactylus* (Amphibia: Leptodactylidae) from Cuba. Proc. Biol. Soc. Washington 70: 209–212.

Eleutherodactylus thomasi. *Eleutherodactylus thomasi* Schwartz 1959: 3. • CUB • Schwartz, A. 1959. A new species of frog of the *Eleutherodactylus ricordi* group from central Cuba. Amer. Mus. Novit. 1926: 1–16.

Eleutherodactylus thorectes. *Eleutherodactylus thorectes* Hedges 1988: 636. • HISP • Hedges, S. B. 1988. A new diminutive frog from Hispaniola (Leptodactylidae: *Eleutherodactylus*). Copeia 1988: 636–641.

Eleutherodactylus toa. *Eleutherodactylus toa* Estrada and Hedges 1991: 140. • CUB • Holotype, CARE (colección de A. R. Estrada) 00438; type-locality, "cabeza del río Yarey, Yateras, Guantánamo Cuba" • Estrada, A. R. and S. B. Hedges. 1991. Nueva especie de *Eleutherodactylus* (Anura: Leptodactylidae) de la región oriental de Cuba. Carib. Jour. Sci. 27: 139–145.

Eleutherodactylus turquinensis. *Eleutherodactylus turquinensis* Barbour and Shreve 1937: 380. • CUB • Barbour, T. and B. Shreve. 1937. Novitates cubanae. Bull. Mus. Comp. Zoöl., Harvard Coll. 80: 377–387.

Eleutherodactylus unicolor. *Eleutherodactylus unicolor* Stejneger 1904: 597. • PRB • Stejneger, L. 1904. The herpetology of Porto Rico. Rept. United States Natl. Mus. 1902: 551–724.

Eleutherodactylus varians. *Hylodes varians* Gundlach and Peters *in* Peters 1864: 390. • CUB • Peters, W. 1864. Über einige neue Säugethiere (*Mormops*, *Macrotus*, *Vesperus*, *Molossus*, *Capromys*), Amphibien (*Platydactylus*, *Otocryptis*, *Euprepes*, *Ungalia*, *Dromicus*, *Tropidonotus*, *Xenodon*, *Hylodes*) und Fische (*Sillago*, *Sebastes*, *Channa*, *Myctophum*, *Carassius*, *Barbus*, *Capoëta*, *Poecilia*, *Sauren-*

chelys, *Leptocephalus*). Monatsber. Königl. Akad. Wiss. Berlin, 1864: 381–399.

Eleutherodactylus varleyi. *Eleutherodactylus varleyi* Dunn 1925: 163. • CUB • Dunn, E. R. 1925. New frogs from Cuba. Occas. Pap. Boston Soc. Nat. Hist. 5: 163–166.

Eleutherodactylus ventrilineatus. *Leptodactylus ventrilineatus* Shreve 1936: 98. • HISP • Shreve, B. 1936. A new *Anolis* and new Amphibia from Haiti. Proc. New England Zoöl. Club 15: 93–99.

Eleutherodactylus warreni. *Eleutherodactylus warreni* Schwartz 1976: 26. • HISP • Schwartz, A. 1976. Variation and relationships of some Hispaniolan frogs (Leptodactylidae, *Eleutherodactylus*) of the *ricordi* group. Bull. Florida State Mus. 21: 1–46.

Eleutherodactylus weinlandi. *Eleutherodactylus weinlandi* Barbour 1914: 246. • HISP • Barbour, T. 1914. A contribution to the zoögeography of the West Indies, with especial reference to amphibians and reptiles. Mem. Mus. Comp. Zoöl., Harvard Coll. 44: 209–359.

Eleutherodactylus wetmorei. *Eleutherodactylus wetmorei* Cochran 1932: 191. • HISP • Cochran, D. M. 1932. A new frog, *Eleutherodactylus wetmorei*, from the Republic of Haiti. Proc. Biol. Soc. Washington 45: 191–193.

Eleutherodactylus wightmanae. *Eleutherodactylus wightmanae* Schmidt 1920: 181. • PRB • Schmidt, K. P. 1920. Contributions to the herpetology of Porto Rico. Ann. New York Acad. Sci. 28: 167–200.

Eleutherodactylus zeus. *Eleutherodactylus zeus* Schwartz 1958: 38. • CUB • Schwartz, A. 1958. Another new large *Eleutherodactylus* (Amphibia: Leptodactylidae) from western Cuba. Proc. Biol. Soc. Washington 71: 37–42.

Eleutherodactylus zugi. *Eleutherodactylus zugi* Schwartz 1958: 127. • CUB • Schwartz, A. 1958. Another new species of *Eleutherodactylus* (Amphibia: Leptodactylidae) from western Cuba. Jour. Washington Acad. Sci. 48: 127–131.

Leptodactylus albilabris. *Cystignathus albilabris* Günther 1859: 217. • PRB • Günther, A. 1859. On the reptiles from St. Croix, West Indies, collected by Messrs. A. and E. Newton. Ann. Mag. Nat. Hist. ser. 3, 4: 209–217. **Plate 1L**.

Leptodactylus dominicensis. *Leptodactlyus dominicensis* Cochran 1923: 184. • HISP • Cochran, D. M. 1923. A new frog of the genus *Leptodactylus*. Jour. Washington Acad. Sci. 13: 184–185.

Leptodactylus fallax. *Leptodactylus fallax* Müller 1926: 200 (substitute name for *Leptodactlyus dominicensis* Müller 1923, not *Leptodactylus dominicensis* Cochran 1923). • LA • Müller, L. 1926. *Leptodactylus fallax* nom. nov. für *Leptodactylus dominicensis* L. Müll. (non Cochran). Zool. Anz. 65: 200. CAAR: Kaiser, H. 1994. No. 583.

Leptodactylus insularum. *Leptodactylus insularum* Barbour 1906: 228. • WC • Barbour, T. 1906. Vertebrata from the savanna of Panama. Pt. IV. Reptilia and Amphibia. Bull. Mus. Comp. Zoöl., Harvard Coll. 46: 224–229.

Leptodactylus validus. *Leptodactylus validus* Garman 1887: 13. • LA • Garman, S. 1887. On West Indian Batrachia in the Museum of Comparative Zoology at Cambridge, Mass. Bull. Essex Inst. 19: 13–16. According to Heyer (1994), West Indian populations previously assigned to *Leptodactylus wagneri* Peters 1862 are, in fact, *L. validus* (*L. wagneri* is restricted to mainland South America).

Family Microhylidae

Gastrophryne carolinensis. *Engystoma carolinense* Holbrook 1836: 83, pl. 2. • BAH, CAY • INT • Holbrook, J. E. 1836. North American Herpetology. [First edition, first version]. Vol. 1. J. Dobson, Philadelphia. CAAR: Nelson, C. E. 1972. No. 120.

Family Ranidae

Rana catesbeiana. *Rana catesbeiana* Shaw 1802: 106. • CUB, HISP, JAM, PRB • INT • Shaw, G. 1802. General Zoology, or Systematic Natural History. Vol. 3. Thomas Davison, London.

Rana grylio. *Rana grylio* Stejneger 1901: 212. • BAH • INT • Stejneger, L. 1901. A new species of bullfrog from Florida and the Gulf Coast. Proc. United States Natl. Mus. 24: 211–215. CAAR: Altig, R. and R. Lohoefener. 1982. No. 286.

Rana sphenocephala. *Rana halecina sphenocephala* Cope 1886: 517. • BAH • INT • Cope, E. D. 1886. Synonymic list of the North American species of *Bufo* and *Rana*, with descriptions of some new species of Batrachia, from specimens in the National Museum. Proc. Amer. Philos. Soc. 23: 514–526.

Class Reptilia, Order Testudines

Family Emydidae

Trachemys decorata. *Pseudemys decorata* Barbour and Carr 1940: 409. • HISP • Barbour, T. and A. F. Carr. 1940. Antillean terrapins. Mem. Mus. Comp. Zoöl., Harvard Coll. 54: 381–415. **Plate 2B**.

Trachemys decussata. *Emys decussata* Gray 1831: 28. • CAY, CUB • Gray, J. E. 1831. Synopsis Reptilium; or Short Descriptions of the Species of Reptiles. Part I. – Cataphracta. Tortoises, Crocodiles, and Enaliosaurians. Treuttel, Wurtz; Sowerby; and Wood, London. CAAR: Seidel, M. E. 1988. No. 440.

Trachemys scripta. *Testudo scripta* Schoepff 1792: 16, pl. 3. • LA • INT • Schoepff, J. D. 1792–1801. Historia Testudinum Iconibus Illustrata. J. J. Palmii, Erlangae.

Trachemys stejnegeri. *Pseudemys stejnegeri* Schmidt 1928: 147. • BAH, HISP, PRB • Schmidt, K. P. 1928. Amphibians and land reptiles of Porto Rico, with a list of those reported from the Virgin Islands. Scientific Survey of Porto Rico and the Virgin Islands. New York Acad. Sci., Sci. Surv. Puerto Rico Virgin Isl. 10: 1–160. CAAR: Seidel, M. E. 1988. No. 441.

Trachemys terrapen. *Testudo terrapen* Lacepède 1788: 129. • BAH, JAM • Lacepède, B.-G.-É. 1788. Histoire Naturelle des Quadrupèdes Ovipares et des Serpens. Vol. 1. Hôtel de Thou, Paris. CAAR: Seidel, M. E. 1988. No. 442.

Family Kinosternidae

Kinosternon scorpioides. *Testudo scorpioides* Linnaeus 1766: 352. • WC • Linnaeus, C. 1766. Systema Naturæ per Regna Tria Naturæ, Secundum Classes, Ordines, Genera, Species cum Characteribus, Differentiis, Synonymis, Locis. Tomus I. Editio duodecima, Reformata. L. Salvii, Holmiae.

Family Pelomedusidae

Pelusios subniger. *Testudo subniger* Lacepède 1788: 175. • LA • INT • Lacepède, B.-G.-É. 1788. Histoire Naturelle des Quadrupèdes Ovipares et des Serpens. Vol. 1. Hôtel de Thou, Paris.

Family Testudinidae

Geochelone carbonaria. *Testudo carbonaria* Spix 1824: 22. • LA, PRB • INT? (this species well may have rafted to the West Indies) • Spix, J. B. von. 1824. Animalia Nova, sive Species Novae Testudinum et Ranarum, quas in Itinere per Brasiliam, Annis MDCCCXVII–MDCCCXX. Vol. 3. F. S. Hübschmanni, Monachii. **Plate 2A**.

ORDER SQUAMATA, SUBORDER AMPHISBAENIA

Family Amphisbaenidae

Amphisbaena bakeri. *Amphisbaena bakeri* Stejneger 1904: 681. • PRB • Stejneger, L. 1904. The herpetology of Porto Rico. Ann. Rept. United States Natl. Mus. 1902: 551–724.

Amphisbaena blanoides. *Cadea blanoides* Stejneger 1916: 85 (substitute name for *Amphisbæna Punctata* Bell 1827, junior homonym of *Amphisbaena punctata* Wied 1825 [=*Leposternon microcephalum* Wagler 1824]). • CUB • Stejneger, L. 1916. Notes on amphisbænian nomenclature. Proc. Biol. Soc. Washington 29: 85. Recently assigned to *Amphisbaena* from *Cadea* (Hedges et al. 1992*b*; also see Hedges, this volume). **Plate 2E**.

Amphisbaena caeca. *Amphisbæna cæca* Cuvier 1829: 73. • PRB • Cuvier, G. L. C. F. D. 1829. Le Règne Animal Distribué d'après son Organisation. Nouvelle Édition. Vol. 2. Déterville, Paris. **Plate 2D**.

Amphisbaena caudalis. *Amphisaena caudalis* Cochran 1928: 58. • HISP • Cochran, D. M. 1928. The herpetological collections made in Haiti and its adjoining islands by Walter J. Eyerdam. Proc. Biol. Soc. Washington 41: 53–59.

Amphisbaena cubana. *Amphisbaena cubana* Gundlach and Peters *in* Peters 1878 (1879): 780. • CUB • Peters, W. 1878 (1879). Über vier neue amerikanische *Amphisbaena*-Arten. Monatsber. Königl. Akad. Wiss. Berlin 1878: 778–781.

Amphisbaena fenestrata. *Diphalus fenestratus* Cope 1861: 76. • PRB • Cope, E. D. 1861. [Some remarks defining the following species of Reptilia Squamata]. Proc. Acad. Nat. Sci. Philadelphia 13: 75–77.

Amphisbaena gonavensis. *Amphisbaena innocens gonavensis* Gans and Alexander 1962: 111. • HISP • Gans, C. and A. A. Alexander. 1962. Studies on amphisbaenids (Amphisbaenia, Reptilia). 2. On the amphisbaenids of the Antilles. Bull. Mus. Comp. Zool., Harvard Univ. 128: 65–158. CAAR: Powell, R. 1992. No. 539.

Amphisbaena innocens. *Amphisbaena innocens* Weinland 1862: 137. • HISP • Weinland, D. F. 1862. Beschreibung und Abbildung von drei neuen Sauriern. (*Embryopus Habichii* und *Amphisbaena innocens* von Haiti, und *Brachymeles Leuckartii* von Neuholland.). Abhandl. Senckenb. Naturf. Ges. 4: 131–143.

Amphisbaena manni. *Amphisbaena manni* Barbour 1914: 318. • HISP • Barbour, T. 1914. A contribution to the zoögeography of the West Indies, with especial reference to the amphibians and reptiles. Mem. Mus. Comp. Zoöl., Harvard Coll. 44: 209–359.

Amphisbaena palirostrata. *Cadea palirostrata* Dickerson 1916: 659. • CUB • Dickerson, M. C. 1916. Description of a new amphisbaenian collected by the late Dr. Charles M. Mead in 1911, on the Isle of Pines, Cuba. Bull. Amer. Mus. Nat. Hist. 35: 659–662. Recently assigned to *Amphisbaena* from *Cadea* (Hedges et al. 1992*b*; also see Hedges, this volume).

Amphisbaena schmidti. *Amphisbaena schmidti* Gans 1964: 3. • PRB • Gans, C. 1964. *Amphisbaena schmidti*, a third species of the genus from Puerto Rico (Amphisbaenia: Reptilia). Breviora 198: 1–11.

Amphisbaena xera. *Amphisbaena xera* Thomas 1966: 7. • PRB • Thomas, R. 1966. Additional notes on the amphisbaenids of Greater Puerto Rico. Breviora 249: 1–23.

SUBORDER SAURIA

Family Anguidae

Celestus agasepsoides. *Diploglossus agasepsoides* Thomas 1971: 2. • HISP • Thomas, R. 1971. A new species of *Diploglossus* (Sauria: Anguidae) from Hispaniola. Occas. Pap. Mus. Zool. Louisiana State

Univ. 40: 1–9. Recently assigned to *Celestus* from *Sauresia* (see Hedges, this volume). **Plate 3A.**

Celestus anelpistus. *Diploglossus anelpistus* Schwartz, Graham, and Duval 1979: 3. • HISP • Schwartz, A., E. D. Graham, Jr., and J. J. Duval. 1979. A new species of *Diploglossus* (Sauria: Anguidae) from Hispaniola. Proc. Biol. Soc. Washington 92: 1–9. Recently assigned to *Celestus* from *Diploglossus* (see Hedges, this volume). CAAR: Henderson, R. W. 1988. No. 424.

Celestus barbouri. *Celestus barbouri* Grant 1940: 101. • JAM • Grant, C. 1940. The herpetology of Jamaica II. The reptiles. Bull. Inst. Jamaica, Sci. Ser. 1: 61–148.

Celestus carraui. *Diploglossus carraui* Inchaústegui, Schwartz, and Henderson 1985: 196. • HISP • Inchaústegui, S. J., A. Schwartz, and R. W. Henderson. 1985. Hispaniolan giant *Diploglossus* (Sauria: Anguidae): Description of a new species and notes on the ecology of *D. warreni*. Amphibia-Reptilia 6: 195–201. Recently assigned to *Celestus* from *Diploglossus* (see Hedges, this volume). CAAR: Henderson, R. W. 1988. No. 425.

Celestus costatus. *Panolopus costatus* Cope 1861 (1862): 494. • HISP, N • Cope, E. D. 1861 (1862). On the genera *Panolopus*, *Centropyx*, *Aristelliger* and *Sphaerodactylus*. Proc. Acad. Nat. Sci. Philadelphia 13: 494–500.

Celestus crusculus. *Diploglossus crusculus* Garman 1887: 22. • CAY, JAM • Garman, S. 1887. On West Indian Geckonidæ and Anguidæ. Bull. Essex Inst. 19: 17–24.

Celestus curtissi. *Celestus curtissi* Grant 1951: 68. • HISP • Grant, C. 1951. The specific characters of the Celesti, with description of a new species of *Celestus* (Sauria, Anguidae). Copeia 1951: 67–69.

Celestus darlingtoni. *Celestus darlingtoni* Cochran 1939: 2. • HISP • Cochran, D. M. 1939. Diagnoses of three new lizards and a frog from the Dominican Republic. Proc. New England Zoöl. Club 18: 1–3.

Celestus duquesneyi. *Celestus duquesneyi* Grant 1940: 157. • JAM • Grant, C. 1940. Notes on the reptiles and amphibians of Jamaica, with diagnoses of new species and subspecies, p. 151–157. *In* Jamaica To-day. Hazell, Watson, and Viney, London and Aylesbury.

Celestus fowleri. *Diploglossus fowleri* Schwartz 1971: 3. • JAM • Schwartz, A. 1971. A new species of bromeliad-inhabiting galliwasp (Sauria: Anguidae) from Jamaica. Breviora 371: 1–10.

Celestus haetianus. *Wetmorena haetiana* Cochran 1927: 91. • HISP • Cochran, D. M. 1927. A new genus of anguid lizards from Haiti. Proc. Biol. Soc. Washington 40: 91–92. Recently assigned to *Celestus* from *Wetmorena* (Hedges et al. 1992b; see also Hedges, this volume).

Celestus hewardii. *Celestus Hewardii* Gray 1845: 118. • JAM • Gray, J. E. 1845. Catalogue of the Specimens of Lizards in the Collection of the British Museum. Trustees of the British Museum, London.

Celestus macrotus. *Celestus macrotus* Thomas and Hedges 1989: 886. • HISP • Thomas, R. and S. B. Hedges. 1989. A new *Celestus* (Sauria: Anguidae) from the Chaine de la Selle of Haiti. Copeia 1989: 886–891.

Celestus marcanoi. *Diploglossus marcanoi* Schwartz and Incháustegui 1976: 242. • HISP • Schwartz, A. and S. J. Incháustegui. 1976. A new species of *Diploglossus* (Reptilia, Lacertilia, Anguidae) from Hispaniola. Jour. Herpetol. 10: 241–246.

Celestus microblepharis. *Diploglossus microblepharis* Underwood 1959: 2. • JAM • Underwood, G. 1959. A new Jamaican galliwasp (Sauria, Anguidae). Breviora 102: 1–13.

Celestus occiduus. *Lacerta Occidua* Shaw 1802: 288. • JAM • Shaw, G. 1802. General Zoology or Systematic Natural History. Vol. 3, part 1. Thomas Davison, London.

Celestus sepsoides. *Sauresia sepsoides* Gray 1852: 282. • HISP • Gray, J. E. 1852. Description of *Sauresia*, a new genus of Scincidae from St. Domingo. Ann. Mag. Nat. Hist., ser. 2, 10: 281–282. Recently assigned to *Celestus* from *Sauresia* (see Hedges, this volume).

Celestus stenurus. *Diploglossus stenurus* Cope 1862 (1863): 188. • HISP • Cope, E. D. 1862 (1863).

Contributions to Neotropical saurology. Proc. Acad. Nat. Sci. Philadelphia 14: 176–188.

Celestus warreni. *Diploglossus warreni* Schwartz 1970: 780. • HISP • Schwartz, A. 1970. A new species of large *Diploglossus* (Sauria: Anguidae) from Hispaniola. Proc. Biol. Soc. Washington 82: 777–787. Recently assigned to *Celestus* from *Diploglossus* (see Hedges, this volume). CAAR: Henderson, R. W. 1988. No. 426.

Diploglossus delasagra. *Seincus* [sic] (*Diploglossus*) *de la sagra* Cocteau 1838: 180. • CUB • Cocteau, J.-T. 1838. *In* J.-T. Cocteau and G. Bibron, 1838–1843. Reptiles, p. (5), 1–142. *In* R. de la Sagra, Historia Física, Política y Natural de la Isla de Cuba. Arthus Bertrand, Paris.

Diploglossus montisserrati. *Diploglossus montisserrati* Underwood 1964: 2. • LA • Underwood, G. 1964. An anguid lizard from the Leeward Islands. Breviora 200: 1–10.

Diploglossus pleii. *Diploglossus Pleii* Duméril and Bibron 1839: 605. • PRB • Duméril, A. M. C. and G. Bibron. 1839. Erpétologie Générale ou Histoire Naturelle Complète des Reptiles. Vol. 5. Fain et Thunot, Paris. **Plate 3B**.

Ophisaurus ventralis. *Anguis ventralis* Linnaeus 1766: 391. • CAY • INT • Linnaeus, C. 1766. Systema Naturæ per Regna Tria Naturæ, Secundum Classes, Ordines, Genera, Species cum Characteribus, Differentiis, Synonymis, Locis. Tomus I. Editio duodecima, Reformata. L. Salvii, Holmiae. CAAR: Holman, J. A. 1971. No. 115.

Family Gekkonidae

Aristelliger barbouri. *Aristelligella barbouri* Noble and Klingel 1932: 4. • BAH • Noble, G. K. and G. C. Klingel. 1932. The reptiles of Great Inagua Island, British West Indies. Amer. Mus. Novit. 549: 1–25. CAAR: Bauer, A. M. and A. P. Russell. 1993. No. 566.

Aristelliger cochranae. *Aristelliger cochranae* Grant 1931: 399. • N • Grant, C. 1931. A new species of *Aristelliger* from Navassa. Jour. Dept. Agric. Puerto Rico 4: 399–400. CAAR: Lynxwiler, J. R. and J. S. Parmerlee, Jr. 1993. No. 567.

Aristelliger expectatus. *Aristelliger expectatus* Cochran 1933: 33. • HISP • Holotype, USNM 75908; type-locality, "Jacmel, Département du Sud-Est, Haiti" • Cochran, D. M. 1933. A new gecko from Haiti, *Aristelliger expectatus*. Proc. Biol. Soc. Washington 46: 33–35. Elevated herein from subspecific status based on: (1) Mertens (1939) relegating *A. expectatus* to subspecific status without comment; (2) differences in size, proportions, pattern, and scalation (see original description of Cochran [1933]); (3) the hiatus between Navassa Island and the main island of Hispaniola, effectively precluding gene exchange; and (4) comments in Powell (1993). CAAR: Lynxwiler, J. R. and J. S. Parmerlee, Jr. 1993. No. 567 (as *Aristelliger cochranae expectatus*).

Aristelliger georgeensis. *Idiodactylus georgeensis* Bocourt 1873: 41. • WC • Bocourt, M.-F. 1873. *In* A. Duméril, M.-F. Bocourt, and F. Mocquard, 1870–1909, Études sur les reptiles, p. i–xiv, 1–1012. *In* Recherches Zoologiques pour servir a l'Histoire de la Faune de l'Amérique Centrale et du Mexique. Mission Scientifique au Mexique et dans l'Amérique Centrale, Recherches Zoologiques. Part 2, sect. 1. Imprimerie Imper., Paris. CAAR: Bauer, A. M. and A. P. Russell. 1993. No. 568.

Aristelliger hechti. *Aristelliger hechti* Schwartz and Crombie 1975: 308. • TC • Schwartz, A. and R. I. Crombie. 1975. A new species of the genus *Aristelliger* (Sauria: Gekkonidae) from the Caicos Islands. Proc. Biol. Soc. Washington 88: 305–313. CAAR: Bauer, A. M. and A. P. Russell. 1993. No. 569.

Aristelliger lar. *Aristelliger lar* Cope 1861 (1862): 497. • HISP • Cope, E. D. 1861 (1862). On the genera *Panolopus*, *Centropyx*, *Aristelliger* and *Sphaerodactylus*. Proc. Acad. Nat. Sci. Philadelphia 13: 494–500. CAAR: Bauer, A. M. and A. P. Russell. 1993. No. 570. **Plate 4A**.

Aristelliger praesignis. *Hemidactylus praesignis* Hallowell 1856: 222. • CAY, JAM, WC • Hallowell, E.

1856. Notes on the reptiles in the collection of the Academy of Natural Sciences in Philad'a. Proc. Acad. Nat. Sci. Philadelphia 8: 221–238. CAAR: Bauer, A. M. and A. P. Russell. 1993. No. 571.

Gonatodes albogularis. *Gymnodactylus albogularis* Duméril and Bibron 1836: 415. • CAY?, CUB, HISP • Duméril, A. M. C. and G. Bibron. 1836. Erpétologie Générale ou Histoire Naturelle Complète des Reptiles. Vol. 3. Libr. Encyclopédique Roret, Paris. **Plate 4B**.

Hemidactylus garnotii. *Hemidactylus Garnotii* Duméril and Bibron 1836: 368. • BAH • INT • Syntypes, MNHN 2318, 2318A, type-locality, "l'Île de Taïti [=Tahiti, French Polynesia]" • Duméril, A. M. C. and G. Bibron. 1836. Erpétologie Générale ou Histoire Naturelle Complète des Reptiles. Vol. 3. Libr. Encyclopédique Roret, Paris.

Hemidactylus haitianus. *Hemidactylus brookii haitianus* Meerwarth 1901: 17. • CUB, HISP, PRB • Syntypes, HZM (Universität Hamburg, Zoologisches Museum, Hamburg) 1500, 2250 (a–b), now destroyed; type-locality, "Haiti, Port-au-Prince" (for HZM 1500, only "Haiti" was given) • Meerwarth, H. 1901. Die westindischen Reptilien und Batrachier des naturhistorischen Museums in Hamburg. Mitt. Naturhist. Mus. Hamburg 18: 1–41. Elevated herein from subspecific status based on; (1) evidence that *H. brookii* in the Eastern Hemisphere constitute a species complex (R. I. Crombie, in. litt., 15 June 1995, to R. W. Henderson); (2) the distinctiveness of West Indian populations (see original description of Meerwarth [1901]); (3) comments in Powell (1993); and (4) previous use of this combination and comments by Powell and Parmerlee (1993). CAAR: Powell, R. and S. A. Maxey. 1990. No. 493 (as *Hemidactylus brookii haitianus*). **Plate 4C**.

Hemidactylus mabouia. *Gecko mabouia* Moreau de Jonnès 1818: 138. • WI • Moreau de Jonnès, A. 1818. Monographie du mabouia des murailles, ou *Gecko Mabouia* des Antilles. Bull. Scient. Soc. Philomath. Paris, ser. 3, 5: 138–139.

Hemidactylus palaichthus. *Hemidactylus palaichthus* Kluge 1969: 39. • LA • Kluge, A. G. 1969. The evolution and geographical origin of the New World *Hemidactylus mabouia-brookii* complex (Gekkonidae, Sauria). Misc. Publ. Mus. Zool. Univ. Michigan 138: 1–78. CAAR: Powell, R. 1990. No. 468.

Hemidactylus turcicus. *Lacerta turcica* Linnaeus 1758: 202. • CUB • INT • Linnaeus, C. 1758. Systema Naturæ per Regna Tria Naturæ, Secundum Classes, Ordines, Genera, Species cum Characteribus, Differentiis, Synonymis, Locis. Tomus I. Editio decima. Reformata. L. Salvii, Holmiae. CAAR: McCoy, C. J. 1970. No. 87.

Phyllodactylus pulcher. *Phyllodactylus pulcher* Gray 1830: 3. • LA • Gray, J. E. 1830. Spicilegia Zoologica. Part 2. Treuttel, Wurtz, London.

Phyllodactylus wirshingi. *Phyllodactylus wirshingi* Kerster and Smith 1955: 229. • HISP, PRB • Kerster, H. W. and H. M. Smith. 1955. The identity of the Puerto Rican species of *Phyllodactylus* (Reptilia: Squamata). Herpetologica 11: 229–232. **Plate 4D**.

Sphaerodactylus altavelensis. *Sphaerodactylus altavelensis* Noble and Hassler 1933: 7. • HISP • Noble, G. K. and W. G. Hassler. 1933. Two new species of frogs, five new species and a new race of lizards from the Dominican Republic. Amer. Mus. Novit. 652: 1–17.

Sphaerodactylus argivus. *Sphaerodactylus argivus* Garman 1888: 103. • CAY • Garman, S. 1888. Reptiles and batrachians from the Caymans and from the Bahamas. Collected by Prof. C. J. Maynard for the Museum of Comparative Zoology at Cambridge, Mass. Bull. Essex Inst. 20: 101–113.

Sphaerodactylus argus. *Sphaerodactylus Argus* Gosse 1850: 347. • BAH, CUB, JAM • Gosse, P. 1850. Description of a new genus and six new species of saurian reptiles. Ann. Mag. Nat. Hist., ser. 2, 6: 334–348.

Sphaerodactylus armasi. *Sphaerodactylus armasi* Schwartz and Garrido 1974: 339. • CUB • Schwartz, A. and O. H. Garrido. 1974. A new Cuban species of *Sphaerodactylus* (Gekkonidae) of the *nigro-*

punctatus complex. Proc. Biol. Soc. Washington 87: 337–343.

Sphaerodactylus armstrongi. *Sphaerodactylus armstrongi* Noble and Hassler 1933: 5. • HISP • Noble, G. K. and W. G. Hassler. 1933. Two new species of frogs, five new species and a new race of lizards from the Dominican Republic. Amer. Mus. Novit. 652: 1–17.

Sphaerodactylus asterulus. *Sphaerodactylus asterulus* Schwartz and Graham 1980: 8. • HISP • Schwartz, A. and E. D. Graham, Jr. 1980. The *shrevei* group of Hispaniolan *Sphaerodactylus* (Reptilia, Gekkonidae). Tulane Studies Zool. Bot. 22: 1–15.

Sphaerodactylus beattyi. *Sphaerodactylus beattyi* Grant 1937: 508. • PRB • Grant, C. 1937. Herpetological notes with new species from the American and British Virgin Islands, 1936. Jour. Dept. Agric. Puerto Rico 21: 503–522.

Sphaerodactylus becki. *Sphaerodactylus becki* Schmidt 1919: 520. • N • Schmidt, K. P. 1919. Descriptions of new amphibians and reptiles from Santo Domingo and Navassa. Bull. Amer. Mus. Nat. Hist. 41: 519–525.

Sphaerodactylus bromeliarum. *Sphaerodactylus bromeliarum* Peters and Schwartz 1972: 395. • CUB • Peters, G. and A. Schwartz. 1972. Ein neuer, Bromelien bewohnender Kugelfingergecko (Gekkonidae: *Sphaerodactylus*) aus Oriente/Cuba. Mitt. Zool. Mus. Berlin 48: 393–399.

Sphaerodactylus caicosensis. *Sphaerodactylus caicosensis* Cochran 1934: 7. • TC • Cochran, D. M. 1934. Herpetological collections from the West Indies made by Dr. Paul Bartsch under the Walter Rathbone Bacon Scholarship, 1928–1930. Smithsonian Misc. Coll. 92(7): 1–48.

Sphaerodactylus callocricus. *Sphaerodactylus callocricus* Schwartz 1976: 66. • HISP • Schwartz, A. 1976. A new species of *Sphaerodactylus* (Sauria, Gekkonidae) from the República Dominicana. Florida Sci. 39: 65–70.

Sphaerodactylus celicara. *Sphaerodactylus celicara* Garrido and Schwartz 1982: 392. • CUB • Garrido, O. H. and A. Schwartz. 1982. A new species of *Sphaerodactylus* (Reptilia: Sauria: Gekkonidae) from eastern Cuba. Proc. Biol. Soc. Washington 95: 392–397.

Sphaerodactylus cinereus. *Sphaerodactylus cinereus* Wagler 1830: 143. • HISP • Wagler, J. G. 1830. Natürliches System der Amphibien, mit vorangehender Classification der Säugethiere und Vögel. Ein Beitrag zur vergleichenden Zoologie. J. G. Cotta, München, Stuttgart, and Tübingen.

Sphaerodactylus clenchi. *Sphaerodactylus clenchi* Shreve 1968: 21. • HISP • Shreve, B. 1968. The *notatus* group of *Sphaerodactylus* (Sauria, Gekkonidae) in Hispaniola. Breviora 280: 1–28.

Sphaerodactylus cochranae. *Sphaerodactylus cochranae* Ruibal 1946: 1. • HISP • Ruibal, R. 1946. A new *Sphaerodactylus* from the Dominican Republic. Amer. Mus. Novit. 1308: 1–4.

Sphaerodactylus copei. *Sphaerodactylus copei* Steindachner 1867: 18. • BAH, HISP • Steindachner, F. 1867. Reptilien, p. 1–98, 3 pl. *In* Reise der Österreichischen Fregatte Novara um die Erde in den Jahren 1857, 1858, 1859 unter den Befehlen des Commodore B. von Wüllerstorf-Urbair. Zoology, Vol. 1, part 3. K. Gerold's Sohn, Wien.

Sphaerodactylus corticola. *Sphaerodactylus corticolus* Garman 1888: 111. • BAH • Garman, S. 1888. Reptiles and batrachians from the Caymans and from the Bahamas. Collected by Prof. C. J. Maynard for the Museum of Comparative Zoology at Cambridge, Mass. Bull. Essex Inst. 20: 101–113.

Sphaerodactylus cricoderus. *Sphaerodactylus cricoderus* Thomas, Hedges, and Garrido 1992: 362. • CUB • Holotype, MNHNCU 238; type-locality, "2.8 km N Uvero, Santiago de Cuba Prov., Cuba, 136 m" • Thomas, R., S. B. Hedges, and O. H. Garrido. 1992. Two new species of *Sphaerodactylus* from eastern Cuba (Sauria: Gekkonidae). Herpetologica 48: 358–367.

Sphaerodactylus cryphius. *Sphaerodactylus cryphius* Thomas and Schwartz 1977: 39. • HISP • Thomas, R. and A. Schwartz. 1977. Three new species of *Sphaerodactylus* (Sauria: Gekkonidae) from Hispaniola. Ann. Carnegie Mus. 46: 33–43.

Sphaerodactylus darlingtoni. *Sphaerodactylus darlingtoni* Shreve 1968: 15. • HISP • Shreve, B. 1968. The *notatus* group of *Sphaerodactylus* (Sauria, Gekkonidae) in Hispaniola. Breviora 280: 1–28.

Sphaerodactylus difficilis. *Sphaerodactylus difficilis* Barbour 1914: 265. • HISP • Barbour, T. 1914. A contribution to the zoögeography of the West Indies, with especial reference to amphibians and reptiles. Mem. Mus. Comp. Zoöl., Harvard Coll. 44: 209–359.

Sphaerodactylus docimus. *Sphaerodactylus docimus* Schwartz and Garrido 1985: 22. • CUB • Schwartz, A. and O. H. Garrido. 1985. The Cuban lizards of the genus *Sphaerodactylus* (Sauria, Gekkonidae). Milwaukee Public Mus. Contrib. Biol. Geol. 62: 1–67.

Sphaerodactylus elasmorhynchus. *Sphaerodactylus elasmorhynchus* Thomas 1966: 1. • HISP • Thomas, R. 1966. A new Hispaniolan gecko. Breviora 253: 1–5.

Sphaerodactylus elegans. *Sphæriodactylus* [sic] *elegans* MacCleay 1834: 12. • CUB, HISP • MacCleay, W. S. 1834. A few remarks tending to illustrate the natural history of two annulose genera, namely *Urania* of Fabricus and *Mygale* of Walckenäer. Proc. Zool. Soc. London 1834: 9–12.

Sphaerodactylus elegantulus. *Sphaerodactylus elegantulus* Barbour 1917: 163. • LA • Barbour, T. 1917. A new Antillean *Sphaerodactylus*. Proc. Biol. Soc. Washington 30: 163–164.

Sphaerodactylus epiurus. *Sphaerodactylus epiurus* Thomas and Hedges 1993: 350. • HISP • Holotype, USNM 317883; type-locality, Loma El Peñón (near García), approximately 5 km airline N Bejucal, 100–290 m, Altagracia Province, Dominican Republic" • Thomas, R. and S. B. Hedges. 1993. A new banded *Sphaerodactlyus* from eastern Hispaniola (Squamata: Gekkonidae). Herpetologica 49: 350–354.

Sphaerodactylus fantasticus. *Sphaerodactylus fantasticus* Duméril and Bibron 1836: 406. • LA • Duméril, A. M. C. and G. Bibron. 1836. Erpétologie Générale ou Histoire Naturelle Complète des Reptiles. Vol. 3. Libr. Encyclopédique Roret, Paris.

Sphaerodactylus gaigeae. *Sphaerodactylus gaigeae* Grant 1932: 31. • PRB • Grant, C. 1932. A new sphaerodactyl from Porto Rico. Jour. Dept. Agric. Puerto Rico 16: 31.

Sphaerodactylus goniorhynchus. *Sphaerodactylus goniorhynchus* Cope 1894 (1895): 440. • JAM • Cope, E. D. 1894 (1895). The Batrachia and Reptilia of the University of Pennsylvania West Indian expedition of 1890 and 1891. Proc. Acad. Nat. Sci. Philadelphia 46: 429–442.

Sphaerodactylus inaguae. *Sphaerodactylus inaguae* Noble and Klingel 1932: 11. • BAH • Noble, G. K. and G. C. Klingel. 1932. The reptiles of Great Inagua Island, British West Indies. Amer. Mus. Novit. 549: 1–25.

Sphaerodactylus intermedius. *Sphaerodactylus intermedius* Barbour and Ramsden 1919: 211. • CUB • Barbour, T. and C. T. Ramsden. 1919. The herpetology of Cuba. Mem. Mus. Comp. Zoöl., Harvard Coll. 47: 69–213.

Sphaerodactylus kirbyi. *Sphaerodactylus kirbyi* Lazell 1994: 2. • LA • Holotype, MCZ 175141; type-locality, "Slope above Friendship Bay, Bequia, Grenadines" • Lazell, J. 1994. A new *Sphaerodactylus* (Sauria: Gekkonidae) from Bequia, Grenada Bank, Lesser Antilles. Breviora 496: 1–20.

Sphaerodactylus klauberi. *Sphaerodactylus klauberi* Grant 1931: 207. • PRB • Grant, C. 1931. The sphaerodactyls of Porto Rico, Culebra and Mona islands. Jour. Dept. Agric. Puerto Rico 15: 199–213.

Sphaerodactylus ladae. *Sphaerodactylus ladae* Thomas and Hedges 1988: 96. • HISP • Thomas, R. and S. B. Hedges. 1988. Two new geckos (*Sphaerodactylus*) from the Sierra Martin Garcia of Hispaniola. Herpetologica 44: 96–104.

Sphaerodactylus lazelli. *Sphaerodactylus lazelli* Shreve 1968: 8. • HISP • Shreve, B. 1968. The *notatus* group of *Sphaerodactylus* (Sauria, Gekkonidae) in Hispaniola. Breviora 280: 1–28.

Sphaerodactylus leucaster. *Sphaerodactylus leucaster* Schwartz 1973: 36. • HISP • Schwartz, A. 1973. A third species of the Hispaniolan *shrevei* group of *Sphaerodactylus* (Sauria, Gekkonidae). Proc. Biol.

Soc. Washington 86: 35–40.

Sphaerodactylus levinsi. *Sphaerodactylus levinsi* Heatwole 1968: 2. • PRB • Heatwole, H. 1968. Herpetogeography of Puerto Rico. V. Description of a new species of *Sphaerodactylus* from Desecheo Island. Breviora 292: 1–6.

Sphaerodactylus macrolepis. *Sphaerodactylus macrolepis* Günther 1859: 215. • LA, PRB • Günther, A. 1859. On the reptiles from St. Croix, West Indies, collected by Messrs. A. and E. Newton. Ann. Mag. Nat. Hist., ser. 3, 4: 209–217. **Plate 4F**.

Sphaerodactylus mariguanae. *Sphaerodactylus mariguanae* Cochran 1934: 9. • BAH, TC • Cochran, D. M. 1934. Herpetological collections from the West Indies made by Dr. Paul Bartsch under the Walter Rathbone Bacon Scholarship, 1928–1930. Smithsonian Misc. Coll. 92(7): 1–48.

Sphaerodactylus microlepis. *Sphaerodactylus microlepis* Reinhardt and Lütken 1862: 278. • LA • Reinhardt, J. and C. F. Lütken. 1862. Bidrag til det vestindiske Öeriges og navnligen til de dansk-vestindiske Öers Herpetologie. Vidensk. Meddel. Naturhist. Foren. Kjøbenhavn 10–18: 153–291.

Sphaerodactylus micropithecus. *Sphaerodactylus micropithecus* Schwartz 1977: 986. • PRB • Schwartz, A. 1977. A new species of *Sphaerodactylus* (Sauria, Gekkonidae) from Isla Monito, West Indies. Proc. Biol. Soc. Washington 90: 985–992. CAAR: Dodd, C. K., Jr. 1985. No. 363.

Sphaerodactylus monensis. *Sphaerodactylus monensis* Meerwarth 1901: 20. • PRB • Meerwarth, H. 1901. Die westindischen Reptilien und Batrachier des naturhistorischen Museums in Hamburg. Mitt. Naturhist. Mus. Hamburg 18: 1–41.

Sphaerodactylus nicholsi. *Sphaerodactylus nicholsi* Grant 1931: 204. • PRB • Grant, C. 1931. The sphaerodactyls of Porto Rico, Culebra and Mona islands. Jour. Dept. Agric. Puerto Rico 15: 199–213.

Sphaerodactylus nigropunctatus. *Sphaerodactylus nigropunctatus* Gray 1845: 168. • BAH, CUB • Gray, J. E. 1845. Catalogue of the Specimens of Lizards in the Collection of the British Museum. Trustees of the British Museum, London.

Sphaerodactylus notatus. *Sphaeriodactylus* [sic] *notatus* Baird 1858 (1859): 254. • BAH, CUB, WC • Baird, S. F. 1858 (1859). Description of new genera and species of North American lizards in the museum of the Smithsonian Institution. Proc. Acad. Nat. Sci. Philadelphia 10: 253–256. CAAR: Schwartz, A. 1970. No. 90.

Sphaerodactylus nycteropus. *Sphaerodactylus nycteropus* Thomas and Schwartz 1977: 41. • HISP • Thomas, R. and A. Schwartz. 1977. Three new species of *Sphaerodactylus* (Sauria: Gekkonidae) from Hispaniola. Ann. Carnegie Mus. 46: 33–43.

Sphaerodactylus ocoae. *Sphaerodactylus ocoae* Schwartz and Thomas 1977: 61. • HISP • Schwartz, A. and R. Thomas. 1977. Two new species of *Sphaerodactylus* (Reptilia, Lacertilia, Gekkonidae) from Hispaniola. Jour. Herpetol. 11: 61–66.

Sphaerodactylus oliveri. *Sphaerodactylus oliveri* Grant 1944: 118. • CUB • Grant, C. 1944. New sphaerodactyls from Cuba and the Isle of Pines. Herpetologica 2: 118–125.

Sphaerodactylus omoglaux. *Sphaerodactylus omoglaux* Thomas 1982: 81. • HISP • Thomas, R. 1982. A new dwarf *Sphaerodactylus* from Haiti (Lacertilia: Gekkonidae). Proc. Biol. Soc. Washington 95: 81–88.

Sphaerodactylus oxyrhinus. *Sphaerodactylus oxyrhinus* Gosse 1850: 347. • JAM • Gosse, P. 1850. Descriptions of a new genus and six new species of saurian reptiles. Ann. Mag. Nat. Hist., ser. 2, 6: 344–348.

Sphaerodactylus parkeri. *Sphaerodactylus parkeri* Grant 1939: 8. • JAM • Grant, C. 1939. Two new sphaerodactyls from Jamaica. Copeia 1939: 7–13.

Sphaerodactylus parthenopion. *Sphaerodactylus parthenopion* Thomas 1965: 117. • PRB • Thomas, R. 1965. A new gecko from the Virgin Islands. Quart. Jour. Florida Acad. Sci. 28: 117–122.

Sphaerodactylus perissodactylius. *Sphaerodactylus perissodactylius* Thomas and Hedges 1988: 101. • HISP • Thomas, R. and S. B. Hedges. 1988. Two new geckos (*Sphaerodactylus*) from the Sierra Martin Garcia of Hispaniola. Herpetologica 44: 96–104.

Sphaerodactylus plummeri. *Sphaerodactylus plummeri* Thomas and Hedges 1992: 289. • HISP • Holotype, USNM 317892; type-locality, "10.6 km N Cabo Rojo on the Alcoa road (2.6 km N Pedernales–Oviedo road), Pedernales Province, Dominican Republic, elevation 80 m" • Thomas, R. and S. B. Hedges. 1992. An unusual new *Sphaerodactylus* from Hispaniola (Squamata: Gekkonidae). Jour. Herpetol. 26: 289–292.

Sphaerodactylus ramsdeni. *Sphaerodactylus ramsdeni* Ruibal 1959: 89. • CUB • Ruibal, R. 1959. A new species of *Sphaerodactylus* from Oriente, Cuba. Herpetologica 15: 89–93.

Sphaerodactylus randi. *Sphaerodactylus notatus randi* Shreve 1968: 5. • HISP • Shreve, B. 1968. The *notatus* group of *Sphaerodactylus* (Sauria, Gekkonidae) in Hispaniola. Breviora 280: 1–28.

Sphaerodactylus rhabdotus. *Sphaerodactylus rhabdotus* Schwartz 1970: 64. • HISP • Schwartz, A. 1970. A new species of gecko (Gekkonidae, *Sphaerodactylus*) from Hispaniola. Jour. Herpetol. 4: 63–67.

Sphaerodactylus richardi. *Sphaerodactylus richardi* Hedges and Garrido 1993: 300. • CUB • Holotype, USNM 325838; type-locality, "8.7 km ESE Playa Girón (Caleta Buena), Matanzas Province, Cuba, sea level" • Hedges, S. B. and O. H. Garrido. 1993. A new species of gecko (*Sphaerodactylus*) from central Cuba. Jour. Herpetol. 27: 300–306.

Sphaerodactylus richardsoni. *Sphaerodactylus Richardsoni* Gray 1845: 168. • JAM • Gray, J. E. 1845. Catalogue of the Specimens of Lizards in the Collection of the British Museum. Trustees of the British Museum, London. **Plate 4E.**

Sphaerodactylus roosevelti. *Sphaerodactylus roosevelti* Grant 1931: 203. • PRB • Grant, C. 1931. The sphaerodactyls of Porto Rico, Culebra and Mona islands. Jour. Dept. Agric. Puerto Rico 15: 199–213.

Sphaerodactylus ruibali. *Sphaerodactylus ruibali* Grant 1959: 53. • CUB • Grant, C. 1959. Another new *Sphaerodactylus* from Guantanamo, Cuba. Herpetologica 15: 53.

Sphaerodactylus sabanus. *Sphaerodactylus sabanus* Cochran 1938: 148. • LA • Cochran, D. M. 1938. Reptiles and amphibians from the Lesser Antilles collected by Dr. S. T. Danforth. Proc. Biol. Soc. Washington 51: 147–155.

Sphaerodactylus samanensis. *Sphaerodactylus samanensis* Cochran 1932: 183. • HISP • Cochran, D. M. 1932. Two new lizards from Hispaniola. Proc. Biol. Soc. Washington 45: 183–187.

Sphaerodactylus savagei. *Sphaerodactylus notatus savagei* Shreve 1968: 7. • HISP • Shreve, B. 1968. The *notatus* group of *Sphaerodactylus* (Sauria, Gekkonidae) in Hispaniola. Breviora 280: 1–28.

Sphaerodactylus scaber. *Sphaerodactylus scaber* Barbour and Ramsden 1919: 126. • CUB • Barbour, T. and C. T. Ramsden. 1919. The herpetology of Cuba. Mem. Mus. Comp. Zoöl., Harvard Coll. 47: 69–213.

Sphaerodactylus schwartzi. *Sphaerodactylus schwartzi* Thomas, Hedges, and Garrido 1992: 359. • CUB • Holotype, NMNHNCU 3438; type-locality, "Loma Redonda, 5 km NW Hatibonico, Guantánamo Prov., Cuba, 100 m" • Thomas, R., S. B. Hedges, and O. H. Garrido. 1992. Two new species of *Sphaerodactylus* from eastern Cuba (Sauria: Gekkonidae). Herpetologica 48: 358–367.

Sphaerodactylus semasiops. *Sphaerodactylus semasiops* Thomas 1975: 183. • JAM • Thomas, R. 1975. The *argus* group of West Indian *Sphaerodactylus* (Sauria: Gekkonidae). Herpetologica 31: 177–195.

Sphaerodactylus shrevei. *Sphaerodactylus shrevei* Lazell 1961: 1. • HISP • Lazell, J. D., Jr. 1961. A new species of *Sphaerodactylus* from northern Haiti. Breviora 139: 1–5.

Sphaerodactylus sommeri. *Sphaerodactylus sommeri* Graham 1981: 363. • HISP • Graham, E. D., Jr. 1981. A new species of lizard (*Sphaerodactylus*) from northwestern Haiti. Jour. Herpetol. 15: 363–366.

Sphaerodactylus sputator. *Lacerta sputator* Sparrman 1784: 164. • LA • Sparrman, A. 1784. *Lacerta sputator* och *Lacerta bimaculata*, två nya Ôdlor från America. Kongl. [Svensk] Vetensk. Acad. Handl., ser. 2, 5: 164–167.

Sphaerodactylus streptophorus. *Sphaerodactylus streptophorus* Thomas and Schwartz 1977: 34. • HISP • Thomas, R. and A. Schwartz. 1977. Three new species of *Sphaerodactylus* (Sauria: Gekkonidae) from Hispaniola. Ann. Carnegie Mus. 46: 33–43.

Sphaerodactylus thompsoni. *Sphaerodactylus thompsoni* Schwartz and Franz 1976: 369. • HISP • Schwartz, A. and R. Franz. 1976. A new species of *Sphaerodactylus* (Sauria: Gekkonidae) from Hispaniola. Proc. Biol. Soc. Washington 88: 367–372.

Sphaerodactylus torrei. *Sphaerodactylus torrei* Barbour 1914: 260. • CUB • Barbour, T. 1914. A contribution to the zoögeography of the West Indies, with especial reference to amphibians and reptiles. Mem. Mus. Comp. Zoöl., Harvard Coll. 44:209–359.

Sphaerodactylus townsendi. *Sphaerodactylus townsendi* Grant 1931: 208. • PRB • Grant, C. 1931. The sphaerodactyls of Porto Rico, Culebra and Mona islands. Jour. Dept. Agric. Puerto Rico 15: 199–213.

Sphaerodactylus underwoodi. *Sphaerodactylus underwoodi* Schwartz 1968: 250. • TC • Schwartz, A. 1968. The geckos (*Sphaerodactylus*) of the southern Bahama Islands. Ann. Carnegie Mus. 39: 227–271.

Sphaerodactylus vincenti. *Sphaerodactylus vincenti* Boulenger 1891: 354. • LA • Boulenger, G. A. 1891. On reptiles, batrachians, and fishes from the lesser West Indies. Proc. Zool. Soc. London 1891: 351–357.

Sphaerodactylus williamsi. *Sphaerodactylus williamsi* Thomas and Schwartz 1983: 96. • HISP • Thomas, R. and A. Schwartz. 1983. Variation in Hispaniolan *Sphaerodactylus* (Sauria: Gekkonidae), p. 86–98. *In* A. G. J. Rhodin and K. Miyata (eds.), Advances in Herpetology and Evolutionary Biology. Essays in Honor of Ernest E. Williams. Mus. Comp. Zool., Harvard Univ., Cambridge, Massachusetts.

Sphaerodactylus zygaena. *Sphaerodactylus zygaena* Schwartz and Thomas 1977: 63. • HISP • Schwartz, A. and R. Thomas. 1977. Two new species of *Sphaerodactylus* (Reptilia, Lacertilia, Gekkonidae) from Hispaniola. Jour. Herpetol. 11: 61–66.

Tarentola americana. *Platydactylus americanus* Gray 1831: 48. • BAH, CUB • Gray, J. E. 1831. A synopsis of the species of the class Reptilia, p. 1–110. *In* E. Griffith and E. Pidgeon, The Class Reptilia arranged by the Baron Cuvier, with Specific Descriptions. *In* E. Griffith and E. Pidgeon, The Animal Kingdom Arranged in Conformity with its Organization, by the Baron Cuvier. Whittaker, Treacher, London. **Plate 4G.**

Thecadactylus rapicauda. *Gekko rapicauda* Houttuyn 1782: 323. • LA, PRB • Houttuyn, M. 1782. Het onderscheid der salamanderen van de haagdissen in ‘t algemeen, en van de gekkos in ‘t byzonder aangetoond. Verh. Zeeuwsch Genoot. Wetensch. Vlissingen, ser. 1, 9: 305–336. **Plate 4H.**

Family Gymnophthalmidae

Bachia heteropus. *Chalcides heteropus* Wiegmann *in* Lichtenstein 1856: 17. • LA • Lichtenstein, H. 1856. Nomenclator Reptilium et Amphibiorum Musei Zoologici Berolinensis. Königl. Akad. Wiss., Berlin. **Plate 2G.**

Gymnophthalmus pleii. *Gymnophthalmus Pleii* Bocourt 1881: 473. • LA • Bocourt, M.-F. 1881. *In* A. Duméril, M.-F. Bocourt, and F. Mocquard, 1870–1909. Études sur les reptiles, p. i–xiv, 1–1012. *In* Recherches Zoologiques pour sevir a l’Histoire de la Faune de l’Amérique Centrale et du Mexique. Mission Scientifique au Mexique et dans l’Amérique Centrale, Recherches Zoologiques. Part 7. Imprimerie Imper., Paris.

Gymnophthalmus underwoodi. *Gymnophthalmus underwoodi* Grant 1958: 228. • LA • Grant, C. 1958. A new *Gymnophthalmus* (Reptilia, Teiidae) from Barbados, B.W.I. Herpetologica 14: 227–228. **Plate 2F.**

Tretioscincus bifasciatus. *Heteropus Bifasciatus* A. Duméril *in* C. Duméril and A. Duméril 1851: 182. • WC • Duméril, C. and A. Duméril. 1851. Catalogue Méthodique de la Collection des Reptiles. Gide et Baudry, Paris. **Plate 2H** (similar species).

Family Iguanidae

Ctenosaura similis. *Iguana* (*Ctenosaura*) *Similis* Gray 1831: 38. • WC • Gray, J. E. 1831. A synopsis of the species of the class Reptilia, p. 1–110. *In* E. Griffith and E. Pidgeon, The Class Reptilia arranged by the Baron Cuvier, with Specific Descriptions. (3), (4). *In* E. Griffith and E. Pidgeon, The Animal Kingdom Arranged in Conformity with its Organization, by the Baron Cuvier. Whittaker, Treacher, London. **Plate 6A.**

Cyclura carinata. *Cyclura carinata* Harlan 1824 (1825): 250. • BAH, TC • Harlan, R. 1824 (1825). Description of two species of Linnæan *Lacerta*, not before described, and construction of the new genus *Cyclura*. Jour. Acad. Nat. Sci. Philadelphia [1st ser.] 4, part 2: 242–251.

Cyclura collei. *Cyclura Collei* Gray 1845: 190. • JAM • Gray, J. E. 1845. Catalogue of the Specimens of Lizards in the Collection of the British Museum. Trustees of the British Museum, London.

Cyclura cornuta. *Lacerta cornuta* Bonnaterre 1789: 40. • HISP, N, PRB • Bonnaterre, P. J. 1789. Erpétologie. *In* Encyclopédie Méthodique. Tableau Encyclopédique et Méthodique des Trois Règnes de la Nature. Panckoucke, Paris.

Cyclura cychlura. *Ig*[*uana*]. *cychlura* Cuvier 1829: 45. • BAH • Cuvier, G. 1829. Le Règne Animal Distribué, d'après son Organisation. Nouvelle Édition [second edition]. Vol. 2. Déterville, Paris.

Cyclura nubila. *Iguana* (*Cyclura*) *Nubila* Gray 1831: 39. • CAY, CUB, PRB • Gray, J. E. 1831. A synopsis of the species of the class Reptilia, p. 1–110. *In* E. Griffith and E. Pidgeon, The Class Reptilia arranged by the Baron Cuvier, with Specific Descriptions. *In* E. Griffith and E. Pidgeon, The Animal Kingdom Arranged in Conformity with its Organization, by the Baron Cuvier. Whittaker, Treacher, London. **Plate 6B.**

Cyclura pinguis. *Cyclura pinguis* Barbour 1917: 100. • PRB • Barbour, T. 1917. Notes on the herpetology of the Virgin Islands. Proc. Biol. Soc. Washington 30: 97–103.

Cyclura ricordii. *Aloponotus ricordii* Duméril and Bibron 1837: 190. • HISP • Duméril, A. M. C. and G. Bibron. 1837. Erpétologie Générale ou Histoire Naturelle Complète des Reptiles. Vol. 4. Libr. Encyclopédique Roret, Paris.

Cyclura rileyi. *Cyclura rileyi* Stejneger 1903: 130. • BAH • Stejneger, L. 1903. A new species of large iguana from the Bahama Islands. Proc. Biol. Soc. Washington 16: 129–132.

Iguana delicatissima. *Iguana delicatissima* Laurenti 1768: 48. • LA • Laurenti, J. N. 1768. Specimen Medicum, Exhibens Synopsin Reptilium Emendatam cum Experimentis circa Venena et Antidota Reptilium Austriacorum, quod Authoritate et Consensu. J. Thomæ Trattnern, Viennae. **Plate 6C.**

Iguana iguana. *Lacerta iguana* Linnaeus 1758: 206. • CAY?, WI • Linnaeus, C. 1758. Systema Naturæ per Regna Tria Naturæ, Secundum Classes, Ordines, Genera, Species cum Characteribus, Differentiis, Synonymis, Locis. Tomus I. Editio decima. Reformata. L. Salvii, Holmiae.

Family Polychrotidae

Anolis acutus. *Anolis acutus* Hallowell 1856: 228. • PRB • Hallowell, E. 1856. Notes on reptiles in the collection of the Academy of Natural Sciences of Philad'a. Proc. Acad. Nat. Sci. Philadelphia 8:

221–238.

Anolis aeneus. *Anolis aeneus* Gray 1840: 114. • LA • Gray, J. E. 1840. Catalogue of the species of reptiles collected in Cuba by W. S. MacLeay, esq.; — with some notes on their habits extracted from his MS. Ann. Mag. Nat. Hist., ser. 1, 5: 108–115.

Anolis ahli. *Anolis ahli* Barbour 1925: 168. • CUB • Barbour, T. 1925. A new Cuban *Anolis*. Occas. Pap. Boston Soc. Nat. Hist. 5: 167–168.

Anolis alayoni. *Anolis alayoni* Estrada and Hedges 1995: 65. • CUB • Holotype, MNHNCU 2746; type-locality, "La Fangosa, Yateras, Guantánamo Prov., Cuba" • Estrada, A. R. and S. B. Hedges. 1995. A new species of *Anolis* (Sauria: Iguanidae) from eastern Cuba. Carib. Jour. Sci. 31: 65–72.

Anolis alfaroi. *Anolis alfaroi* Garrido and Hedges 1992: 27. • CUB • Holotype, MNHNCU 2725; type-locality, "2 km N La Munición, Municipio Yateras, Guantánamo Province, Cuba, 730 m" • Garrido, O. H. and S. B. Hedges. 1992. Three new grass anoles from Cuba (Sauria: Iguanidae). Carib. Jour. Sci. 28: 21–29.

Anolis aliniger. *Anolis chloro-cyanus aliniger* Mertens 1939: 66. • HISP • Mertens, R. 1939. Herpetologische Ergebnisse einer Reise nach der Insel Hispaniola, Westindien. Abhandl. Senckenb. Naturf. Ges. 449: 1–84.

Anolis allisoni. *Anolis allisoni* Barbour 1928: 58. • CUB • Barbour, T. 1928. Reptiles from the Bay Islands. Proc. New England Zoöl. Club 10: 55–61.

Anolis allogus. *Anolis allogus* Barbour and Ramsden 1919: 159. • CUB • Barbour, T. and C. T. Ramsden. 1919. The herpetology of Cuba. Mem. Mus. Comp. Zoöl., Harvard Coll. 47: 69–213.

Anolis altavelensis. *Anolis altavelensis* Noble and Hassler 1933: 9. • HISP • Noble, G. K. and W. G. Hassler. 1933. Two new species of frogs, five new species and a new race of lizards from the Dominican Republic. Amer. Mus. Novit. 652: 1–17.

Anolis alumina. *Anolis alumina* Hertz 1976: 2. • HISP • Hertz, P. E. 1976. *Anolis alumina*, new species of grass anole from the Barahona Peninsula of Hispaniola. Breviora 437: 1–19.

Anolis alutaceus. *Anolis* (*Dracontura*) *alutaceus* Cope 1861: 212. • CUB • Cope, E. D. 1861. Notes and descriptions of anoles. Proc. Acad. Nat. Sci. Philadelphia 13: 208–215.

Anolis anfiloquioi. *Anolis anfiloquioi* Garrido 1980: 17. • CUB • Garrido, O. H. 1980. Revisión del complejo *Anolis alutaceus* (Lacertilia: Iguanidae) y descripción de una nueva especie de Cuba. Poeyana 201: 1–41.

Anolis angusticeps. *Anolis augusticeps* [sic] Hallowell 1856: 228, 237 (where spelled *angusticeps*). • BAH, CUB • Hallowell, E. 1856. Notes on reptiles in the collection of the Academy of Natural Sciences of Philad'a. Proc. Acad. Nat. Sci. Philadelphia 8: 221–238. **Plate 5C**.

Anolis argenteolus. *Anolis* (*Gastrotropis*) *argenteolus* Cope 1861: 213. • CUB • Cope, E. D. 1861. Notes and descriptions of anoles. Proc. Acad. Nat. Sci. Philadelphia 13: 208–215.

Anolis argillaceus. *Anolis* (*Acantholis*) *argillaceus* Cope 1862: 176. • CUB • Cope, E. D. 1862. Contributions to Neotropical saurology. Proc. Acad. Nat. Sci. Philadelphia 14: 176–188.

Anolis armouri. *Audantia armouri* Cochran 1934: 171. • HISP • Cochran, D. M. 1934. Herpetological collections made in Hispaniola by the *Utowana* Expedition. Occas. Pap. Boston Soc. Nat. Hist. 8: 163–188. CAAR: Lenart, L. A., J. S. Parmerlee, Jr., and R. Powell. 1995. No. 607.

Anolis bahorucoensis. *Anolis bahorucoensis* Noble and Hassler 1933: 11. • HISP • Noble, G. K. and W. G. Hassler. 1933. Two new species of frogs, five new species and a new race of lizards from the Dominican Republic. Amer. Mus. Novit. 652: 1–17. CAAR: Sowell, S. P., J. S. Parmerlee, Jr., and R. Powell. 1995. No. 608. **Plate 5E**.

Anolis baleatus. *Eupristis baleatus* Cope 1864: 168. • HISP • Cope, E. D. 1864. Contributions to the herpetology of tropical America. Proc. Acad. Nat. Sci. Philadelphia 16: 166–181.

Anolis baracoae. *Anolis equestris baracoae* Schwartz 1964: 419. • CUB • Schwartz, A. 1964. *Anolis equestris* in Oriente Province, Cuba. Bull. Mus. Comp. Zool., Harvard Univ. 131: 405–428.

Anolis barahonae. *Anolis ricordii barahonae* Williams 1962: 8. • HISP • Williams, E. E. 1962. Notes on Hispaniolan herpetology 6. The giant anoles. Breviora 155: 1–15. CAAR: Calderón, S., S. R. Bowersox, J. S. Parmerlee, Jr., and R. Powell. 1995. No. 609. **Plate 5D**.

Anolis barbatus. *Chamaeleolis barbatus* Garrido 1982: 3. • CUB • Garrido, O. H. 1982. Descripción de una nueva especie cubana de *Chamaeleolis* (Lacertilia: Iguanidae), con notas sobre su comportamiento. Poeyana 236: 1–25. The genus *Chamaeleolis* was synonymized within *Anolis* by Hass et al. (1993).

Anolis barbouri. *Chamaelinorops barbouri* Schmidt 1919: 523. • HISP • Schmidt, K. P. 1919. Descriptions of new amphibians and reptiles from Santo Domingo and Navassa. Bull. Amer. Mus. Nat. Hist. 41: 519–525. The genus *Chamaelinorops* was synonymized within *Anolis* by Hass et al. (1993). **Plate 5B**.

Anolis bartschi. *Deiroptyx bartschi* Cochran 1928: 169. • CUB • Cochran, D. M. 1928. A second species of *Deiroptyx* from Cuba. Proc. Biol. Soc. Washington 41: 169–170.

Anolis bimaculatus. *Lacerta bimaculata* Sparrman 1784: 170. • LA • Sparrman, A. 1784. *Lacerta bimaculata*, en ny Ôdla från America. Kongl. [Svenska] Vetensk. Acad. Handl., ser. 2, 5: 169–171.

Anolis birama. *Anolis birama* Garrido 1990: 158. • CUB • Holotype, MCTH (Museu Carlos de la Torre y Huerta, Holguin, Cuba) 4; type-locality, "las margenes del Río Cauto, Birama, Provincia Granma, Cuba" • Garrido, O. H. 1990. Nueva especie de *Anolis* de la seccion beta (Lacertilia: Iguanidae) para Cuba. Rev. Biol. 4: 157–162.

Anolis bremeri. *Anolis bremeri* Barbour 1914: 288. • CUB • Barbour, T. 1914. A contribution to the zoögeography of the West Indies, with especial reference to amphibians and reptiles. Mem. Mus. Comp. Zoöl., Harvard Coll. 44:209–359.

Anolis brevirostris. *Anolis brevirostris* Bocourt 1870: 11. • HISP • Bocourt, M.-F. 1870. Description de quelques sauriens nouveaux originaires de l'Amérique méridionale. Nouv. Arch. Mus. Hist. Nat. Paris 6: 11–18. CAAR: Moster, J. A., J. S. Parmerlee, Jr., and R. Powell. 1995. No. 610.

Anolis brunneus. *Anolis principalis brunneus* Cope 1894 (1895): 432. • BAH • Cope, E. D. 1894 (1895). The Batrachia and Reptilia of the University of Pennsylvania West Indian expedition of 1890 and 1891. Proc. Acad. Nat. Sci. Philadelphia 46: 429–442.

Anolis caudalis. *Anolis dominicensis caudalis* Cochran 1932: 185. • HISP • Cochran, D. M. 1932. Two new lizards from Hispaniola. Proc. Biol. Soc. Washington 45: 183–187.

Anolis centralis. *Anolis agrillaceus* [sic] *centralis* Peters 1970: 215. • CUB • Peters, G. 1970. Zur Taxonomie und Zoogeographie der kubanischen anolinen Eidechsen (Reptilia, Iguanide). Mitt. Zool. Mus. Berlin 46: 197–234.

Anolis chamaeleonides. *Anolis chamaeleonides* Duméril and Bibron 1837: 168. • CUB • Duméril, A. M. C. and G. Bibron. 1837. Erpétologie Générale ou Histoire Naturelle Complète des Reptiles. Vol. 4. Libr. Encyclopédique Roret, Paris. The genus *Chamaeleolis* was synonymized within *Anolis* by Hass et al. (1993). **Plate 5A**.

Anolis chlorocyanus. *Anolis chloro-cyanus* Duméril and Bibron 1837: 117. • HISP • Duméril, A. M. C. and G. Bibron. 1837. Erpétologie Générale ou Histoire Naturelle Complète des Reptiles. Vol. 4. Libr. Encyclopédique Roret, Paris.

Anolis christophei. *Anolis christophei* Williams 1960: 2. • HISP • Williams, E. E. 1960. Notes on Hispaniolan herpetology. 1. *Anolis christophei*, new species, from the Citadel of King Christophe, Haiti. Breviora 117: 1–7.

Anolis clivicola. *Anolis clivicolus* Barbour and Shreve 1935: 251. • CUB • Barbour, T. and B. Shreve. 1935. Notes on Cuban anoles. Occas. Pap. Boston Soc. Nat. Hist. 8: 249–253.

Anolis coelestinus. *Anolis (Ctenocercus) coelestinus* Cope 1862: 177. • HISP • Cope, E. D. 1862. Contributions to Neotropical saurology. Proc. Acad. Nat. Sci. Philadelphia 14: 176–188.

Anolis concolor. *Anolis* (*Gastrotropis*) *concolor* Cope 1862: 180. • WC • Cope, E. D. 1862. Contributions to Neotropical saurology. Proc. Acad. Nat. Sci. Philadelphia 14: 176–188.

Anolis confusus. *Anolis confusus* Estrada and Garrido 1991: 156. • CUB • Holotype, CARE (colección de A. R. Estrada) 60503; type-locality, "entre Monte Gordo y Vereón, 7 km NNE de Cabo Cruz, Niquero, Granma, Cuba" • Estrada, A. R. and O. H. Garrido. 1991. Dos nuevas especies de *Anolis* (Lacertilia: Iguanidae) de la región oriental de Cuba. Carib. Jour. Sci. 27: 146–161.

Anolis conspersus. *Anolis conspersus* Garman 1887: 273. • CAY • Garman, S. 1887. On the reptiles and batrachians of Grand Cayman. Proc. Amer. Philos. Soc. 24: 273–286. **Plate 5G.**

Anolis cooki. *Anolis cristatellus cooki* Grant 1931: 221. • PRB • Grant, C. 1931. A new species and two new subspecies of the genus *Anolis*. Jour. Dept. Agric. Puerto Rico 15: 219–222. CAAR: Jenssen, T. A. 1990. No. 488.

Anolis cristatellus. *Anolis cristatellus* Duméril and Bibron 1837: 143. • HISP, PRB • Duméril, A. M. C. and G. Bibron. 1837. Erpétologie Générale ou Histoire Naturelle Complète des Reptiles. Vol. 4. Libr. Encyclopédique Roret, Paris.

Anolis cupeyalensis. *Anolis cupeyalensis* Peters 1970: 225. • CUB • Peters, G. 1970. Zur Taxonomie und Zoogeographie der kubanischen anolinen Eidechsen (Reptilia, Iguanide). Mitt. Zool. Mus. Berlin 46: 197–234.

Anolis cuvieri. *Anolis cuvieri* Merrem 1820: 45. • PRB • Merrem, B. 1820. Versuch eines Systems der Amphibien / Tentamen Systematis Amphibiorum. J. C. Kriegeri, Marburgi.

Anolis cyanopleurus. *Anolis* (*Dracontura*) *cyanopleurus* Cope 1861: 211. • CUB • Cope, E. D. 1861. Notes and descriptions of anoles. Proc. Acad. Nat. Sci. Philadelphia 13: 208–215.

Anolis cybotes. *Anolis* (*Anolis*) *cybotes* Cope 1862: 177. • HISP • Cope, E. D. 1862. Contributions to Neotropical saurology. Proc. Acad. Nat. Sci. Philadelphia 14: 176–188. CAAR: Fobes, T. M., J. S. Parmerlee, Jr., and R. Powell. 1993. No. 564.

Anolis darlingtoni. *Xiphocercus darlingtoni* Cochran 1935: 373. • HISP • Cochran, D. M. 1935. New reptiles and amphibians collected in Haiti by P. J. Darlington. Proc. Boston Soc. Nat. Hist. 40: 367–375.

Anolis delafuentei. *Anolis delafuentei* Garrido 1982: 132. • CUB • Garrido, O. H. 1982. Nueva especie de *Anolis* (Lacertilia, Iguanidae) para Cuba. Doñana, Acta Vert. 9: 131–137

Anolis desechensis. *Anolis desechensis* Heatwole 1976: 13. • PRB • Heatwole, H. 1976. Herpetogeography of Puerto Rico. VII. Geographic variation in the *Anolis cristatellus* complex in Puerto Rico and the Virgin Islands. Occas. Pap. Mus. Nat. Hist. Univ. Kansas 46: 1–17.

Anolis distichus. *Anolis distichus* Cope 1861: 208. • BAH, HISP • Cope, E. D. 1861. Notes and descriptions of anoles. Proc. Acad. Nat. Sci. Philadelphia 13: 208–215. CAAR: Schwartz, A. 1971. No. 108.

Anolis dolichocephalus. *Anolis hendersoni dolichocephalus* Williams 1963: 8. • HISP • Williams, E. E. 1963. Notes on Hispaniolan herpetology 8. The forms related to *Anolis hendersoni* Cochran. Breviora 186: 1–13.

Anolis equestris. *Anolis equestris* Merrem 1820: 45. • CUB • Merrem, B. 1820. Versuch eines Systems der Amphibien / Tentamen Systematis Amphibiorum. J. C. Kriegeri, Marburgi.

Anolis ernestwilliamsi. *Anolis ernestwilliamsi* Lazell 1983: 102. • PRB • Lazell, J. D., Jr. 1983. Biogeography of the herpetofauna of the British Virgin Islands, with description of a new anole (Sauria: Iguanidae), p. 99–117. *In* A. G. J. Rhodin and K. Miyata (eds.), Advances in Herpetology and Evolutionary Biology. Essays in Honor of Ernest E. Williams. Mus. Comp. Zool., Harvard Univ., Cambridge, Massachusetts.

Anolis etheridgei. *Anolis etheridgei* Williams 1962: 1 (substitute name for *Anolis darlingtoni* Cochran 1939 preoccupied by *Xiphocercus* (=*Anolis*) *darlingtoni* Cochran 1935). • HISP • Williams, E. E. 1962. Notes on the herpetology of Hispaniola 7. New material of two poorly known anoles: *Anolis monticola* Shreve and *Anolis christophei* Williams. Breviora 164: 1–11.

Anolis eugenegrahami. *Anolis eugenegrahami* Schwartz 1978: 266. • HISP • Schwartz, A. 1978. A new species of aquatic *Anolis* (Sauria, Iguanidae) from Hispaniola. Ann. Carnegie Mus. 47: 261–279.

Anolis evermanni. *Anolis evermanni* Stejneger 1904: 647. • PRB • Stejneger, L. 1904. The herpetology of Porto Rico. Rept. United States Natl. Mus. 1902: 549–724.

Anolis extremus. *Anolis extremus* Garman 1887: 35. • LA • Garman, S. 1887. On West Indian reptiles. Iguanidæ. Bull. Essex Inst. 19: 25–50.

Anolis fairchildi. *Anolis fairchildi* Barbour and Shreve 1935: 357. • BAH • Barbour, T. and B. Shreve. 1935. Concerning some Bahamian reptiles, with notes on the fauna. Proc. Boston Soc. Nat. Hist. 40: 347–365.

Anolis ferreus. *Xiphosurus ferreus* Cope 1864: 168. • LA • Cope, E. D. 1864. Contributions to the herpetology of tropical America. Proc. Acad. Nat. Sci. Philadelphia 16: 166–181.

Anolis fowleri. *Anolis fowleri* Schwartz 1973: 186. • HISP • Schwartz, A. 1973. A new species of montane *Anolis* (Sauria, Iguanidae) from Hispaniola. Ann. Carnegie Mus. 44: 183–195.

Anolis fugitivus. *Anolis fugitivus* Garrido 1975: 28. • CUB • Garrido, O. H. 1975. Distribución y variación del complejo *Anolis cyanopleurus* (Lacertilia: Iguanidae) en Cuba. Poeyana 143: 1–58.

Anolis garmani. *Anolis garmani* Stejneger 1899: 602. • CAY, JAM • Stejneger, L. 1899. A new name for the great crested *Anolis* of Jamaica. Amer. Nat. 33: 601–602. CAAR: Russell, A. P. and A. M. Bauer. 1991. No. 513.

Anolis gingivinus. *Anolis gingivinus* Cope 1864: 170. • LA • Cope, E. D. 1864. Contributions to the herpetology of tropical America. Proc. Acad. Nat. Sci. Philadelphia 16: 166–181.

Anolis grahami. *Anolis Grahami* Gray 1845: 274 (substitute name for *A. punctatus* Gray 1840: 113, not *A. punctatus* Daudin 1802: 84). • JAM • Gray, J. E. 1845. Catalogue of the Specimens of Lizards in the Collection of the British Museum. Trustees of the British Museum, London. CAAR: Russell, A. P. and A. M. Bauer. 1991. No. 514.

Anolis griseus. *Anolis griseus* Garman 1887: 36. • LA • Garman, S. 1887. On West Indian reptiles. Iguanidæ. Bull. Essex Inst. 19: 25–50.

Anolis guafe. *Anolis guafe* Estrada and Garrido 1991: 148. • CUB • Holotype, CARE (colección de A. R. Estrada) 60516; type-locality, "El Guafe (2 km NE de Cabo Cruz), Granma, Cuba" • Estrada, A. R. and O. H. Garrido. 1991. Dos nuevas especies de *Anolis* (Lacertilia: Iguanidae) de la región oriental de Cuba. Carib. Jour. Sci. 27: 146–161.

Anolis guamuhaya. *Chamaeleolis guamuhaya* Garrido, Pérez-Beato, and Moreno 1991: 163. • CUB • Holotype, MNHNCU 500; type-locality, "entre Jibacoa y Topes de Collantes, Sierra de Trinidad (Sierra del Escambray o Macizo de Guamuhaya), Provincia de Sancti Spiritus, a nos 900 m de altura" • Garrido, O. H., O. Pérez-Beato, and L. V. Moreno. 1991. Nueva especie de *Chamaeleolis* (Lacertilia: Iguanidae) para Cuba. Carib. Jour. Sci. 27: 162–168. The genus *Chamaeleolis* was synonymized within *Anolis* by Hass et al. (1993).

Anolis guazuma. *Anolis guazuma* Garrido 1984: 71. • CUB • Garrido, O. H. 1984. Nueva especie de *Anolis* (Lacertilia: Iguanidae) de la Sierra del Turquino, Cuba. Carib. Jour. Sci. 19: 71–76.

Anolis gundlachi. *Anolis Gundlachi* Peters 1876 (1877): 705. • PRB • Peters, W. 1876 (1877). Über eine von Hrn. Viceconsul L. Krug und Dr. J. Gundlach auf der Insel Puertorico gemachte Sammlung von Säugethieren und Amphibien, so wie über die Entwickelung eines Batrachiers, *Hylodes martinicensis* Dum. Bibr., ohne Metamorphose. Monatsber. Königl. Preuss. Akad. Wiss. Berlin, 1876: 703–714.

Anolis haetianus. *Anolis haetianus* Garman 1887: 42. • HISP • Garman, S. 1887. On West Indian reptiles. Iguanidæ. Bull. Essex Inst. 19: 25–50.

Anolis hendersoni. *Anolis hendersoni* Cochran 1923: 225. • HISP • Cochran, D. M. 1923. A new *Anolis* from Haiti. Jour. Washington Acad. Sci. 13: 225–226.

Anolis homolechis. *Xiphosurus homolechis* Cope 1864: 169. • CUB • Cope, E. D. 1864. Contributions to the herpetology of tropical America. Proc. Acad. Nat. Sci. Philadelphia 16: 166–181.

Anolis imias. *Anolis imias* Ruibal and Williams 1961: 237. • CUB • Ruibal, R. and E. E. Williams. 1961. The taxonomy of the *Anolis homolechis* complex of Cuba. Bull. Mus. Comp. Zool., Harvard Univ. 125: 211–246.

Anolis inexpectatus. *Anolis inexpectatus* Garrido and Estrada 1989: 59. • CUB • Garrido, O. H. and A. R. Estrada. 1989. Nueva especies del complejo *Anolis alutaceus* (Lacertilia: Iguanidae) para Cuba. Rev. Biol. 3: 57–66.

Anolis insolitus. *Anolis insolitus* Williams and Rand 1969: 2. • HISP • Williams, E. E. and A. S. Rand. 1969. *Anolis insolitus*, a new dwarf anole of zoogeographic importance from the mountains of the Dominican Republic. Breviora 326: 1–21.

Anolis isolepis. *Anolis isolepis* Cope 1861: 214. • CUB • Cope, E. D. 1861. Notes and descriptions of anoles. Proc. Acad. Nat. Sci. Philadelphia 13: 208–215.

Anolis juangundlachi. *Anolis juangundlachi* Garrido 1975: 34. • CUB • Garrido, O. H. 1975. Distribución y variación del complejo *Anolis cyanopleurus* (Lacertilia: Iguanidae) en Cuba. Poeyana 143: 1–58.

Anolis jubar. *Anolis jubar* Schwartz 1968: 157. • CUB • Schwartz, A. 1968. The Cuban lizards of the *Anolis homolechis* complex. Tulane Studies Zool. 14: 140–184.

Anolis koopmani. *Anolis koopmani* Rand 1961: 1. • HISP • Rand, A. S. 1961. Notes on Hispaniolan herpetology 4. *Anolis koopmani*, new species, from the southwestern peninsula of Haiti. Breviora 137: 1–4.

Anolis krugi. *Anolis Krugi* Peters 1876 (1877): 707. • PRB • Peters, W. 1876 (1877). Über eine von Hrn. Viceconsul L. Krug und Dr. J. Gundlach auf der Insel Puertorico gemachte Sammlung von Säugethieren und Amphibien, so wie über die Entwickelung eines Batrachiers, *Hylodes martinicensis* Dum. Bibr., ohne Metamorphose. Monatsber. Königl. Preuss. Akad. Wiss. Berlin, 1876: 703–714.

Anolis lineatopus. *Anolis lineatopus* Gray 1840: 113. • JAM • Gray, J. E. 1840. Catalogue of the species of reptiles collected in Cuba by W. S. MacLeay, esq.; — with some notes on their habits extracted from his MS. Ann. Mag. Nat. Hist., ser. 1, 5: 108–115.

Anolis lividus. *Anolis lividus* Garman 1887: 43. • LA • Garman, S. 1887. On West Indian reptiles. Iguanidæ. Bull. Essex Inst. 19: 25–50.

Anolis longiceps. *Anolis longiceps* Schmidt 1919: 521. • N • Schmidt, K. P. 1919. Descriptions of new amphibians and reptiles from Santo Domingo and Navassa. Bull. Amer. Mus. Nat. Hist. 41: 519–525.

Anolis longitibialis. *Anolis longitibialis* Noble 1923: 4. • HISP • Noble, G. K. 1923. Four new lizards from Beata Island, Dominican Republic. Amer. Mus. Novit. 64: 1–5.

Anolis loysianus. *Acantholis loysiana* Duméril and Bibron 1837: 100. • CUB • Duméril, A. M. C. and G. Bibron. 1837. Erpétologie Générale ou Histoire Naturelle Complète des Reptiles. Vol. 4. Libr. Encyclopédique Roret, Paris.

Anolis luciae. *Anolis luciae* Garman 1887: 44. • LA • Garman, S. 1887. On West Indian reptiles. Iguanidæ. Bull. Essex Inst. 19: 25–50.

Anolis lucius. *Anolis lucius* Duméril and Bibron 1837: 105. • CUB • Duméril, A. M. C. and G. Bibron.

1837. Erpétologie Générale ou Histoire Naturelle Complète des Reptiles. Vol. 4. Libr. Encyclopédique Roret, Paris.

Anolis luteogularis. *Anolis luteogularis* Noble and Hassler 1935: 113. • CUB • Noble, G. K. and W. G. Hassler. 1935. A new giant *Anolis* from Cuba. Copeia 1935: 113–115.

Anolis macilentus. *Anolis macilentus* Garrido and Hedges 1992: 22. • CUB • Holotype, MNHNCU 2721; type-locality, "Río Pai, Monte Libano, south slope of the Meseta del Guaso, 650 m, Guantánamo Province, Cuba" • Garrido, O. H. and S. B. Hedges. 1992. Three new grass anoles (Sauria: Iguanidae) from Cuba. Carib. Jour. Sci. 28: 21–29.

Anolis marcanoi. *Anolis marcanoi* Williams 1975: 1. • HISP • Williams, E. E. 1975. *Anolis marcanoi* new species: sibling to *Anolis cybotes*: description and field evidence. Breviora 430: 1–9.

Anolis marmoratus. *Anolis marmoratus* Duméril and Bibron 1837: 139. • LA • Duméril, A. M. C. and G. Bibron. 1837. Erpétologie Générale ou Histoire Naturelle Complète des Reptiles. Vol. 4. Libr. Encyclopédique Roret, Paris. **Plate 5F.**

Anolis marron. *Anolis marron* Arnold 1980: 19. • HISP • Arnold, D. L. 1980. Geographic variation in *Anolis brevirostris* (Sauria: Iguanidae) in Hispaniola. Breviora 461: 1–31.

Anolis maynardi. *Anolis maynardi* Garman 1888: 107. • CAY • Garman, S. 1888. Reptiles and batrachians from the Caymans and from the Bahamas, collected by Prof. C. J. Maynard for the Museum of Comparative Zoology at Cambridge, Mass. Bull. Essex Inst. 20: 101–113.

Anolis mestrei. *Anolis mestrei* Barbour and Ramsden 1916: 19. • CUB • Barbour, T. and C. T. Ramsden. 1916. A new *Anolis* from Cuba. Proc. Biol. Soc. Washington 29: 19–20.

Anolis mimus. *Anolis mimus* Schwartz and Thomas 1975: 93 (substitute name for *Anolis montanus* Garrido, preoccupied by *Anolis oculatus montanus* Lazell 1962). • CUB • Schwartz, A. and R. Thomas. 1975. A check-list of West Indian amphibians and reptiles. Carnegie Mus. Nat. Hist. Spec. Publ. 1: 1–216.

Anolis monensis. *Anolis monensis* Stejneger 1904: 646. • PRB • Stejneger, L. 1904. The herpetology of Porto Rico. Rept. United States Natl. Mus. 1902: 549–724.

Anolis monticola. *Anolis monticola* Shreve 1936: 93. • HISP • Shreve, B. 1936. A new *Anolis* and new Amphibia from Haiti. Proc. New England Zoöl. Club 15: 93–99.

Anolis noblei. *Anolis noblei* Barbour and Shreve 1935: 250. • CUB • Barbour, T. and B. Shreve. 1935. Notes on Cuban anoles. Occas. Pap. Boston Soc. Nat. Hist. 8: 249–253.

Anolis nubilis. *Anolis nubilis* Garman 1887: 32. • LA • Garman, S. 1887. On West Indian Reptiles. Iguanidæ. Bull. Essex Inst. 19: 25–50.

Anolis occultus. *Anolis occultus* Williams and Rivero 1965: 4. • PRB • Williams, E. E. and J. A. Rivero. 1965. A new anole (Sauria, Iguanidae) from Puerto Rico. Part I. Description. Breviora 231: 1–9.

Anolis oculatus. *Xiphosurus oculatus* Cope 1879: 274. • LA • Cope, E. D. 1879. Eleventh contribution to the herpetology of tropical America. Proc. Amer. Philos. Soc. 18: 261–277. CAAR: Malhotra, A. and R. S. Thorpe. 1992. No. 540.

Anolis olssoni. *Anolis olssoni* Schmidt 1919: 522. • HISP • Schmidt, K. P. 1919. Descriptions of new amphibians and reptiles from Santo Domingo and Navassa. Bull. Amer. Mus. Nat. Hist. 41: 519–525. CAAR: Smith, J. W., J. S. Parmerlee, Jr., and R. Powell. 1995. No. 611.

Anolis opalinus. *Anolis opalinus* Gosse 1850: 345. • JAM • Gosse, P. 1850. Descriptions of a new genus and six new species of saurian reptiles. Ann. Mag. Nat. Hist., ser. 2, 6: 344–348. CAAR: Russell, A. P. and A. M. Bauer. 1991. No. 515.

Anolis ophiolepis. *Anolis* (*Dracontura*) *ophiolepis* Cope 1861: 211. • CUB • Cope, E. D. 1861. Notes and descriptions of anoles. Proc. Acad. Nat. Sci. Philadelphia 13: 208–215.

Anolis paternus. *Anolis angusticeps paternus* Hardy 1967: 25. • CUB • Hardy, J. D. 1967. Geographic variation in the West Indian lizard, *Anolis angusticeps*, with the description of a new form, *Anolis angusticeps paternus*, subsp. nov., from the Isle of Pines, Cuba (Reptilia: Iguanidae). Carib. Jour. Sci. 6: 23–31.

Anolis pigmaequestris. *Anolis pigmaequestris* Garrido 1975: 4. • CUB • Garrido, O. H. 1975. Nuevos reptiles del archipiélago cubano. Poeyana 141: 1–58.

Anolis pinchoti. *Anolis pinchoti* Cochran 1931: 354. • WC • Cochran, D. M. 1931. A new lizard (*Anolis pinchoti*) from Old Providence Island. Jour. Washington Acad. Sci. 21: 354–355.

Anolis placidus. *Anolis placidus* Hedges and Thomas 1989: 330. • HISP • Hedges, S. B. and R. Thomas. 1989. A new species of *Anolis* (Sauria: Iguanidae) from the Sierra de Neiba, Hispaniola. Herpetologica 45: 330–336.

Anolis poncensis. *Anolis poncensis* Stejneger 1904: 665. • PRB • Stejneger, L. 1904. The herpetology of Porto Rico. Rept. United States Natl. Mus. 1902: 549–724.

Anolis porcatus. *Anolis porcatus* Gray 1840: 112. • CUB, HISP • Gray, J. E. 1840. Catalogue of the species of reptiles collected in Cuba by W. S. MacLeay, esq.; — with some notes on their habits extracted from his MS. Ann. Mag. Nat. Hist., ser. 1, 5: 108–115. CAAR: Powell, R. 1992. No. 541.

Anolis porcus. *Chamaeleolis porcus* Cope 1864: 168. • CUB • Cope, E. D. 1864. Contributions to the herpetology of tropical America. Proc. Acad. Nat. Sci. Philadelphia 16: 166–181. The genus *Chamaeleolis* was synonymized within *Anolis* by Hass et al. (1993).

Anolis pulchellus. *Anolis pulchellus* Duméril and Bibron 1837: 97. • PRB • Duméril, A. M. C. and G. Bibron. 1837. Erpétologie Générale ou Histoire Naturelle Complète des Reptiles. Vol. 4. Libr. Encyclopédique Roret, Paris.

Anolis pumilus. *Anolis pumilus* Garrido 1988: 46. • CUB • Garrido, O. H. 1988. Nueva especies para la ciencia de *Anolis* (Lacertilia: Iguanidae) de Cuba perteneciente al complejo *argillaceus*. Doñana, Acta Vert. 15: 45–57.

Anolis quadriocellifer. *Anolis quadriocellifer* Barbour and Ramsden 1919: 158. • CUB • Barbour, T. and C. T. Ramsden. 1919. The herpetology of Cuba. Mem. Mus. Comp. Zoöl., Harvard Coll. 47: 69–213.

Anolis reconditus. *Anolis reconditus* Underwood and Williams 1959: 44. • JAM • Underwood, G. and E. E. Williams. 1959. The anoline lizards of Jamaica. Bull. Inst. Jamaica Sci. Ser. 9: 1–48.

Anolis richardi. *Anolis richardi* Duméril and Bibron 1837: 141. • LA • Duméril, A. M. C. and G. Bibron. 1837. Erpétologie Générale ou Histoire Naturelle Complète des Reptiles. Vol. 4. Libr. Encyclopédique Roret, Paris.

Anolis ricordii. *Anolis ricordii* Duméril and Bibron 1837: 167. • HISP • Duméril, A. M. C. and G. Bibron. 1837. Erpétologie Générale ou Histoire Naturelle Complète des Reptiles. Vol. 4. Libr. Encyclopédique Roret, Paris.

Anolis rimarum. *Anolis rimarum* Thomas and Schwartz 1967: 19. • HISP • Thomas, R. and A. Schwartz. 1967. The *monticola* group of the lizard genus *Anolis* in Hispaniola. Breviora 261: 1–27.

Anolis roosevelti. *Anolis roosevelti* Grant 1931: 219. • PRB • Grant, C. 1931. A new species and two new sub-species of the genus *Anolis*. Jour. Dept. Agric. Puerto Rico 15: 219–222. CAAR: Dodd, C. K, Jr. and H. W. Campbell. 1982. No. 300.

Anolis roquet. *Lacerta roquet* Lacepède 1788: 1. • LA • Lacepède, B.-G.-É. 1788. Histoire Naturelle des Quadrupèdes Ovipares et des Serpens. Vol. 1. Hôtel de Thou, Paris.

Anolis rubribarbus. *Anolis rubribarbus* Barbour and Ramsden 1919: 156. • CUB • Barbour, T. and C. T. Ramsden. 1919. The herpetology of Cuba. Mem. Mus. Comp. Zoöl., Harvard Coll. 47: 69–213.

Anolis rupinae. *Anolis rupinae* Williams and Webster 1974: 2. • HISP • Williams, E. E. and T. P. Webster. 1974. *Anolis rupinae* new species a syntopic sibling of *A. monticola* Shreve. Breviora 429:

1–22.

Anolis sabanus. *Anolis sabanus* Garman 1887: 39. • LA • Garman, S. 1887. On West Indian reptiles. Iguanidæ. Bull. Essex Inst. 19: 25–50.

Anolis sagrei. *Anolis Sagrei* Cocteau *in* Duméril and Bibron 1837: 149. • BAH, CAY, CUB, JAM, WC • Duméril, A. M. C. and G. Bibron. 1837. Erpétologie Générale ou Histoire Naturelle Complète des Reptiles. Vol. 4. Libr. Encyclopédique Roret, Paris.

Anolis scriptus. *Anolis scriptus* Garman 1887: 28. • BAH, TC • Garman, S. 1887. On West Indian reptiles. Iguanidæ. Bull. Essex Inst. 19: 25–50.

Anolis semilineatus. *Anolis semilineatus* Cope 1864: 171. • HISP • Cope, E. D. 1864. Contributions to the herpetology of tropical America. Proc. Acad. Nat. Sci. Philadelphia 16: 166–181.

Anolis sheplani. *Anolis sheplani* Schwartz 1974: 4. • HISP • Schwartz, A. 1974. A new species of primitive *Anolis* (Sauria, Iguanidae) from the Sierra de Baoruco, Hispaniola. Breviora 423: 1–19.

Anolis shrevei. *Audantia shrevei* Cochran 1939: 2. • HISP • Cochran, D. M. 1939. Diagnoses of three new lizards and a frog from the Dominican Republic. Proc. New England Zoöl. Club 18: 1–3.

Anolis singularis. *Anolis singularis* Williams 1965: 9. • HISP • Williams, E. E. 1965. The species of Hispaniolan green anoles (Sauria, Iguanidae). Breviora 227: 1–16.

Anolis smallwoodi. *Anolis equestris smallwoodi* Schwartz 1964: 412. • CUB • Schwartz, A. 1964. *Anolis equestris* in Oriente Province, Cuba. Bull. Mus. Comp. Zool., Harvard Univ. 131: 405–428.

Anolis smaragdinus. *Anolis smaragdinus* Barbour and Shreve 1935: 355. • BAH • Barbour, T. and B. Shreve. 1935. Concerning some Bahamian reptiles, with notes on the fauna. Proc. Boston Soc. Nat. Hist. 40: 347–365.

Anolis spectrum. *Anolis spectrum* Peters 1863: 136. • CUB • Peters, W. 1863. Über einige neue Arten der Saurier-Gattung *Anolis*. Monatsb. Königl. Akad. Wiss. Berlin: 135–149.

Anolis strahmi. *Anolis strahmi* Schwartz 1979: 11. • HISP • Schwartz, A. 1979. A new species of cybotoid anole (Sauria, Iguanidae) from Hispaniola. Breviora 451: 1–27.

Anolis stratulus. *Anolis striatulus* [sic] Cope 1861 (1862): 209. • PRB • Cope, E. D. 1861 (1862). Notes and descriptions of anoles. Proc. Acad. Nat. Sci. Philadelphia 13: 208–215.

Anolis trinitatis. *Anolis trinitatis* Reinhardt and Lütken 1862: 269. • LA • Reinhardt, J. and C. F. Lütken. 1862. Bidrag til det vestindiske Öeriges og navnligen til de dansk-vestindiske Öers Herpetologie. Vidensk. Meddel. Naturhist. Foren. Kjøbenhavn 10–18: 153–291.

Anolis valencienni. *Xiphocercus valencienni* Duméril and Bibron 1837: 131. • JAM • Duméril, A. M. C. and G. Bibron. 1837. Erpétologie Générale ou Histoire Naturelle Complète des Reptiles. Vol. 4. Libr. Encyclopédique Roret, Paris.

Anolis vanidicus. *Anolis vanidicus* Garrido and Schwartz 1972: 515. • CUB • Garrido, O. H. and A. Schwartz. 1972. The Cuban *Anolis spectrum* complex (Sauria, Iguanidae). Proc. Biol. Soc. Washington 85: 509–521.

Anolis vermiculatus. *Anolis vermiculatus* Cocteau *in* Duméril and Bibron 1837: 128. • CUB • Duméril, A. M. C. and G. Bibron. 1837. Erpétologie Générale ou Histoire Naturelle Complète des Reptiles. Vol. 4. Libr. Encyclopédique Roret, Paris.

Anolis vescus. *Anolis vescus* Garrido and Hedges 1992: 25. • CUB • Holotype, MNHNCU 2729; type-locality, "Palmarito, 4.7 km N Los Calderos, Municipio Imías, Guantánamo Province, 700 meters [Cuba]" • Garrido, O. H. and S. B. Hedges. 1992. Three new grass anoles (Sauria: Iguanidae) from Cuba. Carib. Jour. Sci. 28: 21–29.

Anolis wattsi. *Anolis wattsi* Boulenger 1894: 375. • LA • Boulenger, G. A. 1894. Description of a new *Anolis* from Antigua, West Indies. Ann. Mag. Nat. Hist., ser. 6, 14: 375–376.

Anolis websteri. *Anolis websteri* Arnold 1980: 21. • HISP • Arnold, D. L. 1980. Geographic variation in *Anolis brevirostris* (Sauria: Iguanidae) in Hispaniola. Breviora 461: 1–31.

Anolis whitemani. *Anolis whitemani* Williams 1963: 2. • HISP • Williams, E. E. 1963. *Anolis whitemani*, new species from Hispaniola (Sauria, Iguanidae). Breviora 197: 1–8. CAAR: Powell, R. and D. K. Carr. 1990. No. 462.

Family Scincidae

Mabuya lineolata. *Mabuya lineolata* Noble and Hassler 1933: 16. • HISP • Noble, G. K. and W. G. Hassler. 1933. Two new species of frogs, five new species and a new race of lizards from the Dominican Republic. Amer. Mus. Novit. 652: 1–17. **Plate 2I.**

Mabuya bistriata. *Scincus bistriatus* Spix 1825: 23. • WI • Lectotype, RNH 2512 (designated by Hoogmoed and Gruber 1983); type-locality, "Belém, Pará, Brazil" • Spix, J. B. 1825. Animalia Nova sive Species Novae Lacertarum, quas in Itinere per Brasiliam Annis MDCCCXVII–MDCCCXX. F. S. Hübschmanni, Monachii. Avila-Pires (1995) addressed the status of Amazonian *Mabuya* in what is now considered the *M. mabuya* species complex, but the assignment of Antillean populations remains unresolved.

Family Teiidae

Ameiva ameiva. *Lacerta ameiva* Linnaeus 1758: 202. • LA, WC • Linnaeus, C. 1758. Systema Naturæ per Regna Tria Naturæ, Secundum Classes, Ordines, Genera, Species cum Characteribus, Differentiis, Synonymis, Locis. Tomus I. Editio decima. Reformata. L. Salvii, Holmiae.

Ameiva auberi. *Ameiva auberi* Cocteau 1838: 51. • BAH, CUB • Cocteau, J.-T. 1838 *in* J.-T. Cocteau and G. Bibron. 1838–1843, Reptiles, p. (5), 1–142. *In* R. de la Sagra. 1838. Historia Física, Política y Natural de la Isla de Cuba. Arthus Bertrand, Paris.

Ameiva chrysolaema. *Ameiva chrysolaema* Cope 1868: 127. • HISP • Cope, E. D. 1868. An examination of the Reptilia and Batrachia obtained by the Orton Expedition to Equador and the Upper Amazon, with notes on other species. Proc. Acad. Nat. Sci. Philadelphia 20: 96–140. CAAR: Schell, P. T., J. S. Parmerlee, Jr., and R. Powell. 1993. No. 563.

Ameiva cineracea. *Ameiva cineracea* Barbour and Noble 1915: 453. • LA • Barbour, T. and G. K. Noble. 1915. A revision of the lizards of the genus *Ameiva*. Bull. Mus. Comp. Zoöl., Harvard Univ. 59: 417–479.

Ameiva corax. *Ameiva corax* Censky and Paulson 1992: 187. • LA • Holotype, MCZ 77137; type-locality, "south side of Little Scrub Island, off the northeastern tip of Anguilla" • Censky, E. J. and D. R. Paulson. 1992. Revision of the *Ameiva* (Reptilia: Teiidae) of the Anguilla Bank, West Indies. Ann. Carnegie Mus. 61: 177–195.

Ameiva corvina. *Ameiva corvina* Cope 1861: 312. • LA • Cope, E. D. 1861. On the Reptilia of Sombrero and Bermuda. Proc. Acad. Nat. Sci. Philadelphia 13: 312–314.

Ameiva dorsalis. *Ameiva dorsalis* Gray 1838: 277. • JAM • Gray, J. E. 1838. Catalogue of the slender-tongued saurians, with descriptions of many new genera and species. Ann. Mag. Nat. Hist., ser. 1, 1: 274–283.

Ameiva erythrocephala. *Ameiva erythrocephala* Daudin 1802: 22. • LA • Daudin, F.-M. 1802. Histoire Naturelle, Générale et Particulière des Reptiles. Vol. 3. F. Dufart, Paris.

Ameiva exsul. *Ameiva plei* var. *exsul* Cope 1862: 66. • PRB • Cope, E. D. 1862. Synopsis of the species of *Holcosus* and *Ameiva*, with diagnoses of new West Indian and South American Colubridae. Proc. Acad. Nat. Sci. Philadelphia 14: 60–82.

Ameiva fuscata. *Ameiva fuscata* Garman 1887: 5. • LA • Garman, S. 1887. On the West Indian Teiidæ in the Museum of Comparative Zoology. Bull. Essex Inst. 19: 1–12. CAAR: Malhotra, A. and R. S.

Thorpe. 1995. No. 606.

Ameiva griswoldi. *Ameiva griswoldi* Barbour 1916: 216. • LA • Barbour, T. 1916. Additional notes on West Indian reptiles and amphibians. Proc. Biol. Soc. Washington 29: 215–220.

Ameiva leberi. *Ameiva chrysolaema leberi* Schwartz and Klinikowski 1966: 459. • HISP • Schwartz, A. and R. F. Klinikowski. 1966. The *Ameiva* (Lacertilia, Teiidae) of Hispaniola. II. Geographic variation in *Ameiva chrysolaema* Cope. Bull. Mus. Comp. Zool., Harvard Univ. 133: 425–487.

Ameiva lineolata. *Ameiva lineolata* Duméril and Bibron 1839: 119. • HISP • Duméril, A. M. C. and G. Bibron. 1839. Erpétologie Générale ou Histoire Naturelle Complète des Reptiles. Vol. 5. Libr. Encyclopédique Roret, Paris. **Plate 3D.**

Ameiva major. *Ameiva major* Duméril and Bibron 1839: 117. • LA • Duméril, A. M. C. and G. Bibron. 1839. Erpétologie Générale ou Histoire Naturelle Complète des Reptiles. Vol. 5. Libr. Encyclopédique Roret, Paris.

Ameiva maynardi. *Ameiva maynardi* Garman 1888: 10. • BAH • Garman, S. 1888. Reptiles and batrachians from the Caymans and from the Bahamas, collected by Prof. C. J. Maynard for the Museum of Comparative Zoology at Cambridge, Mass. Bull. Essex Inst. 20: 1–13.

Ameiva plei. *Ameiva plei* Duméril and Bibron 1839: 114. • LA • Duméril, A. M. C. and G. Bibron. 1839. Erpétologie Générale ou Histoire Naturelle Complète des Reptiles. Vol. 5. Libr. Encyclopédique Roret, Paris. **Plate 3C.**

Ameiva pluvianotata. *Ameiva pluvianotata* Garman 1887: 6. • LA • Garman, S. 1887. On the West Indian Teiidæ in the Museum of Comparative Zoology. Bull. Essex Inst. 19: 1–12.

Ameiva polops. *Ameiva polops* Cope 1862: 66. • PRB • Cope, E. D. 1862. Synopsis of the species of *Holcosus* and *Ameiva*, with diagnoses of new West Indian and South American Colubridae. Proc. Acad. Nat. Sci. Philadelphia 14: 60–82. CAAR: Dodd, C. K., Jr. 1980. No. 240.

Ameiva taeniura. *Ameiva taeniura* Cope 1862: 63. • HISP • Cope, E. D. 1862. Synopsis of the species of *Holcosus* and *Ameiva*, with diagnoses of new West Indian and South American Colubridae. Proc. Acad. Nat. Sci. Philadelphia 14: 60–82.

Ameiva wetmorei. *Ameiva wetmorei* Stejneger 1913: 70. • PRB • Stejneger, L. 1913. A new lizard from Porto Rico. Proc. Biol. Soc. Washington 26: 69–71.

Cnemidophorus lemniscatus. *Lacerta lemniscata* Linnaeus 1758: 209. • WC • Linnaeus, C. 1758. Systema Naturæ per Regna Tria Naturæ, Secundum Classes, Ordines, Genera, Species cum Characteribus, Differentiis, Synonymis, Locis. Tomus I. Editio decima. Reformata. L. Salvii, Holmiae. **Plate 3E.**

Cnemidophorus vanzoi. *Ameiva vanzoi* Baskin and Williams 1966: 146. • LA • Baskin, J. N. and E. E. Williams. 1966. The Lesser Antillean *Ameiva* (Sauria, Teiidae). Studies Fauna Curaçao Carib. Isl. 23: 144–176.

Kentropyx borckiana. *Centropyx Borckiana* Peters 1869: 64. • LA • Peters, W. 1869. Über neue Gattungen und Arten von Eidechsen. Monatsb. Königl. Akad. Wiss. Berlin 1869: 57–66. **Plate 3F.**

Family Tropiduridae

Leiocephalus barahonensis. *Leiocephalus barahonensis* Schmidt 1921: 15. • HISP • Schmidt, K. P. 1921. Notes on the herpetology of Santo Domingo. Bull. Amer. Mus. Nat. Hist. 44: 7–20.

Leiocephalus carinatus. *Leiocephalus carinatus* Gray 1827: 208. • BAH, CAY, CUB, WC • Gray, J. E. 1827. A description of a new genus and some new species of saurian reptiles; with a revision of the species of chameleons. Philos. Mag., ser. 2, 2: 207–214. **Plate 3G.**

Leiocephalus cubensis. *Tropidurus* (*Leiolaemus*) *cubensis* Gray 1840: 110. • CUB • Gray, J. E. 1840. Catalogue of the species of reptiles collected in Cuba by W. S. MacLeay, Esq.; with some notes on their habits extracted from his MS. Ann. Mag. Nat. Hist., ser. 1, 5: 108–115.

Leiocephalus endomychus. *Leiocephalus vinculum endomychus* Schwartz 1967: 45. • HISP • Holotype, MCZ 81099; type-locality, "3.4 mi (5.5 km) NE Barrage de Péligre, 1100 ft. (361 meters), Département du Centre, Haiti" • Schwartz, A. 1967. The *Leiocephalus* (Lacertilia, Iguanidae) of Hispaniola. II. The *Leiocephalus personatus* complex. Tulane Studies Zool. 14: 1–53. Elevated from subspecific status by Pregill (1992).

Leiocephalus eremitus. *Liocephalus* [sic] *eremitus* Cope 1868: 122. • N • Cope, E. D. 1868. An examination of the Reptilia and Batrachia obtained by the Orton Expedition to Equador and the Upper Amazon, with notes on other species. Proc. Acad. Nat. Sci. Philadelphia 20: 96–138.

Leiocephalus greenwayi. *Leiocephalus greenwayi* Barbour and Shreve 1935: 358. • BAH • Barbour, T. and B. Shreve. 1935. Concerning some Bahamian reptiles, with notes on the fauna. Proc. Boston Soc. Nat. Hist. 40: 347–365.

Leiocephalus herminieri. *Holotropis herminieri* Duméril and Bibron 1837: 261. • LA • Duméril, A. M. C. and G. Bibron. 1837. Erpétologie Générale ou Histoire Naturelle Complète des Reptiles. Vol. 4. Libr. Encyclopédique Roret, Paris.

Leiocephalus inaguae. *Leiocephalus inaguae* Cochran 1931: 39. • BAH • Cochran, D. M. 1931. New Bahaman reptiles. Jour. Washington Acad. Sci. 21: 39–40.

Leiocephalus loxogrammus. *Leiocephalus loxogrammus* Cope 1887: 437. • BAH • Cope, E. D. 1887. List of the Batrachia and Reptilia of the Bahama Islands. Proc. United States Natl. Mus. 10: 436–439.

Leiocephalus lunatus. *Leiocephalus personatus lunatus* Cochran 1934: 153. • HISP • Cochran, D. M. 1934. A new lizard, *Leiocephalus personatus lunatus*, from the Dominican Republic. Occas. Pap. Boston Soc. Nat. Hist. 8: 153–156.

Leiocephalus macropus. *Liocephalus* [sic] *macropus* Cope 1862: 184. • CUB • Cope, E. D. 1862. Contributions to Neotropical saurology. Proc. Acad. Nat. Sci. Philadelphia 14: 176–188.

Leiocephalus melanochlorus. *Liocephalus* [sic] *melanochlorus* Cope 1862: 184. • HISP • Cope, E. D. 1862. Contributions to Neotropical saurology. Proc. Acad. Nat. Sci. Philadelphia 14: 176–188.

Leiocephalus onaneyi. *Leiocephalus onaneyi* Garrido 1973: 4. • CUB • Garrido, O. H. 1973. Nueva especie de *Leiocephalus* (Lacertilia, Iguanidae) para Cuba. Poeyana 116: 1–19.

Leiocephalus personatus. *Liocephalus* [sic] *personatus* Cope 1862: 182. • HISP • Cope, E. D. 1862. Contributions to Neotropical saurology. Proc. Acad. Nat. Sci. Philadelphia 14: 176–188. **Plate 3H.**

Leiocephalus pratensis. *Hispaniolus pratensis* Cochran 1928: 50. • HISP • Cochran, D. M. 1928. A new genus and species of lizard, *Hispaniolus pratensis*, from the Haitian Republic. Proc. Biol. Soc. Washington 41: 49–51.

Leiocephalus psammodromus. *Liocephalus* [sic] *arenarius* Barbour 1916: 217 (preoccupied by *Steironotus* [=*Leiocephalus*] *arenarius* Tschudi, 1845). • TC • Barbour, T. 1916. Additional notes on West Indian reptiles and amphibians. Proc. Biol. Soc. Washington 29: 215–220.

Leiocephalus punctatus. *Leiocephalus carinatus punctatus* Cochran 1931: 39. • BAH • Cochran, D. M. 1931. New Bahaman reptiles. Jour. Washington Acad. Sci. 21: 39–40.

Leiocephalus raviceps. *Liocephalus* [sic] *raviceps* Cope 1862: 183. • CUB • Cope, E. D. 1862. Contributions to Neotropical saurology. Proc. Acad. Nat. Sci. Philadelphia 14: 176–188.

Leiocephalus rhutidira. *Leiocephalus rhutidira* Schwartz 1979: 273. • HISP • Schwartz, A. 1979. A new species of *Leiocephalus* (Reptilia: Iguanidae) from Hispaniola. Proc. Biol. Soc. Washington 92: 272–279.

Leiocephalus schreibersii. *Pristinotus schreibersii* Gravenhorst 1837 (1838): 739. • HISP • Gravenhorst, J. L. C. 1837 (1838). Beiträge zur genaueren Kenntniss einiger Eidechsengattungen. Nova Acta Acad. Caes. Leop.-Carol. 18: 712–784. CAAR: Schreiber, M. C., J. S. Parmerlee, Jr., and R. Powell. 1995. No. 613.

Leiocephalus semilineatus. *Leiocephalus semilineatus* Dunn 1920: 33. • HISP • Dunn, E. R. 1920. A new lizard from Haiti. Proc. New England Zoöl. Club 7: 33–34.

Leiocephalus stictigaster. *Leiocephalus stictigaster* Schwartz 1959: 121. • CUB • Schwartz, A. 1959. Variation in lizards of the *Leiocephalus cubensis* complex in Cuba and the Isla de Pinos. Bull. Florida State Mus. 4: 97–143.

Leiocephalus vinculum. *Leiocephalus vinculum* Cochran 1928: 54. • HISP • Cochran, D. M. 1928. The herpetological collections made in Haiti and its adjoining islands by Walter J. Eyerdam. Proc. Biol. Soc. Washington 41: 53–59.

Family Xantusiidae

Cricosaura typica. *Cricosaura typica* Gundlach and Peters, *in* Peters 1863: 262. • CUB • Peters, W. 1863. Über eine neue von Hrn. Dr. Gundlach auf Cuba entdeckte Sauriergattung, *Cricosaura typica* Gundlach et Peters. Monatsb. Königl. Akad. Wiss. Berlin 1863: 362–370. CAAR: Crother, B. I. 1988. No. 443. **Plate 5H**.

Suborder Serpentes

Family Boidae

Boa constrictor. *Boa constrictor* Linnaeus 1758: 215. • LA, WC • Linnaeus, C. 1758. Systema Naturæ per Regna Tria Naturæ, Secundum Classes, Ordines, Genera, Species cum Characteribus, Differentiis, Synonymis, Locis. Tomus I. Editio decima. Reformata. L. Salvii, Holmiae. **Plate 6D**.

Corallus hortulanus. *Boa Hortulana* Linnaeus 1758: 215. • LA • Holotype, SMNH Lin. 7; type-locality, "America" • Linnaeus, C. 1758. Systema Naturæ per Regna Tria Naturæ, Secundum Classes, Ordines, Genera, Species cum Characteribus, Differentiis, Synonymis, Locis. Tomus I. Editio decima. Reformata. L. Salvii, Holmiae. CAAR: Henderson, R. W. 1993. No. 576 (as *C. enydris*). McDiarmid et al. (1996) provided reasons for the name change from *C. enydris*. **Plate 6E**.

Epicrates angulifer. *Epicrates angulifer* Bibron 1843: 129, pl. 25. • CUB • Bibron, G. *in* J.-T. Cocteau and G. Bibron. 1838–1843, Reptiles, p. (5), 1–143. *In* R. de la Sagra, Historia Física, Política y Natural de la Isla de Cuba. Arthus Bertrand, Paris.

Epicrates chrysogaster. *Homalochilus chrysogaster* Cope 1871: 557. • BAH, TC • Cope, E. D. 1871. Eighth contribution to the herpetology of tropical America. Proc. Amer. Philos. Soc. 11: 553–559.

Epicrates exsul. *Epicrates exsul* Netting and Goin 1944: 71. • BAH • Netting, M. G. and C. J. Goin. 1944. Another new boa of the genus *Epicrates* from the Bahamas. Ann. Carnegie Mus. 30: 71–76.

Epicrates fordii. *Pelophilus fordii* Günther 1861: 142. • HISP • Günther, A. 1861. On a new species of the family Boidae. Proc. Zool. Soc. London 1861: 142.

Epicrates gracilis. *Chilabothrus gracilis* Fischer 1888: 35. • HISP • Fischer, J. G. 1888. Über eine Kollektion Reptilien und Amphibien von Hayti. Jahrb. Wiss. Anst. Hamburg 5: 23–45. **Plate 6F**.

Epicrates inornatus. *Boa inornata* Reinhardt 1843: 253. • PRB • Reinhardt, J. T. 1843. Beskrivelse af nogle nye Slangearter. Danske Vidensk. Selsk. Afhandl. 10: 233–279.

Epicrates monensis. *Epicrates monensis* Zenneck 1898: 64. • PRB • Zenneck, J. 1898. Die Zeichnung der Boiden. Zeitschr. Wiss. Zool. 64: 1–384.

Epicrates striatus. *Homalochilus striatus* Fischer 1856: 102. • BAH, HISP • Fischer, J. G. 1856. Neue Schlangen des Hamburgischen Naturhistorischen Museums. Abh. Naturwiss. Ver. Hamburg 3: 79–116.

Epicrates subflavus. *Epicrates subflavus* Stejneger 1901: 469 (substitute name for *Epicrates inornatus* Reinhardt [part]). • JAM • Stejneger, L. 1901. A new systematic name for the yellow boa of Jamaica. Proc. United States Natl. Mus. 23: 469–470.

Family Colubridae

Alsophis anomalus. *Zamenis anomalus* Peters 1863: 282. • HISP • Peters, W. 1863. Über einige neue oder weniger bekannte Schlangenarten des zoologischen Museums zu Berlin. Monatsb. Königl. Akad. Wiss. Berlin 1863: 272–289.

Alsophis antiguae. *Alsophis leucomelas antiguae* Parker 1933: 158. • LA • Parker, H. W. 1933. Some amphibians and reptiles from the Lesser Antilles. Ann. Mag. Nat. Hist., ser. 10, 11: 151–158.

Alsophis antillensis. *Psammophis antillensis* Schlegel 1837: 214. • LA • Schlegel, H. 1837. Essai sur la Physionomie des Serpens. Vol. 2, partie descriptive. Arnz, Leiden. **Plate 7D.**

Alsophis ater. *Natrix atra* Gosse 1851: 228. • JAM • Gosse, P. H. 1851. A Naturalist's Sojourn in Jamaica. Longman, Brown, Green and Longmans, London.

Alsophis cantherigerus. *Coluber cantherigerus* Bibron 1843: pl. 27. • CAY, CUB • Bibron, G. *in* J. T. Cocteau and G. Bibron. 1838–1843, Reptiles, p. (5), 1–143. *In* R. de la Sagra, Historia Física, Política y Natural de la Isla de Cuba. Arthus Bertrand, Paris.

Alsophis melanichnus. *Alsophis melanichnus* Cope 1862: 76. • HISP • Cope, E. D. 1862. Synopsis of the species of *Holcosus* and *Ameiva*, with diagnoses of new West Indian and South American Colubridae. Proc. Acad. Nat. Sci. Philadelphia 14: 60–82.

Alsophis portoricensis. *Alsophis portoricensis* Reinhardt and Lütken 1862: 221. • PRB • Reinhardt, J. and C. F. Lütken. 1862. Bidrag til det vestindiske Öeriges og navnligen til de dansk–vestindiske Öers Herpetologie. Vidensk. Meddel. Naturhist. Foren. Kjøbenhavn 10–18: 153–291. **Plate 7E.**

Alsophis rijersmai. *Alsophis rijersmaei* Cope 1869: 154. • LA • Cope, E. D. 1869. Seventh contribution to the herpetology of tropical America. Proc. Amer. Philos. Soc. 11: 147–169.

Alsophis rufiventris. *Dromicus rufiventris* Duméril, Bibron, and Duméril 1854: 688. • LA • Duméril, A. M. C., G. Bibron, and A. H. A. Duméril. 1854. Erpétologie générale ou Histoire naturelle complète des reptiles. Vol. 7. Libr. Encyclopédique Roret, Paris.

Alsophis sanctaecrucis. *Alsophis sancticrucis* Cope 1862: 76. • PRB • Cope, E. D. 1862. Synopsis of the species of *Holcosus* and *Ameiva*, with diagnoses of new West Indian and South American Colubridae. Proc. Acad. Nat. Sci. Philadelphia 14: 60–82.

Alsophis vudii. *Alsophis vudii* Cope 1862: 74. • BAH • Cope, E. D. 1862. Synopsis of the species of *Holcosus* and *Ameiva*, with diagnoses of new West Indian and South American Colubridae. Proc. Acad. Nat. Sci. Philadelphia 14: 60–82.

Antillophis andreae. *Liophis andreae* Reinhardt and Lütken 1862: 214. • CUB • Reinhardt, J. and C. F. Lütken. 1862. Bidrag til det vestindiske Öeriges og navnligen til de dansk–vestindiske Öers Herpetologie. Vidensk. Meddel. Naturhist. Foren. Kjøbenhavn 10–18: 153–291.

Antillophis parvifrons. *Dromicus parvifrons* Cope 1862: 79. • BAH (?), HISP • Cope, E. D. 1862. Synopsis of the species of *Holcosus* and *Ameiva*, with diagnoses of new West Indian and South American Colubridae. Proc. Acad. Nat. Sci. Philadelphia 14: 60–82. **Plate 7F.**

Arrhyton ainictum. *Arrhyton ainictum* Schwartz and Garrido 1981: 216. • CUB • Schwartz, A. and O. H. Garrido. 1981. A review of the Cuban members of the genus *Arrhyton* (Reptilia, Serpentes, Colubridae). Ann. Carnegie Mus. 50: 207–230.

Arrhyton callilaemum. *Natrix callilaema* Gosse 1851: 384. • JAM • Gosse, P. H. 1851. A Naturalist's Sojourn in Jamaica. Longman, Brown, Green and Longmans, London.

Arrhyton dolichura. *Arrhyton dolichurum* Werner 1909: 224. • CUB • Werner, F. 1909. Über neue oder seltene Reptilien des naturhistorischen Museums in Hamburg. I. Schlangen. Mitt. Naturh. Mus. Hamburg 26: 205–247.

Arrhyton exiguum. *Dromicus exiguus* Cope 1862: 79. • PRB • Cope, E. D. 1862. Synopsis of the species of *Holcosus* and *Ameiva*, with diagnoses of new West Indian and South American Colubridae. Proc. Acad. Nat. Sci. Philadelphia 14: 60–82.

Arrhyton funereum. *Alsophis funereus* Cope 1862: 77. • JAM • Cope, E. D. 1862. Synopsis of the species of *Holcosus* and *Ameiva*, with diagnoses of new West Indian and South American Colubridae. Proc. Acad. Nat. Sci. Philadelphia 14: 60–82.

Arrhyton landoi. *Arrhyton vittatum landoi* Schwartz 1965: 109. • CUB • Schwartz, A. 1965. A review of the colubrid snake genus *Arrhyton* with a description of a new subspecies from southern Oriente Province, Cuba. Proc. Biol. Soc. Washington 78: 99–113.

Arrhyton polylepis. *Dromicus polylepis* Buden 1966: 7. • JAM • Buden, D. W. 1966. An evaluation of Jamaican *Dromicus* (Serpentes, Colubridae) with the description of a new species. Breviora 238: 1–10.

Arrhyton procerum. *Arrhyton procerum* Hedges and Garrido 1992: 174. • CUB • Holotype, MNHNCU 3285; type-locality, "11.4 km ESE Playa Girón, Matanzas Prov., Cuba, 0 m" • Hedges, S. B. and O. H. Garrido. 1992. Cuban snakes of the genus *Arrhyton*: two new species and a reconsideration of *A. redimitum* Cope. Herpetologica 48: 168–177.

Arrhyton supernum. *Arrhyton supernum* Hedges and Garrido 1992: 171. • CUB • Holotype, MNHNCU 2704; type-locality, "SW slope of El Yunque de Baracoa, Guantánamo Prov., Cuba, 136 m" • Hedges, S. B. and O. H. Garrido. 1992. Cuban snakes of the genus *Arrhyton*: two new species and a reconsideration of *A. redimitum* Cope. Herpetologica 48: 168–177.

Arrhyton taeniatum. *Arrhyton taeniatum* Günther 1858: 244. • CUB • Günther, A. 1858. Catalogue of the Colubrine Snakes in the Collection of the British Museum. Trustees British Mus., London. **Plate 7G.**

Arrhyton tanyplectum. *Arrhyton tanyplectum* Schwartz and Garrido 1981: 221. • CUB • Schwartz, A. and O. H. Garrido. 1981. A review of the Cuban members of the genus *Arrhyton* (Reptilia, Serpentes, Colubridae). Ann. Carnegie Mus. 50: 207–230.

Arrhyton vittatum. *Cryptodacus vittatus* Gundlach *in* Peters 1861: 1003. • CUB • Peters, W. 1861. Über zwei neue von Hrn. Dr. Gundlach auf Cuba entdeckte Schlangen, *Tropidonotus cubanus* und *Cryptodacus vittatus*. Monatsb. Königl. Akad. Wiss. Berlin 1861: 1001–1004.

Chironius vincenti. *Herpetodryas carinatus* var. *vincenti* Boulenger 1891: 355. • LA • Boulenger, G. A. 1891. On reptiles, batrachians, and fishes from the lesser West Indies. Proc. Zool. Soc. London 1891: 351–357. **Plate 7H** (similar species).

Clelia clelia. *Coluber clelia* Daudin 1803: 330. • LA • Daudin, F.-M. 1803. Histoire Naturelle Générale et Particulière des Reptiles. Vol. 6. F. Dufart, Paris. **Plate 7I.**

Clelia errabunda. *Clelia errabunda* Underwood 1993: 3. • LA • Underwood, G. 1993. A new snake from St. Lucia, West Indies. Bull. Nat. Hist. Mus. (Zool.) 59: 1–9.

Coniophanes andresensis. *Coniophanes fissidens andresensis* Bailey 1937: 4. • WC • Bailey, J. R. 1937. New forms of *Coniophanes* Hallowell, and the status of *Dromicus clavatus*. Occas. Pap. Mus. Zool. Univ. Michigan 362: 1–6. **Plate 8A** (similar species).

Darlingtonia haetiana. *Darlingtonia haetiana* Cochran 1935: 375. • HISP • Cochran, D. M. 1935. New reptiles and amphibians collected in Haiti by P. J. Darlington. Proc. Boston Soc. Nat. Hist. 40: 367–375. CAAR: Schwartz, A. 1986. No. 390. This species may be reassigned to *Arrhyton* (see Hedges and Garrido 1992). **Plate 8B.**

Diadophis punctatus. *Diadophis punctatus* Linnaeus 1766: 376. • CAY • INT • Linnaeus, C. 1766. Systema Naturæ per Regna Tria Naturæ, Secundum Classes, Ordines, Genera, Species cum Characteribus, Differentiis, Synonymis, Locis. Tomus I. Editio duodecima, Reformata. L. Salvii, Holmiae.

Elaphe guttata. *Coluber guttatus* Linnaeus 1766: 385. • BAH (?), CAY • INT • Linnaeus, C. 1766. Systema Naturæ per Regna Tria Naturæ, Secundum Classes, Ordines, Genera, Species cum Characteribus, Differentiis, Synonymis, Locis. Tomus I. Editio duodecima, Reformata. L. Salvii, Holmiae.

Hypsirhynchus ferox. *Hypsirhynchus ferox* Günther 1858: 49. • HISP • Günther, A. C. L. G. 1858. Catalogue of Colubrine Snakes in the Collection of the British Museum. Trustees British Mus, London. CAAR: Sosa, R. A., R. W. Henderson, and R. Powell. 1995. No. 617. **Plate 8C**.

Ialtris agyrtes. *Ialtris aygrtes* Schwartz and Rossman 1976: 90. • HISP • Schwartz, A. and D. A. Rossman. 1976. A review of the Hispaniolan colubrid snake genus *Ialtris*. Stud. Fauna Curaçao Carib. Isl. 50: 76–102. CAAR: Powell, R. and R. W. Henderson. 1994. No. 591.

Ialtris dorsalis. *Philodryas dorsalis* Günther 1858: 126. • HISP • Günther, A. C. L. G. 1858. Catalogue of Colubrine Snakes in the Collection of the British Museum. Trustees British Mus., London. CAAR: Zippel, K. C., J. S. Parmerlee, Jr., and R. Powell. 1994. No. 592. **Plate 8D**.

Ialtris parishi. *Ialtris parishi* Cochran 1932: 189. • HISP • Cochran, D. M. 1932. A new snake, *Ialtris parishi*, from the Republic of Haiti. Proc. Biol. Soc. Washington 45: 189–190. CAAR: Powell, R. and R. W. Henderson. 1994. No. 593.

Liophis cursor. *Coluber cursor* Lacepède 1789: 96. Lacepède, B.-G.-É. 1789. • LA • Histoire Naturelle des Quadrupèdes Ovipares et de Serpens. Vol. 2. Hôtel de Thou, Paris.

Liophis juliae. *Aporophis juliae* Cope 1879: 274. • LA • Cope, E. D. 1879. Eleventh contribution to the herpetology of tropical America. Proc. Amer. Philos. Soc. 18: 261–277. **Plate 8E**.

Liophis melanotus. *Coluber Melanotus* Shaw 1802: 534. • LA • Shaw, G. 1802. General Zoology, or Systematic Natural History. Vol. 3, part 2. Thomas Davison, London.

Liophis ornatus. *Dromicus ornatus* Garman 1887: 281. • LA • Garman, S. 1887. On West Indian reptiles in the Museum of Comparative Zoölogy at Cambridge, Mass. Proc. Amer. Philos. Soc. 24: 278–286.

Liophis perfuscus. *Liophis perfuscus* Cope 1862: 77. • LA • Cope, E. D. 1862. Synopsis of the species of *Holcosus* and *Ameiva*, with diagnoses of new West Indian and South American Colubridae. Proc. Acad. Nat. Sci. Philadelphia 14: 60–82.

Mastigodryas bruesi. *Alsophis bruesi* Barbour 1914: 337. • LA • Barbour, T. 1914. A contribution to the zoögeography of the West Indies, with especial reference to amphibians and reptiles. Mem. Mus. Comp. Zoöl., Harvard Coll. 44:209–359. **Plate 8F**.

Nerodia clarkii. *Regina Clarkii* Baird and Girard 1853: 48. • CUB • Baird, S. F. and C. Girard. 1853. Catalogue of North American Reptiles in the Museum of the Smithsonian Institution. Part I.—Serpents. Smithsonian Inst., Washington.

Pseudoboa neuwiedi. *Scytale neuwiedii* Duméril, Bibron, and Duméril 1854: 1001. • LA • Duméril, A. M. C., G. Bibron, and A. H. A. Duméril. 1854. Erpétologie Générale ou Histoire Naturelle Complète des Reptiles. Vol. 7, part 2. Libr. Encyclopédique Roret, Paris.

Tretanorhinus variabilis. *Tretanorhinus variabilis* Duméril, Bibron, and Duméril 1854: 349. • CAY, CUB • Duméril, A. M. C., G. Bibron, and A. H. A. Duméril. 1854. Erpétologie Générale ou Histoire Naturelle Complète des Reptiles. Vol. 7, part 1. Libr. Encyclopédique Roret, Paris. **Plate 8G**.

Uromacer catesbyi. *Dendrophis catesbyi* Schlegel 1837: 226. • HISP • Schlegel, H. 1837. Essai sur la Physionomie des Serpens. Vol. 2. Arnz, Leiden. CAAR: Schwartz, A. and R. W. Henderson. 1984. No. 356.

Uromacer frenatus. *Ahaetulla frenata* Günther 1865: 94. • HISP • Günther, A. C. L. G. 1865. Fourth account of new species of snakes in the collection of the British Museum. Ann. Mag. Nat. Hist., ser. 3, 15: 89–98. CAAR: Henderson, R. W. and A. Schwartz. 1984. No. 357. **Plate 8H**.

Uromacer oxyrhynchus. *Uromacer oxyrhynchus* Duméril, Bibron, and Duméril 1854: 722. • HISP • Duméril, A. M. C., G. Bibron, and A. H. A. Duméril. 1854. Erpétologie Générale ou Histoire Naturelle Complète des Reptiles. Vol. 7, part. 1. Libr. Encylopédique Roret, Paris. CAAR: Schwartz, A. and R. W. Henderson. 1984. No. 358.

Family Elapidae

Micrurus nigrocinctus. *Elaps nigrocinctus* Girard 1854 (1855): 226. • WC • Syntypes (2), USNM 7347; type-locality, "Taboga Island, Bay of Panama, Panama" • Girard, C. 1854 (1855). Abstract of a report to Lieut. James M. Gilliss, U.S.N., upon the reptiles collected during the U.S.N. Astronomical Expedition to Chili. Proc. Acad. Nat. Sci. Philadelphia 7: 226–227. **Plate 6H**.

Family Leptotyphlopidae

Leptotyphlops asbolepis. *Leptotyphlops asbolepis* Thomas, McDiarmid, and Thompson 1985: 209. • HISP • Thomas, R., R. W. McDiarmid, and F. G. Thompson. 1985. Three new species of thread snakes (Serpentes: Leptotyphlopidae) from Hispaniola. Proc. Biol. Soc. Washington 98: 204–220.

Leptotyphlops bilineata. *Typhlops bilineatus* Schlegel 1839: 36, pl. 12. • LA • Schlegel, H. 1837–1844. Abbildungen neuer oder unvollständig bekannter Amphibien, nach der Natur oder dem Leben entworfen, herausgegeben und mit einem erläuternden Texte begleitet. Arnz & Comp., Düsseldorf.

Leptotyphlops calypso. *Leptotyphlops calypso* Thomas, McDiarmid, and Thompson 1985: 206. • HISP • Thomas, R., R. W. McDiarmid, and F. G. Thompson. 1985. Three new species of thread snakes (Serpentes: Leptotyphlopidae) from Hispaniola. Proc. Biol. Soc. Washington 98: 204–220.

Leptotyphlops columbi. *Leptotyphlops columbi* Klauber 1939: 62. • BAH • Klauber, L. M. 1939. Three new worm snakes of the genus *Leptotyphlops*. Trans. San Diego Soc. Nat. Hist. 9: 59–65. **Plate 7A**.

Leptotyphlops goudotii. *Stenostoma Goudotii* Duméril and Bibron 1844: 330. • WC • Duméril, A. M. C. and G. Bibron. 1844. Erpétologie Générale ou Histoire Naturelle Complète des Reptiles. Vol. 6. Libr. Encyclopédique Roret, Paris.

Leptotyphlops leptipilepta. *Leptotyphlops leptipilepta* Thomas, McDiarmid, and Thompson 1985: 211. • HISP • Thomas, R., R. W. McDiarmid, and F. G. Thompson. 1985. Three new species of thread snakes (Serpentes: Leptotyphlopidae) from Hispaniola. Proc. Biol. Soc. Washington 98: 204–220.

Leptotyphlops pyrites. *Leptotyphlops pyrites* Thomas 1965: 2. • HISP • Thomas, R. 1965. The genus *Leptotyphlops* in the West Indies with description of a new species from Hispaniola (Serpentes, Leptotyphlopidae). Breviora 222:1–12.

Leptotyphlops tenella. *Leptotyphlops tenella* Klauber 1939: 59. • LA • INT? (known only from one record in Antigua) • Klauber, L. M. 1939. Three new worm snakes of the genus *Leptotyphlops*. Trans. San Diego Soc. Nat. Hist. 9: 59–65.

Family Tropidophiidae

Tropidophis canus. *Ungalia cana* Cope 1868: 129. • BAH • Cope, E. D. 1868. An examination of the Reptilia and Batrachia obtained by the Orton Expedition to Equador and the Upper Amazon. Proc. Acad. Nat. Sci. Philadelphia 20: 96–140.

Tropidophis caymanensis. *Tropidophis melanurus caymanensis* Battersby 1938: 558. • CAY • Battersby, J. C. 1938. Some snakes of the genus *Tropidophis*. Ann. Mag. Nat. Hist., ser. 11, 1: 557–560.

Tropidophis feicki. *Tropidophis feicki* Schwartz 1957: 3. • CUB • Schwartz, A. 1957. A new species of boa (genus *Tropidophis*) from western Cuba. Amer. Mus. Novit. 1839: 1–8. **Plate 6G**.

Tropidophis fuscus. *Tropidophis fuscus* Hedges and Garrido 1992: 820. • CUB • Holotype, MNHNCU 2705; type-locality, "Minas Amores (21.7 km NW, 7.7 km SE Baracoa, by road), Guantánamo Province, Cuba, 76 m" • Hedges, S. B. and O. H. Garrido. 1992. A new species of *Tropidophis* from Cuba (Serpentes: Tropidophiidae). Copeia 1992: 820–825.

Tropidophis greenwayi. *Tropidophis pardalis greenwayi* Barbour and Shreve 1936: 2. • TC • Barbour,

T. and B. Shreve. 1936. New races of *Tropidophis* and of *Ameiva* from the Bahamas. Proc. New England Zoöl. Club 40: 347–365.

Tropidophis haetianus. *Ungualia* [sic] *haetiana* Cope 1879: 273. • CUB, HISP, JAM • Cope, E. D. 1879. Eleventh contribution to the herpetology of tropical America. Proc. Amer. Philos. Soc. 18: 261–277.

Tropidophis maculatus. *Leionotus maculatus* Bibron 1843: 127. • CUB • Bibron, G. *in* J.-T. Cocteau and G. Bibron. 1838–1843, Reptiles, p. (5), 1–143. *In* R. de la Sagra, Historia Física, Política y Natural de la Isla de Cuba. Arthus Bertrand, Paris.

Tropidophis melanurus. *Boa melanura* Schlegel 1837: 399. • CUB, N • Schlegel, H. 1837. Essai sur la Physionomie des Serpens. Vol. 2, partie descriptive. Arnz & Comp., Leiden.

Tropidophis nigriventris. *Tropidophis nigriventris* Bailey 1937: 45. • CUB • Bailey, J. R. 1937. A review of some recent *Tropidophis* material. Proc. New England Zoöl. Club 16: 41–52.

Tropidophis pardalis. *Boa pardalis* Gundlach 1840: 359. • CUB • Gundlach, [J.]. 1840. Ueber zwei von mir gesammelte Boen von Cuba. Arch. Naturg., Berlin 6(1): 359–361.

Tropidophis pilsbryi. *Tropidophis maculatus pilsbryi* Bailey 1937: 42. • CUB • Bailey, J. R. 1937. A review of some recent *Tropidophis* material. Proc. New England Zoöl. Club 16: 41–52.

Tropidophis semicinctus. *Ungalia* (*Lionotus*) *maculata* var. *semicincta* Gundlach and Peters *in* W. Peters 1865: 388. • CUB • Peters, W. 1865. Über einige neue Säugethiere (*Mormops*, *Macrotus*, *Vesperus*, *Molossus*, *Capromys*), Amphibien (*Platydactylus*, *Otocryptis*, *Euprepes*, *Ungalia*, *Dromicus*, *Tropidonotus*, *Xenodon*, *Hylodes*) und Fische (*Sillago*, *Sebastes*, *Channa*, *Myctophum*, *Carassius*, *Barbus*, *Capoëta*, *Poecilia*, *Saurenchelys*, *Leptocephalus*). Monatsb. Akad. Wiss. Berlin 1864: 381–399.

Tropidophis wrighti. *Tropidophis wrighti* Stull 1938: 38. • CUB • Stull, O. G. 1938. A revision of the genus *Tropidophis*. Occas. Pap. Mus. Zool. Univ. Michigan 195: 1–49.

Family Typhlopidae

Typhlops biminiensis. *Typhlops biminiensis* Richmond 1955: 2. • BAH, CUB • Richmond, N. D. 1955. The blind snakes (*Typhlops*) of Bimini, Bahama Islands, British West Indies, with description of a new species. Amer. Mus. Novit. 1734: 1–7.

Typhlops capitulatus. *Typhlops capitulatus* Richmond 1964: 2. • HISP • Richmond, N. D. 1964. The blind snakes (*Typhlops*) of Haiti with descriptions of three new species. Breviora 202: 1–12. CAAR: Thomas, R. and R. Powell. 1995. No. 618.

Typhlops catapontus. *Typhlops richardi catapontus* Thomas 1966: 190. • PRB • Holotype, MCZ 77220; type-locality, "Vicinity of the Settlement, Anegada, British Virgin Islands" • Thomas, R. 1966. A reassessment of the Virgin Islands *Typhlops* with the description of two new subspecies. Rev. Biol. Trop. 13: 187–201.

Typhlops caymanensis. *Typhlops caymanensis* Sackett 1940: 1. • CAY • Sackett, J. T. 1940. Preliminary report on results of the West Indies-Guatemala expedition of 1940 for the Academy of Natural Sciences of Philadelphia. Part I.—A new blind snake of the genus *Typhlops*. Notulae Naturae Acad. Nat. Sci. Philadelphia 48: 1–2.

Typhlops dominicanus. *Typhlops dominicana* Stejneger 1904: 687. • LA • Stejneger, L. 1904. The herpetology of Porto Rico. Rept. United States Natl. Mus. 1902: 551–724.

Typhlops epactia. *Typhlops biminiensis epactia* Thomas 1968: 715. • CAY • Thomas, R. 1968. The *Typhlops biminiensis* group of Antillean blindsnakes. Copeia 1968: 713–722.

Typhlops gonavensis. *Typhlops gonavensis* Richmond 1964: 3. • HISP • Richmond, N. D. 1964. The blind snakes (*Typhlops*) of Haiti with descriptions of three new species. Breviora 202: 1–12. CAAR: Thomas, R. and R. Powell. 1995. No. 619.

Typhlops granti. *Typhlops granti* Ruthven and Gaige 1935: 2. • PRB • Ruthven, A. G. and H. T. Gaige. 1935. Observations on *Typhlops* from Puerto Rico and some of the adjacent islands. Occas. Pap. Mus. Zool. Univ. Michigan 307: 1–12.

Typhlops hectus. *Typhlops hectus* Thomas 1974: 12. • HISP • Thomas, R. 1974. A new species of *Typhlops* (Serpentes: Typhlopidae) from Hispaniola. Proc. Biol. Soc. Washington 87: 11–18. CAAR: Thomas, R. and R. Powell. 1992. No. 550.

Typhlops hypomethes. *Typhlops hypomethes* Hedges and Thomas 1991: 452. • PRB • Holotype, USNM 300574; type-locality, "University of Puerto Rico Campus (faculty housing), Río Piedras, San Juan, Puerto Rico" • Hedges, S. B. and R. Thomas. 1991. Cryptic species of snakes (Typhlopidae: *Typhlops*) from the Puerto Rico Bank detected by protein electrophoresis. Herpetologica 47: 448–459.

Typhlops jamaicensis. *Anguis Jamaicensis* Shaw 1802: 588. • JAM • Shaw, G. 1802. General Zoology, or Systematic Natural History. Vol. 3, part 2. Thomas Davison, London.

Typhlops lumbricalis. *Anguis lumbricalis* Linnaeus 1758: 288. • BAH, CUB • Linnaeus, C. 1758. Systema Naturæ per Regna Tria Naturæ, Secundum Classes, Ordines, Genera, Species cum Characteribus, Differentiis, Synonymis, Locis. Tomus I. Editio decima. Reformata. L. Salvii, Holmiae. **Plate 7B.**

Typhlops monastus. *Typhlops monastus* Thomas 1966: 257. • LA • Thomas, R. 1966. Leeward Islands *Typhlops* (Reptilia, Serpentes). Proc. Biol. Soc. Washington 79: 255–265.

Typhlops monensis. *Typhlops monensis* Schmidt 1926: 157. • PRB • Schmidt, K. P. 1926. The amphibians and reptiles of Mona Island, West Indies. Field Mus. Nat. Hist. Ser. Zool. 12: 149–163.

Typhlops platycephalus. *Typhlops platycephalus* Duméril and Bibron 1844: 293. • PRB • Holotype, MNHN 1060; type-locality, "Puerto Rico" • Duméril, A. M. C. and G. Bibron. 1844. Erpétologie Générale ou Histoire Naturelle Complète des Reptiles. Vol. 6. Libr. Encyclopédique Roret, Paris.

Typhlops pusillus. *Typhlops pusillus* Barbour 1914: 323. • HISP • Barbour, T. 1914. A contribution to the zoögeography of the West Indies, with especial reference to the amphibians and reptiles. Mem. Mus. Comp. Zoöl., Harvard Coll. 44: 209–359. CAAR: Thomas, R. and R. Powell. 1994. No. 595.

Typhlops richardi. *Typhlops richardi* Duméril and Bibron 1844: 293. • PRB, TC • Duméril, A. M. C. and G. Bibron. 1844. Erpétologie Générale ou Histoire Naturelle Complète des Reptiles. Vol. 6. Libr. Encyclopédique Roret, Paris.

Typhlops rostellatus. *Typhlops rostellatus* Stejneger 1904: 686. • PRB • Stejneger, L. 1904. The herpetology of Porto Rico. Rept. United States Natl. Mus. 1902: 551–724.

Typhlops schwartzi. *Typhlops schwartzi* Thomas 1989: 409. • HISP • Thomas, R. 1989. The relationships of Antillean *Typhlops* (Serpentes: Typhlopidae) and the description of three new Hispaniolan species, p. 409–432. *In* C. A. Woods (ed.), Biogeography of the West Indies: Past, Present, and Future. Sandhill Crane Press, Gainesville, Florida. CAAR: Thomas, R. and R. Powell. 1994. No. 596.

Typhlops sulcatus. *Typhlops sulcatus* Cope 1868: 128. • HISP, N • Cope, E. D. 1868. An examination of the Reptilia and Batrachia obtained by the Orton Expedition to Equador and the Upper Amazon, with notes on other species. Proc. Acad. Nat. Sci. Philadelphia 20: 96–140. CAAR: Thomas, R. and R. Powell. 1994. No. 597.

Typhlops syntherus. *Typhlops syntherus* Thomas 1965: 436. • HISP • Thomas, R. 1965. A new species of *Typhlops* from the Barahona Peninsula of Hispaniola. Copeia 1965: 436–439. CAAR: White, L. R., J. S. Parmerlee, Jr., and R. Powell. 1992. No. 551.

Typhlops tasymicris. *Typhlops tasymicris* Thomas 1974: 1. • LA • Thomas, R. 1974. A new species of Lesser Antillean *Typhlops* (Serpentes: Typhlopidae). Occas. Pap. Mus. Zool. Louisiana State Univ. 46: 1–5.

Typhlops tetrathyreus. *Typhlops tetrathyreus* Thomas 1989: 414. • HISP • Thomas, R. 1989. The relationships of Antillean *Typhlops* (Serpentes: Typhlopidae) and the description of three new Hispaniolan species, p. 409–432. *In* C. A. Woods (ed.), Biogeography of the West Indies: Past, Present, and Future. Sandhill Crane Press, Gainesville, Florida. CAAR: Thomas, R. and R. Powell. 1994. No. 598.

Typhlops titanops. *Typhlops titanops* Thomas 1989: 415. • HISP • Thomas, R. 1989. The relationships of Antillean *Typhlops* (Serpentes: Typhlopidae) and the description of three new species, p. 409–432. *In* C. A. Woods (ed.), Biogeography of the West Indies: Past, Present, and Future. Sandhill Crane Press, Gainesville, Florida. CAAR: Thomas, R. and R. Powell. 1995. No. 620.

Family Viperidae

Bothrops caribbaea. *Trigonocephalus caribbaea* Garman 1887: 285. • LA • Garman, S. 1887. On West Indian reptiles in the Museum of Comparative Zoology, at Cambridge, Mass. Proc. Amer. Philos. Soc. 24: 278–286. **Plate 7C**.

Bothrops lanceolata. *Coluber lanceolatus* Lacepède 1789: 80. • LA • Lacepède, B.-G.-É. 1789. Histoire Naturelle des Quadrupèdes Ovipares et des Serpens. Vol. 2. Hôtel de Thou, Paris.

Order Crocodylia

Family Alligatoridae

Caiman crocodilus. *Lacerta crocodilus* Linnaeus 1758: 200. • CUB, LA, PRB • INT? (may be an occasional vagrant to the Grenada Bank) • Linnaeus, C. 1758. Systema Naturæ per Regna Tria Naturæ, Secundum Classes, Ordines, Genera, Species cum Characteribus, Differentiis, Synonymis, Locis. Tomus I. Editio decima. Reformata. L. Salvii, Holmiae.

Family Crocodylidae

Crocodylus acutus. *Crocodilus acutus* Cuvier 1807: 55. • CAY, CUB, HISP, JAM • Cuvier, G. 1807. Sur les différentes espèces de crocodiles vivans et sur leurs caractères distinctifs. Ann. Natl. Mus. Hist. Nat. Paris 10: 8–66.

Crocodylus intermedius. *Crocodilus intermedius* Graves 1819: 344. • LA • Graves, M. L. 1819. Sur deux nouvelles espèces de crocodile. Ann. Gen. Sci. Phys. Bruxelles 2: 343–353. CAAR: Thorbjarnarson, J. B. and R. Franz. 1987. No. 406.

Crocodylus rhombifer. *Crocodilus rhombifer* Cuvier 1807: 51. • CUB • Cuvier, G. 1807. Sur les différentes espèces de crocodiles vivans et sur leurs caractères distinctifs. Ann. Natl. Mus. Hist. Nat. Paris 10: 8–66. **Plate 2C**.

Literature Cited

Avila-Pires, T. C. S. 1995. Lizards of Brazilian Amazonia (Reptilia: Squamata). Zool. Verhandl., Nat. Natuurhist. Mus. Leiden 299: 1–706.

Frost, D. R. and R. Etheridge. 1989. A phylogenetic analysis and taxonomy of iguanian lizards (Reptilia: Squamata). Univ. Kansas Mus. Nat. Hist. Misc. Publ. 81: 1–65.

Hass, C. A., S. B. Hedges, and L. R. Maxson. 1993. Molecular insights into the relationships and biogeography of West Indian anoline lizards. Biochem. Syst. Ecol. 21: 97–114.

Hedges, S. B. and O. H. Garrido. 1992. Cuban snakes of the genus *Arrhyton*: two new species and a reconsideration of *A. redimitum* Cope. Herpetologica 48: 168–177.

—, A. R. Estrada, and R. Thomas. 1992*a*. Three new species of *Eleutherodactylus* from eastern Cuba, with notes on vocalizations of other species (Anura: Leptodactylidae). Herpetol. Monogr. 6: 68–83.

—, C. A. Hass, and L. R. Maxson. 1992*b*. Caribbean biogeography: molecular evidence for dispersal in West Indian terrestrial vertebrates. Proc. Natl. Acad. Sci. (USA) 89: 1909–1913.

Heyer, W. R. 1994. Variation within the *Leptodactylus podicipinus-wagneri* complex of frogs (Amphibia: Leptodactylidae). Smithsonian Contrib. Zool. 546: 1–124.

Hoogmoed, M. S. and U. Gruber. 1983. Spix and Wagler type specimens of reptiles and amphibians in the natural history musea in Munich (Germany) and Leiden (The Netherlands). Spixiana, suppl. 9: 319–415.

Kaiser, H., D. M. Green, and M. Schmid. 1994 (1995). Systematics and biogeography of eastern Caribbean frogs (*Eleutherodactylus*: Leptodactylidae), with a description of a new species from Dominica. Canadian Jour. Zool. 72: 2217–2237.

McDiarmid, R. W., T'S. Touré, and J. M. Savage. 1996. The proper name of the tropical American tree boa often referred to as *Corallus enydris* (Serpentes: Boidae). Jour. Herpetol. 30: in press.

Mertens, R. 1939. Herpetologische Ergebnisse einer Reise nach der Insel Hispaniola, Westindien. Abh. Senckenb. Naturf. Ges. 499: 1–84.

Powell, R. 1993. Comments on the taxonomic arrangement of some Hispaniolan amphibians and reptiles. Herpetol. Rev. 24: 135–137.

— and J. S. Parmerlee, Jr. 1993. In the spotlight: *Hemidactylus haitianus* Meerwarth, 1901. Dactylus 2: 54–55.

Pregill, G. K. 1981. Cranial morphology and the evolution of West Indian toads (Salientia: Bufonidae): resurrection of the genus *Peltophryne* Fitzinger. Copeia 1981: 273–285.

—. 1992. Systematics of the West Indian lizard genus *Leiocephalus* (Squamata: Iguania: Tropiduridae). Univ. Kansas Mus. Nat. Hist. Misc. Publ. 84: 1–69.

Schwartz, A. and R. W. Henderson. 1988. West Indian amphibians and reptiles: a checklist. Milwaukee Public Mus. Contrib. Biol. Geol. 74: 1–264.

— and —. 1991. Amphibians and Reptiles of the West Indies: Descriptions, Distributions, and Natural History. Univ. Florida Press, Gainesville.

Powell, R., R. W. Henderson, K. Adler, and H. A. Dundee. 1996. An annotated checklist of West Indian amphibians and reptiles, p. 51–93, color pls. 1–8. *In* R. Powell and R. W. Henderson (eds.), Contributions to West Indian Herpetology: A Tribute to Albert Schwartz. Society for the Study of Amphibians and Reptiles, Ithaca (New York). Contributions to Herpetology, volume 12.

WEST INDIAN AMPHIBIANS AND REPTILES

On the following plates, we have endeavored to provide a glimpse of the diversity that exists among the West Indian herpetofauna by illustrating at least one species in each genus native to the region. We have excluded introduced forms and sea turtles; the latter are widely distributed and have been illustrated frequently in a number of readily available references. When unable to obtain photographs of West Indian specimens, we have used pictures of specimens from adjacent areas, or, when photographs of native species were unavailable, we substituted extralimital forms similar to those found in the Antilles.

Each caption provides available information about the specimen illustrated, credits the photographers, and summarizes aspects of distribution, relationships, and natural history of the genus and the species in question. Specimens subsequently deposited in a museum are indicated.[1]

CONTENTS

Plate 1

Amphibia: Anura
- Bufonidae: *Bufo*[2]
- Dendrobatidae: *Colostethus*
- Hylidae: *Calyptahyla, Hyla, Osteopilus*[2]
- Leptodactylidae: *Eleutherodactylus*,[2] *Leptodactylus*

Plate 2

Reptilia I: Testudines and Crocodylia
- Testudinidae: *Geochelone*
- Emydidae: *Trachemys*
- Crocodylidae: *Crocodylus*

Reptilia II: Squamata – Amphisbaenia and Lacertilia
- Amphisbaenidae: *Amphisbaena*[2]
- Gymnophthalmidae: *Gymnophthalmus, Bachia, Tretioscincus*
- Scincidae: *Mabuya*

Plate 3

Reptilia III: Squamata – Lacertilia (continued)
- Anguidae: *Celestus, Diploglossus*
- Teiidae: *Ameiva*,[2] *Cnemidophorus, Kentropyx*
- Tropiduridae: *Leiocephalus*[2]

Plate 4

Reptilia IV: Squamata – Lacertilia (continued)
- Gekkonidae: *Aristelliger, Gonatodes, Hemidactylus, Phyllodactylus, Sphaerodactylus*,[2] *Tarentola, Thecadactylus*

Plate 5

Reptilia V: Squamata – Lacertilia (continued)
- Polychrotidae: *Anolis*[2]
- Xantusiidae: *Cricosaura*

Plate 6

Reptilia VI: Squamata – Lacertilia (continued)
- Iguanidae: *Ctenosaura, Cyclura, Iguana*

Reptilia VII: Squamata – Serpentes
- Boidae: *Boa, Corallus, Epicrates*
- Tropidophiidae: *Tropidophis*
- Elapidae: *Micrurus*

Plate 7

Reptilia VIII: Squamata – Serpentes (continued)
- Leptotyphlopidae: *Leptotyphlops*
- Typhlopidae: *Typhlops*
- Viperidae: *Bothrops*
- Colubridae: *Alsophis*,[2] *Antillophis, Arrhyton, Chironius, Clelia*

Plate 8

Reptilia IX: Squamata – Serpentes (continued)
- Colubridae (continued): *Coniophanes, Darlingtonia, Hypsirhynchus, Ialtris, Liophis, Mastigodryas, Tretanorhinus, Uromacer*

[1] Acronyms: AMNH (American Museum of Natural History, New York, New York), BWMC (Bobby Witcher Memorial Collection, Avila College, Kansas City, Missouri), CM (Carnegie Museum, Pittsburgh, Pennsylvania), KU (Natural History Museum, University of Kansas, Lawrence, Kansas), MPM (Milwaukee Public Museum, Milwaukee, Wisconsin), MZUSP (Museu de Zoologia, Universidade de São Paulo, Brazil), NMC (Canadian Museum of Nature, Ottawa, Ontario, Canada), and SBH (S. Blair Hedges collection, Pennsylvania State University, University Park, Pennsylvania).

[2] Illustrated by more than one species.

PLATE 1

AMPHIBIA: ANURA

Family Bufonidae

A: ***Bufo peltocephalus*** (formerly *Peltophryne peltocephalus*) from Soroa, Pinar de Río Province, Cuba (200 m) (SBH 172586). *Bufo* is a cosmopolitan genus with 12 species (11 endemic) in the West Indies. This species is widespread on Cuba and occurs in a diverse range of habitats. Food includes ants, beetles, arachnids, gastropods, and crustaceans. Active at night, toads retreat by day under rocks, fallen palm trunks, and even enter the burrows of burrowing owls. Photograph by S. Blair Hedges.

B: ***Bufo marinus***, an adult male from Simla Research Station, Arima Valley, Trinidad. *Bufo marinus* is one of the most widespread anuran species in the world, including a broad West Indian distribution. The species occurs in a wide range of habitats and is common in disturbed areas. Introduced in many countries to control insect pests, it has itself become a pest. Photograph by Hinrich Kaiser.

Family Dendrobatidae

C: ***Colostethus chalcopis***, an adult female from ca. 3 km NE Morne Rouge, Martinique (paratopotype, NMC 33902–9). This is the only representative of this widespread, Neotropical mainland family which is known to occur in the West Indies. This species is endemic to localized elfin woodland and cloud forest on Montagne Pelée, where it was discovered in 1990. Photograph by Hinrich Kaiser.

Family Hylidae

D: ***Calyptahyla crucialis***, adult female (left) and male (right) from a limestone hillock at Marshall's Pen, Mandeville Parish, Jamaica (700 m). The large treefrog in this monotypic genus is endemic to Jamaica, where it occurs in mesic forests. Active by night, frogs retreat to hollows of tree branches by day. Photograph by Jeffrey W. Lang.

E: ***Hyla heilprini*** from 11.2 km S and 1.9 km E Marché Léon (airline distances), Département de la Grand'Anse, Haiti (1360 m) (SBH 104058). Only five species of this genus are native to the West Indies, although others have been introduced. *Hyla heilprini* is endemic to Hispaniola, where it is widespread in mesic areas and often associated with rapidly flowing or torrential streams. Photograph by S. Blair Hedges.

F: ***Osteopilus brunneus***, an adult male at a nest site in the bromeliad *Hohenbergia* sp., Windsor, Trelawney Parish, Jamaica. *Osteopilus* is endemic to the West Indies (Bahamas and Greater Antilles) where three species occur. *Osteopilus brunneus* is endemic to Jamaica, where it is widespread and often associated with bromeliads. Photograph by Rebecca L. Thompson.

G: ***Osteopilus septentrionalis***, an adult male from near Georgetown, Grand Cayman, Cayman Islands. This is the most widespread species of *Osteopilus* in the West Indies, where its range continues to expand. The species is also established in southern Florida, USA. Photograph by Glenn P. Gerber.

Family Leptodactylidae

H: ***Eleutherodactylus armstrongi***, an adult female from near Loma Remigio, Barahona Province, Dominican Republic (1100 m) (BWMC 5657). *Eleutherodactylus* is the most species-rich vertebrate genus in the world, with more than 520 species. Likewise, it has more representatives (at least 139) in the West Indies than any other genus. *Eleutherodactylus armstrongi* is endemic to Hispaniola where it is found in upland pine woods, coffee groves, and broadleaf forests. Diurnal retreats include living bromeliads. Photograph by Robert Powell.

I: ***Eleutherodactylus cooki***, an adult male guarding an egg clutch at San Lorenzo, Puerto Rico. *Eleutherodactylus cooki* inhabitsonly a geographically localized system of caves in southeastern Puerto Rico. Photograph by Rafael L. Joglar.

J: ***Eleutherodactylus schwartzi***, an adult male from Tortola, British Virgin Islands. Males call from near ground level in wooded ravines and from terrestrial bromeliads. Photograph by K. Ovaska, courtesy of The Conservation Agency, Jamestown, Rhode Island, USA.

K: ***Eleutherodactylus martinicensis***, an adult male from the lower Chutes du Carbet, Guadeloupe. This species is widespread in the Lesser Antilles, where it occupies pristine to heavily disturbed habitats. Very abundant where found, it nevertheless has been adversely affected by the recent introduction of *E. johnstonei* on some islands. Photograph by Hinrich Kaiser.

L: ***Leptodactylus albilabris***, an adult female from El Yunque, Puerto Rico (790 m). Five species of *Leptodactylus* occur in the West Indies; four are endemic. *Leptodactylus albilabris* is widespread on Puerto Rico and the Virgin Islands. These frogs occur in a variety of semi-aquatic situations, including muddy drains and gutters in towns. Eggs are deposited in a foam nest under surface debris. Photograph by Rafael L. Joglar.

A, B C

D, E F

G, H I

J, K L

PLATE 2

Reptilia I: Testudines and Crocodylia

Family Testudinidae

A: ***Geochelone carbonaria***, a young male from St. Barthélémy, Lesser Antilles. The genus *Geochelone* is pantropical and *G. carbonaria* has a wide distribution in South America. In the West Indies, the species is widely distributed in the Virgin Islands and Lesser Antilles, where it is highly prized as food by humans. Whether the species is native to the West Indies or not continues to be debated. Photograph by Richard A. Sajdak.

Family Emydidae

B: ***Trachemys decorata***, an adult female from Laguna del Rincón, Barahona Province, Dominican Republic. Six recognized species of *Trachemys* occur only in the Western Hemisphere; five are found in the West Indies and four are endemic to this area. *Trachemys decorata* is endemic to Hispaniola, where it occurs in brackish and freshwater lakes. Photograph by José A. Ottenwalder.

Family Crocodylidae

C: ***Crocodylus rhombifer***, an adult male from Laguna del Tesoro, Zapata Swamp, Cuba. Species of *Crocodylus* are pantropical in distribution and two are native to the West Indies. *Crocodylus rhombifer* occurs on Cuba and, historically, it occurred on Grand Cayman. This species is found primarily in freshwater swamps, where its diet includes fish, turtles, and small mammals. Photograph by José A. Ottenwalder.

Reptilia II: Squamata – Amphisbaenia and Lacertilia

Family Amphisbaenidae

D: ***Amphisbaena caeca*** from the Río Abajo Forest Reserve, Arecibo, Puerto Rico. Ten species of *Amphisbaena* occur in the West Indies and all are endemic to this area. *Amphisbaena caeca* is widely distributed on Puerto Rico and also occurs on two other islands on the Puerto Rico Bank. These worm lizards are found in shaded habitats, where they hide in rotted wood and under rocks and logs. Photograph by Manuel Leal.

E: ***Amphisbaena blanoides*** (formerly *Cadea blanoides*) from San Vicente, Pinar del Río Province, Cuba. This species is found under rocks, manure piles, and planks, and is sometimes plowed up during cultivation. Photograph by William H. Gehrmann.

Family Gymnophthalmidae

F: ***Gymnophthalmus underwoodi*** from ca. 6 km SE Moule, Grand Terre, Guadeloupe (SBH 102275). Eight species of *Gymnophthalmus* occur from Central America to northern Argentina. Two species are found in the West Indies. *Gymnophthalmus underwoodi* is parthenogenetic, diurnally active, and inhabits litter primarily in mesic habitats. Photograph by S. Blair Hedges.

G: ***Bachia heteropus*** from the Arima Valley, Trinidad. Fifteen species of *Bachia* occur in Central and South America, but only *B. heteropus* is found in the West Indies (in addition to Trinidad, Tobago, and northern Venezuela). These lizards are often encountered in soggy cacao and banana trash. Photograph by John C. Murphy.

H: ***Tretioscincus oriximinensis*** from Apiau, Roraima, Brazil (MZUSP, number unknown). Lizards in this genus occur in northern South America. *Tretioscincus oriximinensis* is not found in the West Indies, but superficially resembles the lone West Indian representative, *T. bifasciatus*, which also occurs in northern South America. A single specimen has been taken from Isla de Providencia, but whether the species is established on that island is unknown. Photograph by Laurie J. Vitt.

Family Scincidae

I: ***Mabuya lineolata*** from 10.4 km NW Ca Soleil, Département de l'Artibonite, Haiti (130 m) (SBH 160766). About 85 species of *Mabuya* occur in Asia, Africa, and the Neotropics. Two species are found in the West Indies and only *M. lineolata* is endemic (occurring only on Hispaniola). These diurnally active skinks occupy litter in a variety of habitats. Photograph by S. Blair Hedges.

A

B

C

D

E

F

G

H

I

PLATE 3

REPTILIA III: SQUAMATA – LACERTILIA (continued)

Family Anguidae

A: ***Celestus agasepsoides*** (formerly *Sauresia agasepsoides*), an adult female from 13.7 km E Canoa (airline distance), Barahona Province, Dominican Republic (SBH 160188). Lizards in the genus *Celestus* occur in Central America as well as in the West Indies, where 19 species are endemic. *Celestus agasepsoides* is an attenuate species with an elongated tail (more than twice the snout-vent length). The species occupies xeric habitats on the Barahona Peninsula. Food includes a variety of small arthropods. Photograph by S. Blair Hedges.

B: ***Diploglossus pleii*** from the Río Abajo Forest Reserve, Arecibo, Puerto Rico. About 15 species of *Diploglossus* occur in the Neotropics, including three West Indian endemics from Cuba, Puerto Rico, and Montserrat. *Diploglossus pleii* is confined to Puerto Rico, where it is widespread in a variety of mesic habitats, including heavily human-disturbed situations. Tail autotomy is common. Photograph by Manuel Leal.

Family Teiidae

C: ***Ameiva plei***, an adult male beginning to mount a female; Katouche Bay, Anguilla. This widespread Neotropical genus of about 25 species contains 19 West Indian representatives, of which 18 are endemic. *Ameiva plei* is a diurnal, ground-dwelling active forager endemic to the Anguilla Bank. It occurs in coastal and inland scrub and at some localities competes with hermit crabs (*Paguristes* sp.) for food resources. Photograph by Ellen J. Censky.

D: ***Ameiva lineolata*** from near the military base on Isla Catalina, Dominican Republic. This small species is distinctly xerophilic and is found in open areas, often in disturbed habitats. Photograph by Robert W. Henderson.

E: ***Cnemidophorus lemniscatus***, a juvenile from Cocos Bey, Trinidad. About 40 species of *Cnemidophorus* have been described from North, Central, and South America. Two species occur in the West Indies (*C. vanzoi* is endemic). *Cnemidophorus lemniscatus* has a largely mainland distribution from lower Central America through northern South America. The West Indian distribution is limited to the Western Caribbean. Photograph by John C. Murphy.

F: ***Kentropyx borckiana***, an adult (with tongue protruding from its mouth) from Barbados, Lesser Antilles. Eight species of *Kentropyx* occur in South America, including *K. borckiana*, which also reaches the West Indies, where it is found on Barbados. The species' distribution and natural history on Barbados are virtually unknown. These lizards are unisexual, diurnally active foragers. Photograph by Frank M. Fast.

Family Tropiduridae

G: ***Leiocephalus carinatus*** from the southeastern coast of Little Cayman, Cayman Islands. The genus *Leiocephalus* is endemic to the West Indies, where 23 species are currently known. Thirteen subspecies of *L. carinatus* occur in the Bahama Islands, Cuba, and the Cayman Islands. The species also has become established in the United States. *Leiocephalus carinatus granti* occurs in the Cayman Islands. Photograph by Glenn P. Gerber.

H: ***Leiocephalus personatus***, an adult male from Cayo Monte Chico, Monte Cristi Province, Dominican Republic (about sea level) (BWMC 4854). *Leiocephalus personatus* is endemic to Hispaniola, where it is widely distributed, usually in mesic or semi-mesic habitats, and often in areas characterized by spotty shade. These lizards frequently exploit vacant lots and parks in urban areas. Photograph by Robert Powell.

A

B

C

D

E

F

G

H

PLATE 4

Reptilia IV: Squamata – Lacertilia (continued)

Family Gekkonidae

A: ***Aristelliger lar*** from 3.0 km E Canoa, Barahona Province, Dominican Republic (SBH 160289). Six species of *Aristelliger* are recognized and five are endemic to the West Indies. *Aristelliger lar* has a wide distribution on Hispaniola, where it occurs in a variety of habitats ranging from large trees in broadleaf forests and coconut groves to extremely xeric limestone and karst areas. Usually nocturnal, these geckos occasionally are active by day. Photograph by S. Blair Hedges.

B: ***Gonatodes albogularis***, an adult female from Santiago de Cuba, Santiago de Cuba Province, Cuba (SBH 190348). Fifteen to 20 species of *Gonatodes* occur in the Neotropics, but only *G. albogularis* occurs in the West Indies (in addition to a wide mainland distribution). This gecko is largely active by day. Often associated with buildings or human debris, *G. albogularis* is tolerant of what are frequently considered harsh environments, but also may be found in more moderate coastal areas. Photograph by S. Blair Hedges.

C: ***Hemidactylus haitianus*** from Mano Juan, Isla Saona, Dominican Republic. *Hemidactylus* is a pantropical genus containing about 70–80 species. Four species occur in the West Indies, but some have been introduced to the area. *Hemidactylus haitianus* is endemic to the West Indian islands of Cuba, Hispaniola, and Puerto Rico. These geckos primarily function as human commensals and often are abundant in houses or on walls in densely populated urban areas. As in many other *Hemidactylus*, eggs are capable of withstanding long exposure to salt water and adults are able to survive long periods without food; these characteristics are largely responsible for the tremendous colonizing ability of these geckos. Photograph by Richard A. Sajdak.

D: ***Phyllodactylus wirshingi***, an adult male from Bosque Guánica, Puerto Rico. *Phyllodactylus wirshingi* is endemic to the West Indies, where it occurs on the Puerto Rico Bank and Hispaniola. Although associated primarily with creviced and pocked limestone rocks and boulders, these geckos have adapted to human activities and frequently occupy exposed road cuts or debris (both human and natural), and may be found on paved roads at night (presumably thermoregulating). Photograph by Rafael A. Joglar.

E: ***Sphaerodactylus richardsoni***, an adult female from 0.3 km W Duncans, Trelawny Parish, Jamaica (SBH 161046). About 90 species of these minute geckos occur in the Neotropics, 80 of which are West Indian, and 78 of which are endemic to the region. *Sphaerodactylus richardsoni* has been found in coconut (*Cocos* sp.) plantations under husks and fronds and under rotten logs and piles of limestone rocks in a variety of mesic and semi-mesic situations. Occasionally, these lizards are associated with buildings. Photograph by S. Blair Hedges.

F: ***Sphaerodactylus macrolepis***, an adult male from the Río Abajo Forest Reserve, Arecibo, Puerto Rico. *Sphaerodactylus macrolepis* occurs on the Puerto Rico Bank and on several of the Leeward Islands. Largely scansorial, these diminutive geckos feed on a variety of litter arthropods, mainly collembolans. Photograph by Manuel Leal.

G: ***Tarentola americana*** from Trinidad, Sancti Spíritus Province, Cuba (AMNH 78208). About 15 species of *Tarentola* are known, and all but two (one of which is undescribed) occur in the Eastern Hemisphere. *Tarentola americana* is endemic to the Bahama Islands and Cuba, where it occurs in a variety of habitats, including caves, coconut trash, and buildings. Eggs (clutch size is two) are deposited communally in sites as varied as houses, caves, or under tree bark. Photograph by William H. Gehrmann.

H: ***Thecadactylus rapicauda*** from the Arima Valley, Trinidad. This monotypic genus has a broad distribution in the Neotropics. In the West Indies, this gecko is widespread in the Lesser Antilles and on some of the Virgin Islands. *Thecadactylus rapicauda* occurs in a wide range of habitats, including an association with buildings. Although usually nocturnal, individuals may bask on tree trunks by day. Photograph by John C. Murphy.

A
B
C
D
E
F
G
H

PLATE 5

REPTILIA V: SQUAMATA – LACERTILIA (continued)

Family Polychrotidae

A: ***Anolis chamaeleonides*** (formerly *Chamaeleolis chamaeleonides*), an adult male from San Vicente, Pinar del Río Province, Cuba. More than 200 known species of *Anolis* occur in the Neotropics and as far north as the southern United States. The West Indies are home to 138 described species. *Anolis chamaeleonides* is endemic to Cuba, where it is found in humid, broadleaf forests and coffee groves. Cryptic and slow-moving, these lizards sleep on vines and tree limbs. Photograph by William H. Gehrmann.

B: ***Anolis barbouri*** (formerly *Chamaelinorops barbouri*), an adult male from 21 km off Highway 44 on the road to Guazàra (which is known to many herpetologists as the Green Chapel Road), which turns off the highway 2 km N of Barahona, Barahona Province, Dominican Republic. *Anolis barbouri* is endemic to Hispaniola, where it is found primarily in moist, shaded montane ravines in abundant leaf litter. These lizards are cryptic, sedentary, and apparently do not bask, although they maintain body temperatures well above ambient air and substrate surface temperatures. Photograph by Jonathan B. Losos.

C: ***Anolis angusticeps***, an adult male from near the airport, South Bimini Island, Bahamas. This species is widespread on Cuba and the Bahama Islands and occurs in a variety of habitats, including an association with buildings. Lizards will remain motionless, relying on crypsis, until approached very closely. The diet consists primarily of small insects (e.g., aphids and dipterans). Photograph by Jonathan B. Losos.

D: ***Anolis barahonae***, an adult male from near El Puerto, Barahona Province, Dominican Republic (800 m) (BWMC 5039). This giant anole is found in the Sierra de Baoruco and adjacent lowlands of the Barahona Peninsula. Highly arboreal, this lizard inhabits broadleaf gallery forests and coffee groves. Although largely insectivorous (lepidopteran larvae and orthopterans are prominent dietary components), *A. barahonae* also consumes fruits and seeds. Photograph by Robert Powell.

E: ***Anolis bahorucoensis***, an adult male from 8 km NW Paraíso, Barahona Province, Dominican Republic. A Hispaniolan endemic, this anole usually is found in the deep shade of the forest understory, but may also be common in recently cleared areas where it lives on fallen trees and associated debris. Often foraging in leaf litter, the activity of *A. bahorucoensis* peaks in the late afternoon. Males have a weakly developed dewlap and only rarely display. These small lizards may be prey for other anoles. Photograph by Robert W. Henderson.

F: ***Anolis marmoratus***, an adult male from 4 km WNW on road to Fefé, Basse-Terre, Guadeloupe. This anole is endemic to the Guadeloupean Archipelago, where it is widespread and abundant. Known from xeric habitats to montane rain forests, these lizards are tolerant of a broad range of air and body temperatures without conspicuous thermoregulation. Individuals are commonly found on bushes and trees; males especially are prominent on trunks of trees. Photograph by Jonathan B. Losos.

G: ***Anolis conspersus***, an adult male from near Georgetown, Grand Cayman, Cayman Islands. These scansorial anoles occupy a variety of habitats. Males establish territories in January and February on tree trunks and other above-ground objects. Females and subadults are more secretive. Intermediate between non-basking thermoconformers and extensively basking thermoregulators, *A. conspersus* may aestivate during periods of drought. Photograph by Glenn P. Gerber.

Family Xantusiidae

H: ***Cricosaura typica***, from 2.8 km N Uvero, Santiago de Cuba Province, Cuba (136 m). This monotypic genus has a limited distribution on Cuba, to which it is endemic. Usually found in xeric habitats, these lizards, by day, are secretive under bark or loose soil. Movements often are snake-like. Like some geckos, *C. typica* licks its eye cap. Photograph by S. Blair Hedges.

A

B

C

D

E

F

G

H

PLATE 6

REPTILIA VI: SQUAMATA – LACERTILIA (continued)

Family Iguanidae

A: ***Ctenosaura similis***, an adult male from Cozumel, Quintana Roo, Mexico (KU slide 8907). About ten species in this genus range from northern Mexico to central Panama and also inhabit several offshore islands. Populations of *C. similis* from Isla San Andrés and Isla de Providencia are the only West Indian representatives of this genus. This large iguana is xerophilic, heliophilic, and, although largely ground-dwelling, will climb (especially juveniles). These omnivorous lizards are widely exploited by humans for food. Photograph by William E. Duellman.

B: ***Cyclura nubila***, an adult male from the southeastern coast of Little Cayman, Cayman Islands. Eight species of *Cyclura* are currently recognized and all are endemic to the West Indies. *Cyclura nubila* occurs in Cuba, Puerto Rico, and the Cayman Islands. Often feeding in cultivated areas, these large lizards frequently are hunted by humans for food. Photograph by Glenn P. Gerber.

C: ***Iguana delicatissima***, an adult male from Flamands, St. Barthélémy, Lesser Antilles. Two species of *Iguana* are recognized and both occur in the West Indies, but only *I. delicatissima* is endemic. These lizards occur in a variety of habitats, including some that are heavily disturbed. Active by day in trees or on the ground (where they often are associated with rock piles), iguanas eat carrion and bird eggs as well as vegetable matter. This species hybridizes with *I. iguana* on Les Îles des Saintes (Guadeloupe). As indicated by its specific epithet, and like other iguanas, *I. delicatissima* is exploited by humans for food. Photograph by Mark L. Day.

REPTILIA VII: SQUAMATA – SERPENTES

Family Boidae

D: ***Boa constrictor*** from the Arima Valley, Trinidad. This species ranges from northern Mexico to Argentina. In the West Indies, endemic subspecies occur on Dominica and St. Lucia, and other Antillean populations are found on Providencia, San Andrés, and Sta. Catalina. These snakes occupy a variety of habitats, including many that are disturbed, and may be encountered at ground level or in trees. In the West Indies, the diet includes *Rattus* and *Dasyprocta*. Photograph by John C. Murphy.

E: ***Corallus hortulanus*** (formerly *Corallus enydris*), an adult male from 1.6 km NE Grand Etang, Grenada (460 m) (MPM 27948). Four species of *Corallus* currently are recognized and distributed widely on the Neotropical mainland. Only *C. hortulanus* occurs in the West Indies, with populations on the Grenada Bank and St. Vincent. Almost strictly arboreal, these nocturnally active snakes prey on *Anolis*, birds (rarely), and introduced rodents. A tremendous array of color varieties may be correlated with climatic regimes on Grenada. Photograph by Robert W. Henderson.

F: ***Epicrates gracilis***, a juvenile from near Camp Perrin, Département du Sud, Haiti. *Epicrates* contains ten currently recognized species, nine of which are endemic to the West Indies. *Epicrates gracilis* is the most morphologically specialized member of the genus. This Hispaniolan endemic is not uncommon in mesic habitats, is almost entirely arboreal, and actively forages for anoles, which are its predominant, if not exclusive prey. Photograph by Robert W. Henderson.

Family Tropidophiidae

G: ***Tropidophis feicki*** from San Vicente, Pinar del Río Province, Cuba. Three species of *Tropidophis* occur in South America and 13 in the West Indies. *Tropidophis feicki* is endemic to Cuba, where it occurs in wooded areas. These nocturnally active snakes may be ground-dwelling or arboreal, and are capable of undergoing color changes (light at night and dark by day). Photograph by William H. Gehrmann.

Family Elapidae

H: ***Micrurus nigrocinctus***, an adult male from Atlántida, Honduras (opposite Hog Island). Over 50 species of *Micrurus* occur on mainland America, from the southern United States to central Argentina. Only a single specimen (now lost) of *M. nigrocinctus* has been taken from the West Indies (in 1844 on Isla Providencia); the presence of the species in the West Indies now is questionable. Photograph by William W. Lamar.

A

B

C

D

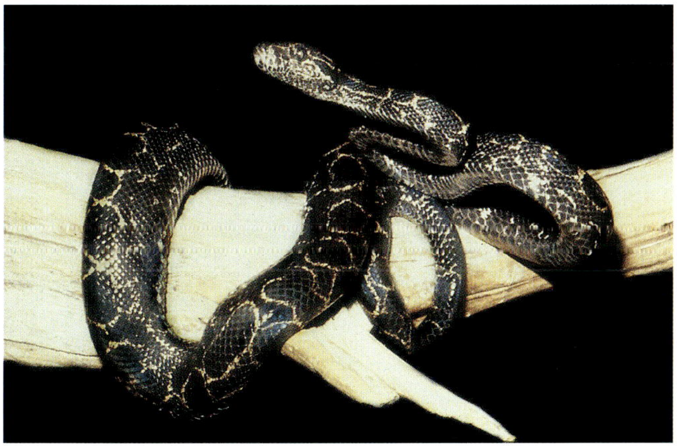

E

F

G

H

PLATE 7

Reptilia VIII: Squamata – Serpentes (continued)

Family Leptotyphlopidae

A: ***Leptotyphlops columbi*** from Little Fortune Hill, San Salvador, Bahama Islands (2 m). The genus *Leptotyphlops* is pantropical. Eight species, of which six are endemic, are known from the West Indies. *Leptotyphlops columbi* is endemic to the Bahamas, where it is encountered in logs. Known locally as the "wood worm," this species is not closely related to any other West Indian member of the genus. Photograph by S. Blair Hedges.

Family Typhlopidae

B: ***Typhlops lumbricalis*** from San Vicente, Pinar del Río Province, Cuba. *Typhlops* is a pantropical genus with 24 endemic species known to occur in the West Indies. *Typhlops lumbricalis* is found in the Bahama Islands and Cuba, where it occupies a variety of habitats. Populations may be particularly dense in disturbed areas such as coppice, sugar cane fields, palm trash, and cornfields. Photograph by William H. Gehrmann.

Family Viperidae

C: ***Bothrops caribbaea*** from the vicinity of Anse la Raye, St. Lucia. More than 30 species of *Bothrops* occur throughout the American tropics, but only two are found in the West Indies (on St. Lucia and Martinique) and both are endemic. *Bothrops caribbaea* is found on St. Lucia, where it is common in disturbed habitats, and especially in piles of coconut trash. These snakes feed primarily on introduced rodents (*Mus* and *Rattus*) and are capable of inflicting lethal bites to humans. Photograph by Richard A. Sajdak.

Family Colubridae

D: ***Alsophis antillensis*** from 0.8 km S Galway's Estate, Montserrat, Lesser Antilles. Species of *Alsophis* occur in western South America and the West Indies; 11 endemic species are known from the latter. *Alsophis antillensis* occurs on Montserrat, the Guadeloupe Archipelago, and Dominica in a wide variety of habitats from arid scrub to rainforests. These snakes are diurnally active, ground-dwelling active foragers whose diets include *Eleutherodactylus* and *Anolis*. Photograph by Richard A. Sajdak.

E: ***Alsophis portoricensis***, an adult exhibiting dorsoventral neck compression (an antipredator response) from the Cambalache Forest Reserve, Arecibo, Puerto Rico. Widely distributed on the Puerto Rico Bank, these snakes are largely ground-dwelling, but will ascend into the crowns of rainforest trees. Although the diet consists largely of *Eleutherodactylus* and *Anolis*, *A. portoricensis* opportunistically will take a wide variety of vertebrate prey. Photograph by Javier A. Rodríguez-Robles.

F: ***Antillophis parvifrons***, an adult male from 8 km ENE of Baní, Peravia Province, Dominican Republic (130 m) (BWMC 5662). This genus is endemic to the West Indies and is comprised of two species. *Antillophis parvifrons* is known from Hispaniola (and many satellite islands) and from one record from Little Inagua in the Bahamas. These snakes are widespread and occupy a wide variety of habitats. Diurnally active foragers, they take a variety of vertebrate prey, but predominantly feed on anoles. Photograph by James D. Forester.

G: ***Arrhyton taeniatum***, a juvenile from 0.2 km E Windmill Beach, United States Naval Station at Guantanamo Bay, Cuba (20 m) (SBH 171002). This genus is endemic to the West Indies and 12 species are currently recognized. *Arrhyton taeniatum* is found on Cuba, where it is probably fossorial. Its diet includes *Amphisbaena* and *Typhlops*. Photograph by S. Blair Hedges.

H: ***Chironius grandisquamis*** from Golfito, Puntarenas Province, Costa Rica. Species of *Chironius* range from Honduras to Argentina and Uruguay. *Chironius grandisquamis* is very similar in appearance to *C. vincenti*, the only West Indian representative, which is endemic to St. Vincent. These diurnally active snakes are ground-dwelling to semiarboreal in rainforest habitats, where they actively forage predominantly or exclusively for anurans. Photography by William W. Lamar.

I: ***Clelia clelia*** from 5.1 km WSW St. Tomas, Izabel, Guatemala (KU 187257). *Clelia* is a widespread genus on the Neotropical mainland (from Mexico to Argentina). Two species occur in the West Indies, one of which is endemic to St. Lucia (*C. errabunda*). *Clelia clelia* has a broad mainland distribution, but occurs only on Grenada in the West Indies (an endemic subspecies has been described). Photograph by Jonathan A. Campbell.

A

B

C

D

E, F

G

H

I

PLATE 8

REPTILIA IX: SQUAMATA – SERPENTES (continued)

Family Colubridae (continued)

A: ***Coniophanes imperialis*** from the Burrell Boom Road at the Belize River ca. 23 km N of Northern Highway, Belize District, Belize (CM, number unknown, one of a series). About 12 species of *Coniophanes* range from extreme southern Texas (United States) to Pacific Ecuador. *Coniophanes imperialis* does not occur in the West Indies, but is grossly similar in appearance to the West Indian endemic *C. andresensis*. The latter is the only West Indian representative of the genus and is endemic to Isla San Andrés. Photograph by Laurie J. Vitt.

B: ***Darlingtonia haetiana***, an adult female from Loma Remigio, Barahona Province, Dominican Republic (1100 m) (BWMC 5662). *Darlingtonia* is a monotypic genus endemic to Hispaniola. This secretive, litter-dwelling snake preys almost exclusively on frogs of the genus *Eleutherodactylus*. Photograph by Robert Powell.

C: ***Hypsirhynchus ferox***, an adult male from 4.3 km NE Oviedo, Pedernales Province, Dominican Republic (30 m) (BWMC 6027). *Hypsirhynchus* is a monotypic genus endemic to Hispaniola. This diurnally active ground-dweller is widespread on Hispaniola, but nowhere is it common. Preying exclusively on lizards, individuals exhibit an ontogenetic shift in diet from *Anolis* to *Ameiva* and *Leiocephalus*. Photograph by Robert Powell.

D: ***Ialtris dorsalis***, an adult female from ca. 3 km N Bois Sec, Département de la Grand'Anse, Haiti (360 m) (SBH 103702). One of three species of *Ialtris*, all of which are endemic to Hispaniola, *I. dorsalis* is the most widely distributed, but, like all members of the genus, is nowhere common. This ground-dweller apparently exploits a wide variety of terrestrial vertebrates as food. Photograph by S. Blair Hedges.

E: ***Liophis juliae*** from 1.6 km E Calibishi, St. Andrew Parish, Dominica. Five species of *Liophis* occur in the West Indies and four are endemic to the area. These snakes, like ground-dwelling *Alsophis*, appear vulnerable to predation by the mongoose. *Liophis juliae* preys predominantly on frogs and is one of the very few West Indian colubrids which occasionally eats insects. Photograph by Richard A. Sajdak.

F: ***Mastigodryas bruesi*** from 3.2 km ENE Layou, St. Patrick Parish, St. Vincent. *Mastigodryas bruesi* is the only member of a predominantly mainland genus which is endemic to the West Indies (St. Vincent and the Grenada Bank). This diurnally active snake is largely ground-dwelling, but will sleep in above-ground vegetation. The diet consists primarily of small frogs and lizards. Photograph by Richard A. Sajdak.

G: ***Tretanorhinus variabilis***, an adult male from Soroa, Pinar del Río Province, Cuba (200 m) (SBH 172473). The only other species in the genus *Tretanorhinus* occurs on the Neotropical mainland. *Tretanorhinus variabilis* is found on Cuba and Grand Cayman. These snakes are largely aquatic and may be associated with freshwater, brackish tidal estuaries, and saltwater marshes. This nocturnal forager preys almost exclusively on fish. Photograph by S. Blair Hedges.

H: ***Uromacer frenatus***, an adult male from 4.3 km NE Oviedo, Pedernales Province, Dominican Republic (30 m) (BWMC 6028). One of three species of *Uromacer*, all endemic to Hispaniola, *U. frenatus* frequents a variety of arboreal habitats, but often will forage on the ground. Virtually all prey is saurian, with an ontogenetic shift in diet from anoles to *Leiocephalus* and *Ameiva*. This species is widely sympatric with *U. catesbyi*. Photograph by Robert Powell.

A B C D E F G H

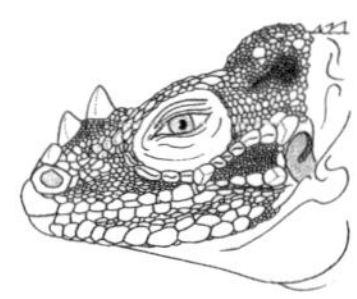

THE ORIGIN OF WEST INDIAN AMPHIBIANS AND REPTILES

S. BLAIR HEDGES[1]

Abstract.—The known West Indian herpetofauna is comprised of 175 species of amphibians (99% endemic) and 457 species of reptiles (93% endemic). Information on distributions, relationships, and times of origin, with emphasis on estimates of divergence time from albumin immunological data, are analyzed in an attempt to understand the origin of the herpetofauna. Seventy-seven independent lineages of West Indian amphibians and reptiles are identified and nearly all (95%) originated in the New World. Of those lineages for which a source area within the New World can be determined, most (79%) show a South American origin, with smaller contributions from Central America (15%) and North America (6%). One very old and diverse lineage, frogs of the genus *Eleutherodactylus*, originated by vicariance or dispersal in the late Cretaceous (70 mya). With one possible exception (the xantusiid lizard *Cricosaura typica*), all other lineages appear to have arrived by dispersal during the Cenozoic, and all but nine lineages in the last half of the Cenozoic (30–0 mya). Most West Indian lineages with multiple endemic species originated in the mid-Tertiary, whereas most lineages with a single endemic species arose in the late Tertiary. Quaternary dispersal is postulated to explain the origin of West Indian populations of mainland species.

The probable explanation for the predominately South American origin of the West Indian herpetofauna is the nearly unidirectional (towards the west-northwest) water current patterns in the Caribbean: water reaching the Greater Antilles originates near the Lesser Antilles and South America. Wide variation in the times of origin for the West Indian lineages does not support the recent suggestion of a mid-Cenozoic landbridge between South America and the Greater Antilles. Dispersal, in most cases, is believed to have occurred by the discharge of organisms attached to flotsam from the mouths of major rivers on the continents and carried by currents to the West Indies. Proto-Antillean vicariance remains a geological possibility, but recent paleocoastline data suggest that a dry land connection between the proto-Antilles and neighboring continents in the late Mesozoic may not have occurred. The widespread invoking of vicariance to explain nearly any distribution of West Indian organism without information on time of origin is seen as a popular trend but one that lacks support.

INTRODUCTION

AMPHIBIANS AND REPTILES MAKE UP ABOUT HALF (632 species) of the terrestrial vertebrates of the West Indies and exhibit the highest levels of endemism. For this reason, they have figured prominently in the literature on Caribbean biogeography. Most recently, attention has focussed on testing two theories for the origin of the West Indian biota. The vicariance model proposes that a proto-Antillean biota existed between North and South America in the late Cretaceous, and later moved eastward with the Caribbean plate during the Cenozoic (Rosen 1975, 1985). The alternative explanation is that overwater dispersal largely has been responsible for the origin of the current West Indian biota (e.g., Matthew 1918; Simpson 1956; Darlington 1957; Williams 1969; Pregill 1981*b*; Briggs 1984). Numerous combinations of these two mechanisms also have been suggested for various groups, although much of this has been speculation, and very few robust phylogenies or established times of origin have been presented.

The geologic history of the Caribbean is complex, but is becoming better understood (Perfit and Williams 1989; Pindell and Barrett 1990; Pitman et

[1] Department of Biology and Institute of Molecular Evolutionary Genetics, Pennsylvania State University, University Park, Pennsylvania 16802, USA.

al. 1993). However, the proposal that a "proto-Antilles" existed between North and South America in the late Cretaceous (70–80 million years ago [mya]) has not changed substantially during the last two decades (e.g., Hedges 1982). Rosen's (1975) vicariance model requires a continuously emergent land mass, otherwise the proto-Antilles could not have provided a link between North and South America and transported terrestrial organisms eastward during the Cenozoic. If the proto-Antilles were not emergent, or if they became submerged at any time during their movement eastward, then the present West Indian biota must owe its origin to dispersal. Unfortunately, details on the emergence or submergence of land in the proto-Antilles are not known (Perfit and Williams 1989), although evidence exists for the complete or nearly complete submergence of Jamaica and the South Island of Hispaniola during the Oligocene (Horsfield 1973; Horsfield and Roobol 1974; reviewed in Buskirk 1985 and Hedges 1989*a*, 1989*b*).

Paleobotanical data suggest that well-developed plant communities, with high-altitude floras, were present in the Oligocene in at least part of the Greater Antilles (Graham and Jarzen 1969; Graham 1993). The oldest vertebrate fossil is a Jurassic pterosaur from Cuba, although the environment likely was shallow-water rather than terrestrial (Colbert 1969). Terrestrial vertebrate fossils from the Tertiary are rare and mostly are from mid-Cenozoic Dominican amber (Poinar 1992). These include frogs, anoline lizards, geckos, mammal hair (presumably rodent), and bird feathers (Rieppel 1980; Böhme 1984; Poinar et al. 1985; Poinar and Cannatella 1987; Poinar 1988, 1992). Fossils of *Anolis*, *Sphaerodactylus*, and *Eleutherodactylus* have been collected at the oldest mine (La Toca) and the age of those fossils has been estimated to be 30–40 my (Lambert et al. 1985; Poinar 1992). However, those older dates, using a somewhat unconventional method (nuclear magnetic resonance) for dating, are tied to a Lower Miocene (20–23 mya) date for Dominican

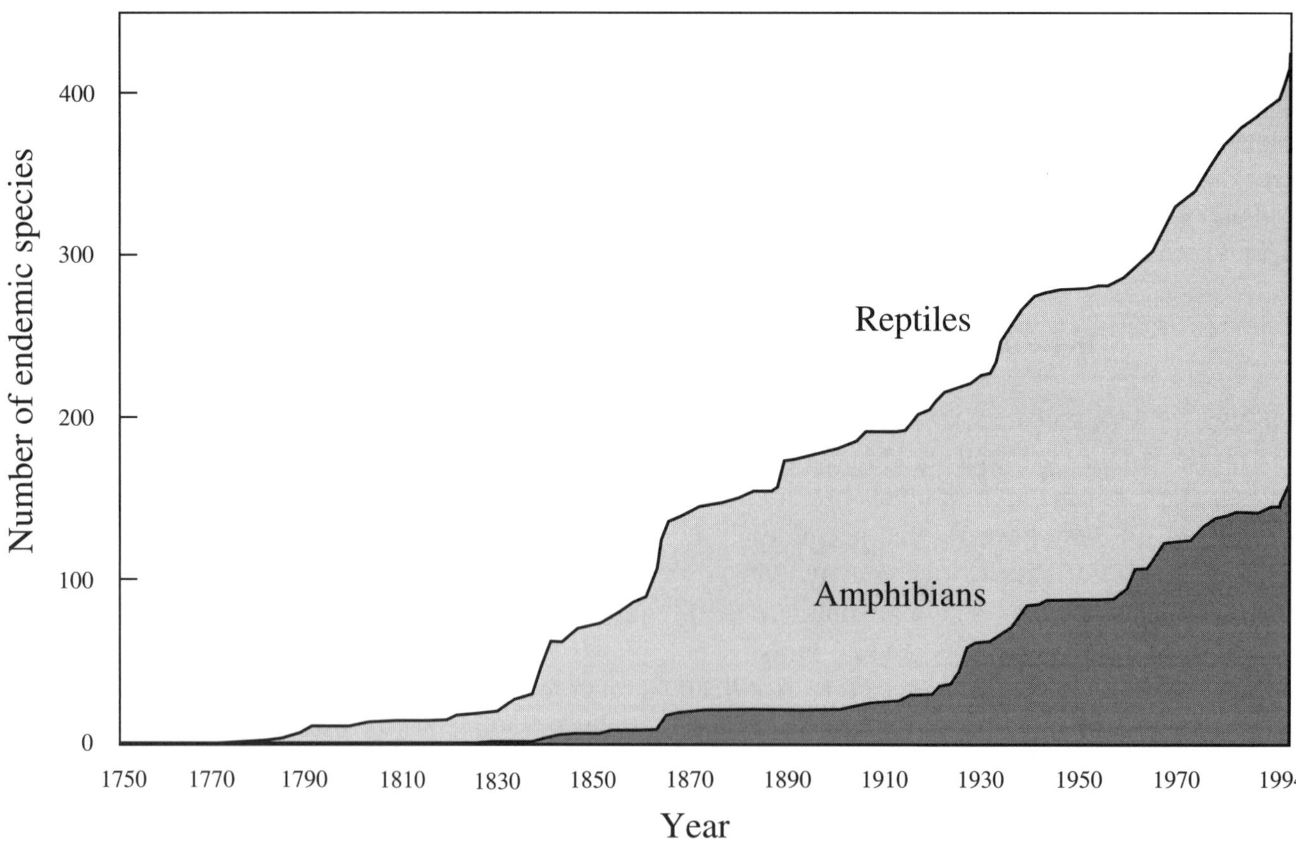

FIGURE 1. Discovery curves of West Indian amphibians and reptiles showing the number of known species at each time interval.

amber based on foraminiferal analysis, and thus a range of 20–40 mya for the La Toca fossils may be more realistic. Miocene fossils of a ground sloth (MacPhee and Ituralde-Vinent 1994) and cichlid fish (Cockerell 1923) also are known. Pleistocene fossils record a diverse vertebrate fauna, much of which disappeared at about the time of human colonization (Pregill and Olson 1981; Morgan and Woods 1986; Woods 1989; Morgan 1993).

Recently, these two hypotheses for the origin of the terrestrial vertebrate fauna — dispersal and vicariance — were tested using estimates of amino acid sequence divergence in the protein serum albumin in diverse groups of West Indian amphibians and reptiles: bufonid, hylid, eleutherodactyline, and leptodactyline frogs; anoline, iguanine, sphaerodactyline, tropidurine, teiid, and anguid lizards; amphisbaenians; and alsophine and tropidophiid snakes (Hedges et al. 1992*b*). Albumin immunological distance (ID) is correlated with time of divergence in vertebrates (Maxson 1992) and this provided the means of estimating times of divergence between West Indian groups and their relatives on the mainland (with new data and those from the literature). All groups examined had lower estimates of divergence than would be predicted by proto-Antillean vicariance, suggesting an origin by overwater dispersal in the mid- to late Cenozoic. Further statistical analyses supported that conclusion (Hedges et al. 1994).

Several possible explanations exist for the finding of a recent origin for these major lineages of the West Indian amphibians and reptiles. The suggestion of an emergent proto-Antillean "isthmus" lacks support from geology and paleontology (Perfit and Williams 1989), in which case an opportunity for vicariance may not have occurred. Another possible explanation was suggested (Hedges et al. 1992*b*): the bolide (meteor or asteroid) impact at the Cretaceous/Tertiary boundary. The proto-Antillean islands were among the closest land areas to the Yucatán impact site (300 km crater; Sharpton et al. 1993) and they would have been devastated by the effects of this impact, including giant tsunamis (tidal waves). Evidence for the bolide impact is substantial (e.g., Hildebrand and Boynton 1990; Maurrasse and Sen 1991; Alvarez et al. 1992; Smit et al. 1992; Swisher et al. 1992; Kring and Boynton 1992; Sharpton et al. 1993), and it was this evidence that helped pinpoint the location of the impact structure. However, whether or not the local and global effects of the impact were sufficient to completely extinguish an ancient proto-Antillean biota that may have existed is speculative.

Analyses of phylogeny and distribution in other West Indian groups also have led to the suggestion of an origin by dispersal. Such groups include scorpions (Armas 1982; Armas and Marcano-Fondeur 1992), freshwater emydid turtles (Seidel 1988), rodents (Woods 1989), bats (Koopman 1989; Breuil and Masson 1991; although see Griffiths and

TABLE 1. Numbers of species of amphibians in the West Indies: total (T), native[1] (N), endemic[2] (E), introduced (I), and percent endemic[3] (%).

Taxon	T	N	E	I	%
Bufonidae					
Bufo [4]	13	12	12	1	100
Dendrobatidae					
Colostethus	1	1	1	0	100
Hylidae					
Hyla	3	1	1	2	100
Osteopilus [4, 5]	9	9	9	0	100
Pseudacris	1	0	0	1	0
Scinax	1	0	0	1	0
Leptodactylidae					
Eleutherodactylus [4]	138	138	138	0	100
Leptodactylus	5	5	3	0	60
Microhylidae					
Gastrophryne	1	0	0	1	0
Ranidae					
Rana	3	0	0	3	0
Total	175	166	164	9	99

[1] Occurs naturally within the West Indies.
[2] Occurs naturally within the West Indies and nowhere else (except as introduced).
[3] Number of endemic species divided by number of native species.
[4] Includes some species currently being described.
[5] Includes *Calyptahyla* and some species previously placed in *Hyla* (see text).

Klingener 1988), and various Lesser Antillean plants and animals (Erard 1991; Lescure et al. 1991). For other groups, vicariance has been implicated. For example, the most basal lineage of xantusiid lizards survives as an eastern Cuba endemic (*Cricosaura*), possibly dating to the proto-Antilles (Hedges et al. 1991; Hedges and Bezy 1993, 1994). Also, the two major subgenera of West Indian *Eleutherodactylus* (*Euhyas* and *Eleutherodactylus*) are separated by a large ID, suggesting a late Cretaceous divergence (Hass and Hedges 1991).

Many groups of West Indian organisms have not yet been investigated, and therefore an origin by vicariance remains a possibility for those groups. Among mammals, the insectivores *Solenodon* and *Nesophontes* are believed to represent an ancient Antillean lineage (MacFadden 1980, 1981), and recently collected fossils establish ground sloths in Cuba in the Miocene (MacPhee and Iturralde-Vinent 1994). Antillean freshwater fishes (Burgess and Franz 1989) and butterflies (Miller and Miller 1989; Johnson 1991) are believed to have originated by both vicariance and dispersal, and Liebherr (1988*a*, 1988*b*) suggested an origin by vicariance for the carabid beetles. The giant endemic Cuban cycad (*Microcycas*) possibly represents an ancient lineage (Jones 1993).

The number of species of West Indian amphibians and reptiles has nearly doubled during the last four decades, largely through the efforts of Albert Schwartz, and the rate of species discovery has not yet declined (Figure 1). Currently 175 species of amphibians (Table 1) and 457 species of reptiles (Table 2) are known from the West Indies; these represent at least 77 independent lineages (see below). Hedges et al. (1992*b*) analyzed ID data for only a subset (13) of those lineages of West Indian amphibians and reptiles. Some minor lineages considered to be the result of recent dispersal were not included in that study, and new data for other lineages have become available since then. Therefore, that analysis is expanded here to include all lineages of West Indian amphibians and reptiles believed to have had independent origins, and, where possible, to (1) estimate the time of origin of each lineage, and (2) identify its mainland source area.

Materials and Methods

This analysis draws from the literature pertaining to the origin of West Indian amphibians and reptiles. Albumin ID data were extracted from published and unpublished studies for the purpose of obtaining estimates of time of divergence. The protein serum albumin has excellent chronological properties (1 ID = approximately 0.6 million years; Maxson 1992), but several sources of error are involved in estimating divergence time from ID data (Hedges et al. 1992*b*, 1994). These include measurement error (± 2 ID; Maxson and Maxson 1979), differences in the reciprocal estimation of ID between two taxa (usually about 10%; Hass and Maxson 1993), unequal rates of change among lineages (Cadle 1988), error in estimating amino acid differences (apparently low; Hass and Maxson 1993), and error involved in calibrating IDs with divergence times estimated by fossils or geology. This last source of error, when estimated for the ID data presented in Hedges et al. (1992*b*), was found not to be substantial and did not affect the conclusions of that study (Hedges et al. 1994). Nonetheless, calibration error will be considered in this analysis (calculation of this error as in Hedges et al. 1994). Also, reciprocal IDs are not available for many species comparisons; this is likely to increase the error in divergence time estimates. However, one advantage of using an ID estimate of divergence time is that it is an estimate of the actual phylogenetic divergence event rather than a minimum or maximum age of that event determined by fossil-based estimates. For example, even if vertebrate fossils in Dominican amber were very accurately dated (now 20–40 mya), this would only establish a minimum time for the arrival of those lineages in the West Indies. The actual time of arrival (or divergence from mainland groups) could be much earlier, which apparently is the case for *Eleutherodactylus*.

The definition of the West Indies used here, and distributional data, are from Schwartz and Henderson (1988, 1991). Taxonomy follows Duellman (1993), King and Burke (1989), Schwartz and Henderson (1991), and Zug (1993), except where noted. All independent lineages of West Indian amphibians and reptiles are discussed. Lineages that clearly are the result of human introduction are mentioned but not treated in the analyses.

Relationships of species within a West Indian lineage, unless they have a bearing on the origin of that lineage, are not considered. For that reason,

TABLE 2. Numbers of species of reptiles in the West Indies: total (T), native (N), endemic (E), introduced (I), and percent endemic (%).

Taxon	T	N	E	I	%
Squamata					
Amphisbaenia					
Amphisbaenidae					
Amphisbaena [1]	14	14	14	0	100
Sauria					
Anguidae					
Ophisaurus	1	0	0	1	0
Celestus	19	19	19	0	100
Diploglossus	3	3	3	0	100
Gymnophthalmidae					
Bachia	1	1	0	0	0
Gymnophthalmus	2	2	1	0	50
Tretioscincus	1	1	0	0	0
Iguanidae					
Anolis [1]	138	138	137	0	99
Ctenosaura	1	1	0	0	0
Cyclura	8	8	8	0	100
Iguana	2	2	1	0	50
Leiocephalus	23	23	23	0	100
Gekkonidae					
Aristelliger	7	7	6	0	86
Gekko	1	0	0	1	0
Gonatodes	1	1	0	0	0
Hemidactylus	5	4	1	1	25
Phyllodactylus	2	2	2	0	100
Sphaerodactylus [1]	80	80	78	0	100
Tarentola [1]	2	2	2	0	100
Thecadactylus	1	1	0	0	0
Scincidae					
Mabuya	2	2	1	0	50
Teiidae					
Ameiva	20	20	19	0	95
Cnemidophorus	2	2	1	0	50
Kentropyx	1	1	0	0	0
Xantusiidae					
Cricosaura	1	1	1	0	100
Serpentes					
Boidae					
Boa	1	1	0	0	0
Corallus	1	1	0	0	0
Epicrates	9	9	9	0	100
Colubridae					
Alsophis	11	11	11	0	100
Antillophis	2	2	2	0	100
Arrhyton	12	12	12	0	100
Chironius	1	1	1	0	100
Clelia	2	2	1	0	50
Coniophanes	1	1	1	0	100
Darlingtonia	1	1	1	0	100
Diadophis	1	0	0	1	0
Elaphe	1	0	0	1	0
Hypsirhynchus	1	1	1	0	100
Ialtris	3	3	3	0	100
Liophis	5	5	4	0	80
Mastigodryas	1	1	1	0	100
Nerodia	1	1	0	0	0
Pseudoboa	1	1	0	0	0
Tretanorhinus	1	1	1	0	100
Uromacer	3	3	3	0	100
Elapidae					
Micrurus	1	1	0	0	0
Leptotyphlopidae					
Leptotyphlops	8	8	6	0	75
Tropidophiidae					
Tropidophis	13	13	13	0	100
Typhlopidae					
Typhlops [1]	24	24	24	0	100
Viperidae					
Bothrops	2	2	2	0	100
Testudines					
Emydidae					
Trachemys	5	4	4	1	100
Kinosternidae					
Kinosternon	1	1	0	0	0
Pelomedusidae					
Pelusios	1	0	0	1	0
Testudinidae					
Geochelone	1	1	0	0	0
Crocodylia					
Alligatoridae					
Caiman	1	0	0	1	0
Crocodylidae					
Crocodylus	3	3	1	0	33
Total	457	449	418	8	93

[1] Includes some species currently being described.

many phylogenetic studies of West Indian amphibians and reptiles are not discussed here. For some groups, IDs are not available between the West Indian lineage and its closest relative on the mainland, and therefore several assumptions have been made concerning times of origin. If IDs are available between that group (e.g., a genus with representatives in the West Indies) and a more distantly related group (e.g., another genus), then it is assumed that the two groups are monophyletic and that the corresponding time estimate is an upper limit for the time of origin in the West Indies. For single species occurring both in the West Indies and on the mainland, a Quaternary (0–1.6 mya) divergence is assumed. Populations within species of amphibians and reptiles that have been examined using microcomplement fixation usually are separated by very low IDs (0–3) concordant with divergence in the Quaternary (e.g., Daugherty et al. 1982; Roberts and Maxson 1986; Hass 1991; Hass and Hedges 1991; Hass et al. 1992), although exceptions to this rule are known, especially in *Leptodactylus* (Maxson and Heyer 1988).

Because the continental source areas are much older than the West Indies and have a much greater diversity of amphibian and reptilian groups, it is assumed (except where noted) that the West Indian lineages originated from mainland source areas and not the reverse. For most lineages, this implicit assumption is supported by additional phylogenetic and distributional data. The primary source areas considered here are Africa and the New World; within the New World, they are North America, Central America, and South America. Neotropical groups occurring in the Lesser Antilles, especially the southern islands, are assumed to have been derived from geographically adjacent South America when no other information is available. South American groups that extend their range into Panamá are treated as South American because the Isthmus of Panama has been emergent only since the Pliocene (Savin and Douglas 1985) indicating recent dispersal northward for those groups.

Only proto-Antillean vicariance (e.g., Rosen 1975) is considered here; intra-Caribbean vicariance may have occurred but is not relevant to the origin of a West Indian group. Geologic models indicate that the proto-Antilles broke apart in the late Cretaceous 70–80 mya (Pindell and Barrett 1990) and therefore vicariance is assumed for groups that arose during that time. Dispersal is assumed for more recent times of origin (i.e., Cenozoic) and for the origin of Lesser Antillean taxa (based on the Eocene origin of that island arc; Pindell and Barrett 1990) when no ID data are available.

RESULTS

Amphibia: Anura

Bufonidae

Bufo marinus (species group) is native to Central and South America and has been introduced widely throughout the West Indies, although it is not established on Cuba (Schwartz and Henderson 1991).

(1) ***Bufo peltocephalus* group.**—This group of toads was recognized at the generic level by Pregill (1981*a*), who resurrected the genus *Peltophryne*. Although evidence for the monophyly of this group was given, no phylogenetic evidence was presented to substantiate its generic distinction from *Bufo*. Recently, an ID of 85 between *Peltophryne* and *B. marinus* (average of *P. guentheri* and *P. peltocephalus* each to *B. marinus*), was presented (Hedges et al. 1992*b*). Because IDs within the virtually global genus *Bufo* range up to 160, and the average ID between African and New World species groups of *Bufo* is 133 ± 14, the *peltocephalus* species group appears to be a radiation derived from within New World *Bufo* and does not warrant generic recognition. Cei (1972) suggested that the affinities of the *Bufo peltocephalus* group are with the *Bufo granulosus* group of South America. The ID data suggest an origin for the *peltocephalus* group in the early Cenozoic (51 ± 5.3 mya) or later by overwater dispersal, probably from South America.

Dendrobatidae

(2) ***Colostethus chalcopis*.**—This species is endemic to Martinique and its affinities within *Colostethus* are not clear (Kaiser et al. 1994*a*). Because *Colostethus* is almost entirely South American in

distribution, and species occur on Trinidad and Tobago (placed by some in a separate genus), *C. chalcopis* almost certainly arrived on Martinique by dispersal from South America. No IDs are available to estimate the time of that dispersal event, although distances among the genera *Colostethus*, *Dendrobates*, and *Phyllobates* (Maxson and Myers 1985) average about 100 (60 mya), suggesting that divergence within *Colostethus* probably occurred subsequently (i.e., during the Cenozoic). The geologic origin of the Lesser Antilles in the Eocene (Pindell and Barrett 1990) places an upper time limit on the origin of *C. chalcopis* (0–45 mya).

Hylidae

Four species have been introduced to the West Indies: *Pseudacris crucifer*, *Hyla cinerea*, *H. squirella*, and *Scinax rubra* (Schwartz and Henderson 1991).

(3) ***Osteopilus***.—Ten native West Indian hylids are currently placed in three genera: *Calyptahyla* (1 species), *Hyla* (5 species), and *Osteopilus* (4 species, including an undescribed Jamaican species). Two of those genera (*Calyptahyla* and *Osteopilus*) are endemic to the West Indies. Hispaniola (4 species) and Jamaica (5 species) are the centers of hylid species diversity, with a single species native to Cuba and the Bahamas Bank. Dunn (1926) proposed that the Hispaniolan and Jamaican species represented independent island radiations but that the West Indian species were "allied," suggesting a single origin for the West Indian hylids. Trueb and Tyler (1974) examined morphological variation in these species and concluded that at least six independent invasions to the West Indies occurred, all from South America via the Lesser Antilles. Data from protein electrophoresis, microcomplement fixation, and DNA sequencing of the West Indian species (Hedges et al. 1992*b*; Hedges, Strafalace, and Maxson, in preparation) suggest that they represent a single monophyletic group (to take the generic name *Osteopilus*), with one exception (*Hyla heilprini*). An albumin ID of 80 (48 my) between *Osteopilus septentrionalis* and the mainland *Osteocephalus taurinus* (Hedges et al. 1992*b*) suggests an early Cenozoic origin by dispersal for this major West Indian lineage. However, few mainland species have been examined and thus a more recent origin is possible. Large IDs (>100) between *Osteopilus septentrionalis* and Holarctic hylids (Maxson and Wilson 1975) indicate that the origin of the West Indian hylids is not from that group. Until more data become available, especially from Central American taxa, the suggestion by Trueb and Tyler (1974) of a South American origin seems most plausible.

(4) ***Hyla heilprini***.—Trueb and Tyler (1974) suggested that the affinities of this species are with the predominantly South American *H. albomarginata* group. Aside from recognizing its distinctiveness among West Indian hylids, molecular data presently are unable to identify the time of entry of this lineage.

Leptodactylidae

(5) ***Eleutherodactylus***.—This is the largest vertebrate genus (>520 species) and is distributed from Arizona and Texas south to Argentina and southeastern Brazil. Two major areas of particularly high species density are the Andes of South America, and the West Indies. All of the 138 West Indian species are endemic to that region and nearly all are restricted to single islands; all but seven species (= 95%) occur in the Greater Antilles (Hedges 1995).

Several studies have addressed higher level relationships in *Eleutherodactylus* (Lynch 1986; Savage 1987; Hedges 1989*b*; Joglar 1989; Hass and Hedges 1991). Five subgenera have been recognized based on evidence from morphological, allozyme, and ID data, three of which occur in the West Indies (Hedges 1989*b*). The subgenus *Euhyas* (82 species) is endemic to the western Caribbean (Cuba, Jamaica, Hispaniola), and *Pelorius* (6 species) is restricted to Hispaniola. The subgenus *Eleutherodactylus* (50 West Indian species) occurs throughout the West Indies and on the mainland (primarily South America). Albumin IDs indicated a close relationship between *Euhyas* and the Central American subgenus *Syrrhophus* (24 species), and between *Pelorius* and West Indian species of the subgenus *Eleutherodactylus* (Hass and Hedges

1991). More extensive sampling of mainland *Eleutherodactylus* and other leptodactylid genera with DNA sequence data has suggested that the subgenus *Eleutherodactylus* is not monophyletic, and that West Indian species of the genus *Eleutherodactylus* (all three subgenera) form a monophyletic group (Youngblood and Hedges, in preparation), although two species in the southern Lesser Antilles possibly have affinities with South American taxa (Kaiser et al. 1994*b*, 1994*c*). The ID data would indicate that *Syrrhophus* also is included in that monophyletic group. Thus, only a single origin is necessary to explain the presence of the genus *Eleutherodactylus* in the West Indies. Major lineages of *Eleutherodactylus* occur in both Central and South America, and the closest relative of the West Indian lineage has not yet been determined. However, the average ID between the two major lineages within the West Indies, 117 (70 mya; *Euhyas* vs. *Eleutherodactylus*; Hass and Hedges 1991), suggests a Mesozoic (Cretaceous) origin for the West Indian lineage, possibly by proto-Antillean vicariance. The calibration error associated with that divergence estimate 70 ± 6.8 mya (= 63.2–76.8 mya) indicates that an origin after the breakup of the proto-Antilles and after the bolide impact (65 mya) cannot be ruled out. Dispersal from Cuba (*Euhyas*) to Central America later in the mid-Cenozoic likely explains the origin of the subgenus *Syrrhophus* (Hedges 1989*b*).

(6) ***Leptodactylus albilabris***.—Heyer (1978) synonymized *L. dominicensis* (Hispaniola) with *L. albilabris* (Puerto Rican Bank); Schwartz and Henderson (1991) continued recognition of these two apparently closely related taxa as distinct species. They are placed in the largely South American *fuscus* group (Heyer 1978), and *L. albilabris* has an ID of 66 (40 mya) to *L. labrosus* (Maxson and Heyer 1988), suggesting dispersal, probably from South America, during the mid-Tertiary.

(7) ***Leptodactylus fallax***.—This Lesser Antillean species is placed in the largely South American *pentadactylus* group (Heyer 1979). It has very low IDs (5–11) to populations of a South American species (*L. stenodema*) that may or may not be its closest relative (Maxson and Heyer 1988), suggesting very recent (3–7 mya) dispersal from South America.

(8) ***Leptodactylus insularum***.—This species of the *ocellatus* group is distributed in Central and northern South America and occurs on Isla de San Andrés and Isla de Providencia. Presumably, those island populations originated by dispersal from Central or South America in the Quaternary.

(9) ***Leptodactylus validus***.—This species of the *melanonotus* group occurs in the southern Lesser Antilles and on Trinidad and Tobago (Heyer 1970, 1994). No ID data are available, but Heyer (1970) suggested that *L. validus* (as *L.* "*wagneri*") arose by dispersal from South America in the Quaternary. A more detailed examination of geographic variation in *L. validus* supports the contention that the Lesser Antillean populations are conspecific with those from Trinidad and Tobago (Heyer 1994).

Microhylidae

A single species of this family, *Gastrophryne carolinensis*, has been introduced in the Bahamas Bank and in the Cayman Islands from North America (Schwartz and Henderson 1991).

Ranidae

One species in this family, *Rana catesbeiana*, has been introduced throughout the West Indies, and two others, *R. grylio* and *R. sphenocephala*, have been introduced in the Bahamas Bank; all are from North America (Schwartz and Henderson 1991).

Reptilia: Crocodylia

Alligatoridae

Caiman crocodilus has been introduced in Cuba and Puerto Rico; this species is native to Central and South America (Schwartz and Henderson 1991).

Crocodylidae

(10) ***Crocodylus acutus***.—This species occurs in North, Central, and South America as well as Cuba, Hispaniola, Jamaica, and the Cayman Islands. Although most abundant in coastal areas, individuals

may swim a considerable distance offshore (Schwartz and Henderson 1991). An ID of 16 (10 mya) between *C. palustris* and *C. acutus* (Hass et al. 1992) establishes an upper time limit for divergences within *C. acutus*, although several species from the New World are believed to be even more closely related to *C. acutus* (Densmore and White 1991) and the origin of the West Indian populations may be much more recent (Quaternary).

(11) ***Crocodylus intermedius***.—This species occurs in northern South America (including Trinidad and Tobago), primarily in large rivers. One record is known from Grenada (Schwartz and Henderson 1991), and is almost certainly the result of recent (Quaternary) dispersal from South America.

(12) ***Crocodylus rhombifer***.—This species is endemic to Cuba (including Isla de Juventud), and occurred in the Cayman Islands until the last century (Morgan et al. 1993). It has an ID of 6 (4 mya) to the Old World species *C. palustris* (Hass et al. 1992), although it is believed to be most closely related to New World species (Densmore and White 1991). These animals most likely arrived by dispersal from Central or South America.

Reptilia: Squamata

Amphisbaenia

Amphisbaenidae

(13) ***Amphisbaena***.—Low IDs among West Indian amphisbaenians and DNA sequence data suggest that the Cuban genus *Cadea* is part of the Antillean *Amphisbaena* radiation and should be synonymized within *Amphisbaena* (Hedges et al. 1992*b*; Hass, Frye, and Hedges, in preparation). Gans (1990) suggested a Mediterranean origin for New World amphisbaenids. However, relationships among African and New World *Amphisbaena* are poorly known, and thus two independent dispersals from Africa, one leading to the South American species and the other to the West Indian species, cannot be ruled out — but is less likely than a single dispersal from Africa to the New World (South America?). Aside from the Antillean taxa, *Amphisbaena* primarily is a South American genus (in the New World), and therefore an ID of 91 (55 mya) between *A. schmidti* (Puerto Rico) and *A. alba* (South America) suggests an early Cenozoic dispersal from South America to the West Indies (Hedges et al. 1992*b*).

Sauria

Anguidae

Ophisaurus ventralis, a species native to the southeastern United States, was introduced in the Cayman Islands (Schwartz and Henderson 1991).

(14) ***Celestus***.—Albumin ID data and DNA sequence data (Hedges et al. 1992*b*; Hass, Maxson, and Hedges, in preparation) indicate that the Hispaniolan endemic genera *Sauresia* and *Wetmorena* are part of a monophyletic radiation of Antillean *Celestus* and should be synonymized within that genus. Also included in *Celestus* are the three Hispaniolan species previously placed in *Diploglossus*. The remaining species of *Celestus* occur in Central America, and although no ID data are available for those species, an ID of 54 (32 mya) between a West Indian species (*haetianus*) and a mainland diploglossine genus (*Ophiodes*) establishes a maximum time (Hedges et al. 1992*b*), and the largest ID among Antillean *Celestus*, 16 (10 mya), establishes a minimum time for their origin. Therefore, the West Indian species of *Celestus* apparently arose by dispersal from Central America in the mid Cenozoic (10 32 mya).

(15) ***Diploglossus***.—Recent immunological and DNA sequence data (Hass, Maxson, and Hedges, in preparation), as noted above, have revealed that the Hispaniolan species of *Diploglossus* belong to *Celestus*, not *Diploglossus*. This finding leaves only three Antillean species with disjunct distributions: *D. delasagra* (Cuba), *D. pleei* (Puerto Rico), and *D. montisserrati* (Montserrat). Whether West Indian *Diploglossus* have affinities with Central or South American species (or both) in the genus is unclear. The absence of this genus on Hispaniola brings up the possibility that the Cuban species was derived from Central America and the Puerto Rican and Lesser Antillean species from South America. No IDs are available between mainland and West In-

dian species, although the ID between the Cuban and Puerto Rican species, 36 (22 mya), suggests a maximum time of origin, if those species were derived separately from mainland taxa (or, conversely, a minimum time if a single origin is indicated).

Gymnophthalmidae

(16) ***Bachia heteropus***.—This species occurs on the Grenada Bank, and is placed in the same subspecies as populations on adjacent Tobago (Thomas 1965*a*; Schwartz and Henderson 1991). The species also occurs in Northern Venezuela, and therefore its origin in the West Indies almost certainly is the result of relatively recent dispersal (Quaternary) from South America.

(17) ***Gymnophthalmus pleei***.—Based on morphology, this Lesser Antillean endemic appears to be most closely related to a species in northern South America, *G. lineatus* (Thomas 1965*a*), suggesting an origin by dispersal from that continent. The geologic origin of the Lesser Antilles in the Eocene (Pindell and Barrett 1990) places an upper time limit on the origin of *G. pleei* (0–45 mya).

(18) ***Gymnophthalmus underwoodi***.—This unisexual species occurs in both the Lesser Antilles (Barbados, St. Vincent, and probably introduced on Guadeloupe) and northern South America (Thomas 1965*a*; Schwartz and Henderson 1991; Vanzolini and Carvalho 1991), suggesting a relatively recent (Quaternary) dispersal from South America.

(19) ***Tretioscincus bifasciatus***.—This species occurs in northern South America and also on Isla de Providencia (Scott and Ayala 1984). The origin of that island population probably was by dispersal (or human introduction) from South America in the Quaternary.

Iguanidae

A vertebra of a large lizard found at an early Miocene site in Puerto Rico was tentatively identified as either a varanid, teiid, or iguanid (MacPhee and Wyss 1990). Until additional material is available allowing more accurate identification, the biogeographic significance of this fossil is uncertain. The reclassification of iguanian lizards proposed by Frost and Etheridge (1989) is not followed here because of the weak resolution of phylogeny obtained by morphological characters (e.g., Frost and Etheridge 1989: figures 8 and 16) and conflicting evidence from published molecular studies (see below).

(20) ***Anolis***.—Lizards of this genus comprise a major fraction of the West Indian herpetofauna and studies on their relationships, biogeography, and origin are numerous. Initially, higher-level groupings within the West Indian species were based largely on karyotypes and osteology (Williams 1976), and since then albumin ID data (e.g., Wyles and Gorman 1980; Shochat and Dessauer 1981; Hass et al. 1993), allozyme data (Burnell and Hedges 1990), and DNA sequence data from the mitochondrial 16S rRNA gene (Hass et al. 1993) have helped to refine those relationships. Abundant evidence now exists for the presence of island radiations such as on Jamaica and Puerto Rico, which formed the basis of the ecomorph concept (Williams 1972, 1983), as well as for many of the series and species groups recognized by Williams (1976). However, the molecular data do not support the major alpha/beta dichotomy based on characteristics of caudal vertebrae, the basal position (and recognition) of the Cuban endemic genus *Chamaeleolis*, or the recognition of the Hispaniolan endemic genus *Chamaelinorops*. For that reason, *Chamaeleolis* and *Chamaelinorops* were synonymized within *Anolis* by Hass et al. (1993). General concordance between albumin ID data and DNA sequence data and discordance with morphology suggests that the generic partitioning of *Anolis* based largely on morphology (Guyer and Savage 1986) was premature (as noted by Williams 1989*b*). The classifications of Williams (1976) and Burnell and Hedges (1990), where informal categories such as series and species groups are used, allow disagreements to occur without affecting non-systematist users of the formal classification.

Because relatively few mainland species were included in those molecular studies, it is not possible at this time to infer the number of independent lineages of West Indian *Anolis* or their closest relatives on the mainland. The IDs among many West Indian species are relatively low, and it is possible, as suggested by Shochat and Dessauer (1981), that a major Antillean radiation exists within the West Indies. At the same time, both Central American affinities for some species and South American affinities for others are indicated by the ID data (summarized in Hass et al. 1993). Thus it would appear that West Indian *Anolis* comprise more than one independent lineage derived from different regions of the mainland.

Although the number of independent West Indian lineages of *Anolis* is not yet known, a maximum time of origin or origins can be estimated from the large amount of ID data now available for the species. Not counting the relatively recent (Pliocene) dispersal from Cuba to Florida leading to the origin of *A. carolinensis* (Buth et al. 1980), IDs between West Indian and mainland species range from 27–60 (Hass et al. 1993), suggesting a mid-Cenozoic (16–36 mya) origin for the West Indian lineages (from Central and/or South America). The oldest fossil West Indian *Anolis*, from Dominican amber (Rieppel 1980), also is mid-Cenozoic (20–40 mya) in age. Roughgarden (1995) recently proposed that the two major groups of anoles in the Lesser Antilles (*bimaculatus* series in the north and *roquet* series in the south) diverged due to plate tectonic separation of the proto-Lesser Antilles in the Mesozoic. He suggested that the islands to the north and south of the Martinique Passage (between Dominica and Martinique) have had separate geological histories. However, the available geological evidence does not support this hypothesis (see Hedges 1995 for a discussion), the distributions of other groups (e.g., *Eleutherodactylus*) show "break points" at different locations in the Lesser Antilles, and the molecular evidence argues strongly against such a vicariance model. Overwater dispersal best explains the distribution of anoles in the Lesser Antilles.

(21) ***Ctenosaura similis***.—This Central American species also occurs on Isla de San Andrés and Isla de Providencia. Presumably, those island populations originated by dispersal from Central America in the Quaternary.

(22) ***Cyclura***.—This endemic West Indian genus is a close relative of *Iguana*, and the ID between those two genera is 20 (mean of *I. iguana* vs. *C. cornuta* and *I. iguana* vs. *C. nubila*; Gorman et al. 1971) suggesting a Miocene (12 mya) origin by dispersal from the mainland (either Central or South America).

(23) ***Iguana delicatissima***.—This Lesser Antillean endemic is a sister species to *I. iguana*. The ID of 7 separating the two species (Gorman et al. 1971) suggests Pliocene (4 mya) dispersal. The source area probably was South America based on geography.

(24–27) ***Iguana iguana***.—At least four separate invasions (dispersals) from the mainland by this species probably occurred, resulting in the following distributions: (24) Swan Islands, (25) Isla de San Andrés and Isla de Providencia, (26) Cayman Islands, and (27) Puerto Rico through the Lesser Antilles. The source area for (24–26) probably was either Central America or South America, whereas the source area for (27) almost certainly was South America. Albumin immunological distances of 0 between Central and South American populations of *I. iguana* (Gorman et al. 1971), while not directly bearing on the West Indian populations, suggests that geographic variation within the species is not great. Presumably, the West Indian populations originated by dispersal in the Quaternary.

(28) ***Leiocephalus***.—A recent parsimony analysis of morphological characters in iguanian lizards placed this endemic West Indian genus of 23 species (Pregill 1992) in its own subfamily ("Leiocephalinae") and as the closest relative of the South American subfamily "Tropidurinae," both in the family "Tropiduridae" (Frost and Etheridge 1989). However, albumin IDs and DNA sequence data (Hedges et al. 1992*b*; Hass, Maxson, and Hedges, in preparation) indicate a closer relationship between *Leiocephalus* and the largely North American genera *Crotaphytus* and *Sceloporus*, placed in other families by Frost and Etheridge ("Crotaphytidae"

and "Phrynosomatidae," respectively). This suggests that the taxonomic rearrangement of Frost and Etheridge (1989) may have been premature. The mean ID between *L. schreibersii* and *Crotaphytus collaris*, 46 (28 mya), suggests a mid-Cenozoic origin by dispersal from North America.

Gekkonidae

One species of *Gekko* (*G. gecko*; Henderson et al. 1993) and one species of *Hemidactylus* (*H. turcicus*) were introduced in the West Indies from the Old World.

(29) **Antillean *Aristelliger*.**—Of the seven species in this genus, six are restricted to the West Indies and the single mainland species occurs in Central America (and nearby islands). Although *Aristelliger* has been considered to be a gekkonine (Kluge 1987), IDs between *Sphaerodactylus* and *Aristelliger* are lower (82.8 ± 3.3) than IDs between *Sphaerodactylus* and several other genera of gekkonines (107.8 ± 10.3; Hass 1991). *Aristelliger* possesses other sphaerodactyline-like traits, such as the absence of cloacal bones and sacs (Kluge 1982) and a single egg per clutch, suggesting that it may be incorrectly classified. Two species within *Aristelliger* (*barbouri* and *cochranae*) appear to be closely related (Hecht 1951, 1952; Bauer and Russell 1993). Otherwise, relationships among the species and to other gekkonids are unclear, therefore making the origin of the West Indian species difficult to infer. For example, the West Indian species may be monophyletic with a Central American origin, or, alternatively, the genus may have been isolated in the West Indies for a long time with the mainland species representing a relatively recent dispersal. In any case, the ID between *Sphaerodactylus* and *Aristelliger*, 83 (50 mya; Hass 1991), sets an upper time limit on the origin of *Aristelliger* and indicates that this event occurred no earlier than early Cenozoic and thus by dispersal. If West Indian species are found to be a monophyletic group with respect to the mainland species (*A. georgeensis*), then a more recent origin by dispersal from Central America would be the most likely scenario for the origin of the West Indian endemics.

(30) ***Aristelliger georgeensis*.**—This species is distributed in Central America and also occurs on Isla de Providencia, Isla de San Andrés, and Isla Santa Catalina. The origin of those island populations probably was by dispersal from Central America in the Quaternary.

(31) ***Gonatodes albogularis*.**—This species is distributed in Central and South America, and the West Indies (Cuba, Cayman Islands, Jamaica, Hispaniola). The Jamaican and Hispaniolan populations are recognized as an endemic subspecies (Schwartz and Henderson 1991), suggesting that at least those populations are not the result of human introductions. The origin of the West Indian populations of this species is most likely due to Quaternary dispersal from Central or South America.

(32) ***Hemidactylus haitianus*.**—Kluge (1969) argued that *H. brookii haitianus* originated by dispersal from the Old World, although human introduction from Africa was suggested by Vanzolini (1978). Powell et al. (this volume) consider this taxon to be a valid species. The origin of the West Indian populations probably occurred during the Quaternary.

(33) ***Hemidactylus mabouia*.**—The two possibilities for the origin of *H. brookii* also apply to this species: either natural dispersal or human introduction from the Old World in the Quaternary (Kluge 1969; Vanzolini 1978).

(34) ***Hemidactylus palaichthus*.**—This species of northern South America also is known from St. Lucia in the Lesser Antilles (Powell 1990; Murphy, this volume). Presumably, these geckos arrived by dispersal from South America at some time during the Quaternary.

(35) ***Phyllodactylus pulcher*.**—This species is endemic to Barbados. The genus occurs in both the Old and New World, including North, Central, and South America. However, this particular species (type species of the genus) has been associated with species from northern South America (Dixon 1962, 1964; Vanzolini 1968), which suggests an origin

from that continent. The geologic origin of the Lesser Antilles in the Eocene (Pindell and Barrett 1990) places an upper time limit on the origin of *P. pulcher* (0 45 mya).

(36) ***Phyllodactylus wirshingi.***—This species is endemic to Hispaniola and Puerto Rico. Schwartz (1979) suggested that the Puerto Rican populations were derived from those on Hispaniola based on scale variation and distribution, although the reverse may have occurred (see below). The species also has been associated with species from the Dutch Leeward Islands and northern South America (Grant and Beatty 1944; Dixon 1962), suggesting an origin from South America.

(37) ***Sphaerodactylus.***—Unlike the two other species-rich genera of West Indian amphibians and reptiles (*Eleutherodactylus* and *Anolis*), most of the species in this genus are West Indian (80 species), with the others occurring in Central and northern South America. Hass (1991) presented a phylogenetic analysis of immunological and protein electrophoretic data for most of the West Indian species, but only one mainland species (*S. molei*) was included. Harris and Kluge (1984) postulated that the mainland species (with only two exceptions) represent a monophyletic group (the *lineolatus* section of Hass 1991). If this suggestions is true, then the determination by Hass (1991) that *S. molei* is basal to the West Indian species suggests that the latter also may be monophyletic (the *sputator* section of Hass 1991). Unfortunately, no IDs are available between species in these two major divisions of *Sphaerodactylus*. However, the primarily South American distributions of other sphaerodactyline genera and the low ID (45; 27 mya) between *Sphaerodactylus* and *Lepidoblepharis* led Hass (1991) to postulate a South American origin for West Indian *Sphaerodactylus* by dispersal in the mid-Cenozoic. This is in agreement with the age, 20–40 mya, of the fossil *Sphaerodactylus* found in Dominican amber (Böhme 1984).

(38) ***Tarentola.***—One species (*T. americana*) is endemic to Cuba and the Bahamas Bank and another undescribed species is endemic to Cuba. All other species in the genus are from the Old World (primarily the Mediterranean region; southern Europe, northern Africa), and dispersal over water from Africa or southern Europe to the Antilles has been suggested (Schwartz 1968). Joger (1984) placed *T. americana* in its own subgenus (*Neotarentola*) within *Tarentola*. In a phylogenetic study of *Tarentola*, Joger (1985), using ID data for other species (none were available for *T. americana* itself), found that *Neotarentola* apparently diverged from the subgenus *Tarentola* about 30 mya, and this date was constrained by an earlier divergence (40 mya) leading to the two pairs of subgenera (Joger 1985: figure 3). According to those data, the origin of West Indian *Tarentola* was by dispersal from North Africa in the mid-Cenozoic (about 30 mya, but within the range 0–40 mya).

(39) ***Thecadactylus rapicauda.***—This widely distributed Neotropical species is found in Central and South America and in the eastern Caribbean (Puerto Rican Bank through the Lesser Antilles). The origin of the West Indian populations, based on geography, would appear to be by dispersal from South America in the Quaternary.

Scincidae

(40) ***Mabuya lineolata.*** The relationship of this Hispaniolan endemic to other congeners is not well known, although Dunn (1935) suggested that a possible affinity with a South American species (*M. guaporicola*) might exist. The time of origin for this species is unknown.

(41) ***Mabuya bistriata.***—This species is distributed in northern South America (north to Panama), and ranges throughout most of the West Indies except for Cuba and the northern Bahamas Bank (Hoogmoed and Gruber 1983). This distribution suggests an origin for the West Indian populations by dispersals (multiple?) from South America in the Quaternary. Dunn and Saxe (1950) suggested that the population on Isla de Providencia may have been derived by human introduction from the southern Lesser Antilles.

Teiidae

(42) ***Ameiva.***—This genus occurs in Central and South America and throughout the West Indies. Whether the West Indian species form a monophyletic group is not known, although ID data suggest that may be the case (Hedges et al. 1992*b*; Hass and Hedges, in preparation). The mean ID between two West Indian species (*A. exsul* and *A. chrysolaema*) and a predominantly mainland species (*A. ameiva*), 60 (36 mya; Hass and Hedges, in preparation) suggests an origin for the West Indian species in the mid-Cenozoic by dispersal from either Central or South America.

(43–45) ***Ameiva ameiva.***—This species occurs from Panamá south through tropical South America. At least three separate dispersals from the mainland by this species probably occurred during the Quaternary, resulting in the following West Indian populations: (43) Swan Islands, (44) Isla de Providencia, both from northern South America (Dunn and Saxe 1950), and (45) the southern Lesser Antilles (St. Vincent and the Grenada Bank), from northern South America.

(46) ***Cnemidophorus lemniscatus.***—This species is distributed in Central and tropical South America and also occurs on Isla de Providencia, Isla de San Andrés, and Isla Santa Catalina. The origin of those island populations probably was by dispersal from Central or South America in the Quaternary.

(47) ***Cnemidophorus vanzoi.***—This species is endemic to the Maria Islands off St. Lucia. Although other species of the genus occur in North, Central, and South America, the location of *C. vanzoi* in the southern Lesser Antilles suggests an origin from South America (Baskin and Williams 1966). The geologic origin of the Lesser Antilles in the Eocene (Pindell and Barrett 1990) places an upper time limit on the origin of *C. vanzoi* (0–45 mya).

(48) ***Kentropyx borckiana.***—This species occurs in northern South America, especially coastal areas (Hoogmoed 1973; Gallagher and Dixon 1992), and also is known from Barbados in the southern Lesser Antilles (Gallagher and Dixon 1992). Presumably, the species colonized Barbados by dispersal from the South American mainland at some time during the Quaternary.

Xantusiidae

(49) ***Cricosaura typica.***—The lizard family Xantusiidae is distributed in a largely disjunct fashion in Central America, southwestern North America, and Cuba, and fossil evidence indicates that it previously occupied a wider distribution, at least in North America (Bezy 1972; Hedges et al. 1991). The single West Indian species, *Cricosaura typica*, is endemic to a small area in eastern Cuba and its origin has been discussed in several phylogenetic studies. Crother et al. (1986) analyzed some published morphological data and concluded that the closest relative of *Cricosaura* is the Central American genus *Lepidophyma*. However, Hedges and Bezy (1993) demonstrated that several key morphological characters in that analysis were incorrectly scored, putting into question that conclusion. DNA sequence data from portions of three mitochondrial genes provide statistical support for *Cricosaura* as the basal lineage within the family (Hedges et al. 1991; Hedges and Bezy 1993, 1994), and this was further supported by chromosome evidence (Hass and Hedges 1992). Indirect evidence from Middle Paleocene (60 mya) xantusiid fossils from North America (Estes 1976) suggests that the lineage leading to *Cricosaura* may have diverged before that time, possibly due to proto-Antillean vicariance (Hedges et al. 1991). However, other explanations are possible. The *Cricosaura* lineage may have dispersed to Cuba in the early Cenozoic after the proto-Antilles broke apart and after the bolid impact at 65 mya; or, alternately, this lineage may have arisen on the mainland and persisted throughout most of the Cenozoic (with dispersal to Cuba occurring at some point during that time), subsequently becoming extinct in all areas except for the small region in eastern Cuba. Although the latter hypothesis is admittedly a considerably more complicated scenario than vicariance, it remains a distinct possibility due to the relictual nature of xantusiid lizard distribution.

Serpentes

Boidae

(50–51) ***Boa constrictor*.**—Three subspecies of *B. constrictor* occur in the West Indies and two are endemic (one each on Dominica and St. Lucia). Otherwise, the species is distributed in Central and South America to about 36°S Latitude. Based on geography and subspecific allocation, two independent origins (in the Quaternary) by dispersal are indicated for the West Indian populations: (50) from Central America, leading to the populations on Isla de San Andrés, Isla de Providencia, and Isla Santa Catalina; and (51) from South America, leading to the populations in the southern Lesser Antilles.

(52) ***Corallus hortulanus*.**—This is a widely distributed species in Central and South America, and it occurs on St. Vincent and the Grenada Bank in the southern Lesser Antilles. A phylogenetic analysis of mitochondrial DNA sequence data from seven widely separated localities indicated that the West Indian populations were derived from the southern (South America) subspecies and not the northern (Central America and northwestern South America) subspecies (Henderson and Hedges 1995). Presumably, dispersal occurred in the Quaternary.

(53) ***Epicrates*.**—Nine of the ten species in this genus are endemic to the West Indies and they are thought to form a monophyletic group (Kluge 1989). No IDs are available among species of *Epicrates*, but an ID of 37 (22 mya) between *Boa constrictor* and *Epicrates cenchria* (Dessauer et al. 1987) sets an upper time limit for evolution within the genus and hence the origin (by dispersal) of West Indian *Epicrates*. This date also is in agreement with an Early Miocene date for the oldest fossil boid in the West Indies, from Puerto Rico (MacPhee and Wyss 1990). This argues against the vicariance model for the origin of West Indian *Epicrates* (Kluge 1988: figure 10) and favors a dispersal model (e.g., Tolson 1987). The mainland species believed to represent the basal lineage within the genus, *E. cenchria*, is widely distributed in Central and South America, and thus the New World source area for the West Indian lineage is not obvious.

Colubridae

Two species, *Diadophis punctatus* and *Elaphe guttata*, have been introduced in the Cayman Islands from North America (Schwartz and Henderson 1991).

Colubrinae

(54) ***Chironius vincenti*.**—This colubrine species is endemic to St. Vincent. Although other species of the genus occur in Central and South America, the location of *C. vincenti* in the southern Lesser Antilles suggests an origin from South America. An ID of 40 (24 mya) between *Trimorphodon biscutatus* and *C. exoletus* (Cadle 1984*c*) places an upper time limit on divergence within *Chironius*; and IDs among congeneric species of colubrid snakes rarely exceed 25 (15 mya) and typically range between 5–15 (3–9 mya) (Cadle 1984*a*, 1984*b*, 1984*c*). This suggests an origin by dispersal from South America in the late Cenozoic (within 0–24 mya, but most likely 0–10 mya).

(55) ***Mastigodryas bruesi*.**—This colubrine species is endemic to St. Vincent and the Grenada Bank. Although other species of the genus occur in Central and South America, the location of *M. bruesi* in the southern Lesser Antilles suggests an origin from South America. An ID of 28 (17 mya) between *Trimorphodon biscutatus* and *M. melanolomus* (Cadle 1984*c*) places an upper time limit on divergence within *Mastigodryas*, and considering the IDs typically encountered among congeneric species of colubrid snakes (5–15), *M. bruesi* likely arose by dispersal from South America in the late Cenozoic (within 0–17 mya, but most likely 0–10 mya).

Natricinae

(56) ***Nerodia clarkii*.**—This North American natricine species inhabits salt marshes and mangrove swamps in the southeastern United States and occurs in similar habitats along the northern coast of Cuba. The Cuban populations are assigned to the same subspecies as those in the Florida Keys. Low IDs (2-3) between species of *Nerodia* and *Thamnophis* (Dowling et al. 1983) suggest that speciation in those genera has been relatively recent and indicates that the origin of Cuban *N. clarkii* almost

certainly was by dispersal from nearby Florida in the Quaternary.

Xenodontinae

(57) ***Clelia clelia***.—This widely distributed xenodontine species is found in Central and South America, and on Grenada (Underwood 1993). Presumably, this snake reached Grenada by dispersal from South America during the Quaternary.

(58) ***Clelia errabunda***.—This species is endemic to St. Lucia and is believed to be most closely related to a clade of three mainland species (*C. clelia*, *C. equatoriana*, and *C. scytalina*) the present ranges of which are in Central and South America (Underwood 1993). No IDs are available between this species and other species within the genus, although an ID of 8 (5 mya) between *C. scytalina* and *Pseudoboa coronata* places an upper limit on divergences within *Clelia* (Cadle 1984*a*) and suggests that the origin of *C. errabunda* was in the Pliocene or Pleistocene (within 5 mya) by dispersal from nearby South America.

(59) ***Liophis cursor* group.**—Four species of the xenodontine genus *Liophis* are endemic to islands of the Lesser Antilles, and appear to form a monophyletic group that is most closely related to the South American (and Grenada) species *L. melanotus* (Maglio 1970). No IDs are available for comparisons among species of *Liophis*, although a mean ID of 30 (18 mya) between *Xenodon severus* and three species of *Liophis* (Cadle 1984*a*) places an upper limit on divergences within *Liophis* and suggests that the origin of this West Indian group was in the late Cenozoic (within 0–18 mya, but most likely 0–10 mya) by dispersal from South America.

(60) ***Liophis melanotus***.—This xenodontine species is widely distributed in South America, including Trinidad and Tobago (Dixon 1989), and a record exists for nearby Grenada, although its validity has been questioned (Henderson 1992; Murphy, this volume). The origin of the Grenada population apparently was by dispersal from South America in the Quaternary.

(61) ***Pseudoboa neuwiedi***.—This primarily South American xenodontine species has been recorded from Grenada in the southern Lesser Antilles, and the validity of this record also has been questioned (Henderson 1992; Murphy, this volume). Presumably, it arrived by dispersal from South America at some time during the Quaternary.

(62) ***Coniophanes andresensis***.—This species is endemic to Isla de San Andrés. Low IDs between *Coniophanes fissidens* and species in closely related genera (*Pliocercus*, *Rhadinea*), 8–15 (5–9 mya) (Cadle 1984*b*), suggest that divergences within *Coniophanes* probably are not greater. These data, and the association of *Coniophanes* with the Central American xenodontine clade (Cadle 1984*b*) indicate that the origin of this island endemic was by dispersal from Central America in the late Cenozoic (0–9 mya).

(63) ***Tretanorhinus variabilis***.—This species is endemic to Cuba and the Cayman Islands. Cadle (1984*b*) identified the genus *Tretanorhinus* as a member of the "Central American xenodontines" (the "Dipsadidae" of Pinou and Dowling 1994). The distribution of the West Indian species in the western Caribbean along with the distribution of the remaining members of the genus in Central America (extending slightly into northwestern South America) indicates that the source area for *T. variabilis* was Central America. Again considering the IDs typically encountered among congeneric species of colubrid snakes (5–15), this West Indian species likely arose by dispersal from Central America in the late Cenozoic. That individuals have been found in saltwater estuaries (Barbour and Amaral 1924) suggests that waif dispersal is not unlikely.

(64) **Alsophines.**—Six of the remaining seven genera are endemic to the Greater Antilles: *Antillophis*, *Arrhyton*, *Darlingtonia*, *Hypsirhynchus*, *Ialtris*, and *Uromacer*. *Alsophis* also occurs in the Lesser Antilles, Galapagos Islands, and western Ecuador. Primarily on the basis of osteological and hemipenial morphology, Maglio (1970) proposed three dispersals from the mainland to account for the origin of

these West Indian xenodontines. However, IDs among species in this assemblage are low (0–27, but most <20 [12 mya]; Cadle 1984*a*; Hedges et al. 1992*b*; Hass and Hedges, in preparation) and are suggestive of a single West Indian radiation. Extensive ID comparisons of Neotropical xenodontines indicated that species in this West Indian group are most closely related to the South American xenodontines (Cadle 1984*a*, 1985). The ID between *Alsophis cantherigerus* and *Philodryas viridissimus* (Cadle 1984*a*), 43 (26 mya), suggests a mid-Cenozoic (12–26 mya) origin for this West Indian assemblage of snakes by dispersal from South America.

Elapidae

(65) ***Micrurus nigrocinctus***.—A single specimen of this species was collected on Isla de Providencia ("Old Providence," the original name for the island) by the "Albatross Expedition in 1884" and deposited in the National Museum of Natural History, but it has since been lost (Dunn and Saxe 1950). Schwartz and Henderson (1988, 1991) did not recognize this species as part of the West Indian herpetofauna. However, the species is widespread in Central America, and occurs on Great Corn Island off the coast of Nicaragua, and therefore little reason exists to doubt the validity of the record for Isla de Providencia, regardless of whether or not a population still is extant on the island. Presumably, this species reached Isla de Providencia by dispersal from Central America.

Leptotyphlopidae

(66) ***Leptotyphlops bilineata* group.**—The five species placed in this group are believed to represent a single West Indian radiation (Thomas et al. 1985). The time of origin for the *L. bilineata* group is unknown. However, differences in hemipenial structure between Old and New World species of *Leptotyphlops* (Branch 1986) indicate some geographic cohesiveness, and thus a New World origin for the *bilineata* group appears to be the most likely explanation.

(67) ***Leptotyphlops columbi***.—This Bahamian endemic (San Salvador Island) represents something of a biogeographic anomaly in that it apparently is not closely related to other West Indian species (Thomas et al. 1985), and no *Leptotyphlops* presently occur in the nearby southeastern United States. Klauber (1939) suggested an association with *L. albifrons*, although Thomas (1965*b*) did not consider it close to that species. The *albifrons* group is primarily South American (Peters and Orejas-Miranda 1970), although one species (*L. goudotii*) extends its range up into Central America. Because the Bahamas Platform was completely submerged at times in the Pliocene or Pleistocene, the origin of *L. columbi* likely was by dispersal in the late Tertiary or in the Quaternary (0–5 mya), probably from South America.

(68) ***Leptotyphlops goudotii***.—This species is distributed in Central and northern South America and also occurs in the Swan Islands, Isla de Providencia, and Isla de San Andrés. The origin of those island populations probably was by dispersal from Central or South America in the Quaternary.

(69) ***Leptotyphlops albifrons***.—An old record exists for this South American species (as *L. tenella*) on Antigua (Thomas 1965*b*; Hoogmoed and Gruber 1983). Apparently, it arrived by dispersal from South America in the Quaternary.

Tropidophiidae

(70) ***Tropidophis***.—Except for three South American species, this is primarily a West Indian genus. Albumin ID data among species in the genus indicate that the West Indian species form a monophyletic group with an ID of 70 (42 mya) between *T. haetianus* (West Indies) and the mainland species *T. paucisquamis* (Hedges et al. 1992*b*; Hass and Hedges, in preparation). These data suggest an origin by dispersal from South America for West Indian *Tropidophis*.

Typhlopidae

(71) ***Typhlops***.—This pan-tropical genus is distributed throughout the West Indies as well as Central and South America. Most New World species are West Indian, and Thomas (1989) concluded that

most West Indian species form a monophyletic group ("major Antillean radiation") whose closest relative is in Africa (*T. caecatus*). The remaining four West Indian species also were considered to form a monophyletic group with affinities to other New World species (Thomas 1989). However, IDs between *T. platycephalus* (Puerto Rico) and other West Indian species (1–44 = 1–26 mya), with lower IDs to *T. biminiensis* than to some species in the "major Antillean radiation," do not support that dichotomy and indicate that West Indian species form a relatively closely related assemblage (Hedges et al. 1992*b*; Hass and Hedges, in preparation). No IDs are available to other New World *Typhlops*, but an ID of 96 (58 mya) between *T. platycephalus* and *T. luzonensis* (Philippines; Hass and Hedges, in preparation) suggests a Cenozoic origin (26–58 mya) for the West Indian species.

Viperidae

(72) ***Bothrops*.**—Two species of this Neotropical genus are endemic to islands in the southern Lesser Antilles. No ID data are available for these species, but IDs between *Bothrops atrox* and other species of *Bothrops* range from 0–21 (0–13 mya) and the reciprocal ID between *B. atrox* and *Crotalus enyo* is 23 (14 mya), placing an upper time limit on divergence within *Bothrops* (Dessauer et al. 1987; Cadle 1992). However, the two West Indian species are most closely related to *B. atrox* (Lazell 1964) or are part of the "polytypic *B. atrox* complex" itself (Gosner 1987). Therefore, the lowest ID between *B. atrox* and another species in the genus, *B. asper* (ID = 7, 4 mya; Cadle 1992), may be a more realistic upper limit on the time of origin for the Antillean species. These data and the distribution of the two species suggest an origin by dispersal from South America in the late Tertiary or Quaternary (0–4 mya).

Reptilia: Testudines

Emydidae

One species of this family, *Trachemys scripta*, has been introduced on Guadeloupe.

(73) ***Trachemys*.**—According to Seidel (1988), this genus includes only a single non-West Indian species, *T. scripta*, although some mainland subspecies of *T. scripta* recently have been elevated to species status (e.g., Seidel 1989; Ernst 1992). A morphological and biochemical analysis of relationships within the genus found that *T. scripta* is polyphyletic, with the North American populations representing a basal lineage within the genus and the Central and South American populations most closely related to *T. decussata* of Cuba (Seidel 1988). He suggested that this may indicate a single invasion of the West Indies from North America (temperate *T. scripta*), radiation within the Greater Antilles, and then dispersal from Cuba to Central America (leading to neotropical *T. scripta*), although he did not rule out an origin for the Antillean radiation by vicariance. Moll and Legler (1971) speculated that *T. scripta* arrived on the Neotropical mainland only recently (Pleistocene) based on life history data, which would support part of that scenario. *Trachemys* can survive in seawater for at least a week (Dunson and Seidel 1986), indicating that over-water dispersal is not unlikely. Pre-Quaternary fossil *Trachemys* are known only from North America and only as far back as the Miocene (Seidel 1988). No evidence exists that *Trachemys* arrived in the West Indies any earlier than the mid-Tertiary, and thus the origin of West Indian *Trachemys* probably was by dispersal from North America in the mid- to late Cenozoic (0–25 mya).

Kinosternidae

(74) ***Kinosternon scorpioides*.**—This is a widely distributed Central and South American species that occurs also on Isla de Cañas and Isla de San Andrés. The origin of those island populations presumably was by dispersal from Central or South America in the Quaternary.

Pelomedusidae

An African species of this family, *Pelusios subniger*, was introduced on Guadeloupe (Schwartz and Henderson 1991).

(75) **Pelomedusids.**—Oligocene fossils of this family are known from Puerto Rico (Wood 1972; Williams 1989*a*; MacPhee and Wyss 1990), although

whether they were saltwater or freshwater turtles is not known with certainty. Presently, the family is restricted to Africa and South America, but it had a much wider distribution in the late Mesozoic and early Cenozoic (Romer 1966) and therefore the source area for this lineage is unknown, but is likely to have been in the New World.

Testudinidae

(76) ***Geochelone carbonaria.***—This widely distributed Neotropical species also occurs in the eastern Caribbean (Virgin Islands through the Lesser Antilles). The species' absence from the Greater Antilles suggests that it arrived in the West Indies by dispersal from South America in the Quaternary. However, literature reports of specimens from the Greater Antilles exist, and the possibility of human introduction to the West Indies has not been ruled out (Censky 1988).

(77) ***Geochelone* sp.**—Although now extinct, giant tortoises occurred throughout the West Indies at least into the Quaternary (Auffenberg 1967). The time of origin and source area for this lineage or lineages is unknown, but was presumably in the New World.

A total of 77 independent lineages of West Indian amphibians and reptiles can be defined (Table 3). Slightly more than half (42/77 = 55%) have species endemic to the West Indies. At least some information pertaining to the time of origin is available for nearly all lineages (73/77 = 95%), and of those, all but one (99%) are in the Cenozoic (65–0 mya) and all but nine in the last half (30–0 mya) of the Cenozoic. Nearly all (73/77 = 95%) lineages appear to have originated in the New World. Of those, a continental source cannot be determined for four (4/73 = 5%), and a distinction between Central America and South America cannot be made for 16 (16/73 = 22%). Of the remaining 53 lineages where a source area within the New World can be identified, most (42/53 = 79%) show a South American origin, eight (8/53 = 15%) indicate a source in Central America, and three (3/53 = 6%) originated in North America. Of the 42 lineages that include endemic West Indian species (Table 4), the proportions change little: 39 (93%) are New World; of those, the continental source cannot be determined for three (3/39 = 8%) and eight (8/39 = 21%) are from either Central or South America; of the rest, most (22/28 = 79%) are from South America, four (4/28 = 14%) are from Central America, and two (2/28 = 7%) are from North America. Of the four lineages for which no data on the time of origin are available (*Hyla heilprini*, *Phyllodactylus wirshingi*, *Mabuya lineolata*, and the *Leptotyphlops bilineata* group), all have congeneric species on mainland Central or South America) and none have highly divergent morphologies that would suggest a long period of isolation. Therefore, all four of those lineages likely also arose in the Cenozoic by dispersal.

DISCUSSION

Overwater Dispersal

The major finding of this analysis is that all but one or two of the 77 independent lineages of amphibians and reptiles present in the West Indies apparently originated by dispersal in the Cenozoic, and the majority of those came from South America. A notable exception is the frog genus *Eleutherodactylus*, which would appear to have originated in the Cretaceous by vicariance (but see below). That one of the most species-rich lineages of terrestrial vertebrates in the West Indies also is one of the oldest, is not surprising, but it presents an unexpected twist in the vicariance-dispersal debate of the last two decades. Whereas dispersal clearly dominates in explaining the origin of nearly all lineages of amphibians and reptiles, vicariance may have been an important mechanism in explaining the origin of nearly one-quarter of the species.

Williams (1989*a*) identified the "central problem in Caribbean biogeography" as the depauperate nature of the Antillean fauna, by which he was referring to higher taxa rather than species diversity. The absence of many groups typical of mainland herpetofaunas has been used as primary evidence to support a dispersal origin for the West Indian fauna. Among amphibians, these include all salamanders and caecilians; and centrolenid, microhylid, pelobatid, pipid, ranid, and rhyno-

TABLE 3. The origin of native West Indian amphibians and reptiles. Each lineage listed is independent in the West Indies and has affinities to a mainland group; those lineages with one or more species endemic to the West Indies are listed in bold. Times of origin indicated in bold are based on albumin ID data calibrated to the vertebrate fossil record (1 ID = 0.6 million years); calibration error estimates are indicated for comparisons believed to indicate approximate time of origin; ranges are given for other estimates; origin in the Quaternary (0–2 mya) is assumed for all West Indian populations of species also occurring on the mainland. Abbreviations: Swan Islands (SI), Isla de San Andrés (ISA), Isla de Providencia (IP), Cayman Islands (CI), Puerto Rico (PR), Lesser Antilles (LA), Isla Santa Catalina (ISC).

Lineage	Number of species	Source area	Time of origin (mya)	Mechanism
(1) ***Bufo peltocephalus* group**	12	South America	**51 ± 5.3**	Dispersal
(2) ***Colostethus chalcopis***	1	South America	0–45	Dispersal
(3) ***Osteopilus***	9	South America?	**48 ± 5.0**	Dispersal
(4) ***Hyla heilprini***	1	South America	?	Dispersal
(5) ***Eleutherodactylus***	138	Central/South America	**70 ± 6.8**	Vicariance
(6) ***Leptodactylus albilabris***	1	South America	**40 ± 4.5**	Dispersal
(7) ***Leptodactylus fallax***	1	South America	**3–7**	Dispersal
(8) *Leptodactylus insularum*	1	Central/South America	0–2	Dispersal
(9) *Leptodactylus validus*	1	South America	0–2	Dispersal
(10) *Crocodylus acutus*	1	New World	0–2	Dispersal
(11) *Crocodylus intermedius*	1	South America	0–2	Dispersal
(12) ***Crocodylus rhombifer***	1	Central/South America	**4 ± 4.0**	Dispersal
(13) ***Amphisbaena***	12	South America	**55 ± 5.6**	Dispersal
(14) ***Celestus***	19	Central America	**10–32**	Dispersal
(15) ***Diploglossus***	3	Central/South America	**22 ± 4.5**	Dispersal
(16) *Bachia heteropus*	1	South America	0–2	Dispersal
(17) ***Gymnophthalmus pleei***	1	South America	0–45	Dispersal
(18) *Gymnophthalmus underwoodi*	1	South America	0–2	Dispersal
(19) *Tretioscincus bifasciatus*	1	South America	0–2	Dispersal
(20) ***Anolis***	138	Central/South America	**16–36**	Dispersal
(21) *Ctenosaura similis*	1	Central America	0–2	Dispersal
(22) ***Cyclura***	8	Central/South America	**12 ± 5.4**	Dispersal
(23) ***Iguana delicatissima***	1	South America	**4 ± 4.0**	Dispersal
(24) *Iguana iguana* (SI)	1 (part)	Central/South America	0–2	Dispersal
(25) *Iguana iguana* (ISA/IP)	1 (part)	Central/South America	0–2	Dispersal
(26) *Iguana iguana* (CI)	1 (part)	Central/South America	0–2	Dispersal
(27) *Iguana iguana* (PR/LA)	1 (part)	South America	0–2	Dispersal
(28) ***Leiocephalus***	22	North America	**24 ± 4.4**	Dispersal
(29) **Antillean *Aristelliger***	6	Central/South America	**0–50**	Dispersal
(30) *Aristelliger georgeensis*	1	Central America	0–2	Dispersal
(31) *Gonatodes albogularis*	1	Central/South America	0–2	Dispersal
(32) ***Hemidactylus haitianus***	1	Africa	0–2	Dispersal
(32) *Hemidactylus mabouia*	1	Africa	0–2	Dispersal
(34) *Hemidactylus palaichthus*	1	South America	0–2	Dispersal
(35) ***Phyllodactylus pulcher***	1	South America	0–45	Dispersal
(36) ***Phyllodactylus wirshingi***	1	South America	?	Dispersal
(37) ***Sphaerodactylus***	78	South America	**27 ± 4.4**	Dispersal
(38) ***Tarentola***	2	Africa	30 (**0–40**)	Dispersal
(39) *Thecadactylus rapicauda*	1	South America	0–2	Dispersal
(40) ***Mabuya lineolata***	1	South America	?	Dispersal
(41) *Mabuya bistriata*	1	South America	0–2	Dispersal
(42) **Antillean *Ameiva***	19	Central/South America	**36 ± 4.4**	Dispersal
(43) *Ameiva ameiva* (SI)	1 (part)	South America	0–2	Dispersal
(44) *Ameiva ameiva* (IP)	1 (part)	South America	0–2	Dispersal

TABLE 3 (continued).

Lineage	Number of species	Source area	Time of origin (mya)	Mechanism
(45) *Ameiva ameiva* (LA)	1 (part)	South America	0–2	Dispersal
(46) *Cnemidophorus lemniscatus*	1	Central/South America	0–2	Dispersal
(47) ***Cnemidophorus vanzoi***	1	South America	0–45	Dispersal
(48) *Kentropyx borckiana*	1	South America	0–2	Dispersal
(49) ***Cricosaura typica***	1	Central America	>60	Vicariance
(50) *Boa constrictor* (ISA/IP/ISC)	1 (part)	Central America	0–2	Dispersal
(51) *Boa constrictor* (LA)	1 (part)	South America	0–2	Dispersal
(52) *Corallus hortulanus*	1	South America	0–2	Dispersal
(53) ***Epicrates***	9	Central/South America	**0–22**	Dispersal
(54) ***Chironius vincenti***	1	South America	**0–10? (0–24)**	Dispersal
(55) ***Mastigodryas bruesi***	1	South America	**0–10? (0–17)**	Dispersal
(56) *Nerodia clarki*	1	North America	0–2	Dispersal
(57) *Clelia clelia*	1	South America	0–2	Dispersal
(58) ***Clelia errabunda***	1	South America	**0–5**	Dispersal
(59) ***Liophis cursor* group**	1	South America	**0–10? (0–18)**	Dispersal
(60) *Liophis melanotus*	1	South America	0–2	Dispersal
(61) *Pseudoboa neuwiedi*	1	South America	0–2	Dispersal
(62) ***Coniophanes andresensis***	1	Central America	**0–9**	Dispersal
(63) ***Tretanorhinus variabilis***	1	Central America	**0–10?**	Dispersal
(64) **Alsophines**	33	South America	**12–26**	Dispersal
(65) *Micrurus nigrocinctus*	1	Central America	0–2	Dispersal
(66) ***Leptotyphlops bilineata* group**	5	New World	?	Dispersal
(67) ***Leptotyphlops columbi***	1	South America?	0–5	Dispersal
(68) *Leptotyphlops goudotii*	1	Central/South America	0–2	Dispersal
(69) *Leptotyphlops albifrons*	1	South America	0–2	Dispersal
(70) ***Tropidophis***	13	South America	**42 ± 4.6**	Dispersal
(71) ***Typhlops***	24	Africa?	**26–58**	Dispersal
(72) ***Bothrops***	2	South America	**0–4**	Dispersal
(73) ***Trachemys***	4	North America	0–25	Dispersal
(74) *Kinosternon scorpioides*	1	Central/South America	0–2	Dispersal
(75) **Pelomedusids**	1	New World	30?	Dispersal
(76) *Geochelone carbonaria*	1	South America?	0–2	Dispersal
(77) ***Geochelone* sp.**	?	New World	0–2?	Dispersal

phrynid frogs, as well as most genera of hylid frogs. Among the "missing" reptiles are most turtles; helodermatid lizards, gerrhonotine and anguine lizards, and most genera of Neotropical iguanid and teiid lizards; anomalepid snakes, elapid snakes (aside from a recent dispersal; lineage 65), and most genera of Neotropical colubrid snakes. Certainly, not all major mainland taxa would be expected to occur in the West Indies, but for large, ecologically diverse islands, at least some are notable absences. Such absences may occur for two reasons: (1) the islands originally were without a fauna, with the present fauna having arisen gradually through dispersal; and (2) the original fauna of the West Indies was a cross section of the mainland fauna (through vicariance) but since has become impoverished due to extinctions. The fossil record of the West Indies, while admittedly poor, has indicated that a slightly more diverse fauna existed before human colonization (Pregill and Olson 1981; Morgan and Woods 1986), but it has not demonstrated that the West Indies ever carried a "cross section" of the mainland fauna (Williams 1989*a*). This analysis provides independent evidence for the interpretation that such a depauperate herpetofauna is the result of dispersal during the Cenozoic.

Recently, a variation on the Cenozoic dispersal model (MacPhee and Iturralde-Vinent 1994) attempted to explain the presence of a Miocene ground sloth fossil in Cuba. They postulated that a land bridge existed between northern South America and the Greater Antilles in the mid-Cenozoic. The underlying geological basis for such a land bridge, the Aves Ridge, has long been recognized and often has been referred to as an "island-arc" (e.g., Malfait and Dinkelman 1972; Pindell and Barrett 1990), although most of the area presently is under deep water (>1000 m). MacPhee and Iturralde-Vinent (1994) speculated that the sharp drop in sealevel during the Oligocene (30 mya), was enough to "create continuous land, or groups of very closely spaced islands from western Cuba through to northern South America." A rising sea level in the late Oligocene (27 mya) and later subsidence of the Aves Ridge caused this opportunity for landbridge dispersal to be very short in geologic time (3 million years).

Whether or not the Aves Ridge, or the proto-Antilles, was a continuous landmass or just a chain of islands (island arc) is not known, although the latter normally is assumed (Donnelly 1989; Perfit and Williams 1989). From a biogeographic standpoint, a direct land connection should have allowed a "cross section" of the mainland fauna to enter (e.g., the Panamanian Isthmus during the last 3 million years) and that is not evident in the fossil or extant faunas, as noted above. In addition, a closer look at the times of origin for the endemic West Indian lineages (Figure 2) does not indicate a cluster or peak in the Oligocene, as would be predicted by such a mid-Cenozoic (30–27 mya) land bridge. Instead, the wide variation in times of origin, with several major lineages appearing earlier and later in the Cenozoic, is in better agreement with an overwater dispersal model.

The times of origin for lineages of West Indian amphibians and reptiles (Table 3; Figure 2) can be arranged, albeit somewhat arbitrarily, into four groups: (I) ancient lineages (e.g., *Eleutherodactylus* and possibly *Cricosaura*), that may have arisen by proto-Antillean vicariance in the late Mesozoic; (II)

TABLE 4. Source areas for lineages of amphibians and reptiles with species endemic to the West Indies.

South America	Central/South America	Central America	North America
Bufo peltocephalus group	*Eleutherodactylus*	*Celestus*	*Leiocephalus*
Colostethus chalcopis	*Crocodylus rhombifer*	*Cricosaura typica*	*Trachemys*
Osteopilus	*Diploglossus*	*Coniophanes andresensis*	
Hyla heilprini	*Aristelliger*	*Tretanorhinus variabilis*	
Leptodactylus albilabris	*Anolis*		
Leptodactylus fallax	*Cyclura*		
Amphisbaena	*Ameiva*		
Gymnophthalmus pleei	*Epicrates*		
Iguana delicatissima			
Phyllodactylus pulcher			
Phyllodactylus wirshingi			
Sphaerodactylus			
Mabuya lineolata			
Cnemidophorus vanzoi		Africa	"New World"
Chironius vincenti			
Mastigodryas bruesi		*Hemidactylus haitianus*	*Leptotyphlops bilineata* group
Clelia errabunda		*Tarentola*	Pelomedusids
Liophis cursor group		*Typhlops*	*Geochelone* sp.
Alsophines			
Leptotyphlops columbi			
Tropidophis			
Bothrops			

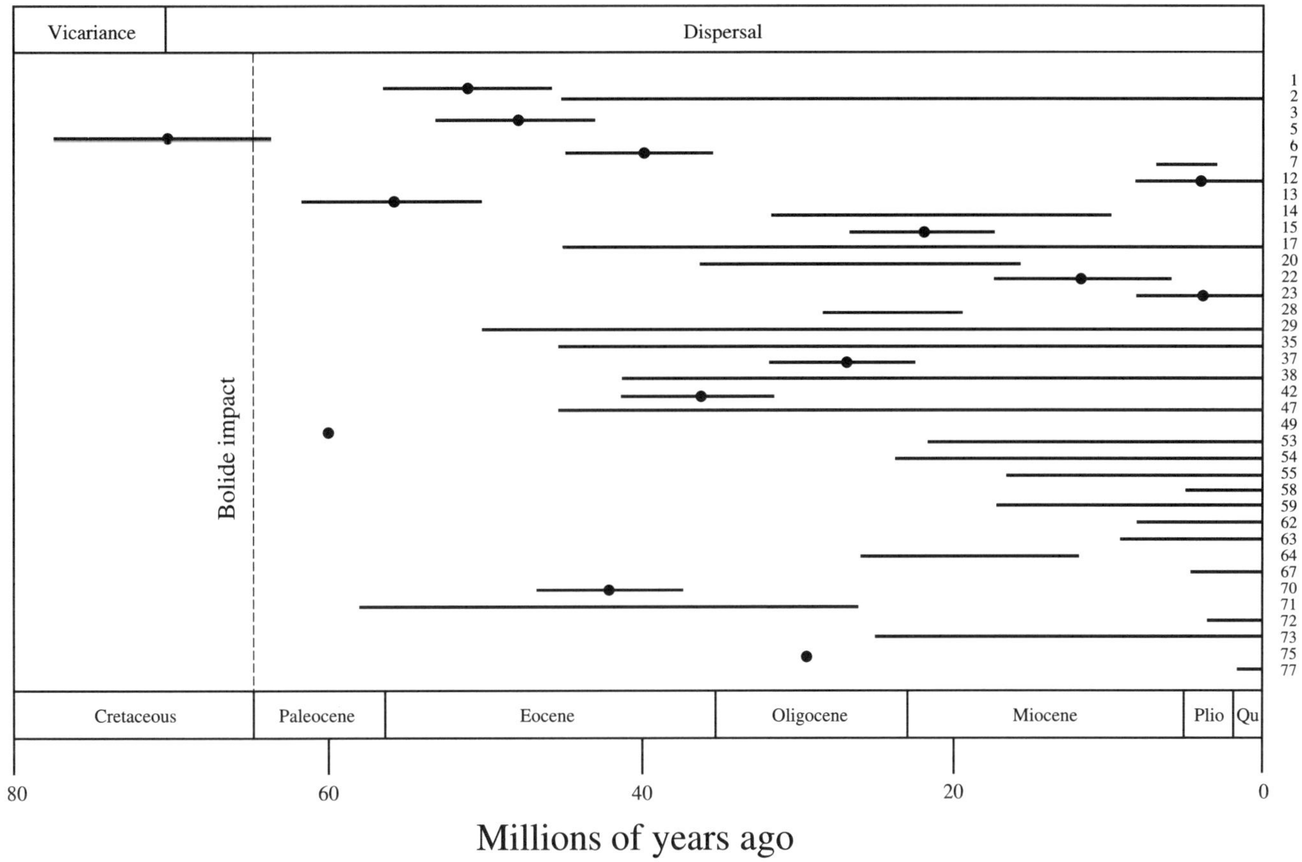

FIGURE 2. Times of origin for the 37 independent lineages of endemic West Indian amphibians and reptiles. The data are from Table 3, with lineage number indicated on right; the 40 non-endemic lineages, all with origin in the Quaternary (0–2 mya), are not shown.

mid-Tertiary colonizers, including most of the large radiations (e.g., *Bufo peltocephalus* group, *Osteopilus*, *Leptodactylus albilabris*, *Amphisbaena*, *Anolis*, *Leiocephalus*, *Sphaerodactylus*, *Ameiva*, *Epicrates*, alsophines, *Tropidophis*, and *Typhlops*; (III) Late Tertiary colonizers, which include mostly single endemic species or small groups of species (e.g., *Leptodactylus fallax*, *Crocodylus rhombifer*, *Cyclura*, *Iguana delicatissima*, *Chironius vincenti*, *Mastigodryas bruesi*, *Liophis cursor* group, *Coniophanes andresensis*, *Tretanorhinus variabilis*, *Bothrops*, and *Trachemys*; and (IV) Quaternary colonizers, which include species present on both the mainland and in the West Indies (all other lineages of known time of origin). The large number of group IV lineages almost certainly is an artifact due to fewer extinctions expected for recent lineages. Also, conspecific populations were assigned to this stage automatically as an initial assumption. Later studies possibly will show some island populations to be considerably more divergent than indicated. The distinction between these four groups facilitates discussion but is artifactual; they are on a time continuum and no reason exists to believe that they represent discrete stages in the evolution of the fauna such as is postulated for the taxon cycle (e.g., Wilson 1961; Ricklefs and Cox 1972).

Water Currents

If no direct land connections existed between the West Indies and the mainland during the Cenozoic, then why do most lineages of amphibians and reptiles show a South American origin? One of the most important considerations in overwater dispersal is the direction of the water currents that carry flotsam from the mouths of large rivers. The current

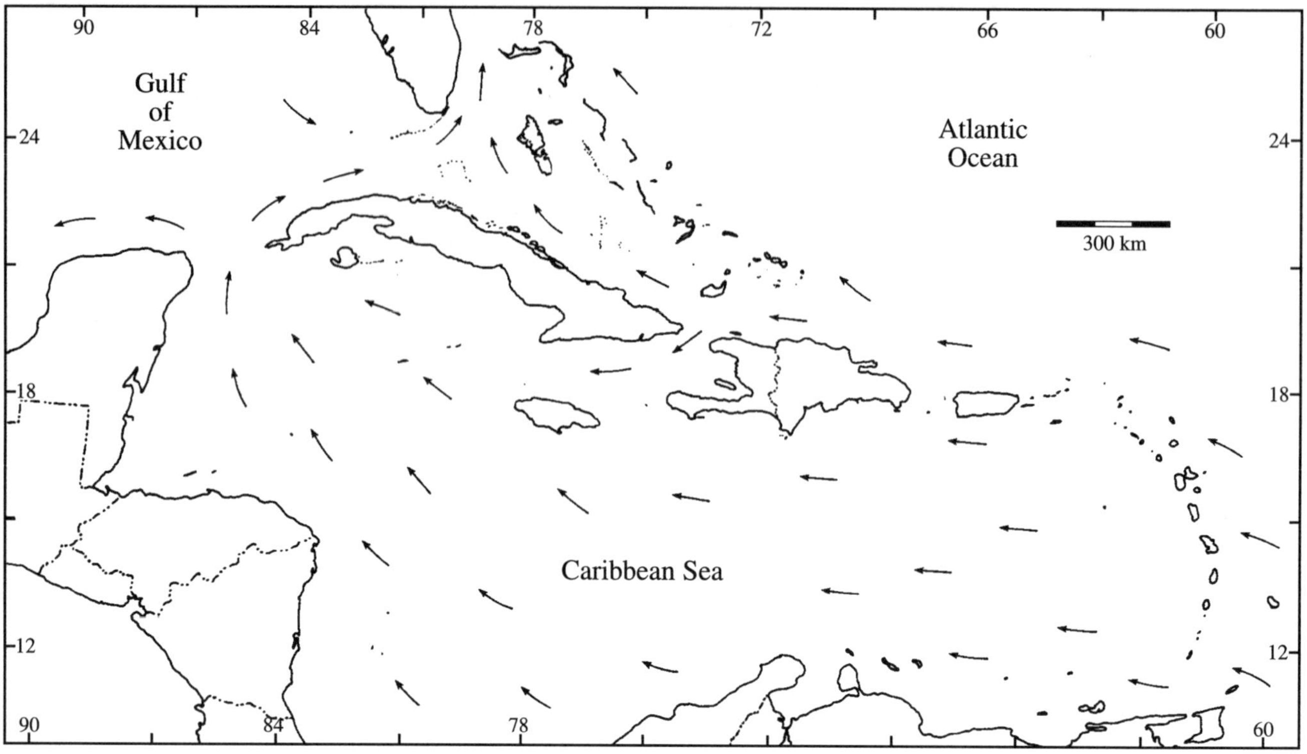

FIGURE 3. The West Indies, showing major water current patterns.

patterns in the West Indies are not random, but flow in a strong westerly and northwesterly pattern (Figure 3), as has been noted earlier (i.e., Guppy 1917; Williams 1969). Because the Caribbean always has been north of the equator during geologic history (Smith et al. 1994), the Coriolis Force would have produced the same clockwise current flow in the past, even while a water connection to the Pacific Ocean was in existence. Such a current pattern agrees with the finding of a strong South American influence in the West Indian herpetofauna in this analysis. The ability of flotsam to carry organisms great distances is well known and already has been discussed for the Caribbean biota (Guppy 1917; King 1962; Heatwole and Levins 1972).

If overwater dispersal has been a significant factor in the origin of the herpetofauna, then two biogeographic "anomalies" are predicted based on this pattern of current flow: (1) that some organisms inhabiting islands in the Western Caribbean will have a greater affinity with South America rather than adjacent Central America; and (2) that some groups inhabiting the southern Lesser Antilles will show a greater affinity with populations from the southeast in the Guianas rather than nearby mainland South America (Figure 4). Both of these patterns are seen in the herpetofauna. For example, *Ameiva ameiva* and possibly *Tretioscincus bifasciatus* for (1), and *Gymnophthalmus underwoodi*, *Kentropyx borckiana*, and *Corallus hortulanus* for (2).

Another interesting distribution pattern is seen in *Tarentola americana* (and the undescribed Cuban sibling species), which colonized the West Indies in the mid-Cenozoic from the Old World. The closest relatives of those species, members of the subgenus *Tarentola*, inhabit islands off the west coast of Africa, including the Cape Verde and Canary Islands (Joger 1984). The distance from those islands to Cuba is 5,300 km (3,800 km to the Lesser Antilles), but this distance would have been about one-third shorter (3,500 km) in the mid-Cenozoic when South America and Africa were closer (Pitman et al. 1993). The Atlantic water current used by Christopher Columbus (North Equatorial Current) was likely the same that brought *Tarentola* to the New World in the mid-Cenozoic (Figure 5). A similar dispersal route was suggested by Kluge (1969) for *Hemidactylus haitianus* and *H. mabouia*, although Vanzolini (1978) argued that human introduction is a more likely explanation (see also Bauer 1993).

Systematic Bias

Recently, a trend has been evident in the literature on Caribbean biogeography to ascribe a vicariance origin for almost any lineage, without knowledge of time of origin (e.g., Crother et al. 1986; Guyer and Savage 1986; Kluge 1989). In some cases, geologic models are modified to fit the phylogenetic or distributional data for a particular group (e.g., Guyer and Savage 1986: 557). For example, a close relationship between Cuba and Jamaica was suggested for a few taxonomic groups, and because this was at odds with the geologic models, the author suggested that "further geological research is necessary" (Kluge 1989: 323). Although possibly true, a more parsimonious explanation is that overwater dispersal is the reason for any particular close biological relationship between Jamaica and Cuba, especially given the complete or nearly complete mid-Cenozoic submergence of Jamaica. Others have suggested that ". . . dispersal should be a last resort for explaining modern distributions and used only after all vicariance possibilities have been considered" (Grimaldi 1988).

The roots of this apparent tendency to reject dispersal in all but the most obvious cases are in the widespread misconceptions throughout the systematic literature that dispersal is "unscientific" and untestable (e.g., Nelson and Platnick 1981: 50; Kluge 1991: 49), and that congruence between phylogenies, and between phylogenies and area relationships, indicates vicariance (Rosen 1978; Humphries 1992). Concordance between phylogenies can just as easily be explained by concordant dispersal (Cadle 1984*c*; Tolson 1987; Page and Lydeard 1994; Hedges et al. 1994), as illustrated by the result of this analysis (concordant dispersals from South America). And concordance between phylogenies and most area relationships also can be explained by dispersal, because organisms are more likely to disperse between geographically closer areas (those areas that usually are geologically more closely related) than between more distant areas (Hedges et al. 1994). Both dispersal and vicariance can be tested with information on time of origin (Hedges et al. 1992*b*, 1994), as has been done in this analysis.

One factor usually overlooked in the vicariance-dispersal debate has been the extent of emergent land in North and South America that presumably connected the proto-Antilles to the mainland in the late Cretaceous (Rosen 1975). New paleocoastline data, however, indicate that much of southern North America and northern South America were below sea level in the late Cretaceous, 90–70 mya (Smith et al. 1994), a problem alluded to earlier by Perfit and Williams (1989: 67). Those data, based on geology and global sea level changes, indicate that even if the proto-Antilles were emergent, they may not have formed a direct land connection to the neighboring continental areas. If the possibility of proto-Antillean vicariance is made less likely by paleocoastline data, then is another explanation possible for the late Cretaceous time of origin for

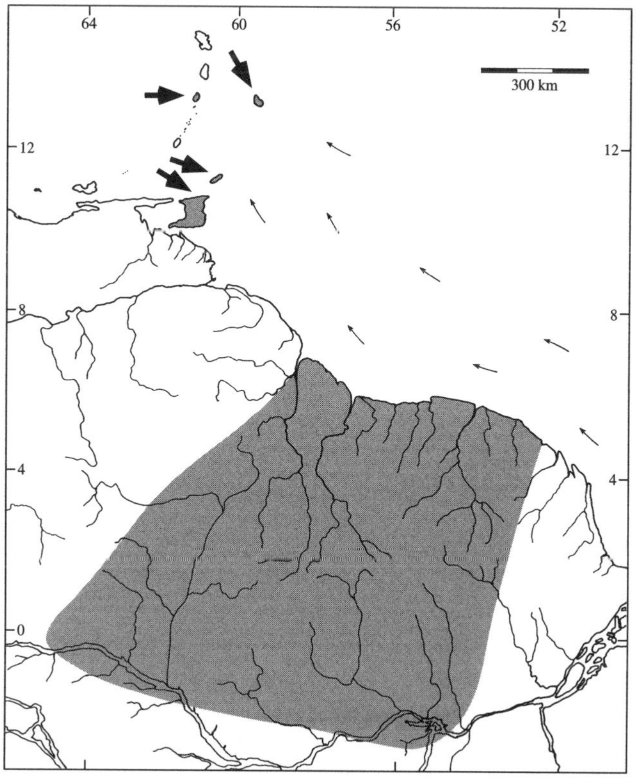

FIGURE 4. Distribution of *Gymnophthalmus underwoodi* in South America (including Trinidad and Tobago) and the Lesser Antilles (Barbados and St. Vincent). The direction of major current patterns is indicated (small arrows).

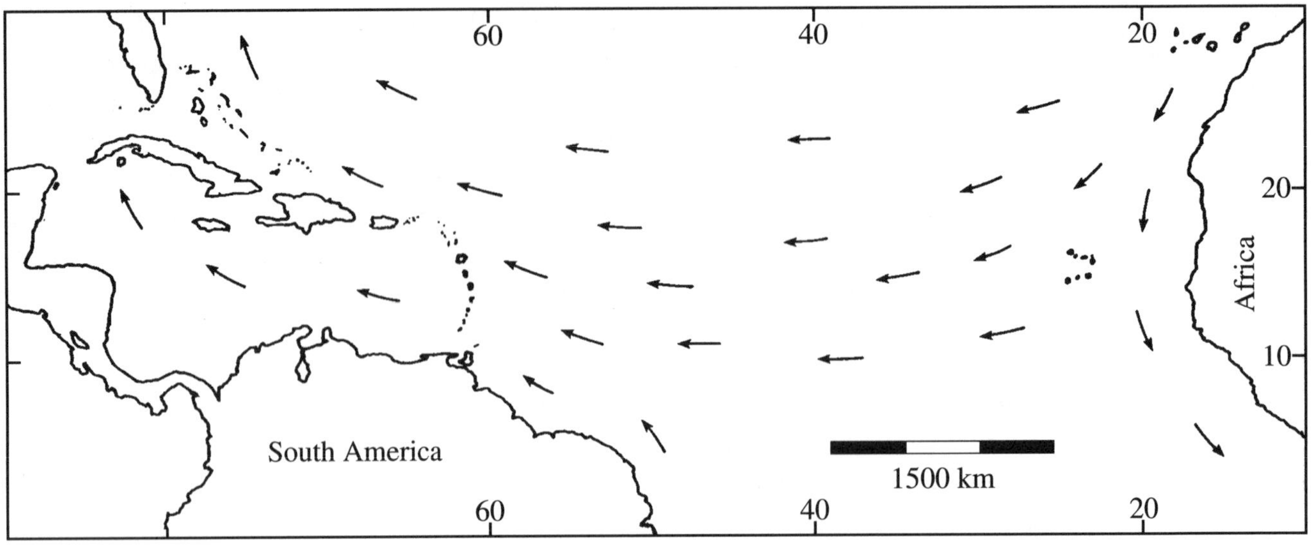

FIGURE 5. The North Equatorial Current flowing from Africa to South America and the West Indies, representing the southern portion of the clockwise pattern of surface currents in the North Atlantic Ocean due to the Coriolis Force. This is the probable mechanism and route for the origin of West Indian geckos of the genus *Tarentola* (and possibly two species of *Hemidactylus*; see text).

Eleutherodactylus (and possibly *Cricosaura*)? Ancient lineages may have arrived in the proto-Antilles by overwater dispersal rather than by vicariance, even in the late Cretaceous. Whereas the current trend in the literature is to accept proto-Antillean vicariance as the "null hypothesis," careful scrutiny of the geologic evidence indicates that such confidence is misplaced.

Predictions

Because most of the information on time of origin in this analysis is based on the chronological properties of serum albumin, the potential error in time estimates must always be considered. Also, the relationships (and number of lineages) of some of the West Indian groups – such as the species-rich genus *Anolis* – have yet to be resolved. In essence, this analysis represents an hypothesis to be tested by additional data. Those data likely will be DNA sequences, although the Tertiary fossil record for the West Indies still has considerable potential for making important contributions (Williams 1989*a*).

For the groups with endemic species on multiple islands in the West Indies, some predictions can be made now that water currents are seen as a probable major factor in the origin of the herpetofauna. Relationships within the West Indies (among islands) also should reflect the prevailing east to west-northwest current flow (Figure 3). In other words, it should be very difficult for an organism to colonize Puerto Rico from Hispaniola (by flotsam) but very easy to do the reverse. Likewise, getting to Hispaniola from either Cuba or Jamaica should be very difficult, but going from Hispaniola to Cuba or Jamaica should be much easier. Colonization of the Lesser Antilles from the Greater Antilles should be much harder than colonization of the Lesser Antilles from South America, as was noted by Baskin and Williams (1966). The direction of water currents allows the Bahamas Platform to be colonized by the Lesser Antilles, Puerto Rico, Hispaniola, and Cuba, but not likely from Jamaica.

Frogs of the genus *Eleutherodactylus* exhibit a pattern of distribution in the West Indies that fits these predictions to some degree. The two major subgenera (*Euhyas* and *Eleutherodactylus*) each have been isolated in the West Indies since the late Mesozoic. However, the western Caribbean group *Euhyas* has not successfully colonized any islands east of Isla Mona (between Hispaniola and Puerto Rico), whereas the primarily eastern subgenus *Eleutherodactylus* occurs throughout the Greater and Lesser Antilles except for Jamaica. In Cuba, the "arboreal" subgenus *Eleutherodactylus* (10 species) clearly is the invader to an island with a large and diverse assemblage of *Euhyas* (Hedges et al. 1992*a*). Likewise, the two species occurring in the southernmost islands of the Lesser Antilles (Grenada and St. Vincent) were derived from lineages in South America (Kaiser et al. 1994*b*).

Hispaniola and Puerto Rico share several species, or have closely related species pairs, that may be the result of relatively recent dispersal from Puerto Rico to Hispaniola (e.g., *Leptodactylus albilabris/L. dominicensis*, *Phyllodactylus wirshingi*, and *Trachemys stejnegeri*). Other cases of interisland relationships may be ascribed to dispersal, but more phylogenetic studies are needed before firm conclusions can be drawn. In general, the prevailing current directions in the West Indies predict that if an "invader" species or lineage can be identified on an island, it most likely arrived from land areas to the east or southeast. For some older groups, intra-Caribbean vicariance also may help to explain distribution patterns (Hedges 1989*b*), although the geologic relationships of the islands still are not well known (Pindell and Barrett 1990).

These predictions pertain to organisms transported on flotsam by water currents. Airborne transport is more complicated, because the prevailing winds from the northeast favor only North America as a source, but nearly all hurricanes follow a path similar to the water currents. Thus, the source area for jetsam (e.g., seeds, many insects, flying vertebrates) depends on which of these two forces (prevailing winds or tropical storms) is more important.

Conclusion

The origin of the diverse West Indian herpetofauna appears to be entirely, or almost entirely, the result of overwater dispersal during the Cenozoic. This agrees with a similar analysis conducted on a subset of the herpetofauna (Hedges et al. 1992*b*). Two possible exceptions are the frog genus *Eleutherodactylus* and the lizard *Cricosaura typica*, which may have arrived in the West Indies in the late Mesozoic, possibly by proto-Antillean vicariance. The largely South American composition of the herpetofauna probably is due to prevailing water current flow from that continent. Because the West Indies have remained north of the Equator since they originated in the Mesozoic, the direction of the currents, due to the Coriolis Force, is inferred to have been the same in the past. The Amazon Basin has the highest levels of terrestrial biodiversity in the World, and most of the region is drained by rivers (e.g., Amazon and Orinoco) that empty into ocean currents that in turn flow through the West Indies. The result is that the West Indies have been in an ideal location for the immigration of a diverse tropical herpetofauna throughout the Cenozoic.

Acknowledgments

I dedicate this paper to the memory of Albert Schwartz, in honor of his enormous contribution to West Indian herpetology spanning four decades. Carla Hass, Robert Henderson, Hinrich Kaiser, Robert Powell, and Richard Thomas offered valuable comments. This work was supported by grants from the National Science Foundation (BSR 8307115 to Richard Highton, and BSR 8906325 and DEB 9123556 to the author).

Literature Cited

Alvarez, W., J. Smit, W. Lowrie, F. Asaro, S. V. Margolis, P. Claeys, M. Kastner, and A. R. Hildebrand. 1992. Proximal impact deposits at the Cretaceous-Tertiary boundary in the Gulf of Mexico: a restudy of DSDP Leg 77 Sites 536 and 540. Geology 20: 697–700.

Armas, L. F. de. 1982. Algunos aspectos zoogeographicos de la escorpionfauna antillana. Poeyana 238: 1–17.

— and E. D. J. Marcano-Fondeur. 1992. New scorpions from the Dominican Republic (Arachnida: Scorpiones). Poeyana 420: 1–36.

Auffenberg, W. 1967. Notes on West Indian tortoises. Herpetologica 31: 34–44.

Barbour, T. and A. do Amaral. 1924. Notes on some Central American snakes. Occas. Pap. Boston Soc. Nat. Hist. 5: 129–132.

Baskin, J. N. and E. E. Williams. 1966. The Lesser Antillean *Ameiva* (Sauria, Teiidae), re-evaluation, zoogeography, and the effects of predation. Stud. Fauna Curaçao Carib. Isl. 23: 143–176.

Bauer, A. M. 1993. African-South American relationships: a perspective from the Reptilia, p. 244–288. *In* P. Goldblatt (ed.), Biological Relationships between Africa and South

America. Yale Univ. Press, New Haven, Connecticut.

— and A. P. Russell. 1993. *Aristelliger*. Cat. Amer. Amphib. Rept.: 565.1–565.4.

Bezy, R. L. 1972. Karyotypic variation and evolution of the lizards in the family Xantusiidae. Contrib. Sci. Nat. Hist. Mus. Los Angeles Co. 227 :1–29.

Böhme, W. 1984. Erstfund eines fossilen Kugelfingergeckos (Sauria: Gekkonidae: Sphaerodactylinae) aus dominikanischem Bernstein (Oligozän von Hispaniola, Antillen). Salamandra 20: 212–220.

Branch, W. R. 1986. Hemipenial morphology of African snakes: a taxonomic review. Part 1. Scolecophidia and Boidae. Jour. Herpetol. 20: 285–299.

Breuil, M. and D. Masson. 1991. Some remarks on Lesser Antillean bat biogeography. Comp. Ren. Séanc. Soc. Biogéogr. 67: 25–39.

Briggs, J. C. 1984. Freshwater fishes and biogeography of Central America and the Antilles. Syst. Zool. 33: 428–434.

Burgess, G. H. and R. Franz. 1989. Zoogeography of the Antillean freshwater fish fauna, p. 263–304. *In* C. A. Woods (ed.), Biogeography of the West Indies: Past, Present, and Future. Sandhill Crane Press, Gainesville, Florida.

Burnell, K. L. and S. B. Hedges. 1990. Relationships of West Indian *Anolis* (Sauria: Iguanidae): an approach using slow-evolving protein loci. Carib. Jour. Sci. 26: 7–30.

Buskirk, R. 1985. Zoogeographic patterns and tectonic history of Jamaica and the northern Caribbean. Jour. Biogeogr. 12: 445–461.

Buth, D. G., G. C. Gorman, and C. S. Lieb. 1980. Genetic divergence between *Anolis carolinensis* and its Cuban progenitor, *Anolis porcatus*. Jour. Herpetol. 14: 279–284.

Cadle, J. S. 1984*a*. Molecular systematics of Neotropical xenodontide snakes. I. South American xenodontines. Herpetologica 40: 8–20.

—. 1984*b*. Molecular systematics of Neotropical xenodontine snakes. II. Central American xenodontines. Herpetologica 40: 21–30.

—. 1984*c*. Molecular systematics of Neotropical xenodontine snakes. III. Overview of xenodontine phylogeny and the history of New World snakes. Copeia 1984: 641–652.

—. 1985. The Neotropical colubrid snake fauna (Serpentes: Colubridae): lineage components and biogeography. Syst. Zool. 34: 1–20.

—. 1988. Phylogenetic relationships among advanced snakes: a molecular perspective. Univ. California Publ. Zool. 119: 1–77.

—. 1992. Phylogenetic relationships among pitvipers: immunological evidence, p. 41–47. *In* J. A. Campbell and E. D. Brodie, Jr. (eds.), Biology of the Pitvipers. Selva, Tyler, Texas.

Cei, J. M. 1972. *Bufo* of South America, p. 82–92. *In* W. F. Blair (ed.), Evolution in the Genus *Bufo*. Univ. Texas Press, Austin.

Censky, E. J. 1988. *Geochelone carbonaria* (Reptila: Testudines) in the West Indies. Florida Sci. 51: 108–114.

Cockerell, T. D. A. 1923. A fossil cichlid fish from the Republic of Haiti. Proc. United States Natl. Mus. 63: 1–3.

Colbert, E. H. 1969. A Jurassic pterosaur from Cuba. Amer. Mus. Novitates 2370: 1–26.

Crother, B. I., M. M. Miyamoto, and W. F. Presch. 1986. Phylogeny and biogeography of the lizard family Xantusiidae. Syst. Zool. 35: 37–45.

Darlington, P. J., Jr. 1957. Zoogeography: The Geographical Distribution of Animals. John Wiley and Sons, New York.

Daugherty, C. H., L. R. Maxson, and B. D. Bell. 1982. Phylogenetic relationships within the New Zealand frog genus *Leiopelma* — immunological evidence. New Zealand Jour. Zool. 9: 239–242.

Densmore, L. D., III and P. S. White. 1991. The systematics and evolution of the Crocodilia as suggested by restriction endonuclease analysis of mitochondrial and nuclear ribosomal DNA. Copeia 1991: 602–615.

Dessauer, H. C., J. E. Cadle, and R. Lawson. 1987. Patterns of snake evolution suggested by their proteins. Fieldiana: Zool. 34: 1–34.

Dixon, J. R. 1962. The leaf-toed geckos, genus *Phyllodactylus*, of northeastern South America. Southwest. Nat. 7: 211–226.

—. 1964. Further data on the geckos (*Phyllodactylus*) of islands of the extreme southern Caribbean. Southwest. Nat. 9: 203–206.

—. 1989. A key and checklist to the Neotropical snake genus *Liophis* with country lists and maps. Smithsonian Herpetol. Info. Serv. 79: 1–28.

Donnelly, T. W. 1989. History of marine barriers and terrestrial connections: Caribbean paleogeographic inference from pelagic sediment analysis, p. 103–118. *In* C. A. Woods (ed.), Biogeography of the West Indies: Past, Present, and Future. Sandhill Crane Press, Gainesville, Florida.

Dowling, H. G., R. Highton, G. C. Maha, and L. R. Maxson. 1983. Biochemical evaluation of colubrid snake phylogeny. Jour. Zool., London 201: 309–329.

Duellman, W. E. 1993. Amphibian Species of the World: Additions and Corrections. Univ. Kansas Mus. Nat. Hist. Spec. Publ. no. 21.

Dunn, E. R. 1926. The frogs of Jamaica. Proc. Boston Soc. Nat. Hist. 38: 11–130.

—. 1935. Notes on American mabuyas. Proc. Acad. Nat. Sci. Philadelphia 87: 533–557.

— and L. H. Saxe, Jr. 1950. Results of the Catherwood-Chaplin West Indies expedition, 1948. Part V. Amphibians and reptiles of San Andrés and Providencia. Proc. Acad. Nat. Sci. Philadelphia 102: 141–165.

Dunson, W. A. and M. E. Seidel. 1986. Salinity tolerance of estuarine and insular emydid turtles (*Pseudemys nelsoni* and *Trachemys decussata*). Jour. Herpetol. 20: 237–245.

Erard, C. 1991. Landbirds of the Lesser Antilles. Comp.

Ren. Séanc. Soc. Biogéogr. 67: 3–23.

Ernst, C. H. 1992. *Trachemys gaigeae*. Cat. Amer. Amphib. Rept.: 538.1–538.4.

Estes, R. 1976. Middle Paleocene lower vertebrates from the Tongue River Formation, southeastern Montana. Jour. Paleontol. 50: 500–520.

Frost, D. R. and R. Etheridge. 1989. A phylogenetic analysis and taxonomy of iguanian lizards (Reptilia: Squamata). Univ. Kansas Mus. Nat. Hist. Misc. Publ. no. 81.

Gallagher, D. S., Jr. and J. R. Dixon. 1992. Taxonomic revision of the South American lizard genus *Kentropyx* Spix (Sauria: Teiidae). Boll. Mus. Reg. Sci. Nat. Torino 10: 125–171.

Gans, C. 1990. Patterns in amphisbaenian biogeography: a preliminary analysis, p. 133–143. *In* G. Peters and R. Hutterer (eds.), Vertebrates in the Tropics. Museum Alexander Koenig, Bonn.

Gorman, G. C., A. C. Wilson, and M. Nakanishi. 1971. A biochemical approach towards the study of reptilian phylogeny: evolution of serum albumin and lactic dehydrogenase. Syst. Zool. 20: 167–185.

Gosner, K. L. 1987. Observations on Lesser Antillean pit vipers. Jour. Herpetol. 21: 78–80.

Graham, A. 1993. Contribution toward a Tertiary palynostratigraphy for Jamaica: the status of Tertiary paleobotanical studies in northern Latin America and preliminary analysis of the Guys Hill Member (Chapelton formation, middle Eocene) of Jamaica, p. 443–461. *In* R. M. Wright and E. Robinson (eds.), Biostratigraphy of Jamaica. Geol. Soc. Amer., Boulder, Colorado.

— and D. M. Jarzen. 1969. Studies in Neotropical paleobotany. I. The Oligocene communities of Puerto Rico. Ann. Missouri Bot. Garden 56: 308–357.

Grant, C. and H. A. Beatty. 1944. Herpetological notes on St. Croix, Virgin Islands. Herpetologica 2: 110–113.

Griffiths, T. A. and D. Klingener. 1988. On the distribution of Greater Antillean bats. Biotropica 20: 240–251.

Grimaldi, D. A. 1988. Relicts in the Drosophilidae (Diptera), p. 183–213. *In* J. K. Liebherr (ed.), Zoogeography of Caribbean Insects. Cornell Univ. Press, Ithaca, New York.

Guppy, H. B. 1917. Plants, Seeds, and Currents in the West Indies and Azores. Williams and Norgate, London.

Guyer, C. and J. M. Savage. 1986. Cladistic relationships among anoles (Sauria: Iguanidae). Syst. Zool. 35: 509–531.

Harris, D. M. and A. G. Kluge. 1984. The *Sphaerodactylus* (Sauria: Gekkonidae) of Middle America. Occas. Pap. Mus. Zool. Univ. Michigan 706: 1–59.

Hass, C. A. 1991. Evolution and biogeography of West Indian *Sphaerodactylus* (Sauria: Gekkonidae): a molecular approach. Jour. Zool., London 225: 525–561.

— and S. B. Hedges. 1991. Albumin evolution in West Indian frogs of the genus *Eleutherodactylus* (Leptodactylidae): Caribbean biogeography and a calibration of the albumin immunological clock. Jour. Zool., London 225: 413–426.

— and —. 1992. Karyotype of the Cuban lizard *Cricosaura typica* and its implications for xantusiid phylogeny. Copeia 1992: 563–565.

—, —, and L. R. Maxson. 1993. Molecular insights into the relationships and biogeography of West Indian anoline lizards. Biochem. Syst. Ecol. 21: 97–114.

—, M. A. Hoffman, L. D. Densmore, III, and L. R. Maxson. 1992. Crocodilian evolution: insights from immunological data. Molec. Phylogen. Evol. 1: 193–201.

— and L. R. Maxson. 1993. Nonreciprocity in immunological studies: a reply. Syst. Biol. 42: 210–215.

Heatwole, H. and R. Levins. 1972. Biogeography of the Puerto Rican Bank: flotsam transport of terrestrial animals. Ecology 53: 112–117.

Hecht, M. K. 1951. Fossil lizards of the West Indian genus *Aristelliger* (Gekkonidae). Amer. Mus. Novitates 1538: 1–33.

—. 1952. Natural selection in the lizard genus *Aristelliger*. Evolution 6: 112–124.

Hedges, S. B. 1982. Caribbean biogeography: implications of the recent plate tectonic studies. Syst. Zool. 31: 518–522.

—. 1989*a*. An island radiation: allozyme evolution in Jamaican frogs of the genus *Eleutherodactylus* (Anura, Leptodactylidae). Carib. Jour. Sci. 25: 123–147.

—. 1989*b*. Evolution and biogeography of West Indian frogs of the genus *Eleutherodactylus*: slow-evolving loci and the major groups, p. 305–370. *In* C. A. Woods (ed.), Biogeography of the West Indies: Past, Present, and Future. Sandhill Crane Press, Gainesville, Florida.

—. 1995. Analysis of West Indian amphibian distributions. *In* W. E. Duellman (ed.), Regional Patterns of Amphibian Distribution: A Global Perspective. Univ. Kansas Mus. Nat. Hist. Spec. Publ.: in press.

— and R. L. Bezy. 1993. Phylogeny of xantusiid lizards: concern for data and analysis. Molec. Phylogen. Evol. 2: 76–87.

— and —. 1994. Xantusiid lizards and phylogenetic inference. Molec. Phylogen. Evol. 3: 275–278.

—, —, and L. R. Maxson. 1991. Phylogenetic relationships and biogeography of xantusiid lizards inferred from mitochondrial DNA sequences. Molec. Biol. Evol. 8: 767–780.

—, A. R. Estrada, and R. Thomas. 1992*a*. Three new species of *Eleutherodactylus* from Eastern Cuba, with notes on vocalizations of other species (Anura: Leptodactylidae). Herpetol. Monogr. 6: 68–83.

—, C. A. Hass, and L. R. Maxson. 1992*b*. Caribbean biogeography: molecular evidence for dispersal in West Indian terrestrial vertebrates. Proc. Natl. Acad. Sci. (USA) 89: 1909–1913.

—, —, and —. 1994. Towards a biogeography of the Caribbean. Cladistics 10: 43–55.

Henderson, R. W. 1992. Consequences of predator introductions and habitat destruction on amphibians and reptiles in the post-Columbus West Indies. Carib. Jour. Sci. 28: 1–10.

—, A. Delatte, and T. J. McCarthy. 1993. *Gekko gecko* (Sauria: Gekkonidae) established on Martinique, French West Indies. Carib. Jour. Sci. 29: 128–129.

— and S. B. Hedges. 1995. Origin of West Indian populations of the geographically widespread boa *Corallus enydris* inferred from mitochondrial DNA sequences. Molec. Phylogen. Evol. 4: 88–92.

Heyer, W. R. 1970. Studies on the frogs of the genus *Leptodactylus* (Amphibia: Leptodactylidae). VI. Biosystematics of the *melanonotus* group. Los Angeles Co. Mus. Contr. Sci. 191: 1–48.

—. 1978. Systematics of the *fuscus* group of the frog genus *Leptodactylus* (Amphibia, Leptodactylidae). Nat. Hist. Mus. Los Angeles Co. Sci. Bull. 29: 1–85.

—. 1979. Systematics of the *pentadactylus* group of the frog genus *Leptodactylus* (Amphibia, Leptodactylidae). Smithsonian Contr. Zool. 301: 1–43.

—. 1994. Variation within the *Leptodactylus podicipinus-wagneri* complex of frogs (Amphibia: Leptodactylidae). Smithsonian Contr. Zool. 546: 1–124.

Hildebrand, A. R. and W. V. Boynton. 1990. Proximal Cretaceous-Tertiary boundary impact deposits in the Caribbean. Science 248: 843–847.

Hoogmoed, M. S. 1973. Notes on the Herpetofauna of Surinam IV. The Lizards and Amphisbaenians of Surinam. W. Junk, The Hague.

— and U. Gruber. 1983. Spix and Wagler type specimens of reptiles and amphibians in the Natural History Musea in Munich (Germany) and Leiden (The Netherlands). Spixiana 9 (suppl.): 319–415.

Horsfield, W. T. 1973. Late Tertiary and Quaternary crustal movements in Jamaica. Jour. Geol. Soc. Jamaica 13: 6–13.

— and M. J. Roobol. 1974. A tectonic model for the evolution of Jamaica. Jour. Geol. Soc. Jamaica 14: 31–38.

Humphries, C. J. 1992. Cladistic biogeography, p. 137–159. *In* P. L. Forey et al. (eds.), Cladistics: A Practical Course in Systematics. Clarendon Press, Oxford.

Joger, U. 1984. Taxonomische Revision der Gattung *Tarentola* (Reptilia: Gekkonidae). Bonn. Zool. Beitr. 35: 129–174.

—. 1985. The African gekkonine radiation — preliminary phylogenetic results, based on quantitative immunological comparisons of serum albumins, p. 479–494. *In* K. L. Schuchmann (ed.), Proceedings of the International Symposium on African Vertebrates: Systematics, Phylogeny, and Evolutionary Ecology. Zoologisches Forschungsinstitut und Museum Alexander Koenig, Bonn.

Joglar, R. L. 1989. Phylogenetic relationships of the West Indian frogs of the genus *Eleutherodactylus*: a morphological analysis, p. 371–408. *In* C. A. Woods (ed.), Biogeography of the West Indies: Past, Present, and Future. Sandhill Crane Press, Gainesville, Florida.

Johnson, K. 1991. Cladistics and the biogeography of two trans-Caribbean hairstreak butterflies: *Nesiostrymon* and *Terra* (Lepidoptera, Lycaenidae). Amer. Mus. Novit. 3011: 1–43.

Jones, D. L. 1993. Cycads of the World. Smithsonian Press, Washington, D. C.

Kaiser, H., L. A. Coloma, and H. M. Gray. 1994*a*. A new species of *Colostethus* (Anura: Dendrobatidae) from Martinique, French Antilles. Herpetologica 50: 23–32.

—, J. D. Hardy, Jr., and D. M. Green. 1994*b*. Taxonomic status of Caribbean and South American frogs currently ascribed to *Eleutherodactylus urichi* (Anura: Leptodactylidae). Copeia 1994: 780–796.

—, T. F. Sharbel, and D. M. Green. 1994*c*. Systematics and biogeography of eastern Caribbean *Eleutherodactylus* (Anura: Leptodactylidae): evidence from allozymes. Amphibia-Reptilia 15: 375–394.

King, F. W. 1962. The occurrence of rafts for dispersal of land animals into the West Indies. Quart. Jour. Florida Acad. Sci. 25: 45–52.

— and R. L. Burke (eds.). 1989. Crocodilian, Tuatara, and Turtle Species of the World. Assoc. Syst. Coll., Washington, D. C.

Klauber, L. M. 1939. Three new worm snakes of the genus *Leptotyphlops*. Trans. San Diego Soc. Nat. Hist. 9: 59–65.

Kluge, A. G. 1969. The evolution and geographical origin of the New World *Hemidactylus mabouia - brooki* complex (Gekkonidae, Sauria). Mus. Zool. Univ. Michigan Misc. Publ. 138: 1–78.

—. 1982. Cloacal bones and sacs as evidence of gekkonoid lizard relationships. Herpetologica 38: 348–355.

—. 1987. Cladistic relationships in the Gekkonoidea (Squamata, Sauria). Mus. Zool. Univ. Michigan Misc. Publ.173: 1–54.

—. 1988. Parsimony in vicariance biogeography: a quantitative method and a Greater Antillean example. Syst. Zool. 37: 315–328.

—. 1989. A concern for evidence and a phylogenetic hypothesis of relationships among *Epicrates* (Boidae, Serpentes). Syst. Zool. 38: 7–25.

—. 1991. Boine snake phylogeny and research cycles. Mus. Zool. Univ. Michigan Misc. Publ. 178: 1–58.

Koopman, K. F. 1989. A review and analysis of the bats of the West Indies, p. 635–644. *In* C. A. Woods (ed.), Biogeography of the West Indies: Past, Present, and Future. Sandhill Crane Press, Gainesville, Florida.

Kring, D. A. and W. V. Boynton. 1992. Petrogenesis of an augite-bearing melt rock in the Chicxulub structure and its relationship to K/T impact spherules in Haiti. Nature 358: 141–144.

Lambert, J. B., J. S. Frye, and G. O. Poinar, Jr. 1985. Amber from the Dominican Republic: analysis of nuclear magnetic resonance spectroscopy. Archaeometry 27: 43–51.

Lazell, J. D., Jr. 1964. The Lesser Antillean representatives of *Bothrops* and *Constrictor*. Bull. Mus. Comp. Zool., Harvard Univ., 132: 245–273.

Lescure, J., J. Jeremie, W. Lourenco, J. P. Mauries, J. Pierre, C. Sastre, and J. M. Thibaud. 1991. Biogeography and insularity: the example of the Lesser Antilles. Comp. Ren. Séanc. Soc. Biogéogr. 67: 41–59.

Liebherr, J. K. 1988*a*. Biogeographic patterns of West Indian *Platynus* carabid beetles (Coleoptera), p. 121–152. *In* J. K. Liebherr (ed.), Zoogeography of Carribean Insects. Cornell Univ. Press, Ithaca, New York.

—. 1988*b*. General patterns in West Indian insects, and graphical biogeographic analysis of some circum-Caribbean *Platynus* beetles. Syst. Zool. 37: 385–409.

Lynch, J. D. 1986. The definition of the Middle American clade of *Eleutherodactylus* based on jaw musculature (Amphibia: Leptodactylidae). Herpetologica 42: 248–258.

MacFadden, B. 1980. Rafting mammals or drifting islands? Biogeography of the Greater Antillean insectivores *Nesophontes* and *Solenodon*. Jour. Biogeogr. 7: 11–22.

—. 1981. Comments on Pregill's appraisal of historical biogeography of Caribbean vertebrates: vicariance, dispersal, or both? Syst. Zool. 30: 370–372.

MacPhee, R. D. E. and M. A. Ituralde-Vinent. 1994. First Tertiary land mammal from Greater Antilles: an early Miocene sloth (Xenarthra, Megalonchidae) from Cuba. Amer. Mus. Novit. 3094: 1–13.

— and A. R. Wyss. 1990. Oligo-Miocene vertebrates from Puerto Rico, with a catalog of localities. Amer. Mus. Novit. 2965: 1–45.

Maglio, V. J. 1970. West Indian xenodontine colubrid snakes: their probable origin, phylogeny, and zoogeography. Bull. Mus. Comp. Zool., Harvard Univ., 141: 1–54.

Malfait, B. T. and M. G. Dinkelman. 1972. Circum-Caribbean tectonic and igneous activity and the evolution of the Caribbean plate. Bull. Geol. Soc. Amer. 83: 251–272.

Matthew, W. D. 1918. Affinities and origin of the Antillean mammals. Bull. Geol. Soc. Amer. 29: 657–666.

Maurrasse, F. J.-M. R. and G. Sen. 1991. Impacts, tsunamis and the Haitian Cretaceous-Tertiary boundary layer. Science 252: 1690–1693.

Maxson, L. R. 1992. Tempo and pattern in anuran speciation and phylogeny: an albumin perspective, p. 41–57. *In* K. Adler (ed.), Herpetology: Current Research on the Biology of Amphibians and Reptiles. Proceedings of the First World Congress of Herpetology. Soc. Study Amphib. Rept., Oxford, Ohio. Contrib. Herpetol., vol. 9.

— and W. R. Heyer. 1988. Molecular systematics of the frog genus *Leptodactylus* (Amphibia: Leptodactylidae). Fieldiana: Zool. (new ser.) 41: 1–13.

— and R. D. Maxson. 1979. Comparative albumin and biochemical evolution in plethodontid salamanders. Evolution 33: 1057–1062.

— and C. W. Myers. 1985. Albumin evolution in tropical poison frogs (Dendrobatidae): a preliminary report. Biotropica 17: 50–56.

— and A. C. Wilson. 1975. Albumin evolution and organismal evolution in tree frogs (Hylidae). Syst. Zool. 24: 1–15.

Miller, L. D. and J. Y. Miller. 1989. The biogeography of West Indian butterflies (Lepidoptera: Papilionoidea, Hesperioidea): a vicariance model, p. 229–262. *In* C. A. Woods (ed.), Biogeography of the West Indies: Past, Present, and Future. Sandhill Crane Press, Gainesville, Florida.

Moll, E. O. and J. M. Legler. 1971. The life history of a Neotropical slider turtle, *Pseudemys scripta* (Schoepff) in Panama. Nat. Hist. Mus. Los Angeles Co. Sci. Bull. 11: 1–102.

Morgan, G. S. 1993. Quaternary land vertebrates of Jamaica, p. 417–442. *In* R. M. Wright and E. Robinson (eds.), Biostratigraphy of Jamaica. Geol. Soc. Amer., Boulder, Colorado.

—, R. Franz, and R. I. Crombie. 1993. The Cuban crocodile, *Crocodylus rhombifer*, from Late Quaternary fossil deposits on Grand Cayman. Carib. Jour. Sci. 29: 153–164.

— and C. A. Woods. 1986. Extinction and the zoogeography of West Indian land mammals. Biol. Jour. Linnean Soc. 28: 167–203.

Nelson, G. and N. I. Platnick. 1981. Systematics and Biogeography: Cladistics and Vicariance. Columbia Univ. Press, New York.

Page, R. D. M. and C. Lydeard. 1994. Towards a cladistic biogeography of the Caribbean. Cladistics 10: 21–41.

Perfit, M. R. and E. E. Williams. 1989. Geological constraints and biological retrodictions in the evolution of the Caribbean Sea and its islands, p. 47–102. *In* C. A. Woods (ed.), Biogeography of the West Indies: Past, Present, and Future. Sandhill Crane Press, Gainesville, Florida.

Peters, J. A. and B. Orejas-Miranda. 1970. Catalogue of the Neotropical Squamata. Part 1. Snakes. Bull. United States Natl. Mus. 297: 1–347.

Pindell, J. and S. F. Barrett. 1990. Geological evolution of the Caribbean region: a plate tectonic perspective, p. 405–432. *In* G. Dengo and J. E. Case (eds.), The Geology of North America. Vol. H. The Caribbean Region. Geol. Soc. Amer., Boulder, Colorado.

Pinou, T. and H. G. Dowling. 1994. The phylogenetic relationships of the Central American snake *Tretanorhinus*: data from morphology and karyology. Amphibia-Reptilia 15: 297–305.

Pitman, W. C., III, S. Cande, J. LaBrecque, and J. Pindell. 1993. Separation of Africa from South America, p. 15–36. *In* P. Goldblatt (ed.), Biological Relationships between Africa and South America. Yale Univ. Press, New Haven, Connecticut.

Poinar, G. O., Jr. 1988. Hair in Dominican amber: evidence for Tertiary land mammals in the Antilles. Experientia 44: 88–89.

—. 1992. Life in Amber. Stanford Univ. Press, Stanford, California.

— and D. C. Cannatella. 1987. An Upper Eocene frog from the Dominican Republic and its implication for Caribbean biogeography. Science 237: 1215–1216.

—, K. Warheit, and J. Brodzinsky. 1985. A fossil feather in Dominican amber. Internatl. Res. Comm. Sys., Med. Sci. 13: 927.

Powell, R. 1990. *Hemidactylus palaichthus*. Cat. Amer. Amphib. Rept.: 468.1.

Pregill, G. K. 1981*a*. Cranial morphology and the evolution of West Indian toads (Salientia: Bufonidae): resurrection of the genus *Peltophryne* Fitzinger. Copeia 1981: 273–285.

—. 1981*b*. An appraisal of the vicariance hypothesis of Caribbean biogeography and its application to West Indian terrestrial vertebrates. Syst. Zool. 30: 147–155.

—. 1992. Systematics of the West Indian lizard genus *Leiocephalus* (Squamata: Iguania: Tropiduridae). Univ. Kansas Mus. Nat. Hist. Misc. Publ. no. 84.

— and S. L. Olson. 1981. Zoogeography of West Indian vertebrates in relation to Pleistocene climatic cycles. Ann. Rev. Ecol. Syst. 12: 75–98.

Ricklefs, R. E. and G. W. Cox. 1972. Taxon cycles in the West Indian avifauna. Amer. Nat. 106: 195–219.

Rieppel, O. 1980. Green anole in Dominican amber. Nature 286: 486–487.

Roberts, J. D. and L. R. Maxson. 1986. Phylogenetic relationships in the genus *Limnodynastes* (Anura: Myobatrachidae): a molecular perspective. Australian Jour. Zool. 34: 561–573.

Romer, A. S. 1966. Vertebrate Paleontology. Univ. Chicago Press, Chicago, Illinois.

Rosen, D. E. 1975 (1976). A vicariance model of Caribbean biogeography. Syst. Zool. 24: 431–464.

—. 1978. Vicariant patterns and historical explanation in biogeography. Syst. Zool. 27: 159–188.

—. 1985. Geological hierarchies and biogeographic congruence in the Caribbean. Ann. Missouri Bot. Garden 72: 636–659.

Roughgarden, J. 1995. Anolis Lizards of the Caribbean: Ecology, Evolution, and Plate Tectonics. Oxford Univ. Press, New York.

Savage, J. M. 1987. Systematics and distribution of the Mexican and Central American rainfrogs of the *Eleutherodactylus gollmeri* group (Amphibia: Leptodactylidae). Fieldiana: Zool. (new ser.) 33: 1–57.

Savin, S. M. and R. G. Douglas. 1985. Sea level, climate, and the Central American Land Bridge, p. 303–324. *In* F. G. Stehli and S. D. Webb (eds.), The Great American Biotic Interchange. Plenum Press, New York.

Schwartz, A. 1968. Geographic variation in the New World gekkonid lizard *Tarentola americana* Gray. Proc. Biol. Soc. Washington 81: 123–142.

—. 1979. The status of Greater Antillean *Phyllodactylus* (Reptilia, Gekkonidae). Jour. Herpetol. 13: 419–426.

— and R.W. Henderson. 1988. West Indian amphibians and reptiles: a checklist. Milwaukee Public Mus. Contrib. Biol. Geol. no. 74.

— and —. 1991. Amphibians and Reptiles of the West Indies: Descriptions, Distributions, and Natural History. Univ. Florida Press, Gainesville.

Scott, N. J., Jr. and S. C. Ayala. 1984. *Tretioscincus bifasciatus*. Herpetol. Rev. 15: 21.

Seidel, M. E. 1988. Revision of the West Indian emydid turtles (Testudines). Amer. Mus. Novit. 2918: 1–41.

—. 1989. *Trachemys dorbigni*. Cat. Amer. Amphib. Rept.: 486.1–486.3.

Sharpton, V. L., K. Burke, A. C.-Zanoguera, S. A. Hall, D. S. Lee, L. E. Marin, G. S.-Reynoso, J. M. Q.-Muñeton, P. D. Spudis, J. U.-Fucugauchi. 1993. Chicxulub multiring impact basin: size and other characteristics derived from gravity analysis. Science 261: 1564–1567.

Shochat, D. and H. C. Dessauer. 1981. Comparative immunological study of albumins of *Anolis* lizards of the Caribbean islands. Comp. Biochem. Physiol. 68A: 67–73.

Simpson, G. G. 1956. Zoogeography of West Indian land mammals. Amer. Mus. Novit. 1759: 1–28.

Smit, J. A. Montanari, N. H. M. Swinburne, W. Alvarez, A. R. Hildebrand, S. V. Margolis, P. Claeys, W. Lowrie, and F. Asaro. 1992. Tektite-bearing, deep-water classic unit at the Cretaceous-Tertiary boundary in northeastern Mexico. Geology 20: 99–103.

Smith, A. G., D. G. Smith, and B. M. Funnell. 1994. Atlas of Mesozoic and Cenozoic Coastlines. Cambridge Univ. Press, Cambridge.

Swisher, C. C., J. M. G. Nishimura, A. Montanari, S. V. Margolis, P. Claeys, W. Alvarez, P. Renne, E. C. Pardo, F. J.-M. R. Maurrasse, G. H. Curtis, J. Smit, and M. O. McWilliams. 1992. Coeval 40AR/39AR ages of 65.0 million years ago from Chicxulub crater melt rock and Cretaceous-Tertiary boundary tektites. Science 257: 954–958.

Thomas, R. 1965*a*. The smaller teiid lizards (*Gymnophthalmus* and *Bachia*) of the southeastern Caribbean. Proc. Biol. Soc. Washington 78: 141–154.

—. 1965*b*. The genus *Leptotyphlops* in the West Indies with description of a new species from Hispaniola (Serpentes, Leptotyphlopidae). Breviora 222: 1–12.

—. 1989. The relationships of Antillean *Typhlops* (Serpentes: Typhlopidae) and the description of three new Hispaniolan species, p. 409–432. *In* C. A. Woods (ed.), Biogeography of the West Indies: Past, Present, and Future. Sandhill Crane Press, Gainesville, Florida.

—, R. W. McDiarmid, and F. G. Thompson. 1985. Three new species of thread snakes (Serpentes: Leptotyphlopidae) from Hispaniola. Proc. Biol. Soc. Washington 98: 204–220.

Tolson, P. J. 1987. Phylogenetics of the boid snake genus

Epicrates and Caribbean vicariance theory. Occas. Pap. Mus. Zool. Univ. Michigan 715: 1–68.

Trueb, L. and M. J. Tyler. 1974. Systematics and evolution of Greater Antillean hylid frogs. Occas. Pap. Mus. Nat. Hist. Univ. Kansas 24: 1–60.

Underwood, G. 1993. A new snake from St Lucia, West Indies. Bull. Nat. Hist. Mus., Zool. 59: 1–9.

Vanzolini, P. E. 1968. Geography of the South American Gekkonidae (Sauria). Arq. Zool., Sao Paulo 17: 85–112.

—. 1978. On South American *Hemidactylus*. Pap. Avul. Zool., Sao Paulo 31: 307–343.

— and C. M. Carvalho. 1991. Two sibling and sympatric species of *Gymnophthalmus* in Roraima, Brasil (Sauria, Teiidae). Pap. Avul. Zool., Sao Paulo 37: 173–226.

Williams, E. E. 1969. The ecology of colonization as seen in the zoogeography of anoline lizards on small islands. Quart. Rev. Biol. 44: 345–389.

—. 1972. The origin of faunas: evolution of lizard congeners in a complex island fauna — a trial analysis. Evol. Biol. 6: 47–89.

—. 1976. West Indian anoles: a taxonomic and evolutionary summary. I. Introduction and a species list. Breviora 440: 1–21.

—. 1983. Ecomorphs, faunas, island size, and diverse end points in islands radiations of *Anolis*, p. 326–370. *In* R. B. Huey, E. R. Pianka, and T. W. Schoener (eds.), Lizard Ecology: Studies of a Model Organism. Harvard Univ. Press, Cambridge, Massachusetts.

—. 1989*a*. Old problems and new opportunities in West Indian biogeography, p. 1–46. *In* C. A. Woods (ed.), Biogeography of the West Indies: Past, Present, and Future. Sandhill Crane Press, Gainesville, Florida.

—. 1989*b*. A critique of Guyer and Savage (1986): cladistic relationships among anoles (Sauria: Iguanidae): are the data available to reclassify the anoles?, p. 433–478. *In* C. A. Woods (ed.), Biogeography of the West Indies: Past, Present, and Future. Sandhill Crane Press, Gainesville, Florida.

Wilson, E. O. 1961. The nature of the taxon cycle in the Melanesian ant fauna. Amer. Nat. 95: 169–193.

Wood, R. C. 1972. A fossil pelomedusid turtle from Puerto Rico. Breviora 392: 1–13.

Woods, C. A. 1989. The biogeography of the West Indian rodents, p. 741–798. *In* C. A. Woods (ed.), Biogeography of the West Indies: Past, Present, and Future. Sandhill Crane Press, Gainesville, Florida.

Wyles, J. S. and G. C. Gorman. 1980. The albumin immunological and Nei electrophoretic distance correlation: a calibration for the saurian genus *Anolis* (Iguanidae). Copeia 1980: 66–71.

Zug, G. R. 1993. Herpetology. Academic Press, San Diego, California.

Resumen.—La herpetofauna antillana comprende unas 175 especies de anfibios (99% endémico) y 457 especies de reptiles (93% endémico). Se analiza información sobre la distribución, relaciones y época de origen, con énfasis en estimados de epocas de divergencia a partir de datos moleculares, en un intento por entender el origen de la herpetofauna. Son identificados sententa y siete linajes independientes de anfibios y reptiles y casi todos (95%) fueron originados en el Nuevo Mundo. En aquellos linajes donde se puede identificar un area fuente dentro del Nuevo Mundo, la mayoria (79%) muestra un origen sudamericano, con contribuciones mas pequenas desde centroamerica (15%) y norteamerica (6%). Un linaje muy viejo y diverso, las ranas del genero *Eleutherodactylus*, fue originado por vicarianza o dispersión en el Cretaceo Tardío (70 mya). Con una posible excepción (el lagarto xantúsiido *Cricosaura typica*), todos los otros linajes parecen haber arrivado por dispersión durante el Cenozoico y todos, exceptuando nueve linajes, en la ultima mitad del Cenozoico (30–0 mya). La mayoría de los linajes antillanos con múltiples especies endémicas fueron originados en el Terciario Medio mientras que la mayoria de los linajes con una sola especie endémica se originaron en el Terciario Tardío. Se postula la dispersión durante El Cuarternario para explicar el origen de las poblaciones antillanas de especies continentales.

La explicación probable del origen predominantemente sudamericano de la herpetofauna antillana, es el patrón casi unidireccional (hacia el oeste-noroeste) de las corrientes marinas en el Caribe: el agua que alcanza las Antillas Mayores se origina cerca de las Antillas Menores y Sur America. La amplia variación en las epocas de origen de los linajes antillianos no sustenta la reciente sugerencia de un puente de tierra durante el Cenozoico Medio entre Sud America y las Antillas Mayores. Se cree que, en la mayoria de los casos, debe haber ocurrido la dispersion por la descarga de organismos viajando en balsas desde las desembocaduras de los ríos mas grandes en el continente y arrastrados por las corrientes hacia las Antillas. La vicarianza de las Proto Antillas queda como una posibilidad geológica, pero datos recientes de las paleolinea de costas sugieren que puede que no halla ocurrido una coneccion de tierra seca entre las Proto Antillas y las tierras continentales vecinas en el Mesozoico Tardío. La extendida invocación de la vicarianza para explicar casi cualquier distribución de organismos en las Antillas sin información acerca de la epoca de origen es visto como una tendencia popular pero carente de sustentación.

Résumé.—La faune herpétologique connue des Antilles comporte 175 espèces d'amphibiens (98% endémiques) et 457 espèces de reptiles (93% endémiques). Les données sur la répartition, les rapports entre espèces, l'origine (notamment à l'aide d'évaluations du temps de divergence au moyen de données immunologiques sur l'albumine) ont été analysées pour essayer de comprendre l'origine de la faune herpétologique. Soixante-dix-sept lignées indépendantes d'amphibiens et de reptiles des Antilles ont été identifiées et presque toutes (95%) proviennent du Nouveau Monde. Lorsque l'origine même de ces lignées au Nouveau Monde peut être déterminée, il s'avère que la plupart (81%) origine d'Amérique du Sud et que les autres, à savoir 15% et 6%, proviennent respectivement d'Amérique Centrale et d'Amérique du Nord. Une lignée très ancienne et diversifiée de grenouilles du genre *Eleutherodactylus* doit son origine à la vicariance ou à la dispersion de la fin du Crétacé (70 millions d'années). À une seule exception près (le Xanthusidé *Cricosaura typica*), toutes les autre lignées sont arrivées par dispersion pendant le Cénozoïque et toutes sauf neuf, à la fin du Cénozoïque (30–0 millions d'années). La plupart des lignées des Antilles comportent plusieurs espèces endémiques dont l'origine remonte au milieu de l'ère tertiaire alors que la plupart des lignées à une seule espèce endémique datent de la fin de l'ère tertiaire. La dispersion du quaternaire est émise comme hypothèse pour expliquer l'origine des populations antillaises d'espèces que l'on retrouve sur le continent.

L'origine majeure d'Amérique du Sud de la faune herpétologique des Antilles est probablement due au sens quasi unidirectionnel (ouest/nord-ouest) des courants marins dans la mer des Caraïbes: l'eau qui baigne les grandes Antilles provient des petites Antilles et d'Amérique du Sud. Les grandes variations au titre de l'origine des lignées antillaises ne permettent pas de confirmer l'hypothèse récente d'un pont continental entre l'Amérique du Sud et les grandes Antilles, vers la moitié du Cénozoïque. La dispersion, dans la plupart des cas, semble s'être produit par l'évacuation d'organismes attachés aux épaves dans les embouchures des principales rivières continentales, transportés ensuite par les courants vers les Antilles. La vicariance proto-antillaise reste une possibilité géologique mais des données paléo-côtières récentes donnent à penser qu'il n'y a pas eu de pont en terre ferme entre les proto-Antilles et les continents voisins à la fin du Mésozoïque. L'invocation de la thèse de la vicariance pour expliquer pratiquement toutes les distributions d'organismes antillais sans données sur leur origine temporelle semble être aussi répandue qu'infondée.

Hedges, S. B. 1996. The origin of West Indian amphibians and reptiles, p. 95–128. *In* R. Powell and R. W. Henderson (eds.), Contributions to West Indian Herpetology: A Tribute to Albert Schwartz. Society for the Study of Amphibians and Reptiles, Ithaca (New York). Contributions to Herpetology, volume 12.

Systematics and Biogeography of Eastern Caribbean *Eleutherodactylus* (Anura: Leptodactylidae): Consensus from a Multidisciplinary Approach

Hinrich Kaiser[1]

Abstract.—An investigation of the systematic relationships among Eastern Caribbean frogs, genus *Eleutherodactylus*, was conducted by characterizing each species using five independently-derived data sets (external morphology, osteology, morphometrics, allozymes, chromosomes). In each case, phenetic and phylogenetic algorithms produced largely congruent hypotheses of relationship that differed only in minor topological details. According to the consensus found, Eastern Caribbean *Eleutherodactylus* have at least two biogeographic origins, with northern species [topology: ((*E. barlagnei*, *E. pinchoni*) (*E. johnstonei* (*E. martinicensis*, *E. amplinympha*)))] forming a monophyletic subgroup in an "Antillean" clade, and with southern species [topology: (*E. urichi* (*E.* cf. *terraebolivaris* (*E. euphronides*, *E. shrevei*)))] belonging to one or more South American clades. *Eleutherodactylus* has biogeographically enveloped the Caribbean, and the diverse radiations from the Greater Antilles and South America meet in the southern Lesser Antilles.

Introduction

The neotropical frog genus *Eleutherodactylus*, with over 500 described species (Duellman 1993) the most speciose vertebrate genus, is an excellent example of an explosive radiation. Although species diversity has grown immensely during this radiation, combinations of phenotypic and genotypic characteristics have evolved or coevolved based on a relatively homogeneous morphotype (e.g., Lynch 1971), but with few apparent constraints. As a result, *Eleutherodactylus* can present highly complex systematic and taxonomic problems (e.g., Hedges 1989; Kaiser et al. 1994*a*), and whereas every year several new species of *Eleutherodactylus* are described, others may be synonymized (e.g., Lynch 1992; Lynch and La Marca 1993). Considering such complexity, it may seem surprising that only a few studies (e.g., Miyamoto 1983) in which balanced character sets derived from a variety of techniques were used towards systematic or biogeographic ends.

The Eastern Caribbean (encompassing the Lesser Antilles, Trinidad, and Tobago) is a biogeographic region where elements of the Greater Antillean biota meet and intermingle with elements from South America (e.g., Koopman 1989; Miller and Miller 1989; Kaiser et al. 1994*a*, 1994*b*, 1994*c* [1995]; Hedges, this volume; Murphy, this volume). The potential biogeographic connection between these biotas has been explored for a variety of organisms (see papers in Woods 1989 and their references), but the data for poor cross-saltwater dispersers such as frogs remain incomplete (see Kaiser et al. 1994*b*). Furthermore, the opposing views of origin for Eastern Caribbean *Eleutherodactylus* presented, for example, by Schwartz (1967) and Lescure (1983) may have indicated that the diverse Greater Antillean and South American radiations of this speciose genus meet somewhere in the Eastern Caribbean.

[1] Institut für Humangenetik, Universität Würzburg, Biozentrum am Hubland, 97074 Würzburg, Germany, *and* Redpath Museum, McGill University, Montréal, Québec, Canada H3A 2K6.

In this study, I used several data sets for Eastern Caribbean *Eleutherodactylus*, in part from previously published work, so that hypotheses of relationships could be tested repeatedly for their robustness. Given these hypotheses, I used the emerging biogeographic implications to take issue with questions about dispersal and species origin of these frogs.

Materials and Methods

Four types of data were obtained largely from literature sources: morphometric data (Kaiser 1993), allozyme data (Kaiser et al. 1994*b*), and data for external morphology and osteology (Kaiser et al. 1994*c* [1995]). Techniques for specimen and data collection as well as specimen deposition were provided in these studies. Data sets were analyzed separately in order to permit sequential and comparative testing of alternately-created hypotheses. Characters and character states used for the morphological and osteological data set are listed in Kaiser et al. (1994*c* [1995]). For their analysis, *Eleutherodactylus antillensis* and *E. coqui* were used as representatives of Greater Antillean taxa, *E.* cf. *terraebolivaris*[1] was used as a representative of South American taxa, and *E. fitzingeri* was used as a possible outgroup. Discriminant scores calculated by multiple discriminant function analysis (MDA: Hair et al. 1992; Kaiser et al. 1994*a*) from ten standard measurements (head width—HW, eye diameter—ED, tympanum diameter—TD, tympanum-eye distance—TE, interorbital distance—IOD, snout length—SL, internarial distance—ID, tympanum-naris distance—TN, snout-vent length—SVL, and tibia length—TL) were available from 1008 *Eleutherodactylus* specimens. Data for individual species (Analysis I) as well as for species clustered by geography (Analysis II) were plotted. From the original data matrix I also calculated Mahalanobis distances between presumptive taxa and clustered these with the UPGMA algorithm (Sneath and Sokal 1973). Discriminant analyses were done on a Macintosh computer using Systat 5.2 software, and the UPGMA phenogram was hand-calculated. Data from starch

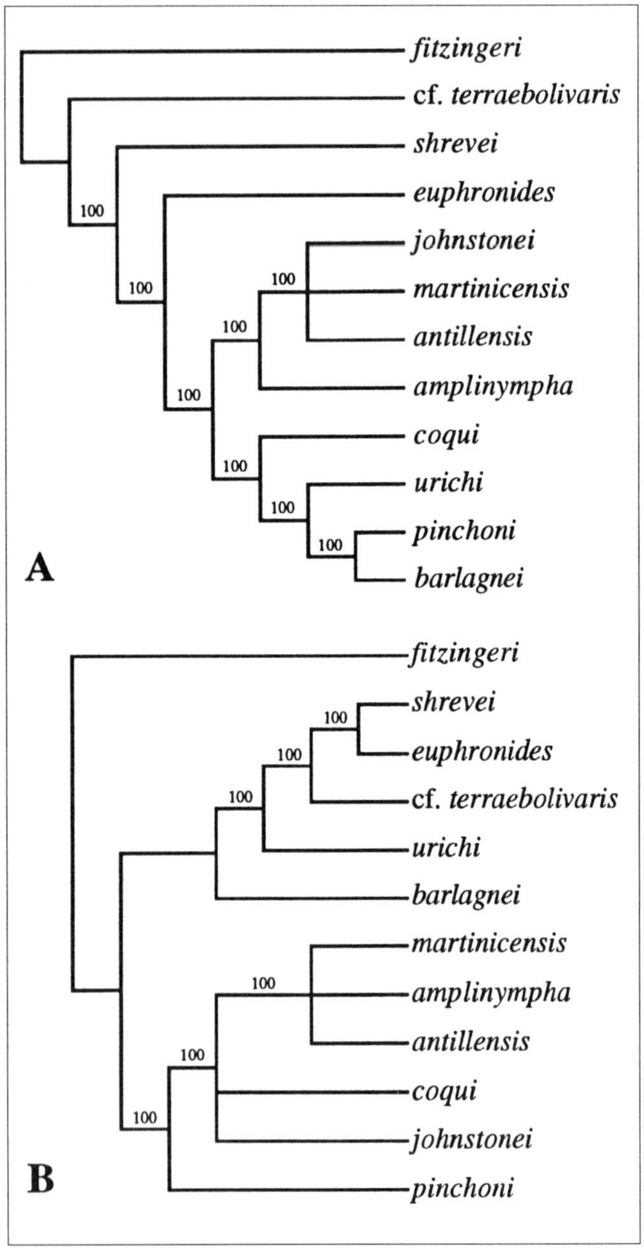

FIGURE 1. Majority-rule consensus trees of (A) 56 external morphological and (B) 86 osteological characters for selected *Eleutherodactylus* from the Eastern Caribbean, the Greater Antilles, and Central America.

[1] *Eleutherodactylus terraebolivaris* was described by Rivero (1961) from Venezuela. Some *Eleutherodactylus* populations on the island of Tobago have hitherto been considered conspecific with *E. terraebolivaris*. However, whereas these frogs are clearly of South American origin, Hardy (1982) indicated that their taxonomy warrants a closer investigation. Pending the results from such a study, I opted for a conservative approach in calling the problematic Tobagonian frogs *E.* cf. *terraebolivaris*.

gel electrophoresis were available for 22 allozyme loci (n = 277 specimens; Kaiser et al. 1994*b*), and the allele frequency values were used in phenetic and cladistic algorithms to produce relationship hypotheses. The fifth data set consisted of basic chromosome data obtained directly through the methodology of Schmid (1978), or from the works of DeWeese (1976), Bogart (1991), and Kuramoto (1991). At least five representative metaphases were obtained and counted for males and females of each species. Phylogenetic analysis was separately conducted for discrete data (external morphology, osteology, allozymes) using unweighted and unordered data matrices in PAUP 3.1.1 (Swofford 1993). Throughout, ACCTRAN optimization was used, and searches were conducted with options as used by Kaiser et al. (1994*b*) and bootstrapping (1000 repetitions). Characters that were not analyzed cladistically (morphometric and chromosome data) were subsequently mapped onto the most parsimonious topologies, with exploratory branch swapping conducted manually using MacClade 3.01 (Maddison and Maddison 1993).

RESULTS

External Morphology

Phylogenetic analysis produced three most parsimonious trees of length 106 steps with a consistency index (CI) of 0.509. The majority-rule consensus tree (Figure 1A) was fully resolved except for the trichotomy of *Eleutherodactylus antillensis*, *E. johnstonei*, and *E. martinicensis*. This tree contained one major clade with two subclades, one containing the just-mentioned trichotomy and *E. amplinympha*, the other the remaining northern Eastern Caribbean species and *E. urichi*. Only one sister group relationship was apparent, for *E. barlagnei* and *E. pinchoni*. The remaining three taxa nested further out from the major clade towards the outgroup in the sequence *E. euphronides*, *E. shrevei*, and *E.* cf. *terraebolivaris*.

Osteology

Analysis of osteological data resulted in two most parsimonious trees of length 184 steps with a CI of 0.462. The majority-rule consensus tree (Figure 1B) consisted of a single major clade comprising two subclades. One subclade contained all northern Eastern Caribbean species with the exception of *Eleutherodactylus barlagnei*. The overall arrangement is equivocal, however, as indicated by the presence of two polytomies, one with *E. antillensis*, *E. martinicensis*, and *E. amplinympha*, the other incorporating that polytomy, plus *E. coqui* and *E. johnstonei*. The second subclade grouped all southern species, with *E. barlagnei* nested on the outside of that clade. Again only one sister group relationship was resolved, here for *E. euphronides* and *E. shrevei*.

Morphometrics

The MDA for all species (Analysis I) showed that both size and shape were instrumental in characterizing Eastern Caribbean *Eleutherodactylus*. Loadings for discriminant function (DF) 1 were all positive and for DF3 all negative (Table 1); such sign homogeneity identified these as size-determined loadings. The loadings with the greatest discrimi-

TABLE 1. Discriminant loadings from two multiple discriminant function analyses of ten length measurements from Eastern Caribbean *Eleutherodactylus*. Characters with the relatively greatest discriminating power for each discriminant function (DF) are marked with asterisks (*). Cutoff values were assigned arbitrarily at 0.400 (DF1), + or – 0.100 (DF2), and –0.400 (DF3) in Analysis I, and at 0.400 (DF1) and 0.200 (DF2) in Analysis II. Abbreviations of measurements are explained in the text (facing page).

Measurement	Analysis I			Analysis II	
	DF 1	DF 2	DF 3	DF 1	DF 2
log HW	0.342	0.020	–0.544*	0.281	0.079
log ED	0.485*	0.062	–0.441*	0.432*	–0.082
log EN	0.273	–0.296*	–0.340	0.259	0.237*
log TD	0.261	0.385*	–0.230	0.220	–0.037
log TE	0.253	0.204*	–0.486*	0.185	0.137
log IOD	0.451*	–0.019	–0.324	0.416*	0.173
log SL	0.388	0.185*	–0.419*	0.319	0.128
log ID	0.481*	0.077	–0.418*	0.417*	–0.106
log SVL	0.305	0.008	–0.354	0.268	0.267*
log TIB	0.560*	–0.013	–0.344	0.507*	0.159

nating power (ranked in order from greatest to smallest contribution) were TIB, ED, ID, and IOD for DF1, and HW, TE, ED, SL, and ID for DF3 (Table 1). Ranked as before, the variables with the greatest discriminating power in DF2, indicative of shape, were TD, EN, TE, and SL (Table 1). Species assignments based on individual posterior probabilities showed that the classification using the calculated discriminant function was 65.3% correct over all 983 classified specimens (Table 2). Specimens of *E. martinicensis* (35.0%) were the least well classified (Table 2), with all other specimens classified correctly an average of 71.1% of the time. Much of the poor classification for specimens of *E. martinicensis* was due to incorrect grouping with *E. johnstonei* and *E. amplinympha* (both 22.3%; Table 2). The converse effect was also observed: 14.1% of *E. johnstonei* specimens and 17.2% of *E. amplinympha* specimens were statistically misaligned with *E. martinicensis* (Table 2).

Discriminant score (DS) plots of DS1 against DS2 (Figure 2) showed that differences of size (DS1) and shape (DS2) existed between species, but that graphic representation was insufficient to clearly separate species or species groups. The southern species *Eleutherodactylus euphronides*, *E. shrevei*, and *E.* cf. *terraebolivaris* were well separated from northern species along DS1, whereas *E. urichi* placed in a position between northern and southern taxa (Figure 2A). The nine taxa could be divided arbitrarily into five clusters based on their morphometric characteristics and their geographic positions and closer to northern taxa (Figure 2A). The two extreme positions were occupied by *E.* cf. *terraebolivaris* at the maximum values of DS1, and by *E. barlagnei* at the highest values of DS2. The other taxa lay in intermediate positions with respect to DS values. With increasing distance from the position of *E.* cf. *terraebolivaris* these were the clusters of: (1) *E. shrevei* and *E. euphronides*; (2) *E. urichi* by itself; and (3) *E. amplinympha*, *E. johnstonei*, *E. martinicensis,* and *E. pinchoni* (Figure 2A).

An MDA of two sets of species (Analysis II) resulted in loadings similar to those for the analysis of all species: DF1 was a size-determined function, whereas DF2 was shape-determined. The greatest discriminating power in DF1, ranked as above, were made again by TIB and ED, followed by ID and IOD (Table 1). The main contributions to shape (DF2) were very different when comparing species sets; the only outstanding contributions were made by SVL and EN (Table 1). Group assignments (Table 3)

TABLE 2. Group assignments for 983 specimens of *Eleutherodactylus* from the Eastern Caribbean from a multiple discriminant function analysis (MDA) of 10 metric characters. Rows are MDA predictions, columns are actual population groupings (i.e., of 157 MART specimens, 55 are correctly classified by the MDA, whereas 35 are incorrectly classified statistically as JHN). Differences between MDA groupings tested significant at $P \leq 0.001$ (Pearson chi-square). Species codes are *E. martinicensis* (MART), *E. johnstonei* (JHN), *E. barlagnei* (BAR), *E. euphronides* (EUP), *E. shrevei* (SHR), *E. urichi* (URI), *E. pinchoni* (PIN), *E.* cf. *terraebolivaris* (TER), and *E. amplinympha* (AMP). Percent classification success >10% is given in parentheses.

Group	MART	JHN	BAR	EUP	SHR	URI	PIN	TER	AMP	Total
MART	55 (35.0)	35 (22.3)	9	4	5	8	6	0	35 (22.3)	157
JHN	73 (14.1)	359 (69.4)	10	0	2	6	41	0	26	517
BAR	1	1	31 (91.2)	0	0	1	0	0	0	34
EUP	3	0	0	56 (66.7)	19 (22.6)	2	0	3	1	84
SHR	1	0	0	6 (14.3)	31 (73.8)	1	0	3	0	42
URI	1	1	1	0	1	18 (72.0)	2	1	0	25
PIN	1	2	1	0	0	0	25 (78.1)	0	0	29
TER	0	0	0	2	2	1	0	33 (86.8)	0	38
AMP	10 (17.2)	2	2	1	1	3	4	0	34 (59.7)	57
Total	145	400	54	69	61	40	78	40	96	983 (65.3)

from individual discriminant probabilities strongly (90.6%) supported the presence of two distinctive species groups: a northern group consisting of *Eleutherodactylus amplinympha*, *E. barlagnei*, *E. johnstonei*, *E. martinicensis*, *E. pinchoni*; and a southern group, consisting of *E. euphronides*, *E. shrevei*, and *E. urichi*, more closely aligned with the South American *E.* cf. *terraebolivaris* than with any northern species. Northern specimens were grouped correctly 93.7% of the time, whereas southern specimens were assigned correctly in 87.4% of the cases. A discriminant score plot of DS1 against DS2 (Figure 2B) showed that species groups clearly could be separated along the size axis (DS1) but only partially along the shape axis (DS2). In both analyses, discrimination based on size (DS1) was more successful than that based on shape (DS2).

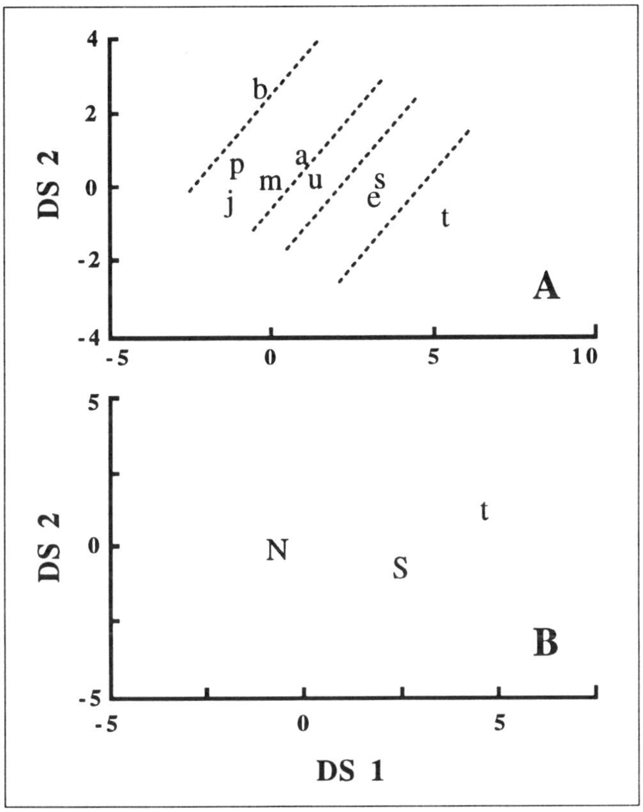

FIGURE 2. Discriminant score (DS) plots from a multiple discriminant function analysis (MDA) of ten measurements from 1008 Eastern Caribbean *Eleutherodactylus*: (A) Plot of scores from Analysis I (see text). Dashed lines are arbitrary parallel dividers of morphometric groupings. (B) Plot of scores from Analysis II (see text). The horizontal axis (DS1) reflects size characteristics, whereas the vertical axis (DS2) distinguishes mainly by shape. In both analyses, size is of greater importance to discriminating these taxa than is shape. Letters denote the centers of score distribution for northern species (N), southern species (S), *E. amplinympha* (a), *E. barlagnei* (b), *E. euphronides* (e), *E. johnstonei* (j), *E. martinicensis* (m), *E. pinchoni* (p), *E. shrevei* (s), *E.* cf. *terraebolivaris* (t), and *E. urichi* (u).

TABLE 3. Group assignments for 1002 specimens of northern and southern Eastern Caribbean *Eleutherodactylus* from a multiple discriminant function analysis (MDA) of 10 metric characters. Rows are MDA predictions, columns are actual population groupings (i.e., of 813 N specimens, 762 are correctly classified by the MDA, while 50 are incorrectly classified statistically as S). Differences between groupings tested significant at $P \leq 0.001$ (Pearson chi-square). Species codes are northern species (N), southern species (S), and *E.* cf. *terraebolivaris* (TER). Percent classification success is given in parentheses.

Group	N	S	TER	Total
N	762 (93.7)	50	1	813
S	9	132 (87.4)	10	151
TER	0	5	33 (86.8)	38
Total	771	187	44	1002 (90.6)

The phenetic clustering of the Mahalanobis distances resembled the above arrangement closely (Figure 3). Two major clusters were identified, one composed of the southern Eastern Caribbean taxa *Eleutherodactylus euphronides* and *E. shrevei*, as well as *E.* cf. *terraebolivaris*. The second cluster included all of the northern Eastern Caribbean taxa, but also *E. urichi*. The clusters of *E. euphronides* with *E. shrevei*, and of *E. amplinympha* with *E. martinicensis* showed the highest degrees of similarity between taxa.

Allozymes

Of 22 resolved loci only one (GTDH) was monomorphic. Clusters in the UPGMA tree (Figure 4A) showed that the *Eleutherodactylus* species from the Eastern Caribbean assorted into a northern and a southern group. The five northern species grouped into two clusters, with the following topologies: (*E. barlagnei*, *E. pinchoni*) and (*E. johnstonei* (*E. martinicensis*, *E. amplinympha*)). In the southern group, *E. euphronides* and *E. shrevei* clustered together, with *E.* cf. *terraebolivaris* clustered more distantly to that group (Figure 4A). *Eleutherodactylus urichi* was situated even farther distant from the cluster with these three species at a genetic distance >1, making it inappropriate to deduce relationships. Three Puerto Rican species placed more closely to the northern Eastern Caribbean species than to the southern species; and the Hispaniolan representative in this analysis placed most closely to the northern Eastern Caribbean group.

Two major clades were identified by the cladistic analysis of allozyme data, one containing northern Eastern Caribbean taxa, the other containing those with presumed South American affinities (Figure 4B). Majority-rule values were quite high (90–100%), but the evidence from bootstrapping was less strong (50–74%). Nevertheless, monophyly of northern taxa and the topology of (*Eleutherodactylus johnstonei* (*E. martinicensis*, *E. amplinympha*)) were supported in a clear majority of analyses. In the southern clade, the triad of ((*E. euphronides*, *E. shrevei*) *E.* cf. *terraebolivaris*) was not unequivocally supported, making the arrangement somewhat tenuous.

Chromosomes

The northern Eastern Caribbean species of *Eleutherodactylus* (*E. amplinympha*, *E. barlagnei*, *E. johnstonei*, *E. martinicensis*, *E. pinchoni*) can be grouped together based on a common diploid chromosome number of $2n = 28$, whereas southern species (*E. euphronides*, *E. shrevei*, *E. urichi*) have $2n = 32$, and *E.* cf. *terraebolivaris* has $2n = 36$. *Eleutherodactylus fitzingeri* has a chromosome number of $2n = 22$. Other cytogenetic characteristics, such as the position of nucleolar organizer regions or the amount of constitutive heterochromatin, are highly similar among the northern species. These characteristics differ significantly between northern and southern species, as well as among the southern species. The karyotypes of *E. euphronides* and *E. shrevei* are very similar, but show considerable differences compared to those of *E. urichi* and *E.* cf. *terraebolivaris*.

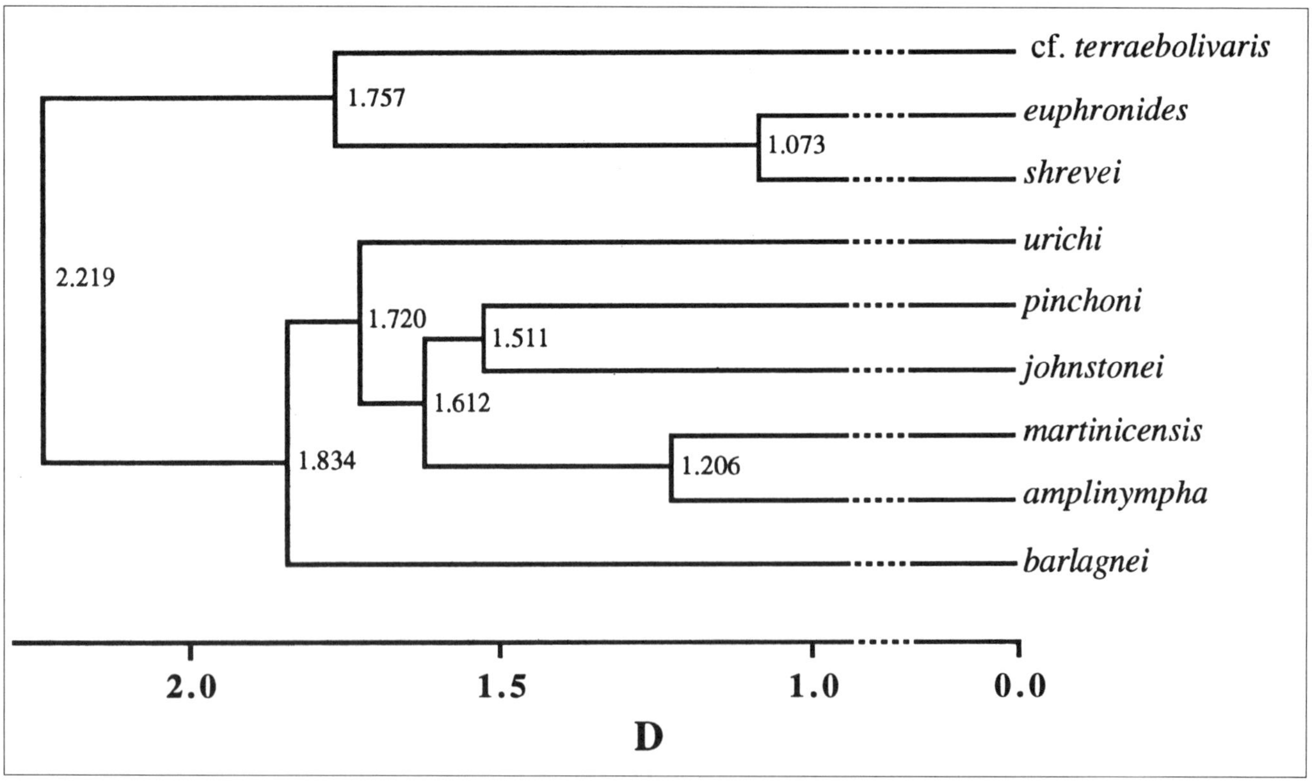

FIGURE 3. UPGMA phenogram of Mahalanobis distances from a multiple discriminant function analysis (MDA) of ten measurements from 1008 Eastern Caribbean *Eleutherodactylus*.

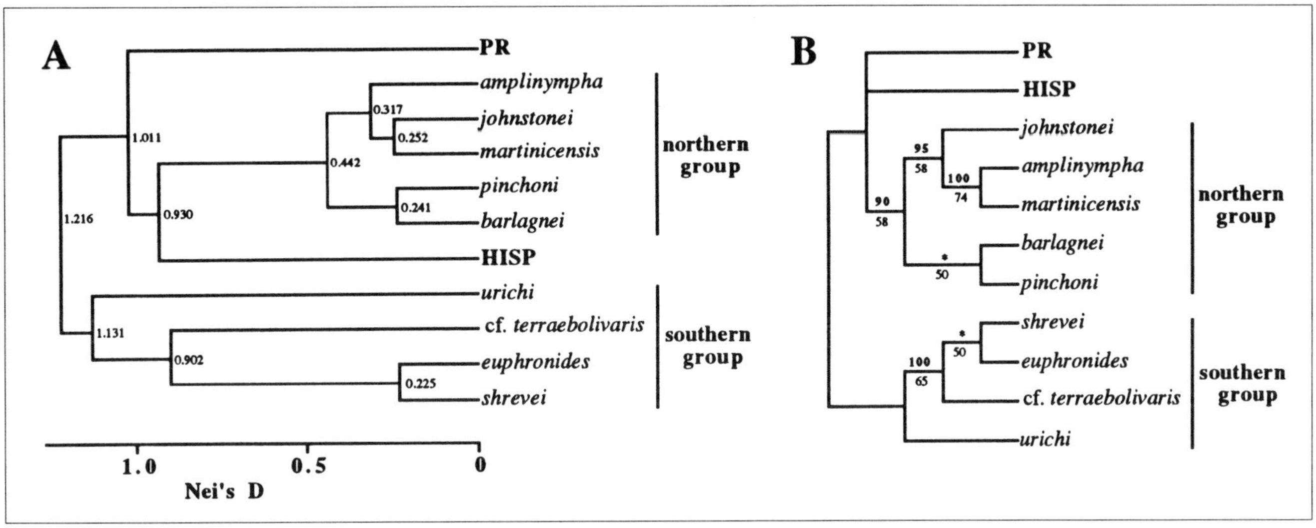

FIGURE 4. Results from an analysis of allozyme data for 277 Eastern Caribbean, Puerto Rican (PR), and Hispaniolan (HISP) *Eleutherodactylus* (figures modified from Kaiser et al. 1994*b*): (A) UPGMA phenogram of Nei's genetic distances. Relationships at D > 1 are not conclusive. (B) Majority-rule consensus tree of a phylogenetic analysis using the same allozyme data as in (A). Boldface numbers are majority-rule values, plain numbers are bootstrapping values for 1000 repetitions. Asterisks (*) denote polytomies in the majority-rule results.

DISCUSSION

General Comments

The above results show individually and in consensus that Eastern Caribbean *Eleutherodactylus* have evolved from two or more ancestral lineages and that these lineages have different geographic origins. They also show that the sequential analysis of several data sets may allow the resolution of some systematic problems that might not have been resolved using only one specific data set, one specific type of analysis, or even a single data matrix derived from several data sets. Whereas some groupings were supported repeatedly in several branching arrangements, some were only produced once. Such rare patterns are of importance to detailed *post hoc* checks of character transformations. They may have remained obscured in studies using fewer data sets or a consensus data matrix.

Phylogenetic Hypotheses

The topology of the majority-rule consensus tree of external morphological data (Figure 1A) seems to be the most unreliable hypothesis for several reasons. Very few of the arrangements are reproduced in subsequent analyses, many of the characteristics included in the data set are known to have high probabilities of homoplasy, and their polarities have not been resolved (see Kaiser et al. 1994*c* [1995]). The arrangement found using osteological data produced two distinct clades, and only *Eleutherodactylus barlagnei* placed with the southern taxa, outside its geographic range (Figure 1B). The placement of *E. euphronides* and *E. shrevei* as sister taxa was supported by all subsequent analyses and is by far the most reliable grouping found in this study. Relationships among northern taxa were poorly resolved. Allozyme data provided the identical arrangement for the southern group, although *E. barlagnei* placed with the northern species in this analysis. Greater and Lesser Antillean taxa formed an unresolved trichotomy.

Phenetic Hypotheses

The phenetic analyses provided some solid evidence for the presence of two different major groups. The discriminant score plots (Figure 2) show that differentiation at the species level can be difficult using morphometric data, although the difference between species groups was pronounced. In both types of analysis, more effective discrimination of taxa was achieved along the size-dependent axis (DS1) rather than the shape-dependent axis (DS2). This reflects the "common" morphotype from which the present *Eleutherodactylus*-diversity likely was derived, and stresses the potential difficulties in conducting cursory morphological investigations of these frogs. The UPGMA phenogram of Mahalanobis distances (Figure 3) shows that a very close morphometric relationship exists between *E. martinicensis* and *E. amplinympha* in addition to *E. euphronides* and *E. shrevei*, but the placement of *E. barlagnei* and *E. urichi* is incongruent with any of the other analyses. The phenogram calculated from genetic distances (Figure 4A) supports some of the arrangements from morphometric data, although with some rearrangements. That the distances calculated using allozyme data created a "crown" of taxa more distant from the base of the phenogram than did morphometric distances is noteworthy.

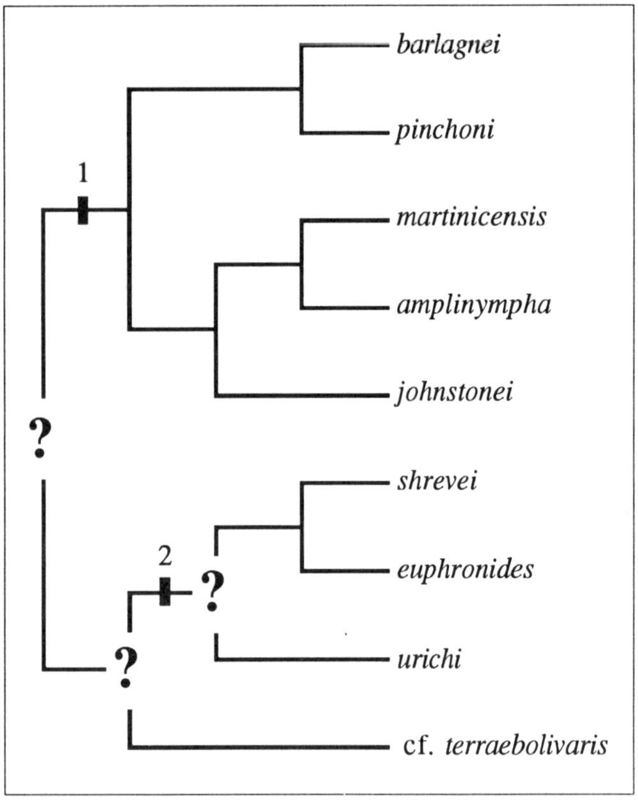

FIGURE 5. Hypothesis of relationships for Eastern Caribbean *Eleutherodactylus* from a consensus of five independently derived data sets. This is the tree with the highest congruence between data sets. Two distict groups can be identified, one containing a monophyletic northern species assemblage, the other containing the four southern Eastern Caribbean taxa. Relationships of the latter with each other and with the northern Eastern Caribbean taxa are as yet uncertain because of a lack of presumably South American ingroup taxa (indicated by "?"). Character state 1 denotes the acquisition of a chromosome number of $2n = 28$, state 2 the acquisition of a $2n = 32$ karyotype.

Chromosomal Information

Considering the karyotypes of some other Antillean *Eleutherodactylus* (e.g., Bogart 1970, 1981, 1991; Bogart and Hedges 1994), it is unlikely that the common diploid number of northern Eastern Caribbean taxa was derived convergently. Assumption of a single common derivation is most parsimonious, particularly in view of the remote yet clustered geographic position of these species. Although the karyotypes of *E. euphronides* and *E. shrevei* are exceedingly similar, their chromosome morphology is very distinct from that of *E. urichi*, despite an identical chromosome number ($2n = 32$). The karyotype of *E.* cf. *terraebolivaris* contains $2n = 36$ telocentric chromosomes. Among southern Eastern Caribbean taxa, no affinities beyond the relationship of *E. euphronides* with *E. shrevei* can be deduced.

Consensus

Based on the arrangements from the different analyses, a single hypothesis of relationships emerges (Figure 5). The monophyly of the five northern Eastern Caribbean taxa is supported by a variety of characteristics, among which vocalization patterns (multi-note calls) and chromosome number ($2n = 28$) stand out as synapomorphies. That the northern Lesser Antillean group is not lacking any ingroup taxa because of recent surveys is very unlikely (Kaiser 1993; Kaiser and Henderson 1994), and therefore this topology should remain stable. This is the *martinicensis* group (*martinicensis* series, *auriculatus* group) of Hedges (1989), with the addition of the recently described *Eleutherodactylus amplinympha*. The topologies for the southern Eastern Caribbean species included in the preceding analyses are not stable, with the exception of a very close relationship of *E. euphronides* and *E. shrevei*. Although phenetic analyses produced clusters for these four taxa, albeit including large morphometric and genetic distances, chromosome morphology and several other characteristics are not supportive of a close relationship of *E. urichi* with the other two taxa. Likewise, the position of *E.* cf. *terraebolivaris* as part of a southern clade is very tenuous, and its placement requires further analysis with additional South American taxa. As a consequence, to suggest possible South American species group affinities would be premature.

Biogeography

The above consensus tree (Figure 5) is indicative of two main centers of *Eleutherodactylus*-endemism (Figure 6), one in the central Lesser Antilles (Dominica, Guadeloupe and its archipelago of smaller islands, and Martinique), the other in close proximity to South America at the southern end of the volcanic arc (Grenada, St. Vincent, Tobago, and Trinidad). Islands peripheral to these regions have received their *Eleutherodactylus* populations most likely as the result of chance dispersal (e.g., via hurricane debris, pumice floats, human-mitigated introductions), and some of these instances have been documented (e.g., Censky 1989; Henderson et al. 1992; Kaiser 1992). The geological history of Trinidad and Tobago, whose mountain ranges form an extension of the Península de Paria of Venezuela, indicates close ties to a northern South American fauna (e.g., Mertens 1969; Hardy 1983; Murphy, this volume). This fauna is represented by several frogs of the genera *Hyla* and *Leptodactylus*, which are widespread in South America, and a species of glass frog, *Hyalinobatrachium orientalis*, exists on Tobago. In comparison with the Lesser Antilles, the area of northern South America that includes Trinidad and Tobago, has a relatively poor species diversity in *Eleutherodactylus*. Although this may be an artifact of taxonomy, it may nonetheless explain why *E. urichi* is the only *Eleutherodactylus* species endemic to Trinidad and Tobago.

The most probable biogeographic hypothesis links the southern Eastern Caribbean islands to the South American mainland by the tremendous freshwater discharge from the Orinoco delta (Gade 1961). The force of running water frequently detaches vegetation floats from the river's banks and carries them out into the Atlantic currents. In this region, prevalent winds and currents can direct such a temporary island into the Caribbean Sea. A highly speculative series of chance events might thus trap a small population of frogs on a floating island, transport this tiny biota out to sea, and set it adrift to be swept past one of the southern Lesser Antillean islands. However, to assume that safe passage results in a safe landing would be incorrect, particularly since the windward coasts of the Lesser Antilles are rugged and inundated with saltwater spray. A further requirement for a successful colonization event would be that the float drifts just past an island and gets caught in a back eddy current which deposits it

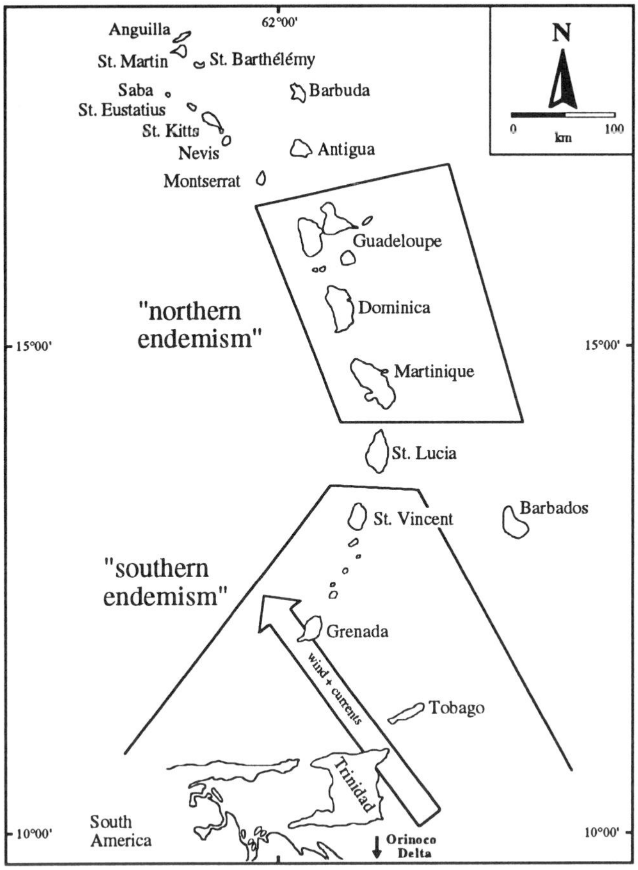

FIGURE 6. Map of the Eastern Caribbean indicating two centers of endemism, one for the northern Eastern Caribbean taxa, the other for the southern taxa. The southern center of endemism is not closed because biogeographic and systematic relationshiops are not yet fully resolved. The prevalent wind and current direction are indicated. A hypothesis that winds and curents are able to transport "habitat arks" into the south ern Eastern Caribbean seems reasonable, and as a result, dispersal from South America (e.g., the Orinoco River delta) is one of the most important biogeographic mechanisms.

near an outflow of freshwater to permit the crossing of frogs. Although this example in its entirety is conjecture, some anecdotal evidence exists which confirms at least individual parts (e.g., Hardy 1983).

Conclusion

I cannot claim to have completely solved the systematic and biogeographic puzzle presented by Eastern Caribbean *Eleutherodactylus*. What I have presented is an accumulation of research and thought, begun in the 1960s by the Lesser Antillean work of Albert Schwartz, which now points towards a very likely hypothesis. More importantly, I have tested a methodology that is more diverse and more comprehensive than ever before used on *Eleutherodactylus* and which may become very useful where single data sets or analyses fail to resolve relationships.

Acknowledgements

This study is derived in part from my dissertation work at McGill University, Montréal. For constructive criticisms and discussions at every stage of the way, I thank G. Bell, M. W. Caldwell, R. L. Carroll, R. L. Holmes, D. L. Kramer, L. A. Lowcock, J. Rasmussen, H. M. Reiswig, D. A. Roff, T. F. Sharbel, K. Sittmann, and most especially, D. M. Green. During this endeavour, my studies were supported by scholarships and grants from the Hans-Krüger-Stiftung and the Studienstiftung des Deutschen Volkes, the Metcalfe Foundation, and the Faculty of Science at McGill University. Their support is gratefully acknowledged. I am indebted to my field assistants, C. A. George, H. M. Gray, G. Schäfer, T. F. Sharbel, H. H. Schwarten, D. von Stösser, and A. Werres, whose tireless good spirits in the search for tiny frogs made all the difference. Heartfelt thanks to C. M. Dwyer, who made many useful comments on this manuscript throughout its development and, with calm composure, endured several early versions of the presentation from which it results. Special thanks to R. W. Henderson and R. Powell for inviting me to the Schwartz symposium, and to M. Schmid and the experimental cytogenetics team for making chromosome work so enjoyable. During completion of this research, I was supported by in-house funding at the Institut für Humangenetik, Universität Würzburg, and by a postdoctoral fellowship from Boehringer Ingelheim Fonds.

Literature Cited

Bogart, J. P. 1970. Los cromosomas de anfibios del genero *Eleutherodactylus*. Actas IV Congreso Latinoamer. Zool. 1: 65–78.

—. 1981. Chromosome studies in *Sminthillus* from Cuba and *Eleutherodactylus* from Cuba and Puerto Rico (Anura: Leptodactylidae). Life Sci. Contr., Royal Ontario Mus. 129: 1–22.

—. 1991. The influence of life history on karyotypic evolution in frogs, p. 233–258. *In* D. M. Green and S. K. Sessions (eds.), Amphibian Cytogenetics and Evolution. Academic Press, San Diego, California.

— and S. B. Hedges. 1995. Rapid chromosome evolution in Jamaican frogs of the genus *Eleutherodactylus* (Leptodactylidae). Jour. Zool. (London) 235: 9–31.

Censky, E. J. 1989. *Eleutherodactylus johnstonei* (Salientia: Leptodactylidae) from Anguilla, West Indies. Carib. Jour. Sci. 25: 229–230.

DeWeese, J. E. 1976. The karyotypes of Middle American frogs of the genus *Eleutherodactylus* (Anura: Leptodactylidae): a case study of the significance of the karyologic method. Doctoral dissertation, Univ. Southern California, Los Angeles.

Duellman, W. E. 1993. Amphibian species of the world: additions and corrections. Univ. Kansas Mus. Nat. Hist. Spec. Publ. 21: 1–372.

Gade, H. G. 1961. On some oceanographic observations in the southeastern Caribbean Sea and the adjacent Atlantic Ocean with special reference to the influence of the Orinoco River. Bol. Inst. Oceanográfico, Univ. Oriente (Venezuela) 1: 287–342.

Hair, J. F., Jr., R. E. Anderson, R. L. Tatham, and W. C. Black. 1992. Multivariate Data Analysis. 3rd ed. Macmillan, New York.

Hardy, J. D., Jr. 1982. Biogeography of Tobago, West Indies, with special reference to amphibians and reptiles: a review. Bull. Maryland Herpetol. Soc. 18: 37–142.

—. 1983. Tobago—the forgotten island. Nat. Mag. 4: 8–22, 41–43, 50.

Hedges, S. B. 1989. Evolution and biogeography of West Indian frogs of the genus *Eleutherodactylus*: slow-evolving loci and the major groups, p. 305–370. *In* C. A. Woods (ed.), Biogeography of the West Indies Past, Present, and Future. Sandhill Crane Press, Gainesville, Florida.

Henderson, R. W., J. Daudin, G. T. Haas, and T. J. McCarthy. 1992. Significant distribution records for some amphibians and reptiles in the Lesser Antilles. Carib. Jour. Sci. 28: 101–103.

Kaiser, H. 1992. The trade-mediated introduction of *Eleutherodactylus martinicensis* (Anura: Leptodactylidae) on St. Barthélémy, French Antilles, and its implications for Lesser Antillean biogeography. Jour. Herpetol. 26: 264–273.

—. 1993. Systematics and biogeography of Eastern Caribbean frogs. Doctoral dissertation, McGill Univ., Montréal.

— and R. W. Henderson. 1994 (1995). The conservation status of Lesser Antillean frogs. Herpetol. Nat. Hist. 2: 41–56.

—, J. D. Hardy, Jr., and D. M. Green. 1994*a*. The taxonomic status of Caribbean and South American frogs currently ascribed to *Eleutherodactylus urichi* (Anura: Leptodactylidae). Copeia 1994: 780–796.

—, T. F. Sharbel, and D. M. Green. 1994*b*. Systematics and biogeography of Eastern Caribbean *Eleutherodactylus* (Anura: Leptodactylidae): evidence from allozymes. Amphibia-Reptilia 15: 375–394.

—, D. M. Green, and M. Schmid. 1994*c* [1995]. Systematics and biogeography of Eastern Caribbean frogs of the genus *Eleutherodactylus* (Anura: Leptodactylidae) with the description of a new species from Dominica. Canadian Jour. Zool. 72: 2217–2237.

Koopman, K. F. 1989. A review and analysis of the bats of the West Indies, p. 635–644. *In* C. A. Woods (ed.), Biogeography of the West Indies Past, Present, and Future. Sandhill Crane Press, Gainesville, Florida.

Kuramoto, M. 1990. A list of chromosome numbers of anuran amphibians. Bull. Fukuoka Univ. Educ. 39: 83–127.

Lescure, J. 1983. Introductions passives et actives des reptiles et d'amphibiens dans les Antilles et les Guyanes. Comp. Rend. Soc. Biogéogr. 59: 59–70.

Lynch, J. D. 1971. Evolutionary relationships, osteology, and zoogeography of leptodactyloid frogs. Univ. Kansas Mus. Nat. Hist. Misc. Publ. 53: 1–238.

—. 1992. Two new species of *Eleutherodactylus* from southwestern Colombia and the proposal of a new species group (Amphibia: Leptodactylidae). Jour. Herpetol. 26: 53–59.

— and E. La Marca. 1993. Synonymy and variation in *Eleutherodactylus bicumulus* (Peters) from northern Venezuela, with a description of a new species (Amphibia: Leptodactylidae). Carib. Jour. Sci. 29: 133–146.

MacArthur, R. H. and E. O. Wilson. 1967. The theory of island biogeography. Monogr. Pop. Biol. 1: 1–203.

Maddison, W. P. and D. R. Maddison. 1992. MacClade: Analysis of Phylogeny and Character Evolution. Version 3.01. Sinauer Assoc., Sunderland, Massachusetts.

Mertens, R. 1969. Herpetologische Beobachtungen auf der Insel Tobago. Salamandra 5: 63–70.

Miller, L. D. and J. Y. Miller. 1989. The biogeography of West Indian butterflies (Lepidoptera: Papilionoidea, Hesperioidea): a vicariance model, p. 229–262. *In* C. A. Woods (ed.), Biogeography of the West Indies Past, Present, and Future. Sandhill Crane Press, Gainesville, Florida.

Miyamoto, M. M. 1983. Frogs of the *Eleutherodactylus rugulosus* group: a cladistic study of allozyme, morphological and karyological data. Syst. Zool. 32: 109–124.

Rivero, J. A. 1961. Salientia of Venezuela. Bull. Mus. Comp. Zool., Harvard Univ. 126: 1–207.

Schmid, M. 1978. Chromosome banding in Amphibia. I. Constitutive heterochromatin and nucleolus organizer regions in *Bufo* and *Hyla*. Chromosoma 66: 361–388.

Schwartz, A. 1967. Frogs of the genus *Eleutherodactylus* in the Lesser Antilles. Stud. Fauna Curaçao Carib. Isl. 24: 1–62.

Sneath, P. H. A. and R. R. Sokal. 1973. Numerical Taxonomy. W. H. Freeman, San Francisco, California.

Swofford, D. L. 1993. PAUP: Phylogenetic Analysis using Parsimony, Version 3.1.1. Computer program distributed by the Illinois Natural History Survey, Champaign.

Woods, C. A. (ed.). 1989. Biogeography of the West Indies Past, Present, and Future. Sandhill Crane Press, Gainesville, Florida.

Resumen.—Una investigación de las relaciones sistemáticas entre las ranas del género *Eleutherodactylus* del Caribe Oriental fue conducida caracterizando cada especie usando cinco conjuntos de datos derivados independientemente (morfología externa, osteología, morfometría, alozimas, cromosomas). En cada caso, algoritmos fenéticos y filogenéticos produjeron hipótesis de relaciones grandemente congruentes que difirieron solamente en detalles topológicos menores. De acuerdo al consenso encontrado, los *Eleutherodactylus* del Caribe Oriental tienen al menos dos orígenes biogeográficos, con especies septentrionales [topología: ((*E. barlagnei*, *E. pinchoni*) (*E. johnstonei* (*E. martinicensis*, *E. amplinympha*)))] formando un subgrupo monofilético en un linaje "antillano" y con especies meridionales [topología: (*E. urichi*) (*E.* cf. *terraebolivaris* (*E. euphronides*, *E. shrevei*)))] que pertenecen a uno o más linajes suramericanos. El género *Eleutherodactylus* ha cubierto biogeográficamente el Caribe, y las diversas radiaciones de las Antillas Mayores se encuentran con las de América del Sur en las Antillas Menores.

Résumé.—Une étude des relations systématiques entre les grenouilles du genre *Eleutherodactylus* dans l'Est des Antilles a été entreprise en caractérisant chaque espèce à l'aide de cinq ensembles de données dérivées séparément (morphologie extérieure, ostéologie, morphométrie, allozymes et chromosomes). Dans chaque cas, des algorithmes phénétiques et phylogénétiques ont produit des hypothèses largement congruentes qui ne diffèrent que sur des détails topologiques mineurs. Selon le consensus obtenu, l'*Eleutherodactylus* de l'Est des Antilles possède au moins deux origines biogéographiques, les espèces du nord [topologie: ((*E. barlagnei*, *E. pinchoni*) (*E. johnstonei* (*E. martinicensis*, *E. amplinympha*)))] formant un sous-groupe monophylétique dans un clade "antillais" et les espèces du sud [topologie: (*E. urichi* (*E.* cf. *terraebolivaris* (*E. euphronides*, *E. shrevei*)))] appartenant à un ou plusieurs clades d'Amérique du Sud. *Eleutherodactylus* a biogégraphiquement colonisé les Antilles et les différentes radiations des Grandes Antilles et d'Amérique du Sud convergent dans les Petites Antilles du sud.

Kaiser, H. 1996. Systematics and biogeography of Eastern Caribbean *Eleutherodactylus* (Anura: Leptodactylidae): consensus from a multidisciplinary approach, p. 129–140. *In* R. Powell and R. W. Henderson (eds.), Contributions to West Indian Herpetology: A Tribute to Albert Schwartz. Society for the Study of Amphibians and Reptiles, Ithaca (New York). Contributions to Herpetology, volume 12.

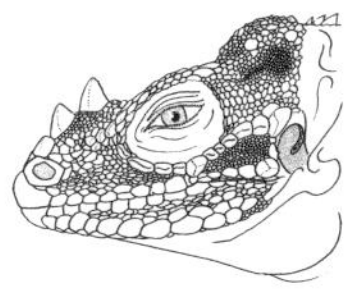

THE RELATIONSHIPS OF THE HISPANIOLAN FROGS OF THE SUBGENUS *PELORIUS* (*ELEUTHERODACTYLUS*: LEPTODACTYLIDAE)

JOHN D. LYNCH[1]

Abstract.—Two phenetic groups are evident using crania of frogs of the subgenus *Pelorius* of the genus *Eleutherodactylus*, and these phenetic groups correspond to groups defined using external characters and electromorphs. However, when these characters (cranial and otherwise) are subjected to rigorous outgroup comparison, most of the differences between groups are found to be uninformative of relationships. The *inoptatus* group (*E. chlorophenax*, *E. inoptatus*, *E. nortoni*) share one synapomorphy and the *ruthae* group (*E. hypostenor*, *E. parapelates*, *E. ruthae*) share two others, but relationships within groups are not resolved and no synapomorphy supports the subgenus. A single apparent synapomorphy links the *inoptatus* group with part of the subgenus *Euhyas*.

INTRODUCTION

WITH MORE THAN 500 SPECIES CURRENTLY RECOGnized, the leptodactylid frog genus *Eleutherodactylus* is the most speciose genus of vertebrates and one of the richest vertebrate elements of the West Indies. The known richness is due in no small measure to the work in alpha taxonomy by Albert Schwartz. At present (Hedges 1989), *Eleutherodactylus* is subdivided into five subgenera of very different sizes and geographic extensions. The subgenus *Craugastor* contains approximately 75 species and is distributed primarily in Mexico and Central America (Texas to Ecuador, with a single Venezuelan species). The subgenus *Syrrhophus* contains 24 species and is distributed from Texas to Guatemala. The subgenus *Euhyas* contains approximately 80 species and is distributed naturally in the Greater Antilles and the island of Mona (introduced in Florida, Louisiana, and Veracruz; Schwartz 1974). The subgenus *Pelorius* contains six species found on Hispaniola and the subgenus *Eleutherodactylus* contains about 325 species found in the West Indies (all islands except Jamaica), South America, and lower Central America.

Although my focus for more than 25 years has been mainland species and their relationships, I have a long-standing interest in the relationships of all of the frogs of the tribe Eleutherodactylini. In the past 10 years, I have renewed my attention to osteological features of the genus in the hopes of discovering monophyletic groups within *Eleutherodactylus*. My intent is not to criticize Hedges' (1989) subgeneric arrangement here but to focus on one particular cluster of species that Hedges (1989) named as the subgenus *Pelorius*.

The subgenus *Pelorius* contains six species confined to Hispaniola, including four endemic to the "South Island" (Schwartz and Henderson 1991). Having studied the skulls of some 230 species of mainland *Eleutherodactylus* in the past dozen years, I sought to extend this data set into the Antilles in order to test some of my conjectures about polarities as well as to test the hypotheses advanced by Hedges (1989) and Joglar (1989).

A Brief History of Species of *Pelorius*

The subgenus *Pelorius* comprises six species of *Eleutherodactylus* with a complex taxonomic his-

[1] School of Biological Sciences, The University of Nebraska, Lincoln, Nebraska 68588-0118, USA.

tory. Only eight species names have been applied to frogs of the subgenus *Pelorius*. Tschudi (1838) named *Cornufer unicolor*, but that name became lost in the literature of the Australasian world (as a genus of ranids currently called *Platymantis*) later in the 19th century. Barbour (1914) named the same (or a similar, see below) species as *Leptodactylus inoptatus* and, soon thereafter, Schmidt (1921) transferred it to *Eleutherodactylus*. Noble (1923) named the smaller (but still large, relative to most Antillean *Eleutherodactylus*) *E. ruthae* and reported that it was closely related to the larger *E. inoptatus*, but did not provide reasons for thinking so. Although skeptical, Cochran (1941) agreed with Noble and cited size, coloration, and toe morphology as evidence to support her view. Cochran (1956) named *E. beebei* from Guyana and noted that it was like *E. inoptatus* in some features (large size and presence of a dermal horn on the upper eyelid), but she also compared it to broad-headed eleutherodactyline frogs from South America.

Schwartz (1965) reviewed *Eleutherodactylus inoptatus* and *E. ruthae* and described three additional subspecies of the latter and another species, *E. hypostenor*. He explicitly rejected Shreve and Williams' (1963) assignment of *E. inoptatus* and *E. ruthae* to the *auriculatus* group and cited, as evidence, the presence of a double vocal sac in *E. ruthae* rather than the single subgular sac in the *auriculatus* group and the form of the vomerine tooth row (a short series that is rather long and not patch-like, unlike those seen in frogs of the *auriculatus* group). Zweifel (1967) showed that the lectotype of *Cornufer unicolor* was not a ranid and suggested that it was an older name for *E. inoptatus* (but see next paragraph). Rather than precipitate a nomenclatural nightmare, Zweifel sought to suppress Tschudi's names. The Commission of Zoological Nomenclature acted after a protracted delay (Anonymous 1978) and removed *Cornufer* Tschudi and *unicolor* Tschudi from further consideration for eleutherodactyline frogs. Lynch (1976) pointed out that *E. beebei* was the same as *E. inoptatus*, that the locality data were probably confused, that several errors could be found in Cochran's description, and that frogs of the *inoptatus* group were unlike other Antillean *Eleutherodactylus* because the first finger was longer than the second. Schwartz (1976) named two additional large species (*E. chlorophenax* and *E. nortoni*) that he thought to be related to *E. inoptatus*. Hedges and Thomas (1987) named *E. parapelates*, completing the list of taxa, and pointed out that all six species (including *E. ruthae*) have a single internal vocal sac. As noted by Hedges and Thomas (1987), *E. chlorophenax*, *E. inoptatus*, and *E. nortoni* are larger frogs (60–90 mm snout-vent length [SVL]) having more-or-less distinctive conical tubercles on the upper eyelid than the three smaller species (45–55 mm SVL) which lack eyelid tubercles and which share a cornified pad at the tip of the snout (hypothesized by those authors to be used in construction of the below-ground nests and calling sites).

Although Duméril and Bibron (1841) and Tschudi (1838) had noted a frontoparietal depression, most authors dealing with the largest three species have failed to detect frontoparietal crests (which produce the depression). The depression is readily visible in Hedges and Thomas' (1987) photograph of *Eleutherodactylus chlorophenax* and less so in their photographs of *E. inoptatus* and *E. nortoni*. Zweifel (1967) noted the depression (and cranial crests) and the transverse fold of skin crossing it, but did not encounter the two features jointly in *E. inoptatus*. Schwartz (1976) explicitly described the transverse skin fold in *E. chlorophenax* and *E. nortoni*, but described the frontoparietal as ". . . a pair of longitudinal folds between the eyes . . ." and ". . . a depressed area between the eyes . . ." in the two species, respectively. *Cornufer unicolor* Tschudi is not an available name (Anonymous 1978), but is probably an older name for *E. nortoni* whose disk size and ear size match those of the lectotype. Cochran (1956), in describing *E. beebei*, emphasized ". . . the presence of smooth bony ridges forming a 'U' between and behind the eyes . . ." and although she contrasted *E. beebei* with *E. inoptatus*, she failed to note the presence of such crests in *E. inoptatus* (nor did she mention crests in her 1941 description).

The current views on relationships had their genesis with Noble's (1923) description of

Eleutherodactylus ruthae and his assertion that it was most nearly related to the larger *E. inoptatus*. To the degree possible, Noble and, later, Cochran (1941) apparcntly used the color patterns and the relatively large size of the frogs as evidence that they were related. Cochran (1941: 25) said ". . . *inoptatus*, does not seem to have a close relative there [Hispaniola], although its vomerine teeth are similar to those of *ruthae* and *schmidti*." In the same publication, Cochran (1941: 33) said of *E. inoptatus* and *E. ruthae*, in justifying Noble's opinion that they were related, ". . . essential color pattern, in the presence of a dorsolateral fold and a subsidary diagonal row of glands on the sides, and in the relative proportions of the toes[,] the two species are similar." In describing *E. beebei*, Cochran (1956) noted that it resembled *E. inoptatus* in having dermal horns on the upper eyelids and being large in size.

The descriptions of four additional species between 1965 and 1987 did not alter the view that these species form a species group and Schwartz (1965) emphasized that point by representing *Eleutherodactylus hypostenor* as intermediate between the two extremes (*E. inoptatus* and *E. ruthae*). Aside from a brief inclusion of some of these large frogs in the

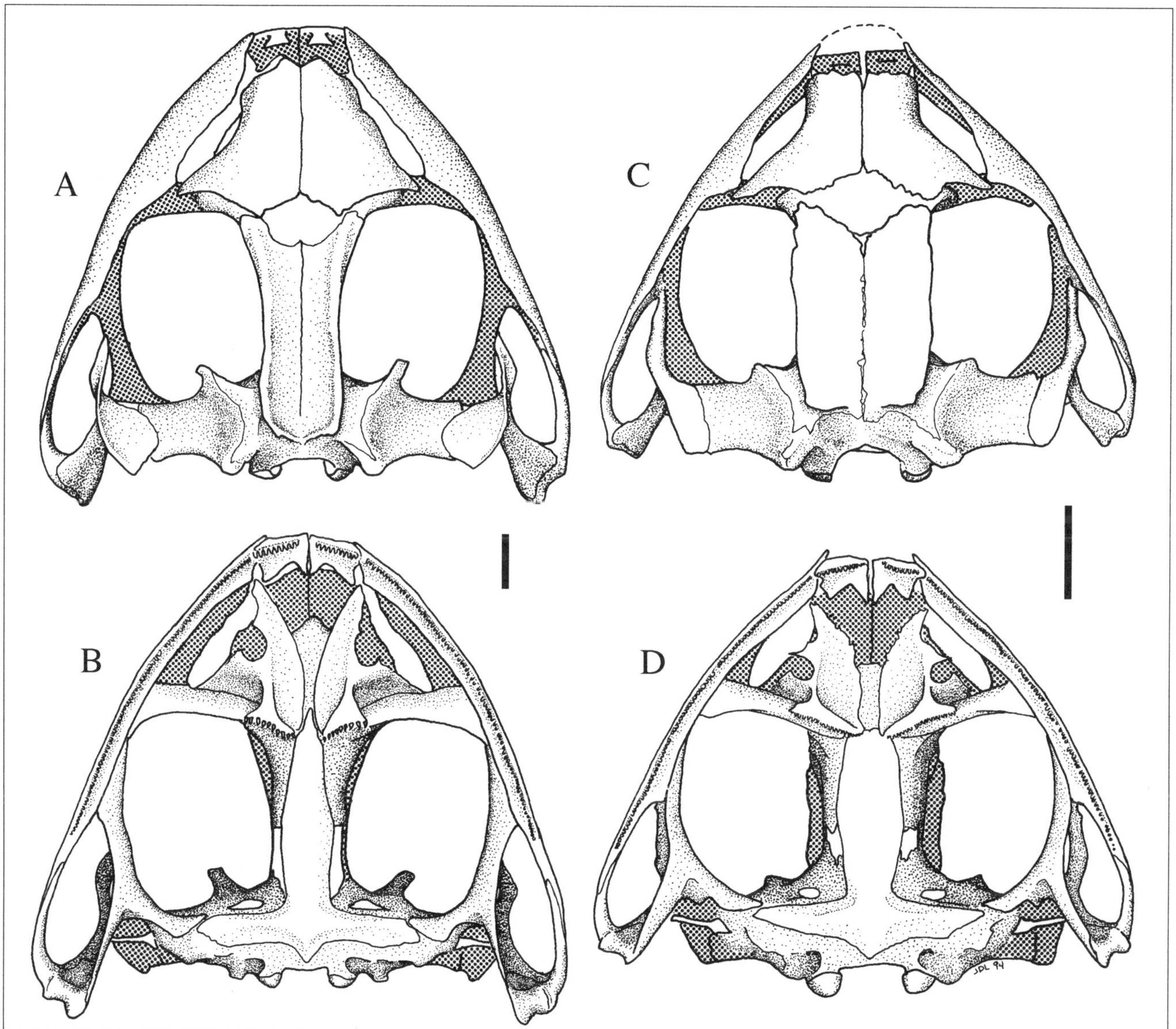

FIGURE 1. Dorsal and ventral views of crania of (A-B) *Eleutherodactylus chlorophenax* (UF 56797) and (C-D) *E. ruthae* (KU: AS v4237). Scales equal 3 mm.

auriculatus group (*E. inoptatus* and *E. ruthae*; Shreve and Williams 1963), these frogs have been treated as a single phenetic unit of Antillean *Eleutherodactylus*, explicitly so since Schwartz (1965).

Two analyses of Antillean *Eleutherodactylus* by Hedges (1989) and Joglar (1989) agreed that these six species constituted a monophyletic subset of the genus. Neither author identified the "essential color pattern" (cited and used by Cochran 1941) or dorsolateral folds as characteristics useful in diagnosing *Pelorius*. However, these authors identified as synapomorphies two features of dubious value (first finger longer than second and body size large, 50–90 mm SVL). Joglar also cited head width (HW >40% SVL) as a synapomorphy. As I argue below, these characteristics may serve in phenetics but not in cladistics, because outgroup comparison quickly establishes that each is a plesiomorphy and therefore uninformative of relationships. Hedges identified three allelic characters as synapomorphies, but none is sufficient to diagnose *Pelorius* and, if derived, the allelic characters are incompatible with one another.

Neither Hedges (1989) nor Joglar (1989) mentioned one character cited by Cochran (1941), the relative proportions of the toes, but she was not very specific about the nature of the character either. The six species of *Pelorius* are peculiar among Antillean *Eleutherodactylus* because the fifth toe is longer than the third when each is adpressed against the fourth. The fifth toe extends distally to a point between the distal and penultimate subarticular tubercle of Toe IV, whereas the third reaches only the penultimate subarticular tubercle. Most frogs of the subgenus *Euhyas* have the fifth toe shorter than the third and

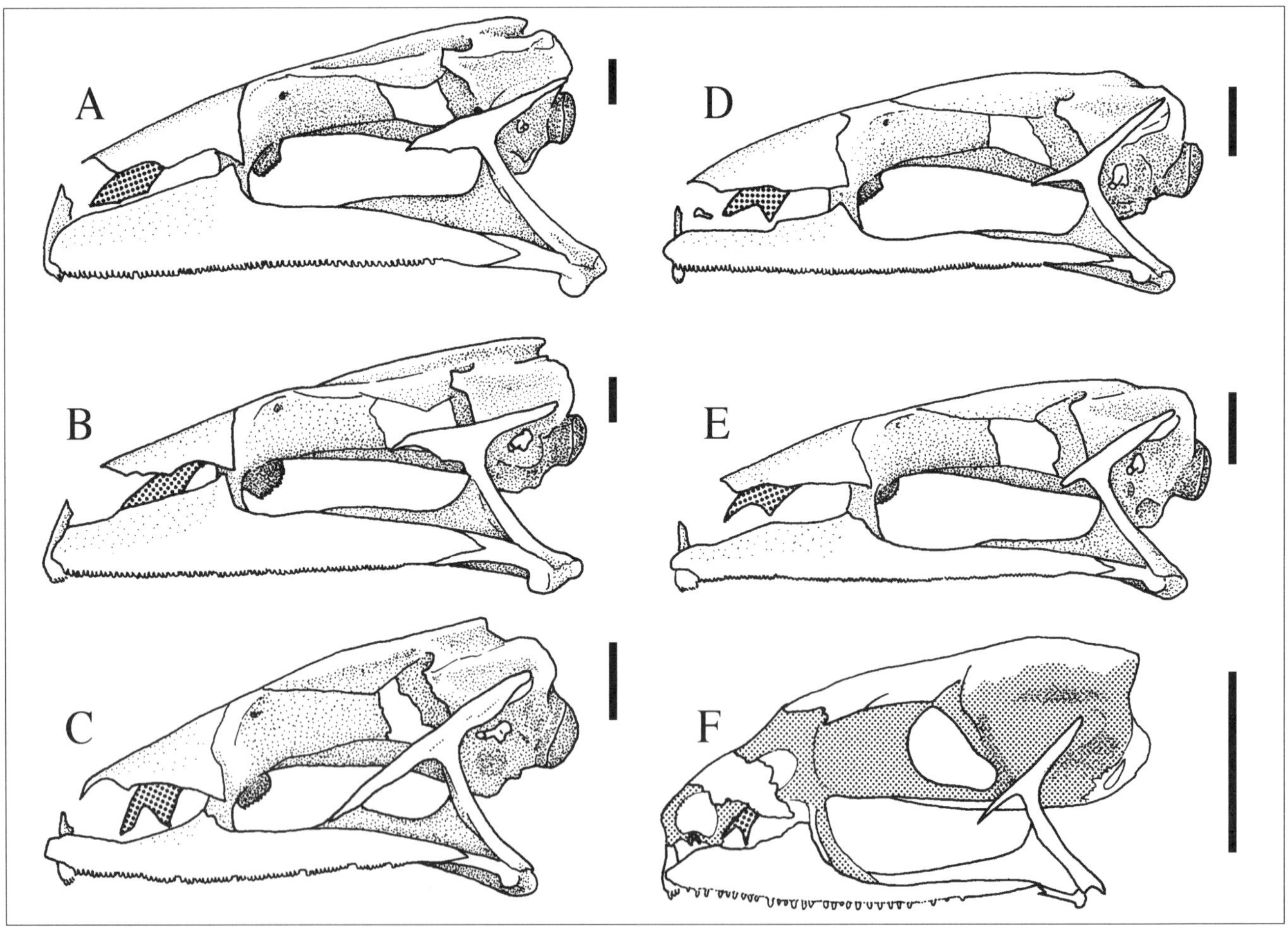

FIGURE 2. Lateral views of crania of frogs of the subgenus *Pelorius*: (A) *chlorophenax* (UF 56797), (B) *nortoni* (USNM 257746), (C) *ruthae* (KU: AS V4237), (D) *hypostenor* (KU 79768), (E) *parapelates* (USNM 257718), and (F) *ruthae* (KU: AS X2908) juvenile 13.2 mm SVL. Scales equal 2 mm.

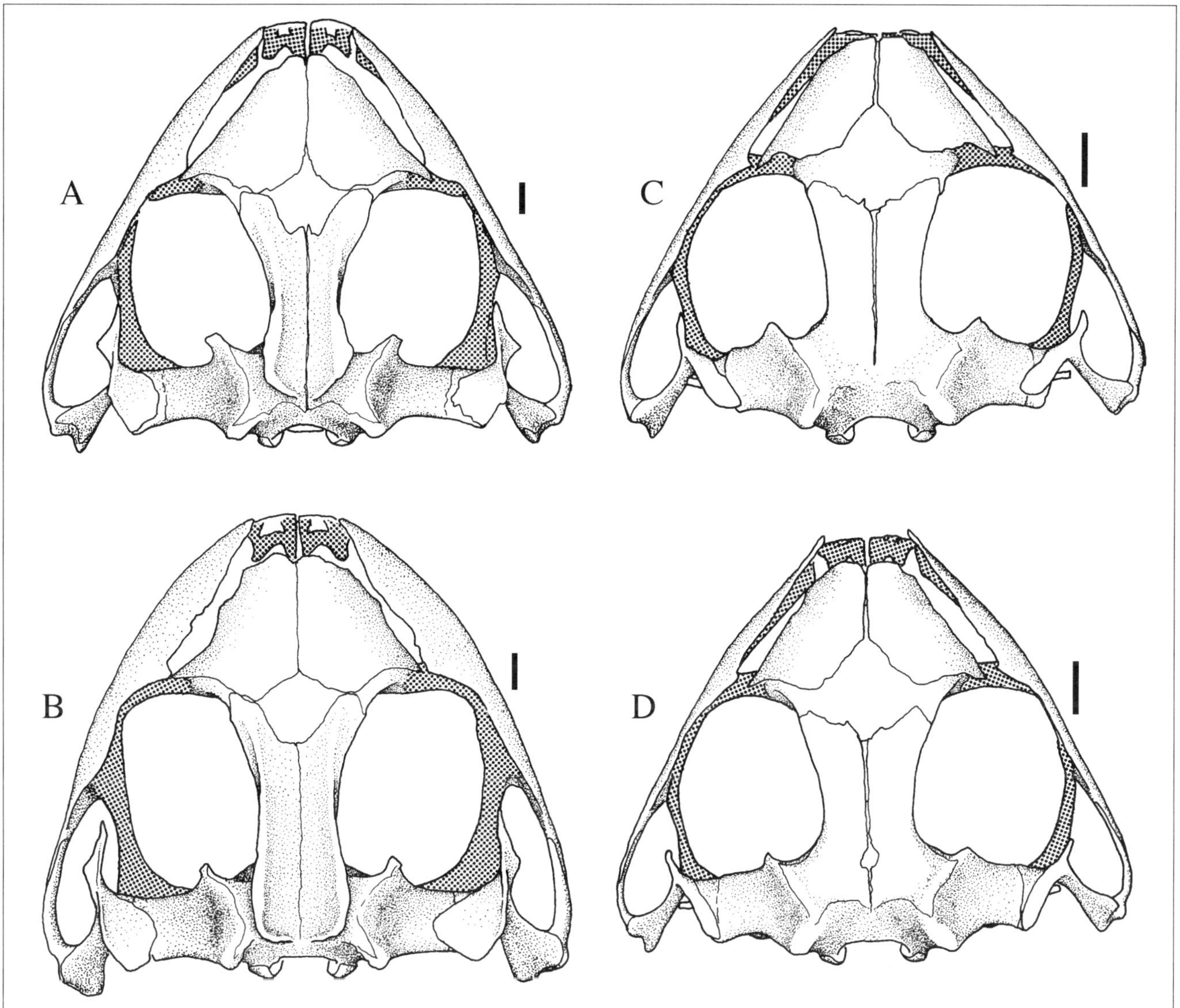

FIGURE 3. Dorsal views of crania of (A) *Eleutherodactylus inoptatus* (KU 280103), (B) *E. nortoni* (USNM 257746), (C) *E. hypostenor* (KU 79768), and (D) *E. parapelates* (USNM 257718). Scales equal 2 mm.

most Antillean species of the subgenus *Eleutherodactylus* have the fifth toe reaching to the distal border of the distal subarticular tubercle of Toe IV.

The current state of affairs as concerns *Pelorius* is that we do not have very satisfactory evidence that the group is monophyletic. What we have is evidence (phenetic) that these frogs do not very much look like other Antillean *Eleutherodactylus.* As I noted 25 years ago (Lynch 1971), *E. inoptatus* and *E. ruthae* are easily distinguished on the basis of their skulls (Figures 1–2). Furthermore, their skulls are readily distinguished from the skulls of any other Antillean *Eleutherodactylus* (Joglar 1989; Lynch 1971). This study was undertaken because I had the opportunity to examine skulls of all six species of *Pelorius* and wanted to test the hypothesis that *Pelorius* is monophyletic.

MATERIALS AND METHODS

Skeletons were prepared either by flensing and washing in warm 25% bleach (Chlorox) for 5–10 min followed by rinsing in a cold water bath for 12 h or as cleared and doubly stained skeletons as described by Dingerkus and Uhler (1977). Flensed skulls are stored either as dry skulls or in alcohol.

Several of these frogs are rare. A single adult was studied for each of four species, but available series of *Eleutherodactylus inoptatus* and *E. ruthae* included adult females, adult males, and juveniles. The following crania (or skeletons) were examined: *E. chlorophenax* (UF 56797), *E. hypostenor* (KU 79763), *E. inoptatus* (KU 280103, 280328, 280339, 280341, 280351, 280353, AS X2356), *E. nortoni* (USNM 257746), *E. parapelates* (USNM 257718), *E. ruthae aporostegus* (KU, AS V47099, V48961, X2908), *E. ruthae bothroboans* (KU, AS V4236–37), and *E. ruthae ruthae* (KU, AS V1039). The series of skulls was supplemented by judicious dissections of preserved specimens.

Characters were polarized using outgroup comparisons. For any character having two or more conditions in the ingroup (*Pelorius*), the condition(s) found in the outgroup is primitive for the ingroup, whereas a condition unique to the ingroup (not found in the outgroup) is derived for the ingroup. Observing that a state is found both in the ingroup and in the outgroup means: (1) the state is primitive for the monophyletic ingroup, (2) the state is derived and the ingroup is not monophyletic, or (3) we do not know enough about what is homologous. Because I was concerned that some additional West Indian taxa might be near relatives of the frogs included in *Pelorius*, I used approximately 50 species

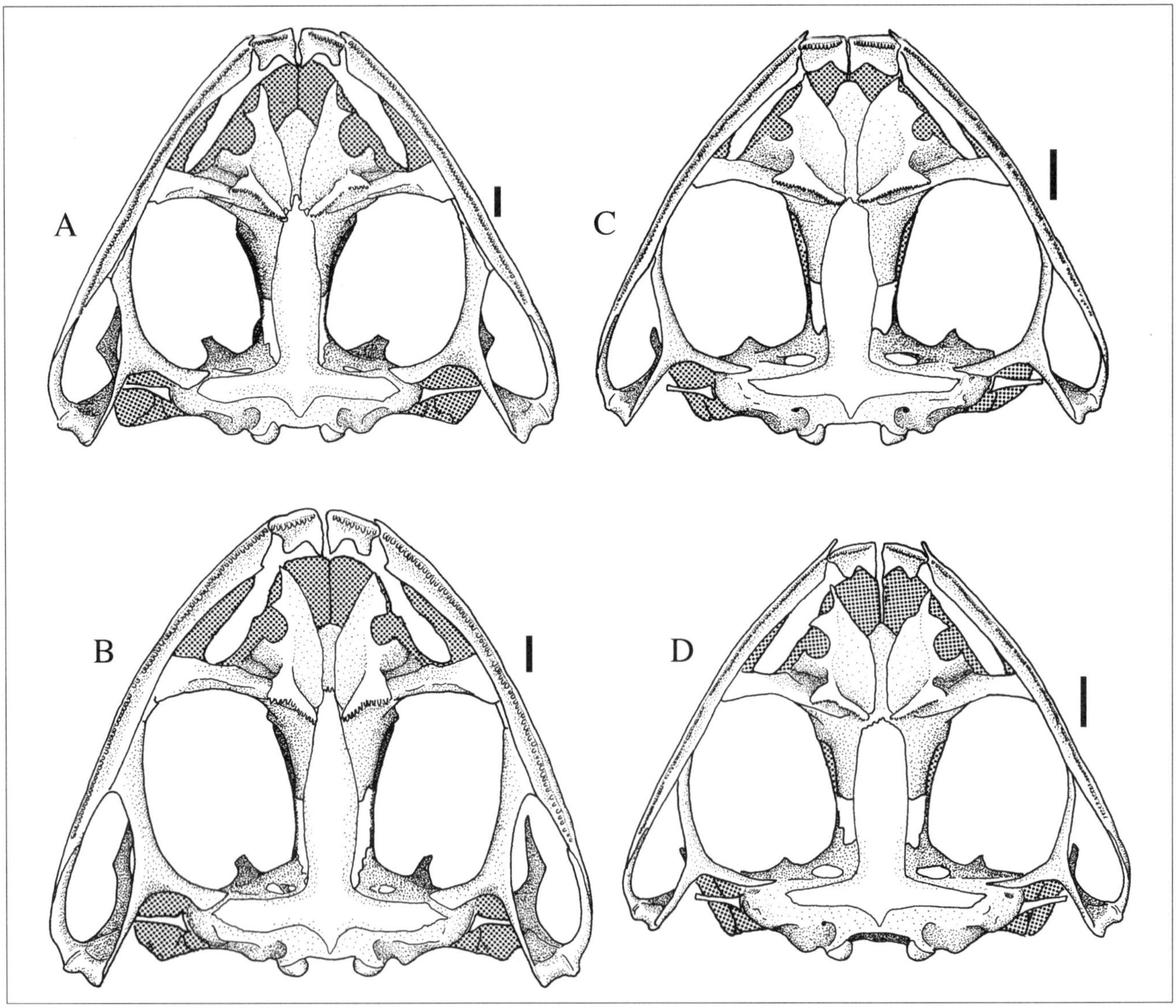

FIGURE 4. Ventral views of crania of (A) *Eleutherodactylus inoptatus* (KU 280103), (B) *E. nortoni* (USNM 257746), (C) *E. hypostenor* (KU 79768), and (D) *E. parapelates* (USNM 257718). Scales equal 2 mm.

of the subgenus *Craugastor* and 150 species of the subgenus *Eleutherodactylus* (Colombia, Ecuador, and Peru) as my outgroup (Appendix I).

Thc usc of a large outgroup avoids artificial polarization of variable traits within the ingroup. Using a single outgroup species obviates the possibility of encountering more than a single condition and maximizes the possibility of polarizing variable characters. Because the outgroup method depends on unique ingroup states being derived, one must exercise some caution to avoid inflating the value of a data matrix.

Hedges (1989) recognized an *inoptatus* species group and a *ruthae* species group within his subgenus *Pelorius*. Joglar (1989) recognized a more inclusive *inoptatus* group (=*Pelorius*). Joglar's usage of *inoptatus* group is in the same sense as previous usages (Hedges and Thomas 1987; Lynch 1976; Schwartz 1965, 1976). To distinguish between usages, I will use *inoptatus* group and *ruthae* group in the sense of Hedges (1989) and use *Pelorius* as a synonym for Joglar's (1989) *inoptatus* group.

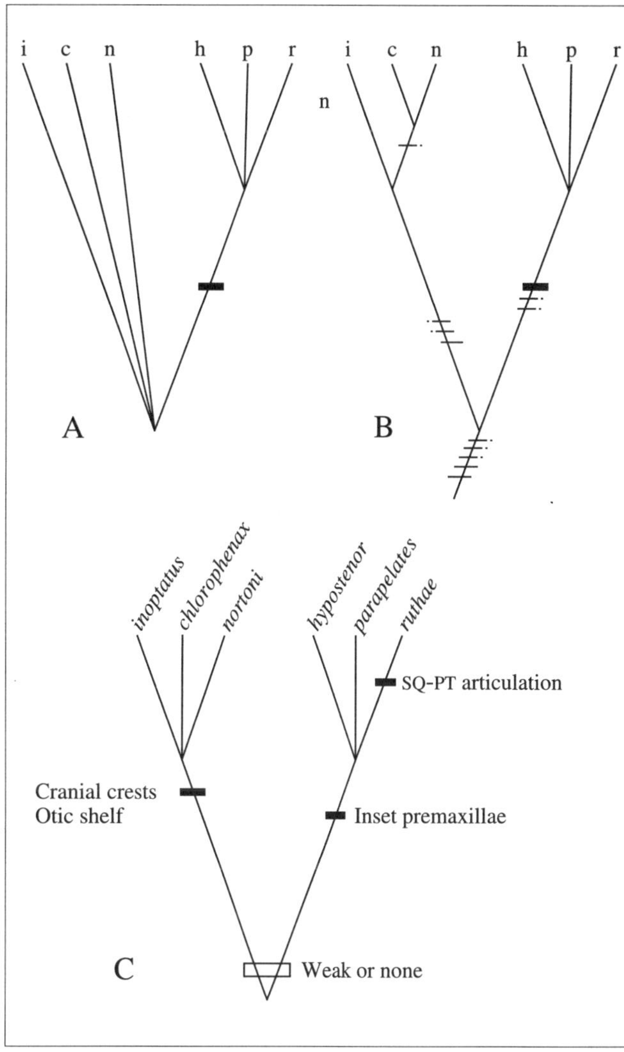

FIGURE 5. Cladograms for frogs of the subgenus *Pelorius*: (A) Hedges and Thomas' 1987 hypothesis, (B) Hedges' 1989 hypothesis, (C) proposed here. Synapomorphies are identified with tics: the heavy bar refers to the four traits given by Hedges and Thomas (1987); thin bars with a dot are biochemical traits from Hedges (1989) and Table 1; thin bars not labeled are morphological traits given by Hedges (1989).

CRANIA

I had expected two cranial modes among the six species and in some respects found just that. The crania are broader than long, not or only slightly clcvatcd posteriorly, and have full complements of bones. No fusions of bones are seen. The frontoparietals are in broad median contact, covering the fontanelles (Figure 1).

***inoptatus* group** (Figures 1A-B, 2A-B, 3A-B, 4A-B) Crania of members of the *inoptatus* group (*chlorophenax*, *inoptatus*, and *nortoni*) are nearly indistinguishable. Each species has cranial crests, although Joglar (1989) claimed that cranial crests are lacking in all Antillean taxa, and a curious bony flange extending anteriorly from the epiotic eminence (Figures 1A-B, 3A-B). The braincase is slightly wider anteriorly than posteriorly and the cranial crests are smooth and appropriately described as U-shaped (as did Cochran 1956). The frontoparietals contact the nasals by way of an anterolateral flange which rests on the sphenethmoid and a posterior process of the nasal (Figures 1A, 3A-B). The nasals are large elements, broadly in contact on the midline. The cristae paroticae are relatively slender. The epiotic eminences are prominent and a bony flange lies between the *m. levator posterior longus* and the *m. pterygoideus* (also visible in lateral views, Figure 2A-B). Each species has an extensive otic shelf.

The zygomatic process of the squamosal is blade-like (tip pointed, process on ventral border) and is nearly as long as the otic process (Figure 2A-B). The maxilla is relatively deep and nearly contacts the nasal. The premaxillae are terminal in position (visible even in lateral view) and the alary processes are deflected slightly posteriorly (Figure 2A-B).

The palatine shelf of the premaxilla is dissected but shallow relative to those of many Andean species. The maxilla lacks a pterygoid process. The vomers are large bones, considerably longer than wide, narrowly separated medially, and bearing large vomerine odontophores that lie well posterior to the choanae (Figures 1B, 4A-B). The vomerine tooth rows are arched (median portion posterior to lateral portion) and extend laterally to the level of the middle of the choanae. The width of the vomerine tooth row is greater in *Eleutherodactylus inoptatus* than in *E. chlorophenax* or *E. nortoni*. The median ends of the neopalatines lie dorsal to the posteromedial portions of the vomers. The cultriform process of the parasphenoid is long, extending anteriorly between the neopalatines. The anterior portion of the cultriform process narrows and is point-like (Figures 1B, 4A-B). The parasphenoid alae are deflected slightly posteriorly and are in contact with the median rami of the pterygoids. The pterygoids are massive bones with long anterior rami (extending nearly to the *plani antorbitalia*).

***ruthae* group** (Figures 1C-D, 2C-F, 3C-D, 4C-D)
In contrast to the crania of frogs of the *inoptatus* group, the crania of frogs of the *ruthae* group are heterogeneous. The cranium of *Eleutherodactylus ruthae* is very different from those of *E. hypostenor* and *E. parapelates*. The three species of the *ruthae* group lack cranial crests and epiotic flanges (Figures 1C, 3C-D) and have short cultriform processes of the parasphenoid (Figures 1D, 4C-D). The frontoparietals do not meet the nasals, the maxilla is shallow, and the zygomatic ramus of the squamosal is slender. These animals also have inset premaxillae, reflecting the connective tissue pad and mentioned by Hedges (1989) as a synapomorphy.

In *Eleutherodactylus ruthae*, the nasals are constricted in their anterior one-third but in broad median contact (Figure 1C). The nasals do not contact the frontoparietals although they have the posterior process. The frontoparietals are indented anteriorly (i.e., have anterolateral processes extending anteriorly over the sphenethmoid) but have approximately parallel borders as seen from above. The frontoparietals do not bear crests but extend laterally over the posteromedian edges of the orbits. The cristae paroticae are somewhat stockier than those of frogs of the *inoptatus* group and the epiotic eminences developed equally, but they lack the bony flange extending between the *m. levator posterior longus* and the *m. pterygoideus*. The otic shelf of the squamosal is narrow unlike the broad, medially-extended otic shelf seen in frogs of the *inoptatus* group and its axis is more nearly parallel to the sagittal plane (Figure 1D).

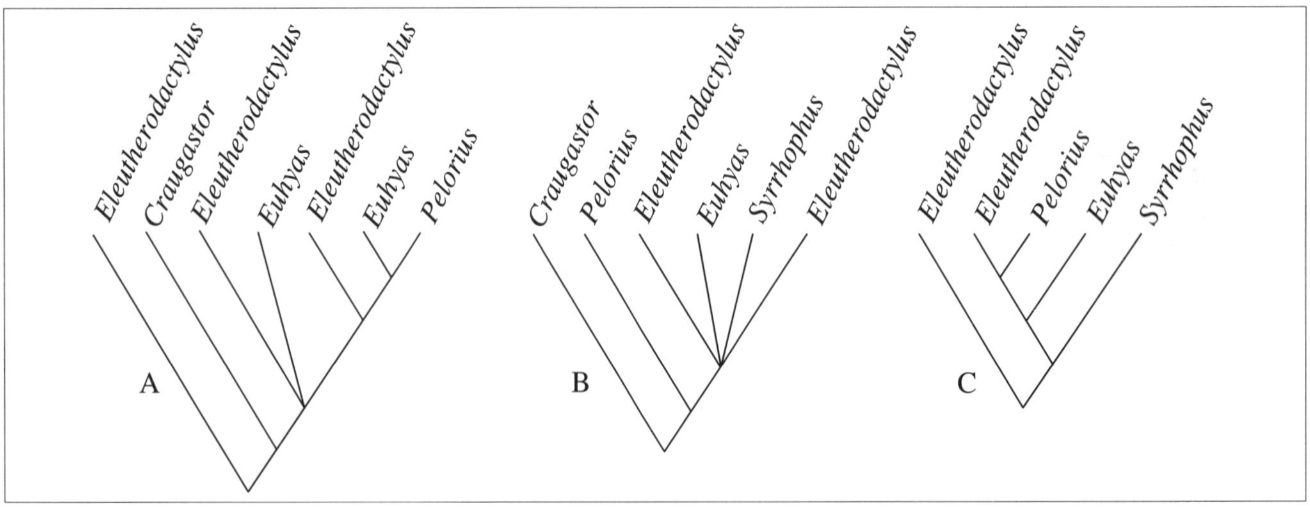

FIGURE 6. Cladograms relating *Pelorius* to other groups and subgenera of *Eleutherodactylus*: (A) Hedges (1989); (B) Joglar (1989); (C) Joglar (1989).

The zygomatic process of the squamosal is round in cross-section, longer than the otic ramus, and sutured to the dorsal surface of the anterior ramus of the pterygoid (Figure 2C). The *pars facialis* of the maxilla is shallow anteriorly, exposing the palate to lateral view. The anterior tip of the maxilla extends anteriorly to the anteriormost extent of the premaxilla as a toothless flange (Figures 1C-D, 2C). The premaxillae, as viewed laterally, are inset relative to the maxillae. The alary processes of the premaxillae are short and directed dorsally.

The palatal shelf of the premaxilla is relatively broad and not deeply dissected. The palatal shelf of the maxilla develops a modest pterygoid process. The vomers are large and nearly as broad as long. The vomers are separated medially and the vomerine odontophores lie well posterior to the choanae and lie ventral to the anteromedial portions of the neopalatines. The vomerine tooth row is broad and extends laterally to the level of the lateral margin of the choanae. The tooth row slants posteromedially but is not arched. The cultriform process of the parasphenoid is only slightly narrower anteriorly than at its midpoint and is relatively blunt anteriorly, being barely overlapped by the posteromedial extensions of the vomers (Figure 1D). The parasphenoid alae taper laterally, are perpendicular to the cultriform process, and touch the median rami of the pterygoids. The anterior rami of the pterygoids extend about three-fourths the length of the orbit.

Eleutherodactylus hypostenor and *E. parapelates* differ in several respects from *E. ruthae* and their crania are described jointly. The nasals are not constricted anteriorly and are in median contact for less than half their lengths. Their posterior edges do not display obvious posteriorly-directed processes and the nasals do not contact the frontoparietals (Figure 3C-D). The frontoparietals have concave anterior borders with suggestions of anterolateral processes and are broader anteriorly than in mid-orbit. The cristae paroticae are relatively stocky with elevated epiotic eminences. The anterior process of the epiotic eminence is angular but does not bear a bony flange. The otic ramus of the squamosal has a narrow shelf and is not parallel to the sagittal plane (Figure 3C-D).

TABLE 1. Distribution of electromorph states for six biochemical characters (from Hedges 1989).

Electromorph		Species[1] C	N	I	H	P	R	Hedges' conjecture
ICD-1	p5	X	X	O	X	X	X	synapomorphy *Pelorius*
LGl	al	X	X	X	X	O	X	synapomorphy *Pelorius*
PGM	o	O	O	X	X	X	X	synapomorphy *Pelorius*
PGM	q	X	X	O	O	O	O	synapomorphy C + N[2]
CK	j2	X	X	X	O	O	O	synapomorphy C + N + I[2]
ACP	b	O	O	O	X	X	X	synapomorphy H + P + R[2]
CK	j1A	O	O	O	X	O	X	synapomorphy H + P + R[3]
PT-3	l1	X	X	X	O	O	O	synapomorphy C + N + I[2]

[1] Species: C = *chlorophenax*, N = *nortoni*, I = *inoptatus*, H = *hypostenor*, P = *parapelates*, R = *ruthae*.
[2] Congruent with Hedges' cladogram.
[3] Most parsimoniously viewed as a synapomorphy for H + R.

The zygomatic rami of the squamosals are short (shortest in *Eleutherodactylus parapelates*) and thin (not round in cross-section) but not blade-like. The *pars facialis* of the maxilla is shallow and widely separated from the nasal (Figure 2D-E). The anterior tip of the maxilla is toothless and extends anteriorly to the premaxillae which are inset and have dorsally-directed alary processes (Figure 2D-E).

The palatal shelf of the premaxilla is relatively deep but is not deeply dissected. The maxilla does not bear a pterygoid process. The vomers are large bones, narrowly separated medially, and bear large vomerine odontophores well posteriad to the choanae. The tooth row is more arched in *Eleutherodactylus parapelates* than in *E. hypostenor*, the latter closely resembling that of *E. ruthae*. The lateral extent of the tooth row reaches the level of the middle of the choanae. The vomers underlie the median extensions of the neopalatines. The cultriform processes of the parasphenoids are short (do not extend between the vomers) and are not narrowed appreciably anteriorly (Figure 4C-D). The parasphenoid alae are perpendicular to the cultriform process and taper laterally. The median rami of the pterygoids are thin compared to those of *E. ruthae* (or frogs of the *inoptatus* group) and are narrowly separated from the parasphenoid alae. The pterygoids are slender bones and the anterior rami extend about halfway to the *plani antorbitalia*.

Summary of Crania

The phenetic approach can be continued using the crania to recognize two groups, one for *chlorophenax*, *inoptatus*, and *nortoni* (large nasals in broad median contact and contacting the frontoparietals which bear U-shaped cranial crests, thin cristae paroticae with a bony flange on the anterior end of the epiotic eminence, broad otic shelf, blade-like zygomatic rami of the squamosals, deep maxillae, terminal premaxillae having posteriorly directed alary processes, arched vomerine odontophores separated posteromedially by narrow cultriform process of parasphenoid, parasphenoid alae deflected posteriorly), and a second group for *hypostenor*, *parapelates*, and the distinctive *ruthae* (smaller nasals not contacting frontoparietals which lack cranial crests, stocky cristae paroticae lacking flanges on epiotic eminences, narrow otic shelf, shallow maxillae with toothless terminal portion, inset premaxillae having dorsally directed alary processes, vomerine odontophores usually not arched, not separated by cultriform process of parasphenoid which is broad and short, parasphenoid alae of parasphenoid perpendicular to cultriform process, tapering laterally).

Polarization of Characters

Characters Supporting Monophyly of *Pelorius*

Although the cited features (osteological as well as external) testify to the distinctness of these two phenetic groups, the features need not be informative. To be informative, features must be shared and must be arguably derived (otherwise such characters fail to discriminate among competing hypotheses). Both Hedges (1989) and Joglar (1989) argued that the long first finger was derived and synapomorphic and that large size was derived and synapomorphic for *Pelorius*. Lynch (1976) detailed the distribution of the long first finger condition but, for present purposes, it is sufficient to note that this condition is seen in most *Craugastor* and many *Eleutherodactylus*. Thus, this character state must be considered primitive even though all other Antillean species have the first finger shorter than the second. Use of large body size, defined or not, similarly is not informative because large species of *Craugastor* are known (e.g., *E. bufoniformis*, *E. cheiroplethus*). Joglar (1989) also cited head width (more than 40% SVL) as a synapomorphy, and broad heads also were mentioned descriptively by Hedges (1989).

Head width, expressed as HW/SVL, was used by Lynch (1975) as apomorphic for what he considered a monophyletic group even though he noted that certain species (*Eleutherodactylus anomalus*, *E. lymani*) were intermediate between what he called broad-headed and narrow-headed. Subsequently, Lynch (1986) argued that broad heads had evolved twice in mainland taxa and Savage (1987) showed considerable ontogenetic variation within *E. laticeps* (a species of the subgenus *Craugastor* having as broad a head as any species of *Pelorius*).

Individually, Hedges' (1989) biochemical characters are insufficient to characterize *Pelorius* because one or more species lacks each of the supposedly derived conditions (Table 1). Treated as a combination, one might salvage a diagnosis but combinations are not characters. Hedges (1989), like Joglar (1989), employed "standard" parsimony methods to discover support for particular nodes of the cladogram. The weakness of such an approach is that it generates fully polythetic hypotheses which are then treated as though they were objectively monophyletic. The approach I advocate is the minority position within the systematic community, but has the advantage of allowing rejection of hypotheses. Furthermore, Hedges' biochemical characters have only been surveyed for two species in the outgroup (one specimen each). The failure to observe the "derived" condition in the outgroup is best explained as a sampling error.

The only other characteristics that might qualify as a synapomorphy for *Pelorius* are the size and shape of the vomerine odontophores (short but not patch-like) and relative toe length. The description of the odontophores is in reference to the long odontophores of frogs of the *ricordii* group (*Euhyas sensu* Hedges 1989) and the clumps of teeth (patch-like) seen in most members of the *auriculatus* group. Each condition is seen in frogs of the subgenus

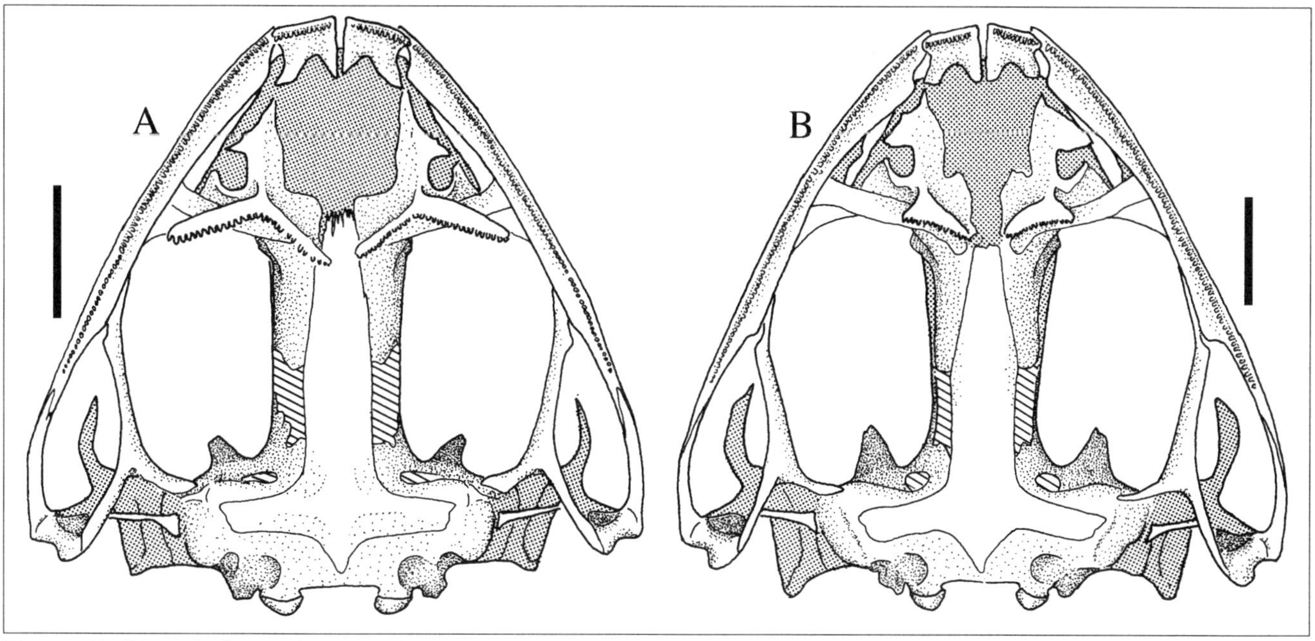

FIGURE 7. Ventral views of crania of frogs of the subgenus *Euhyas*: (A) *Eleutherodactylus dimidiatus* (KU 203386); (B) *E. glandulifer* (KU 278757). Scales equal 3 mm.

Craugastor and, therefore, each condition must be viewed as primitive relative to the hypothesis that *Pelorius* forms a monophyletic group.

Toe length is more complicated, because using all other leptodactylids as the outgroup for *Eleutherodactylus* (in which three states or conditions occur), the primitive condition is to have the fifth toe shorter than the third. Discovering that the other two conditions are derived does not tell us which of those is derived relative to the other or if they are derived independently. Using *Craugastor* as the outgroup for the West Indian eleutherodactyline fauna, the condition seen in *Pelorius* (fifth toe slightly longer than third) is plesiomorphic because the *alfredi* group of *Craugastor* has the same condition as is seen in *Pelorius* but other *Craugastor* have the primitive condition (toe V shorter than III).

Cochran (1941) cited dorsolateral folds as a characteristic for the group, and such folds are seen in all six species (Hedges and Thomas 1987; Schwartz and Henderson 1991) but not in other Antillean *Eleutherodactylus*. However, such folds are observed frequently in mainland taxa (e.g., *E. achatinus*, *E. devillei*, *E. noblei*).

The failure to discover an unequivocal synapomorphy for frogs of the subgenus *Pelorius* does not represent evidence that the group is paraphyletic or primitive — it only reveals that the group is not demonstrably monophyletic.

Less Inclusive Characters

The other features cited above represent features which might serve to support part of a phylogenetic hypothesis. The external features are: (1) presence of a dermal horn on upper eyelid, (2) cornified snout tip, (3) U-shaped cranial crests and the three characteristics cited by Hedges and Thomas (1987), (4) chevrons on shanks, (5) laterally extended vocal sacs, and (6) lack of sexual dimorphism in size. Of these, (4) may be apomorphic but has not been subjected to outgroup comparison; (5) occurs as well in frogs of the *nigrovittatus* group (Lynch 1989), and (6) is seen in *Eleutherodactylus scolodiscus* (Lynch and Burrowes 1990).

Dermal horns are uncommon features evident in many mainland *Eleutherodactylus* (e.g., *E. appendiculatus*, *E. cerastes*, *E. cruentus*), but could not serve as a synapomorphy for the *inoptatus* group (*sensu* Hedges). The cornified snout tip of the *ruthae* group is not evident in the outgroup (200 mainland species) although a fleshy snout tip is known for frogs of the *nigrovittatus* group (Lynch 1989). This feature is a synapomorphy for *E. hypostenor*, *E. parapelates*, and *E. ruthae*. The presence of cranial crests in such mainland species as *E. biporcatus*, *E. curtipes*, *E. galdi*, *E. sulcatus*, and *E. surdus* would seem to argue against U-shaped cranial crests being synapomorphic, but the smooth crests forming a U-shaped pattern is not duplicated in the outgroup of mainland species nor in any other *Eleutherodactylus* known to me. This particular form of cranial crests is a synapomorphy for *E. chlorophenax*, *E. inoptatus*, and *E. nortoni*.

The inset premaxillae (and edentulous anterior tip of the maxilla) is an osteological character reflecting the presence of the connective tissue pad (or cornified pad) cited by Hedges and Thomas (1987). Because the external and the internal characters are completely correlated, to consider them as independent characters would be an error. The epiotic flange seen in the *inoptatus* group (*sensu* Hedges) is not seen in the outgroup and cannot be conceived as being the same character as the U-shaped frontoparietal crests.

Most species of *Craugastor* have contact between the nasals and frontoparietals (e.g., *Eleutherodactylus fitzingeri*, *E. merendonensis*), but many other species do not (e.g., *E. anomalus*, *E. mexicanus*). Therefore, the two conditions of the articulation between the nasals and frontoparietals must each be treated as primitive. Three conditions are seen for the otic shelf in frogs of the subgenus *Pelorius* (broad in the *inoptatus* group, narrow in the *ruthae* group; not parallel to sagittal plane in *E. hypostenor* and *E. parapelates*). Broad or narrow states are seen in the subgenera *Craugastor* and *Eleutherodactylus* (Lynch 1971), as are parallel and non-parallel axes (Lynch 1971). The shape of the otic shelf in the *inoptatus* group *sensu* Hedges is very distinctive, but what appears to be the same condition occurs as well in *E. biporcatus*. Both long and tapering (e.g., *E. gollmeri*, *E. rostralis*) and short and blunt (*E. greggi*, *E. mexicanus*) cultriform processes of the parasphenoid occur in species of the subgenus *Craugastor*, as do perpendicular versus posteriorly deflected parasphenoid alae. The laterally tapering alae of the *ruthae* group may be derived.

The suture between the zygomatic ramus of the squamosal and the pterygoid of *Eleutherodactylus ruthae* is not seen in the outgroup nor is the shape of its skull roof (parallel sided frontoparietals), but each of these is unshared (autapomorphic) and uninformative of relationships. In very small *E. ruthae* (Figure 2F), the zygomatic ramus is short and not in contact with the pterygoid, and very much resembles the condition seen in *E. hypostenor*, providing ontogenetic "confirmation" of one transformation.

HYPOTHESES ABOUT RELATIONSHIPS

Within *Pelorius*

Hedges and Thomas (1987) provided a verbal account of relationships within *Pelorius* (Figure 5A) that is partially congruent with the cladogram presented by Hedges (1989). The latter (Figure 5B) is more completely resolved and provides topological support for recognizing two species groups. Joglar (1989) did not comment on relationships within *Pelorius*.

The 1987 hypothesis was supported by morphology (snout with cornified tip, no sexual dimorphism in size, lateral extensions of vocal sac, and crural chevrons), and Hedges and Thomas (1987) were reluctant to claim that *Eleutherodactylus chlorophenax* and *E. nortoni* shared large disks as a synapomorphy, although Hedges (1989) united them on the strength of an electromorph. The 1989 cladogram contained these morphological features as well as the claim that large size (50–90 mm SVL) and a long first finger were synapomorphies of *Pelorius*, whereas very large size (66–90 mm SVL) was a synapomorphy for the *inoptatus* group *sensu* Hedges.

Although Hedges (1989) claims to have several biochemical synapomorphies for *Pelorius* and its subgroups (Table 1), an alternate view is that his biochemical characters are largely incompatible, support only parts of his cladogram, and provide further resolution (the j1A state of CK is shared only by *Eleutherodactylus hypostenor* and *E. ruthae*, but Hedges treated it as a synapomorphy, not of this pair of species but of the *ruthae* group).

Ignoring the biochemical characters for which rigorous outgroup comparisons have not been done, outgroup comparisons reduce these lists to: (a) snout with cornified tip and (b) crural chevrons. The osteological synapomorphies cited above include: (1) U-shaped cranial crests, (2) bony epiotic flange, and (3) inset premaxillae. As mentioned above, (3) must be considered only the internal reflection of (a). These four characteristics generate a partially resolved cladogram consistent with Hedges' (1989) classification, but do not resolve the relationships within either group nor is there evidence that *Pelorius* is monophyletic (Figure 5C).

Relationships Relative to Other Groups

Hedges (1989) and Joglar (1989) believed, mistakenly in my view, that they had evidence that *Pelorius* is monophyletic but differed in the identification of the nearest relatives of *Pelorius*. Hedges' hypothesis (Figure 6A) did not include *Syrrhophus* and represented *Eleutherodactylus* and *Euhyas* as paraphyletic to *Pelorius* with *Pelorius* being most closely related to Jamaican *Euhyas*.

Joglar presented two incompatible hypotheses but did not comment on their incompatibility. In one case (Figure 6B, the one discussed by Joglar), *Pelorius* is seen as the sister group to *Euhyas* + *Syrrhophus* + the Antillean *Eleutherodactylus* + 13 species of the *unistrigatus* group (part of his "outgroup"), whereas the other hypothesis treats *Pelorius* and the Antillean *Eleutherodactylus* as sister groups (Figure 6C). The three hypotheses are incompatible with one another although Hedges' and one of Joglar's (Figure 6C) may share a component (*Pelorius* + *Eleutherodactylus* [part] + *Euhyas* [part]). Inspecting their character lists and appendices allows one to discover that the "synapomorphies" will not pass muster in a rigorous outgroup comparison and are more in keeping with phenetic rather than cladistic traditions.

I have not adequately surveyed the Antillean *Eleutherodactylus* (*sensu lato*) to offer a competing hypothesis, but do have some data that appear instructive. The epiotic flange cited for *Pelorius* also occurs in some *Euhyas* (Figure 7), including at least most of the species from Jamaica (Flores 1984). If that character is derived, it follows that the *inoptatus* group must be more closely related to some *Euhyas* than to the *ruthae* group. Countering this claim is the observation that the vomerine architecture of *Pelorius* is relatively constant and intermediate between the extremes seen in frogs of the subgenus *Euhyas* (Figure 7). Hedges (1989) chose to argue that while long tooth rows might be informative, short rows had evolved repeatedly within *Euhyas*. In contrast, Joglar's *ricordii* group is made up only of species with "long" and "very long" (equal to Figure 7A) tooth rows, and those with "short" tooth rows (as in Figure 7B) were assigned to the *auriculatus* group. My view of group assignments is closer to Hedges' than to Joglar's, except that I am sympathetic to Joglar's attention to vomerine anatomy (*E. lentus* and *E. richmondi* are Puerto Rican species of *Euhyas* rather than species of the subgenus *Eleutherodactylus* with long vomerine toothrows). Although *Pelorius* appears not to be monophyletic, I suggest that use of the name continue until a sufficient sample of Antillean *Eleutherodactylus* has been studied and the broad outlines of the relationships among the various groups have been formulated based on shared derived characteristics.

ACKNOWLEDGMENTS

David Auth (Florida Natural History Museum, UF), William Duellman (Natural History Museum, University of Kansas, KU), Ronald Heyer (National Museum of Natural History, USNM), and the late Albert Schwartz permitted me to prepare skeletons (or skulls) of material in their care.

LITERATURE CITED

Anonymous. 1978. Opinion 1104/Relative precedence of *Cornufer* Tschudi, 1838, and *Platymantis* Günther, 1858 (Amphibia Salientia). Bull. Zool. Nomencl. 34: 222–233.

Barbour, T. 1914. A contribution to the zoogeography of the West Indies, with especial reference to amphibians and reptiles. Mem. Mus. Comp. Zoöl., Harvard Univ. 44: 209–359.

Cochran, D. M. 1941. The herpetology of Hispaniola. U. S. Natl. Mus. Bull. 177: vii, 1–398.

—. 1956. A new species of frog from Kartabo, British Guiana. Zoologica, N. Y. 41: 11–12.

Dingerkus, G. and L. D. Uhler. 1977. Enzyme clearing of alcian blue stained whole small vertebrates for demonstration of cartilage. Stain Technol. 52: 229–232.

Duméril, A. M. C. and G. Bibron. 1841. Erpétologie Générale ou Histoire Naturelle Complète des Reptiles. Vol. 8. Libr. Encylopedique Roret, Paris.

Flores, G. 1984. Comparative Osteology, Relationships, and Evolution in Jamaican Frogs of the Genus *Eleutherodactylus*. Unpubl. B. A. Thesis, Harvard University.

Hedges, S. B. 1989. Evolution and biogeography of West Indian frogs of the genus *Eleutherodactylus*: slow-evolving loci and the major groups, p. 305–369. *In* C. A. Woods (ed.), Biogeography of the West Indies Past, Present, and Future. Sandhill Crane Press, Gainesville, Florida.

— and R. Thomas. 1987. A new burrowing frog from Hispaniola with comments on the *inoptatus* group of the genus *Eleutherodactylus* (Anura: Leptodactylidae). Herpetologica 43: 269–279.

Joglar, R. L. 1989. Phylogenetic relationships of the West Indian frogs of the genus *Eleutherodactylus*: a morphological analysis, p. 371–408. *In* C. A. Woods (ed.), Biogeography of the West Indies Past, Present, and Future. Sandhill Crane Press, Gainesville, Florida.

Lynch, J. D. 1971. Evolutionary relationships, osteology, and zoogeography of leptodactyloid frogs. Univ. Kansas Mus. Nat. Hist. Misc. Publ. 53: 1–238.

—. 1975. A review of the broad-headed eleutherodactyline frogs of South America (Leptodactylidae). Occas. Pap. Mus. Nat. Hist. Univ. Kansas 38: 1–46.

—. 1976. The species groups of the South American frogs of the genus *Eleutherodactylus* (Leptodactylidae). Occas. Pap. Mus. Nat. Hist. Univ. Kansas 61: 1–24.

—. 1986. The definition of the Middle American clade of *Eleutherodactylus* based on jaw musculature (Amphibia: Leptodactylidae). Herpetologica 42: 248–258.

—. 1989. Intrageneric relationships of mainland *Eleutherodactylus* (Leptodactylidae). I. A review of the frogs assigned to the *Eleutherodactylus discoidalis* species group. Contr. Biol. Geol. Milwaukee Public Mus. 79: 1–25.

— and P. A. Burrowes. 1990. The frogs of the genus *Eleutherodactylus* at the La Planada Reserve in southwestern Colombia with descriptions of eight new species. Occas. Pap. Mus. Nat. Hist. Univ. Kansas 136: 1–31.

Noble, G. K. 1923. Six new batrachians from the Dominican Republic. Amer. Mus. Novit. 61: 1–6.

Savage, J. M. 1987. Systematics and distribution of the Mexican and Central American rainfrogs of the *Eleutherodactylus gollmeri* group (Amphibia: Leptodactylidae). Fieldiana, Zoology (new ser.) 33: 1–57.

Schmidt, K. P. 1921. Notes on the herpetology of Santo Domingo. Bull. Amer. Mus. Nat. Hist. 44: 7–20.

Schwartz, A. 1965. Variation and natural history of *Eleutherodactylus ruthae* on Hispaniola. Bull. Mus. Comp. Zool., Harvard Univ. 132: 479–508.

—. 1974. *Eleutherodactylus planirostris*. Cat. Amer. Amphib. Rept.: 154.1–154.4.

—. 1976. Two new species of Hispaniolan *Eleutherodactylus* (Leptodactylidae). Herpetologica 32: 163–171.

— and R. W. Henderson. 1991. Amphibians and Reptiles of the West Indies: Descriptions, Distributions, and Natural History. Univ. Florida Press, Gainesville, Florida.

Shreve, B. and E. E. Williams. 1963. Part II. The frogs, p. 302–342. *In* E. E. Williams, B. Shreve, and P. S. Humphrey, The herpetology of the Port-au-Prince region and Gonave Island, Haiti. Parts I–II. Bull. Mus. Comp. Zool., Harvard Univ. 129: 291–342.

Tschudi, J. J. 1838. Classification der Batrachier, mit Berucksichtigung der fossilien Thiere dieser Abtheilung der Reptilien. Petitpierre, Neuchatel (reprint 1967, Soc. Study Amphib. Rept.).

Zweifel, R. G. 1967. Identity of the frog *Cornufer unicolor* and application of the generic name *Cornufer*. Copeia 1967: 117–121.

APPENDIX. The members of the outgroup employed in the analysis of the relationships of *Pelorius* (and *Euhyas*) are listed below in alphabetical order within the subgenera *Craugastor* and *Eleutherodactylus*.

Craugastor: *andi, angelicus, anomalus, augusti, azueroensis, berkenbuschii, biporcatus, bocourti, bransfordii, brocchi, bufoniformis, cheiroplethus, crassidigitus, daryi, decoratus, emcelae, escoces, fitzingeri, fleischmanni, gollmeri, greggi, laticeps, lineatus, longirostris, matudai, maussi, melanostictus, merendonensis, mexicanus, milesi, necerus, noblei, occidentalis, podiciferus, punctariolus, pygmaeus, raniformis, rhodopis, rostralis, rugulosus, sartori, spatulatus, stejnegerianus, talamancae, taurus, uno, vocalis, xucanebi*, and *zygodactylus*.

Eleutherodactylus: *acerus, achatinus, actites, acuminatus, acutirostris, altae, altamazonicus, anolirex, appendiculatus, atratus, babax, bacchus, baryecuus, bellona, bicolor, bicumulus, boconoensis, bogotensis, boulengeri, bromeliaceus, buckleyi, cacao, cadenai, cajamarcensis, calcarulatus, caprifer, carmelitae, caryophyllaceus, celator, cerasinus, cerastes, chalceus, chloronotus, colodactylus, condor, conspicillatus, cornutus, cremnobates, crenunguis, cristinae, croceoinguinis, crucifer, cruentus, cruralis, cryophilius, cryptomelas, curtipes, danae, delicatus, devillei, diadematus, diaphonus, diastema, discoidalis, dolops, duellmani, elassodiscus, elegans, erythropleura, fenestratus, gaigei, galdi, ginesi, glandulosus, gracilis, hectus, imitatrix, incanus, incomptus, ingeri, insignitus, jaimei, labiosus, lacrimosus, lancinii, lanthanites, latens, leoni, leptolophus, leucopus, loustes, luteolateralis, lymani, lynchi, malkini, mantipus, martiae, megalops, modipeplus, myersi, nervicus, nicefori, nigrovittatus, nyctophylax, obmutescens, ockendeni, ocreatus, orcesi, orestes, ornatissimus, orpacobates, palmeri, pardalis, peraticus, percultus, permixtus, philipi, phoxocephalus, platydactylus*, *prolatus*, *prolixodiscus*, *proserpens*, *pseudoacuminatus*, *pugnax*, *quaquaversus*, *quinquagesimus*, *racemus*, *repens*, *rhabdolaemus*, *ridens*, *riveti*, *roseus*, *rubicundus*, *ruidus*, *ruizi*, *ruthveni*, *sanctaemartae*, *satagius*, *sernai*, *simoterus*, *spilogaster*, *spinosus*, *stenodiscus*, *sulcatus*, *supernatis*, *surdus*, *taeniatus*, *tamsitti*, *tayrona*, *tenebrionis*, *terraebolivaris*, *thectopternus*, *thymalopsoides*, *thymelensis*, *toftae*, *trachyblepharis*, *trepidotus*, *unistrigatus*, *uranobates*, *variabilis*, *ventrimarmoratus*, *versicolor*, *vertebralis*, *vicarius*, *vidua*, *walkeri*, and *w-nigrum*.

Resumen.—Se encuentra dos grupos féneticos usando los cráneos de las ranas del subgénero *Pelorius* del género *Eleutherodactylus* y los grupos son iqual que grupos definidos usando carácteristicas externas y de electroforesis. Sin embargo, cuando se examine las carácteristicas por un método objectivo de grupos ajenos, se encuentra que la mayoría de las carácteres no son informativas de los parentescos. Las especies del grupo *inoptatus* (*E. chlorophenax*, *E. inoptatus*, *E. nortoni*) comparten una sinapomorfia y las especies del grupo *ruthae* (*E. hypostenor*, *E. parapelates*, *E. ruthae*) comparten dos más pero los parentescos dentros los grupos no son resueltos y no se encuentra una sinapomorfia para el subgénero. A más de esto, parece que hay una sinapomorfia para un grupo incluyendo el grupo *inoptatus* y una parte del subgénero *Euhyas*.

Résumé.—Deux groupes phénétiques se dégagent de l'étude des crânes de grenouilles du sous-genre *Pelorius* (genre *Eleutherodactylus*); ses groupes phénétiques correspondent à des groupes définis au moyen de caractéristiques externes et d'électromorphes. Toutefois, lorsque ces caractéristiques (crâniennes et autres) sont soumises à de rigoureuses comparaisons hors groupe, la plupart des différences n'éclairent nullement les rapports inter-groupes. Le groupe *inoptatus* (*E. Chlorophenax*, *E. inoptatus*, *E. nortoni*) partage une synapomorphie et le groupe *ruthae* (*E. hypostenor*, *E. parapelates*, *E. ruthae*) en partage deux autres mais les rapports au sein des groupes ne sont pas élucidés et aucune synapomorphie ne permet d'expliquer le sous-genre. Une seule synapomorphie apparente permet d'établir un lien entre le groupe *inoptatus* et une partie du sous-genre *Euhyas*.

Lynch, J. D. 1996. The relationships of the Hispaniolan frogs of the subgenus *Pelorius* (*Eleutherodactylus*: Leptodactylidae), p. 141–155. *In* R. Powell and R. W. Henderson (eds.), Contributions to West Indian Herpetology: A Tribute to Albert Schwartz. Society for the Study of Amphibians and Reptiles, Ithaca (New York). Contributions to Herpetology, volume 12.

A KARYOLOGICAL PERSPECTIVE ON THE MONOPHYLY OF THE HYLID GENUS *OSTEOPILUS*

KAREN ANDERSON[1]

Abstract.—The hylid genus *Osteopilus* currently is represented by three species: *Osteopilus brunneus* from Jamaica, *O. dominicensis* from Hispaniola, and *O. septentrionalis* from Cuba, the Bahamas, and southern Florida. Karyological evaluations of these treefrogs indicate that *O. dominicensis* and *O. septentrionalis* are closely related; both species have 12 homomorphic metacentric or submetacentric chromosome pairs of similar morphology, C-bands in centromeric and most telomeric regions, and a nucleolar organizer region (NOR) on Chromosome 9. The *O. septentrionalis* karyotype can be derived from that of *O. dominicensis* by an inversion in the short arm of Chromosome 9 and small modulations in heterochromatin. *Osteopilus brunneus* is unusual among hylids in having 17 chromosome pairs, 10 of which are telocentric; the NOR is on the telocentric Chromosome 17. This karyotype can be derived from that of *O. dominicensis* by six fission events, one subsequent fusion event, and an inversion in the derived Chromosome 17. Telomere sequence patterns support a fission scenario; none of the karyotypes show interstitial or centromeric telomere signal. Out-group comparisons do not support a close relationship between *Osteopilus* and *Osteocephalus* or the West Indian hylids *Hyla marianae* and *H. wilderi*.

INTRODUCTION

THE GENUS *OSTEOPILUS* IS REPRESENTED BY THREE species of casque-headed hylids generally endemic to the West Indies: *O. brunneus* from Jamaica, *O. dominicensis* from Hispaniola, and *O. septentrionalis* from Cuba, the Bahamas, and southern Florida. The genus is characterized most notably by a suite of cranial features and a supplementary apical element of the intermandibularis muscle (Trueb and Tyler 1974).

The relationship of these three species to other West Indian hylid species remains unclear. Dunn (1926) suggested that *Osteopilus brunneus* gave rise in Jamaica to *Hyla marianae*, *H. wilderi*, and *Calyptahyla crucialis*, and that *O. dominicensis* gave rise in Hispaniola to *H. heilprini*, *H. pulchrilineata*, and *H. vasta*. Trueb and Tyler (1974) argued that the species in *Osteopilus* are closely related, and proposed that the extant West Indian hylid fauna may represent up to six separate invasions from South America, with species in *Osteopilus* allied with those of *Osteocephalus*. In addition, they suggested that *H. marianae* and *H. wilderi* are closely related, but not allied with other West Indian hylids. Albumin microcomplement fixation data support, in part, the Dunn hypothesis of paraphyly: a single invasion of a South American stock with subsequent speciation on island masses (Maxson 1992) and a more recent divergence of *O. dominicensis* and *O. brunneus* from an *O. septentrionalis*-like stock (Hedges et al. 1992).

Each hypothesis carries different karyological implications. If *Osteopilus* is monophyletic, the karyotypes of *O. brunneus*, *O. dominicensis*, and *O. septentrionalis* would be similar or, at least, easily derived from one another. In the Trueb and Tyler (1974) hypothesis, all three of these karyotypes would be similar to those in *Osteocephalus* and the karyotypes of *Hyla marianae* and *H. wilderi* would be similar to one another and distinct from those in *Osteopilus*. If, on the other hand, *Osteopilus* is

[1] Department of Biology, Hofstra University, Hempstead, New York 11550, USA.

paraphyletic, the species in this genus would have primitive karyotypes and the karyotypes in each island assemblage would be easily derived from one another.

The present study focuses primarily on chromosome evolution within *Osteopilus* and suggests likely derivations among these karyotypes. Out-group comparisons are restricted by the recent lack of availability of specimens from Hispaniola; however, karyological comparisons with *Hyla marianae* and *H. wilderi* and with *Osteocephalus taurinus* allow for a preliminary test of the paraphyly of *Osteopilus*.

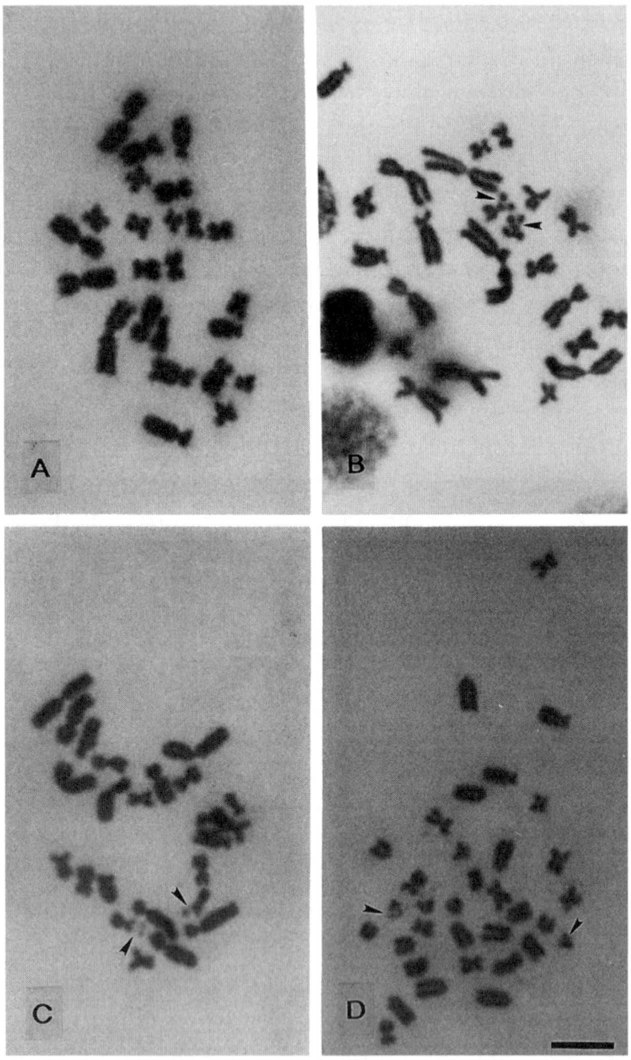

FIGURE 1. Giemsa-stained metaphase plates of: (A and B) *Osteopilus dominicensis*, (C) *O. septentrionalis*, and (D) *O. brunneus*. Nucleolar organizer regions (NORs) appear as gaps in the chromosome arms (arrowheads). In *O. dominicensis*, the NORs are not evident in mid-metaphase plates (A), but are clearly evident in early-metaphase plates (B). Bar = 10 μm.

MATERIALS AND METHODS

Collection Sites and Specimens

Two specimens of *Osteopilus dominicensis* were collected from the Dominican Republic, six specimens of *O. brunneus* were collected from Jamaica, and a total of 10 specimens of *O. septentrionalis* was collected from Bimini, Cuba, and southern Florida. Out-group specimens include one *Osteocephalus taurinus* from French Guiana, one *Hyla marianae* from Jamaica, and two *H. wilderi* from Jamaica (see Appendix).

All specimens were maintained in the laboratory for several weeks prior to sacrifice. With the exception of two *Osteopilus brunneus* (Karen Anderson Collection [KA] 1260 and 1268), all vouchers were placed in 10% buffered formalin, stored in 70% ethanol, and later deposited in the National Museum of Natural History (USNM). Specimens numbered KA 1260 and KA 1268 were skeletonized in a dermestid beetle colony and currently are in a collection held by Herndon G. Dowling.

Preparation and Banding Procedures

Frogs were inoculated with 0.2 ml of 0.5% colchicine (Sigma) three or 36 hours prior to sacrifice. Preparation of metaphase chromosomes from intestinal mucosa followed standard procedures. Several plates were stained with Giemsa (Harleco #620), photographed, and subsequently C-banded and silver stained. C-band procedures followed two rounds of the BSG technique (Sumner 1972). Silver stain procedures followed the AgAS technique (Goodpasture and Bloom 1975) with some modification (Anderson and Moler 1986). Ten to 30 metaphase plates were analyzed for each species. Chromosome measurements were made with vernier calipers from photographs, and idiograms were generated from relative length calculations.

Fluorescent in situ Hybridization

Unstained metaphase plates were hybridized under low stringency with a biotinylated DNA probe (Oncor #P5097B5) for human telomere sequence. Procedures followed ONCOR instructions. Hybridization was detected by two amplification rounds of FITC-

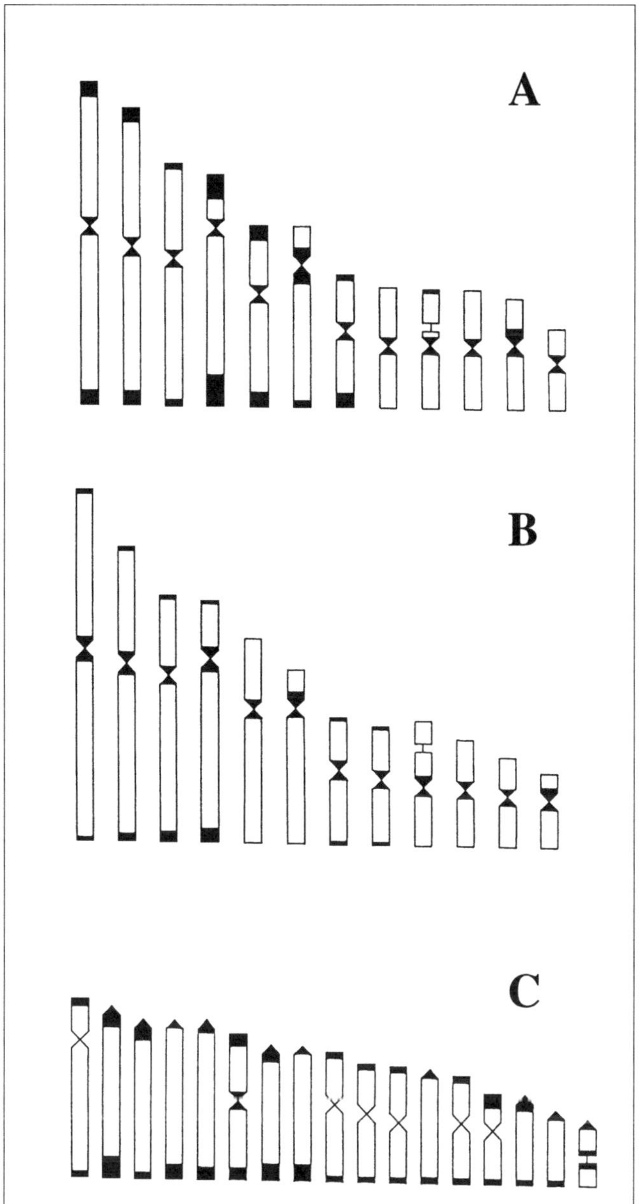

FIGURE 2. Idiograms of: (A) *Osteopilus dominicensis*, (B) *O. septentrionalis*, and (C) *O. brunneus*. Solid regions indicate C-band positive heterochromatin; nucleolar organizer regions are indicated by gaps in the chromosome arm.

conjugated avidin. Slides were mounted in propidium iodide and stored in a dark chamber for four days prior to examination.

Photography

All metaphase plates were photographed with an Olympus BH-2 microscope with fluorescent capability. Giemsa and silver stained plates were photographed with Kodak TMAX 100 film and were printed on Agfa Rapitone P1–2 paper. C-banded plates were photographed with Kodak Tech Pan film and were printed on Kodabrome II RC paper. Hybridization slides were excited at 460 nm (BH2-DMBG filter system), photographed on Kodak EKTAR 1000 and TMAX films, and printed on Kodabrome II RC paper.

RESULTS

Karyotypes of *Osteopilus*

The karyotypes of *Osteopilus dominicensis* and *O. septentrionalis* are very similar in general chromosome morphology (Figures 1 and 2). Each has 12 homomorphic pairs ($2n = 24$, FN = 24), with the first six pairs being distinctly larger than the remaining six pairs. With the exception of the subtelocentric chromosomes 4 and 6, all pairs are metacentric or submetacentric. In both species, the nucleolar organizer region (NOR) is located on the short arm of Chromosome 9 (Figure 1) and has little or no C-band positive heterochromatin at the NOR junctions; however, the NOR, represented as a gap in the idiograms, is moderately C-band positive. C-bands are present at all centromeric regions and at several telomeric sites (Figure 3); the centromeric C-band on Chromosome 6 extends well into the chromosome arms and the telomeric C-band on Chromosome 4 in *O. dominicensis* is prominent. These variations in C-band pattern most probably represent local amplification of heterochromatin.

In the two female specimens of *Osteopilus dominicensis*, no variation exists in C-band patterns or placement of the NOR on the small arm. In contrast, among the 10 specimens of *O. septentrionalis*, variation is evident in both the number and placement of the NOR. The one male specimen from Cuba has an NOR placed approximately two-thirds distal to the centromere (Figure 4A); this position represents the NOR on the idiogram for the species (Figure 2B). The one male and two female specimens from Bimini have an NOR placed median on the short arm (Figure 4B). In Florida, two males have a terminal NOR pair, one male has a two-thirds distal pair, one male has one terminal and one two-thirds distal pair, and one female has only one two-thirds distal

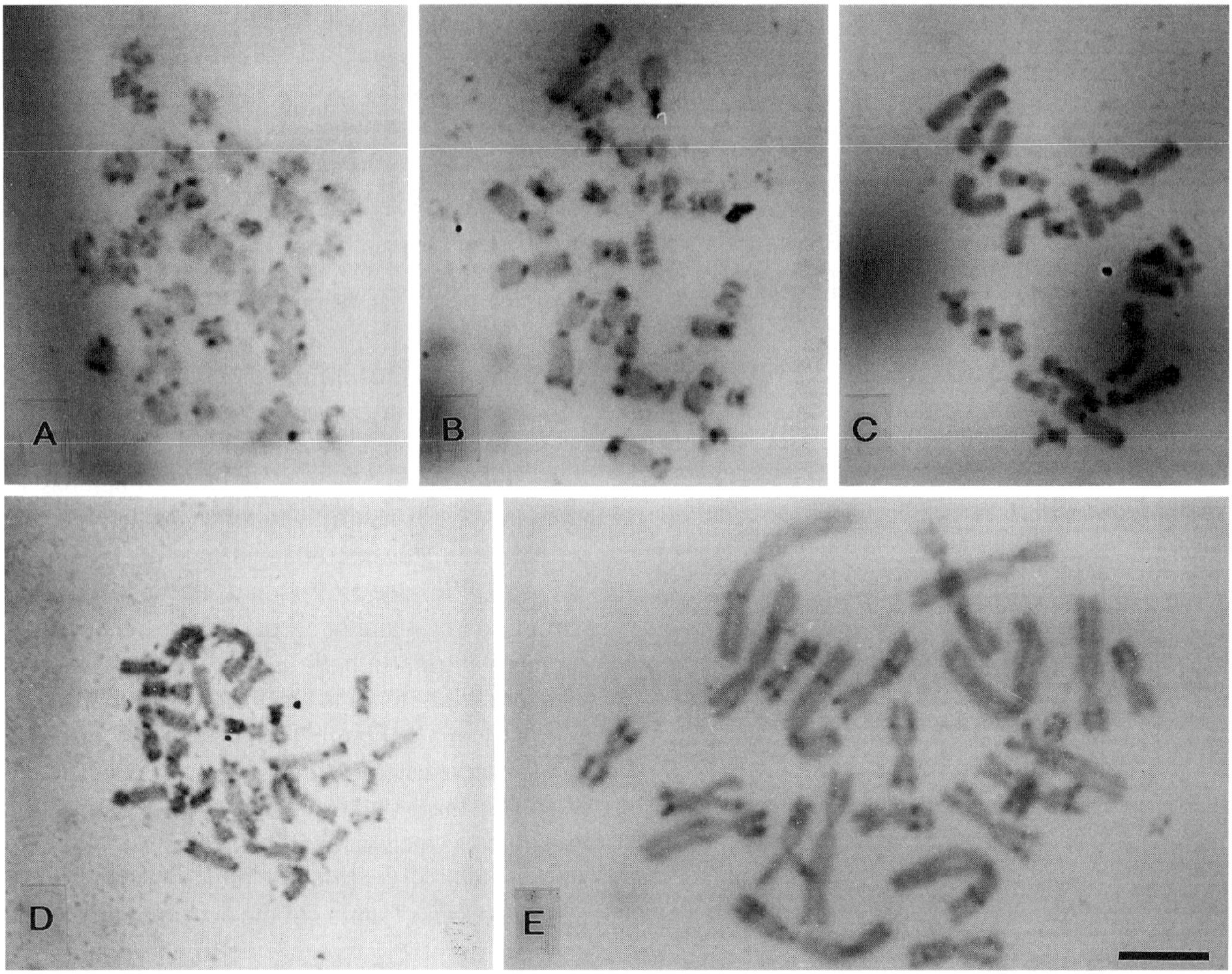

FIGURE 3. C-band patterns in mid-metaphase chromosomes of: (A) *Osteopilus brunneus*, (B) *O. dominicensis* (same plate as in Figure 1A), (C) *O. septentrionalis* (same plate as in Figure 1C), (D) *Hyla wilderi*, and (E) *Osteocephalus taurinus* (same plate as in Figure 8D and 8E). Bar = 10 µm.

NOR-bearing chromosome (Figure 4C). In all cases, the number of silver-stained nucleoli in interphase nuclei agree with the number of NORs visible in Giemsa-stained metaphase plates. The variation seen in the placement of the NOR on the short arm of Chromosome 9 suggests that several paracentric inversions have occurred on this chromosome and that specific inversions may be fixed in the isolated populations on Bimini and Cuba. However, the present study has only small samples from these islands and the possibility of other NOR variants from these regions cannot be excluded.

Finally, one female from Dade County, Florida, has a clear translocation between chromosomes 6 and 9 (Figure 4D). In Giemsa-stained plates, the "homologues" of chromosomes 6 and 9 do not show the same general morphology (Figure 5A). In the Chromosome 6 pair, the atypical homologue has a secondary constriction indicating a presumptive NOR. This homologue has moderate C-band positive heterochromatin in the secondary constriction (Figure 5B) and is silver-stain positive (Figure 5C). Likewise, the atypical homologue of the Chromosome 9 pair (Figure 5A) has the C-band pattern of Chromosome 6 (Figure 5B) and is silver-stain negative (Figure 5C). A translocation with breakpoints near the centromere on the long arms of chromosomes 6 and 9 would result in the patterns seen.

The karyotype of *Osteopilus brunneus* is distinct from those of *O. dominicensis* and *O. septen-*

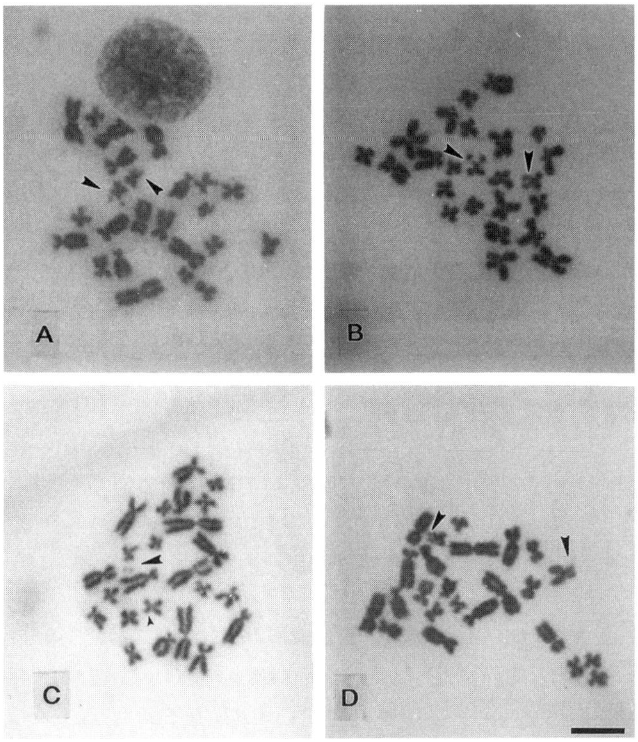

FIGURE 4. Giemsa-stained metaphase chromosomes of *Osteopilus septentrionalis* from: (A) Cuba; (B) Bimini; (C) St. Lucie County, Florida; and (D) Dade County, Florida. The specimen from Dade County has a reciprocal translocation between chromosomes 6 and 9. Large arrowheads indicate the location of the nucleolar organizer region (NOR) on Chromosome 9; the small arrowhead indicates the NOR-negative homologue of Chromosome 9. Bar = 10 μm.

trionalis (Figures 1 and 2). This species has 17 homomorphic pairs ($2n = 34$, FN = 24), with no clear division between larger and smaller chromosomes. Ten pairs are telocentric and the remaining seven pairs are metacentric or submetacentric. The NOR appears in a median position on the telocentric Chromosome 17 (Figure 1D); it is C-band positive and is bordered by a small amount of heterochromatin. All of the pairs have C-bands at telomeric sites (Figure 3A). Among the metacentric and submetacentric pairs, only Chromosome 6 has a C-band at the centromeric region. Without the aid of additional chromosomal markers, distinguishing certain pairs from one another is difficult, if not impossible.

Five of the six specimens of *Osteopilus brunneus* have silver-positive NORs on both homologues of Chromosome 17. The slides of one specimen were destroyed after C-band procedures and, therefore, could not be silver-stained; however, secondary constrictions are present on Chromosome 17 in these Giemsa-stained plates. Four specimens (3 males and 1 female) have an additional NOR on both homologues of Chromosome 16 (Figure 6). This NOR is subtelocentric and not easily seen in Giemsa-stained preparations (Figure 6A); it is easily seen in silver-stained plates (Figure 6B). Individuals with the Chromosome 16 NORs have up to four silver-positive nucleoli in interphase nuclei (Figure 6C); the one female specimen lacking a NOR on Chromosome 16 has up to two nucleoli in interphase nuclei.

Fluorescent in situ hybridization (FISH) with a human telomere probe was performed on chromosomes from one female *Osteopilus septentrionalis* and one male and one female *O. brunneus* (Figure 7). Metaphase plates from both species show hybridization signals at the telomeric regions only. No evidence suggests a signal at the centromeric region of Chromosome 6 in *O. brunneus*. The differences

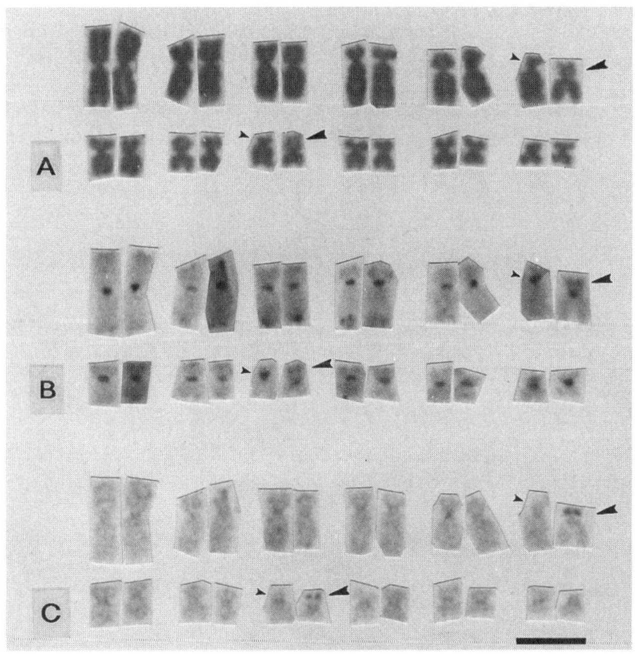

FIGURE 5. The karyotype from the metaphase plate shown in Figure 1D: (A) Giemsa-stained chromosomes; (B) the same chromosomes C-banded; and (C) the same chromosomes silver-stained. This specimen of *Osteopilus septentrionalis* has a reciprocal translocation between the long arms of chromosomes 6 and 9, resulting in the shift of the entire small and a portion of the long arm. Large arrowheads indicate the chromosome segment bearing a nucleolar organizer region (= short arm of Chromosome 9); small arrowheads indicate the segment for Chromosome 6. Bar = 10 μm.

in signal intensity at telomeric regions reflect varying focal planes and the conversion of color to black and white images.

Karyotypes of *Hyla marianae* and *H. wilderi*

The karyotypes of *Hyla marianae* and *H. wilderi* differ from one another in several aspects (Figures 8 and 9). *Hyla marianae* has 12 homomorphic chromosome pairs ($2n = 24$, FN = 24), with the first five pairs distinctly larger than the remaining seven pairs; all pairs are metacentric or submetacentric. No secondary constrictions are apparent in Giemsa-stained plates (Figure 8A) and neither C-band nor silver stain information was obtained. *Hyla wilderi* has 14 homomorphic pairs ($2n = 28$, FN = 26), three of which are distinctly larger than the remaining pairs. Chromosomes 7 and 10 are telocentric; all of the other pairs are metacentric or submetacentric. C-band positive heterochromatin is present on all telomeric and most centromeric sites; chromosomes 6 and 8 have interstitial C-bands, and the long arm of Chromosome 12 is mainly C-band positive (Figure 3D).

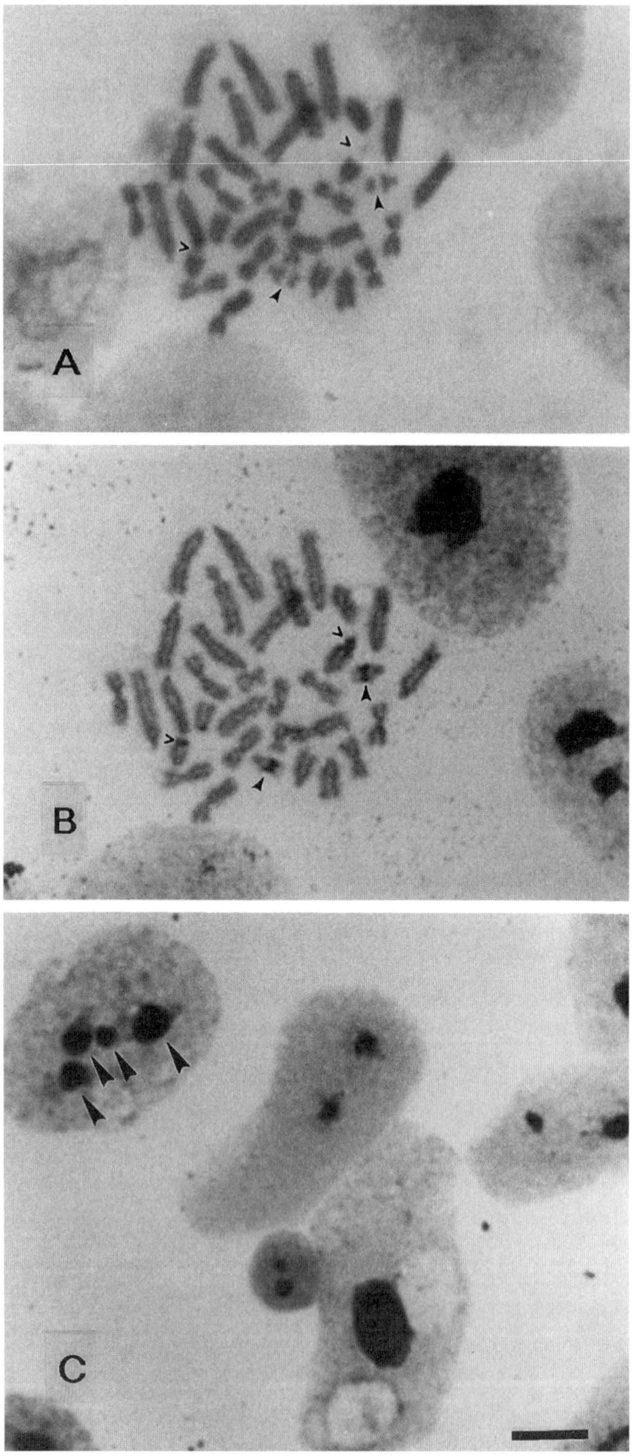

FIGURE 6. The chromosomes from a specimen of *Osteopilus brunneus* with four nucleolar organizer regions: (A) Giemsa-stained plate, showing a secondary constriction in Chromosome 17 (solid arrowheads) and the region of the supplemental NOR at the telomere of Chromosome 16 (open arrowheads); (B) same plate, silver-stained: secondary constrictions now stain black from silver deposition, and nucleoli within nuclei also stain black; and (C) silver-stained interphase nuclei from the same specimen: nuclei have up to four silver-positive nucleoli (solid arrowheads) — as each cell matures, the nucleoli associate within the nucleus and appear as one large nucleolus. Bar = 10 µm.

Karyotype of *Osteocephalus taurinus*

The karyotype of *Osteocephalus taurinus* is represented by 12 homomorphic pairs ($2n = 24$, FN = 24) that have no distinct size groupings (Figures 8 and 9). Chromosomes are metacentric or submetacentric and a NOR is present on the short arm of both homologues of Chromosome 12 (Figures 8D and 8E). C-band patterns differ from those seen in the karyotypes of *Osteopilus* and *Hyla wilderi* (Figure 3E). Most C-bands in the *Osteocephalus taurinus* karyotype are interstitial; few centromeric or telomeric bands are evident. In addition, two fragile sites (silver-negative secondary constrictions) exist within heterochromatic regions in chromosomes 6 and 9. Although no genome size values were determined, it is important to note that mid-metaphase chromosomes in *O. taurinus* are twice as large as those found in *Osteopilus* (see Figure 3) or *H. marianae* and *H. wilderi* (see Figure 8). This difference in size is not evident in idiograms drawn from relative length values (Figure 9).

DISCUSSION

The herpetofauna of the West Indies has few representatives of the family Hylidae. Only nine species occur among the three major islands of Cuba, Hispaniola, and Jamaica, and each species, with the exception of *Osteopilus septentrionalis*, is endemic to one island mass. *Osteopilus septentrionalis* occurs in Cuba, the Bahamas, southern Florida, the Puerto Rican Bank, and St.-Martin (Schwartz and Thomas 1975; Powell et al. 1992). This current distribution presumably represents recent immigrations or introductions from Cuba. Hispaniola and Jamaica each has four species of hylids: *Hyla heilprini*, *H. pulchrilineata*, *H. vasta*, and *O. dominicensis* on Hispaniola and *Calyptahyla crucialis*, *H. marianae*, *H. wilderi*, and *O. brunneus* in Jamaica.

Dunn (1926) reviewed the distribution and life histories of the West Indian hylids and recognized that the three species of *Osteopilus* were "closely allied." He also recognized some "resemblance" among the remaining West Indian hylids. However, based largely on external morphological features, breeding sites, and tadpole morphology, he suggested that the three species of *Hyla* on Hispaniola are derivatives of *O. dominicensis* and that the two species of *Hyla* and *Calyptahyla crucialis* on Jamaica are derivatives of *O. brunneus*.

Trueb and Tyler (1974) provided a more extensive review of the morphology and ecology of the West Indian hylids. They also recognized that *Osteopilus brunneus*, *O. dominicensis*, and *O. septentrionalis* are closely related, placed these species into the genus *Osteopilus*, and noted that *O. dominicensis* and *O. septentrionalis* are most similar in osteological features. They suggested that the West Indian hylid fauna may represent up to six separate invasions from South America. By this scenario, one stock invaded the West Indies and evolved the throat musculature unique to the genus; subsequent isolation of this stock on Cuba, Hispaniola, and Jamaica gave rise to the three recognized species in *Osteopilus*; the ancestral stock in South America may have given rise to *Osteocephalus*. In addition, one hylid stock from South America may have given rise to *Hyla marianae* and a more derived *H. wilderi*. Each of the remaining hylid species would represent a separate lineage from a South American ancestor.

More recent molecular analyses using albumin microcomplement fixation techniques dispute the monophyly of *Osteopilus* (Hedges et al. 1992; Maxson 1992). To summarize the points relevant to the present discussion: *O. brunneus* and *O. do-*

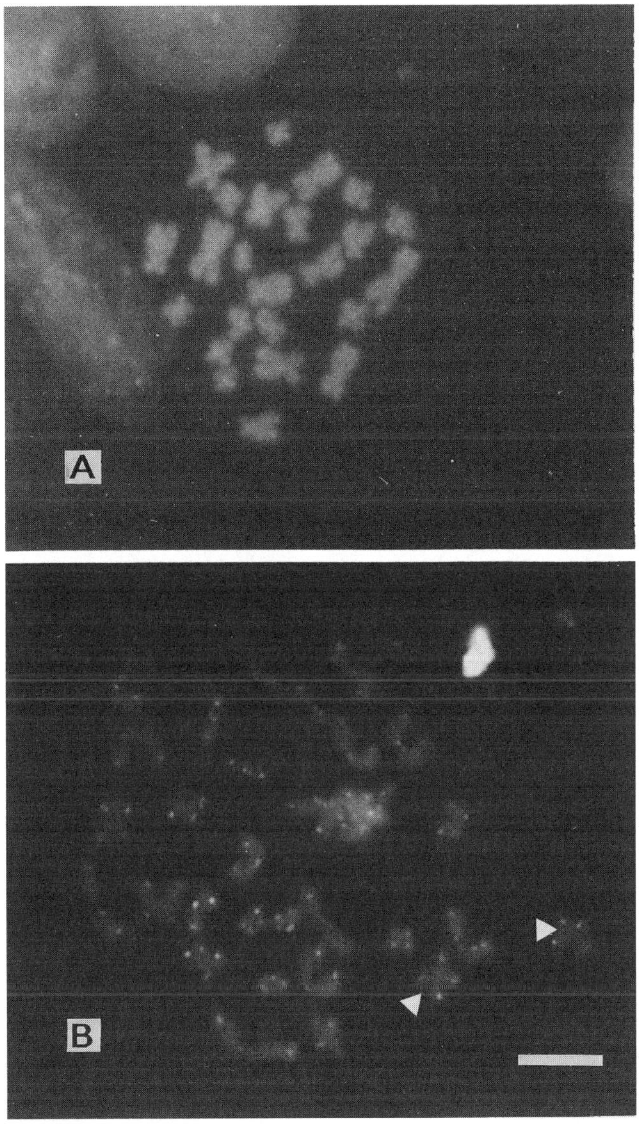

FIGURE 7. Fluorescent in situ hybridization with a human telomere probe on chromosomes from: (A) *Osteopilus septentrionalis,* and (B) *O. brunneus.* Telomere signal is present only at telomeres. The white arrowheads indicate the centromeric regions of the proposed metacentric fusion chromosomes in *O. brunneus*. Bar = 10 μm.

minicensis have apparently diverged more recently from one another than from *O. septentrionalis* (Hedges et al. 1992), *O. septentrionalis* diverged early from an ancestral stock, *Hyla pulchrilineata* and *O. dominicensis* share a common ancestor, and *Calyptahyla crucialis*, *H. marianae*, and *O. brunneus* share a common ancestor (Maxson 1992). These findings generally agree with those of Dunn (1926); namely that hylid evolution has occurred in isolation on each island mass.

Chromosomal studies can provide an independent perspective on relationships among closely allied species. Although no one has suggested that chromosome evolution functions as a molecular clock, an ancestral karyotype puts some constraints on the types of chromosomal rearrangements that may occur. For example, an ancestral karyotype that contains only metacentric or submetacentric chromosomes could sustain fission, but not fusion, events, and an ancestral genome size could not double without some evidence of polyploidy or considerable amplification of heterochromatin.

Chromosomal studies of the West Indian hylids have been confined to determinations of chromosome number and general chromosome morphology for *Osteopilus brunneus* and *O. septentrionalis*

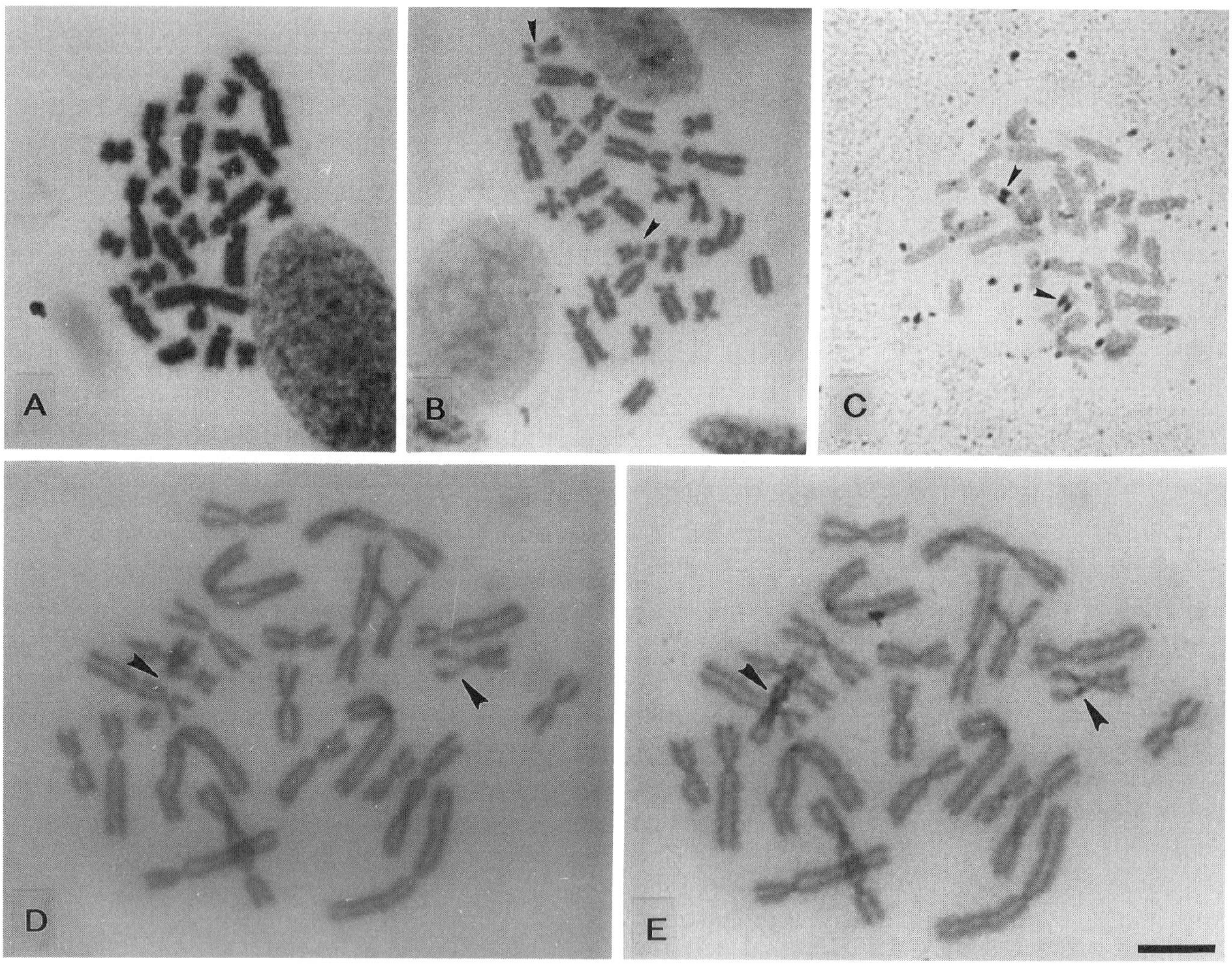

FIGURE 8. Mid-metaphase chromosomes of out-group hylids to *Osteopilus*: (A) Giemsa-stained plate from *Hyla marianae*, no nucleolar organizer regions (NOR) are evident; (B) Giemsa-stained plate from *H. wilderi* with NOR gaps (arrowheads) in both homologues of Chromosome 9; (C) silver-stained plate from *H. wilderi* with silver deposited only in the gaps of Chromosome 9; (D) Giemsa-stained plate from *Osteocephalus taurinus* with NOR gaps in both homologues of Chromosome 12; and (E) silver-stained plate as in D (and Figure 3E) with silver deposited in the gaps of Chromosome 12 (the difference in NOR size is consistent in all chromosome preparations from this individual frog). Bar = 10 μm.

(Duellman and Cole 1965; Goin et al. 1978). The limited number of species analyzed and the limited nature of these chromosomal studies have restricted the use of this karyological information in discussions of West Indian hylid relationships. However, Goin et al. (1978) presented a possible derivation of the *O. brunneus* karyotype from that of *O. septentrionalis* by fissions of the five largest chromosomes in *O. septentrionalis* to form the ten telocentric chromosomes in *O. brunneus*.

The present study includes a more extensive evaluation of the karyotypes of the species currently allocated to *Osteopilus*. *Osteopilus dominicensis* and *O. septentrionalis* have similar karyotypes. The karyotype of *O. septentrionalis* can be derived easily from that of *O. dominicensis* by local modulation of heterochromatic material at centromeric and telomeric sites, and a paracentric inversion on the small arm of Chromosome 9 (Figure 10). The location of such an inversion cannot be determined by the parameters of this study; no interstitial C-bands (other than within the NOR) mark possible breakpoints. Little chromosome evolution apparently has occurred since these two species diverged. The *O. brunneus* karyotype can be derived from that of *O. dominicensis* by a series of fission events. The simplest derivation would involve five fissions of metacentric and submetacentric chromosome pairs in *O. dominicensis* and result in 10 telocentric pairs in *O. brunneus*. Such a scenario is not compatible with the chromosome morphologies presented here. Rather, six fission events with a subsequent fusion event in the *O. dominicensis* karyotype would result in a close approximation of the *O. brunneus* karyotype (Figure 10). The metacentric Chromosome 6 of *O. brunneus* would be the product of the fusion of the short arms of chromosomes 4 and 5 of *O. dominicensis*. In this derivation, Chromosome 9 of *O. dominicensis* would undergo a fission and result in chromosomes 16 and 17 of *O. brunneus*. That the NOR found on Chromosome 16 in many specimens of *O. brunneus* originated from an inversion in Chromosome 9 of *O. dominicensis* is possible, and that by unequal cross-over, this NOR is being lost on some Chromosome 16 homologues. That the proposed fusion chromosome in *O. brunneus* is the only metacentric or submetacentric pair with a C-band positive centromeric region is interesting to note. Differences in C-band patterns between the two species may represent whole arm inversions

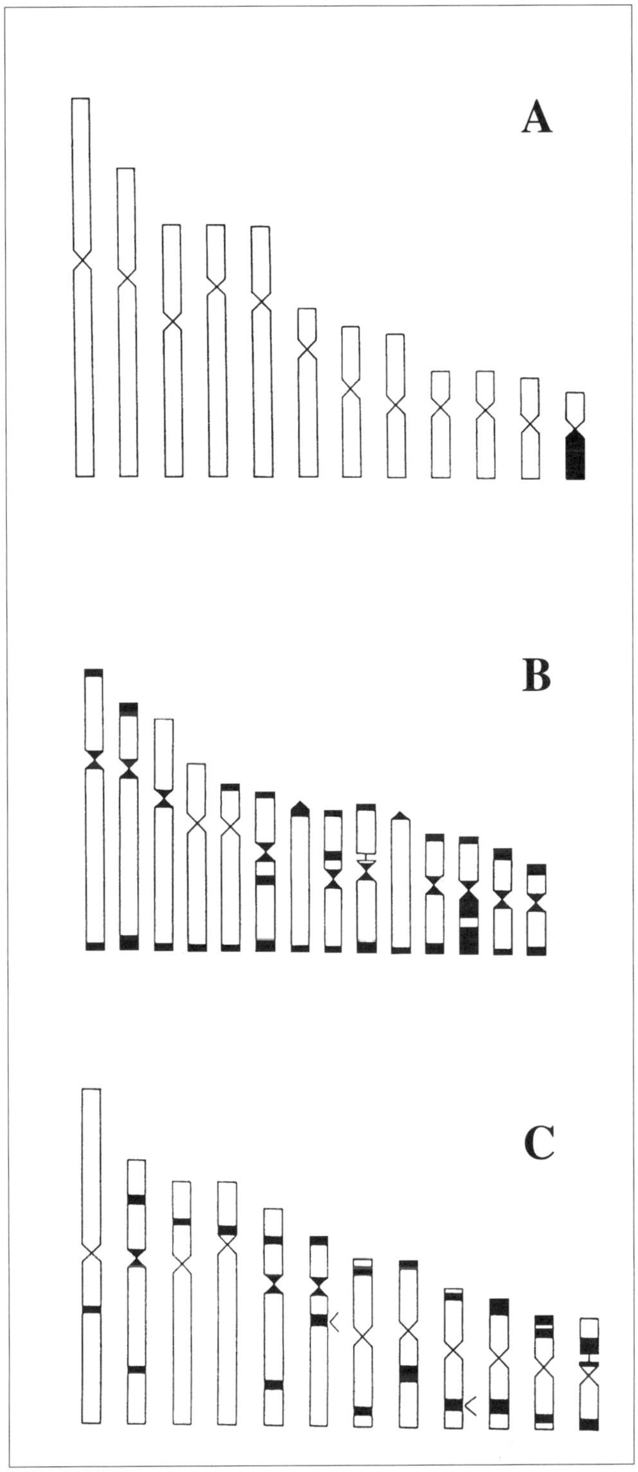

FIGURE 9. Idiograms of: (A) *Hyla marianae*, (B) *H. wilderi*, and (C) *Osteocephalus taurinus*. Codes follow Figure 2; open arrowheads indicate the location of fragile sites. Chromosomes from *H. marianae* were not C-banded or silver-stained.

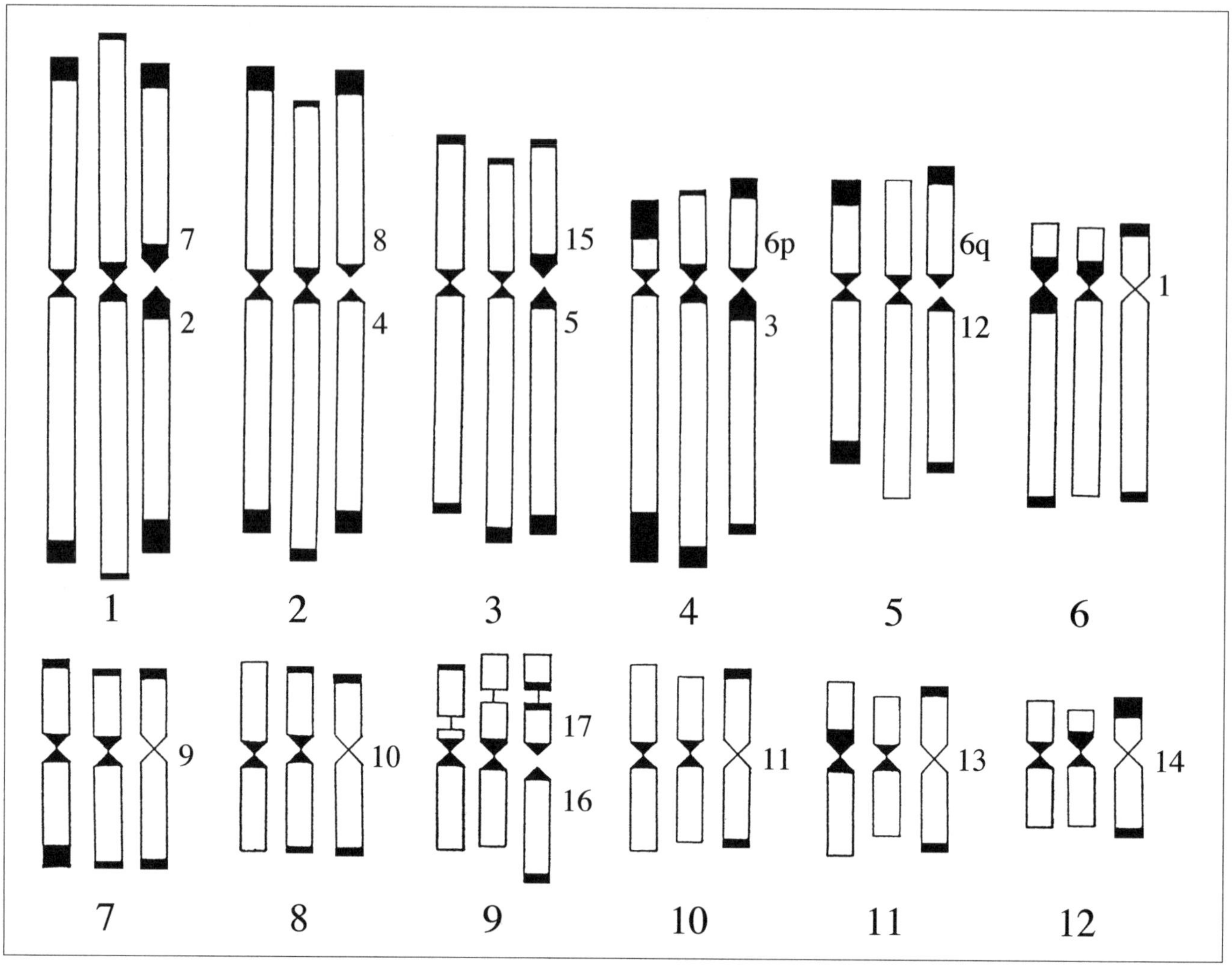

FIGURE 10. A comparison of the idiograms from *Osteopilus dominicensis* (left), *O. septentrionalis* (center), and *O. brunneus* (right). Numbers at the bottom refer to chromosome designations for both *O. dominicensis* and *O. septentrionalis*; numbers at the side refer to chromosome designations for *O. brunneus*. Abbreviations: p = short arm, and q = long arm. The *O. brunneus* idiogram is arranged to conform to that of the other two species.

and/or local modulation of heterochromatic sequences.

Telomere sequence patterns, as determined by FISH, are consistent with the proposed polarity of fission events in *Osteopilus*. If 10 telocentric chromosomes had fused to form the metacentric and submetacentric chromosomes seen in the karyotypes of *O. dominicensis* and *O. septentrionalis*, one might expect telomeric sequences to be present in centromeric regions. No hybridization signal was seen in any centromeric region in *O. septentrionalis* (Figure 7). Likewise, hybridization results did not support the proposed fusion for Chromosome 6 in *O. brunneus*; the centromeric region of this chromosome did not have detectable telomeric sequence. However, the lack of detectable telomeric DNA at the centromeric regions does not conclusively exclude possible fusions; telomeric sequences may be lost in this type of rearrangement.

Karyological data are consistent with a monophyletic origin for *Osteopilus* and the close relationship between *O. dominicensis* and *O. septentrionalis* proposed by Trueb and Tyler (1974). However, chromosomal results do not support a close relationship between *Osteopilus* and *Osteocephalus*. In addition, the *Hyla wilderi* karyotype is not easily derived from an *Osteopilus dominicensis*-like form; chromosome morphologies and C-band patterns are quite different. However, the location of the NOR in the karyotypes of both species is similar; the NOR is found on the short arm of a small metacentric chromosomc and is proximal to the centromere. Most likely, *H. wilderi* represents a separate lineage from a South American hylid stock.

The lack of information on the chromosomes of *Hyla marianae* limits any analysis of the possible derivation of the *H. wilderi* karyotype to considerations of chromosome morphology only. That *H. wilderi* does have a derived karyotype may be assumed by its unusual chromosome number; it is the only *Hyla* with a $2n = 28$ complement. The two telocentric chromosomes in *H. wilderi* may be the result of a fission of a metacentric chromosome in an ancestral karyotype; *H. marianae* does not have such a chromosome of appropriate size. In fact, more than simple "arm-switching" would be necessary to derive the *H. wilderi* karyotype from a *H. marianae*-like form.

The high level of variation in the placement of the NOR on Chromosome 9 in *Osteopilus septentrionalis* and the high frequency of an additional NOR in *O. brunneus* was surprising. With a larger sample size, such variation also may be present in the *O. dominicensis* karyotype.

Clearly, karyological studies need to be expanded to include all West Indian hylids. Such studies would add insight into thc relationships of the West Indian species of *Hyla*, and possibly suggest their South American affinities. Based on the karyological data presented here, assignment of any West Indian species of *Hyla* to the genus *Osteopilus* or reassignment of the species in *Osteopilus* to *Hyla* would seem premature.

ACKNOWLEDGMENTS

I sincerely thank the many people who donated or assisted in the collection of specimens for this study: Ronald Crombie, Herndon Dowling, Blair Hedges, Julian Lee, John Morrissey, Carleton Phillips, Theodora Pinou, Wayne Van Devender, Robert Powell, Peter Warny, and Steve Williams. This research was supported in part by a Faculty Research and Development Grant from Hofstra University.

LITERATURE CITED

Anderson, K. and P. E. Moler. 1986. Natural hybrids of the Pine Barrens treefrog, *Hyla andersonii* with *H. cinerea* and *H. femoralis* (Anura: Hylidae): morphological and chromosomal evidence. Copeia 1986: 70–76.

Duellman, W. E. and C. J. Cole. 1965. Studies of chromosomes of some anuran amphibians (Hylidae and Centrolenidae). Syst. Zool. 14: 139–143.

Dunn, E. R. 1926. The frogs of Jamaica. Proc. Boston Soc. Nat. Hist. 38: 111–130.

Goin, C. J., O. B. Goin, and G. R. Zug. 1978. Introduction to Herpetology. 3rd ed. W. H. Freeman & Co., San Francisco.

Goodpasture, C. and S. E. Bloom. 1975. Visualization of nucleolar organizer regions in mammalian chromosomes using silver staining. Chromosoma (Berlin) 53: 37–50.

Hedges, S. B., C. A. Hass, and L. R. Maxson. 1992. Caribbean biogeography: molecular evidence for dispersal in West Indian terrestrial vertebrates. Proc. Natl. Acad. Sci. USA 89: 1909–1913.

Maxson, L. R. 1992. Tempo and pattern in anuran speciation and phylogeny: an albumin perspective, p. 41–57. *In* K. Adler (ed.), Herpetology: Current Research on the Biology of Amphibians and Reptiles. Soc. Study Amphib. Rept., Oxford, Ohio, Contrib. Herpetol. no. 9.

Powell, R., R. J. Passaro, and R. W. Henderson. 1992. Noteworthy herpetological records from Saint [*sic*] Maarten, Netherlands Antilles. Carib. Jour. Sci. 28: 234–235.

Schwartz, A. and R. Thomas. 1975. A check-list of West Indian amphibians and reptiles. Carnegie Mus. Nat. Hist. Spec. Publ. 1: 1–216.

Sumner, A. T. 1972. A simple technique for demonstrating centromeric heterochromatin. Exp. Cell Res. 75: 304–306.

Trueb, L. and M. J. Tyler. 1974. Systematics and evolution of the Greater Antillean hylid frogs. Occas. Pap. Mus. Nat. Hist. Univ. Kansas 24: 1–60.

APPENDIX. Specimens examined: *Hyla marianae* — Jamaica, Trelawney Parish (USNM 343981, female). *Hyla wilderi* — Jamaica, Quick Step (USNM 343982, female); Trelawney Parish (USNM 343983, fcmalc). *Ostcoccphalus taurinus* French Guiana, Sinnamary (USNM 343984, female). *Osteopilus brunneus* — Jamaica, Trelawney Parish (KA 1260 and USNM 343985, females; USNM 343986–7, males); St. Andrew Parish (KA 1268 and USNM 343988, males). *Osteopilus dominicensis* — Dominican Republic, La Vega Province (USNM 343989, female); Santo Domingo (USNM 343990, male). *Osteopilus septentrionalis* — Bimini (USNM 343991, male; USNM 343992–3, females); Cuba, Guantanamo Province (USNM 343994, male); Florida, Dade County (USNM 343995–8, males; USNM 343999, female); St. Lucie County (USNM 344000, female).

Resumen.—El género hílido *Osteopilus* está representado actualmente por tres especies: *O. brunneus* de Jamaica, *O. dominicensis* de La Española, y *O. septentrionalis* de Cuba, las Islas Bahamas, y el sur de la Florida. Los estudios cariológicos de estos sapos arbóreos indican que *O. dominicensis* y *O. septentrionalis* están relacionados íntimamente; ambas especies poseen 12 pares de cromosomas homomórficos metacéntricos o submetacéntricos de morfología similar, bandas C en las regiones centroméricas y en la mayoría de las regiones teloméricas, y una región del organizador nucleolar (RON) en el cromosoma 9. El cariotipo de *O. septentrionalis* puede ser derivado del de *O. dominicensis* por una inversión en el brazo corto del cromosoma 9 y modulaciones pequeñas en heterocromatina. *Osteopilus brunneus* es extraño entre los hílidos porque tiene 17 pares de cromosomas, 10 de los cuales son telocéntricos; el ron está en el cromosoma telocéntrico 9. Este cariotipo puede ser derivado del de *O. dominicensis* mediante seis eventos de fisión, un evento subsiguiente de fusión, y una inversión en el cromosoma derivado 17. Los patrones de secuencias teloméricas apoyan un escenario de fisión; ninguno de los cariotipos muestra una señal telomérica intersticial o centromérica. Las comparaciones con grupos externos no apoyan una relación íntima entre *Osteopilus* y *Osteocephalus* o los hílidos de las Antillas *Hyla marianae* y *H. wilderi*.

Résumé.—L'Hylidé *Osteopilus* compte trois représentants: *Osteopilus brunneus* de la Jamaïque, *O. dominicensis* d'Hispaniola et *O. septentrionalis* de Cuba et des Bahamas. L'étude du caryotype de ces grenouilles indique que *O. dominicensis* et *O. septentrionalis* sont étroitement apparentées; les deux espèces comptent 12 paires de chromosomes homomorphiques métacentriques ou submétacentriques de morphologic comparable, des bandes C dans les régions centromériques et dans la plupart des régions télomériques et une région nucléolaire sur le chromosome 9. Le caryotype de *O. septentrionalis* peut être dérivé de celui de *O. dominicensis* en inversant le bras court du chromosome 9 et en procédant à de petites modulations de l'hétérochromatine. *Osteopilus brunneus* se distingue des autres Hylidés dans la mesure où il compte 17 paires de chromosomes dont 10 sont télocentriques; la région nucléolaire est située sur le chromosome télocentrique 17. Ce caryotype peut être dérivé de celui de *O. dominicensis* par fissiparité, fusion ou inversion sur le chromosome 17 dérivé. L'organisation des séquences télomères confirme le scénario de fissiparité; aucun des caryotypes n'affiche de signal télomère centromérique ou interstitiel. Les comparaisons hors groupe ne permettent pas de voir d'étroites ressemblances entre *Osteopilus* et *Osteocephalus* ou entre les Hylidés des Antilles, *Hyla marianae* et *H. wilderi*.

Anderson, K. 1996. A karyological perspective on the monophyly of the hylid genus *Osteopilus*, p. 157–168. *In* R. Powell and R. W. Henderson (eds.), Contributions to West Indian Herpetology: A Tribute to Albert Schwartz. Society for the Study of Amphibians and Reptiles, Ithaca (New York). Contributions to Herpetology, volume 12.

Current Status of Biogeography of the West Indian Turtles in the Genus *Trachemys* (Emydidae)

Michael E. Seidel[1]

Abstract.—Natural populations of *Trachemys* in the West Indies occur primarily in the Greater Antilles. Their origin is most likely the result of dispersal from continental Mesoamerica, followed by intra-Caribbean dispersal and vicariance. Records of sliders in the Lesser Antilles appear to represent human introductions. Recent additions to the known distribution of *Trachemys* in the central Bahamas (Great Bahama Bank) and the discovery of fossils on San Salvador raise the strong possibility that *Trachemys*, if not presently, at least once was part of the natural fauna of that region.

Introduction

All of the freshwater or pond turtles native to the West Indies are sliders of the genus *Trachemys* (Emydidae). Their primary distribution is the Greater Antilles where four species occur: *T. terrapen* on Jamaica, *T. decussata decussata* and *T. d. angusta* on Cuba (including Isla de la Juventud), *T. decorata* and *T. stejnegeri vicina* on Hispaniola, and *T. s. stejnegeri* on Puerto Rico. *Trachemys s. stejnegeri* also has been reported on Vieques and Culebra islands, 30 km east of Puerto Rico. A third subspecies of *T. stejnegeri*, *T. s. malonei*, is found on Great Inagua Island, southern Bahamas (Seidel 1988). In a recent revision of the West Indian *Trachemys*, Seidel (1988) presented several theories on the origin and status of these turtles for all islands on which they were known to occur. Since that review, *Trachemys* has been reported from several additional West Indian islands, and the first record of a fossil *Trachemys* outside the Greater Antilles was documented. The present chapter is an update and summary of these recent discoveries as they pertain to evolutionary origins and taxonomy.

The four species of *Trachemys* in the West Indies are ecologically similar to the widespread continental species, *T. scripta*. They are opportunistic omnivores with well-developed basking habits. They utilize a wide variety of aquatic habitats, but often become terrestrial during dry periods. Clearly the Greater Antilles, with their rivers and relatively large bodies of freshwater, support most of the natural populations of these turtles. Islands of the Bahamas and Lesser Antilles are comparatively small and aquatic habitats are often seasonal. As with much of the West Indian fauna, the zoogeographic origin of the freshwater turtles remains uncertain.

Biogeography of *Trachemys*

Following the vicariance model of Rosen (1975), an ancestral form of *Trachemys* may have occurred on a proto-Antillean plate which became separated from North America during the Tertiary Period. Subsequent intra-Caribbean vicariance (late Tertiary) may have resulted in the isolation and speciation of *Trachemys* on Jamaica, Cuba, Hispaniola, and Puerto Rico. Seidel (1988) proposed a phylogeny for the West Indian *Trachemys* based on morphological and biochemical data (Figure 1). If Jamaica broke off from the plate early, this would explain why *T. terrapen* appears to be the most distinct (divergent) of the endemic West Indian species. The most recent geographic separations (perhaps late Pliocene) may have occurred between Puerto Rico

[1] Department of Biological Sciences, Marshall University, Huntington, West Virginia 25755, USA.

and Hispaniola and between the two paleoislands of Hispaniola. This would explain the presence of *T. stejnegeri* on both of the two former islands and the close relationship that exists between *T. stejnegeri* and *T. decorata*, both of which occur on Hispaniola.

Several problems arise in accepting a vicariance theory for the origin of *Trachemys* in the West Indies. Seidel and Jackson (1990) indicated that the oldest stratigraphically fixed fossil of *Trachemys* is *T. inflata*, reported from the Upper Miocene of Florida (Weaver and Robertson 1967). Separation of the proto-Antillean plate apparently occurred no later than the Eocene, at least 20 million years earlier (Rosen 1975). Tertiary fossils of terrestrial vertebrates are virtually absent from the West Indies, and Pregill (1981) pointed out that if the Caribbean plate was separated from the continent by the late Eocene, currently shared mainland and West Indian genera (such as *Trachemys*) must have evolved in parallel throughout the Cenozoic. This seems very unlikely. Furthermore, vicariance is not consistent with the observation that species diversity in the West Indies is depauperate compared to the continent (Williams 1989). A specific problem with the vicariance model is that it does not explain the close relationship between *T. decussata* on Cuba and *T. scripta* in continental Mesoamerica (Neotropical *T. scripta*, Figure 1). However, it could be argued that Cuban turtles dispersed back to nuclear Central America, thus reintroducing "Antillean" genes into continental populations. Legler (1990) suggested this by stating that parts of Central and South America may have been colonized by slider turtles from the Antilles. Alternative theories on the origin of *Trachemys* in the Greater Antilles involve multiple invasions (dispersals) from Mesoamerica, with Jamaican *Trachemys* representing early colonization and Cuban *Trachemys* a more recent colonization. In any event, the freshwater turtles in the Greater Antilles are likely a para- or polyphyletic assemblage and their origin has involved early dispersals followed by intra-Caribbean vicariance and dispersal. Mitochondrial DNA work currently in progress (Starkey et al. 1995) may help resolve some of these questions.

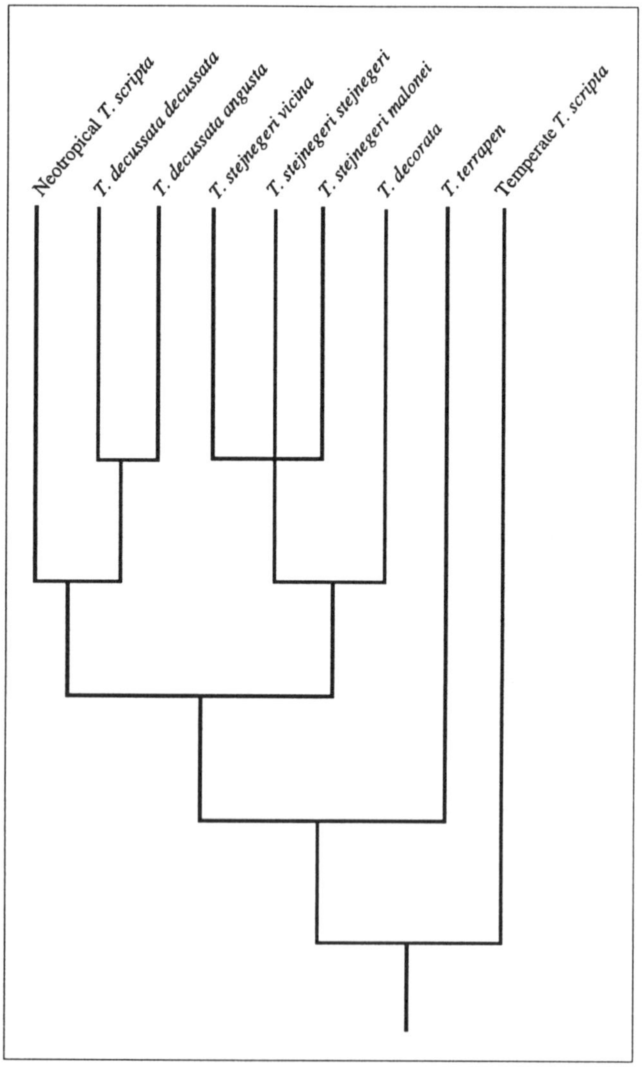

FIGURE 1. A theory on the phylogeny of *Trachemys* in the West Indies (modified from Seidel 1988) based on cladistic analysis of morphological and biochemical data.

In contrast to the Greater Antilles, vicariance does not offer an explanation for the occurrence of *Trachemys* in the southern Bahamas, Cayman Islands, and Lesser Antilles. Natural dispersal or human introduction have apparently resulted in the presence of these turtles on small islands which presumably have never been connected to other land masses. The only recorded population of *Trachemys* in the southern Bahamas is *T. stejnegeri malonei* on Great Inagua. Because it is morphologically distinct, this endemic subspecies most likely originated by natural dispersal from nearby Hispaniola, where *T. s. vicina* occurs. Seidel (1988) briefly mentioned a record of *T. stejnegeri* (UF 49423) from Pine Cay (Caicos Islands) in the southern Bahamas archipelago, but attributed it to human introduction

(David Auth, personal communication). Shell fragments of *Trachemys* have recently been recovered from archeological sites over 700 years old on Middle Caicos and Grand Turk (W. F. Keegan, personal communication). A well-established population of *T. decussata* occurs on the small island of Grand Cayman, south of Cuba. Whether this represents an early human introduction of turtles from western Cuba (*T. d. angusta*) or natural dispersal from that island is uncertain. Among a diversity of fossilized reptile bones which have been recovered from "cow wells" on Grand Cayman, remains of *Trachemys* are conspicuously absent (G. Morgan and R. Franz, personal communication). *Trachemys* have apparently been transported from Grand Cayman to Cayman Brac, 150 km northeast, where a small population may exist (Seidel 1988).

Five years ago only two records of *Trachemys* were known from the Lesser Antilles. Schwartz and Thomas (1975) and Schwartz and Henderson (1991) reported introductions of the red-eared slider, *T. scripta elegans*, at two localities on Guadeloupe. On nearby Marie-Galante, an established population of *T. s. stejnegeri* has apparently been introduced from Puerto Rico (Seidel 1988; Schwartz and Henderson 1991). Recently, Fritz (1991) reported *Trachemys* from Dominica, the major island just to the south of Guadeloupe and Marie-Galante. Twelve specimens were shipped to Germany and examined by Fritz at the zoological park in Stuttgart. Based on his description, these turtles appear to be *T. stejnegeri* and were probably introduced on Dominica from Marie-Galante. An additional Lesser Antillean record consists of bone fragments (neural and nuchal; catalog number 4700474, zooarcheological collection, Florida Museum of Natural History) recovered from an archeological site on Saba Island, 200 km northwest of Guadeloupe. These bones, estimated to be less than 1300 years old, were found among human food remains. Despite these records for *Trachemys* in the Lesser Antilles, freshwater turtles probably were never part of the natural fauna of the region.

No reports are known of *Trachemys* in the northern Bahamas, which includes Great Abaco and Grand Bahama. These islands, located on the Little Bahama Bank, are separated from central Bahamian islands by deep water, and probably were never connected to them. In contrast, numerous *Trachemys* records exist for the central Bahamas, situated on the Great Bahama Bank which includes the major islands Eleuthera, Andros, New Providence, Cat, Long, and Great Exuma (Figure 2). Seidel (1988) reported *T. terrapen* on Andros, Cat, Eleuthera, and New Providence, but suggested they all represent human introductions. Franz et al. (1993) have recently reported a population of *Trachemys* on Great

TABLE 1. Records for *Trachemys* in the Bahaman Archipelago.

Island	Species	Material	Source
1. Cat	*T. terrapen*	Extant	Barbour 1935
2. Eleuthera	*T. terrapen*	Extant	Schwartz and Thomas 1975
3. Andros	*T. terrapen* x *stejnegeri*	Extant	Schwartz and Thomas 1975
4. New Providence	*T. terrapen* x *stejnegeri*	Extant	Campbell 1978
5. Exuma	*T. terrapen* x *stejnegeri*	Extant	Franz et al. 1993
6. Long	*Trachemys* sp.	Extant, sighting only	Craton 1968
7. San Salvador	*Trachemys* sp.	Fossil, archeological	Olson et al. 1990; Berman 1994
8. Great Inagua	*T. stejnegeri malonei*	Extant	Barbour and Carr 1938
9. Pine Cay	*Trachemys* sp.	Extant, single specimen	UF 49423
10. Middle Caicos	*Trachemys* sp.	Archeological	Keegan (pers. comm.)
11. Grand Turk	*Trachemys* sp.	Archeological	Keegan (pers. comm.)

Exuma. Residents in the area mention that turtles also occur (and have so for at least 50 years) in ponds on smaller nearby islands including Little Exuma, Darby Island, and Hog Cay. Examination of four fluid-preserved specimens (UF 81579–82) from Great Exuma indicated a blend of characters that are shared by *T. terrapen*, *T. stejnegeri*, and *T. decussata*. However, the turtles appear to be most similar to *T. terrapen* which is native to Cat Island, less than 100 km north. Craton (1968) made brief mention of freshwater turtles (*Trachemys*?) on Long Island, less than 50 km east of Great Exuma. Seidel (1988) dismissed this reference, partially based on a discussion with Albert Schwartz who made field observations on Long Island, but never encountered *Trachemys*. However, Schwartz also did not encounter *Trachemys* in the Exumas. Because *Trachemys* can be quite secretive, especially during dry seasons when they burrow under substrate, their presence easily could be overlooked.

Although no geologic evidence exists to suggest that any of the Bahamas have been connected to other land masses during the Cenozoic (Morgan 1989), islands of the Great Bahama Bank likely were interconnected during the last or Wisconsin glaciation (late Pleistocene, about 17,000 BP). They presumably formed a single paleoisland larger than Puerto Rico, Jamaica, or Hispaniola (Olson and

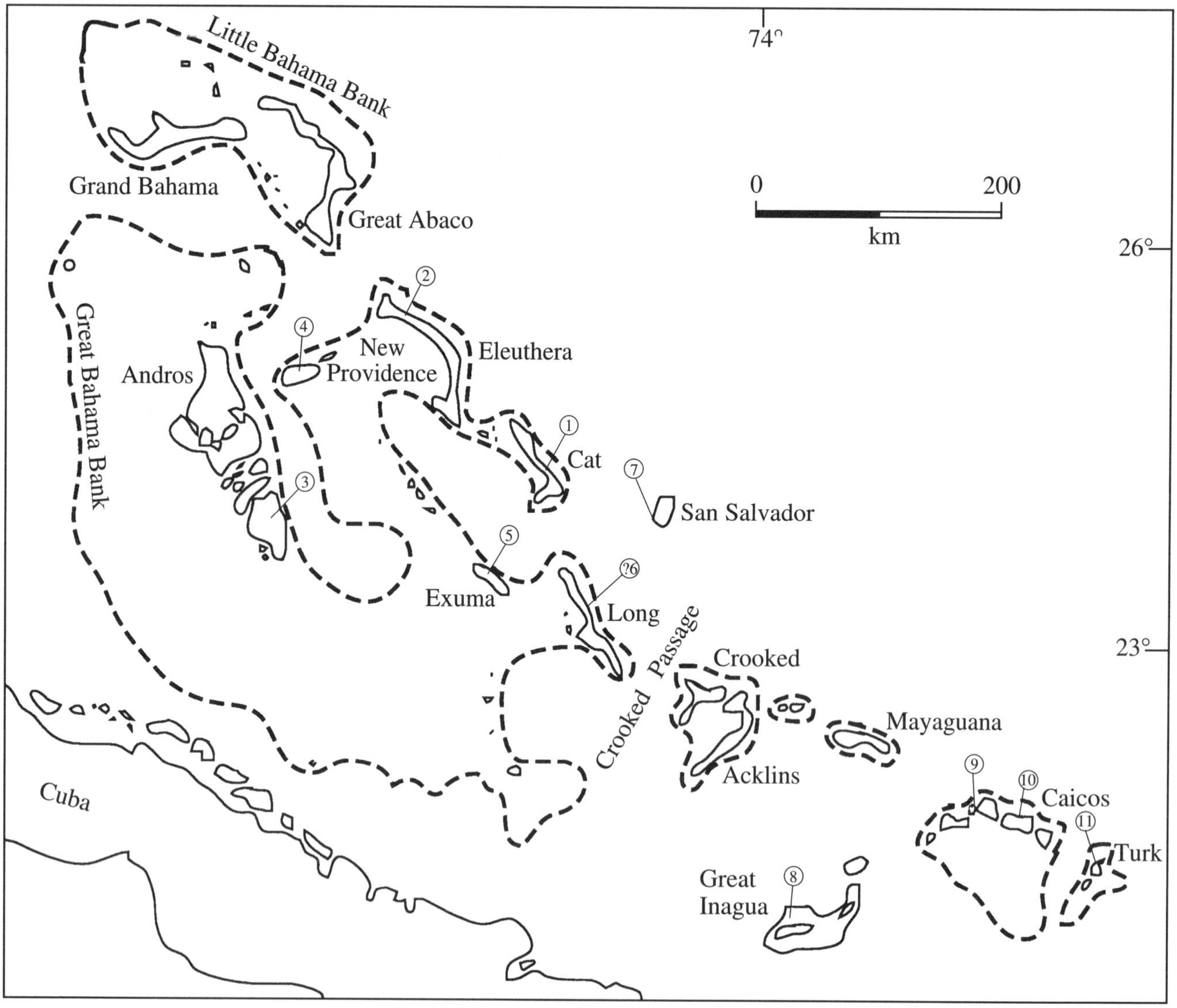

FIGURE 2. Map of the Bahamas showing the banks (broken lines) and islands. Records for *Trachemys* are numbered and listed in Table 1.

Pregill 1982). With the addition of the Exumas, *Trachemys* has now been reported from all six of the large islands on the Great Bahama Bank (Table 1 and Figure 2). Today, *T. terrapen* is the predominant slider in the central Bahamas, although on some islands it has hybridized with other, introduced species of *Trachemys* (Seidel and Adkins 1987). Between 6,000–14,000 BP, sea levels rose rapidly and reached current levels in the early Holocene. This event apparently fragmented the large paleoisland. If an ancestral form of *T. terrapen* was present on the paleoisland during the Pleistocene, this vicariance event could have isolated populations of *Trachemys* on the six major islands of the Great Bahama Bank.

Without question, the most significant recent discovery of *Trachemys* in the Bahamas is fossil fragments from San Salvador. Olson et al. (1990) reported appendicular and plastral elements of an emydid turtle (apparently *Trachemys*) recovered from late Pleistocene deposits. The bones are well mineralized, and represent at least two individuals less than 200 mm carapace length. This strongly suggests the presence of sliders in the central Bahamas prior to the arrival of humans. San Salvador lies just to the east of the Great Bahama Bank and presumably has been isolated from the rest of the archipelago throughout its history. Nevertheless, late Pleistocene dispersals between San Salvador and the nearby paleoisland of the bank easily could have occurred. In spite of its relatively small size (156 km^2), San Salvador has an extensive system of inland lakes. Although they are now saline, during glacial periods these lakes may have held fresh water and provided suitable habitat for *Trachemys*. Olson et al. (1990) stated that freshwater turtles probably flourished in the large inland lakes of San Salvador and may have disappeared as a result of increased salinity following the last eustatic rise in sea level. During this period, populations of *Trachemys* may have been able to survive elsewhere in the Bahamas, perhaps on islands with higher elevation and more stable freshwater (i.e., Cat Island, 67 m). The extinction of freshwater turtles on San Salvador may have occurred later. Recently, Berman (1994) reported shell fragments from an archeological site on San Salvador which is less than 1,200 years old.

CONCLUSION

Systematic evidence strongly indicates that *Trachemys* in the central Bahamas, especially populations on Eleuthera and Cat Island, are taxonomically indistinct from *T. terrapen* on Jamaica (Seidel 1988). Considering the geographic positions of Jamaica and the central Bahamas, a single model (dispersal or vicariance) which would explain their common origin is difficult to imagine. Therefore, the interpretation that extant *Trachemys* in the Bahamas are the result of human introduction remains a reasonable postulate. Nevertheless, evidence now suggests that pond sliders (*Trachemys*?) were at least once part of the natural Bahamian biota.

ACKNOWLEDGMENTS

I thank U. Fritz, J. Iverson, R. Franz, G. Pregill, M. Berman, W. Keegan, E. Wing, and D. Auth for information on locality records. I thank V. Crager for typing the manuscript.

LITERATURE CITED

Barbour, T. 1935. A new *Pseudemys* from Cat Island, Bahamas. Occas. Pap. Boston Soc. Nat. Hist. 8: 205–206.

— and A. F. Carr. 1938. Another Bahamian freshwater tortoise. Proc. New England Zool. Club 17: 75–76.

Berman, M. J. 1994. Preliminary report on a vertebrate assemblage excavated from the Three Dog Site, San Salvador, Bahamas. Unpubl. report in Proc. Fifth Nat. Hist. Bahamas Conf.

Campbell, D. G. 1978. The Ephemeral Islands, a Natural History of the Bahamas. Macmillan Co., New York.

Craton, M. 1968. A History of the Bahamas. Collins Press, London.

Franz, R., C. K. Dodd, and D. W. Buden. 1993. Distributional records of amphibians and reptiles from the Exuma Islands, Bahamas, including the first reports of a freshwater turtle and an introduced gecko. Carib. Jour. Sci. 29: 165–173.

Fritz, U. 1991. Contribution to the knowledge of the Hispaniolan slider, *Trachemys decorata* (Barbour and Carr, 1940). Sauria 1991(1): 11–14.

Legler, J. M. 1990. The genus *Pseudemys* in Mesoamerica:

taxonomy, distribution, and origins, p. 82–105. *In* J. W. Gibbons (ed.), Life History and Ecology of the Slider Turtle. Smithsonian Inst. Press, Washington, D. C.

Morgan, G. S. 1989. Fossil Chiroptera and Rodentia from the Bahamas, and the historical biogeography of the Bahamian mammal fauna, p. 685–740. *In* C. A. Woods (ed.), Biogeography of the West Indies Past, Present, and Future. Sandhill Crane Press, Gainesville, Florida.

Olson, S. L., and G. K. Pregill. 1982. Fossil vertebrates from the Bahamas, p. 1–7. *In* S. L. Olson (ed.), Fossil Vertebrates from the Bahamas. Smithsonian Inst. Press, Washington, D. C.

—, G. K. Pregill, and W. B. Hilgartner. 1990. Studies on fossil and extant vertebrates from San Salvador (Watling's) Island, Bahamas. Smithsonian Contr. Zool. 508: 1–13.

Pregill, G. 1981. An appraisal of the vicariance hypothesis of Caribbean biogeography and its application to West Indian terrestrial vertebrates. Syst. Zool. 30: 147–155.

Rosen, D. E. 1975. A vicariance model of Caribbean biogeography. Syst. Zool. 24: 431–464.

Schwartz, A. and R. W. Henderson. 1991. Amphibians and Reptiles of the West Indies: Descriptions, Distributions, and Natural History. Univ. Florida Press, Gainesville.

— and R. Thomas. 1975. A checklist of West Indian amphibians and reptiles. Carnegie Mus. Nat. Hist. Spec. Publ. no. 1.

Seidel, M. E. 1988. Revision of the West Indian emydid turtles. Amer. Mus. Novitates 2918: 1–41.

— and M. D. Adkins. 1987. Biochemical comparisons among West Indian *Trachemys* (Emydidae: Testudines). Copeia 1987: 485–489.

— and D. R. Jackson. 1990. Evolution and fossil relationships of slider turtles, p. 68–73. *In* J. W. Gibbons (ed.), Life History and Ecology of the Slider Turtle. Smithsonian Inst. Press, Washington, D. C.

Starkey, D. E., R. W. Guthrie, and S. K. Davis. 1995. Molecular systematics and biogeography of the turtle genus *Trachemys*. Abstracts. 75th Ann. Meet. Amer. Soc. Ichthyol. Herpetol., Edmonton, Alberta.

Weaver, W. G. and J. S. Robertson. 1967. A re-evaluation of fossil turtles of the *Chrysemys scripta* group. Tulane Stud. Geol. 5: 53–66.

Williams, E. E. 1989. Old problems and new opportunities in West Indian biogeography, p. 1–46. *In* C. A. Woods (ed.), Biogeography of the West Indies Past, Present, and Future. Sandhill Crane Press, Gainesville, Florida.

Resumen.—Poblaciónes naturales de *Trachemys* in las Antillas ocurren principalmente en las Gran Antillas. Su origen es probablemente el resultado del disperso del medio america continental, seguido por el disperso y fragmentación (vicariance) dentro del Caribe. Registros de tortugas de las Antillas Menores parecen representar introducciónes humanos. Reciente añadidura a la distribución conocida de *Trachemys* en las Bahamas Centrales (el bajio grande de las Bahamas) y el descubrimiento de fósiles en San Salvador susitan la firme posibilidad que *Trachemys*, si no actualmente, por lo menos una vez, fue parte de la fauna natural de esa región.

Résumé.—Les populations naturelles de *Trachemys* dans les Antilles se trouvent essentiellement dans les Grandes Antilles. Leur origine résulte vraisemblablement de la dispersion provenant de la Mesoamérique continentale suivie de dispersions et vicariances intra-antillaises. Les mentions de tortues du genre *Trachemys* dans les Petites Antilles semblent être dues à des introductions humaines. Des additions récentes à la distribution connue des *Trachemys* dans le centre des Bahamas (grand banc des Bahamas) et la découverte de fossiles à San Salvador semblent indiquer que *Trachemys* a déjà fait partie de la faune naturelle de cette région.

Seidel, M. E. 1996. Current status of biogeography of the West Indian turtles in the genus *Trachemys* (Emydidae), p. 169–174. *In* R. Powell and R. W. Henderson (eds.), Contributions to West Indian Herpetology: A Tribute to Albert Schwartz. Society for the Study of Amphibians and Reptiles, Ithaca (New York). Contributions to Herpetology, volume 12.

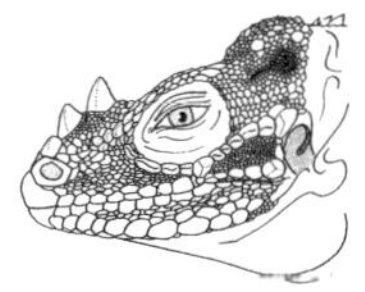

Relationships among West Indian Geckos of the Genus *Sphaerodactylus*: A Preliminary Analysis of Mitochondrial 16s Ribosomal RNA Sequences

Carla Ann Hass[1]

Abstract.—A portion of the mitochondrial 16s ribosomal RNA gene was sequenced in 28 species of West Indian *Sphaerodactylus* and two sphaerodactyline outgroup taxa, *Gonatodes* and *Lepidoblepharis*. These data were used to examine the phylogenetic relationships among the major lineages within this group, and to elucidate the placement of the *ramsdeni* group, a morphologically unique assemblage of three Cuban species (*S. ramsdeni*, *S. cricoderus*, and *S. schwartzi*). Both neighbor-joining (NJ) and maximum parsimony analyses significantly support the recognition of the following species groups: *ramsdeni* (P_C [confidence values for interior-branch test in NJ analysis] = 95%), *nigropunctatus* (P_C = 96%), and *scaber* (P_C = 99%). Also, a close relationship between the Cuban *nigropunctatus* and *scaber* groups, an association which has not been suggested previously, is strongly supported by the NJ analyses (P_C = 95%). Representatives of three species groups (*shrevei*, *richardsoni*, and *thompsoni*) comprise a strongly supported cluster (P_C = 99%). Other associations suggested by a previous molecular study are supported moderately. The eleven species representing the *argus* series cluster (P_C = 80%), as do the two members of the *cinereus* group, *S. cinereus* and *S. elegans* (P_C = 73%). All three Lesser Antillean species examined, members of both the *sputator* and *fantasticus* series, cluster (P_C = 77%); this relationship was indicated by immunological data, but was not apparent in an allozyme study. At higher levels, these data do not support the recognition of the *cinereus* series as a monophyletic group. Instead, these data suggest that the *argus* series is derived from within that group (P_C = 85%). The affinities of the *ramsdeni* group remain unresolved, although it does not appear to be closely related to any other species group of West Indian *Sphaerodactylus*.

Introduction

Approximately 80 species of *Sphaerodactylus* are known to occur in the West Indies (Schwartz and Henderson 1991; Thomas and Hedges 1992, 1993; Thomas et al. 1992; Hedges and Garrido 1993; Lazell 1994; Powell et al., this volume). The degree of endemism in this group is high, with few species found on more than one island and many species restricted to small ranges within an island. Early morphological work on this group resulted in the definition of several species groups (Thomas and Schwartz 1966, 1983; Shreve 1968; Thomas 1975; Schwartz and Graham 1980; Schwartz and Garrido 1981, 1985; Schwartz 1983), but offered little insight into how these groups are related.

In order to examine higher-level relationships among West Indian *Sphaerodactylus*, a study using two molecular techniques, starch-gel electrophoresis of proteins and immunological comparisons of serum albumin, was carried out with about half of the species of *Sphaerodactylus* that occur in the West Indies (Hass 1991). Based upon those data, the Greater Antillean *Sphaerodactylus* were partitioned into two main groups, the *cinereus* series and the *argus* series; species from the Lesser Antilles are members of two other series (*sputator* and *fantasticus*). The *cinereus*, *argus*, and *sputator* series com-

[1] Department of Biology and Institute of Molecular Evolutionary Genetics, The Pennsylvania State University, University Park, Pennsylvania 16802, USA.

prise the *sputator* subsection, whereas the Lesser Antillean species in the *fantasticus* series were placed in the *fantasticus* subsection.

In that revised classification, based primarily upon the electrophoretic data, the majority of Cuban species were placed in the *cinereus* series. Although no morphological characters appear to define this group, its members tend to have large body size and to show strong sexual dichromatism as adults. This series is further partitioned into seven species groups (Hass 1991). Those molecular data did not support recognition of the *nigropunctatus* group of Schwartz and Garrido (1985). Instead, *Sphaerodactylus nigropunctatus*, *S. torrei*, *S. docimus*, and *S. intermedius* were placed in the *nigropunctatus* group, whereas *S. cinereus* (a Hispaniolan species) and *S. elegans* were recognized as the *cinereus* group. *Sphaerodactylus oliveri* was placed in this series, based upon immunological comparisons, as a member of the *scaber* group; no other species in this Cuban group (Hedges and Garrido 1993) were available for comparison.

The remaining members of the *cinereus* series occur outside of Cuba. In Jamaica, two species, *Sphaerodactylus parkeri* and *S. richardsoni*, comprise the *richardsoni* group (Grant 1939). The ranges of these species are restricted to coastal areas on the southern and northern regions of the island, respectively. In Hispaniola, three additional *cinereus* series groups occur. The *shrevei* group consists of four species that are found in xeric habitats (Schwartz and Graham 1980); the three species examined are very similar at the molecular level (Hass 1991). A single species, *S. thompsoni* (Schwartz and Franz 1976) represents the *thompsoni* group. The *copei* group, represented on Hispaniola by *S. copei*, also has representatives on Navassa and the Islas de la Bahia, Honduras (Schwartz and Garrido 1981; Hass 1991).

All of the remaining species in the Greater Antilles have been placed in the *argus* series (Hass 1991). This series is not as morphologically diverse (particularly in scale characters) as the *cinereus* series, although its members do show a greater range of body size. The Cuban species, *Sphaerodactylus notatus* (as well as *S. bromeliarum* and *S. celicara*), is closely related to a group of Hispaniolan species, some of which originally were described as subspecies of *S. notatus* (Shreve 1968), and is placed in the *notatus* group of the *notatus* subseries (Hass 1991). The other Cuban members of this series, *S. ruibali* and *S. armasi*, remain unassigned to subseries. Based on morphology, these two species had been allied to the *nigropunctatus* group (Schwartz and Garrido 1985), but they were placed in the *argus* series as a result of the molecular study (Hass 1991).

The remaining Jamaican species were placed in the *argus* subseries as members of the *argus* species group except for *Sphaerodactylus goniorhynchus*. The molecular data suggest that this species is closely related to that group. Its cryptic coloration and differences in external morphology, probably a result of its leaf-litter habitat (see below), led to its placements in its own species group (Hass 1991). All of the Puerto Rico Bank species are members of the *argus* series. Based upon the electrophoretic data, they do not appear to be a monophyletic group, with three members allocated to the *notatus* subseries (*S. gaigeae*, *S. klauberi*, and *S. macrolepis*), and the remaining eight unassigned to subseries.

The majority of the 49 species in the *argus* series — 28 species — are found on Hispaniola and the islands of the Bahamas Bank (some species occur in Cuba, Jamaica, and Puerto Rico, as discussed above). All of the Hispaniolan species that belong to the *argus* series are in the *notatus* subseries; only three species from the southern Bahamas Bank (*Sphaerodactylus corticola*, *S. mariguanae*, and *S. caicosensis*) could not be assigned to subseries (Schwartz 1968; Hass 1991).

One species from the Lesser Antilles, *Sphaerodactylus sputator*, was placed in the *sputator* series. The two molecular data sets provided equivocal information on this species (Hass 1991). Whereas the electrophoretic data suggested a closer relationship to the *argus* series, the immunological data indicated that it might be more closely related to other Lesser Antillean species. However, this was based on similarity of cross-reaction to a panel of antisera from species in the Greater Antilles; an antiserum against the serum albumin from a species in the Lesser Antilles was not available (Hass 1991). The other Lesser Antillean species examined were placed in the *fantasticus* series of the *fantasticus* subsection, outside of the other West Indian species.

Three West Indian species (*Sphaerodactylus gilvitorques*, *S. elasmorhynchus*, and *S. ramsdeni*) could not be allocated to subsection. None of the three was available for molecular analysis and two (*S. gilvitorques* and *S. elasmorhynchus*) are known only from one specimen (Thomas 1966; Thomas and Schwartz 1974). However, *S. ramsdeni* was known from multiple specimens but occurs only in a small area in eastern Cuba (Schwartz and Henderson 1991). Two closely related species, *S. cricoderus* and *S. schwartzi*, also from eastern Cuba, recently have been described, and all three have been placed in the *ramsdeni* group (Thomas et al. 1992).

Species in the *ramsdeni* group share a combination of morphological characters that are unique among Cuban *Sphaerodactylus* (Schwartz and Garrido 1985; Thomas et al. 1992). In scalation, they are similar to members of the *nigropunctatus* group, which have small, usually granular, dorsal scales. However, most members of that group show strong sexual dichromatism as adults, whereas adults of the *ramsdeni* group species show little color or pattern differentiation. The members of the *notatus* group usually exhibit weaker sexual dichromatism, but their dorsal scales are very different from those of the *ramsdeni* group. The members of the *scaber* group have large, swollen dorsal scales and a middorsal zone of small, granular scales, characters not seen in the *ramsdeni* group. The species in the *ramsdeni* group also have very short snouts and more than one internasal scale (Thomas et al. 1992), whereas other Cuban species have a modal number of one internasal (Schwartz and Henderson 1991).

Although they could not ally *Sphaerodactylus ramsdeni* with any other *Sphaerodactylus* based upon the morphological data, Schwartz and Garrido (1985) advanced several hypotheses for the evolutionary history of this species. One suggestion was that *S. ramsdeni* may represent a remnant of an old lineage that has persisted only in relatively remote areas. They also proposed that *S. ramsdeni* may be closely related to the Jamaican species *S. goniorhynchus*. Both are forest-dwelling animals and have dark coloration and a similar color pattern, particularly lighter dorsoventral stripes in the sacral region. However, Schwartz and Garrido (1985) also discussed the possibility that this similarity in coloration might be the result of convergence, with selection for dark coloration in forest-dwelling species.

This idea of convergence in morphology due to similar ecologies has been examined in other groups of West Indian vertebrates, particularly among the members of the lizard genus *Anolis*. In 1972, Williams coined the term "ecomorph" to describe this convergence, defining it as "species with the same structural habitat/niche, similar in morphology and behavior, but not necessarily close phyletically." Since then, this concept has been examined extensively in *Anolis* (e.g., Williams 1983; Losos 1992). A number of ecomorphs also have been defined within the West Indian members of the frog genus *Eleutherodactylus* (Hedges 1989; Hedges et al. 1992). The ecomorph concept in *Sphaerodactylus* was discussed by Thomas et al. (1992), who indicated that the members of the *ramsdeni* group represent the forest ecomorph in *Sphaerodactylus* and suggested that this ecomorph also has arisen multiple times within the *argus* series.

Members of the *ramsdeni* group were not included in Hass (1991) because tissues were not available. I since have obtained DNA sequence data from a portion of a mitochondrial gene, 16S ribosomal RNA from 28 species of West Indian *Sphaerodactylus*, including the three members of the *ramsdeni* group. With these data, the relationships of the species within that group, the idea that these species represent the forest ecomorph, and the relationships among the major lineages of West Indian *Sphaerodactylus* can be examined.

Materials and Methods

The animals used in this study and their collecting localities are listed in Appendix A. In both the field and laboratory, lizards were sacrificed by cryothermy, blood samples taken, and the animal either frozen whole or tissues (heart, liver, kidney, a piece of intestine, and a portion of the tail) removed and stored at –75°C. A phenol/chloroform DNA extraction was done using either a small piece of liver, tissues (listed above), approximately 10 µl of red blood cells, or about 50 µl of tissue homogenate from a previous electrophoretic study (Hass 1991).

A portion of the mitochondrial 16S ribosomal gene corresponding to sites 2607–3055 in the human mitochondrial genome (Anderson et al. 1981) was amplified (Hedges et al. 1991) using specific primers (Hass et al. 1993) from 28 species of West Indian *Sphaerodactylus*, and one representative each from the sphaerodactyline genera *Gonatodes* and *Lepidoblepharis*. These two species were included as outgroup taxa. The tribe Sphaerodactylini was considered to be monophyletic by Kluge (1987) and

TABLE 1. Aligned DNA sequence data for a 430 bp segment of the mitochondrial 16S rRNA gene from 28 species of West Indian *Sphaerodactylus*, and two mainland sphaerodactyline genera. Gaps (–) were inserted to increase sequence similarity and ambiguities are denoted by "?". A solid line above a segment of sequence indicates a region not used in the phylogenetic analyses because it is highly variable among taxa and site homology could not be determined. Alignments shown for these variable regions may not be optimal.

Species	Sequence	60
ramsdeni	GTATGAATGGCTAAATGAGACCCCGACTCTCTCCTACGACCAATCAATGAAACTGATCTC	
schwartzi	...................G................T.......................	
cricoderus	.......C...........G....A..........GT.....................C.	
altavelensis	............GC.....GA..AA..........TT.GTT..........C.......T	
armstrongi	............GC.....GA..AA..........T....T..........C........	
cryphius	.............C.....GA..GA..........TT..............C......CT	
darlingtoni	............GC.....GA..AA..........TT.G............C...C...T	
notatus	.......C...C.C.....GA..AA..........TT...T..........C.......T	
goniorhynchus	.............C.....GA..AA..........TT...T...................	
semasiops	............GC.....GA..AA..........TT...T..........C........	
gaigeae	.............C.....GA..AA..........TT..............C......C.	
klauberi	............GC.....GA..AAT.........TT..............C........	
nicholsi	............GC.....GA..AA..........TT..............C.......T	
macrolepis	.............C.....GA..AA..........TT..............C......C.	
elegans	...................G...AA..........................C.......T	
cinereus	...................G....A..........CT..............C.......T	
intermedius	C..........AC......G....A...........T..............C...C...T	
nigropunctatus	.......C...AC......G....A..........CT..............C...C...T	
torrei	...........ATC..........A..........CT...T..........C.......G	
oliveri	........?..C.C..........A...........T...T..........C.......T	
richardi	...........C.C..........A...........T..............C.......T	
richardsoni	.......C...........GA.A.C..........T...............C.......T	
leucaster	.......C...........GA.AAA..........T...............C.......T	
thompsoni	.......C...........GA.AGA..........T...............C........	
copei	...................G....A..........GT...T..........C.......T	
elegantulus	.......C...........G...AA...........AT.............C........	
fantasticus	...................G...AA...........AT..T..........C........	
sputator	...................G...AA...........A...T..........C........	
Gonatodes	.......C.C.C.--....G....A.........CTT.G...C........C......C.	
Lepidoblepharis	.......C...........G...TAC....T.....T.G..CG........C.......T	

contains four other genera (*Coleodactylus*, *Gonatodes*, *Lepidoblepharis*, and *Pseudogonatodes*), all of which are appropriate to use as outgroups in this study (Hass 1991). This double-stranded DNA was isolated in a low-melting-point gel, the band excised, and this DNA used in asymmetrical PCR, with one primer diluted to 1:100 (Gyllensten and Erlich 1988), to generate single-stranded template for dideoxy-sequencing with Taq DNA polymerase (sequencing grade; Promega). Both strands were sequenced for all species except *S. oliveri*; for that species only the L strand could be amplified in asymmetrical PCR.

The autoradiographs were digitized using the Gelin Program (S. W. Schaeffer, Pennsylvania State University) and alignments were done by eye using the multisequence editing program ESEE (Cabot and Beckenbach 1989). These aligned sequences were analyzed using the MEGA software package for analysis of molecular data (Kumar et al. 1994). Both neighbor-joining (NJ; Saitou and Nei 1987) and maximum parsimony (MP; Fitch 1971) analyses

TABLE 1 (continued).

Species	Sequence 120
ramsdeni	CCAGTACAAAAGCTGGAATACCCACATAAGACGAGAAGACCCTGTGGAGCTTAAAATATT
schwartzi	....
cricoderus	................G.C....C...................................C..C
altavelensis	.T..................AA.C............................T.......
armstrongi	T..............AG...AAAC............................T......A
cryphius	.T............A.....AA.C....................................
darlingtoni	.T............A.....AA.C....................................
notatus	.T............A.G..CTA.C............................T.....CC
goniorhynchus	.T..............G...AAA...................................C.A.
semasiops	................C..TAAAC............................C......A
gaigeae	...............AG....AAC...................................AC
klauberi	T.........G....AG...AGAC........A..........................CC
nicholsi	T..............AG...ATAC........A..........................CC
macrolepis	....................AAAC...................................A
elegans	.T............A......A.C...................................C
cinereus	.T............A.....AA.T...................................A
intermedius	....................AT......................................
nigropunctatus	....................A.......................................
torrei	................CG..AA.C....................................
oliveri	.............C......AA.C...................................C
richardi	.............C......AA.C...................................A
richardsoni	.T............A.....TA.C................................C.C.
leucaster	.T............A.....AA.C..................................C.
thompsoni	.T............A......A.C..................................C.
copei	.T............A.G...AA.C..................................G.
elegantulus	...............AG....AAC...................................A
fantasticus	....A
sputator	.T............A.....TA.C....................................
Gonatodes	................G...A.....................C...........G..CAA
Lepidoblepharis	...................TAA.C............................C...C.AC

were performed using this program. Analyses were done both with complete deletion (sites with a gap or missing data in any one taxon were removed from the analysis) and pairwise deletion (sites with gaps or missing data were removed only when calculating pairwise distances) options to determine if the insertion of gaps and the presence of missing data had any affect on the tree topologies.

Two types of statistical analyses were performed for the NJ trees. The bootstrap method (Felsenstein 1985) was used to assess the statistical significance of the clusters obtained. Much controversy has surrounded the interpretation of this test, and simulation studies indicate the bootstrap confidence value (P_B) may be a conservative estimate of the significance of a cluster (Zarkikh and Li 1992*a*, 1992*b*; Hillis and Bull 1993; Felsenstein and Kishino 1993; Sitnikova et al. 1995). Therefore, the standard error test for interior branch lengths (Rzhetsky and Nei 1993) also was performed to calculate the confidence value (P_C) of each interior branch for trees generated in the NJ analyses. Recently, computer

TABLE 1 (continued). See caption on page 178.

Species	Sequence 180
ramsdeni	A ATACTACCCACAAGCCCGCCCAA-CTATGCA--GCACCCAACC-AAAAT-ACTATACA
schwartzi	..G......A.....C.A.CAAA.C..GCAA...CA....AT.A.............C...
cricoderus	..C............A..AA...GC..ACG.C..CA.C...CC..............GG.G
altavelensis	.GGC....A..CA----.A..A.CC..ATAA.G.CC...GAT..T...........GC...
armstrongi	..C.........A----.AC.A.CC..ACCA.G.CC...A.C...............C...
cryphius	...C....A..CA----AAC.A..C..ATCA.G.CC...AAC.............G....
darlingtoni	..GC....AT.CA----.AC.A.CC..ATCA.G.CC...T.C..............GC...
notatus	..GC....A....----.ACTA.CC..A.CA.G.CTT..G.C.A..G.........GCC..
goniorhynchus	.........TA..---GG.C...CC..A.AA.G.ACT..A.C...C............G..
semasiops	........G...A---.A.TTTACC..C.C....CC...A.CG.....C..........T.
gaigeae	........A....----.AC.A.CC..GCCA.G.CC...A.C..................
klauberi	........A..C.----..T.AACC..ACAA.G.CT...A.C..................
nicholsi	........A....---.ATTTT.CC..ACCA.G.CC...A.C..................
macrolepis	G.CC....A...A--CA.AA...CG.TCTCA.G.CC...A...TA...........GC...
elegans	........A...G-CC..AC.A.CC.AAGCA.G.CC...A.C...............A...
cinereus	..G.....A....---.AT..AGCC.-------..----A.C.--............C...
intermedius	..G.G....G...--A..ACTAACC.AATC-.G.CC...AAT.TA...C......C.C...
nigropunctatus	..G.GT..TA...--AT.ACTAACC.AATC-.G.CC...AAT.TA...C......C.C...
torrei	..C.G....A...--AG..C...CC..A.CA.G.CC...A-GCAA...C......C.....
oliveri	....G....T...--AA.TA.AACC.AAC.A.G.CC...AAG..A...G......C.C...
richardi	....G........--AA.TA.AACC.AA-.A.G.CC...AAG..A...T......C.C...
richardsoni	G.C.CTC.G..TG--CA..CAA.CCAAC.CA.G.AC...A.C.......C..G..CGC.T.
leucaster	G.C.GT..G..CG--CATAAAA.CC.ACCAA.G.CC...AAC.....GC..A.CC.C.T.
thompsoni	G..CC..G.ATC.-CCA..CAG.CC.AACAA.G.AC...A.C..A..GC..GG.C.C...
copei	T.......A.A..--CAA.TA.GCC.AACA-.G.CC....A...A.-..........G..
elegantulus	..C.G...AT...--CAGTAAA.TC.AACAA..CAC....A......T.......C.C...
fantasticus	..G.....ATTTA-.TA..AAA.CC.AACAA.GCAC...A.C..T..C.........C...
sputator	T.G.....A...A--CATTAAA.CC.AATAA.GCAC...A.C..T..T.........C...
Gonatodes	-.CC..C.---------------.-------..---.G.CC.T.---T..G..GCCTC
Lepidoblepharis	-.C.G...---------------.-------..---A..C..A.---TA.GCATGGT.

simulations, without the assumption of a molecular clock, have shown that P_C values provide more reliable estimates of interior branch significance than do P_B values. Although the P_C values may be slight overestimates when calculated for a tree estimated from actual data, the P_C value and the value obtained when this overestimate is corrected (P'_C) are approximately the same for values greater than 90% (Sitnikova et al. 1995).

Because the stochastic variation in nucleotide substitutions is not taken into account when constructing MP trees, determining an appropriate statistical test to evaluate the significance of groups is difficult (Kumar et al. 1993). Therefore, the population of MP trees obtained was used to generate a consensus tree indicating the percentage of MP trees in which a particular cluster appears. Because the criterion for choosing the most likely tree in a parsimony analysis is to choose the shortest MP tree(s), this percentage reflects the support for each cluster in the population of MP trees. These values do not reflect statistical significance.

TABLE 1 (continued).

Species	Sequence	240
ramsdeni	TTTTTAGTTGGGGCGACTGCGGAACCAAACCTAACTTCCACGA--CACCTATGGCACTCA	
schwartzi	..................A....GA.....A....C....T.......GG.C--...CA.	
cricoderus	..................A............G........T.T..T..A.GC--...CAG	
altavelensis	....C.............A...........AA...............AGGCA--T....C	
armstrongi	..................A...........AG...............TAGGG--.CT.TC	
cryphius	....C.............A............A...............AGACA--...C.C	
darlingtoni	..................A............A...............AGACA-----C.C	
notatus	..................A...........AA...............-AGGG-C.....C	
goniorhynchus	..............................AA................AAGG-C..-C.C	
semasiops	.........................A....AA........T......TAGGG-C...C.G	
gaigeae	..................A...........AA...............AAAGA-C..-C.C	
klauberi	..................A...........AA................AAGA-C..-C.C	
nicholsi	..............................AA........T.......AAGA-C..-C.C	
macrolepis	..................A.....T.....AA................AGG--C..AC.C	
elegans	..................A......A.....................TGGGCACTC.A.C	
cinereus	....T....GAAC...C..	
intermedius	.....G...........C.......A...TTA.........TTTT...-GGC-C....AG	
nigropunctatus	.....G...........C.......A...TAA.........TT.T...-GGC-C....AG	
torrei	.....G...........CA......T..GT.C.........T.ATTTT-AGC-C.T.C.G	
oliveri	........................T.....TA................-..A-C..TAA.	
richardi	........................T.....T.................-..A-C..TAA.	
richardsoni	....C.......................G..C................-GGA-C.....C	
leucaster	.........................T...A.C.G..............-GGA-C...-.C	
thompsoni	............................G..C.G.............A-GGA-.....A.	
copei	........................TT....GC................-AGG-C..AAT.	
elegantulus	........................TA....AA...............A-GGG-C.G.CA.	
fantasticus	........................AA...TAA........T.C..ACA-GGA-C...AA.	
sputator	.........................A....AA........T.C....A-GGA-A...CA.	
Gonatodes	.C..C.............A.....AA...TAA..........C..--------A...AA.	
Lepidoblepharis	C.................A......A....AA..........C...T.-GGA-C...C..	

RESULTS

For these 30 species of sphaerodactyline geckos, 430 base pairs (bp) were sequenced (Table 1). Two regions were too variable to be aligned with confidence across all the taxa. The less variable regions showed strong sequence length conservation and the majority of gaps inserted to increase sequence similarity (170/207 across all taxa) were a single nucleotide. Therefore, alignment by eye is justified with these portions of the data set. However, within the variable regions only closely related species (as defined below) could be aligned with any confidence, and in some instances considerable length variation existed even between these taxa. After these variable regions were excluded, the remaining 372 sites were included in the phylogenetic analyses; of these, 150 sites were variable and 103 were informative under parsimony. This data set is very similar to that obtained for 25 species of West Indian anoline lizards. In that analysis, 352 of the 442 bp sequenced could be aligned and the variable regions that were

TABLE 1 (continued). See caption on page 178.

Species	Sequence	300
ramsdeni	ACCATACCA--AGGTCAACAAACCAAC-ACACCCATT-----CACCCAGTACC-ACTGAT	
schwartzi	C...C...-......T.................AT..T.............AG.......	
cricoderus	C.A.C...-...........C.....T........................A........	
altavelensis	C..TAG.GG......TG............A.TAAC................AT.......	
armstrongi	C..TCC...........................AC................AT......C	
cryphius	T.T.CC.A-.................G..T..TAC................AT......C	
darlingtoni	C..C-----.................T....A.AC................A.......C	
notatus	C...C..A-................G.....T.AT.........T....C..T.G....C	
goniorhynchus	C.T.CC.A-...................G....AC................AT.......	
semasiops	C..C....C...............-...G....AC................AT......C	
gaigeae	.T..GC..................T........AC................AT......C	
klauberi	.T.CCC..C...............T.T.C.C..AT...............TAT......C	
nicholsi	GT..AC..................T.T......AC................TT......C	
macrolepis	C..TCCA.CCA................C..G.TAC................AT......C	
elegans	CAATA.---....................T.T.AC................A........	
cinereus	GTTCAC....................AGC....AC................AT.......	
intermedius	G..-ACA.C.................-.T..T.AC..............C.AT.G....C	
nigropunctatus	G..-ACA.C.................-.T....AC..............C.AT.G....C	
torrei	G..-ACA.C.................-.T..T.AC..............C.G..G....G	
oliveri	TTTTATA.C................G..T....AC..T.............AT.......	
richardi	TTTTACA.C.................T.T....AC..T.............AT.......	
richardsoni	CT.CAT.A.......T............G...ATC.........T......AT......C	
leucaster	GT.CAC.A.......T............G...A.C................A....C...	
thompsoni	CT.CCC.A....................GT.AA.C................AT......C	
copei	C.T.ACTA-.......................ATT.-..............AT.......	
elegantulus	C..TAC.ACC................A......AC..............C.AT.G.....	
fantasticus	.T.CAC.ACC.....T..........A..T.AA.C..............C..T.G....G	
sputator	TT.C.G...CC.....................TATA.............C..T.G.....	
Gonatodes	TTACAG.--....C.-....GG....T.GT.ATA...AACT...........A.......	
Lepidoblepharis	T..TATAA-.....CACC..CG.......T.A..C..AACTC........C.AT......	

excluded from the analyses were in the same position and were almost the same length as the variable regions in this study (Hass et al. 1993). This suggests a similar secondary structure (or perhaps tertiary) in this region for these two groups; however, because the entire 16S gene has not been sequenced in these taxa, I cannot speculate about the nature of this structure.

The NJ algorithm allows the use of different distance measures that account for the phenomenon of multiple substitutions at a single nucleotide site between taxa, the deviation from equal frequencies of the four nucleotides, and/or a bias in transitional to transversional substitutions that is sometimes seen in mitochondrial data (Brown et al. 1982). For this data set, the base composition was comparable to that found in other vertebrate mitochondrial genes, with approximately an equal ratio of A + T (52.7%) to G + C (47.3%), with base frequencies of 32.0% (A), 20.7% (T), 26.9% (C) and 20.4% (G). A slight transition/transversion bias existed, with the ratio ranging from 3.250 (between *Sphaerodactylus*

TABLE 1 (continued).

Species	Sequence 360
ramsdeni	CAACGAACCAAGTTACCCCAGGGATAACAGCGCCATCCTCTTCAAGAGCCCATATCGCCA
schwartzi	....T.C........
cricoderus	....
altavelensis	....
armstrongi	....
cryphius	.................................T..........................
darlingtoni	....
notatus	....
goniorhynchus	....C........
semasiops	....C........
gaigeae	....C........
klauberi	....C........
nicholsi	....
macrolepis	....C........
elegans	.....G..
cinereus	....
intermedius	......................................C.....................
nigropunctatus	......................................C.....................
torrei	....C........
oliveri	....TT................
richardi	....TT................
richardsoni	....
leucaster	....
thompsoni	....T.................
copei	....
elegantulus	....C........
fantasticus	....C.....A..
sputator	....
Gonatodes	T..................G......................AT.......C........
Lepidoblepharis	T...T........C........

notatus and *S. nicholsi*) to 0.412 (*S. torrei* and *S. fantasticus*); only 28/435 values were greater than 2.0. The Jukes-Cantor distance (Jukes and Cantor 1969) is the most appropriate to use under the above conditions (Kumar et al. 1993). However, distances that correct for deviation of nucleotide frequencies from 25% (Tajima and Nei 1984) and to correct for a transition bias (Kimura 1980) also were calculated and used with the NJ algorithm to see if any affect on tree topology and/or the significance of the clusters was evident.

An NJ tree constructed from these data (Figure 1) using the Jukes-Cantor distance with a complete deletion of all gaps and missing sites (resulting in 330 sites in the analysis). Both P_C values and P_B values are shown; as predicted by the conservative nature of the test, the P_B values are consistently lower. The two outgroup taxa, *Gonatodes* and *Lepidoblepharis*, cluster. The group containing all the West Indian *Sphaerodactylus* is strongly supported in both the interior-branch and bootstrap tests (P_C = 99%; P_B = 97%).

TABLE 1 (continued). See caption on page 178.

Species	Sequence	420
ramsdeni	AGAGGGTTTACGACCTCGATGTTGGATCAGGACCCCCGGGTGGTGCAACCGCTACCAGTG	
schwartzi	.................................TT..A...................A..	
cricoderus	....???????????	
altavelensis	.....................................A.....C...C.........A..	
armstrongi	.....................................A...................AC.	
cryphius	.....................................C...................AA.	
darlingtoni	.....................................A...................A..	
notatus	..................................?..A...................A..	
goniorhynchus	.................................A...A...................AC.	
semasiops	.....................................A...................AC.	
gaigeae	.....................................C.........G.........AC.	
klauberi	.....................................A...................AC.	
nicholsi	.....................................A.............?.....AC.	
macrolepis	.................................G...A...................AC.	
elegans	.................................TT..A...................AC.	
cinereus	.................................T...A.............?.....A..	
intermedius	..G............................G.....A...................AA.	
nigropunctatus	..G............................G.....A...................AA.	
torrei	...............................G.....A.............?....TAA.	
oliveri	.....................................A...................AC.	
richardi	.....................................A...................AC.	
richardsoni	.....................................A...................AA.	
leucaster	.....................................A...................AC.	
thompsoni	.............................CC......A...................A..	
copei	.....................................A..C..........-.G...A..	
elegantulus	.....................................A...................A..	
fantasticus	.....................................A...................AC.	
sputator	.....................................A...................AC.	
Gonatodes	.....................................AAC.......G.......T.A..	
Lepidoblepharis	...............................G.....AA........G-.A....T.AC.	

Within West Indian *Sphaerodactylus*, the cluster containing the Cuban species *S. ramsdeni*, *S. schwartzi*, and *S. cricoderus* is supported significantly (P_C = 95%) and a sister species relationship is suggested for *S. ramsdeni* and *S. schwartzi*. This group is basal to the remaining West Indian species, but this position is not strongly supported by either statistical test. The three Lesser Antillean species examined (*S. elegantulus*, *S. fantasticus*, and *S. sputator*) cluster (P_C = 77%), and the alliance of *S. fantasticus* and *S. sputator* is strongly supported.

TABLE 1 (continued).

Species	Sequence 430
ramsdeni	G-TTCGTTTG
schwartzi	....
cricoderus	?.????????
altavelensis	....
armstrongi	....
cryphius	....
darlingtoni	....
notatus	....
goniorhynchus	....
semasiops	....
gaigeae	....
klauberi	....
nicholsi	....
macrolepis	....
elegans	....
cinereus	....
intermedius	....
nigropunctatus	....
torrei	....
oliveri	....
richardi	....
richardsoni	....
leucaster	....
thompsoni	....
copei	....
elegantulus	....
fantasticus	....
sputator	....
Gonatodes	......????
Lepidoblepharis	.G........

The next cluster in the tree contains five Cuban species and is supported statistically (P_C = 95%). The two subgroups within this group are supported above the 95% level by both statistical tests; only one other group in the tree is supported at that level. *Sphaerodactylus oliveri* and *S. richardi*, both members of the *scaber* group (Hedges and Garrido 1993), are closely related; these species show little sequence differentiation (the Jukes-Cantor distance is 0.0124). The remaining three species — *S. torrei*, *S. intermedius*, and *S. nigropunctatus* — form a second group, with *S. intermedius* and *S. nigropunctatus* as sister taxa.

The remaining species form an assemblage that is moderately supported (P_C = 85%); within this assemblage are three groups. One group, consisting of the Hispaniolan species *Sphaerodactylus copei*, *S. elegans* (also found in Cuba and Florida), and *S. cinereus*, is supported, as is the association between *S. elegans* and *S. cinereus*. The group containing the Hispaniolan species *S. thompsoni* and *S. leucaster*, and the Jamaican *S. richardsoni*, is significant in both analyses (P_C = 99%; P_B = 98%). The remaining taxa cluster as a group that is moderately supported by the interior-branch test (P_C = 80%). All of the species in that group are members of the *argus* series (Hass 1991). This group includes taxa from all of the Greater Antilles and is split into two lineages that are moderately supported (P_C = 79% and 80%); one lineage contains taxa from Hispaniola (*S. altavelensis*, *S. cryphius*, and *S. darlingtoni*) and Cuba (*S. notatus*), the other has taxa from Puerto Rico (*S. gaigeae*, *S. klauberi*, and *S. nicholsi*), Jamaica (*S. argus* and *S. goniorhynchus*), and Hispaniola (*S. armstrongi*).

An NJ analysis using a distance (Tajima-Nei) that gives a better estimate of actual nucleotide substitutions in the case of unequal nucleotide frequencies gives a tree that is virtually identical to that from the previous analysis. The only differences are minor rearrangements in the cluster containing species from Jamaica, Puerto Rico, Hispaniola, and Cuba (*argus* series). This indicates that the deviation of base frequencies from 25% in this data set does not have an impact on the estimation of nucleotide substitutions between taxa.

An analysis that accounts for a bias towards transitional substitutions (Kimura 2-parameter model) also was employed in an NJ analysis. This does have some effect on tree topology. However, none of the branches that are affected show alternative topologies with strong statistical support. The moderately supported (P_C = 85%) group of five Cuban species clusters with the group of Lesser Antillean species, but this branch is not well supported (P_C = 46%). The group containing three Hispaniolan species (*Sphaerodactylus cinereus*, *S. copei*, and *S. elegans*) clusters with the group containing *S. thompsoni*, *S. richardsoni*, and *S. leucaster*, but this also is not well supported (P_C = 31%). Again, some minor rearrangements exist in the moderately supported (P_C = 86%) group of species from the *argus* series. Whereas none of these differences from the Jukes-Cantor tree is statistically significant, they do indicate that the slight transitional bias seen in these data does have an affect on data analyses.

Neighbor-joining analyses also were performed using a pairwise, rather than complete, deletion (372 sites analyzed). The only major changes in the tree topology were: (1) the dissociation of the assemblage, moderately supported in the other NJ analyses (P_C = 78–85%), containing the group of species from across the Greater Antilles (*argus* series), the cluster of *Sphaerodactylus thompsoni*, *S. richardsoni*, and *S. leucaster*, and the group of the three Hispaniolan species, *S. copei*, *S. elegans*, and *S. cinereus*; and (2) the association of these three Hispaniolan species with the *ramsdeni* group (P_C = 76%).

An MP analysis, in which gaps and missing data were excluded, was performed (Figure 2). Twenty-one MPTs were found, each of 376 steps. The topology is similar to that obtained in NJ analyses and most clusters appeared in all trees, except for some within the *argus* series. This analysis also does not show the *cinereus* series as a monophyletic group, with the cluster of *Sphaerodactylus thompsoni*/*S.*

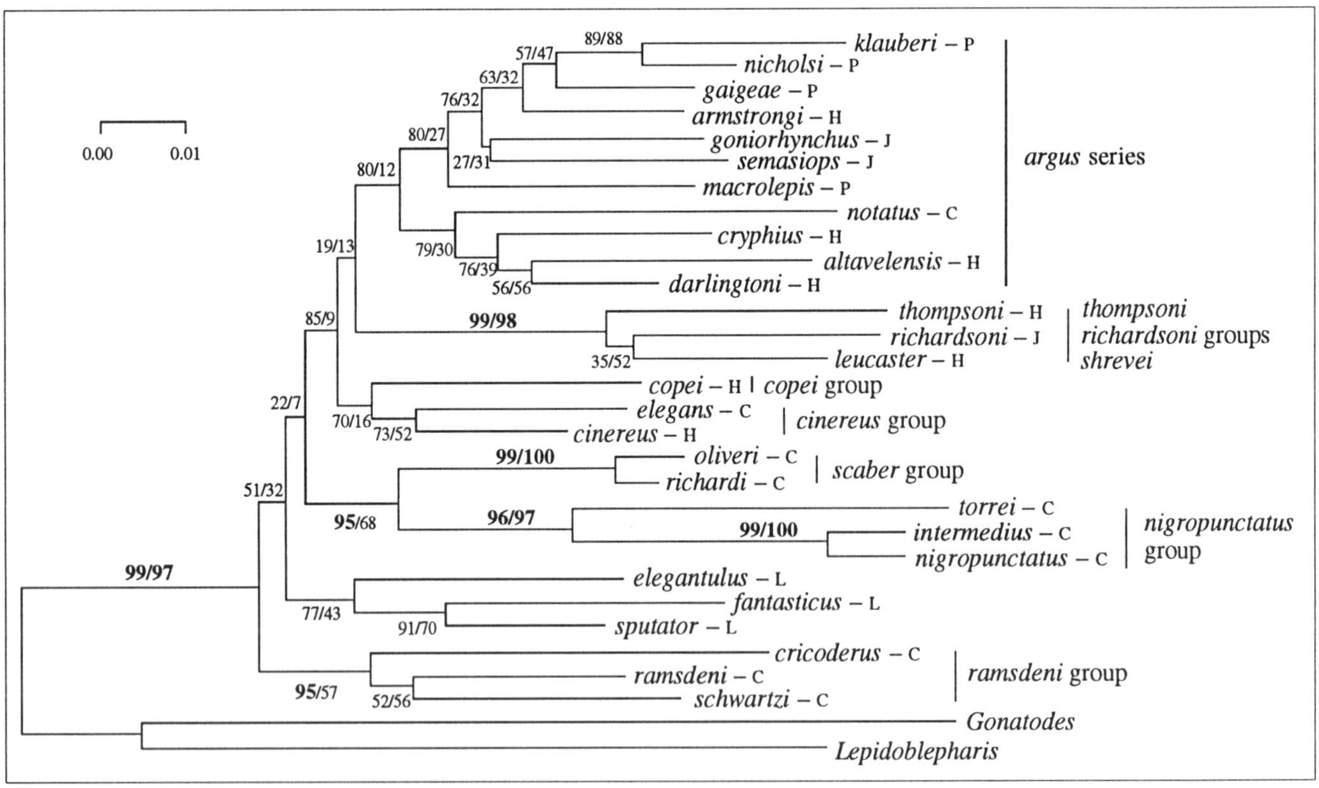

FIGURE 1. Phylogenetic tree constructed using 372 aligned sites from the mitochondrial 16S rRNA gene and rooted with two mainland sphaerodactyline genera, *Gonatodes* and *Lepidoblepharis*. This tree was constructed using the neighbor-joining algorithm with the Jukes-Cantor distance (all sites with missing data and gaps were excluded, resulting in 330 sites). The numbers on the branches are P_C values/P_B values (2000 replicates; Hedges 1992). Bold figures indicate significant support. The scale bar indicates a Jukes-Cantor distance of 0.01. Island designations are as follows: C = Cuba; H = Hispaniola; J = Jamaica; P = Puerto Rico; L = Lesser Antilles.

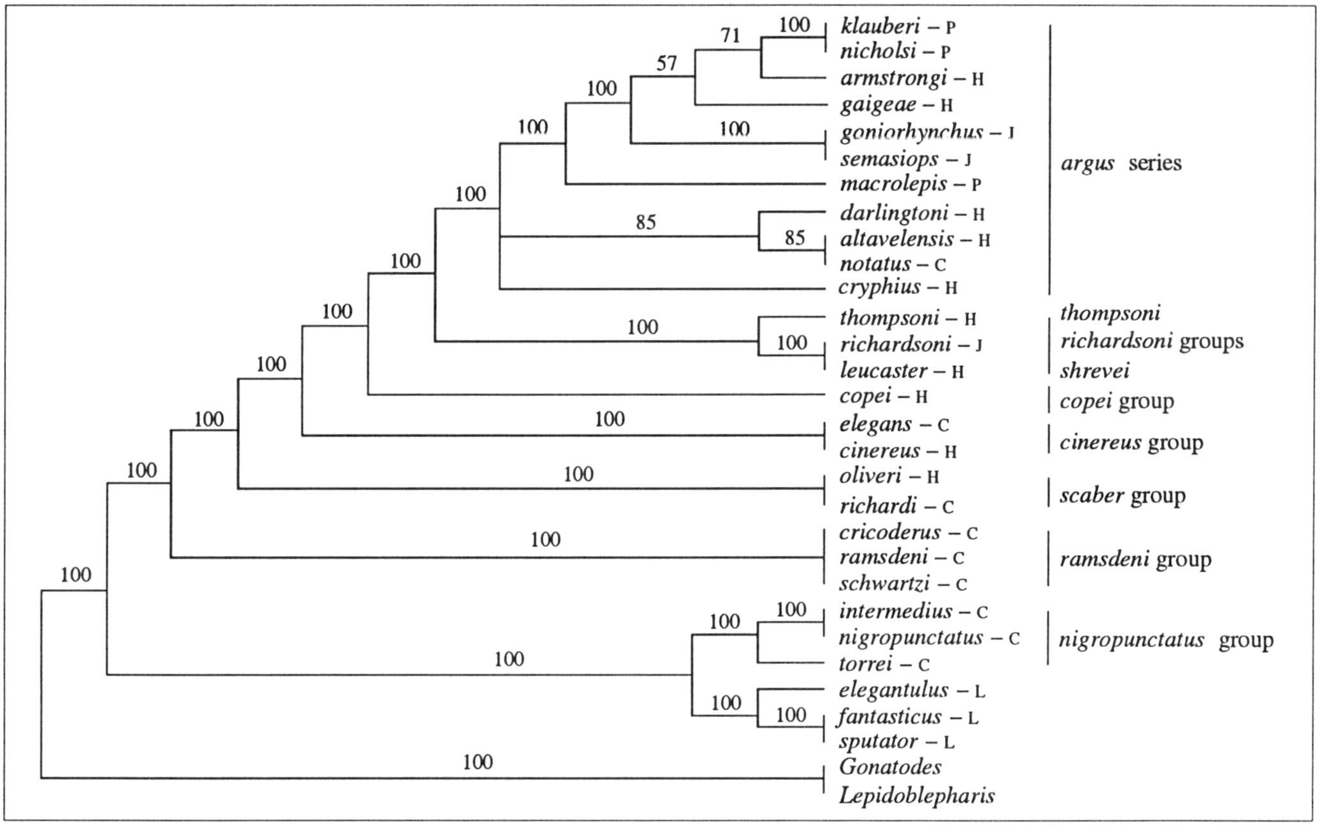

FIGURE 2. Maximum parsimony analysis of the same data as in Figure 1. This tree was constructed through a heuristic search, which resulted in 21 most parsimonious trees (MPTs), each with a length of 376 steps. The values on the branches are the percentage of MPTs in which the cluster was found. Island designations are as in Figure 1.

richardsoni/*S. leucaster* as the sister group of the *argus* series. *Sphaerodactylus copei* no longer clusters with the Hispaniolan species *S. elegans* and *S. cinereus*. Also, the group of five Cuban species is not found. *Sphaerodactylus oliveri* and *S. richardi* form a separate group, while *S. intermedius*, *S. nigropunctatus,* and *S. torrei* cluster with the Lesser Antillean species. The group containing *S. ramsdeni* is not basal to the remaining West Indian species.

DISCUSSION

Relationships within Antillean *Sphaerodactylus*

When examining the phylogenetic trees generated from this data set, a number of clusters are strongly supported across all the analyses. For the most part, these correspond to the species groups and subseries identified in Hass (1991). These sequence data also provide some evidence for the relationships among these groups.

These sequence data strongly support ($P_C \geq 95\%$ in all NJ analyses; appears in all MPTs) the recognition of the *ramsdeni* group as proposed by Thomas et al. (1992) based upon morphology; this group was not included in previous molecular studies (Hass 1991). However, these data do not unequivocally resolve relationships within this group. The NJ analyses with complete deletion place *Sphaerodactylus ramsdeni* and *S. schwartzi* as sister taxa, whereas using the pairwise deletion allies *S. ramsdeni* to *S. cricoderus*, but neither of these arrangements is strongly supported. The third possible arrangement, with *S. schwartzi* and *S. cricoderus* as sister species, was hypothesized by Thomas et al. (1992) based upon morphological characters, but does not appear in any analysis of these sequence data.

Another group that is strongly supported across all analyses is comprised of two Hispaniolan species, *Sphaerodactylus leucaster* and *S. thompsoni*, and *S. richardsoni* from Jamaica. In a previous molecular study, these species were placed in three different species groups within the *richardsoni* subseries of the *cinereus* series based upon both allozyme and immunological data (Hass 1991). Although no unique morphological characters appear to define this subseries, all these species are similar ecologically, inhabiting relatively xeric areas on or near the coast (*S. rhabdotus* occurs at sea level in the Valle de Neiba, Dominican Republic) and are moderate to large in size (Schwartz and Henderson 1991). The other species that have been placed in the *richardsoni* subseries, members of the *scaber* group (Hass 1991; Hedges and Garrido 1993), do not cluster with these species in any of the analyses of the DNA sequence data (see below).

Five Cuban species also form a significantly supported group ($P_C = 95\%$) in the NJ analyses (this cluster was not seen in the MP analysis). Three of these species — *Sphaerodactylus torrei*, *S. intermedius*, and *S. nigropunctatus* — which form a strongly supported cluster in both analyses, were placed in the *nigropunctatus* group by Schwartz and Garrido (1985). These species share a number of morphological characteristics, including similar scale morphology, ontogeny of color pattern, and strong sexual dichromatism. The previous molecular data also supported the close relationship of *S. nigropunctatus* and *S. torrei*; *S. intermedius* was not available for that study (Hass 1991). The latter species is unique in that it is the only member of this group that occurs in western Cuba (in Habana and Matanzas Provinces); the closest population of *S. nigropunctatus* is on Cayo Coco in the Archipiélago de Camagüey. *Sphaerodactylus intermedius* and *S. nigropunctatus* have very similar DNA sequences (the Jukes-Cantor distance between them is 0.018), and this indicates that *S. intermedius* may have originated as a result of allopatric speciation after a disjunction in a continuous coastal range of the common ancestor of the two species.

The remaining two species in this group of Cuban species, *Sphaerodactylus oliveri* and *S. richardi* (cluster at $P_C = 99\%$), are very different morphologically from members of the *nigropunctatus* group. *Sphaerodactylus richardi* was described recently and that paper also examined the relationships of the *scaber* group (Hedges and Garrido 1993). All the members of that group (*S. oliveri*, *S. richardi*, *S. scaber*, and *S. storeyae*), found in central Cuba, Isla de la Juventud, and the Archipiélago de los Canarreos, have a sexually dimorphic color pattern as adults, and large dorsal scales with a middorsal zone of small, granular scales. Previous molecular information is available for only one species, *S. oliveri*, and the immunological data indicated that it was allied to members of the *richardsoni* subseries (see above); this animal was not included in the electrophoretic study (Hass 1991). However, the sequence data do not place these species within that group, but instead suggest a relationship with the Cuban *nigropunctatus* group. Although the scale morphology of these two groups is very different (Schwartz and Garrido 1985; Hedges and Garrido 1993), they do share color pattern similarities including strong sexual dichromatism, with banded juveniles (and females in some species) and generally unicolor males (no males are banded as adults); this pattern dichromatism and ontogeny also is seen in some members of the *richardsoni* subseries. Because of their boss-like scales, members of the *scaber* group at one time were considered to be related to the Hispaniolan *S. copei* (Schwartz and Garrido 1981). These DNA sequence data do not support that placement.

None of the other groups present in the NJ tree are supported at significant P_C or P_B values. However, a number of groups do appear in all of the trees (NJ and MP analyses). One of the most intriguing associations is the cluster of three species from the Lesser Antilles. Previous molecular analyses (Hass 1991) had placed *Sphaerodactylus fantasticus* and *S. sabanus* in the *fantasticus* subsection, which is restricted to the Lesser Antilles. However, *S. sputator*, found on several islands in the northern Lesser Antilles, was placed in its own series as a member of the *sputator* subsection, which included species from the Greater Antilles and Bahamas Bank. The sequence data support a sister species relationship between *S. fantasticus* and *S. sputator* ($P_C = 91\%$; cluster seen in 100% of the MPTs), with *S. elegantulus*, found on the Antigua Bank, as the third member of this group. The immunological data also

TABLE 2. Classification of the genus *Sphaerodactylus*. This scheme is based primarily upon the electrophoretic data presented by Hass (1991), with species that were not placed in that study on the basis of morphology (see Hass 1991 for detailed discussion). Some taxa were reallocated based upon the data presented in this study (see text). Abbreviations are as follows: B = Bahamas, C = Cuba, H = Hispaniola, J = Jamaica, L = Lesser Antilles, N = Navassa, P = Puerto Rican Bank, Y = Cayman Islands.

Genus *Sphaerodactylus* Wagler 1830
♦ *sputator* section – West Indies (81 species)
- • *cinereus* subsection
  - + *cinereus* series
    - – *richardsoni* subseries
      - - *richardsoni* species group
        - *parkeri* (J), *richardsoni* (J)
      - - *shrevei* species group
        - *asterulus* (H), *leucaster* (H), *rhabdotus* (H), *shrevei* (H)
      - - *thompsoni* species group
        - *thompsoni* (H)
    - – *nigropunctatus* subseries
      - - *nigropunctatus* species group
        - *docimus* (C), *intermedius* (C), *nigropunctatus* (C), *torrei* (C)
      - - *scaber* species group
        - *oliveri* (C), *richardi* (C), *scaber* (C), *storeyae* (C)
    - – *cinereus* subseries
      - - *cinereus* species group
        - *cinereus* (H), *elegans* (C)
    - – *copei* subseries
      - - *copei* species group
        - *becki* (N), *copei* (H), *rosaurae* (Honduras)
  - + *argus* series
    - – *argus* subseries
      - - *argus* species group
        - *argivus* (Y), *argus* (J), *oxyrhinus* (J), *semasiops* (J)
      - - *goniorhynchus* species group
        - *goniorhynchus* (J)
    - – *notatus* subseries
      - - *notatus* species group
        - *bromeliarum* (C), *celicara* (C), *clenchi* (H), *cryphius* (H), *darlingtoni* (H), *difficilis* (H), *inaguae* (B), *ladae* (H), *lazelli* (H), *notatus* (C), *ocoae* (H), *omoglaux* (H), *plummeri* (H), *randi* (H), *savagei* (H), *sommeri* (H), *underwoodi* (B)
      - - unassigned to species group
        - *altavelensis* (H), *armstrongi* (H), *callocricus* (H), *cochranae* (H), *epiurus* (H), *gaigeae* (P), *klauberi* (P), *macrolepis* (P), *nycteropus* (H), *perissodactylius* (H), *samanensis* (H), *streptophorus* (H), *williamsi* (H), *zygaena* (H)
    - – unassigned to subseries
      - *armasi* (C), *beattyi* (P), *caicosensis* (B), *corticola* (B), *levinsi* (P), *mariguanae* (B), *micropithecus* (P), *monensis* (P), *nicholsi* (P), *parthenopion* (P), *roosevelti* (P), *ruibali* (C), *townsendi* (P)
- • *sputator* subsection
  - *sputator* (L)
- • *fantasticus* subsection
  - + *fantasticus* series
    - *elegantulus* (L), *fantasticus* (L), *kirbyi* (L), *microlepis* (L), *sabanus* (L), *vincenti* (L)
- • *ramsdeni* subsection
  - + *ramsdeni* series
    - - *ramsdeni* species group
      - *cricoderus* (C), *ramsdeni* (C), *schwartzi* (C)
- • unassigned to subsection
  - *gilvitorques* (J), *elasmorhynchus* (H)

♦ *lineolatus* section – Central and South America (10 species)

suggested that *S. sputator* is as different from the Greater Antillean and Bahamian species as are *S. fantasticus* and *S. sabanus* (Hass 1991). *Sphaerodactylus sputator* possesses a zone of middorsal granules, which is present in other Lesser Antillean species (King 1962), but is not found in members of the *argus* series. Therefore, the Lesser Antillean species apparently form a group, but a more complete understanding of this group requires further study and the addition of the three remaining Lesser Antillean species to the data set. The MP analysis indicates that the closest relative of this group is the *nigropunctatus* group, but this cluster was not found in any of the NJ analyses and, based on geography, seems less likely that the association of the *scaber* and *nigropunctatus* groups.

The species *Sphaerodactylus cinereus* and *S. elegans* cluster in all analyses, although this group is not significantly supported in the NJ analyses. These two species comprise the *cinereus* group (Hass 1991), but the relationship of this group to other West Indian species has not been resolved. The Hispaniolan species *S. copei* clusters with the *cinereus* group in the NJ analyses, but this association is not statistically significant. *Sphaerodactylus copei* was placed in the *copei* subseries of the *cinereus* series with two other species (*S. becki* and *S. rosaurae*); all three species in this subseries have both hair-bearing and knob-like scale organs (Schwartz and Garrido 1981).

The remaining 11 species examined were all members of the *argus* series (Hass 1991), which contains about 50 species. This group is supported moderately in all of the NJ analyses (P_C = 79–86%) and appears in all of the MPTs. However, the relationships of the species within this series vary among analyses, although one cluster, containing *Sphaerodactlyus klauberi*, *S. nicholsi*, *S. gaigeae*, and *S. armstrongi*, is always present. Whereas the electrophoretic data placed *S. klauberi* and *S. gaigeae* as sister taxa (Hass 1991), the sequence data consistently cluster *S. nicholsi* and *S. klauberi*. The two members of the *argus* subseries, *S. argus* and *S. goniorhynchus*, consistently cluster, but this is not well supported (P_C = 27%). The transition/transversion ratio is higher (between 2.0 and 3.0) for a number of comparisons between members of the *argus* series, indicating that these species may have diverged more recently and therefore sequence data from a more rapidly evolving gene or region may be more informative for investigating relationships within this group.

The classification presented by Hass (1991) can be examined in light of these sequence data. Table 2 shows that classification (including species), with modifications based upon the sequence data presented here. In that classification, the majority of West Indian species were placed in the *sputator* subsection, which was further divided into three series; *argus*, *cinereus*, and *sputator*. As discussed, all of the species placed in the *argus* series cluster in both NJ and MP analyses and that series has been retained with no modification.

The recognition of the *cinereus* series, containing the *richardsoni*, *nigropunctatus*, *cinereus*, and *copei* subseries (Hass 1991), was not supported by any analysis of the DNA sequence data, although the composition of these subseries, with the exception of the transfer of the *scaber* group to the *nigropunctatus* subseries (see above), was supported by the sequence data. Most analyses placed the *richardsoni* subseries (excluding the *scaber* group) as the closest relative of the *argus* series. However, whereas this was seen in all of the MPTs, it is not well supported in the NJ analyses. In the Kimura 2-parameter NJ analysis, the *richardsoni* subseries clusters (P_C = 31%) with *Sphaerodactylus copei* and the *cinereus* group, and this cluster is supported as the sister taxon to the *argus* series (P_C = 78%). Because of the overall lack of resolution in higher level relationships provided by the sequence data, and the support for this series by the electrophoretic data, the *cinereus* series has been retained pending further investigation.

Two of the Lesser Antillean species examined, *Sphaerodactylus fantasticus* and *S. elegantulus*, are members of the *fantasticus* subsection. However, the sequence data suggest that *S. sputator* also is a member of this subsection. Because of the conflict between the electrophoretic and sequence data sets, *S. sputator* has been moved to its own subsection. Likewise, the three members of the *ramsdeni* group form a distinct cluster with no clear affinities to any other groups of Antillean *Sphaerodactylus*. Therefore, they are classified as the *ramsdeni* subsection.

As the DNA sequence data largely support the species groups and subseries suggested by the analyses of the electrophoretic and, to a lesser extent, the immunological data, the relationships among the higher taxonomic groups (series and subsections) within West Indian *Sphaerodactylus* remain to be explored. This lack of resolution most likely is due to the short length of the region included in the analyses (372 sites; 150 of which were variable) relative to the number of taxa in the study (Lecointre et al. 1994).

Relationships: *Sphaerodactylus ramsdeni* Group Schwartz and Garrido (1985) posed a number of possible evolutionary histories for *Sphaerodactylus ramsdeni* and the other members of the *ramsdeni* group; these data allow the examination of some of these hypotheses. These data demonstrate that *S. ramsdeni* is not closely related to the Jamaican species *S. goniorhynchus*. Schwartz and Garrido (1985) had proposed this because of the proximity of the islands and the brown, cryptic color pattern that is seen in both species. Also, the members of the *ramsdeni* group and *S. goniorhynchus* have more than one internasal, whereas the modal number in other Cuban species is one (Schwartz and Henderson 1991). *Sphaerodactylus goniorhynchus* clusters with the other Jamaican species examined, *S. semasiops*, in the MP analysis. These two species cluster in two of the NJ analyses (Jukes-Cantor and Kimura distances with complete deletion), although this relationship is not well supported in those analyses (P_C = 27–30%). This relationship of *S. goniorhynchus* to the other Jamaican species in the *argus* group also was detected through immunological comparisons of serum albumin (Hass 1991). *Sphaerodactylus goniorhynchus* appears to be an example of a forest ecomorph that has been derived from, in this case, a group of species that primarily are lineate, all as juveniles and some as adults (Thomas 1975).

The analyses of these sequence data also do not suggest that the members of the *ramsdeni* group are closely related to any other Cuban groups of *Sphaerodactylus*. Only in the NJ analysis using the Jukes- Cantor distance with a pairwise deletion was the *ramsdeni* group allied to another group of West Indian species with some support (P_C = 76%). That group contains the Hispaniolan species *S. cinereus*, *S. copei*, and *S. elegans*. *Sphaerodactylus elegans* does occur in Cuba, although it is more abundant in the western part of Hispaniola (Schwartz and Henderson 1991). The species also does not show sexual dichromatism in adults (both males and females are unicolor), although juveniles are strongly banded, as in *S. cinereus*. Thus, while a relationship between the *ramsdeni* group and *cinereus* group cannot be ruled out, these data do not provide strong support.

The third possibility that Schwartz and Garrido (1985) presented was that *Sphaerodactylus ramsdeni* represents an old lineage of *Sphaerodactylus* with no close relatives among the extant West Indian species. In all the NJ analyses with complete deletion, the *ramsdeni* group occupies the basal position among West Indian *Sphaerodactylus*, even outside of the Lesser Antillean species, which were the most basal group in the analyses of allozyme data (Hass 1991). However, this placement is not well supported (P_C = 51–64%). In the MP analysis, this group is not basal and, as mentioned above, the *ramsdeni* group clusters with a group of Hispaniolan species in one NJ analysis. All species of *Sphaerodactylus* in the West Indies were not examined in this study, and therefore whether the *ramsdeni* group has strong affinities to any other West Indian species is uncertain, although the present analysis suggests that is unlikely. Also, from these data I cannot estimate the age of divergence of the *ramsdeni* group from other West Indian or mainland species to determine if it is indeed an ancient lineage. In light of these findings, the applicability of the ecomorph concept to the members of the *ramsdeni* group cannot yet be tested.

ACKNOWLEDGMENTS

I thank Robert Henderson and Robert Powell for inviting me to participate in this tribute to Albert Schwartz. Blair Hedges, Richard Thomas, and colleagues contributed significantly to this study through extensive field work on Hispaniola and Cuba. Ronald Crombie and Guido Pereira also provided some specimens. Blair Hedges kindly reviewed the manuscript. Linda Maxson provided some facilities and Marion Emerson assisted technically. Collecting and export permits were provided by J. Novo and G. Silva (Cuba), E. T. Lindberg and T. Price (Guantanamo Bay United States Naval Station, Cuba), S. J. and Y. A. Incháustegui (Dominican Republic), J. Fifi (Guadeloupe), G. Hermatin, E. Magny, P. Paryski, and F. Sergil (Haiti), P. Fairbairn and A. Haynes (Jamaica), E. Cardona (Puerto Rico), B. S. Nisbett (St. Kitts and Nevis), and J. L. Robinson (Antigua). Financial support was provided by the National Science Foundation (USA) (grants BSR 83-07115 to R. Highton and BSR-8906325 and DEB-9123556 to S. B. Hedges). All experimental protocols with animal tissue were approved by The Pennsylvania State University Institutional Animal Care and Use Committee, approval code number 1418. The 16S rRNA sequences presented here have been deposited in the European Bioinformatics Institute Nucleotide Sequence Database (EMBL), accession numbers X86034–X86063.

LITERATURE CITED

Anderson, S., A. T. Bankier, B. G. Barrell, M. H. L. DeBruijn, A. R. Coulson, J. Drouin, I. C. Eperon, D. P. Nierlich, B. A. Roe, F. Sanger, P. H. Schreier, A. J. H. Smith, R. Staden, and I. G. Young. 1981. Sequence and organization of the human mitochondrial genome. Nature (London) 290: 457–465.

Brown, W. M., E. M. Prager, A. Wang, and A. C. Wilson. 1982. Mitochondrial DNA sequences of primates: tempo and mode of evolution. Jour. Molec. Evol. 18: 225–239.

Cabot, E. L. and A. T. Bechkenbach. 1989. Simultaneous editing of multiple nucleic acid and protein sequences with ESEE. Comput. Appl. Biosci. 5: 233–234.

Felsenstein, J. 1985. Confidence limits on phylogenies: an approach using the bootstrap. Evolution 39: 783–791.

— and H. Kishino. 1993. Is there something wrong with the bootstrap on phylogenies? A reply to Hillis and Bull. Syst. Biol. 42: 193–200.

Fitch, W. M. 1971. Towards defining the course of evolution: minimum change for a specific tree topology. Syst. Zool. 20: 406–416.

Grant, C. 1939. Two new sphaerodactyls from Jamaica. Copeia 1939: 7–13.

Gyllensten, U. B. and H. A. Erlich. 1988. Generation of single-stranded DNA by the polymerase chain reaction and its implications to direct sequencing of the HLA DQ-alpha locus. Proc. Natl. Acad. Sci. USA 85: 7652–7656.

Hass, C. A. 1991. Evolution and biogeography of West Indian *Sphaerodactylus* (Sauria: Gekkonidae): a molecular approach. Jour. Zool. (London) 225: 525–561.

—, S. B. Hedges, and L. R. Maxson. 1993. Molecular insights into the relationships and biogeography of West Indian anoline lizards. Biochem. Syst. Ecol. 21: 97–114.

Hedges, S. B. 1989. Evolution and biogeography of West Indian frogs of the genus *Eleutherodactylus*: slow-evolving loci and the major groups, p. 305–370. *In* C. A. Woods (ed.), Biogeography of the West Indies: Past, Present and Future. Sandhill Crane Press, Gainesville, Florida.

—. 1992. The number of replications needed for accurate estimation of the bootstrap P value in phylogenetic studies. Molec. Biol. Evol. 9: 366–369.

—, R. L. Bezy, and L. R. Maxson. 1991. Phylogenetic relationships and biogeography of xantusiid lizards, inferred from mitochondrial DNA sequences. Molec. Biol. Evol. 8: 767–780.

—, A. R. Estrada, and R. Thomas. 1992. Three new species of *Eleutherodactylus* from eastern Cuba, with notes on vocalizations of other species (Anura: Leptodactylidae). Herpetol. Monogr. 6: 68–83.

— and O. H. Garrido. 1993. A new species of gecko (*Sphaerodactylus*) from central Cuba. Jour. Herpetol. 27: 300–306.

Hillis, D. M. and J. J. Bull. 1993. An empirical test of bootstrapping as a method for assessing confidence in phylogenetic analysis. Syst. Biol. 42: 182–192.

Jukes, T. H. and C. R. Cantor. 1969. Evolution of protein molecules, p. 21–132. *In* H. N. Munroe (ed.), Mammalian Protein Metabolism. Academic Press, New York.

Kimura, M. 1980. A simple method for estimating evolutionary rate of base substitutions through comparative studies of nucleotide sequences. Jour. Molec. Evol. 16: 111–120.

King, W. 1962 Systematics of Lesser Antillean lizards of the genus *Sphaerodactylus*. Bull. Florida State Mus. Biol. Sci. 7: 1–52.

Kluge, A. G. 1987. Cladistic relationships in the Gekkonidae (Squamata, Sauria). Misc. Publ. Mus. Zool., Univ. Michigan 173: 1–53.

Kumar, S., K. Tamura, and M. Nei. 1993. MEGA: Molecular Evolutionary Genetics Analysis, version 1.0. Pennsylvania State Univ., University Park.

—, —, and —. 1994. MEGA: molecular evolutionary genetics analysis software for microcomputers. Comput. Appl. Biosci. 10: 189–191.

Lazell, J. 1994. A new *Sphaerodactylus* (Sauria: Gekkonidae) from Bequia, Grenada Bank, Lesser Antilles. Breviora 496: 1–20.

Lecointre, G., H. Philippe, J. L. V. Lê, and H. Le Guyader. 1994. How many nucleotides are required to resolve a phylogenetic problem? The use of a new statistical method applicable to available sequences. Molec. Phylogen. Evol. 3: 292–309.

Losos, J. B. 1992. The evolution of convergent structure in Caribbean *Anolis* communities. Syst. Biol. 41: 403–420.

Rzhetsky, A. and M. Nei. 1992. A simple method for estimating and testing minimum-evolution trees. Molec. Biol. Evol. 9: 945–967.

Saitou, N. and M. Nei. 1987. The neighbor-joining method: a new method for reconstructing phylogenetic trees. Molec. Biol. Evol. 4: 406–425.

Schwartz, A. 1968. The geckos (*Sphaerodactylus*) of the southern Bahama Islands. Ann. Carnegie Mus. Nat. Hist. 39: 227–271.

—. 1983. The *difficilis* complex of *Sphaerodactylus* (Sauria, Gekkonidae) of Hispaniola. Pt. 1. *Sphaerodactylus difficilis, S. clenchi,* and *S. lazelli*. Bull. Carnegie Mus. Nat. Hist. 22: 5–30.

— and R. Franz. 1976. A new species of *Sphaerodactylus* (Sauria: Gekkonidae) from Hispaniola. Proc. Biol. Soc. Washington 88: 367–372.

— and O. H. Garrido. 1981. Las salamanquitas cubanos del género *Sphaerodactylus* (Sauria: Gekkonidae). 1. El grupo *copei*. Poeyana 230: 1–27.

— and —. 1985. The Cuban lizards of the genus *Sphaerodactylus* (Sauria, Gekkonidae). Milwaukee Public Mus. Contrib. Biol. Geol. 62: 1–67.

— and E. D. Graham, Jr. 1980. The *shrevei* group of Hispaniolan *Sphaerodactylus* (Reptilia, Gekkonidae). Tulane Studies Zool. Bot. 22: 1–15.

— and R. W. Henderson. 1991. Amphibians and Reptiles of the West Indies: Descriptions, Distributions, and Natural History. Univ. Florida Press, Gainesville, Florida.

Shreve, B. 1968. The *notatus* group of *Sphaerodactylus* (Sauria: Gekkonidae) in Hispaniola. Breviora 280: 1–28.

Sitnikova, T., A. Rzhetsky, and M. Nei. 1995. Interior-branch and bootstrap tests of phylogenetic trees. Molec. Biol. Evol. 12: 319–333.

Tajima, F. and M. Nei. 1984. Estimation of evolutionary distance between nucleotide sequences. Molec. Biol. Evol. 1: 269–285.

Thomas, R. 1966. A new Hispaniolan gecko. Breviora 253: 1–5.

—. 1975. The *argus* group of West Indian *Sphaerodactylus* (Sauria: Gekkonidae). Herpetologica 31: 177–195.

— and S. B. Hedges. 1992. An unusual new *Sphaerodactylus* from Hispaniola (Squamata: Gekkonidae). Jour. Herpetol. 26: 289–292.

— and —. 1993. A new banded *Sphaerodactylus* from eastern Hispaniola (Squamata: Gekkonidae). Herpetologica 49: 350–354.

—, —, and O. H. Garrido. 1992. Two new species of *Sphaerodactylus* from eastern Cuba (Squamata: Gekkonidae). Herpetologica 48: 358–367.

— and A. Schwartz. 1966. *Sphaerodactylus* (Gekkonidae) in the Greater Puerto Rico Region. Bull. Florida State Mus. Biol. Sci. 10: 193–260.

— and —. 1974. The status of *Sphaerodactylus gilvitorques* Cope and of *Sphaerodactylus nigropunctatus* Gray (Sauria: Gekkonidae). Jour. Herpetol. 8: 353–358.

— and —. 1983. The *difficilis* complex of *Sphaerodactylus* (Sauria: Gekkonidae) of Hispaniola. Pt. 2. *Sphaerodactylus savagei*, *S. cochranae*, *S. darlingtoni*, *S. armstrongi*, *S. streptophorus*, and conclusions. Bull. Carnegie Mus. Nat. Hist. 22: 31–60.

Williams, E. E. 1972. The origin of faunas. Evolution of lizard congeners in a complex island fauna: a trial analysis. Evol. Biol. 6: 47–89.

—. 1983. Ecomorphs, faunas, island size and diverse end points in island radiation of *Anolis*, p. 326–370. *In* R. B. Huey, E. R. Pianka, and T. W. Schoener (eds.), Lizard Ecology: Studies of a Model Organism. Harvard Univ. Press, Cambridge, Massachusetts.

Zarkikh, A. and W.-H. Li. 1992*a*. Statistical properties of bootstrap estimation of phylogenetic variability from nucleotide sequences. I. Four taxa with a molecular clock. Molec. Biol. Evol. 9: 1119–1147.

— and —. 1992*b*. Statistical properties of bootstrap estimation of phylogenetic variability from nucleotide sequences. II. Four taxa without a molecular clock. Jour. Molec. Evol. 35: 356–366.

Appendix A. Sample localities and specimen numbers of *Sphaerodactylus* used in this study (SBH = S. Blair Hedges frozen tissue collection, Pennsylvania State University; RIC = Ronald I. Crombie field collection).

Cuba: *S. cricoderus* – Santiago de Cuba: 1.5 km WSW La Tabla, 545 m (SBH 190623); *S. elegans* – Pinar del Río, Soroa, 155 m (SBH 191529); *S. intermedius* – La Habana: Narigon (near Puerto Escondido), 0 m (SBH 191484); *S. nigropunctatus* – Guantanamo Bay United States Naval Station, E of Bay at horse corral, 20 m (SBH 161962); *S. oliveri* – Sancti Spíritus: Manacal, 136 m (SBH 172853); *S. ramsdeni* – Santiago de Cuba: Isabelica (at the top of Gran Piedra), 1150 m (SBH 190305); *S. richardi* – Matanzas: 11.4 km ESE Playa Girón, 0 m (SBH 191501); *S. schwartzi* – Guantánamo: Loma Redonda, 100 m (SBH 190959); *S. torrei* – Santiago de Cuba: Santiago (near city zoo and Hotel Leningrado) (SBH 190641).

United States: *S. notatus* – Florida. Monroe County: No Name Key (SBH 160824–29).

Hispaniola: *S. altavelensis* – Dominican Republic. Barahona: Barahona (on the grounds of Hotel Guarocuya), 0 m (SBH 102744); *S. armstrongi* – Dominican Republic. Barahona: 11.3 km S Barahona (SBH 102589); *S. cinereus* – Haiti. l'Ouest: 10 km SE Fond Parisien (SBH 160501); *S. copei* – Haiti. Sud'Est: 9.5 km E Jacmel, 10 m (SBH 104281); *S. cryphius* – Dominican Republic. Independencia: 5 km SE La Florida, 90 m (SBH 102792); *S. darlingtoni* – Dominican Republic. Barahona: 0.3 km N, 13.8 km E Canoa (airline distance), 550 m (SBH 160160); *S. leucaster* – Dominican Republic. Barahona: 3.0 km E Canoa (SBH 102669); *S. thompsoni* – Dominican Republic. Pedernales: 6.4 km SW Las Mercedes, 61 m (SBH 102584).

Jamaica. *S. goniorhynchus* – St. Mary: 1.3 mi W Little Bay (SBH 190706); *S. richardsoni* – Trelawny: 0.3 km W Duncans, 80 m (SBH 161043); *S. semasiops* – Trelawny: 10.6 km S Kinloss (SBH 161357).

Puerto Rico: *S. gaigeae* – Isla Vieques (SBH 161776); *S. klauberi* – 9.0 km S Isabela, 200 m (SBH 172214); *S. macrolepis* – Isla Verde (near San Juan), 0 m (SBH 172613); *S. nicholsi* – Cabo Rojo (vicinity of lighthouse on cape) (SBH 161312).

Lesser Antilles: *S. elegantulus* – Antigua: Top Side Hill (SBH 192763); *S. fantasticus* – Guadeloupe. Basse-Terre: 3 km W Pointe de la Grande Anse, 5 m (SBH 102250); *S. sputator* – St. Martin. 2.7 km N Orleans (SBH 172774).

Other taxa: *Lepidoblepharis xanthostigma* – (RIC 051869). *Gonatodes vittatus* – Venezuela (SBH 161802).

Resumen.—Un segmento de arn del gene mitocondrial ribosomal 16S fue secuenciado en 28 especies de *Sphaerodactylus* de las Antillas y dos taxones esferodactilinos externos, *Gonatodes* y *Lepidoblepharis*. Estos datos fueron utilizados para examinar las relaciones filogenéticas entre los linajes principales de este grupo y para dilucidar la posición del grupo *ramsdeni*, un grupo morfológicamente único de tres especies cubanas (*S. ramsdeni*, *S. cricoderus*, y *S. schwartzi*). Tanto los análisis de método de agrupamiento por el vecino más cercano (AVC), como los de parsimonia máxima (PM) apoyan significativamente el reconocimiento de los siguientes grupos de especies: *ramsdeni* (P_C [valores de confiabilidad para las ramas internas en un análisis de AVC] = 95%), *nigropunctatus* (P_C = 96%), y *scaber* (P_C = 99%). Además, una relación íntima entre los grupos cubanos *nigropunctatus* y *scaber*, una asociación que no había sido sugerida previamente, es apoyada fuertemente por los análisis de AVC (P_C = 95%). Los representantes de tres grupos de especies (*shrevei*, *richardsoni*, y *thompsoni*) forman un conglomerado apoyado fuertemente (P_C = 99%). Otras asociaciones sugeridas por un estudio molecular anterior son apoyadas moderadamente. Las once especies que representan la serie *argus* (P_C = 80%) se agrupan, al igual que lo hacen los dos miembros del grupo *cinereus*, *S. cinereus* y *S. elegans* (P_C = 73%). Las tres especies de las Antillas Menores estudiadas, miembros de la serie *sputator* y de la serie *fantasticus*, se agrupan (P_C = 77%); esta relación fue indicada por datos immunológicos, pero no fue aparente en un estudio de alozimas. A niveles superiores, estos datos no apoyan el reconocimiento de la serie *cinereus* como un grupo monofilético. En cambio, estos datos sugieren que la serie *argus* se deriva de ese grupo (P_C = 85%). Las afinidades del grupo *ramsdeni* continúan sin resolverse, aunque este grupo no parece estar más íntimamente relacionado a ningún otro grupo de los *Sphaerodactylus* de las Antillas.

Résumé.—Une portion du gène ARN ribosomique 16S des mitochondries a été séquencée chez 28 espèces de *Sphaerodactylus* des Antilles et chez deux espèces hors groupe, *Gonatodes* et *Lepidoblepharis*. Les données recueillies ont permis d'examiner les rapports phylogénétiques entre les différentes lignées au sein de ce groupe et de percer le mystère de la place du groupe *ramsdeni*, assemblage morphologiquement unique de trois espèces cubaines (*S. ramsdeni*, *S. cricoderus* et *S. schwartzi*). Les analyses de parcimonie maximale et de fusion entre voisins (FV) confirment de façon marquée la reconnaissance des groupes d'espèces suivants: *ramsdeni* (P_c [degré de certitude pour le test intrabranches dans l'analyse FV] = 95%), *nigropunctatus* (P_c = 96%) et *scaber* (P_c = 99%). Les analyses FV (P_c = 95%) mettent également en lumière les rapports très étroits qui existent entre *nigropunctatus* de Cuba et les groupes *scaber*, association qui n'avait jamais été soupçonnée auparavant. Les représentants des trois groupes d'espèces (*shrevei*, *richardsoni* et *thompsoni*) forment un agrégat important (P_c = 99%). Les autres associations mises en lumière par une étude moléculaire antérieure sont moyennement confirmées. Les onze espèces qui représentent la série *argus* forment un agrégat (P_c = 80%), comme les deux membres du groupe *cinereus*, *S. cinereus* et *S. elegans* (P_c = 73%). Les trois espèces des Petites Antilles examinées, membres des séries *sputator* et *fantasticus*, se regroupent (P_c = 77%); ce rapport était apparent dans les données immunologiques mais pas dans l'étude des allozymes. À des niveaux supérieurs, ces données ne permettent pas de considérer la série *cinereus* comme un groupe monophylétique. Au contraire, ces données donnent à penser que la série *argus* est dérivée de ce groupe (P_c = 85%). Les affinités du groupe *ramsdeni* restent mystérieuses, mais ne semblent pas avoir de rapports très étroits avec les autres groupes d'espèces de *Sphaerodactylus* des Antilles.

Hass, C. A. 1996. Relationships among West Indian geckos of the genus *Sphaerodactylus*: a preliminary analysis of mintochondrial 16S ribosomal RNA sequences, p. 175–194. *In* R. Powell and R. W. Henderson (eds.), Contributions to West Indian Herpetology: A Tribute to Albert Schwartz. Society for the Study of Amphibians and Reptiles, Ithaca (New York). Contributions to Herpetology, volume 12.

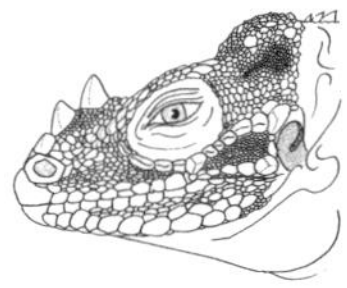

CAREENING ISLAND AND THE GOAT ISLANDS: EVIDENCE FOR THE ARID-INSULAR INVASION WAVE THEORY OF DICHOPATRIC SPECIATION IN JAMAICA

JAMES LAZELL[1]

Abstract.—Compared to the other Greater Antilles, the Jamaica Bank is little fragmented by historically dramatic sea level fluctuations spanning some 140 m. Opportunities for dichopatric speciation on satellite islands are ostensibly best on the arid-insular lands of Portland Bight, such as the Goat Islands. Here, forms which evolve in isolation in simple, xeric ecosystems may be preadapted to succeed competitively when able to colonize the mainland, especially if there is a long-term trend towards climatic deterioration. Dichopatric speciation always renders parent species "paraphyletic" and, to the extent that ecotypic forms are produced, provides an opportunity for polyphyly. *Anolis* lizards of the *grahami* complex are exemplars. Relictual, montane, mesic country forms are predicted.

INTRODUCTION

THE JAMAICA BANK, CONTINUOUS LAND AT GLACIAL maximum, was mid-sized among the Greater Antilles; it was much larger than any of the three Cayman Banks, Mona, or Navassa, but much smaller than Cuba or Hispaniola. The present main island of Jamaica is about the same size as Puerto Rico (11,526 km^2 and 8,860 km^2, respectively), but the Puerto Rico Bank is almost twice the size of the Jamaica Bank: about 25,000 km^2 versus 15,000 km^2, respectively. Sea level fluctuations certainly extending back into the Pliocene and associated with glaciations span some 140 m (Dawson 1992). High stand sea level, such as that during the Sangamon Interglacial, about 120 thousand years ago (ka), fragmented the larger banks into veritable archipelagos, but had little effect on Jamaica (Dawson 1992; Figure 1). At low stand during glacial maxima, land areas closely approximated the maximum areas of each bank. Sea level fluctuations thus provided a cyclic reiteration of many islands as potential evolutionary theaters followed by terrestrial continuity allowing overland movement of species to and from previous island sites. Rand (1969), in a cogent analysis of the effect of sealevel changes on *Anolis* speciation, demonstrated that within-bank diversity is high in the Greater Antilles, like Jamaica, and results largely from self-contained radiations. However, he did not propose a model for intra-island speciation.

The Jamaica Bank is the simplest example of the intra-island radiation phenomenon, but it has the fewest satellite islands. Nevertheless, Jamaica has 11 currently recognized taxa of *Anolis* lizards, of which ten are likely to constitute a monophyletic group endemic to their bank (i.e., all except *A. sagrei*; Underwood and Williams 1959). Eight of these forms belong to the *lineatopus* and *grahami* complexes, are very closely related (Underwood and Williams 1959), and must have arisen from a common ancestor on the Jamaica Bank (Hedges and Burnell 1990).

The basic idea that dichopatric, xeric-adapted species evolved on the Portland Bight islands during interglacials and then invaded the dry mainland south coast of Jamaica at glacial maxima was pro-

[1] The Conservation Agency, 6 Swinburne Street, Jamestown, Rhode Island 02835, USA.

posed to account for the evolution of *Anolis reconditus*, a relictual species of the Blue Mountains (Lazell 1966). A corollary to this theory is that an overall, or more-or-less continuous, drying trend — climatic deterioration — has occurred over at least the last several million years.

The concept of climatic deterioration, a general if ragged trend towards cooler and drier global climates, beginning at least in the Miocene, is entrenched in geological literature (e.g., Brooks 1949; Frakes 1980; Kerr 1989). Of course, pluvial punctuations have occurred; cyclic glaciations have caused swings in weather; and controversy exists over just where and when conditions were relatively wet or dry (Flint 1935; Pregill and Olson 1981). The reality of Holocene desertification in the northern tropics is not debatable in my opinion (Pielou 1992: 311). In any case, repeated periods of relative drought would enable my model to function, albeit more slowly, even if no long-term drying trend can be proven.

The alternative to my view of speciation, suggested by Underwood and Williams (1959: 23), is ecological isolation within a continuous land area. Because of the abundance and overland dispersal abilities of anoles, I must reject this hypothesis.

I was long taunted by the near-absence of herpetological records from the Goat Islands in Portland Bight. Celebrated in the literature for their once-abundant *Iguana* (=*Cyclura*) *collei* (Grant 1940, and references therein), only one specimen with precise data appears to be extant: National Museum of Natural History (USNM) 103688, collected by Gardner Lynn in 1932 on Little Goat. The various other specimens cited by Grant (1940: 71) are attributed only to an unspecified "Goat Island" or "Goat Islands." The boa, *Epicrates subflavus*, is mentioned by Grant (1940), but no specimen seems to exist. In view of the fact that Thomas Barbour of the Museum of Comparative Zoology (MCZ) visited at least one of these islands, and that at least Little Goat was a site of considerable United States Navy activity during World War II (Calderhead 1985; United States Navy 1946, 1947), the paucity of records seemed nearly incredible. Further, Careening Island often is shown as a rather large place on

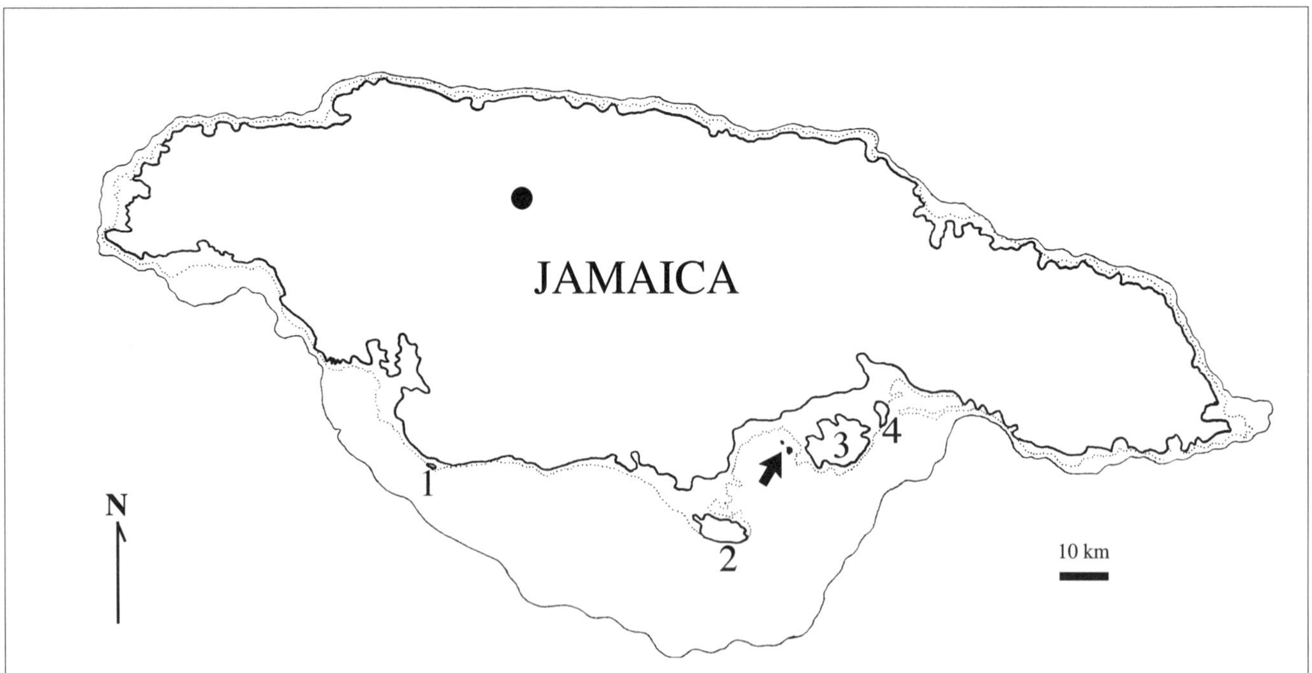

FIGURE 1. Jamaica: the bold line indicates the Sangamon high sealevel, ca. 120 ka; the fine line marks the greater Jamaica Bank, corresponding to glacial maximum low sea level (e.g., ca. 20 ka); the dotted line represents the present coastline. The arrow indicates the Goat Islands (Careening Island is too small to indicate, see Figure 2), the solid circle marks the area of Ram Goat Cave in the Cockpits. 1 = Great Pedro Bluff; 2 = Portland Ridge; 3 = Hellshire; 4 = Port Henderson Hills.

some maps (Texaco 1964), yet no herpetological records from there are known. My theory of *Anolis* evolution is scarcely tenable if no *Anolis* occurred on the islands during the present interglacial.

Materials and Methods

On 24 February 1993 I instigated expeditions to Careening and both Goat Islands that were led, in fact, by George Proctor and Edwin Duffus. Six people hunted for animals and plants. On 1 March 1995 we returned to both Goat Islands. I took notes on the catch, general aspects of vegetation and geology, and the extent of human and exotic animal disturbance.

On 27 February 1993, George Proctor led a party of five to Ram Goat Cave, at about 500 m, in the Cockpit Country, 5.1 km by road north of Burnt Hill, Trelawny Parish (Figure 1). Several of us returned on 25 February and 5 March 1995. Hunting, catching, and recording proceeded as before.

I made detailed color notes on all lizards in life, both before and after capture, so as to document color change. All specimens were fixed by injection with one part formalin (37 percent formaldehyde in water, buffered) in 15 parts of 70% isopropyl alcohol. Specimens were stored in 70% isopropanol for several days until switched to 70% ethanol at MCZ.

Results

We found lizards, including *Anolis*, on Careening and both Goat Islands (Table 1). However, *Anolis* were strangely scarce there compared to similar habitats elsewhere. At Ram Goat Cave we found two species of *Anolis*: *A. garmani* (MCZ A-11597), not found on any island, and a distinctive form not among those currently granted taxonomic recognition in Jamaica (MCZ A-11598-603, A-15089–92, and A-15126–31), but seemingly allied to *A. grahami*. The possibility that this anole might be a relictual mesic country species evolutionarily analogous to *A. reconditus*, combined with finding two *grahami*-group species on the Goat Islands, inspired the present work.

The islands are shown in Figure 2. Careening Island, locally called "Crane Island," is a modest mangrove with a tiny patch of sandy ground, about 40 m^2, above high tide, but clearly frequently overwashed. I do not believe chthonic or strictly terrestrial reptiles could survive there. *Anolis lineatopus* was spotted and collected in the single buttonwood (*Conocarpus erectus*, Laguncularíaceae); its arboreality ameliorates its survival potential. A couple of seaside mallows (*Thespesia populnea*, Malvaceae) were also found. No other woody terrestrials were encountered, but all three mangrove species are present.

Little Goat Island was clear-cut during World War II (United States Navy 1946). Levelled terraces, pits, bunkers, and extensive concrete roads and platforms attest to massive modification. *Sphaerodactylus argus* was the only reptile we commonly encountered. *Anolis opalinus* was present and color notes include a male displaying, but these lizards were so scarce and shy we could catch only one, a juvenile. We encountered only two individuals of *Anolis grahami* in over an hour on the island, and collected just one.

The vegetation was extremely degraded; the enormous goat population must have severely retarded regrowth. I counted over 100 goats in one herd congregated on a large concrete pad on the north shore. These goats are evidently heavily subsidized. Water and food containers were numerous and appeared recently supplied. The goats obviously are of economic importance and this fact alone would make ecological restoration expensive.

TABLE 1. Lizard records for Careening and the Goat Islands, Jamaica. The Museum of Comparative Zoology at Harvard University (MCZ) numbers are field tags because accessioning was incomplete at the time of writing.

Species	Careening	Little Goat	Great Goat
Anolis lineatopus	A-11586	—	—
Anolis opalinus	—	A-15108	A-11592–3
Anolis grahami	—	A-11590	A-15109
Iguana (=Cyclura) collei	—	103688*	—
Sphaerodactylus argus	—	A-11587–9	A-11591

* National Museum of Natural History, Washington (USNM).

Enormous effort would be required to restore Little Goat to a semblance of naturalness. Nevertheless, some plant species important in the diet of iguanas, like *Picrodendron baccatum* (Euphorbiaceae), were common and fruiting (seemingly ignored by goats).

Great Goat Island is in better ecological condition than Little Goat and probably escaped clear-cutting. This island is higher and far more rugged. *Sphaerodactylus argus*, *Anolis opalinus*, and *A. grahami* are common, but no other reptiles were encountered. Goats are present and numerous, but not so dense as on Little Goat, although also apparently fed and watered. Potential iguana food species — *Picrodendron* and several *Capparis* spp. (Capparaceae) — are common.

Other islands in the extensive mangrove swamps flanking southwestern Hellshire may be worthy of biological exploration.

The *Anolis lineatopus* of Careening Island fits the nominate subspecies well. The pattern is unusually bold but in typical shades of light and dark gray-brown. The snout was yellowish in life. The throat fan was pale yellow distally with a bright peach-colored center. The fan faded to near-white proximally and had pale yellow to near-white scales. Generally, *A. l. lineatopus* is the most abundant anole in xeric habitats of southern Jamaica, and its ab-

FIGURE 2. Careening (1), Little Goat (2), and Great Goat (3) islands. The first, bold contour represents about 20 m elevation, corresponding to the Sangamon high sea level. Other contours on Great Goat are at about 60 m and 90 m. Mangrove swamp is indicated. This map was modified from Jamaica Government, 1973, Middlesex, Preliminary Edition, Survey Department: 87B, with ground truthing 24 February 1993.

sence from the Goat Islands seems very peculiar. However, Crombie et al. (1983) found *A. grahami* dominant over *A. lineatopus* in disturbed habitats on Cabarita Island off Jamaica's north coast.

Anolis grahami on both Goat Islands seem also to fit the nominate form well. One individual was green (MCZ A-15109, Great Goat), but the other (MCZ A-11590, Little Goat) was brilliant sky-blue, a common ground color in southern Jamaica. Both had purple tail bases. The adult males collected had deep bright orange throat fans with bright yellow borders and scales. Other individuals seen varied from blue to green. Color changes were to an overall dark bluish-gray and, additionally, to a heavily reticulated pattern of blue-gray and pale greenish-yellow-gray. This pattern was reminiscent of *A. conspersus*.

Anolis opalinus has been little noticed in terms of geographic variation; no subspecies have been described. This may be an oversight. Both Goat Island populations appeared similar to each other and those of the coastal mainland. MCZ A-11592, our largest, an adult male, was 55 mm snout-vent length (SVL) after fixation and two days in spirits. Goat Island *A. opalinus* were largely gray with only a hint of greenish on their tail bases. Their lateral stripes were prominent, light gray, and tinted with a little bluish or greenish. Their throat fans were as described by Underwood and Williams (1959): bright orange with a yellow border and center spot. In the uplands of Jamaica I have found *A. opalinus* to be quite mossy greenish (*contra* Grant, 1940: 94–95): MCZ A-11594, Green Hills, Portland Parish. Throat fans may lack the central yellow spot: MCZ A-11583–4, Muirton, Portland Parish, and MCZ A-11607, Christiana, Manchester Parish.

Sphaerodactylus argus from the Goat Islands seem to fit the nominate form. They were very dark brown with bold ash-white ocellae. Subcaudal coloration varied from rosy red through pink and terra cotta to mottled brick red and slate gray.

Anoles from the Ram Goat Cave, Cockpits, are believed to be related to *grahami*-group anoles because the pericardium is very darkly pigmented; the ventrals are smooth and flat to slightly swollen; and the lizards chirp when handled. Other scale characters include a well-defined double row of enlarged middorsals; the interparietal about half the size of the supraocular disk; and high lamellae counts: 24–28 lamellae under phalanges III–IV of the fourth toe.

When first encountered, most males were sooty blackish-brown. Extensive, bold, lighter gray-brown, subcircular dots occurred in rough longitudinal rows, especially prominent anteriorly. The throat fan was bright orange with brilliant chartreuse scales and border. The underside of the tail, elevated in display, was blotched with brick-red and rust orange.

Upon capture, the color changed to a lighter, dull gray-brown. Larger males (e.g. MCZ A-15089–90 and A-15126) can be olive, or dull mossy green, or largely bluish-gray. The spots turned ash-white with dark borders and became notably bolder and even more striking than before. The eyelids turned ashy. The fan scales turned sooty. The color of the fan border faded to lemon-yellow on the posterior fan base. This color shaded into the pale blue-green of the abdomen. The tail turned dull blue-green, except for the tip which remained rusty.

Females were very similar, but the fan was small and entirely chartreuse. Their middorsal stripes were prominent and narrowly broken with transverse sooty lines. Some females had warm-brown snouts and developed blue-green head tints. Several females (e.g., MCZ A 11601–2, 45 and 43 mm SVL, respectively) contain shelled eggs.

I would proceed with caution in declaring this anole currently nameless, especially in view of the available names currently in the synonymy of *Anolis grahami* (*punctatissimus* Hallowell seems especially apt).

Scenarios

My thesis is that much, possibly most, of the evolutionary diversification within Jamaica results from the juxtaposition of a high, wet, windward set of forested mountain ranges (John Crow, Blue, Main Ridge, and Manchester Plateau) and a relatively dry, flat, southern, rain-shadowed coastal plain varying in size by a factor of three because of sea level fluctuations and variably studded with islands (Figure

1). This is the evolutionary theater in which my scenarios play.

Invoking the same processes of speciation by geographic isolation used previously to explain the existence of a montane relic (Lazell 1966), I will develop a model I believe has most everything that commonly occurs in real evolution within the species group: peripheral isolates, parapatric subspeciation, cladogenesis, anagenesis, paraphyly, polyphyly, classic character divergence, and histories both recoverable and forever unrecoverable. With modest tinkering my model can provide secondary subspeciation, laddered or bridged clades, and dichopatric character divergence. At a simplistic level, Hoelzer and Melnick (1994) have described some of these standard, typical phenomena and noted the havoc they wreak upon cladistic theoretical dogma.

This scenario plays on background of fairly continuous climatic deterioration: drying. This progressive drying trend is especially demonstrable in coastal lowlands. Wet forest — mesic habitat — remains in the upland interior of the main island, although it retreats from the lowlands and diminishes in area. I do not envision Jamaica's forest ever naturally fragmenting significantly into disjunct mesic enclaves. The time span envisioned is from glacial maximum through an interglacial to a second glacial maximum; all that can fit into less than 200,000 years (Dawson 1992).

The scenario (rather simplistically diagrammed in Figure 3) begins at a glacial maximum with all Jamaica Bank lands united by low sea level, overall conditions rather mesic, and a ubiquitous small anole, hereafter called "ancestor," showing weak, modal, clinal geographic variation. In the wettest interior habitats all individual ancestor anoles are green or greenish and have a bicolored throat fan (principal ground color and border color). In the most dry, remote, south coast lowlands all ancestor anoles are gray and have a tricolor throat fan (ground, border, and central spot colors). Over 90% of Jamaica, without any real ecological zonation and generally mesic conditions, ancestor anoles are greenish-gray and have a weak wash of slightly differentiated central color in their basically bicolor throat fans. Not even the most ravenous splitter would describe the two ends of this cline as subspecies; this is just very typical, slight geographic variation in a widespread monotypic form.

Comes the interglacial, stranding two ancestor populations, one on each of the Goat Islands. Climatic deterioration sharpens the ecological zonation of Jamaica, expanding near-uniform arid, xeric lowlands, reducing mesic upland, and narrowing the ecotone between them. Selection pressure for differentiated ecotypes on the mainland increases. Competition with other species shifts ancestor anoles toward smaller size and more rupicolous (less generalized) habits. Ancestor anoles of southern, mainland Jamaica lose ground to other species and begin to disappear from this area (although this has not yet occurred in the present interglacial).

Meantime, island anoles find themselves in a situation of solitude. Through a series of stochastic events, no other congener has survived on the stranded lands of what are now the Goat Islands. Through ecological release, island anoles evolve larger size and generalized, but strongly arboreal, habits; and avian predators exert a strong selection for palid color, but island anoles retain their tricolor throat fans (this also has not yet occurred in the present interglacial).

Evolution in the island anoles is parallel. The same phenotypes, basically, are selected for: big, arboreal, and pale. Some of the same fortuitous mutations appear and the two genotypes evolve with much in common, but the evolution of each is utterly independent of the other.

By the end of the interglacial, three distinct sorts have been derived from the ancestor and are easily recognized phenotypically. On the mainland, the ancestor has become a rupicolous dwarf sharply divided into two parapatric subspecies, a green form with a bicolor fan (hereafter "bigreen") confined to the interior uplands, and a gray form with a tricolor fan (hereafter "trigray") confined to the north coast. A zone of intergradation unites these two, but most of the species' range is occupied by one or the other. These variants are no longer two ends of a gradual cline.

The third is the island form: big, arboreal, and very pale. To the human observer, the two island

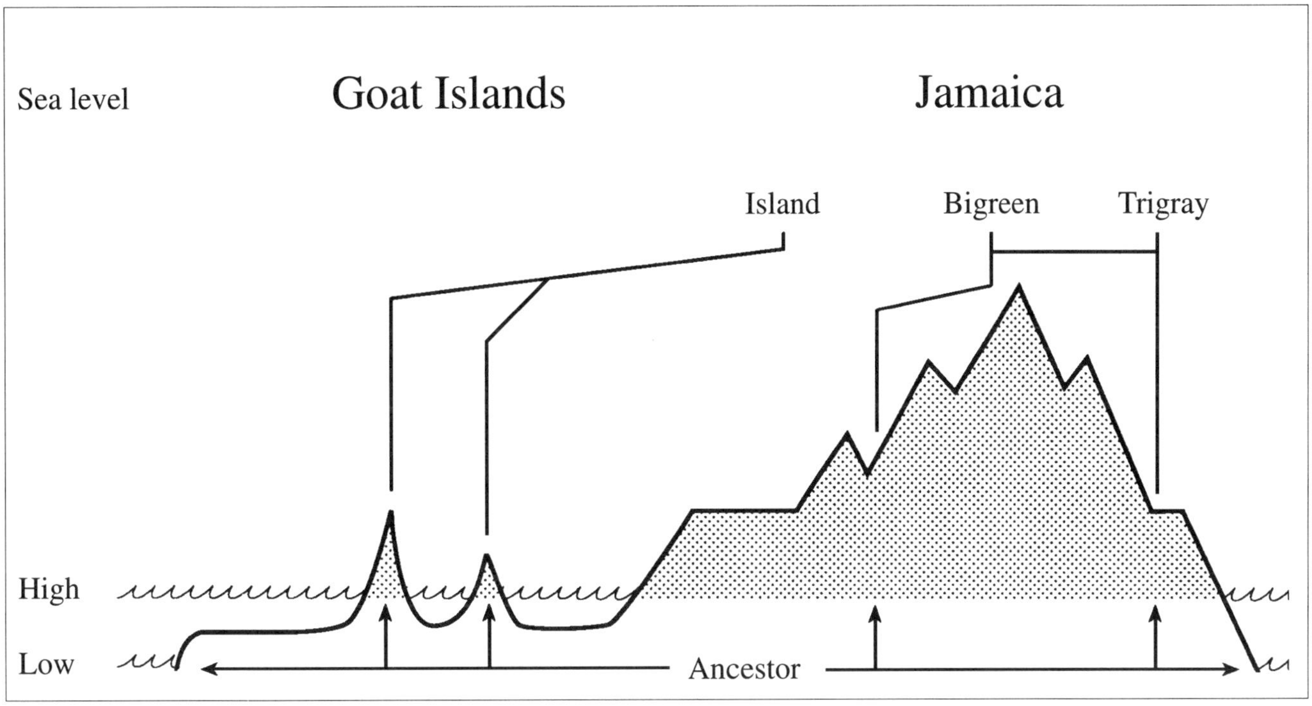

FIGURE 3. A diagramatic view of interglacial dichopatric speciation followed by glacial maximum invasion of the Jamaican mainland. See scenarios in text.

populations look nearly identical. No thought of taxonomic separation *inter se* would be entertained at this stage. However, the two island populations have evolved in complete isolation from each other; their resemblances are purely the result of parallel evolution. To name an island taxon at this stage would be to manufacture a polyphyletic fiction.

Comes the glacial maximum. Sea level drops, and all Jamaica is reunited. The first thing that happens to our anoles is that the two island populations meet in a frenzy of male combat and indiscriminate copulation. As the falling sea spreads vast new lands before them, the mongrel horde colonizes rapidly. No mainland anole is so well equipped to take over new, expanding, arid terrain. The new island anole is a genetic combination of its two island parent stocks. The Little Goat individuals, forged by brutal selection on their tiny islet, are the biggest, the most physiologically adept at surviving aridity, and the palest: a set of extremes giving their descendants a sweeping advantage. The Great Goat stock, however, was far more populous on its vastly larger islet. In the initial mixing of these stocks, Great Goat females so outnumber Little Goat females that, when the dust settles several generations later, only Great Goat mitochondrial DNA (mtDNA) remains in island descendants. This bit of history cannot be recovered. All that eventually remains of the Great Goat contribution to the new, conquering, island anole is mtDNA. All the rest of the genotype — everything that matters evolutionarily — has come from Little Goat.

As the new island anole (polyphyletic, but now indisputably unitary) expands onto the mainland, it eventually meets the descendants of ancestor anole. The initial contacts are between the island form and bigreen. These two are now so different in appearance and behavior that few attempts at hybridization occur. The handful of hybrid offspring produced fail to prosper. The failure of introgression to succeed constitutes wasted reproductive effort for those parents that attempted it, and, therefore, a *de facto* selection pressure for any characters (colors, behavior) that mitigate hybridization. Classic character divergence results, with the ancestor derivative becoming even smaller and more rupicolous, and the island derivative becoming even larger and more arboreal. By the time island animals meet trigray, with whom they still share a tricolor fan, introgression will be out of the question.

At the end of this scenario two full, biological, unequivocal species exist, one with two subspecies. The ancestor species has been rendered paraphyletic and the island derivative is polyphyletic. To name this composite insular derivative now necessarily acknowledges a polyphyletic reality.

One can easily modify the scenario to include occasional waif dispersal events that ladder and bridge the two island clades during their interglacial isolation. Similarly, one need not postulate the complete retreat of the ancestor from the mainland south coast. If it remained there, one could invoke occasional waif dispersers colonizing its habitat and producing a "nudge effect" for dichopatric character divergence (Williams 1969; Lazell 1972: 102; Goodyear and Lazell 1986).

Finally, one need not invoke such dramatic differentiation. Then, perhaps, when the invading island anoles meet ancestor anoles, introgression might succeed and a zone of intergradation might develop between them. Then we would have three parapatric, interbreeding subspecies exhibiting both primary and secondary intergradation *sensu* Mayr (1942: 99).

REALITY

Do examples of these phenomena exist? I believe I see them all the time. *Anolis opalinus* on Jamaica today is my exemplar of ancestor, just beginning the process with modest, clinal, geographic variation. *Anolis cristatellus wileyae* on Tortola is another example. The highest upland, inland variants are greenish and have yellow bellies (University of Texas at Austin GP-316, 325–9; MCZ A-11932–6: Sage Mountain, Tortola, elevation over 500 m). Coastal individuals are usually quite gray with near-white bellies (MCZ A-11937–41: East End, Tortola). Over most of Tortola, *A. c. wileyae* varies modally between these extremes, but on satellite Beef Island all individuals are at the gray, white-bellied extreme (MCZ Z-22957–9, 158922).

Frogs of the genus *Eleutherodactylus* on Jamaica also may provide examples. With *E. cundalli* playing the analog of *Anolis lineatopus*, *E. cavernicola* would then be an arid-insular species set to invade from Portland Point. Although relationships are unresolved (Crombie 1977, 1986; Hass and Hedges 1991), *E. grabhami* might be the corresponding mesic relic analogous to *A. reconditus* or *A. lineatopus neckeri*; *E. sisyphodemus* might be an alternative.

Crombie (1986) suggested that speciation of *Eleutherodactylus* occurred in mesic enclaves, much as Underwood and Williams (1959) postulated for *Anolis*. Hedges and Burnell (1990) inclined towards an arid-insular dichopatric speciation scenario for the Jamaican anoles, but Hass and Hedges (1991) do not commit themselves with respect to *Eleutherodactylus* — a far more speciose, complex, and possibly much older group on this bank. Thus, *Eleutherodactylus* may provide opportunities to test both scenarios as molecular clock data develop.

The Jamaican dwarf boas, genus *Tropidophis*, are exemplary, with *T. haetianus stejnegeri* in the mesic interior and north coast, *T. h. jamaicensis* along the drier southern coast, and disjunct *T. h. stullae* out on peninsular Portland Ridge (Tolson and Henderson 1993).

Basically, this scenario was first devised for the Jamaican *Anolis* of the *lineatopus* group (Lazell 1966). Today we have *A. l. neckeri* and *A. l. merope* playing the roles of my "bigreen" and "trigray," respectively, while nominate *A. l. lineatopus* pushes *A. l. neckeri* from the dry south. Whether *A. l. lineatopus* and *A. l. neckeri* are full species producing rare, maladapted hybrids, or subspecies showing secondary intergradation, is presently arguable, but I incline toward the former view (Lazell 1993).

In the *lineatopus* group scenario, *Anolis lineatopus ahenobarbus* and *A. reconditus* are seen as relics in the mesic uplands of older ancestral stocks that gave way to, for example, the invasion wave producing *A. l. neckeri* and *A. l. merope*. That *A. reconditus* seems fully immune to introgression from any other form is notable (but this may not have been tested against *A. l. ahenobarbus*), whereas *A. l. ahenobarbus* intergrades (secondarily?) with *A. l. lineatopus* (Underwood and Williams 1959). This situation is not surprising: evolutionary rates in potential reproductive isolating mechanisms vary enormously, as they do in any other sorts of characters. Reproductive isolation is needed for any clade to exist independently, but the age of a clade has very little, if anything, to do with its distinctiveness or ability to remain independent. Clades diverge and amalgamate almost regardless of their relative ages.

On the Puerto Rico Bank, *Anolis cristatellus wileyae* today is fragmented into nearly a hundred populations on the islands east of Puerto Rico itself, and actually intergrades with nominate *A. c. cristatellus* in extreme eastern Puerto Rico (Heatwole 1976); these are ecotypic subspecies: *A. c. wileyae* in the arid lowlands, *A. c. cristatellus* on the mesic north coast. *Anolis gundlachi* is a presumed relic in the uplands, predating the wave that brought onshore the ancestral stock of modern *A. c. cristatellus*.

Anolis ernestwilliamsi is a presumed autochthon on Carrot Rock, the present interglacial's product of dichopatric speciation (Lazell 1983). However, it has managed to colonize adjacent Peter Island (MacLean 1982), and seems to be thereon providing the celebrated nudge effect (Williams 1969). Because *A. ernestwilliamsi* is very large and can tolerate and thrive in the most desolately arid conditions, it may do very well in competition with *A. cristatellus wileyae*. In any case, it renders not just *A. cristatellus* "paraphyletic," but the subspecies *A. c. wileyae* paraphyletic too.

In Les Îles des Saints off southern Guadeloupe, *Anolis marmoratus terraealtae* and *A. m. caryae* are intergrading subspecies united at low sea levels. However, they are ecotypes too: *A. m. terraealtae* (despite its name) holds the low, dry land whereas *A. m. caryae* retains the more mesophilic. Is *A. m. terraealtae*, occurring on several islands, polyphyletic — the product of what Mayr (1994) describes as parallelophyly? In Les Îles de le Petite Terre, Guadeloupe, interglacial high sea level have stranded *A. m. chrysops* on two different islands (Lazell 1964). Are we here seeing polyphyletic parallelophyly? Mitochondrial DNA might now reveal the truth.

An interesting side-scenario could develop on some island banks with arid-insular dichopatric speciation potentials on more than one flank. For example, Puerto Rico has not just the eastern, Virgin theater, but (like Jamaica) coastal cays off the xeric south side as well. Has dichopatric speciation occurred here? Of course, *Anolis cooki* represents the arid counterpart of *A. cristatellus*. *Anolis cooki*, however, seems to be faring poorly in competition with *A. cristatellus* (Jenssen et al. 1984), although its status as a distinct species is fully demonstrated. Heatwole (1976, and references cited therein) presents evidence that *A. cooki* may be closer to a common ancestral stock than is *A. cristatellus*. If this is the case, then *A. cooki* would be a relic in southwestern Puerto Rico and on the tiny Punta Aguila coastal cay, and a case of the arid-insular form being reversed, as it were, losing ground to a mainland (but apparently younger) relative (Marcellini et al. 1985).

CONCLUSIONS

If repeated cycles of interglacial, dichopatric, arid-insular speciation followed by glacial maximum invasions displacing mesic-adapted mainland forms have occurred, then two predictions follow: (1) of course, various populations on the arid insular land areas will show degrees of speciation; and (2) relictual, mesic-adapted taxa will persist in upland habitats. Both predictions seem fully borne out.

At Portland Point, the potentially most disjunct, fairly large arid-insular land mass off Jamaica, we get our degrees of speciation. *Eleutherodactylus cavernicola* and *Tropidophis haetianus stullae* are recognized taxa at different levels. The "*coxi*" variant of *Anolis l. lineatopus* is not well enough differentiated to formally qualify (Underwood and Williams 1959), but awaits the coming separation of Portland Point by rising sealevel. On the Goat Islands, *A. opalinus* is certainly not yet a taxon separable from the mainland populations, but the individuals are big and may be diverging behaviorally from mainland norms. I found them to be very much trunk-ground anoles, not at all trunk-crown anoles as stated by Hedges and Burnell (1990). And sea level is rising.

Inland, mesic-adapted relics are many. Just within *Anolis*, *A. reconditus* and the spotted, Ram

Goat Cave anoles noticed herein are exemplary. *Anolis lineatopus neckeri* currently may be approaching relictual status. I predict more taxa like these will be discovered.

I see two distinct scenarios for speciation on the Jamaica Bank: (1) arid-insular invasion waves (Lazell 1966; Hedges and Burnell 1990); or (2) mesic enclave refugia (Underwood and Williams 1959; Crombie 1986). Separating examples should be easy. Forms confined to arid-insular areas should be young and those confined to mesic enclaves old in the first scenario. The exact opposite, old arid-insular forms with younger counterparts in mesic enclaves would support the latter scenario. We await only the definitive molecular clock data.

Real evolution frequently involves paraphyly, polyphyly, and parallelophyly. The symplesiomorphy of reproductive continuity may amalgamate and obliterate clades regardless of their ages. Dogmatic philosophies that attempt to codify "relationship" while ignoring evolutionary realities are not of practical value for comprehending the real world — the world of lizards.

Acknowledgments

I am indebted to George Proctor and Edwin Duffus for their leadership in our efforts. Numi Goodyear, Glenn Mitchell, Wenhua Lu, Ann Sutton, Dan Freitas, Hillrie Quin, and Derek Campbell assisted in many ways. Proctor and Goodyear provided botanical data, including names and relative edibility (for iguanas) of plants encountered. Robert Henderson went far beyond the call of editorial duty to supply me, in the field, with a number of critical references suggested by perspicacious reviewers.

Literature Cited

Brooks, C. P. E. 1949. Climate Through the Ages: TheStudy of Climatic Factors and Their Variations. McGraw-Hill Book Co., New York.

Calderhead, W. L. 1985. Caribbean, the, U. S. Naval bases, 1940–, p. 62–65. *In* P. E. Coletta (ed.), United States Navy and Marine Corps Bases, Overseas. Greenwood Press, Westport, Connecticut.

Crombie, R. I. 1977. A new species of frog of the genus *Eleutherodactylus* (Amphibia: Leptodactylidae) from the Cockpit Country of Jamaica. Proc. Biol. Soc. Washington 90: 194–204.

—. 1986. Another new forest-dwelling frog (Leptodactylidae: *Eleutherodactylus*) from the Cockpit Country of Jamaica. Trans. San Diego Soc. Nat. Hist. 12: 145–153.

—, D. W. Steadman, and J. C. Barber. 1983. A preliminary survey of the vertebrates of Cabarita Island, St. Mary Parish, Jamaica. Atoll Res. Bull. 280: 1–15.

Dawson, A. G. 1992. Ice Age Earth. Routledge, London and New York.

Flint, R. F. 1935. Climates of the Pleistocene: a review. Amer. Jour. Sci. 29: 381–385.

Frakes, L. A. 1980. Climates Throughout Geologic Time. Elsevier/North-Holland, Inc., New York.

Goodyear, N. C. and J. Lazell. 1986. Relationships of the silver rice rat *Oryzomys argentatus* (Rodentia: Muridae). Postilla 198: 1–7.

Grant, C. 1940. The herpetology of Jamaica II. The reptiles. Bull. Inst. Jamaica Sci. Ser. 1: 61–148.

Hass, C. A. and S. B. Hedges. 1991. Albumin evolution in West Indian frogs of the genus *Eleutherodactylus* (Leptodactylidae): Caribbean biogeography and a calibration of the albumin immunological clock. Jour. Zool. 225: 413–426.

Heatwole, H. 1976. Herpetology of Puerto Rico. VII. Geographic variation in the *Anolis cristatellus* complex in Puerto Rico and the Virgin Islands. Occas. Pap. Mus. Nat. Hist., Univ. Kansas 46: 1–18.

Hedges, S. B. and K. L. Burnell. 1990. The Jamaican radiation of *Anolis* (Sauria: Iguanidae): an analysis of relationships and biogeography using sequential electrophoresis. Carib. Jour. Sci. 26: 31–44.

Hoelzer, G. A. and D. J. Melnick. 1994. Patterns of speciation and limits to phylogenetic resolution. Trends Ecol. Evol. 9: 104–107.

Jenssen, T., D. Marcellini, C. Pague, and L. Jenssen. 1984. Competitive interference between the Puerto Rican lizards, *Anolis cooki* and *A. cristatellus*. Copeia 1984: 855–862.

Kerr, R. A. 1989. Did the roof of the world start an Ice Age? Science 244: 1441–1442.

Lazell, J. 1964. The anoles (Sauria: Iguanidae) of the Guadeloupeen archipelago. Bull. Mus. Comp. Zool., Harvard Univ. 131: 359–401.

—. 1966. Studies on *Anolis reconditus* Underwood and Williams. Bull. Inst. Jamaica Sci. Ser. 18: 1–15.

—. 1972. The anoles (Sauria: Iguanidae) of the Lesser Antilles. Bull. Mus. Comp. Zool., Harvard Univ. 143: 1–115.

—. 1983. Biogeography of the herpetofauna of the British Virgin Islands with description of a new anole (Sauria: Iguanidae), p. 99–117. *In* A. Rhodin and K. Miyata (eds.), Advances in Herpetology and Evolutionary Biology: Essays in Honor of Ernest E. Williams. Mus. Comp. Zool., Harvard Univ., Cambridge, Massachusetts.

—. 1993. Geographic distribution: *Anolis lineatopus lineatopus* (Liguanea bush anole). Herpetol. Rev. 24: 108.

Maclean, W. P. 1982. Reptiles and Amphibians of the Virgin Islands. MacMillan and Co., London.

Marcellini, D., T. Jenssen, and C. Pague. 1985. Distribution of the lizard *Anolis cooki*, with comments on its possible future extinction. Herpetol. Rev. 16: 99–102.

Mayr, E. 1942. Systematics and the Origin of Species. Columbia Univ. Press, New York.

—. 1994. Cladistics and convergence. Trends Ecol. Evol. 9: 149–150.

Pielou, E. C. 1992. After the Ice Age. Univ. Chicago Press, Chicago.

Pregill, G. K. and S. L. Olson. 1981. Zoogeography of West Indian vertebrates in relation to Pleistocene climatic cycles. Ann. Rev. Ecol. Syst. 12: 75–98.

Rand, A. S. 1969. Competitive exclusion among anoles (Sauria: Iguanidae) on small islands in the West Indies. Breviora 319: 1–16.

Texaco. 1964. Jamaica. Rand McNally, Chicago.

Tolson, P. J. and R. W. Henderson. 1993. The Natural History of West Indian Boas. R & A Publ., Ltd., Taunton, Somerset, U. K.

Underwood, G. and E. E. Williams. 1959. The anoline lizards of Jamaica. Bull. Inst. Jamaica Sci. Ser. 9: 1–48.

United States Navy. 1946. Caribbean Sea Frontier. Caribbean Commands, vol. II. U. S. Naval Administrative History of World War II, vol. 165. United States Navy Dept. Libr., Washington, D. C.

—. 1947. Building the Navy's Bases in World War II. History of the Bureau of Yards and Docks and the Civil Engineer Corps 1940–1946, vol. II. United States Govt. Print. Off., Washington, D. C.

Williams, E. E. 1969. The ecology of colonization as seen in the zoogeography of anoline lizards on small islands. Quart. Rev. Biol. 44: 345–389.

Resumen.—Comparado con las demás Antillas Mayores, el Banco de Jamaica está poco fragmentado por fluctuaciones históricamente dramáticas en el nivel del mar que llegan a los 140 m. Las oportunidades para la especiación dicopátrida en islas satélites son ostensiblemente mejores en las tierras insulares áridas de Portland Bight, como las Islas Goat. Aquí, formas que evolucionan aisladamente en ecosistemas xéricos simples pueden estar preadaptadas a tener éxito competitivo cuando son capaces de colonizar la isla de Jamaica, especialmente si existe una tendencia desde hace mucho tiempo hacia el deterioro del clima. La especiación dicopátrida siempre causa especies parentales "parafiléticas" y, a la medida que se producen formas ecotípicas, se provee una oportunidad para la polifilia. Las lagartijas del complejo de *Anolis grahami* son un buen ejemplo de este patrón. Se predice que se producirán poblaciones relictuales montanas de terrenos húmedos.

Résumé.—Comparé aux autres Grandes Antilles, le banc Jamaïque est peu fragmenté par les fluctuations spectaculaires du niveau de la mer dont l'éventail couvre quelques 140 mètres. Les occasions de spéciation dichopatrique sur les îles satellites sont ostensiblement meilleures sur les terrains arido-insulaires de Portland Bight, comme les Goat Islands. Dans ces îles, les formes isolées qui évoluent dans des écosystèmes secs simples sont peut-être préadaptées à la survie si elles sont capables de coloniser le continent, notamment s'il y a une tendance à long terme vers la détérioration climatique. La spéciation dichopatrique rend toujours les espèces jumelles "paraphylétiques" et dans la mesure où des formes écotypiques sont produites, elle fournit l'occasion de polyphylie. Les *Anolis* du complexe de *grahami* sont à ce titre exemplaires. Des formes mésiques subalpines relictuelles sont prévues.

Lazell, J. 1996. Careening Island and the Goat Islands: evidence for the arid-insular invasion wave theory of dichopatric speciation in Jamaica, p. 195–205. *In* R. Powell and R. W. Henderson (eds.), Contributions to West Indian Herpetology: A Tribute to Albert Schwartz. Society for the Study of Amphibians and Reptiles, Ithaca (New York). Contributions to Herpetology, volume 12.

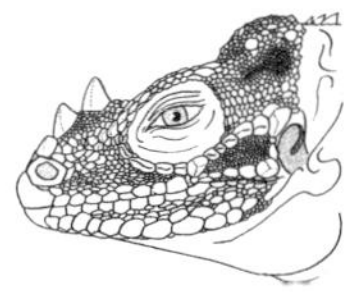

Crossing Bond's Line: The Herpetofaunal Exchange between the Eastern Caribbean and Mainland South America

John C. Murphy[1]

Abstract.—The herpetofauna of Trinidad and Tobago is almost entirely continental and distinct from the fauna of the Lesser Antilles. Bond's Line, a salt water gap of 140 km between Grenada and Tobago, separates the two faunas. The Trinidad and Tobago herpetofauna includes as many as 129 species and subspecies and, of these, only five (3.8%) are of Lesser Antillean origin. The one frog (*Eleutherodactylus johnstonei*) and four lizards (*Anolis aeneus*, *A. extremus*, *A. richardi*, and *A. trinitatis*) have probably had minimal negative impact on the native herpetofauna, and they appear restricted to urban areas and other human-modified environments. Conversely, 19 extant taxa from Trinidad-Tobago-mainland South America have been reported in the Lesser Antilles. Several of these are represented by specimens with erroneous locality data, others are one-time waifs that failed to establish a population, others represent older dispersals and current taxonomy fails to note their distinctiveness, and some have dispersed with human assistance. Thus, Bond's Line must be considered a formidable barrier for amphibians and reptiles, and the taxa that have successfully overcome the barrier from the north have mostly done so with the aid of humans.

Introduction

Travel guides and geography texts consider the islands of Trinidad and Tobago part of the West Indies. However, geologically, floristically, and faunistically these islands are continental. This position has been supported by numerous authors over the past 150 years (DeVerteuil 1858; Boettger 1895; Darlington 1957; Lescure 1980; Hardy 1982; Williams 1989). Beard (1944) may have said it best: "The mountains of Tobago constitute the last and most easterly link in the great coastal cordillera of Venezuela. Tobago, the Northern Range of Trinidad, and the Paria Peninsula of Venezuela are the three ridges of related structure arranged parallel and *en echelon*."

Trinidad (4769 km^2) is three times larger than Guadeloupe, the largest island in the Lesser Antilles. Tobago (300 km^2) would rank sixth in size in the Lesser Antilles. Both Trinidad and Tobago were almost completely covered with rain forests and seasonal forests in pre-Columbian times (Beard 1944, 1946). For a large portion of the past 140,000 years these islands have been attached by dry land to main land South America due to sea level fluctuations (based upon my interpretation of data in Shackleton 1987 and discussion in Dawson 1992). Since their colonization by humans, the habitat on both islands has been radically altered. Today the vegetation is a mosaic of agriculture, derived and natural savanna, and secondary and primary forest.

The West Indian ornithologist, James Bond, recognized the dramatic change in bird species that occurred between Tobago and Grenada, and Lack (1976) labeled the 140 km-wide ocean barrier "Bond's Line." Lescure (1980) noted the uniqueness of the Lesser Antillean herpetofauna from the Greater Antillean and the Trinidad-Tobago-mainland South American faunas, but ignored the important role humans have played in the distribution of those species that have successfully crossed Bond's Line.

[1] Division of Amphibians and Reptiles, Field Museum of Natural History, Chicago, Illinois 60614, USA.

Herein I examine extant amphibian and reptilian species that have been reported to have traversed Bond's Line. The literature, museum collections, and my 114 days of field work (from 1982–1994) on Trinidad and Tobago suggest that the islands maximally hold a herpetofauna of 129 species and subspecies representing 32 families (Table 1). This number includes species introduced by humans, waifs from the flooding Orinoco, museum material with possibly erroneous locality data, marine turtles, and undescribed taxa. The 129 taxa (119 on Trinidad, 69 on Tobago, 56 shared) can be categorized into eight distributional patterns (modified from Hoogmoed 1979). Records of South American taxa invading the Lesser Antilles are based on Schwartz and Henderson (1985, 1988, 1991), who listed 81 taxa in 18 families, exclusive of sea turtles, in the Lesser Antilles. Plainly, the herpetofauna of Trinidad and Tobago is continental in origin and contains relatively few species from the Lesser Antillean fauna (Table 2).

Moving Southward Across Bond's Line

Five taxa showing a Lesser Antillean distribution include one frog (*Eleutherodactylus johnstonei*) and four saurians (*Anolis aeneus*, *A. extremus*, *A. richardi*, and *A. trinitatis*) that have made their way to Trinidad and Tobago. Some have been intentionally introduced by humans, whereas others may have arrived by more serendipitous routes.

Eleutherodactylus johnstonei was first reported from the Port of Spain dock area by Kenny (1979), who considered the introduction deliberate; the species' presence on Trinidad is documented with a specimen (USNM 244420). No evidence exists to suggest an expansion of its range in agricultural or natural habitats in Trinidad; my field work has not produced any additional specimens. Concern regarding successful competition with native species is justified on some islands (Kaiser and Henderson 1994 [1995]). Due to its isolation in an urban center, this frog does not appear to be a problem on Trinidad, at least for the present.

Anolis aeneus is native to the Grenada Bank and has been introduced into Guyana, Trinidad, and the

TABLE 1. Amphibian and reptilian taxa occurring on Trinidad (T) and Tobago (To). An "x" denotes the taxon's presence on an island. A "?" notes that a taxon's presence is based on questionable literature records or museum specimens with possibly erroneous data. A "?+" denotes possible extirpation. Distributional patterns (D) are modified from Hoogmoed (1979): W = Widespread, A = Amazonian, LGE = Lowland Guiana Endemic, CCR = Caribbean Coastal Range, OB = Orinoco Basin, Cos = Cosmopolitan, LA = Lesser Antilles, UK = Unknown, TE = Trinidad Endemic, ToE = Tobago Endemic, TTE = Trinidad and Tobago Endemic. 1 = found on Chacachacare Island, off Trinidad's northwest peninsula.

Species	T	To	D
Bufo beebei	x		OB
Bufo marinus	x	x	W
Hyalinobatrachium orientalis tobagoensis		x	CCR/ToE
Mannophryne olmonae		x	CCR/ToE
Mannophryne trinitatis	x		CCR/TE
Flectonotus fitzgeraldi	x	x	CCR
Hyla boans	x		W
Hyla crepitans	x	x	W
Hyla geographica	x		A
Hyla microcephala misera	x		OB
Hyla minuscula	x		OB
Hyla minuta	x	x	A
Hyla punctata	x		A
Phrynohyas venulosa	x	x	W
Phyllodytes auratus	x		CCR/TE
Phyllomedusa trinitatis	x		CCR
Scinax rubra	x	x	W
Sphaenorhynchus lacteus	x		A
Adenomera hylaedactyla	x		A
Eleutherodactylus johnstonei	x		LA
Eleutherodactylus cf. *rozei*		x	CCR/ToE
Eleutherodactylus terraebolivaris		x	CCR
Eleutherodactylus urichi	x	x	CCR
Leptodactylus bolivianus	x	?	W
Leptodactylus fuscus	x	x	W
Leptodactylus knudseni	x	?	A
Leptodactylus macrosternum	x		A
Leptodactylus nesiotus	x		OB
Leptodactylus validus	x	x	CCR
Lithodytes lineatus	x		A
Physalaemus pustulosus	x	x	W
Elachistocleis ovalis	x		A
Elachistocleis surinamensis	x		W
Pipa pipa	x		A
Pseudis paradoxa caribensis	x		OB
Rana palmipes	x	?	A
Typhlonectes sp.	?		OB
Chelus fimbriatus	x		A
Phrynops gibbus	x		A
Caretta caretta	x	x	Cos

Species (continued)	T	To	D
Chelonia mydas	x	x	Cos
Eretomochelys imbricata imbricata	x	x	Cos
Lepidochelys olivacea	x		Cos
Dermochelys coriacea	x	x	Cos
Rhinoclemmys punctularia punctularia	x	?	A
Kinosternon scorpioides scorpioides	x		A
Podocnemis expansa	?	?	A
Podocnemis unifilis	?	?	A
Geochelone carbonaria	?	?	A
Geochelone denticulata	x	?	A
Caiman crocodilus crocodilus	x	x	A
Crocodylus acutus	?	?	W
Crocodylus intermedius	?	?	OB
Amphisbaena alba	x		A
Amphisbaena fuliginosa fuliginosa	x		LGE
Gonatodes albogularis	?	?	W
Gonatodes ceciliae	x		CCR
Gonatodes humeralis	x	x	A
Gonatodes ocellatus		x	CCR/ToE
Gonatodes vittatus	x	x	CCR
Gymnodactylus geckoides	?		UK
Hemidactylus mabouia	x	x	Cos
Hemidactylus palaichthus	x	x	LGE
Sphaerodactylus molei	x	x	CCR
Thecadactylus rapicauda	x	x	W
Anolis aeneus	x		LA
Anolis chrysolepis planiceps	x		LGE
Anolis extremus	x		LA
Anolis cf. *lemurinus*	?		W
Anolis richardi		x	LA
Anolis trinitatis	?+		LA
Polychrus marmoratus	x	x	A
Iguana iguana	x	x	W
Tropidurus plica	x		A
Mabuya bistriata	x	x	W
Ameiva ameiva	x	x	W
Bachia flavescens		x	LGE
Bachia heteropus alleni		x	CCR/ToE
Bachia heteropus trinitatis	x		CCR
Cnemidophorus lemniscatus	x	x	W
Gymnophthalmus speciosus	1		W
Gymnophthalmus underwoodi	x		OB
Kentropyx striatus	x	?	OB
Proctoporus shrevei	x		CCR/TE
Tupinambis teguixin	x	x	A
Helminthophis sp.	x		CCR/TE
Boa constrictor constrictor	x	x	W
Corallus hortulanus cookii	x	x	W
Epicrates cenchria maurus	x	x	A
Eunectes murinus	x		A
Atractus trilineatus	x	x	LGE
Atractus univittatus		x	CCR
Chironius carinatus	x		W
Clelia clelia clelia	x		W
Dipsas variegata trinitatus	x		CCR
Drymarchon corais corais	x	x	A
Erythrolamprus aesculapii aesculapii	x		W
Erythrolamprus ocellatus		x	CCR/ToE
Helicops angulatus	x		A
Hydrops triangularis neglectus	x		LGE
Imantodes cenchoa cenchoa	x	x	W
Leptodeira annulata ashmeadi	x	x	W
Leptophis ahaetulla coeruleodorsus	x	x	OB
Leptophis riveti	x		CCR
Liophis cobella cobella	x		LGE
Liophis melanotus nesos	x	x	CCR
Liophis reginae ssp.		x	CCR/ToE
Liophis reginae zweifeli	x		CCR
Mastigodryas boddaerti boddaerti	x		A
Mastigodryas boddaerti dunni		x	CCR/ToE
Ninia atrata	x	x	W
Oxybelis aeneus	x	x	W
Oxyrhopus petola petola	x	x	CCR
Pseudoboa neuwiedii	x	x	W
Pseustes poecilonotus polylepis	x		A
Pseustes sulphureus sulphureus	x		A
Sibon nebulata nebulata	x	x	W
Siphlophis cervinus	x		W
Spilotes pullatus pullatus	x	x	W
Tantilla melanocephala	x	x	W
Thamnodynastes sp.	x		LGE
Tripanurgos compressus	x		W
Micrurus lemniscatus diutius	x		LGE
Micrurus psyches circinalis	x		CCR
Leptotyphlops albifrons	x	?	A
Typhlops brongersmianus	x		A
Typhlops trinitatus	x	x	CCR
Bothrops asper	x		W
Lachesis muta muta	x		A

TABLE 2. The eight distributional patterns exhibited by the 129 taxa of Trinidad and Tobago amphibians and reptiles (abbreviations are as in Table 1).

Distributional pattern	Number of taxa (%)
Widespread (W)	28 (22%)
Amazonian (A)	38 (29%)
Caribbean Coastal Range (CCR)	28 (22%)
Lowland Guiana Endemics (LGE)	14 (11%)
Lesser Antilles (LA)	5 (3.8%)
Cosmopolitan (Cos)	6 (4.7%)
Orinoco Basin (OB)	5 (3.8%)
Unknown (UK)	5 (3.8%)
Total taxa	129 (100%)

following Trinidadian satellite islands: Gasparee, Gasparillo, Little Centipede, Carrera, Caledonia, Lenagan, and Nelson Island (Boos 1977, 1984, 1990). The majority of the literature records and museum specimens document urban or agricultural areas, suggesting the species has not invaded forest. Three specimens (FMNH 176926–28) from Spring Hill Farm, the Arima Valley (now the Asa Wright Nature Centre) indicate that the species may have had an opportunity to colonize this secondary forest area (locality data may be erroneous), but extensive field work at this location by me and others has failed to produce additional specimens. *Anolis aeneus* has impacted the population of another introduced species, *A. trinitatis*. Males of the former are larger and more aggressive than *A. trinitatis* males and *aeneus* x *trinitatis* hybrids have been documented by Gorman and colleagues (Gorman and Boos 1972; Gorman and Licht 1975; Gorman et al. 1971).

Anolis extremus is native to Barbados and has been introduced on Bermuda, St. Lucia, Trinidad, Huevos Island (off Trinidad's northwest peninsula), and Caracas, Venezuela. Boos (1977) reported introducing six pairs of *A. extremus* on Huevos Island in 1965, and noted the species' continued presence in 1976. A series of five specimens (USNM 244439–42) collected along Wrightson Road opposite the Holiday Inn in Port of Spain in 1982 documents its presence on that island. No lizards were found during a search of the Holiday Inn parking lot and surrounding area in 1993 and, therefore, data were inconclusive in regard to the species' continuing presence there. At present no evidence exists to suggest that this lizard has expanded into other urban, agricultural, or natural areas.

Anolis richardi is native to the Grenada Bank and has been introduced on Tobago and in Suriname (Hoogmoed 1981). Lazell (1972) stated that *A. richardi* is widespread on Tobago and suggested that it was not a recent introduction; however, Yang et al. (1974) considered it to be a recent introduction to Tobago. Gorman and Kim (1978) provided four reasons why this species should be considered introduced on Tobago: (1) the Tobago population is indistinguishable from the Grenada population; (2) it is a giant anole on an island by itself and giant species usually occur within a suite of *Anolis* species; (3) on Tobago it appears associated with disturbed habitats and is absent from the Main Ridge's primary forest; and (4) the Tobago population shows reduced allelic polymorphism and heterozygosity from the Grenada population. Hardy (1982), however, noted remains of this species from the Crusoe Cave system and hypothesized that the remains could be 1600–2000 years old. Another piece of evidence suggesting *A. richardi* has a long history on Tobago is the larger body size exhibited by the Tobago population (Lazell 1972).

Field observations and museum material support Gorman and Kim's (1978) suggestion that this species is absent from primary forest. The large size of *Anolis richardi* (Tobago males reach 140 mm snout-vent length, females 77 mm) makes it readily observable. Furthermore, the species appears to be absent from the 90 ha island of Little Tobago, an island undoubtedly connected to Tobago at glacial maxima. A 40 m-deep channel currently separates Tobago and Little Tobago (Underwood 1962) suggesting that the two islands were connected for most of the last 140,000 years, but likely have been separated for the past 10,000 years (based upon my interpretation of data in Shackleton 1987 and discussion in Dawson 1992). Little Tobago has a relatively undisturbed forest and has been visited frequently by naturalists. Dinsmore (1970) lived there for nine months and did not encounter this species. Surely, if it occurred there, he would have noticed it. He did report the less obvious, smaller (48 mm), tree-trunk dwelling, endemic gecko *Gonatodes ocellatus,* which inhabits more mesic, shaded microhabits.

Anolis trinitatis is native to St. Vincent and Chateaubelair Island and was introduced to Trinidad, despite what its scientific name implies. Lazell (1972) suggested that *A. trinitatis* reached Trinidad via breadfruit saplings brought from St. Vincent. Based on the literature and museum specimens, the species' distribution on Trinidad is decidedly urban. Gorman and Boos (1972) showed that female *A. trinitatis* mate with male *A. aeneus* and found the resulting hybrids to be sterile. My field work failed to find *A. trinitatis* or *trinitatis* x *aeneus* hybrids.

MOVING NORTHWARD ACROSS BOND'S LINE

As might be expected from the relative diversities of the herpetofaunas, a greater number of extant South American species have colonized the Lesser Antilles than have northern species moved south. Three anurans, one tortoise, two crocodilians, seven saurians, and six serpents are reported to have reached the Lesser Antilles from the mainland (Schwartz and Henderson 1991).

Bufo marinus is undoubtedly the most successful mainland colonist in the islands, spread with human help for the purposes of pest control. The species' plasticity in habitat preference and diet, tolerance of salt water, toxic skin, and high reproductive potential make it an outstanding predator-resistant "weed" species (Zug and Zug 1979). The hylid frog *Scinax rubra* has established populations in St. Lucia and is a well-known human commensal (Schwartz and Henderson 1991). That this frog reached this island with human help rather than by chance over-water dispersal seems probable. Lescure (1980) considered two mainland leptodactylids, *Eleutherodactylus urichi* and *Leptodactylus wagneri*, to have invaded the Lesser Antilles. However, the Lesser Antillean populations of *E. urichi* have been shown to be composed of two separate species, *E. euphronides* and *E. shrevei* (Kaiser et al. 1994). Furthermore, Kaiser et al. (1994) recognized *E. urichi* as a Trinidad and Tobago endemic and considered the mainland populations previously identified as *E. urichi* to be misidentifications.

Heyer (1994) resurrected Garman's taxon *Leptodactylus validus* (formerly the Trinidad, Tobago, and Windward Island populations of *L. wagneri*). This species' distribution, as currently understood, includes Trinidad, Tobago, and the Lesser Antillean islands of St. Vincent, Bequia, and Grenada, a distribution unique among Trinidadian frogs. *Leptodactylus validus* inhabits both open habitats and forest on Trinidad and Tobago (Kenny 1969), but is associated more with disturbed environments in the Lesser Antilles (Schwartz and Henderson 1991; Kaiser and Henderson 1994 [1995]). Therefore, *L. validus* is more likely to have moved from Trinidad and Tobago to the Lesser Antilles than the reverse.

One South American tortoise, *Geochelone carbonaria*, has invaded the Lesser Antilles and its island distribution has been much discussed (Hardy 1982; Censky 1988; Lazell 1993). Whereas *G. carbonaria* has been reported from Trinidad (DeVertueil 1858) and Tobago (Hardy 1982 reported feral animals), only *G. denticulata* appears to be extant on Trinidad and the presence of a natural tortoise population on Tobago is uncertain. All references to *G. carbonaria* on Trinidad and nearby islets appear to refer to animals with questionable data (USNM 15882) or to feral animals (Boos and Quesnel 1994), and the species' historical status remains undetermined. However, *G. carbonaria* has somehow dispersed into the Lesser Antilles; humans may well be at least partially responsible, for this tortoise has been, and still is, valued as food (DeVerteuil 1858). Garman (1887) wrote about a tortoise on Trinidad: "It feeds readily in captivity and is kept about the houses and carried from place to place much as the more common domestic animals." To complicate the picture, Lazell (1993) reported tortoise remains similar to, if not conspecific with, *G. carbonaria* from undated Pleistocene deposits on Anguilla, near the northern end of the Windward island arc. Thus, over-water dispersal for this tortoise cannot be dismissed.

Two crocodilians, *Caiman crocodilus* and *Crocodylus intermedius*, native to mainland South America, have been recorded in the Lesser Antilles. The occurrence of *C. intermedius* is based on a single waif (Barbour 1914) and the account of *C. crocodilus* in Schwartz and Henderson (1991) is based on a single historical report (Groome 1970; R. W. Henderson, personal communication). Thus, no crocodilians are known to have established populations in the Lesser Antilles.

Two mainland gekkonids have dispersed into the Lesser Antilles. *Hemidactylus palaichthus* is a lowland Guiana endemic (Hoogmoed 1979) that has found its way into the island arc. The morphological and ecological similarity of this lizard to its cosmopolitan relative *H. mabouia* suggests that *H. palaichthus* may be more widely distributed in the

Lesser Antilles than is currently recognized. Its presence on Trinidad and Tobago proper is highly questionable, but its presence on satellites of both islands has been verified. The species' presence on two islets off Saint Lucia has been confirmed, but it does not occur on St. Lucia proper. The widespread gecko, *Thecadactylus rapicauda*, is present in the Lesser Antilles, and its close association with humans makes it a likely candidate for dispersal via human traffic; Pregill et al. (1988), however, noted fossil remains of this species from Late Holocene deposits on Antigua. Lescure (1980) considered four South American lizards, *Ameiva ameiva*, *Anolis richardi* (discussed previously), *Bachia heteropus alleni*, and *Gymnophthalmus underwoodi*, to have invaded the Lesser Antilles. *Ameiva ameiva* is a widespread species complex in need of biochemical analysis, but its origin appears to be mainland South America, and its distribution in the Lesser Antilles is relatively well known (Schwartz and Henderson 1991) and unlikely to be recent. *Bachia heteropus alleni* occurs on Tobago, Little Tobago, and in the Windward Islands of Grenada and Bequia, an unusual distribution paralleling that of *Leptodactylus validus*. Thus, *B. heteropus alleni* is most likely a Tobago endemic which has dispersed into the Lesser Antilles. Human transport of adults or eggs in soil surrounding food or ornamental plants seems a likely mode of transport for this leaf-litter dwelling microteiid. *Gymnophthalmus underwoodi*, a unisexual species formed from the hybridization of *G. cryptus* and *G. speciosus* on mainland South America (Cole et al. 1993), appears to have dispersed subsequently into the Lesser Antilles. Human involvement in the dispersal of *G. underwoodi* seems likely considering its association with habitats such as lawns and gardens.

Iguana iguana has successfully colonized the island arc. Underwood (1962), Lescure (1980), and others have suggested that this was accomplished with human assistance due to this species' value as food. However, Lazell (1973) reviewed the evidence and concluded that geographic variation in Lesser Antillean *I. iguana* populations rules out human-mediated dispersal. The widespread skink, *Mabuya bistriata* (previously *Mabuya mabouya mabouya*, see Hoogmoed and Gruber 1983) is undoubtedly a species complex that probably originated on the mainland. Island populations may eventually be shown to be endemic taxa. Thus, seven mainland saurians (including *Thecadactylus*) have crossed Bond's Line to colonize the Lesser Antilles, and three of these (*Ameiva ameiva*, *Iguana iguana*, and *Mabuya bistriata*) may well represent ancient dispersal events.

Six mainland snake species have been reported in the Lesser Antilles: two boids (*Boa constrictor*, *Corallus hortulanus*), three colubrids (*Clelia clelia*, *Liophis melanotus*, and *Pseudoboa neuwiedii*), and one leptotyphlopid (*Leptotyphlops albifrons* [formerly *L. tenella*, see Hoogmoed and Gruber 1983]). All of these species occur on Trinidad and all except *C. clelia* and *Leptotyphlops albifrons* also occur on Tobago. Three of these snakes (*Liophis melanotus*, *P. neuwiedii*, and *Leptotyphlops albifrons*) are known from only one specimen in the Lesser Antilles and all likely have erroneous locality data (Dixon 1981; Schwartz and Henderson 1991; R. W. Henderson, pers. comm.). The Grenadian specimens of *Liophis melanotus* and *P. neuweidii* were collected by or presented to the Museum of Comparative Zoology at Harvard University by the same person, Peter Gallinan, in 1880 (R. W. Henderson, personal communication). Whereas these snakes may have had viable populations on Grenada and were extirpated by the mongoose, erroneous locality data appear more likely. Underwood (1993) described the St. Lucia population of *Clelia* as the endemic species *C. errabunda*, and suggested that records for Dominica were in error. Thus, *C. clelia* is most likely restricted to Grenada in the Lesser Antilles.

Two widespread boids have populations in the Windward Islands. One of these, *Boa constrictor*, appears to have resulted from an old dispersal event because island populations have differentiated from the mainland populations (Lazell 1964); Price and Russo (1991) have suggested that Lesser Antillean populations be granted specific status. Populations of *Corallus hortulanus* have recently been compared using mtDNA sequences from seven geographically distant localities (Henderson and Hedges 1995) and the resultant data support a monophyletic ancestry and two separate taxa, a more northern form, the range of which includes Trinidad and Panama (*C. h. cooki*), and a more southern form in the Guianas, Amazonia, and the Atlantic rainforest (*C. h. hortulanus*). Unexpectedly, the Lesser Antillean populations clustered with the Guiana sample rather than the more geographically proximate Trinidad sample. The authors inferred a recent Guianan-Amazonian origin of the Antillean island populations. This evidence could be interpreted to support a human-mediated dispersal (although unlikely) of *C. hortulanus* into the Windward Islands or an overwater dispersal via the South Equatorial Current, which sweeps up the northern coast of South America toward the Lesser Antilles. This discovery may have implications for other South American taxa in the Lesser Antilles which await further investigation.

Discussion

Elements of the Lesser Antillean herpetofauna inhabiting Tobago comprise two species (2.9%) of the island's herpetofauna; and on Trinidad they number at most five species (4.2%) of that island's herpetofauna. Lesser Antillean taxa invading Trinidad and Tobago appear to have had a minimal negative impact on the native fauna and are mostly restricted to human-modified habitats.

Northward dispersal of mainland taxa into the island arc is much more substantial than the reverse. Extant taxa from mainland South America, Trinidad, and Tobago have invaded the Lesser Antilles at a relatively huge rate (three anurans, one tortoise, two crocodilians, seven saurians, six serpents). Of these 19 reported taxa, 14 have apparently been successful in establishing viable populations; and of these, ten have been able to extend their distribution north of St. Vincent, a distance of more than 313 km from the South American mainland.

Other South American or Middle American species have previously colonized, radiated, and specialized in the island arc to become Lesser Antillean endemics (e.g., *Eleutherodactylus*, *Anolis*, *Sphaerodactylus*, *Amphisbaena*, *Epicrates*, *Leptotyphlops*, *Typhlops*, *Bothrops*). An interesting question is why so few extant mainland species occur in the Lesser Antilles. A partial answer is that Bond's Line, the 140 km-wide saltwater gap, appears to be a formidable barrier to amphibians and reptiles. Yet the Orinoco River floods annually and carries a substantial amount of fresh water, debris, and probably amphibians and reptiles into the Caribbean. Human traffic and trade between islands and the mainland is considerable, and so are opportunities for dispersal. Still, only 14 species from the rich mainland fauna have been able to colonize the Lesser Antilles. Similarly, the Lesser Antillean fauna, composed of so many species well-suited for dispersal over water, has contributed little to the fauna of Trinidad and Tobago. In large part this is probably due to the northward direction of the Orinoco current, pushing water away from the mainland, and from Bond's Line.

The exact role humans have played in bringing amphibians and reptiles across Bond's Line is difficult, if not impossible, to document. Circumstantial evidence suggests that this role has been major for southward dispersal. The *Eleutherodactylus johnstonei* population in the vicinity of the Port of Spain dock is clearly the result of human activity. Similarly, the urban distributions of *Anolis aeneus*, *A. extremus*, and *A. trinitatis* speak of human interventions; and, Boos (1977) directly documented the establishment of *A. extremus* on Huevos Island. An interesting note is that the Lesser Antilles hold a rich radiation of 16 species of *Anolis*, whereas Trinidad contains a single species of mainland anole, *Anolis chrysolepis planiceps* (although a single specimen of the Middle American and trans-Andean *Anolis* cf. *lemurinus* was collected in the middle of the last century). Tobago contains no native or mainland anoles.

The role of humans in the northward dispersal, with the exception of *Bufo marinus*, is not well documented. The use of *B. marinus* in sugar cane pest control during the late 19th and early 20th century is well-known and its dispersal via humans has been documented extensively in the literature (see Lewis 1989; Zug and Zug 1979). *Geochelone carbonaria* and *Iguana iguana* are valuable human food items and as such are likely candidates for dispersal by humans, but fossil material and geographic variation argue otherwise. The two mainland geckos, *Hemidactylus palaichthus* and *Thecadactylus rapicauda*, frequent buildings, act largely as human commensals, and therefore are possible hitchhikers on human transport. Yet, the rarity and unusual distribution of *H. palaichthus* make this mode of dispersal somewhat implausible, and the Antigua fossil remains of *Thecadactylus* suggest its presence may be older than that of human Caribbean populations. *Gymnophthalmus underwoodi* is parthenogenetic, and a single individual can found a new colony. Its close association with lawns and gardens makes it a likely stowaway in ornamental plants; and *Bachia heteropus alleni* could have easily dispersed by the same route. Although an unlikely scenario, the two mainland snakes (*Clelia clelia*, *Corallus hortulanus*) with established populations have large body sizes and may have been transported by Caribbean/Amazonian peoples for rodent control, or more probably for their symbolic importance in Amerindian culture; however, chance over-water transport via the South Equatorial Current cannot be ruled out for *C. hortulanus*. Thus, Bond's Line continues to be an effective barrier separating the South American herpetofauna from that of the Lesser Antilles.

Acknowledgments

I thank R. W. Henderson and W. W. Lamar for useful comments and moral support during this investigation; H. Kaiser (Redpath Museum, McGill University) and an anonymous reviewer for useful critical comments; A. Resetar and H. Voris (Field Museum of Natural History) for laboratory space and support; R. Crombie, W. R. Heyer, R. McDiarmid, R. P. Reynolds, and A. Wynne (United States National Museum) for the loan of specimens; and H. Boos and J. Seyjagat (Emperor Valley Zoo) and J. D. Hardy, Jr. for stimulating conversation and logistical support. J. D. Lazell, Jr. offered numerous useful critical comments and insights. C. James, D. Garcia, H. Nelson, and B. Ramdial of the Trinidad and Tobago Forestry Department provided permits and useful information. Asa Wright Nature Centre employees J. and C. Price, E. and J. Rooks, R. Hernandez, and R. Quiaman provided a variety of support and lodging on Trinidad; P. and C. Turpin and the Blue Water Inn provided lodging on Tobago. K. and S. Murphy and A. A. Stern helped catch anoles in the Holiday Inn parking lot and provided excellent field support. Students D. Alderson, K. Kramer, and C. Spangler, and Chicago Herpetological Society members S. Barten, M. Bruns, M. Dloogatch, G. Hermann, R. and D. Humbert, M. Miller, and R. Shepstone assisted with field work; as did H. Brittin, J. and R. Downie, B. Montgomery, M. Read, J. Read, D. Reznick, and the insect collecting crew from the Cincinnati Zoo. A. Mills and K. Nelson translated Lescure's article, and R. Tsaliagos proof-read the ms (remaining errors are mine alone).

Literature Cited

Barbour, T. 1914. A contribution to the zoogeography of the West Indies, with especial reference to amphibians and reptiles. Mem. Mus. Comp. Zoöl., Harvard Univ. 44: 209–395.

Beard, J. S. 1944. The natural vegetation of the island of Tobago, British West Indies. Ecol. Monogr. 14: 137–163.

—. 1946. The natural vegetation of Trinidad. Oxford Forest Mem. no. 2.

Boettger, O. 1895 (1896). A contribution to the herpetological fauna of the island of Tobago. Jour. Trinidad Field Nat. Club 2: 145–146.

Boos, H. E. A. 1977. *Iguana* – relict of the dinosaur age (Part 2). Trinidad Nat. 1(11): 24–30.

—. 1984. A consideration of the terrestrial reptile fauna on some offshore islands northwest of Trinidad. Living World, Jour. Trinidad Tobago Field Nat. Club 1983–1984: 19–26.

—. 1990. Additions to the terrestrial fauna of the offshore islands northwest of Trinidad. Living World, Jour. Trinidad Tobago Field Nat. Club 1989–1990: 9.

— and V. C. Quesnel. 1994. Additional notes on the fauna recorded for the offshore islands northwest of Trinidad. Living World, Jour. Trinidad Tobago Field Nat. Club 1993–1994: 6–7.

Censky, E. 1988. *Geochelone carbonaria* (Reptilia: Testudines) in the West Indies. Florida Sci. 51: 108–114.

Cole, C. J., H. C. Dessauer, and A. L. Markezich. 1993. Missing link found: the second ancestor of *Gymnophthalmus underwoodi* (Squamata: Teiidae), a South American unisexual lizard of hybrid origin. Amer. Mus. Novit. 3055: 1–13.

Darlington, P. J., Jr. 1957. Zoogeography, the Geographical

Distribution of Animals. John Wiley and Sons, New York.

Dawson, A. G. 1992. Ice Age Earth, Late Quaternary Geology and Climate. Routledge, New York.

DeVerteuil, L. A. A. G. 1858. Trinidad: Its Geography, Natural Resources, Administration, Present Condition, and Prospects. Ward and Lock, London.

Dinsmore, J. J. 1969 (1970). Reptiles of Little Tobago Island, West Indies. Quart. Jour. Florida Acad. Sci. 32: 307–309.

Dixon, J. R. 1981. The neotropical colubrid snake genus *Liophis*: The eastern Caribbean complex. Copeia 1981: 296–304.

Garman, S. 1887. On West Indian reptiles in the Museum of Comparative Zoology, at Cambridge, Mass. Proc. Amer. Philos. Soc. 24: 278–286.

Gorman, G. C. and J. O. Boos. 1972. Extinction of a local population of *Anolis* lizard through competition with a congener. Syst. Zool. 21: 440–441.

— and Y. J. Kim. 1978. The genetics of colonization: loss of variability among introduced populations of *Anolis* lizards (Reptilia, Lacertilia, Iguanidae). Jour. Herpetol. 12: 47–51.

— and P. Licht. 1975. Differences between the reproductive cycles of sympatric *Anolis* lizards on Trinidad. Copeia 1975: 332–337.

—, P. Licht, H. C. Dessauer, and J. O. Boos. 1971. Reproductive failure among the hybridizing *Anolis* lizards of Trinidad. Syst. Zool. 20: 1–18.

Groome, J. R. 1970. A Natural History of the Island of Grenada, W. I. Caribbean Printers, Trinidad.

Hardy, J. D., Jr. 1982. Biogeography of Tobago, West Indies, with special reference to amphibians and reptiles: a review. Bull. Maryland Herpetol. Soc. 18: 37–142.

Henderson, R. W. and S. B. Hedges. 1995. Origin of West Indian populations of the geographically widespread boa *Corallus enydris* inferred from mitochondrial DNA sequences. Molec. Phylog. Evol. 4: 88–92.

Heyer, W. R. 1994. Variation within the *Leptodactylus podicipinus-wagneri* complex of frogs (Amphibia: Leptodactylidae). Smithsonian Contrib. Zool. 546: 1–124.

Hoogmoed, M. S. 1979. The herpetofauna of the Guiana region, p. 217–240. *In* W. E. Duellman (ed.), The South American Herpetofauna: Its Origin, Evolution, and Dispersal. Mus. Nat. Hist., Univ. Kansas Monogr. no. 7.

—. 1981. Introduced species of reptiles in Suriname. Notes on the herpetofauna of Suriname VIII. Amphibia-Reptilia 1: 277–285.

— and U. Gruber. 1983. Spix and Wagler type specimens of reptiles and amphibians in the Natural History Musea in Munich (Germany) and Leiden (The Netherlands). Spixiana suppl. 9: 319–415.

Kaiser, H., J. D. Hardy, Jr., and D. M. Green. 1994. Taxonomic status of Caribbean and South American frogs currently ascribed to *Eleutherodactylus urichi* (Anura: Leptodactylidae). Copeia 1994: 780–796.

— and R. W. Henderson. 1994 (1995). The conservation status of Lesser Antillean Frogs. Herpetol. Nat. Hist. 2: 41–56.

Kenny, J. S. 1969. The Amphibia of Trinidad. Studies Fauna Curaçao Carib. Isl. 29: 1–78.

—. 1979. Some recent colonizations. Jour. Trinidad Tobago Field Nat. Club 78–79: 27.

Lack, D. 1976. Island biology illustrated by the land birds of Jamaica. Stud. Ecol. 3: 1–445.

Lazell, J. D. 1964. The Lesser Antillean representatives of *Bothrops* and *Constrictor*. Bull. Mus. Comp. Zool., Harvard Univ. 132: 245–273.

—. 1972. The anoles (Sauria, Iguanidae) of the Lesser Antilles. Bull. Mus. Comp. Zool., Harvard Univ. 143: 1–115.

—. 1973. The lizard genus *Iguana* in the Lesser Antilles. Bull. Mus. Comp. Zool., Harvard Univ. 145: 1–28.

—. 1993. Tortoise, cf. *Geochelone carbonaria*, from the Pleistocene of Anguilla, Northern Lesser Antilles. Jour. Herpetol. 27: 485–486.

Lescure, J. 1980. Le peuplement en reptiles et amphibiens des Petites Antilles. Bull. Soc. Zool. France 112: 327–342.

Lewis, S. 1989. Cane Toads, an Unnatural History. Dolphin/Doubleday, New York.

Pregill, G. K., D. W. Steadman, S. L. Olson, and F. V. Grady. 1988. Late Holocene fossil vertebrates from Burma Quarry, Antigua, Lesser Antilles. Smithsonian Contr. Zool. 463: 1–27.

Price, R. M. and P. Russo. 1991. Revisionary comments on the genus *Boa* with the description of a new subspecies of *Boa constrictor* from Peru. The Snake 23: 29–35.

Schwartz, A. and R. W. Henderson. 1985. A Guide to the Identification of the Amphibians and Reptiles of the West Indies exclusive of Hispaniola. Milwaukee Public Mus., Milwaukee, Wisconsin.

— and —. 1988. West Indian amphibians and reptiles: a checklist. Milwaukee Public Mus. Contr. Biol. Geol. 74: 1–264.

— and —. 1991. Amphibians and Reptiles of the West Indies: Descriptions, Distributions, and Natural History. Univ. Florida Press, Gainesville.

Shackleton, N. J. 1987. Oxygen isotopes, ice volume and sea level. Quaternary Sci. Rev. 6: 183–190.

Underwood, G. 1962. Reptiles of the Eastern Caribbean. Carib. Affairs, new ser. 1: 1-192.

—. 1993. A new snake for St. Lucia, West Indies. Bull. Nat. Hist. Mus. (Zool.) 59: 1–9.

Williams, E. E. 1989. Old problems and new opportunities in West Indian biogeography, p. 1–46. *In* C. A. Woods (ed.), Biogeography of the West Indies: Past, Present, and Future. Sandhill Crane Press, Gainesville, Florida.

Yang, S. Y., M. Soulé, and G. C. Gorman. 1974. *Anolis* lizards of the eastern Caribbean: a case study in evolution. 1. Genetic relationships, phylogeny and colonization sequence of the *roquet* group. Syst. Zool. 23: 387–399.

Zug, G. R. and P. B. Zug. 1979. The marine toad, *Bufo marinus*. A natural history resume of native populations. Smithsonian Contrib. Zool. 284: iv + 58 p.

Resumen.—La herpetofauna de Trinidad y Tobago es casi completamente de origen continental y es muy diferente a la fauna de las Antillas Menores. La l ínea de Bond, una brecha de agua salada de 140 km entre Granada y Tobago, separa ambas faunas. La herpetofauna de Trinidad y Tobago incluye hasta unas 129 especies y subespecies y sólo cinco (3.8%) de éstas tienen su origen en las Antillas Menores. La única rana (*Eleutherodactylus johnstonei*) y las cuatro lagartijas (*Anolis aeneus*, *A. extremus*, *A. richardi*, y *A. trinitatis*) parecen estar restringidas a áreas urbanas y otros ambientes modificados por el hombre y probablemente han tenido un impacto negativo mínimo en la herpetofauna nativa. Por el contrario, 19 taxones aún vivientes en Trinidad y Tobago y en el continente suramericano han sido también identificados en las Antillas Menores. Varios de estos taxones están representados por taxones cuyos datos sobre localización son erróneos, otros son animales que se dispersaron al azar en una sola ocasión y que no establecieron una población, otros están representados por especies que se dispersaron hace tiempo y la taxonom ía actual no reconoce cuán distintivas son estas especies, y algunos se han dispersado con la asistencia del hombre. Por lo tanto, la l ínea de Bond debe ser considerada una barrera formidable para los anfibios y reptiles, y los taxones septentrionales que han cruzado exitosamente la barrera lo han logrado principalmente con la ayuda del hombre.

Résumé.—La faune herpétologique de Trinité et Tobago est presque entièrement continentale et se distingue de la faune des Petites Antilles. Bond's Line, fossé d'eau salée de 140 km entre la Grenade et Tobago, sépare les deux faunes. La faune herpétologique de Trinité et Tobago compte 129 espèces et sous-espèces dont cinq seulement (3,8 %) trouvent leur origine dans les Petites Antilles. Une grenouille (*Eleutherodactylus johnstonei*) et quatre lézards (*Anolis aeneus*, *A. extremus*, *A. richardi* et *A. trinitatis*) ont probablement eu un impact négatif minimum sur la faune herpétologique autochtone et semblent se limiter aux zones urbaines ou à d'autres environnements modifiés par l'être humain. Inversement, 19 taxons encore existants du continent-Trinité-Tobago sud-américain ont été signalés dans les Petites Antilles. Plusieurs d'entre eux sont représentés par des spécimens pour lesquels les données en matière de localité sont erronées; d'autres sont des spécimens abandonnés qui n'ont pu établir de population, d'autres encore résultent d'anciennes dispersions et la taxonomie actuelle ne permet pas de noter leur caractère distinct. Enfin certains se sont dispersés avec l'aide de l'être humain. Par conséquent, Bond's Line doit être considérée comme une imposante barrière pour les amphibiens et reptiles et les taxons qui ont réussi à franchir cette barrière en provenance du nord l'ont probablement fait avec l'aide de l'homme.

Murphy, J. C. 1996. Crossing Bond's Line: the herpetofaunal exchange between the eastern Caribbean and mainland South America, p. 207–216. *In* R. Powell and R. W. Henderson (eds.), Contributions to West Indian Herpetology: A Tribute to Albert Schwartz. Society for the Study of Amphibians and Reptiles, Ithaca (New York). Contributions to Herpetology, volume 12.

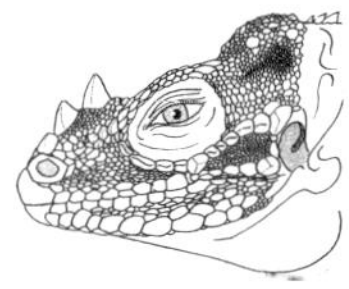

The Role of Amphibians and Reptiles in a West Indian Rain Forest Food Web

Douglas P. Reagan[1]

Abstract.—Amphibians and reptiles occupy all trophic levels in the food web of the rain forest near the El Verde field station, Puerto Rico. They are conspicuous and abundant in aboveground webs and are important consumers in litter and subterranean food webs. Recent investigations have shown that anoline lizards (*Anolis* spp.) and leptodactylid frogs (*Eleutherodactylus* spp.) play a prominent role in structuring the trophic relationships among invertebrate predators and their prey. Species belonging to both genera forage throughout the vertical extent of the forest. Detailed food web studies show that amphibians and reptiles are major consumers of invertebrates in separate compartments of the food web; *Anolis* dominates the diurnal web and *Eleutherodactylus* dominates the nocturnal web. Detailed analyses of prey to the species level showed that only a 13% overlap exists in prey taxa between anoles and frogs. These vertebrate genera are also prey for large invertebrate predators including tarantulas, crab spiders, tailless whip-scorpions, and centipedes. Detailed data on the feeding relationships document cross-predation among lizards and frog species, and between these groups and the large invertebrates. Canopy anoles also exert control over food web structure and function by suppressing the abundance of large arthropods, including spiders.

Introduction

AMPHIBIANS AND REPTILES ARE CONSPICUOUS AND abundant in terrestrial ecosystems on the islands of the West Indies (Moermond 1979; Williams 1969, 1983; Stewart and Pough 1983), yet despite the number of biologists who have worked and are working in the region, little is known about the natural history of 95% of the described species (Schwartz and Henderson 1991). Food web relationships have been described for some groups (e.g., anoline lizards; Reagan 1986; Roughgarden 1995), but the roles of most amphibians and reptiles in ecosystem processes have not been investigated in any detail.

In 1980, biologists in the Terrestrial Ecology Division of the Center for Energy and Environment Research (CEER) at the University of Puerto Rico initiated an investigation of cycling and transport processes in the rain forest near the El Verde field station, Puerto Rico. The study was designed to build on the substantial body of existing information on the effects of radiation stress on the rain forest ecosystem, including populations, mineral cycles, and metabolism (Odum and Pigeon 1970). Whereas animal populations had been investigated in the forest prior to 1980, the structure of the animal community, in particular the food web organization, had not been described. Because of the fundamental importance of feeding relationships as the basis for energy and nutrient cycling in the animal community, studies focused on defining the feeding relationships among animal species in the forest. Food web investigations have continued at El Verde as part of the Luquillo Long Term Experimental Research Program, funded by the National Science Foundation (USA), and have added important information concerning the effect of large scale disturbances (e.g., hurricanes) to the trophic structure of the animal community at El Verde.

Inspection of the list of species occurring at El Verde (Table 1) shows that widespread and speciose

[1] Terrestrial Ecology Division, University of Puerto Rico, San Juan, Puerto Rico 00936, USA.

genera of amphibians and reptiles in terrestrial ecosystems in the West Indies were represented in the forest at El Verde. Whereas no single location can be considered truly representative of the West Indies, evaluation of the food web of the forest at El Verde provides a basis for understanding the role of these groups in a West Indian ecosystem, representative at least of moist tropical forest systems throughout the region.

My role in the original CEER studies was to investigate the feeding relationships involving amphibians and reptiles. Evaluation of the food web drew upon past studies, studies conducted by CEER scientists, and on the work of visiting scientists investigating various animal groups in the forest. In particular, the long-term studies by Margaret Stewart, Larry Woolbright, and their colleagues have provided detailed information on leptodactylid frogs (*Eleutherodactylus*) at El Verde.

My goal herein is to report on findings made by me and others on the food web relationships of amphibians and reptiles in the forest at El Verde. Specific objectives are to: (1) describe the various roles of amphibians and reptiles in the general trophic structure of the food web at El Verde; (2) discuss the influence of these groups on general food web structure and function; and (3) demonstrate how studies of amphibians and reptiles have elucidated the functional interrelationships within the forest food web.

THE FOREST

The rain forest near El Verde possesses climatic, topographic, and vegetation characteristics similar to many insular rain forest communities in the West Indies. Mean monthly temperatures are 21–24°C, annual rainfall is approximately 3.5 m, and a mild dry season occurs from January to mid-April (Brown et al. 1983; McDowell and Estrada Pinto 1988). A perennial stream, the Quebrada Sonadora, and several smaller streams traverse the area along steep rocky channels. General topography is rough, with deeply dissected drainages and steep slopes.

The El Verde research area of the forest is located at elevations between 250 and 500 m on the eastern slopes of the Luquillo Mountains in eastern Puerto Rico. The forest at this elevation falls within the subtropical wet forest life zone (Ewel and Whitmore 1973) and is dominated by tabonuco (*Dacryodes excelsa*), sierra palm (*Prestoea montana*), and lesser known species (*Buchenavia capitata*, *Inga fagifolia*, and *Guarea trichilioides*). Canopy height is 20–24 m and, unlike most mainland rain forests, no well-defined subcanopy is present. Although some tree species are deciduous, the forest is evergreen and the canopy remains closed

TABLE 1. The trophic level and food of amphibian and reptile genera found in the forest at El Verde, Puerto Rico (Rivero 1978; Reagan et al. 1982; Schwartz and Henderson 1991; Stewart and Woolbright, in press; Thomas and Kessler, in press).

Trophic level Genera	Food	Habitat
Primary consumers		
Amphibia		
Bufo (larvae)	Aquatic vegetation	Aquatic
Leptodactylus (larvae)	Aquatic vegetation	Aquatic
Omnivore		
Anolis cuvieri	Plants and animals	Canopy
Secondary consumers		
Amphibia		
Bufo (adults)	Insects	Ground
Leptodactylus (adults)	Invertebrates	Ground-aquatic
Eleutherodactylus	Invertebrates and vertebrates	Ground-canopy
Reptilia		
Amphisbaena	Invertebrates	Soil, litter
Diploglossus	Arthropods	Litter
Sphaerodactylus	Invertebrates	Litter
Anolis	Invertebrates	Ground-canopy
Typhlops	Invertebrates	Ground, litter
Tertiary consumers (top predators)		
Reptilia		
Arrhyton	Lizards, frog eggs	Ground, litter
Alsophis	Frogs and lizards	Ground-canopy
Epicrates	Vertebrates	Ground-arboreal

throughout the year. A few tree species have well-developed buttresses. Bromeliads and lianas are conspicuous elements of the vegetation.

The animal community of the forest contains fewer species of most taxonomic groups than occur in comparable mainland communities. Only 40 bird species have been reported from the forest at El Verde over several decades of observation (Waide, in press). No large mammalian herbivores (e.g., tapirs, capybaras) or carnivores (e.g., ocelots, jaguars) comparable to those that occur in mainland forests are found at El Verde. However, leptodactylid frogs and anoline lizards achieve extremely high population densities (Reagan and Waide, in press).

Food Web Structure

Trophic Levels

Amphibians and reptiles occur at all trophic levels in the food web of the forest at El Verde. Table 1 lists the genera of amphibians and reptiles inhabiting the forest, their food, and the trophic level occupied by each. Snakes, hawks, and owls fill the role of top predators. The largest species are the red-tailed hawk (*Buteo jamaicensis*), Puerto Rican screech owl (*Otus nudipes*), and Puerto Rican boa (*Epicrates inornatus*).

By comparing the trophic status of amphibian and reptilian genera from the forest with the trophic status of amphibians and reptiles for all of Puerto Rico (Table 2), the similarity in roles of amphibians and reptiles in the El Verde and island-wide food webs is apparent. From a functional perspective, the herpetofauna of Puerto Rico is representative of the other Greater Antilles in that amphibians and reptiles occupy all trophic levels. Primary consumers and omnivores include amphibian larvae, large lizards (*Cyclura*, *Ameiva*), and turtles (*Trachemys*). The giant anole (*Anolis cuvieri*), which inhabits the canopy of the forest at El Verde, also consumes plant and animal material (Reagan et al. 1982).

TABLE 2. The trophic level and food of amphibian and reptilian genera of Puerto Rico (Rivero 1978; Reagan et al. 1982; Schwartz and Henderson 1991; Stewart and Woolbright, in press; Thomas and Kessler, in press).

Trophic level Genera	Food	Habitat
Primary Consumers		
Amphibia		
Bufo (larvae)	Aquatic vegetation	Aquatic
Leptodactylus (larvae)	Aquatic vegetation	Aquatic
Reptilia		
Cyclura	Plant material	Ground
Omnivores		
Reptilia		
Trachemys	Aquatic plant and animal matter	Pools, lagoons
Ameiva	Plant and animal material	Ground
Anolis	Plants and animals	Canopy
Secondary consumers		
Amphibia		
Bufo	Invertebrates	Ground
Leptodactylus	Invertebrates	Ground-aquatic
Eleutherodactylus	Invertebrates and vertebrates	Ground-canopy
Reptilia		
Amphisbaena	Invertebrates	Soil, litter
Diploglossus	Invertebrates	Litter
Mabuya	Invertebrates	Ground, litter
Hemidactylus	Arthropods	Buildings
Phyllodactylus	Arthropods	Ground
Sphaerodactylus	Invertebrates	Litter
Ameiva	Invertebrates	Ground
Anolis	Invertebrates	Ground-canopy
Typhlops	Invertebrates	Ground, litter
Tertiary consumers (top predators)		
Reptilia		
Arrhyton	Lizards, frog eggs	Ground, litter
Alsophis	Frogs and lizards	Ground-canopy
Epicrates	Vertebrates	Ground-arboreal

Stratification and Microhabitats

Amphibians and reptiles in the forest at El Verde and in wet forests throughout the West Indies occur in all strata of the three-dimensional habitat provided by the forest. At El Verde, amphisbaenian species in the genus *Amphisbaena* and snakes in the genus *Typhlops* burrow in the soil. The lizard genera

Diploglossus and *Sphaerodactylus* are semi-fossorial, burrowing through litter and beneath surface debris. The amphibian genera *Bufo* and *Leptodactylus* have aquatic larvae that live in pools and puddles; adults are confined to the forest floor and are usually found near streams.

Above ground, anoline lizards occur from ground level to the forest canopy, but species within this genus are generally stratified (Reagan 1992). *Anolis gundlachi* is a trunk-ground predator that occurs at and near ground level. *Anolis stratulus* inhabits the forest canopy, generally at heights from 10–24 m above ground level. *Anolis evermanni* is a generalist that occurs at all levels in the forest and is the only species that regularly forages along rocky stream channels. The giant anole, *A. cuvieri*, inhabits the forest canopy as an adult, but juveniles are frequently found in shrubby vegetation near ground level. *Anolis krugi* generally occupies large openings in the forest and along roads, but may occasionally be found in treefall gaps. *Anolis occultus* is rare in the forest at El Verde, but has been found along streams and in forest openings created by Hurricane Hugo (D. P. Reagan, personal observation).

The leptodactylid frog, *Eleutherodactylus coqui*, occurs primarily on vegetation in the lower few meters of the forest. During the day, individuals retreat to the litter or other hiding places in the vegetation to avoid desiccation and predators. When conditions are favorable, many individuals climb to the canopy where they forage. In the brief period before dawn, these individuals launch themselves into the air and parachute to ground level (Stewart 1985, 1993). Several other species of *Eleutherodactylus* inhabit the forest, but at considerably lower densities than *E. coqui*. *Eleutherodactylus wightmanae* inhabits herbaceous vegetation near the forest floor, *E. hedricki* places its nests in tree holes well above ground level, and *E. portoricensis* occurs in forest habitat similar to the more abundant *E. coqui* (Rivero 1978; Stewart and Woolbright, in press). The large frog, *E. karlschmidti*, has been reported occurring along streams in the forest, but now appears to be absent from this habitat and is possibly extinct (R. Thomas and R. Joglar, personal communication).

The Puerto Rican boa, *Epicrates inornatus*, is considered arboreal, but occurs at and near ground level in the forest (Reagan 1984). Individuals frequently bask on rocks or low vines in forest openings near stream channels, but have been observed 3–4 m above ground level on tree branches and vines. In the limestone karst area of western Puerto Rico, the species occurs on rocky ledges and vines at the mouth of caves where it feeds on emerging bats (Rodriguez and Reagan 1984). Other species of *Epicrates* occur in similar habitats on other West Indian islands (Hardy 1957).

The colubrid snake, *Alsophis portoricensis*, is generally terrestrial but may sometimes be found in trees (Rivero 1978). Observations made by me and others (J. Gillingham, personal communication) in the forest at El Verde indicate that this species inhabits the forest canopy, where it rests on branches and in cavities and forages for anoles. While often observed near ground level in openings, in areas of closed forest this species occurs primarily in the branches and vines of the forest canopy.

Abundance

Eleutherodactylus coqui is the most abundant amphibian species in the forest at El Verde. Like other members of the genus *Eleutherodactylus* in the West Indies, it deposits its eggs in terrestrial sites and the young undergo direct development to the adult form within the egg (Townsend and Stewart 1985). This life history feature permits the species to become distributed throughout the forest. Stewart and Pough (1983) and Stewart and Woolbright (in press) report mean densities of 2,900 adults/ha, and a maximum density of 24,800 individuals/ha.

West Indian anoles also attain high population densities. Turner and Gist (1970) estimated the population density of *Anolis gundlachi* from the forest at El Verde at 800 individuals/acre (2,000 individuals/ha), Ruibal and Philibosian (1974) estimated the density of *A. acutus* on St. Croix as 2,000–5,600 individuals/ha, and Gorman and Harwood (1977) estimated up to 20,000 *A. pulchellus*/ha in grassy lowland areas of Puerto Rico.

Using multiple mark and recapture methods, I obtained mean estimates of 21,333–25,870 individuals/ha of the canopy anole, *Anolis stratulus*, from the forest at El Verde (Reagan 1992). These high densities are made possible because individual lizards are small (adult body mass of 1.5–2.5 g), prefer perches of small diameter (< 5 cm), such as small branches found in forest canopy, and occupy ellipsoidal home ranges generally 6–7 m in diameter layered within the 10–14 m level of the forest canopy (Reagan 1992). The highest of these densities (> 2 individuals/m^2) are from Puerto Rico and are the highest reported for any lizard species. They are several times greater than those that have been estimated for most mainland lizards.

Population density estimates of other amphibians and reptiles that inhabit the forest at El Verde document the lower densities of other species. Stewart and Woolbright (in press) calculated maximum densities of 800 *Eleutherodactylus portoricensis*, 400 *E. wightmanae*, 400 *E. antillensis*, 200 *E. hedricki*, and 100 *E. richmondi*/ha, based on nocturnal transect counts from the forest at El Verde. Observations made during field surveys for litter invertebrates and limited field searches in small areas of ground litter indicated that at least some areas of the forest may contain densities of *Sphaerodactylus* spp. comparable to those of anoles (W. Pfieffer and R. Thomas, personal communication), but quantitative estimates have not been made.

Food Web Processes

Energy and Nutrient Cycling

Because amphibians and reptiles are abundant in the forest, they are important as consumers. The feeding rates of frogs and anoline lizards have been calculated for the forest at El Verde (Table 3). Whereas the size of individual prey items is small, the abundance and ubiquitous distribution of anoles and frogs in the forest habitat is significant. Although comparable data on abundance and feeding rates are not available for most other insectivores and predators, their numbers appear to be substantially lower than those estimated for frogs and lizards at El Verde. The combined feeding rates of the three anoline and one frog species of more than half a million prey items/day/ha indicate that they are important in the flow of nutrients and energy through the animal community.

The feeding rates of predators on amphibians and reptiles is not known, but numerous observations over more than a decade at El Verde suggest substantial predation. Tarantulas (*Avicularia*), crab spiders (*Stasina*) (Formanowicz et al. 1981), and screech owls prey largely on frogs, and lizard cuckoos (*Saurothera vieilloti*), pearly-eyed thrashers (*Margarops fuscatus*), red-tailed hawks (*Buteo jamaicensis*), and colubrid snakes (*Alsophis portoricensis*) feed regularly on anoles (Reagan and Waide, in press).

Compartments

One of the challenges of understanding food web processes is understanding if the potential feeding relationships from theoretical food webs, such as compartmentalization, occur in nature. Pimm and Lawton (1980) examined two extreme hypotheses concerning food web compartments: (1) the reticulate hypothesis which states that species interactions will be homogeneous throughout the food web, and (2) the compartmented hypothesis which states that only species within particular compartments will interact. Pimm (1982) re-examined these hypotheses and concluded that compartments (subwebs) could exist on the basis of habitat divisions, but found no evidence that compartments would occur in the same habitat.

TABLE 3. Feeding rates of the most abundant amphibian (*Eleutherodactylus*) and reptilian (*Anolis*) species found in the forest at El Verde, Puerto Rico.

Species	Prey items per day per ha	Source
Amphibia		
E. coqui	114,000	Stewart & Woolbright (in press)
Reptilia		
A. stratulus	340,000	Reagan (1984)
A. gundlachi	50,800	Reagan (in press)
A. evermanni	51,900	Reagan (in press)

Within-habitat compartmentalization was investigated for the above-ground food web of the forest at El Verde (Reagan et al., in press). The study focused on evaluating the predators and prey of the two most abundant groups of secondary and tertiary consumers: anoline lizards, which are diurnal, and frogs, which are nocturnal. The major anoline predators are diurnal birds and the colubrid snake, *Alsophis portoricensis*, which is also active diurnally. Major frog predators, such as the Puerto Rican screech owl, crab spider, tarantula, and tailless whip-scorpion, *Phrynus longipes*, are nocturnal.

Results of frog and anole gut content analyses showed that both groups consumed a variety of prey, but were predominantly insectivorous. However, detailed taxonomic identification of prey (genus/species level) revealed that frogs and anoles had a dietary overlap of only 13%. Additional data on potential prey items were obtained from invertebrate trapping studies. The number of invertebrates collected in day and night samples were comparable, but taxonomic overlap at the species level was only 12%. Within taxonomic groups species demonstrated a strong tendency to be either nocturnal or diurnal (e.g., Blattodea, 7 at night, 0 by day; Orthoptera, 8 at night, 0 by day; dipteran suborders Brachycera and Cyclorrapha [except Phoridae], 28 at night, 114 by day) (Reagan et al., in press). Not surprisingly, the taxa found to be most active at night were also found most commonly as prey items by frogs. Numerically, ants were the most common prey eaten by *Anolis stratulus*, comprising nearly 60% of all items consumed. Homoptera (chiefly planthoppers), Diptera, Lepidoptera, Coleoptera, Orthoptera, Acarina, and Araneida were also consumed (Reagan 1995).

Dial (1992) conducted intensive investigations of foraging and habitat use by *Anolis stratulus* and *A. evermanni* in the canopy of the rain forest in the Luquillo Mountains. Through a series of experiments and observations, he showed that anoles in the canopy suppressed the abundance of large arthropods and spiders. He concluded that anoles play a role in shaping the diurnal subweb, and possibly affecting overall food web structure. *Anolis* effects on arthropod abundance have been previously noted on St. Eustatius (Netherland Antilles) (Pacala and Roughgarden 1984).

The data on feeding relationships from primary consumers through top predators indicate a trophic structure subdivided into day and night compartments. Although the ultimate energy base is plants, the animal community of the forest is clearly compartmented. As abundant intermediate consumers, amphibians (frogs) and reptiles (anoles) play key roles in the transfer of nutrients and energy in the nocturnal and diurnal webs, respectively.

Food Loops

Another principle of food web organization that theoretically exists, but is controversial in nature, is the food loop (reciprocal predation). A food loop is a cyclic feeding interaction between two or more species. For example, species A eats species B and B eats A, or species A eats B, B eats C, and C eats A. Many workers believe that although food loops may exist, they are unimportant in food web structure because of the dynamic constraints imposed by reciprocal predation; if species A only eats species B and vice versa, entropy would render this arrangement unlikely (Pimm 1982; Pimm and Lawton 1982; Lawton and Warren 1988). However, the assumptions made by these authors are simplistic and do not consider situations with multiple interconnected loops. Close inspection of the feeding relationships at El Verde showed that reciprocal predation often involves multiple combinations of species and the adults of one species consuming the juvenile stages or eggs of another.

The food web at El Verde contains many food loops, and many of these loops involve amphibians and reptiles (Reagan et al., in press). Figure 1 summarizes feeding relationships among three lizard, one frog, and four invertebrate predators. A variety of loops exists among these species, including one instance of A eats A (predation by male *Eleutherodactylus coqui* on eggs); eight instances of A eats B, B eats A; two loops where A eats B, B eats C, and C eats A; and one loop where A eats B, B eats C, C eats D, and D eats A. Of the 23 feeding interactions illustrated, 18 have been observed in the food web at El Verde, and the rest are considered probable. Inter-

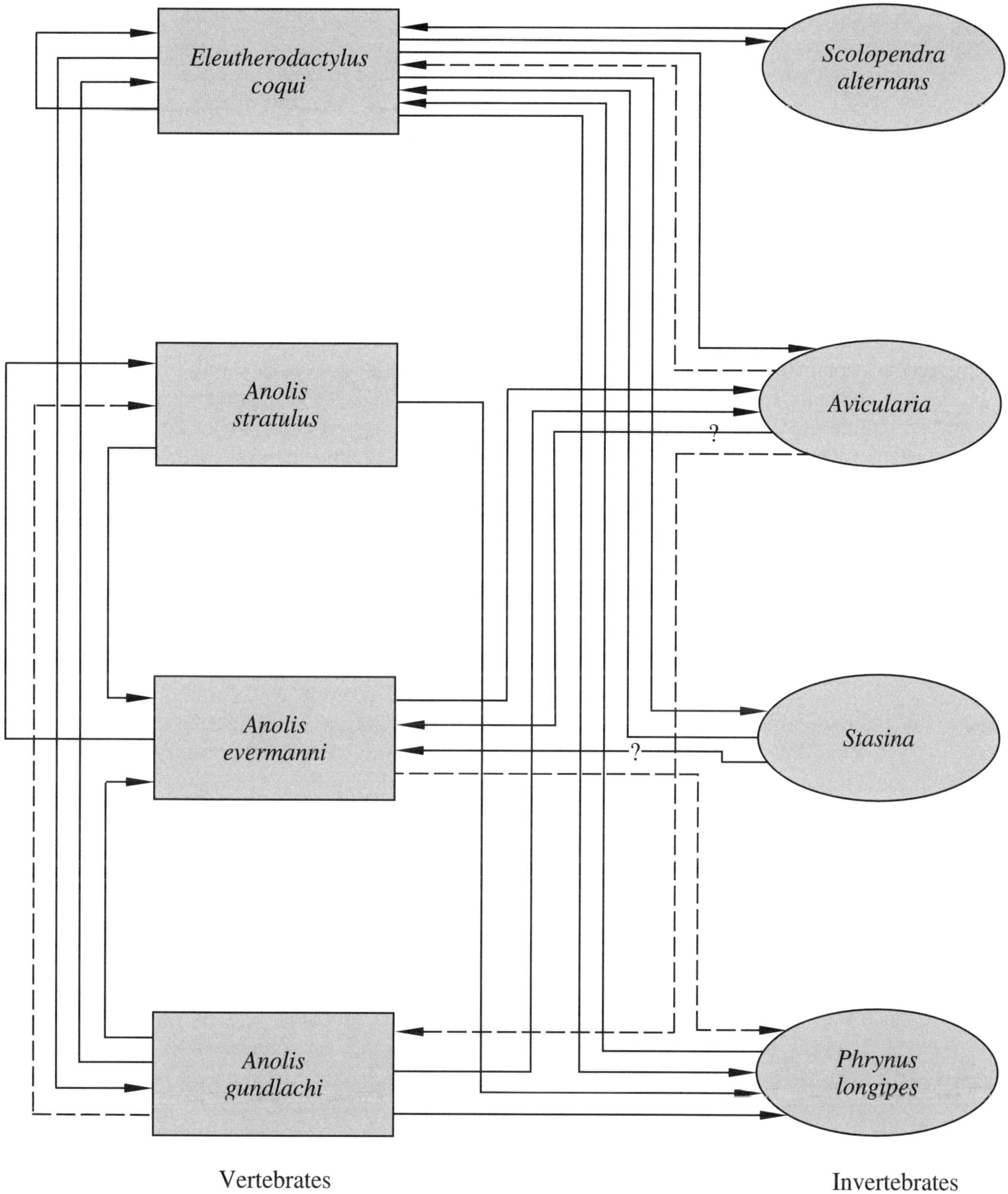

FIGURE 1. Feeding interactions among one amphibian (*Eleutherodactylus coqui*), three reptilian (*Anolis stratulus*, *A. evermanni*, *A. gundlachi*), and four invertebrate species in the El Verde, Puerto Rico food web. Invertebrates include a centipede (*Scolopendra alternans*), tailless whip-scorpion (*Phrynus longipes*), and two spiders: the crab spider (genus *Stasina*) and the tarantula (genus *Avicularia*) (modified from Reagan et al., in press). Solid lines indicate documented relationships; dashed lines mark probable pathways.

actions in these loops center on amphibians and reptiles; no feeding interactions have been observed or are expected among the invertebrate predators in this grouping.

The importance of food loops in the food web at El Verde may be that they provide multiple alternative pathways for the flow of energy among trophic levels, thus preserving food web structure and function during episodic disturbances such as hurricanes. Reciprocal predation may serve the additional functions of controlling predators by consuming the young of predatory species and of providing feedback control of intraspecific competition (e.g., through consumption of the eggs of competing males). While the dynamics and significance of these relationships remain poorly known, their existence in nature is no longer debatable. As indicated by Figure 1, amphibians (*Eleutherodactylus coqui*) and reptiles (*Anolis* spp.) are key components in the food loops of the forested communities in the West Indies.

Summary

Analysis of the food web for the rain forest at El Verde, Puerto Rico illustrates the central role of amphibians and reptiles in food web structure and function. Amphibians and reptiles occur at all trophic levels and occupy all strata, including below-ground, litter, ground level, aquatic (pools), and canopy. Amphibians and reptiles are extremely abundant. Frogs at El Verde attain one of the highest densities known for any amphibian species (Stewart and Woolbright, in press). The population density of *Anolis stratulus*, a species which occupies the three dimensional structural habitat of the forest canopy, is the highest-known population density of any lizard species. The abundance of these groups throughout the West Indies is in contrast to mainland forests where mammalian species play a more prominent role in the animal community (Gentry 1990).

Amphibians and reptiles play a central role in energy and nutrient flow through the animal community of the forest. They are abundant small predators, preying chiefly on insects and small invertebrates, and are in turn consumed by top predators such as owls, hawks, and snakes. The abundance and importance of some groups, such as anoline lizards, may be the result of successful competition with insectivorous birds (Wright 1981), or it may be a consequence of reduced predation pressure (Waide and Reagan 1983).

Anoline lizards and frogs are central in the diurnal and nocturnal subwebs, respectively. Direct observations of predation on these groups and analyses of their gut contents have made it possible to determine the degree of compartmentalization that is characteristic of the food web at El Verde. Documentation of day-night compartmentalization of this food web suggests that similar compartmentalization exists in other ecosystems, particularly tropical forests where conditions of temperature and humidity are suitable for nocturnally active species at all trophic levels.

The food web of the forest at El Verde is characterized by multiple food loops, many of which involve amphibians and reptiles. This appears to be the result of the superabundance of small poikilothermic vertebrates (anoline lizards and leptodactylid frogs) and large invertebrate predators (tarantulas, tailless whip-scorpions, centipedes). McCormick and Polis (1982) surveyed the literature and recorded over 200 instances of arthropod predation on vertebrates and stated their belief in the existence of food loops in terrestrial ecosystems. Our results support their assertion. Whereas loops are a common feature of planktonic food webs (Sprules 1972; Williams 1980), their presence in terrestrial webs has not been widely documented. Our results from studies at El Verde suggest that food loops may be widespread in similar forested areas throughout the West Indies, because the types of animals involved in the food loops at El Verde are widely distributed and are likely to have similar food habits throughout their ranges.

Amphibians and reptiles occur at all levels in West Indian food webs and are important in food web structure and function. Elucidation of feeding relationships involving amphibians and reptiles demonstrated the existence of food web compartments and food loops in the forest food web. These studies have contributed significantly to a more thorough understanding of the particular food web of the forest and to the understanding of the basic principles of food web organization.

ACKNOWLEDGMENTS

I thank R. Garrison, M. Stewart, R. Thomas, R. Waide, M. Willig, L. Woolbright, and the other many investigators who have collected data on food web relationships for the animal community of the forest at El Verde, Puerto Rico. R. Garrison analyzed the gut contents of anoles and frogs which provided the basis for many of the feeding relationships in the El Verde food web. M. Stewart reviewed and made many helpful comments on an earlier draft of this manuscript. Much of the research reported herein was supported by the Office of Health and Environmental Research, United States Department of Energy under contract no. HA 0203010 to the Center for Energy and Environmental Research of the University of Puerto Rico. Portions of this investigation were performed under grant BSR-8702336 from the National Science Foundation (USA) to the Center for Energy and Environment Research and the Institute for Tropical Forestry as part of the Long Term Ecological Research Program in the Luquillo Experimental Forest. Additional support was provided by the University of Puerto Rico.

LITERATURE CITED

Brown, S., A. E. Lugo, S. Silander, and L. Liegel. 1983. Research history and opportunities in the Luquillo Experimental Forest. Forest Serv., Southern Forest Exp. Sta., Gen. Tech. Rept. SO-44. New Orleans, Louisiana.

Dial, R. 1992. A Food Web for a Tropical Rain Forest: The Canopy View from *Anolis*. Ph.D. Dissertation, Stanford Univ., Stanford, California.

Ewel, J. J. and J. L. Whitmore. 1973. The Ecological Life Zones of Puerto Rico and the U.S. Virgin Islands. Research paper ITF-18, United States Forest Serv., Inst. Tropical Forestry, Río Piedras, Puerto Rico.

Formanowicz, D. R., Jr., M. M. Stewart, F. H. Pough, and P. F. Brussard. 1981. Predation by giant crab spiders on the Puerto Rican frog *Eleutherodactylus coqui*. Herpetologica 37: 125–129.

Gentry, A. H. (ed.). 1990. Four Neotropical Rainforests. Yale Univ. Press, New Haven, Connecticut.

Gorman, G. C. and R. Harwood. 1977. Notes on population density, vagility, and activity patterns of the Puerto Rican grass lizard, *Anolis pulchellus* (Reptilia, Lacertilia, Iguanidae). Jour. Herpetol. 11: 363–368.

Hardy, J. D. 1957. Bat predation by the Cuban boa, *Epicrates angulifer*. Copeia 1957: 151–152.

Lawton, J. H. and P. H. Warren. 1988. Static and dynamic explanations for patterns in food webs. Trends Ecol. Evol. 3: 242–245.

McCormick, S. and G. A. Polis. 1982. Arthropods that prey on vertebrates. Biol. Rev. Cambridge Philos. Soc. 57: 29–58.

McDowell, W. H. and A. Estrada-Pinto. 1988. Rainfall at the El Verde Field Station 1964–1986. CEER-T-228, Center for Energy and Environment Research, Río Piedras, Puerto Rico.

Moermond, T. C. 1979. The influence of habitat structure on *Anolis* foraging behavior. Behaviour 70: 147–167.

Odum, H. T. and R. F. Pigeon. 1970. A Tropical Rain Forest: A Study of Irradiation and Ecology at El Verde, Puerto Rico. United States Atomic Energy Comm., Oak Ridge, Tennessee.

Pacala, S. and J. Roughgarden. 1984. Control of arthropod abundance by *Anolis* lizards on St. Eustatius (Neth. Antilles). Oecologia 64: 160–162.

Pimm, S. L. 1982. Food Webs. Chapman and Hall, London.

— and J. H. Lawton. 1980. Are food webs divided into compartments? Jour. Anim. Ecol. 49: 879–898.

Reagan, D. P. 1984. Ecology of the Puerto Rican boa (*Epicrates inornatus*) in the Luquillo Mountains of Puerto Rico. Carib. Jour. Sci. 20: 119–127.

—. 1986. Foraging behavior of *Anolis stratulus* in a Puerto Rican rain forest. Biotropica 18: 157–160.

—. 1992. Congeneric species distribution and abundance in a three-dimensional habitat: the rain forest anoles of Puerto Rico. Copeia 1992: 392–403.

—. 1995. Lizard ecology in the canopy of an island rain forest, p. 149–164. *In* M. D. Lowman and N. Nadkarni (eds.), Forest Canopies. Academic Press, San Diego, California.

—, G. R. Camilo, and R. B. Waide. In press. The community food web: major properties and patterns of organization. *In* D. P. Reagan and R. B. Waide (eds.), The Food Web of a Tropical Rain Forest. Univ. Chicago Press, Chicago, Illinois.

—, R. W. Garrison, J. E. Martinez, R. B. Waide, and C. P. Zucca. 1982. Tropical Rain Forest Cycling and Transport Program: Phase I Report. CEER-T-17. Center for Energy and Environment Research, San Juan, Puerto Rico.

— and R. B. Waide (eds.). In press. The Food Web of a Tropical Rain Forest. Univ. Chicago Press, Chicago, Illinois.

Rivero, J. A. 1978. Los anfibios y reptiles de Puerto Rico. Ed. Universitario, Río Piedras, Puerto Rico.

Rodriguez, G. A. and D. P. Reagan. 1984. Bat predation by the Puerto Rican boa, *Epicrates inornatus*. Copeia 1984:

219–220.

Roughgarden, J. 1995. *Anolis* Lizards of the Caribbean: Ecology, Evolution, and Plate Tectonics. Oxford Univ. Press, Oxford.

Ruibal, R. and R. Philibosian. 1974. The population ecology of the lizard *Anolis acutus*. Ecology 55: 525–537.

Schwartz, A. and R. W. Henderson. 1991. Amphibians and Reptiles of the West Indies: Descriptions, Distributions, and Natural History. Univ. Florida Press, Gainesville, Florida.

Sprules, W. G. 1972. Effects of size-selective predation and food competition on high altitude zooplankton communities. Ecology 53: 375–386.

Stewart, M. M. 1985. Arboreal habitat use and parachuting by a subtropical forest frog. Jour. Herpetol. 19: 391–401.

—. 1993. Frequent flyers. Nat. Hist. 102: 42–49.

— and F. H. Pough. 1983. Population density of tropical forest frogs: relation to retreat sites. Science 221: 570–572.

— and L. L. Woolbright. In press. Amphibians. *In* D. P. Reagan and R. B. Waide (eds.), The Food Web of a Tropical Rain Forest. Univ. Chicago Press, Chicago, Illinois.

Thomas, R. and A. G. Kessler. In press. Non-anoline reptiles. *In* D. P. Reagan and R. B. Waide (eds.), The Food Web of a Tropical Rain Forest. Univ. Chicago Press, Chicago, Illinois.

Townsend, D. S. and M. M. Stewart. 1985. Direct development in *Eleutherodactylus coqui* (Anura: Leptodactylidae): a staging table. Copeia 1985: 423–436.

Turner, F. B. and C. S. Gist. 1970. Observations of lizards and tree frogs in an irradiated Puerto Rican forest, Ch. E-2. *In* H. T. Odum and R. F. Pigeon, (eds.), A Tropical Rain Forest. Atomic Energy Comm., Oak Ridge, Tennessee.

Waide, R. B. In press. Birds. *In* D. P. Reagan and R. B. Waide (eds.), The Food Web of a Tropical Rain Forest. Univ. Chicago Press, Chicago, Illinois.

— and D. P. Reagan. 1983. Competition between West Indian anoles and birds. Amer. Nat. 121: 133–138.

Williams, E. E. 1969. The ecology of colonization as seen in the zoogeography of anoline lizards on small islands. Quart. Rev. Biol. 44: 345–389.

—. 1983. Ecomorphs, faunas, island size, and diverse endpoints in island radiations of *Anolis*, p. 326–370. *In* R. B. Huey, E. R. Pianka, and T. W. Schoener (eds.), Lizard Ecology: Studies of a Model Organism. Harvard Univ. Press, Cambridge, Massachusetts.

Williams, E. H. 1980. Disjunct distributions of two aquatic predators. Limnol. Oceanogr. 25: 999–1006.

Wright, S. J. 1981. Extinction-mediated competition: the *Anolis* lizards and insectivorous birds of the West Indies. Amer. Nat. 117: 181–192.

Resumen.—Los anfibios y reptiles ocupan todos los niveles tróficos de la red alimentaria del bosque lluvioso cerca de la Estación de Campo de El Verde, Puerto Rico. Éstos son conspicuos y abundantes en redes alimenticias sobre el suelo y son consumidores importantes en redes alimentarias subterráneas y de la hojarasca. Investigaciones recientes han demostrado que las lagartijas anolinas (*Anolis* spp.) y ranas leptodactílidas (*Eleutherodactylus* spp.) tienen un rol prominente en la estructuración de las relaciones tróficas entre los depredadores invertebrados y sus presas. Las especies de ambos géneros forrajean por toda la extensión vertical del bosque. Los estudios detallados de redes alimenticias muestran que los anfibios y reptiles son importantes consumidores de invertebrados en compartimientos separados de la red alimentaria; *Anolis* domina la red diurna y *Eleutherodactylus* domina la red nocturna. Los análisis detallados de las presas a nivel de especie demostraron que solamente existe un 13% de coincidencia entre las lagartijas anolinas y las ranas. Estos dos géneros de vertebrados también son presas de invertebrados grandes como tarántulas, cangrejos araña, vinagrillos sin cola, y ciempiés. Los datos precisos de las relaciones alimenticias documentan una depredación mutua entre especies de lagartijas y ranas, y entre estos dos grupos e invertebrados grandes. Las lagartijas del dosel del bosque también ejercen control sobre la estructura y función de la red alimentaria al disminuir la abundancia de artrópodos grandes, incluyendo arañas.

Résumé.—Les amphibiens et reptiles occupent tous les réseaux trophiques de la chaîne alimentaire de la forêt tropicale près d'El Verde à Porto Rico. Ils sont apparents et abondants dans les réseaux de surface et nombreux dans la litière et les réseaux souterrains. Des études récentes ont démontré que les *Anolis* (*Anolis* spp.) et les grenouilles Leptodactylidés (*Eleutherodactylus* spp.) jouaient un rôle proéminent dans la structure des relations trophiques entre les prédateurs invertébrés et leurs proies. Les espèces appartenant aux deux genres se retrouvent dans toute l'extension verticale de la forêt. Les études détaillées des chaînes alimentaires montrent que les amphibiens et les reptiles sont d'importants consommateurs d'invertébrés dans des compartiments séparés de la chaîne; *Anolis* domine la chaîne diurne et *Eleutherodactylus* domine la chaîne nocturne. Des analyses détaillées du rapport proies-espèces indiquent que 13% seulement des proies sont capturées à la fois par les *Anolis* et les grenouilles. Ces vertébrés sont également les proies de gros prédateurs invertébrés parmi lesquels figurent les tarentules, les araignées-crabes, les scorpions et les chilopodes. Des données détaillées sur les rapports alimentaires illustrent également la prédation entre les lézards et les grenouilles ainsi qu'entre ces groupes et les gros invertébrés. Les *Anolis* du couvert forestier exercent également un certain contrôle sur la structure de la chaîne alimentaire en freinant la prolifération des gros arthropodes, araignées comprises.

Reagan, D. P. 1996. The role of amphibians and reptiles in a West Indian rain forest food web, p. 217–227. *In* R. Powell and R. W. Henderson (eds.), Contributions to West Indian Herpetology: A Tribute to Albert Schwartz. Society for the Study of Amphibians and Reptiles, Ithaca (New York). Contributions to Herpetology, volume 12.

Patterns of Parental Care in Frogs of the Genus *Eleutherodactylus*

Daniel S. Townsend[1]

Abstract.—Among a suite of adaptations for terrestrial reproduction, frogs of the genus *Eleutherodactylus* exhibit several forms of post-fertilization parental care. One Puerto Rican species, *E. jasperi*, exhibits ovoviviparity; other species exhibit parental attendance of eggs or hatchlings at terrestrial oviposition sites. Of approximately 525 named species in this large Neotropical genus, documented observations of the presence or absence of parental attendants at nest sites were collected for 59 species. The largest data set exists for West Indian species, which constitute about one fourth of the genus. Twenty-four species exhibited female attendance, 20 species exhibited male attendance, two species exhibited amphisexual attendance (either males or females were observed to attend clutches), and 13 species exhibited no attendance. Female attendance was strongly associated with ground oviposition sites. Female attendance also was predominant in mainland species and on the West Indian island of Jamaica. Male attendance was not associated with any particular type of oviposition site. However, calling from a nest site and/or the occurrence of multiple clutches in a nest site has been reported for nine species, all of them with male attendance. Male attendance was strongly associated geographically with the eastern West Indies, especially Puerto Rico and Hispaniola.

Introduction

Frogs of the genus *Eleutherodactylus* constitute the largest genus of vertebrates and a dominant component of the Neotropical herpetofauna. At present, approximately 525 species in the genus have been named (J. D. Lynch, personal communication); these include members of the synonymized genera *Ladailadne*, *Sminthillus*, *Hylactophryne*, *Tomodactylus*, and *Syrrhophus* (Hedges 1989). The range of the genus extends from South America through Middle America into the southwestern United States and the West Indies. Although a majority of species (approximately 385) are mainland, and are especially numerous in the northern Andes, slightly more than one fourth of the genus (*circa* 140 species) is West Indian. With few exceptions, West Indian species are endemic to single islands, suggesting that little natural (i.e., non-anthropogenic) inter-island dispersal has occurred subsequent to insular speciation events.

Despite their ubiquitous presence in Neotropical habitats and wide geographic distribution, information on reproduction, ecology, and behavior of these frogs is unavailable for most species. The available information suggests a highly derived suite of adaptations in a group of anurans that has left the ancestral aquatic habitat and achieved tremendous success as terrestrial amphibians.

Frogs in the genus *Eleutherodactylus* are characterized by the direct development of large, heavily-yolked eggs laid in a variety of terrestrial sites (Townsend and Stewart 1985). In concert with large egg size, clutch sizes are small (in proportion to body size) by comparison to most aquatic-breeding frogs (Duellman and Trueb 1986 [1985]).

Internal fertilization is known in two Puerto Rican species, *Eleutherodactylus jasperi* (Drewry and Jones 1976; Wake 1978) and *E. coqui* (Townsend et al. 1981), and possibly a third Puerto Rican species, *E. portoricensis* (D. S. Townsend and M.

[1] Department of Biology, University of Scranton, Scranton, Pennsylvania 18510, USA.

M. Stewart, unpublished data), as well as a Costa Rican species, *E. angelicus* (Hayes 1985). *Eleutherodactylus jasperi* exhibits ovoviviparity, females giving birth to froglets that undergo complete development within modified oviducts (Wake 1978).

Another adaptive strategy exhibited by these frogs involves the provision of parental care to eggs and hatchlings in terrestrial nest sites. Within the genus, species may exhibit female, male, and no parental care. Amphisexual care (*sensu* Simon 1983) has been documented for at least one species. Parental care may involve nest preparation and burying of eggs during oviposition, short-term attendance of a newly laid clutch, or prolonged attendance of eggs and even hatchlings. A brief review of parental care in *Eleutherodactylus* was presented by Wells (1981).

In this chapter, I review parental care in *Eleutherodactylus*, and assess the relationship of parental care patterns to the biogeography, phylogeny (especially in the West Indies), and some aspects of the ecology of these frogs. While the occurrence of ovoviviparity in the Puerto Rican species *E. jasperi* certainly satisfies the definition of parental care, I have restricted the major focus of this review to the "voluntary" association of parents with their clutches, and hence do not consider *E. jasperi* further in the main analysis.

The functions of parental care in frogs and toads, with regard to how parents increase survival of offspring, have been studied in relatively few species (Wells 1981; Duellman and Trueb 1986 [1985]). For terrestrial-breeding species, principal functions include reduction of desiccation and protection from predation, including cannibalism. Very few studies have addressed the functions or adaptive significance of parental care in *Eleutherodactylus* (Wake 1978; Townsend et al. 1984; Diesel et al. 1995).

Methods

I considered both published records and unpublished observations of parental care. Because most parental care in frogs of the genus *Eleutherodactylus* consists of attendance of eggs at a nest site, the observation of an adult frog sitting on or immediately adjacent to a clutch could be considered adequate evidence of parental care. Because single observations for species may render a high risk of a Type I Error (especially in deciding that clutches without attendants lack parental care), I created two levels of data in the analysis.

Species for which at least two observations existed of an adult attending a clutch were considered as reliable accounts of parental care, with the exception that behavioral observations of parental fidelity or tenacity (e.g., repeated observations of an adult brooding eggs, an adult that returned to the eggs if removed, or an adult that refused to abandon the eggs even when disturbed) in the case of a single observation were considered valid evidence of parental care. Because a parent may be periodically absent from its nest even in a species which exhibits a high level of parental care (e.g., *E. coqui*; Townsend et al. 1984), I set a sample size of three observations of eggs without an attendant to decide that a species lacked parental attendance. Species that met these sampling criteria I have dubbed "Good" species in this review.

"Marginal" species were those for which only one documented observation existed of an attendant parent the sex of which was determined, or for which one to two observations were available of clutches without attendant parents.

Results and Discussion

I obtained records of eggs or parental care observations for 70 species of *Eleutherodactylus* (excluding the live-bearing *E. jasperi*). Thirty-three species satisfied the sampling criteria of "Good" species (Table 1) and 26 species were "Marginal" (Table 2). For the remaining 11 species (Table 3), accounts failed to state whether or not an adult was present with a clutch, the sex of an attendant parent was undetermined, or accounts conflicted.

TABLE 1 (opposite page). Parental attendance patterns of frogs in the genus *Eleutherodactylus* that met the sampling criteria for "good species" (see Methods). Subgeneric assignments, in parentheses, follow S. B. Hedges (1989; personal communication) for West Indian species, and J. D. Lynch (personal communication) for mainland species.

Species	Distribution	Source
Female attendance		
E. (*Craugastor*) *angelicus*	Costa Rica	Hayes (1985)
E. (*Craugastor*) *decoratus*	Mexico	Bogert (1969)
E. (*Craugastor*) *fitzingeri*	Nicaragua to Colombia	Dunn (1931), Lynch and Myers (1983)
E. (*Craugastor*) *maussi*	Venezuela	Heatwole (1962)
E. (*Eleutherodactylus*) *affinis*	Colombia	J. D. Lynch (pers. comm.)
E. (*Eleutherodactylus*) *caryophyllaceus*	Costa Rica to Colombia	Myers (1969)
E. (*Eleutherodactylus*) *lymani*	Ecuador to Peru	J. D. Lynch (1969; pers. comm.)
E. (*Eleutherodactylus*) *terraebolivaris*	Venezuela	Heatwole (1963)
E. (*Euhyas*) *cundalli*	Jamaica	Diesel et al. (1995)
E. (*Euhyas*) *glaucoreius*	Jamaica	D. S. Townsend (unpubl. data)
E. (*Euhyas*) *nubicola*	Jamaica	Lynn (1942), D. S. Townsend (unpubl. data)
E. (*Euhyas*) *pantoni*	Jamaica	Lynn (1940), D. S. Townsend (unpubl. data)
E. (*Euhyas*) *pentasyringos*	Jamaica	Lynn and Dent (1943)
E. (*Euhyas*) *ventrilineatus*	Hispaniola South Island	R. Franz and D. S. Townsend (unpubl. data)
Male attendance		
E. (*Eleutherodactylus*) *cooki*	Puerto Rico	Grant (1932), Drewry (1971), Rivero (1978), R. Thomas (pers. comm.)
E. (*Eleutherodactylus*) *coqui*	Puerto Rico	Townsend et al. (1984)
E. (*Eleutherodactylus*) *guantanamera*	Cuba	Hedges et al. (1992), Estrada (1990)
E. (*Eleutherodactylus*) *hedricki*	Puerto Rico	Rivero (1963), Drewry (1971), Rivero (1978), D. S. Townsend and M. M. Stewart (unpubl. data)
E. (*Eleutherodactylus*) *portoricensis*	Puerto Rico	D. S. Townsend and M. M. Stewart (unpubl. data)
E. (*Eleutherodactylus*) *wightmanae*	Puerto Rico	D. S. Townsend and M. M. Stewart (unpubl. data)
E. (*Euhyas*) *amadeus*	Hispaniola South Island	Hedges et al. (1987), R. Franz and D. S. Townsend (unpubl. data)
E. (*Euhyas*) *bakeri*	Hispaniola South Island	S. B. Hedges (personal communication), R. Franz and D. S. Townsend (unpubl. data)
E. (*Euhyas*) *glandulifer*	Hispaniola South Island	S. B. Hedges (pers. comm.), R. Franz and D. S. Townsend (unpubl. data)
E. (*Pelorius*) *inoptatus*	Hispaniola North and South islands	S. B. Hedges (pers. comm.)
Amphisexual attendance		
E. (*Eleutherodactylus*) *johnstonei*	Lesser Antilles	Bayley (1950), K. Ovaska (pers. comm.), G. Bourne (pers. comm.)
E. (*Euhyas*) *alticola*	Jamaica	Lynn (1937), D. S. Townsend (unpubl. data), S. B. Hedges (pers. comm.), M. M. Stewart (pers. comm.)
No attendance		
E. (*Eleutherodactylus*) *cruentus*	Costa Rica to Panama	C. W. Myers (1969; pers. comm.)
E. (*Eleutherodactylus*) *curtipes*	Colombia to Ecuador	J. D. Lynch (1981; pers. comm.)
E. (*Eleutherodactylus*) *guentheri*	Brazil	Lynn and Lutz (1946)
E. (*Eleutherodactylus*) *tayrona*	Colombia	Ruthven (1915), Lynch and Ruiz-Carranza (1985), J. D. Lynch (pers. comm.)
E. (*Eleutherodactylus*) *vicarius*	Colombia	Lynch and Ruiz-Carranza (1983), J. D. Lynch (pers. comm.)
E. (*Euhyas*) *planirostris*	Cuba	Goin (1947), Stewart (1979)
E. (*Syrrhophus*) *marnocki*	Southwestern United States	Jameson (1954)

TABLE 2. Parental attendance patterns of frogs in the genus *Eleutherodactylus* that did not meet sampling criteria for "marginal species," but had at least one documented observation of parental attendance or lack thereof (see Methods). Subgeneric assignments, in parentheses, follow Hedges (1989, personal communication) for West Indian species and J. D. Lynch (personal communication) for mainland species.

Species	Distribution	Source
Female attendance		
E. (*Craugastor*) *anomalus*	Colombia to Ecuador	Lynch and Myers (1983)
E. (*Craugastor*) *psephosypharus*	Guatemala to Belize	Campbell et al. (1994)
E. (*Craugastor*) *zygodactylus*	Colombia	J. D. Lynch (pers. comm.)
E. (*Eleutherodactylus*) *nicefori*	Colombia	J. D. Lynch (pers. comm.)
E. (*Eleutherodactylus*) *percultus*	Ecuador	J. D. Lynch (pers. comm.)
E. (*Eleutherodactylus*) *simoterus*	Colombia	J. D. Lynch (pers. comm.)
E. (*Euhyas*) *varleyi*	Cuba	Estrada (1987)
E. (*Euhyas*) *atkinsi*	Cuba	Novo et al. (1987)
E. (*Euhyas*) *dimidiatus*	Cuba	Estrada (1987)
E. new sp. P	Bolivia	O. de la Riva and J. D. Lynch (pers. comm.)
Male attendance		
E. (*Craugastor*) *augusti*	Mexico to southwestern United States	Jameson (1950)
E. (*Eleutherodactylus*) *abbotti*	Hispaniola North and South islands	S. B. Hedges (pers. comm.)
E. (*Eleutherodactylus*) *auriculatoides*	Hispaniola North Island	S. B. Hedges (pers. comm.)
E. (*Eleutherodactylus*) *diastema*	Costa Rica to Panama	Dunn (1937)
E. (*Eleutherodactylus*) *montanus*	Hispaniola North Island	S. B. Hedges (pers. comm.)
E. (*Eleutherodactylus*) *unicolor*	Puerto Rico	L. L. Woolbright (pers. comm.)
E. (*Eleutherodactylus*) new sp. A	Lesser Antilles	Kaiser et al. (1994 [1995]), H. Kaiser (pers. comm.)
E. (*Euhyas*) *intermedius*	Cuba	Estrada (1992)
E. (*Euhyas*) new sp. TE	Cuba	A. R. Estrada (pers. comm.)
E. cf. *ventrilineatus*	Hispaniola South Island	R. Franz and D. S. Townsend (unpubl. data)
No attendance		
E. (*Craugastor*) *rugulosus*	Mexico to Guatemala	J. A. Campbell (pers. comm.)
E. (*Eleutherodactylus*) *chloronotus*	Ecuador	J. D. Lynch (pers. comm.)
E. (*Eleutherodactylus*) *prolixodiscus*	Colombia	J. D. Lynch (pers. comm.)
E. (*Eleutherodactylus*) *unistrigatus*	Colombia to Ecuador	J. D. Lynch (1981, pers. comm.)
E. (*Euhyas*) *gossei*	Jamaica	Dunn (1926), Stewart (1979)
E. (*Syrrhophus*) *cystignathoides*	Mexico to southwestern United States	Hayes-Odum (1990)

Of the 59 "Good" and "Marginal" species (Tables 1 and 2), 46 exhibited parental attendance of eggs and/or hatchlings, while 13 species exhibited no parental attendance. Of species that exhibited parental attendance, 24 showed female attendance, 20 showed male attendance, and 2 species exhibited amphisexual attendance.

Amphisexual attendance (*sensu* Simon 1983) is the condition in which either males or females attend individual clutches. This condition has been observed in the Jamaican species *Eleutherodactylus alticola* (Lynn 1937; D. S. Townsend, unpublished data) and in the Lesser Antillean species *E. johnstonei* (Bayley 1950; K. Ovaska, personal communication; G. Bourne, personal communication). For *E. johnstonei*, the attending parent was most often male, while for *E. alticola*, females were most often found attending clutches. In both species, however, the secondary sex was observed as an attendant on more than one occasion.

Besides egg attendance, two other parental behaviors have been observed in *Eleutherodactylus*.

Two mainland species have been observed to bury their eggs during oviposition in shallow holes or depressions along streams. Females of *E. angelicus* apparently remain with and guard the buried eggs for at least a few days after oviposition (Hayes 1985), whereas females of *E. rugulosus* appear to abandon the buried clutch soon after oviposition (J. A. Campbell, personal communication). Diesel et al. (1995) reported females carrying hatchlings in the Jamaican species *E. cundalli*, a behavior that has been independently observed in the field (R. Thompson, personal communication) and the laboratory (M. M. Stewart, personal communication). Females of a related Jamaican species, *E. glaucoreius*, have been observed with hatchlings on their backs in nest sites (D. S. Townsend, unpublished data).

Geographic Patterns

Of approximately 385 mainland species of *Eleutherodactylus*, breeding records existed for 28 species (n_{good} = 14), about 7% of the total (Tables 1 and 2). Fifteen mainland species exhibited female attendance (n_{good} = 8), whereas 11 species exhibited no parental attendance (n_{good} = 6). Male parental attendance has been described for only two mainland species, both based on single records (Table 2). Jameson (1950) reported finding a male *E. augusti* under a rock with eggs and speculated that the male may have urinated on the eggs to keep them from desiccating. Dunn (1937: 163) reported taking a male *E. diastema* "with a clutch of 11 eggs in a bromeliad."

Of approximately 140 West Indian species of *Eleutherodactylus*, breeding records existed for 31 species (n_{good} = 19) or about 22% of the total (Tables 1 and 2). Male parental care was strongly associated with the eastern Caribbean. Six Puerto Rican species and one new species from the Lesser Antilles (new species A of Kaiser et al. 1994 [1995]) exhibited male attendance of eggs (Table 4). No known female-caring species are known from the Puerto Rican bank or the Lesser Antilles. The only exception was *E. johnstonei*, a Lesser Antillean species that has been observed to have either males or females attending clutches, although males were the most common attendants (G. Bourne, personal communication).

In contrast, on Jamaica, female attendance was the predominant condition (Table 4). Females have been found to care for eggs in five species (all "Good" species; Table 1), including the observation of attendance and hatchling transport by female *Eleutherodactylus cundalli* (Diesel et al. 1995). No male-caring species are known from Jamaica.

TABLE 3. Species in the genus *Eleutherodactylus* for which eggs or parental behavior have been observed, but information is unclear as to their parental care status. Subgeneric assignments, in parentheses, follow S. B. Hedges (1989; personal communication) for West Indian species, and J. D. Lynch (personal communication) for mainland species.

Species	Distribution	Source
E. (Craugastor) crassidigitatus	Costa Rica to Panama	Taylor (1952), Lynch and Myers (1983)
E. (Craugastor) spatulatus	Mexico	J. D. Lynch (pers. comm.)
E. (Eleutherodactylus) antillensis	Puerto Rico	Castillo and Mayorga (1984), D. S. Townsend and M. M. Stewart (unpubl. data)
E. (Eleutherodactylus) fowleri	Hispaniola South Island	Schwartz (1973)
E. (Eleutherodactylus) melacara	Cuba	Hedges et al. (1992)
E. (Eleutherodactylus) nasutus	Brazil	Lynn and Lutz (1947)
E. (Eleutherodactylus) parvus	Brazil	Lutz (1944)
E. (Eleutherodactylus) richmondi	Puerto Rico	Drewry (1971)
E. (Eleutherodactylus) urichi	Trinidad	Kenny (1969)
E. (Euhyas) jamaicensis	Jamaica	Dunn (1926), Lynn and Dent (1943), D. S. Townsend (unpubl. data)
E. (Pelorius) ruthae	Hispaniola North and South islands	Schwartz (1965)

Whereas *E. alticola* apparently provides amphisexual care, females were found most frequently as attendants (Lynn 1940; Townsend, unpublished data). Only one Jamaican species, *E. gossei*, appears to lack parental attendance (Table 4).

Hispaniola has the highest diversity of *Eleutherodactylus* in the West Indies, with 48 currently recognized species. Of nine species for which records of parental attendance existed (n_{good} = 5), eight exhibited male parental care. Only one species, *E. ventrilineatus*, exhibited female egg attendance (Table 4).

The island of Hispaniola is composed of two paleoislands, a larger North Island and a smaller South Island, that are joined by an east-west low elevation plain, the Cul de Sac in Haiti which is continuous with the Valle de Neiba in the Dominican Republic. The North Island is generally considered to have been part of the proto-Antilles and to have been emergent for a much longer portion of the Cenozoic than the South Island, which may have been submergent until the early Miocene (Hedges 1989). More importantly, the South Island was disjunct from the North Island until the late Miocene (approximately 10 myBP) when it collided with the North Island. The herpetofauna of Hispaniola therefore represents separately-evolved geographic clusters that only had the chance to exchange taxa relatively recently.

In examining the nine Hispaniolan species for which parental attendance is known, the single species with female parental care was restricted to the South Island. Of eight species with male parental attendance, two were strictly North Island, four were strictly South Island, and the remaining two species were found on both paleoislands (Tables 1 and 2).

TABLE 4. Geographic patterns of parental care in West Indian frogs of the genus *Eleutherodactylus*. For table entries that contain slashes, the first number represents "Good" species, the second number "Marginal" species under sampling criteria (see Methods). Abbreviations: Cuba (C), Jamaica (J), Hispaniola (H), and Puerto Rico (P). "P" includes Puerto Rico, the Puerto Rican Bank, and the Lesser Antilles.

Parental care	C	J	H	P
Female care	0/3	5/0	1/0	0
Male care	1/2	0	4/4	5/2
Amphisexual care	0	1/0	0	1
No care	1/0	0/1	0	0

The least complete data set for parental care in *Eleutherodactylus* was for Cuba. Of seven species for which records exist (n_{good} = 2), all but one (*E. planirostris*) have been observed by Alberto Estrada and his colleagues. Three species may have female attendance (all are "Marginal" records). Three species appear to practice male attendance, including *E. guantanamera*, one of the two species with sample sizes greater than one (Table 4). The other, *E. planirostris*, lacks parental care (Goin 1947; Stewart 1979).

Phylogenetic Patterns

Mainland species of *Eleutherodactylus*, comprising nearly three-quarters of the genus (approximately 385 species), have been regarded as constituting several species groups (e.g., Savage 1987) or subgenera (e.g., Hedges 1989). These include the subgenera *Craugastor* (with about 75 species), *Syrrhophus* (with about 24 species), and *Eleutherodactylus*. Given the relatively few mainland species for which records of parental care exist (28 of 385 species), a phylogenetic analysis of mainland *Eleutherodactylus* would be premature.

Recently, Hedges (1989) suggested a classification of West Indian *Eleutherodactylus* into three subgenera: *Eleutherodactylus*, *Euhyas*, and *Pelorius*. Although disagreement exists on the assignment of some species within Hedges' scheme (J. D. Lynch, personal communication), it serves as a useful means for a preliminary examination of phylogenetic associations of parental care (Table 5). The subgenus *Pelorius* is small (6 species) and contains only one species, *E. inoptatus*, for which reliable data on parental care were available.

All nine West Indian species that exhibited female parental attendance (n_{good} = 6, $n_{marginal}$ = 3) occur in the subgenus *Euhyas* (Table 5). The Jamaican species, *Eleutherodactylus alticola*, that exhibited amphisexual care predominately by females is also placed in *Euhyas* (Hedges 1989). Additionally, three (J. D. Lynch, personal communication)

TABLE 5. Phylogenetic patterns of parental care in West Indian frogs of the genus *Eleutherodactylus*. For table entries that contain slashes, the first number represents "Good" species, the second number "Marginal" species under sampling criteria (see Methods). Abbreviations: *Eleutherodactylus* (EL), *Euhyas* (EU), and *Pelorius* (PE).

Parental care	Subgenus		
	EL	EU	PE
Female care	0	6/3	0
Male care	6/5	3/2	1
Amphisexual care	1	1	0
No Care	0	1/1	0

to five (Hedges 1989) *Euhyas* species exhibited male attendance, and two lacked parental care (Table 5).

Within the West Indian subgenus *Eleutherodactylus* are species with male attendance, amphisexual attendance (*E. johnstonei*), and no attendance (Table 5). No instances of female attendance are known within this subgenus in the West Indies, although several mainland species with female attendance are placed in this subgenus (Tables 1 and 2).

Ecological Patterns

Although all *Eleutherodactylus* deposit non-aquatic eggs, they use a broad spectrum of oviposition sites in the terrestrial environment. To examine the relationship between oviposition site and parental attendance behavior, I divided oviposition site data into three major categories: ground nesting, litter nesting, and arboreal nesting. Ground nesters included species that laid their eggs under rocks, in holes in soil or sand, or in subterranean burrows. Litter nesters were species that laid their eggs within the leaf litter layer on the ground. Arboreal nesting included the widest range of sites, including bromeliads and treeholes but also sites used by species like *E. cooki* that lay eggs on the sides of boulders in grottoes (Rivero 1978). I also examined the use of bromeliads as a subcategory of arboreal nesters.

Of 43 species that exhibited parental attendance ($n_{good} = 26$), female attendance and amphisexual attendance were strongly associated with ground nests (Table 6). Only two of the 23 female-caring species, *Eleutherodactylus caryophyllaceus* (Myers 1969) and *E. percultus* (J. D. Lynch, personal communication), have been observed to attend eggs at arboreal sites and only the latter species used bromeliads as nest sites. In contrast, species with male attendance were fairly equally distributed among all three oviposition site types (Table 6). However, arboreal species with male attendance were predominately bromeliad users, the exceptions being two Puerto Rican species, *E. coqui* (Townsend et al. 1984) and *E. hedricki* (Rivero 1963; D. S. Townsend and M. M. Stewart, unpublished data). *Eleutherodactylus coqui* is an adaptable species that uses various elevated sites, including palm axils, hanging dead leaves and palm fronds, and even artificial elevated substrates (Townsend 1989), but frequently uses ground litter nest sites as well. *Eleutherodactylus hedricki* is a treehole nesting species (Rivero 1963).

Species that lack parental care showed a tendency to use ground nest sites, although four of the 15 species were found to use arboreal sites, three of them in bromeliads (Table 6).

Male parental attendance may be more likely to evolve if eggs are deposited in nest sites that are close to or coincident with regular calling sites (Wells 1981). Males thereby could continue calling to attract additional females while still providing some degree of protection or care to an existing clutch. Seven of the 19 species that exhibited male attendance (*Eleutherodactylus bakeri*, *E. cooki*, *E.*

TABLE 6. Relationship between parental attendance patterns and oviposition sites in frogs of the genus *Eleutherodactylus*. For table entries that contain slashes, the first number represents "Good" species, the second number "Marginal" species under sampling criteria (see Methods). Abbreviations: ground nesting (G), litter nesting (L), and arboreal nesting (bromeliads) (A [B]).

Parental care	Oviposition site			
	G	L	A	(B)
Female care	13/6	0/2	1/1	(0/1)
Male care	3/3	3/2	4/3	(2/3)
Amphisexual care	2	0	0	—
No care	5/5	1/0	2/2	(1/2)

coqui, *E. hedricki*, *E. intermedius*, *E. unicolor*, and *E.* new sp. TE) have been observed to call from nest sites. Furthermore, attendance of multiple clutches at a single nest site has been observed only in male-caring species, including *E. amadeus*, *E. bakeri*, *E. cooki*, *E. coqui*, *E. guantanamera*, *E. hedricki*, and *E.* new sp. TE. No records exist of multiple clutches for any species with female attendance or amphisexual attendance. Multiple clutches in a single nest cavity have been observed for an Andean species, *E. curtipes*, that lacks parental attendance (Lynch 1981).

Internal Fertilization

Two Puerto Rican species, *Eleutherodactylus jasperi* and *E. coqui*, are known to exhibit internal fertilization, and circumstantial or anecdotal evidence suggests that *E. portoricensis* and *E. angelicus* do so as well (Wake 1978; Townsend et al. 1981; Hayes 1985; D. S. Townsend and M. M. Stewart, unpublished data).

In analyses of mating systems, internal fertilization is generally considered to select against male parental investment, because females may be able to achieve multiple matings and hence males may have lower certainty of paternity and some probability of being cuckolded. From that perspective, that both *Eleutherodactylus coqui* and *E. portoricensis* exhibited prolonged male attendance is striking. However, at least in the case of these two species, the timing of internal fertilization yields a very low chance for the occurrence of cuckoldry.

In *Eleutherodactylus coqui*, for which the behavior and timing of fertilization and subsequent oviposition have been documented, oviposition occurs within a few hours (4–5 hr at most) of insemination (Townsend and Stewart 1986). Furthermore, the sequence of ovulation, insemination, and oviposition appears to be relatively "hard-wired," such that once a female has begun the process, she is unable to voluntarily halt or prolong it. This is suggested by the observation that females who are disturbed while mating, and placed in containers or other foreign environments, lay an entire ovulated clutch in the same time frame as females left undisturbed in their natural nest sites. Under these circumstances, it would be virtually impossible for a female, once inseminated, to abandon the male parent and seek a second male which would be cuckolded. That a female would have sufficient time to leave the breeding site, enter a separate site, and lay her eggs is highly unlikely. Observations suggest that females that reject potential mates do so early in the courtship sequence, well before ovulation occurs.

Given the tremendous success of frogs in the genus *Eleutherodactylus* as terrestrial amphibians, that internal fertilization occurs in other species is likely, and may be widespread in the genus. We currently have no basis for determining how widespread this trait may be, or whether it represents an ancestral condition for the genus. Until additional species have been observed, accounting for any influence of fertilization mode on the evolution of parental investment patterns in these frogs is impossible.

Summary

The strongest geographic pattern of parental care in the West Indies corresponds to an apparent dichotomy in the sex of the attending parent. The eastern islands of the Puerto Rican bank and the Lesser Antilles are dominated by male attendance, and harbor no known female-caring species. Jamaica, in the western Caribbean, presents the opposite situation, with no known male-caring species. Hispaniola, geographically intermediate to Puerto Rico and Jamaica, displays a mixture of male-caring and female-caring species. Whereas the sole female-caring species is a South Island endemic, male attendance occurs in North Island, South Island, and pan-Hispaniolan species. Cuba, which might shed light on any geographic hypothesis, is too poorly studied.

The geographic pattern in the West Indies is paralleled by a phylogenetic pattern with respect to the sex of the attending parent. For "Good" species that met the sampling criteria, male attendance is the predominant condition in the subgenus *Eleutherodactylus* of Puerto Rico and Cuba. Female attendance predominates in the subgenus

Euhyas, especially in Jamaica. Hispaniola presents a confusing picture, with one *Euhyas* species with female attendance, and three *Euhyas* species that exhibit male attendance.

Ecology, life history, and mating system data are known for too few species of *Eleutherodactylus* to provide a useful basis for addressing functional questions about the evolution of parental care in these frogs.

ACKNOWLEDGMENTS

For access to unpublished data and/or manuscripts, I am indebted to Blair Hedges, John Lynch, Richard Franz, Alberto Estrada, Jonathan Campbell, Hinrich Kaiser, and Larry Woolbright. Richard Franz kindly provided access to specimens in the Florida Museum of Natural History, University of Florida; specimens were collected during field work in Haiti funded by a grant to Charles Woods from the USAID and the MacArthur Foundation. For helpful discussion of ideas and critical review of the manuscript, I thank Martha Crump, Blair Hedges, John Lynch, and Margaret Stewart. I am grateful for funding of my field work from the United States National Science Foundation, the National Geographic Society, and the I. W. Killam Foundation (Dalhousie University, Halifax, Nova Scotia). This review was prepared during a sabbatical leave granted me by the University of Scranton.

LITERATURE CITED

Bayley, I. 1950. The whistling frogs of Barbados. Jour. Barbados Mus. Hist. Soc. 17: 161–169.

Bogert, C. M. 1969. The eggs and hatchlings of the Mexican leptodactylid frog *Eleutherodactylus decoratus* Taylor. Amer. Mus. Novit. 2376: 1–9.

Campbell, J. A., J. M. Savage, and J. R. Meyers. 1994. A new species of *Eleutherodactylus* (Anura: Leptodactylidae) of the *rugulosus* group from Guatemala and Belize. Herpetologica 50: 412–419.

Castillo, J. A. and H. M. Mayorga. 1984. Distribucion, Habitos, Ecologilos, Reproduccion y Embriologia Externa de *Eleutherodactylus antillensis* (Anura). Lic. Thesis, Univ. Panama, Panama City.

Diesel, R., G. Bäurle, and P. Vogel. 1995. Cave breeding and froglet transport: a novel pattern of anuran brood care in the Jamaican frog, *Eleutherodactylus cundalli*. Copeia 1995: 354–360.

Drewry, G. E. 1971. The role of amphibians in the ecology of Puerto Rican rain forest, p. 16–54. *In* R. G. Clements, G. E. Drewry, and R. J. Lavigne (eds.), The Rain Forest Project Annual Report. Puerto Rico Nuclear Center, San Juan.

— and K. L. Jones. 1976. A new ovoviviparous frog, *Eleutherodactylus jasperi* (Amphibia, Anura, Leptodactylidae), from Puerto Rico. Jour. Herpetol. 10: 161–165.

Duellman, W. E. and L. Trueb. 1986 (1985). Biology of Amphibians. McGraw-Hill, New York.

Dunn, E. R. 1926. The frogs of Jamaica. Proc. Boston Soc. Nat. Hist. 38: 111–130.

—. 1931. The amphibians of Barro Colorado Island. Occas. Pap. Boston Soc. Nat. Hist. 5: 403–421.

—. 1937. The amphibian and reptilian fauna of bromeliads in Costa Rica and Panama. Copeia 1937: 163–167.

Estrada, A. R. 1987. Los nidos terrestres de dos especies de anfibios cubanos del genero *Eleutherodactylus* (Anura: Leptodactylidae). Poeyana 352: 1–9.

—. 1990. Las puestas de *Eleutherodactylus varians* (Gundlach y Peters). Rev. Biol. 4: 163–167.

—. 1992. Comentarios sobre *Eleutherodcactylus intermedius* Barbour et Shreve, 1937 de la region oriental de Cuba. Comm. Brev. Zool.: 12–14.

Goin, C. J. 1947. Studies on the life history of *Eleutherodactylus ricordii planirostris* (Cope) in Florida. Univ. Florida Studies, Biol. Sci. ser. 4: 1–66.

Grant, C. 1932. A new frog from Puerto Rico. Jour. Dept. Agric. Puerto Rico 16: 145–148.

Hayes, M. P. 1985. Nest structure and attendance in the stream-dwelling frog, *Eleutherodactylus angelicus*. Jour. Herpetol. 19: 168–169.

Hayes-Odum, L. A. 1990. Observations on reproduction and embryonic development in *Syrrhophus cystignathoides campi* (Anura: Leptodactylidae). Southwest. Nat. 35: 358–361.

Heatwole, H. 1962. Contributions to the natural history of *Eleutherodactylus cornutus maussi*. Stahlia Misc. Pap. 2: 1–7.

—. 1963. Contribucion a la historia natural de *Eleutherodactylus terraebolivaris* (Anura). Acta Biol. Venezuelica 3: 301–313.

Hedges, S. B. 1989. Evolution and biogeography of West Indian frogs of the genus *Eleutherodactylus*: slow-evolving loci and the major groups, p. 305–370. *In* C. A. Woods (ed.), Biogeography of the West Indies: Past, Present, and Future. Sandhill Crane Press, Gainesville, Florida.

—, A. R. Estrada, and R. Thomas. 1992. Three new species of *Eleutherodactylus* from eastern Cuba, with notes on vocalizations of other species (Anura: Leptodactylidae). Herpetol. Monogr. 6: 68–83.

—, R. Thomas, and R. Franz. 1987. A new species of *Eleutherodactylus* (Anura, Leptodactylidae) from the Massif de la Hotte, Haiti. Copeia 1987: 943–949.

Jameson, D. L. 1950. The development of *Eleutherodactylus latrans*. Copeia 1950: 44–46.

—. 1954. Social patterns in the leptodactylid frogs *Syrrhophus* and *Eleutherodactylus*. Copeia 1954: 36–38.

Kaiser, H., D. M. Green, and M. Schmid. 1994 (1995). Systematics and biogeography of Eastern Caribbean frogs (*Eleutherodactylus*: Leptodactylidae), with the description of a new species from Dominica. Canadian Jour. Zool. 72: 2217–2237.

Kenny, J. S. 1969. The Amphibia of Trinidad. Studies Fauna Curaçao Carib. Isl. 108: 1–78.

Lutz, B. 1944. The hatching of *Eleutherodactylus parvus*. Copeia 1944: 64.

Lynch, J. D. 1969. Taxonomic notes on Ecuadorian frogs (Leptodactylidae: *Eleutherodactylus*). Herpetologica 25: 262–274.

—. 1981. Leptodactylid frogs of the genus *Eleutherodactylus* in the Andes of northern Ecuador and adjacent Colombia. Univ. Kansas Mus. Nat. Hist. Misc. Publ. 72: 1–46.

— and C. W. Myers. 1983. Frogs of the *fitzingeri* group of *Eleutherodactylus* in eastern Panama and Chocoan South America (Leptodactylidae). Bull. Amer. Mus. Nat. Hist. 175: 481–572.

— and P. M. Ruiz-Carranza. 1983. New frogs of the genus *Eleutherodactylus* from the Andes of southern Colombia. Trans. Kansas Acad. Sci. 86: 99–112.

— and —. 1985. A synopsis of the frogs of the genus *Eleutherodactylus* from the Sierra Nevada de Santa Marta, Colombia. Occas. Pap. Mus. Zool., Univ. Michigan 711: 1–59.

Lynn, W. G. 1937. Two new frogs from Jamaica. Herpetologica 1: 88–90.

—. 1940. The amphibians of Jamaica, p. 3–60. *In* W. G. Lynn and C. Grant, The Herpetology of Jamaica. Bull. Inst. Jamaica, Sci. Ser. 1. Kingston.

—. 1942. The embryology of *Eleutherodactylus nubicola*, an anuran which has no tadpole stage. Carnegie Inst. Washington, Contrib. Embryol. 30: 27–62.

— and J. N. Dent. 1943. Notes on Jamaican amphibians. Copeia 1943: 234–242.

— and B. Lutz. 1946. The development of *Eleutherodactylus guentheri* Stdnr. 1864. Bol. Mus. Nac., nova ser. 71: 1–46.

— and —. 1947. The development of *Eleutherodactylus nasutus* Lutz. Bol. Mus. Nac., nova ser. 79: 1–13.

Myers, C. W. 1969. The ecological geography of cloud forest in Panama. Amer. Mus. Novit. 2396: 1–52.

Novo, J., A. R. Estrada, and L. V. Moreno. 1987. Apuntes sobre la reproduccion de *Eleutherodactylus atkinsi* (Anura: Leptodactylidae). Misc. Zool. (Acad. Cien. Cuba) 29: 2–3.

Rivero, J. A. 1963. *Eleutherodactylus hedricki*, a new species of frog from Puerto Rico (Salientia, Leptodactylidae). Breviora 185: 1–7.

—. 1978. Los Anfibios y Reptiles de Puerto Rico. Univ. Puerto Rico, San Juan.

Ruthven, A. G. 1915. The breeding habits of *Hylodes cruentus* Peters. Occ. Pap. Mus. Zool. Univ. Michigan 1: 1–6.

Savage, J. M. 1987. Systematics and distribution of the Mexican and Central American rainfrogs of the *Eleutherodactylus gollmeri* group (Amphibia: Leptodactylidae). Fieldiana Zool., new ser. 33: 1–57.

Schwartz, A. 1965. Variation and natural history of *Eleutherodactylus ruthae* on Hispaniola. Bull. Mus. Comp. Zool., Harvard Univ. 132: 479–508.

—. 1973. Six new species of *Eleutherodactylus* (Anura, Leptodactylidae) from Hispaniola. Jour. Herpetol. 7: 249–273.

Simon, M. P. 1983. The ecology of parental care in a terrestrial breeding frog from New Guinea. Behav. Ecol. Sociobiol. 14:61–67.

Stewart, M. M. 1979. The role of introduced species in a Jamaican frog community, p. 113–146. *In* H. Wolda (ed.), Actas del IV Simposium Internacional de Ecologia Tropical. Tomo 1. Panama City, Panama.

Taylor, E. H. 1952. A review of the frogs and toads of Costa Rica. Univ. Kansas Sci. Bull. 35: 577–923.

Townsend, D. S. 1989. The consequences of microhabitat choice for male reproductive success in a tropical frog (*Eleutherodactylus coqui*). Herpetologica 45: 451–458.

— and M. M. Stewart. 1985. Direct development in *Eleutherodactylus coqui* (Anura: Leptodactylidae): a staging table. Copeia 1985: 423–436.

— and —. 1986. Courtship and mating behavior of a Puerto Rican frog, *Eleutherodactylus coqui*. Herpetologica 42: 165–170.

—, —, and F. H. Pough. 1984. Male parental care and its adaptive significance in a Neotropical frog. Anim. Behav. 32: 421–431.

—, —, —, and P. F. Brussard. 1981. Internal fertilization in an oviparous frog. Science 212: 469–471.

Wake, M. H. 1978. The reproductive biology of *Eleutherodactylus jasperi* (Amphibia, Anura, Leptodactylidae), with comments on the evolution of live-bearing systems. Jour. Herpetol. 12: 121–133.

Wells, K. D. 1981. Parental behavior of male and female frogs, p. 184–197. *In* R. D. Alexander and D. W. Tinkle (eds.), Natural Selection and Social Behavior: Recent Research and New Theory. Chiron Press, New York.

APPENDIX. Specimens examined: Florida Museum of Natural History (University of Florida): *Eleutherodactylus amadeus*, UFFSM 60497–60505; *E. bakeri*, UFFSM 60456–60459, 60558; *E. glandulifer*, UFFSM 60451–60452; *E. ventrilineatus*, UFFSM 604006, 69138; *E.* cf. *ventrilineatus*, UFFSM 69137.

Resumen.—Entre una serie de adaptaciones para la reproducción terrestre, las ranas del género *Eleutherodactylus* exhiben varias formas de cuido parental postfecundación. Una especie de Puerto Rico, *E. jasperi*, es ovovivípara; otras especies exhiben cuido parental de los huevos o las crías en lugares terrestres de oviposición. De aproximadamente 525 especies descritas en este gran género neotropical, se coleccionaron observaciones documentadas de la presencia o ausencia de asistencia parental en los lugares de anidaje para 59 especies. Existe una mayor cantidad de datos para las especies de las Antillas, que constituyen casi una cuarta parte del género. Veinticuatro especies tienen cuido femenino, 20 mostraron asistencia por los machos, dos especies tuvieron cuido anfisexual (o los machos o las hembras fueron observados atendiendo las camadas) y 13 especies no exhibieron cuido parental. El cuido por las hembras estuvo fuertemente asociado con lugares de oviposición en el suelo. La asistencia por las hembras también fue predominate en especies del continente y de la isla antillana de Jamaica. El cuido por los machos no estuvo asociado con ningún tipo particular de lugar de oviposición. Sin embargo, cantar desde un lugar de anidaje y/o la ocurrencia de varias camadas en un mismo lugar de anidaje han sido reportados para nueve especies, todas ellas con asistencia por los machos. El cuido por los machos estuvo fuertemente asociado geográficamente con las Antillas Orientales, especialmente Puerto Rico y La Española.

Résumé.—Parmi la série d'adaptations qui ont conduit les grenouilles du genre *Eleutherodactylus* à se reproduire sur la terre ferme figurent également plusieurs formes de soins parentaux après la fécondation. Une espèce porto-ricaine, *E. jasperi*, est ovovivipare; d'autres espèces s'occupent des oeufs ou des têtards aux sites d'ovoposition terrestre. Des observations documentées de la présence ou de l'absence de soin parental aux sites de reproduction ont été effectuées pour 59 espèces de ce genre néotropical qui compte environ 525 espèces connues. Le groupe de données le plus important concerne les espèces antillaises qui forment environ un quart du genre. Chez vingt-quatre espèces, la femelle s'occupe des oeufs/têtards, chez 20 espèces ce sont les mâles et chez deux espèces ce sont soit les mâles soit les femelles. Treize espèces ne s'occupent pas des oeufs/têtards. Lorsque ce sont les femelles qui s'occupent des oeufs/têtards, on a noté un rapport très étroit avec les sites d'ovoposition terrestre. La présence des femelles domine chez les espèces continentales et sur l'île de la Jamaïque. La présence des mâles n'a aucun rapport avec le site d'ovoposition. Toutefois, les coassements provenant d'un site de reproduction et(ou) l'occurrence de multiples pontes dans un site de reproduction ont été signalés pour 9 espèces chez qui la présence des mâles a été observée. Dans les Antilles orientales, la présence des mâles dépend étroitement de la géographie, notamment à Porto Rico et à Hispaniola.

Townsend, D. S. 1996. Patterns of parental care in frogs of the genus *Eleutherodactylus*, p. 229–239. *In* R. Powell and R. W. Henderson (eds.), Contributions to West Indian Herpetology: A Tribute to Albert Schwartz. Society for the Study of Amphibians and Reptiles, Ithaca (New York). Contributions to Herpetology, volume 12.

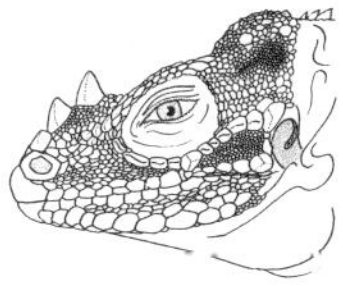

MATE SELECTION IN THE PUERTO RICAN FROG, *ELEUTHERODACTYLUS COQUI*

PAMELA T. LOPEZ[1]

Abstract.—Spectral and temporal features of the advertisement call ("co qui"), as well as male phenotype (body size, physical condition, and fat reserves), varied extensively in a population of *Eleutherodactylus coqui* at El Verde, Puerto Rico. Mate selection in *E. coqui* was investigated by addressing the following two questions: (1) Upon what attribute of the male advertisement call is female choice based?, and (2) What is the adaptive significance of this preference? Both laboratory two-choice trials and observations of natural courtship events indicated female preference for relatively high call rates, but not for call dominant frequencies. Neither size-assortative mating or large-male advantage regarding mating success were observed. However, all mated males called at a rate that was greater than the populational average and the average for unmated males. Eighty-three percent of the observed matings occurred prior to 2130 h. No significant association between male phenotype and calling effort (a male's average call rate within one night) was found, and no significant difference in phenotype existed between males which had recently acquired a clutch of eggs and the general population of calling males. Therefore, females of *E. coqui* cannot reliably determine male phenotype by assessing call rate. Alternatively, female preference for high call rate in this species may be based in the locatability of males in the complex acoustic environment of the Puerto Rican rainforest. Forty-four percent of the males studied were calling from sites with no vegetative cover and maximal sound propagation, thereby presumably enhancing localization by females.

INTRODUCTION

THE FUNCTIONAL SIGNIFICANCE OF THE ANURAN advertisement call as a premating isolating mechanism has been documented extensively (e.g., reviews by Blair 1964; Littlejohn 1977). Studies which showed that females discriminate conspecific males from heterospecifics on the basis of vocalizations alone (e.g., Loftus-Hills and Littlejohn 1971; Schwartz 1987) indicated that the potential costs to reproductive success associated with mismatings are high.

Male vocalizations may also provide acoustic cues with which females can assess male quality, specifically in prolonged breeders, where territory or oviposition site quality is not judged directly by females. Female preference, as determined in two- or four-choice laboratory trials, has been documented for the following call parameters: call frequency (Ryan 1980*a*, 1983; Gerhardt 1982, 1987; Doherty and Gerhardt 1984; Forester and Czarnowsky 1985; Robertson 1986; Forester and Harrison 1987; Morris and Yoon 1989), intensity (Fellers 1979; Forester and Czarnowsky 1985), call rate (Whitney and Krebs 1975; Gerhardt 1982, 1987; Sullivan 1982, 1983, 1985; Forester and Czarnowsky 1985; Klump and Gerhardt 1987; Arak 1988), pulse repetition rate (Gerhardt 1982; Klump and Gerhardt 1987; Arak 1988), and call complexity (Ryan 1980b; Schwartz 1986). The association between call frequency and male body size has been shown to affect male mating success in species where females prefer larger males (Ryan 1980*a*; Arak 1983, 1988; Morris and Yoon 1989). However, the adaptive significance of female preference for the majority of these very specific call attributes has not yet been addressed.

[1] Department of Biology, Pacific University, Forest Grove, Oregon 97116, USA.

Herein I review the major findings regarding mate selection in a population of the frog *Eleutherodactylus coqui* (Leptodactylidae) located at the El Verde field station in the Luquillo Mountains of northeastern Puerto Rico. The population and site have been described in Lopez (1990) and Lopez and Narins (1991). Field work was conducted during the months of July through September 1985–1988. Two specific questions addressed in this chapter are: (1) Upon what attribute of the male advertisement call is female choice in *E. coqui* based?, and (2) What is the adaptive significance of this preference? A considerable amount of research has been conducted on *E. coqui* since the mid-1970s, much of which served as the basis for questions concerning mate choice in this species. Among the salient features of the behavioral ecology and auditory physiology of *E. coqui* is information concerning the function and perception of the male advertisement call, and the parental care behavior shown by males.

Males produce a two-note advertisement call ("co qui"), each note having a unique communicative function (Narins and Capranica 1976, 1978). Of significance to this review is the "qui" note, which functions in female attraction; the peripheral auditory system of females being most sensitive to frequencies that comprise the "qui" note (Narins and Capranica 1976, 1978). Variation in this note presumably forms the basis of intersexual selection. Specific to the study population discussed herein, a considerable degree of intrapopulational, temperature-independent variation exists in the temporal and spectral attributes of both call notes (Lopez and Narins 1991). Male snout-vent length (SVL) and body mass are significantly, albeit weakly, associated with "co" and "qui" dominant frequency, but not with call rate (Lopez and Narins 1991). Moreover, no large male advantage or size-assortative mating is known to exist in this study population of *E. coqui* (Lopez and Narins 1991; Townsend and Stewart 1994).

Females initiate courtship with calling males by making physical contact. Males then lead females to the oviposition site where internal fertilization occurs (Townsend et al. 1981). Following oviposition, the males brood the eggs for an extended period of time (Townsend et al. 1984). Egg brooding is a period of relaxed foraging, as indicated by a high frequency of empty stomachs in brooding males; additionally, males lose an average of 20.9% of their body mass while brooding (Townsend 1986). Furthermore, fat body size in males is significantly reduced during the brooding period (Townsend 1986). Of critical importance to female reproductive success is the fact that hatching success is variable and directly related to the quality of paternal care (Townsend 1986, 1989; Townsend and Stewart 1994). This suggests that female mate selection in *Eleutherodactylus coqui* should be directed toward males of relatively superior phentoype, specifically regarding those characteristics that affect paternal quality.

Materials and Methods

Female Choice Trials

Two-choice trials of female preference for call dominant frequency and rate were conducted at the El Verde field station in an arena (2.5 x 2 m) located in a sound-absorbing room. All trials were performed during the active period of *Eleutherodactylus coqui* (between 1900 and 0430 h) and under ambient conditions of temperature and humidity; dim red lighting was used to minimize disturbance to females. In order to accurately record the location of females during trials, a grid was painted on the floor of the arena. Stimuli were delivered with two Uher 4200 tape recorders and broadcast from two miniature loudspeakers covered with black cloth, each located on the ground 1 m from the center of the arena and facing each other.

Stimuli used in female choice trials were generated with a portable animal sound synthesizer (Narins and Capranica 1978) and were representative of the natural range of variation of call dominant frequency and rate documented for the study population in 1985. Stimuli of "standard" calls ("co" dominant frequency = 1425 Hz, "qui" dominant frequency = 2430–2600 Hz, "co qui" rate = 21/min) were characteristic of average males in the population, while stimuli of "low" ("co" dominant frequency = 1330 Hz, "qui" dominant frequency = 2285–2460 Hz, "co qui" rate = 12/min) and "high" ("co" dominant frequency = 1520 Hz, "qui" dominant frequency = 2575–2740 Hz, "co qui" rate = 31/min) calls represented the observed maxima and minima for these call attributes. Duration of the "co" note (98 ms), "qui" note (124 ms), and internote interval (142 ms) were representative of the averages of the study population and held constant for all stimulus types. In all trials only one call attribute, either call rate or call dominant frequency, varied. In frequency preference trials, one speaker broadcast standard calls while the other broadcast calls of either low or high dominant frequency; call rate was held constant at the standard value of 21 calls per minute. In rate preference trials, stimulus pairs were either standard versus high rate or standard versus low rate; the dominant frequency of each call note was held constant at the standard values. In all trials speaker output was equalized to standard intensity levels ("co" = 67 dB SPL, "qui" = 85 dB SPL) and randomized.

Gravid females used in these trials were collected within 24 hours of experimentation from the forest surrounding the study area at El Verde. Each female was allowed a 15-minute acclimation period prior to a trial, during which the female, held in a plastic bag, was placed in the arena and exposed to ambient environmental conditions and the stimuli. The female was then removed from the bag and placed on the dampened floor of the arena so that she was facing neither speaker. The movements and orientation of each test female while exposed to the stimuli were observed for 15 minutes. A "choice" was recorded when the female entered a circle of radius 26.5 cm (5% of the total area of the arena) around either speaker.

Field Observations of Male Calling Effort

Detailed accounts of the calling activity and calling effort (average number of "co qui" calls emitted per minute) during a single night were recorded for 36 males (hereafter referred to as "callers"). After an actively calling male was located, a minimum of 25 "co qui" calls were recorded using a cassette recorder (Sony TC-D5M) and a directional shotgun microphone with windscreen (AKG CE-8, with an SE5E-10 power module) held directly in front of the male at a distance of 1 m; these calls were later spectrally and temporally analyzed using a real-time sonagraph (Kay DSP 5500). Each male was then monitored for a period of four hours or until calling activity ceased, during which the number of "co qui" and "co" notes produced each minute during a 5-min period every 12 minutes was recorded. The number of "qui" notes was not quantified because males do not emit this note in the absence of "co" notes during their advertisement calling behavior. The amount of vegetative cover while calling (using the scheme of Narins and Hurley 1982) was recorded for each of these 36 males, as well as for 68 additional males whose vocalizations were recorded and analyzed during another phase of this study.

Male Phenotype

At the end of the observation period of male calling effort, each male was measured for SVL and body mass, toe-clipped with a unique number, and frozen for later analysis of fat reserves. Twelve males that had acquired a clutch of eggs not more than 24 h before they were located ("brooders") were likewise measured for SVL and body mass, toe-clipped, and frozen.

The fat content of fat bodies and carcasses (body minus the stomach, intestines, and fat bodies) was analyzed for the 36 callers and the 12 brooders defined above using petroleum ether lipid extractions. Following dissection and freeze-drying at –60°C to

TABLE 1. Female preference behavior in *Eleutherodactylus coqui* for call dominant frequency ($n = 51$) and call rate ($n = 29$). In all trials, females were exposed to standard calls versus calls representing one frequency or rate extreme for the population.

Extreme	Number of choices		P
	Standard	Extreme	
Low frequency	19	13	NS
High frequency	11	8	NS
Low rate	10	1	<0.01
High rate	4	14	<0.025

a constant weight, fat bodies and carcasses were soaked in petroleum ether (20:1 solvent to solute ratio) for 24 hours. The dissolved lipids and solvent were then vacuum-filtered, the solvent was evaporated off for 48 hours, and dry carcass and fat body lipids were then weighed.

RESULTS

Female Choice Trials

Overall, females showed a distinct preference for one of the two stimuli in 70% of the two-choice trials that were performed. In such cases females typically moved rather quickly toward the preferred stimulus and often jumped onto the speaker delivering that stimulus. In 30% of the trials, females either did not move away from the center of the arena or they hopped randomly about the arena.

Females showed no preference for either the low or high frequency calls over the standard call for this population ($\chi^2 = 1.13$, d.f. = 1, NS; $\chi^2 = 0.47$, d.f. = 1, NS, respectively) (Table 1). In contrast, females showed a strong preference for calls of relatively higher rate in trials of standard versus low rate ($\chi^2 = 7.36$, d.f. = 1, $P < 0.01$) and standard versus high rate ($\chi^2 = 5.56$, d.f. = 1, $P < 0.025$) (Table 1).

Field Observations of Male Calling Effort

Cumulative sum analysis (Woodward and Goldsmith 1964) indicated that call rate during a single night was not constant for each of the 38 males. One to five significant changes in call rate occurred within a night (see Figure 1 for samples of calling patterns), with the largest set (40%) of males showing only a single significant change in rate. Maximum call rate was attained by each male before 2300 h. Prior to 2300 h, the majority of males (58%) called at a constant rate that was greater than or equal

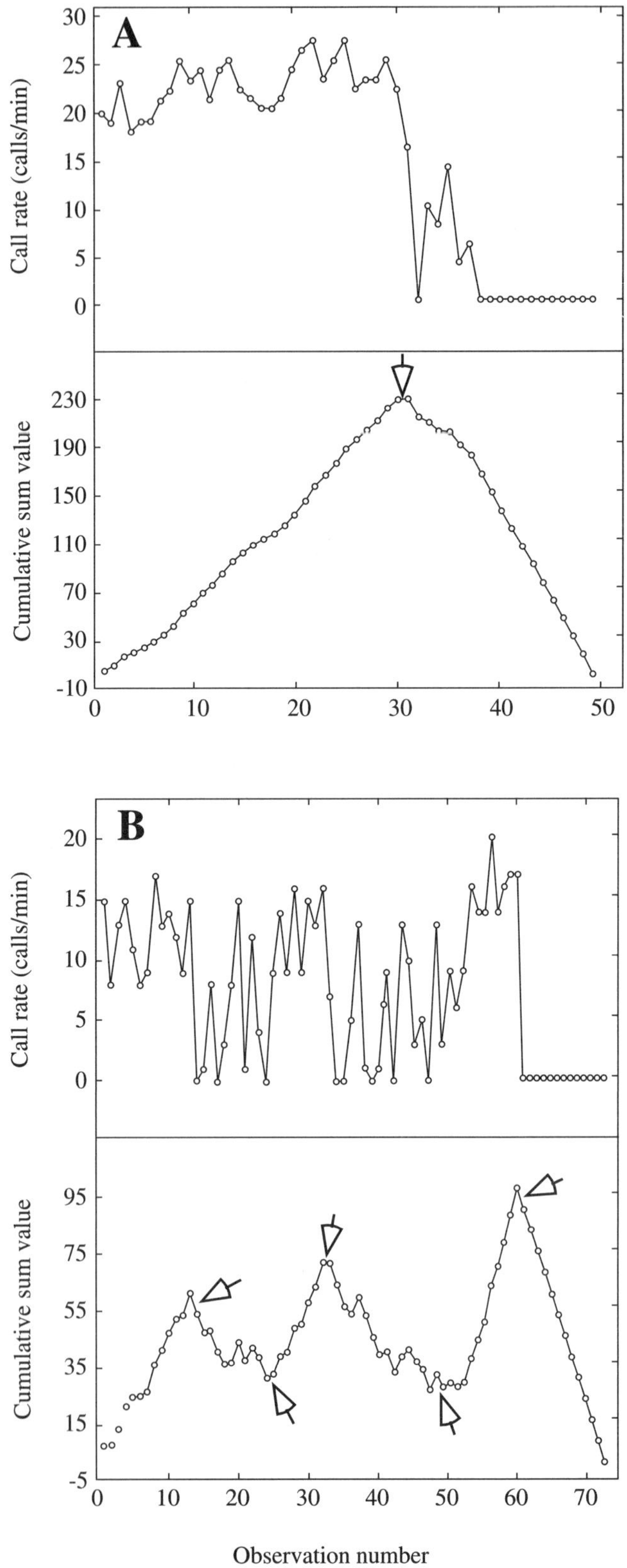

FIGURE 1. Representatives of two of the six calling patterns observed for *Eleutherodactylus coqui*. In each couplet, call rate data are shown in the upper plot and cumulative sum values in the lower plot. Occurrences of significant changes in call rate are indicated with arrows. (A) One significant change in call rate (male number 0204, $n = 49$: overall mean call rate = 14.9, range = 0–27, $s^2 = 105.06$). (B) Five significant changes in call rate (male number 0012, $n = 73$: overall mean call rate = 7.49, range = 0–20, $s^2 = 40.36$). Figure taken, in part, from Lopez and Narins (1991).

to their average rate. Forty-four percent of the males called from sites that had no immediate surrounding vegetation, and only 5% of the males called while completely surrounded by vegetation. Calling males showed a significant bias toward using call sites that were completely exposed ($\chi^2 = 46.38$, d.f. = 4, $P < 0.001$).

Six instances of courtship initiation, all of which occurred prior to 2130 h, were observed while monitoring male calling effort; these males are hereafter referred to as "mated" males. The average call rate, average "co" and "qui" dominant frequency, cover, SVL, and body mass were determined for each mated male (prior to courtship initiation, after which call rate increases and call level decreases) as well as for 43 unmated males that were studied during the same year. Mated and unmated males did not differ significantly with respect to amount of cover or call dominant frequency (Student's t-test, all $P > 0.05$). However, the average call rate of mated males (mean ± SE: 19.4 ± 1.6) was significantly greater than that of unmated males (13.9 ± 0.7) (Mann-Whitney $U = 562.0$, $P < 0.025$). Moreover, the average call rate of each of five of the six mated males was greater than the populational average call rate of 14.8 calls/min (Figure 2).

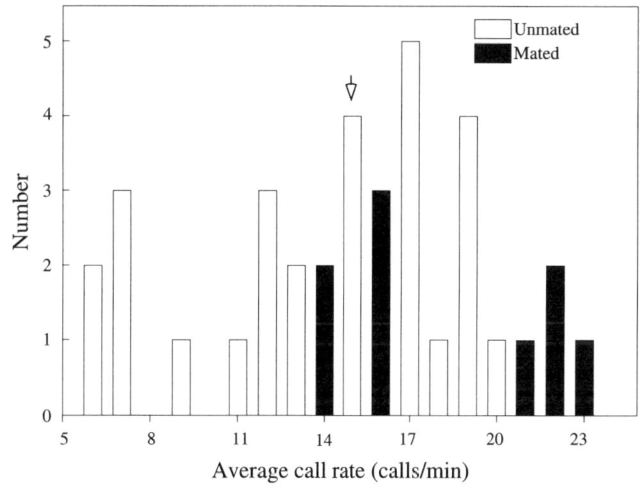

FIGURE 2. Comparison of average call rate for mated ($n = 6$, solid bars) and unmated ($n = 32$, open bars) males of *Eleutherodactylus coqui*. The arrow indicates the overall populational average call rate. Figure taken from Lopez and Narins (1991).

TABLE 2. Calling behavior of mated males of *Eleutherodactylus coqui* during two stages of courtship. "Pre-courtship" is the calling period prior to courtship initiation by the female (mean duration = 78.6 min, range: 16–184 min); "pre-nest courtship" is the calling period during which the male leads the female into the nesting site (mean duration = 15.4 min, range: 7–25 min); "n" represents the number of one-minute observation periods during which calling activity was monitored. Call rate data shown are average values. Significance levels were derived from Mann-Whitney U-tests.

Male number	Pre-courtship		Pre-nest courtship		P
	Call rate	n	Call rate	n	
0024	21.9	26	26.0	8	0.0284
0044	22.7	80	23.4	5	0.0946
0130	21.5	10	24.7	13	0.0881
0203	21.4	25	25.0	3	0.0075
0310	13.6	36	18.9	9	0.0107

Data on male call rate were collected for five of the six instances of courtship initiation. In each of these cases, male call rate increased during movement of the male and female to the nesting site (Table 2). A significant change in call rate was observed for three of the males, with the remaining two approaching significance (Mann-Whitney U-test). After moving an average distance of 124.5 cm (range: 31–343 cm) from the male's calling site into the nesting site, male call rate typically decreased. Males then began to produce very low level calls with a single "co" note and multiple "qui" notes (Figure 3).

Male Phenotype

Mated and unmated males did not differ significantly with respect to SVL or body mass (Student's t-test, all $P > 0.05$). For callers and brooders combined ($n = 48$), the mean amount of carcass fat was 36.45 mg (12.20–73.20 mg, $s^2 = 183.87$), which equals 6.85% of dry carcass mass (2.06–10.84%, $s^2 = 4.44$). Fat bodies were sufficiently large to be detected and weighed in 56% of these males.

For the 36 callers, absolute fat and percent fat (the ratio of carcass fat mass to dry carcass mass) were not significantly associated with call rate (Pearson product-moment correlations, all $P > 0.05$).

Moreover, regressions of average call rate on these two variables were not significant (Figure 4). Callers were not significantly different from brooders with respect to SVL, wet mass, gut mass, carcass dry mass, carcass fat mass, or percent fat (Table 3).

DISCUSSION

Both field and laboratory observations of female choice behavior in *Eleutherodactylus coqui* strongly suggest that the best correlate of male mating success is the rate at which a male calls. Contrary to the majority of anuran species that have been studied (e.g., Doherty and Gerhardt 1984; Robertson 1990), females of *E. coqui* showed no preference for call dominant frequency. Because only a weak association existed between male SVL or mass and "qui" dominant frequency, females would gain little information regarding male body size by assessing spectral qualitites of the advertisement call prior to courtship initiation.

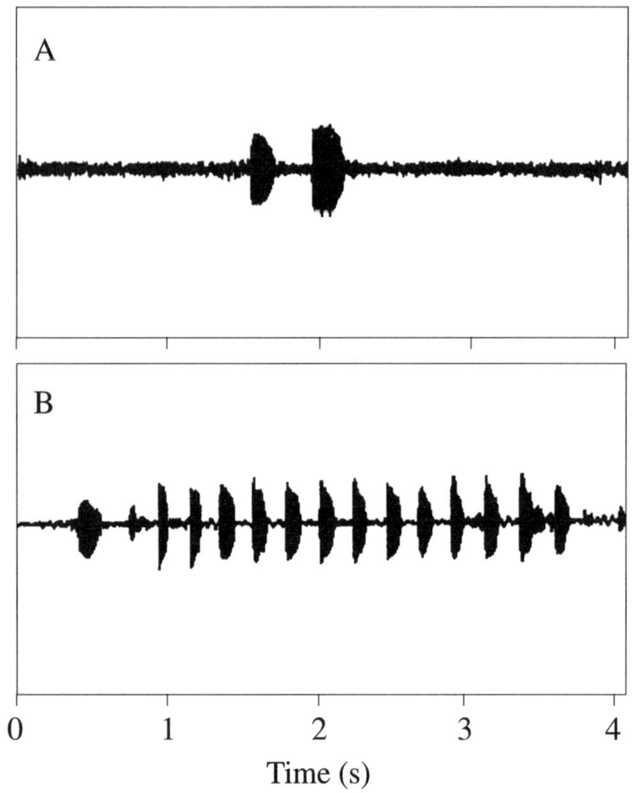

FIGURE 3. Natural advertisement and courtship calls of *Eleutherodactylus coqui*. (A) Oscillogram of a representative "co qui" advertisement call, recorded at 25.0°C ("co" dominant frequency = 1522 Hz, "co" duration = 100.41 ms, internote interval = 138.31 ms; "qui" dominant frequency = 2354 Hz, "qui" duration = 130.67 ms). (B) Oscillogram of a representative courtship call (one "co" note and 14 "qui" notes) produced by a male while calling in the nest with a female. Temperature during recording: 25.0°C. Thirty-nine courtship calls were recorded from this male, with the number of "qui" notes ranging from 1–14 (mean = 4.53). Spectral and temporal features specific to this call are as follows: "co" dominant frequency = 1440 Hz, "co" duration = 93.75 ms; mean "qui" dominant frequency = 1831.4 Hz, range of "qui" dominant frequencies = 1520–2240 Hz, mean "qui" duration = 58.92 ms, range of "qui" note durations = 18.75–68.75 ms, total call duration = 1993.0 ms.

Instead, females of *Eleutherodactylus coqui* showed a significant preference for calls produced at a relatively high rate, even though no significant association exists between male body size and this call attribute. This preference appears to be the norm among those anuran species that have been studied to date (see review by Gerhardt 1988). Although the call rate of males of *E. coqui* is typically highly variable within a single night, with from one to five significant changes in rate, all males reach their maximum call rate prior to 2300 h. This is particularly noteworthy because the timing of male calling effort qualitatively matches the temporal availability of receptive females within a night as predicted by Woolbright (1985). Therefore, based on both two-choice laboratory trials, natural observations of female mate selection, and the predicted availability of receptive females, I predict that male mating success can be maximized by calling relatively rapidly and early in the night.

The fact that males increase their call rate while leading females toward and into the nesting site may be interpreted as further support of female preference for this call attribute. Additionally, the similarity between the courtship call and the aggressive call of males of *Eleutherodactylus coqui* is noteworthy. Although five types of aggressive calls were documented by Stewart and Rand (1992), one type is a multi-note call and is structurally very similar to the courtship call. Males use the aggressive call most commonly at dawn and dusk when they enter or leave, respectively, their retreat sites. High quality retreat sites are known to be a limiting resource for *E. coqui*, and aggressive calls may be used to advertise occupied retreats; additionally, males may

TABLE 3. Comparison of body size, gut mass, and fat content for callers (n = 36) and early brooders (n = 12) of *Eleutherodactylus coqui*. Values represent averages ± 1 SE. No significant differences were found for any variable (Student's t-tests, all $P > 0.05$).

Measurement	Callers	Brooders
SVL (mm)	35.70 ± 0.25	35.90 ± 0.47
Wet mass (g)	2.78 ± 0.06	2.80 ± 0.10
Gut mass (g)	0.24 ± 0.02	0.24 ± 0.35
Carcass dry mass (g)	0.53 ± 0.01	0.53 ± 0.02
Carcass fat mass (g)	0.04 ± 0.00	0.03 ± 0.00
% fat	7.24 ± 0.31	5.68 ± 0.72

use retreats as nesting sites (Stewart and Rand 1992). Therefore, because of the similarity between the courtship and aggressive calls, the courtship call may serve a dual function in this species: for female attraction and as a territorial signal to potential interlopers.

From the female perspective of optimizing lifetime reproductive success, the adaptive significance of the preference for high call rate in *Eleutherodactylus coqui* should be based in the accrual of high quality paternal care. Paternal care is obligatory, males experience significant weight loss and reduction in fat body size during this period of minimal foraging, and hatching success is highly variable and directly related to the quality of paternal care in this species (Townsend 1986, 1989; Townsend and Stewart 1994). Reliable assessment of paternal quality, specifically energy reserves, prior to fertilization should therefore be favored by natural selection. Call rate was not significantly associated with energy reserves, nor were brooders (males that presumably had been calling at a relatively high rate) significantly different from the general population of malcs with rcgard to fat stores. Therefore, female preference for high call rate in *E. coqui* cannot be explained as a mechanism whereby females are accruing high quality paternal care by selecting males with relatively large energy reserves.

Alternatively, female preference for high call rate in *Eleutherodactylus coqui* may be based in the localization of males that are calling in a very complex and noisy acoustic environment. Zelick and

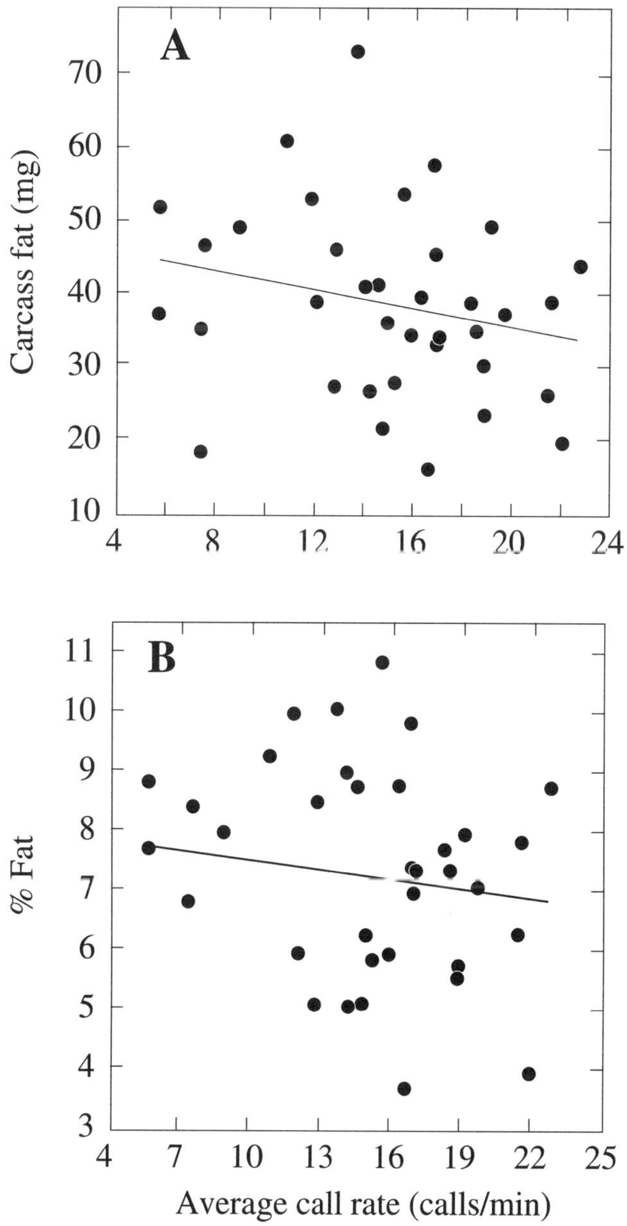

FIGURE 4. (A) The relationship between average call rate (calls/min) within a night and absolute carcass fat (mg) in males of *Eleutherodactylus coqui*. Average call rate was a poor predictor of carcass fat (r^2 = 0.052), and the regression of the two variables was not significant ($y = 48.01–0.63x$, $t = -1.37$, d.f. = 34, $P > 0.05$). (B) The association between average call rate (calls/min) within a night and percent fat in males of *E. coqui*. Average call rate did not explain a significant amount of the variance in fat stored in the carcass (r^2 = 0.02), and the regression of the two variables was not significant ($y = 8.03– 0.05x$, $t = -0.77$, d.f. = 34, $P > 0.05$).

Narins (1983, 1985) have demonstrated that males of *E. coqui* minimize call overlap with neighboring males by adjusting the timing of their calls; presumably this behavior enhances their ability to attract females. Moreover, Narins (1982) and Parker (1983) suggested that males whose calls are of high energy content (rate or intensity) should be more easily detected and located by females. Furthermore, the distribution of call site choice by males in this study was significantly skewed toward locations with no immediate surrounding vegetation and therefore maximal call propagation. Although the potential gain of genetic advantages (*sensu* Zahavi 1975; Hamilton and Zuk 1982) associated with mating with specific males has yet to be investigated in *E. coqui*, ease of male localization by females appears to be the most parsimonious explanation for the strong preference for high call rate in this species.

Acknowledgments

Many thanks to those who provided invaluable field assistance and humor during the course of this research, particularly E. Alkaslassy, A. Estrada-Pinto, J. Howland, L. Wolfenbarger, and N. Simmons. Support during various phases of the research and manuscript preparation was provided by the Organismic Biology Fund, a University Fellowship and a Dissertation Year Fellowship (University of California at Los Angeles), and NIH grant NS19725 (to P. M. Narins).

Literature Cited

Arak, A. 1983. Male-male competition and mate choice in anuran amphibians, p. 181–210. *In* P. Bateson (ed.), Mate Choice. Cambridge Univ. Press, Cambridge.

—. 1988. Female mate selection in the natterjack toad: active choice or passive attraction? Behav. Ecol. Sociobiol. 22: 317–327.

Blair, W. F. 1964. Isolating mechanisms and interspecies interactions in anuran amphibians. Quart. Rev. Biol. 39: 334–344.

Doherty, J. A. and H. C. Gerhardt. 1984. Evolutionary and neurobiological implications of selective phonotaxis in the spring peeper (*Hyla crucifer*). Anim. Behav. 32: 875–881.

Fellers, G. M. 1979. Mate selection in the gray treefrog, *Hyla versicolor*. Copeia 1979: 286–290.

Forester, D. C. and R. Czarnowsky. 1985. Sexual selection in the spring peeper, *Hyla crucifer* (Amphibia, Anura): role of the advertisement call. Behaviour 92: 112–127.

— and W. K. Harrison. 1987. The significance of antiphonal vocalisation by the spring peeper, *Pseudacris crucifer* (Amphibia, Anura). Behaviour 103: 1–15.

Gerhardt, H. C. 1982. Sound pattern recognition in some North American treefrogs (Anura, Hylidae): implications for mate choice. Amer. Zool. 22: 581–595.

—. 1987. Evolutionary and neurobiological implications of selective phonotaxis in the green treefrog, *Hyla cinerea*. Anim. Behav. 35: 1479–1489.

—. 1988. Acoustic properties used in call recognition by frogs and toads, p. 455–483. *In* B. Fritzsch, W. Wilczynski, M. Ryan, T. Hetherington, and W. Walkowiak (eds.), The Evolution of the Amphibian Auditory System. John Wiley, New York.

Hamilton, W. D. and M. Zuk. 1982. Heritable true fitness and bright birds: a role for parasites? Science 218: 384–387.

Klump, G. M. and H. C. Gerhardt. 1987. Use of non-arbitrary criteria in mate choice by female gray treefrogs. Nature, London 326: 286–288.

Littlejohn, M. J. 1977. Long-range acoustic communication in anurans: an integrated and evolutionary approach, p. 263–294. *In* D. H. Taylor and S. I. Guttman (eds.), The Reproductive Biology of Amphibians. Plenum Press, New York.

Loftus-Hills, J. J. and M. J. Littlejohn. 1971. Pulse repetition rate as the basis for mating call discrimination by two sympatric species of *Hyla*. Copeia 1971: 154–156.

Lopez, P. T. 1990. Mate Choice in the Neotropical Anuran, *Eleutherodactylus coqui*. Unpubl. Ph.D. dissertation, Univ. California, Los Angeles.

— and P. M. Narins. 1991. Mate choice in the Neotropical frog, *Eleutherodactylus coqui*. Anim. Behav. 41: 757–772.

Morris, M. R. and S. L. Yoon. 1989. A mechanism for female choice of large males in the treefrog *Hyla chrysoscelis*. Behav. Ecol. Sociobiol. 25: 65–71.

Narins, P. M. 1982. Effects of masking noise on evoked calling in the Puerto Rican coqui (Anura: Leptodactylidae). Jour. Comp. Physiol. 147: 439–446.

— and R. R. Capranica. 1976. Sexual differences in the auditory system of the tree frog *Eleutherodactylus coqui*. Science 192: 378–380.

— and —. 1978. Communicative significance of the two-note call of the treefrog *Eleutherodactylus coqui*. Jour. Comp. Physiol. 127: 1–9.

— and D. D. Hurley. 1982. The relationship between call intensity and function in the Puerto Rican coqui (Anura: Leptodactylidae). Herpetologica 38: 287–295.

Parker, G. A. 1983. Mate quality and mating decisions, p. 141–166. *In* P. Bateson (ed.), Mate Choice. Cambridge

Univ. Press, Cambridge.

Robertson, J. G. M. 1986. Female choice, male strategies and the role of vocalizations in the Australian frog *Uperolia rugosa*. Anim. Behav. 34: 773–784.

—. 1990. Female choice increases fertilization success in the Australian frog, *Uperolia rugosa*. Anim. Behav. 39: 639–645.

Ryan, M. J. 1980*a*. Female mate choice in a Neotropical frog. Science 209: 523–525.

—. 1980*b*. The reproductive behavior of the bullfrog (*Rana catesbeiana*). Copeia 1980: 108–114.

—. 1983. Sexual selection and communication in a Neotropical frog, *Physalaemus pustulosus*. Evolution 37: 261–272.

Schwartz, J. J. 1986. Male calling behavior and female choice in the Neotropical treefrog *Hyla microcephala*. Ethology 73: 116–127.

—. 1987. The importance of spectral and temporal properties in species and call recognition in a Neotropical treefrog with a complex vocal repertoire. Anim. Behav. 35: 340–347.

Stewart, M. M. and A. S. Rand. 1992. Diel variation in the use of aggressive calls by the frog *Eleutherodactylus coqui*. Herpetologica 48: 49–56.

Sullivan, B. K. 1982. Sexual selection in Woodhouse's toad (*Bufo woodhousei*). I. Chorus organization. Anim. Behav. 30: 680–686.

—. 1983. Sexual selection in Woodhouse's toad (*Bufo woodhousei*) II. Female choice. Anim. Behav. 31: 1011–1017.

—. 1985. Sexual selection and mating system variation in anuran amphibians of the Arizona-Sonora desert. Great Basin Nat. 45: 688–696.

Townsend, D. S. 1986. The costs of male parental care and its evolution in a Neotropical frog. Behav. Ecol. Sociobiol. 19: 187–195.

—. 1989. The consequences of microhabitat choice for male reproductive success in a tropical frog (*Eleutherodactylus coqui*). Herpetologica 45: 451–458.

— and M. M. Stewart. 1994. Reproductive ecology of the Puerto Rican frog *Eleutherodactylus coqui*. Jour. Herpetol. 28: 34–40.

—, —, and F. H. Pough. 1984. Male parental care and its adaptive significance in a Neotropical frog. Anim. Behav. 32: 421–431.

—, —, —, and P. F. Brussard. 1981. Internal fertilization in an oviparous frog. Science 212: 469–471.

Whitney, C. L. and J. R. Krebs. 1975. Spacing and call timing in Pacific treefrogs, *Hyla regilla*. Canadian Jour. Zool. 53: 1519–1527.

Woodward, R. H. and P. L. Goldsmith. 1964. Cumulative Sum Tests: Theory and Practice. Oliver and Boyd, Edinburgh.

Woolbright, L. L. 1985. Patterns of nocturnal movement and calling by the tropical frog, *Eleutherodactylus coqui*. Herpetologica 41: 1–9.

Zahavi, A. 1975. Mate selection — a selection for a handicap. Jour. Theor. Biol. 53: 205–214.

Zelick, R. D. and P. M. Narins. 1983. Intensity discrimination and the precision of call timing in two species of Neotropical treefrogs. Jour. Comp. Physiol. 153: 403–412.

— and —. 1985. Characterization of the advertisement call oscillator in the frog *Eleutherodactylus coqui*. Jour. Comp. Physiol. 156: 223–229.

Resumen.—Las características espectrales y temporales del canto de anuncio ("co qui"), al igual que el fenotipo del macho (tamaño del cuerpo, condición física, y reservas de grasa), variaron grandemente en una población de la rana *Eleutherodactylus coqui* de El Verde, Puerto Rico. Se investigó la selección de pareja respondiendo las siguientes dos preguntas: (1) ¿en qué atributo del canto de anuncio del macho se basa la selección de la hembra?, (2) ¿cuál es el significado de esta preferencia en términos de adaptación? Los experimentos de dos alternativas de elección efectuados en el laboratorio, así como observaciones de eventos naturales de cortejo, indicaron que las hembras preferían las razones de canto relativamente altas, no las frecuencias de canto dominantes. No se observó ni apareamiento de acuerdo a tamaño, ni ventaja de los machos grandes en cuanto a éxito de apareamiento. Sin embargo, todos los machos que se aparearon tuvieron una razón de canto superior al promedio de los machos que no se aparearon y al promedio de la población. El ochenta y tres por ciento de los apareamientos observados ocurrió antes de las 2130 horas. No se encontró ninguna asociación significativa entre el fenotipo del macho y el esfuerzo de canto (el promedio de la razón de canto del macho en una noche), y no hubo ninguna diferencia significativa en fenotipo entre machos que obtuvieron recientemente una camada de huevos y la población general de machos cantores. Por lo tanto, las hembras de *E. coqui* no pueden determinar con confiabilidad el fenotipo del macho estudiando la razón de canto. Alternativamente, la preferencia de las hembras de esta especie por una razón de canto alta puede estar basada en la capacidad de localizar a los machos en los ambientes acústicamente complejos del bosque lluvioso de Puerto Rico. El cuarenta y cuatro por ciento de los machos estudiados estaba cantando desde lugares que no estaban cubiertos por vegetación y que tenían una propagación de sonido máxima, presumiblemente aumentando así la probabilidad de que las hembras los pudieran localizar.

Résumé.—Les caractéristiques spectrales et temporales du cri ("co qui"), de même que le phénotype mâle (taille corporelle, condition physique et réserves de graisse), varie considérablement dans la population d'*Eleutherodactylus coqui* de El Verde (Porto Rico). La sélection des partenaires chez *E. coqui* a été étudiée dans le but de répondre aux deux questions suivantes: (1) Sur quel attribut du cri du mâle repose le choix de la femelle? et (2) Quelle est l'intérêt de cette préférence en termes d'adaptation? Les essais à deux choix en laboratoire et les observations de parades nuptiales naturelles indiquent que les femelle préfèrent les cris relativement aigus aux fréquences dominantes. Nous n'avons observé ni d'homogamie en fonction de la taille ni de préférence pour les gros mâles. Toutefois, tous les mâles qui se sont accouplés avaient un registre supérieur à celui de la population moyenne et de la moyenne des mâles qui ne se sont pas accouplés. Quatre-vingt-trois pour cent des accouplements observés ont eu lieu avant 21 h 30. Aucun lien significatif entre le phénotype du mâle et les cris (moyenne des cris poussés par le mâle en une nuit) n'a été observé. De même, aucune différence significative n'a été observée entre le phénotype des mâles qui venaient d'acquérir un ensemble d'oeufs et celui de la population mâle générale. Par conséquent, les femelles *E. coqui* ne peuvent déterminer avec exactitude le phénotype du mâle en évaluant les cris. Par contre, la préférence des femelles pour les cris aigus est peut-être due à la localisation des mâles dans l'environnement acoustique complexe de la forêt tropicale porto-ricaine. Les cris de quarante-quatre pour cent des mâles étudiés provenaient de sites sans couverture végétale qui bénéficiaient d'une propagation maximale du son, ce qui améliorait vraisemblablement les possibilités de localisation par les femelles.

Lopez, P. T. 1996. Mate selection in the Puerto Rican frog, *Eleutherodactylus coqui*, p. 241–250. *In* R. Powell and R. W. Henderson (eds.), Contributions to West Indian Herpetology: A Tribute to Albert Schwartz. Society for the Study of Amphibians and Reptiles, Ithaca (New York). Contributions to Herpetology, volume 12.

BIOLOGY OF THE PUERTO RICAN CAVE-DWELLING FROG, *ELEUTHERODACTYLUS COOKI*, AND SOME RECOMMENDATIONS FOR CONSERVATION

RAFAEL L. JOGLAR,[1,2] PATRICIA A. BURROWES,[2] AND NEFTALÍ RIOS[1]

Abstract.—With the exception of *Eleutherodactylus coqui*, little is known about population fluctuations and reproductive biology of *Eleutherodactylus* in Puerto Rico. *Eleutherodactylus cooki* inhabits a system of caves on the southeastern part of the island. During a 33-month study, we determined that reproductive activity and population density decrease during the winter months, when air temperatures and precipitation decrease. Most egg clutches were in rock depressions; fewer were in rock crevices and on flat surfaces. Most of the clutches were attended by males that attend 1–3 clutches from different females. We consider that *E. cooki* could be at risk and is in need of further monitoring in order to determine its status. An important step should be the protection of its habitat, all of which is privately owned.

INTRODUCTION

ON 24 JANUARY 1932, CHAPMAN GRANT COLLECTED an adult frog that he later described as the most elusive species he discovered. He named the species *Eleutherodactylus cooki* in honor of Melville T. Cook. Grant (1932) eloquently described the habitat and behavior of the species:

> This is the most romantic species on the Island. It inhabits the "guajonales" of the Panduras Mountains, and is known locally as a "guajone" [sic]. A "guajonal" is a place where wild bamboo grows, but here used to designate a mountain gorge tumbled full of granite boulders from bungalow to grand piano size. One can hear from the surface a most melodious note coming from the depths, a sweet liquid pe-pe-pe-pe-pe-pe-pe- resounds from the gloomy caves, echoes, re-echoes and is repeated by other "guajones." "The "guajone" [sic] is only a voice. No one has ever seen one," the natives say. In the day, with a flash light, one can crawl down one, two, three tiers of jumbled boulders to the hidden stream bed in disintegrating granite. The "guajones" sing, but it is impossible to locate them by ear. The flash light and a slender twig will serve to locate and dislodge them from deep cracks less than half an inch wide under husks of exfoliating granite, or from the damp earth where earth and boulder meet. It took me three all day trips to secure one specimen. It was on the third trip that I discovered their hiding places and then several escaped after being pried out. One might as well try to bribe a mountaineer to catch a ghost as a "guajone" [sic]. I tried it; money is no object.

Eleutherodactylus cooki is known in Puerto Rico as "guajón" and/or "demonio" of Puerto Rico (Rivero 1978). This species is a member of the West Indian subset of the *unistrigatus* group of *Eleutherodactylus* (Joglar 1989). This petricolous species (Schwartz and Henderson 1985) inhabits crevices and grottos known as "guajonales" along the Sierra de Panduras region in southeastern Puerto Rico (Figure 1) and west to the San Lorenzo–Patillas road (Schwartz and Thomas 1975; Rivero 1978).

With the exception of the common coqui (*Eleutherodactylus coqui*), little is known about the population fluctuations and reproductive biology of *Eleutherodactylus* in Puerto Rico. Despite the fact that *E. cooki* was reported to be declining in some

[1] Departamento de Biología, Universidad de Puerto Rico, Recinto de Río Piedras, Apartado 23360, San Juan, Puerto Rico 00931–3360, USA.

[2] Natural History Museum, The University of Kansas, Lawrence, Kansas 66045–2454, USA.

areas (Moreno 1991) and it was proposed as a candidate for protection as a threatened species (Drewry 1986), no status surveys have been conducted on this species in recent years. The purposes of our research were to: (1) study several aspects of the biology of *E. cooki*, including distribution, population fluctuations, and reproduction; and (2) make recommendations for the species' conservation and management.

Methods and Materials

Distribution was determined from collection data from the following museums: American Museum of Natural History; Carnegie Museum of Natural History; Museum of Natural History, The University of Kansas; Museum of Comparative Zoology, Harvard University; Museum of Zoology, University of Michigan; Universidad de Puerto Rico, Recinto de Mayagüez; National Museum of Natural History, Washington; and Museo de Biología, Universidad de Puerto Rico, Recinto de Río Piedras. Historic localities were visited and the presence of rock boulders ("guajonales" = caves) and frogs ("guajones") were verified in the field. All localities were visited by day. Acoustic surveys and visual search for individual frogs were performed within the caves.

One locality, km 16.1 on Route 181, Barrio Espino, San Lorenzo, Puerto Rico (18°05'00"S, 65°59'30"W) was chosen for long-term population studies. At this locality, two cave systems were sampled monthly from October 1991 through June 1994, for a total of 33 months. The caves were visited between 1100 and 1400 h. In each of these visits, the same area was sampled and frogs were counted and categorized by age and sex. Because of the complexity of the caves, the area was not measured, and thus, population counts represent only an estimate of the relative abundance of the species at a particular time. These estimates are used as indicators of natural population fluctuations for *Eleutherodactylus cooki*.

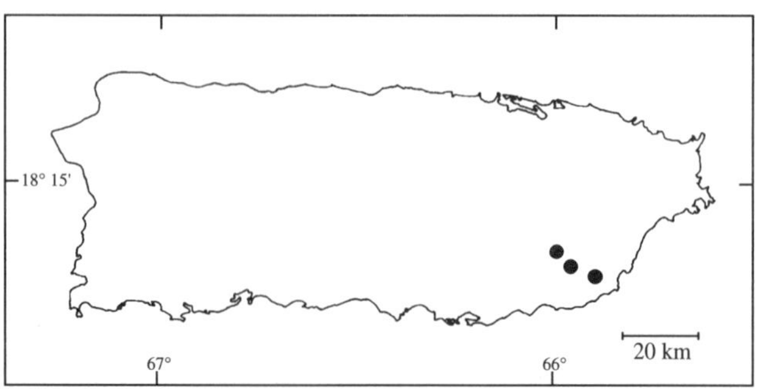

FIGURE 1. Map of locality records in Puerto Rico for *Eleutherodactylus cooki*.

Snout-vent length (SVL) of animals was measured following Woolbright (1989). Estimates of the relative abundance of the species included individuals of all age and sex categories. These categories were defined as: (1) adults (females, > 46 mm; males, > 40 mm); (2) subadults (females, 30–45 mm; males, 30–39 mm); and (3) juveniles (< 29 mm). When egg clutches were found, the following data were taken: (1) microhabitat; (2) presence or absence of guarding frog; (3) sex and coloration of guarding frog; (4) number of eggs per clutch; (5) egg diameter; (6) presence or absence of differences in size and stage of development among eggs within egg mass; and (7), if eggs hatched, number and SVL of hatchlings. Townsend and Stewart's (1985) staging table for *Eleutherodactylus coqui* was used to determine developmental stages of eggs.

Preliminary work on nocturnal behavior was initiated. During five nights caves and their surroundings were surveyed from dusk to dawn. Air temperature and relative humidity were measured in and outside of caves on each visit with a Fisher NIST Certified Hygrometer/Thermometer. Monthly precipitation was recorded from February 1993 to June 1994 using a RainWise Automatic Electronic Rain Gauge. The relationship between activity patterns of *Eleutherodactylus cooki* and climate was determined by calculating Pearson product-moment correlation coefficients. A regression analysis on the abundance of frogs over time was performed to determine if the population was declining.

For comparative purposes we present data on *Eleutherodactylus coqui*, a better-known relative of *E. cooki*. Unless otherwise stated, the data on *E. coqui* are from our ongoing study at 661 m above sea level in El Yunque, Sierra de Luquillo, northeastern Puerto Rico.

TABLE 1. Size and reproductive biology of *Eleutherodactylus cooki* and *E. coqui*. Unless indicated otherwise, data include mean (range); *n*. All measurements are in mm. Data from Townsend et al. (1984) are indicated by an asterisk (*); data from Townsend and Stewart (1994) are marked by a double asterisk (**).

Parameters	*E. cooki*	*E. coqui*
SVL		
Males	43.1 (39.6–47.4); 19	36.4 (30.0–42.5); 435
Females	48.9 (45.0–54.0); 19	46.4 (38.8–57.7); 103
Parental care		
Attended egg clutches (%)	85.0; 58	98.2; 616*
Non-attended egg clutches	15.0	1.8*
Egg clutch microhabitat (%)	Depressions 49.2; 61	Bromeliads 72.1; 165
	Crevices 23.0	Palm fronds 23.0
	Flat surfaces 19.7	Tree cavities 4.8
	Other 8.1	Other 0
Number of eggs per clutch	16 (4–34); 66	25.2 (10–40); 119
Multiple clutches (%)	27.6; 58	5.0; 119*
Double clutches	22.4	4.7*
Triple clutches	5.2	0.3*
Egg diameter	5.2 (4.4–6.3); 54	4.5 (3.5–5.5); 45
SVL (hatchlings)	8.7 (8.1–10.3); 19	6.9 (6.0–7.9); 16
Hatching success (%)	64.9 (61.5–68.4); 2	60.0 (0–100.0); 109**

RESULTS

Size and Coloration

In comparison with all other *Eleutherodactylus* in Puerto Rico, *E. cooki* is characterized by having: (1) largest digital disk diameter to SVL ratio; (2) second largest eye diameter to SVL ratio; and (3) largest tibia length to SVL ratio (Joglar 1981). These characteristics are considered adaptations to rupiculous habits (Joglar 1983) and are shared by other West Indian cave-dwelling species (Hedges 1989). As in most species of *Eleutherodactylus*, females of *E. cooki* are larger than males (Table 1). The coloration of *E. cooki* is dark brown by day or when under white light, but pale when photographed by flash. Only males have yellow throats; in some males, the yellow color extends onto the abdominal and dorsolateral regions.

Distribution

Eleutherodactylus cooki is known only from Southeastern Puerto Rico, in the Cuchilla de Pandura region and between San Lorenzo and Patillas. Altitudinal distribution is 91–303 m (Figure 1; Appendix A).

Calls

The call of *Eleutherodactylus cooki* consists of 3–7 similar notes (Drewry and Rand 1983; R. L. Joglar, personal observations). Vocal activity from dusk to dawn reaches a maximum around 2100 h and then decreases until dawn (Drewry and Rand 1983). However, we observed that vocal activity was more intense by day than by night. Several differences were observed between *E. cooki* and *E. coqui* regarding calling behavior. Males of *E. cooki* often continue to call while attending single or double egg clutches. These calling males are usually separated from their egg clutch by distances of less than 40 cm. The yellow throat in males is associated with the vocal sac and other areas of the body that are inflated when the frog calls; perhaps this coloration is involved with male-male or male-female communication. We never heard sounds produced by females.

Parental Care

Parental care is performed exclusively by males, and may consist of egg attendance, egg brooding, and occasional defense of egg clutches. Most egg clutches observed were attended by males (Table 1). Unattended clutches (15%) probably represent situations in which males temporarily abandoned clutches in response to our presence. One male that abandoned his clutch returned to it in less than 45 min. Two guarding males attacked us and bit our fingers while we were measuring eggs in their clutches. Egg-brooding behavior in *Eleutherodactylus cooki* resembles that which has been described for *E. coqui* (Townsend et al. 1984). Males of *E. cooki* usually brood one or more clutches by covering eggs with their bodies. However, unlike *E. coqui*, in which males usually cover most of the clutch with their bodies (Townsend et al. 1984; R. L. Joglar, personal observations), *E. cooki* can cover only part of the clutch, because the eggs usually are deposited in a single layer.

Reproductive Activity

The number of egg clutches per month was used as an indicator of reproductive activity. Reproductive activity in *Eleutherodactylus cooki* was significantly correlated with precipitation ($r = 0.7021$; $n = 17$) and air temperature ($r = 0.5657$; $n = 24$). We observed a distinct seasonal pattern. Reproductive activity decreased in winter when precipitation and air temperature also decreased (Figure 2). No egg clutches were found during January through March 1993, which were the driest and coldest months. An atypical, long dry period extended from November 1993 through May 1994; no egg clutches were found during that time.

Most egg clutches (72.1%) were in protected microhabitats (Table 1). Shallow depressions and small crevices in the rocks probably reduce evaporative water loss and the possibility of predation of eggs and guarding males. *Eleutherodactylus cooki* may prefer these microhabitats to flat rock surfaces (19.7%).

More than one fourth (27.6%) of the egg masses was considered multiple egg clutches (Table 1). Obvious differences in sizes and stages of development of eggs permitted the identification of one to three distinct clutches among egg masses. Most multiple clutches (68.7%) were contiguous egg masses. Five of the 16 multiple clutches were separated by 1.5–85.0 mm ($\bar{x} = 40.3$ mm).

Two egg clutches were observed both before and after they hatched. One had 19 recently laid eggs and produced 13 hatchlings 22 days later. The other clutch had 13 eggs and produced eight hatchlings 29 days later. Percent hatching success (initial number of eggs per number of eggs hatched x 100) was similar in both clutches (61.5% and 68.4%) (Table 1). From these observations, we estimate that developmental time (time between oviposition and hatching) in *Eleutherodactylus cooki* is 20–29 days. In all observed clutches, hatchlings remained together as a group in the nest for several days, then they dispersed. This behavior resembles that of *E. coqui* (Townsend et al. 1981, 1984; R. L. Joglar, personal observations).

Four hatchlings from two clutches were captured in order to observe color and pattern. A dorsal pattern of dark brown inverted parentheses [)(] was present in all four hatchlings. Two had supratympanic lines, pale spots on a dark background in the dorsolateral region, and bars on fore- and hindlimbs.

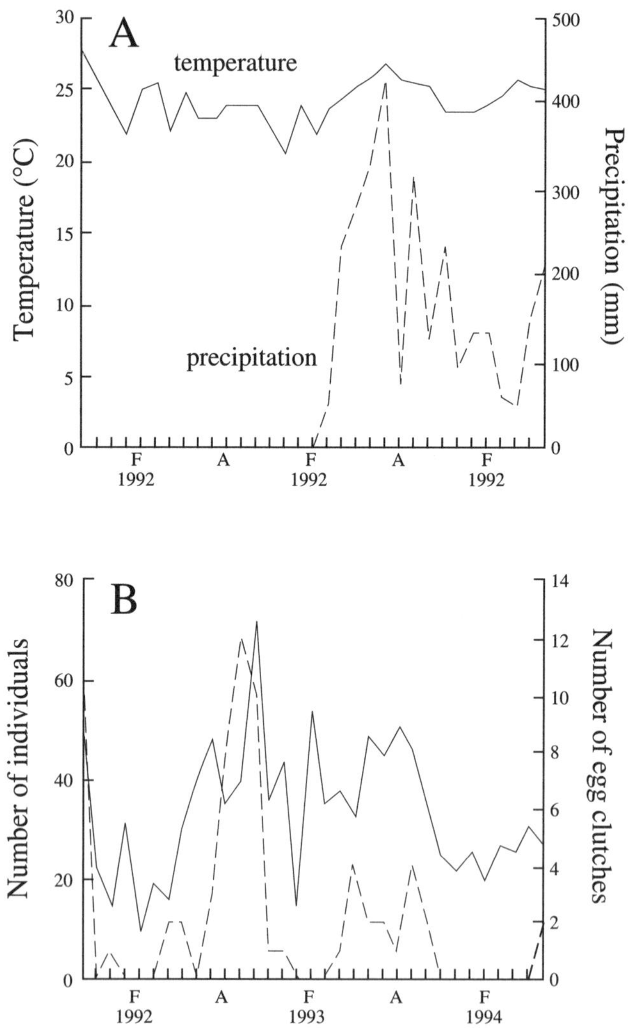

FIGURE 2. A. Air temperature inside of caves (solid line) and precipitation (dashed line) at study site. B. Population fluctuations (solid line) and number of egg clutches per month (dashed line) of *Eleutherodactylus cooki*.

Population Fluctuations

During our study, population fluctuations of *Eleutherodactylus cooki* were associated with precipitation and air temperature (Figure 2). Relative abundance decreased during winter months when precipitation and air temperature decreased. We did not find a significant linear relationship between number of individuals and time ($y = 33.5 + 0.015\,x$; $r^2 = 0$; P = 0.953).

Diel Activity

Diurnal activity of *Eleutherodactylus cooki* occurs only inside caves. Areas of the caves receive different amounts of sunlight; no individuals were observed in the illuminated areas of caves. By day we observed males: (1) calling; (2) brooding eggs; (3) attending egg clutches without brooding eggs; and (4) calling while attending egg clutches. Because males do not call outside of caves, we assume amplexus takes place during the day inside caves. Diurnal activity within caves is not limited to rock substrate. During dry conditions, individuals commonly were in contact with soil and in rock crevices, some of which contained soil. Individuals also were observed on roots and vines inside caves.

We observed that many *Eleutherodactylus cooki* leave the caves at dusk and return before dawn. Males attending egg clutches remain within the caves. Females are seen more easily at dusk while leaving caves or at night outside of caves than inside caves during the day. The following additional observations were made at night: (1) individuals moving at least 20.7 m away from the cave entrance; (2) individuals perching outside of caves; (3) individuals feeding outside of caves; (4) males calling inside caves (no males were heard calling outside caves); and (5), on dry nights, individuals adopting a water-conserving posture outside of caves.

Parasites

Examination of preserved and live specimens revealed that these frogs are parasitized by the tick *Ornithodoros talaje*, of which 1–13 ($\bar{x} = 7.0$) were found on a sample of five live males. The effect of this parasite on *Eleutherodactylus cooki* is not known.

Discussion

Sexual Dimorphism

If *Eleutherodactylus karlschmidti* is, indeed, extinct (as suggested by Joglar and Burrowes, this volume), *E. cooki* is the largest species of the genus in Puerto Rico (Joglar 1981). However, sexual dimorphism in size is less obvious in *E. cooki* than in *E. coqui* (Table 1). Snout-vent lengths of female *E. cooki* are 13.5% greater than those of males. In contrast, *E. coqui* females from the population that we are studying currently are 27.5% larger than males, and Woolbright (1989) found that in a different locality they can be 29% larger than males. Sexual size dimorphism in *E. coqui* may be maintained by energetic constraints associated with male reproductive behavior that reduce the growth of males relative to females (Woolbright 1989). Thus, following Woolbright's hypothesis, one can infer that, if *E. cooki* males are not much smaller than females, they must have lower energy expenditure associated with mating and parental care than do males of *E. coqui*. Environmental and ecological conditions provided by caves (lower risk of predation, less evaporative water loss, and possibly better acoustics for mating calls) could decrease parental energy expenditure in *E. cooki* and allow the species to allocate more resources to growth. *Eleutherodactylus cooki* may exhibit another form of sexual dimorphism that has not been reported for other species of *Eleutherodactylus* in Puerto Rico. Yellow throats occur in calling and attending males but are absent in gravid females; thus, *E. cooki* seems to be the only species of *Eleutherodactylus* in Puerto Rico that exhibits sexual dimorphism in color.

Multiple Clutches

Males of *Eleutherodactylus coqui* markedly reduce their calling activity while attending egg clutches (Townsend et al. 1984; Townsend and Stewart 1994; R. L. Joglar and P. A. Burrowes, personal observations); in doing so, they reduce their opportunity to mate. By continuing to call while attending egg clutches, males of *E. cooki* reduce paternal investment, i.e., the extent to which parental care of individual offspring reduces parents' residual reproductive value (Clutton-Brock 1991), and possibly mate more often. This could explain the difference in numbers of multiple clutches in these two species — 4.8 times more double clutches and 19 times more triple clutches in *E. cooki* than in *E. coqui*.

Different Females for Multiple Clutches?
According to Rivero (1978), egg clutches of *Eleutherodactylus cooki* contain up to 59 eggs, probably from several females. Based on our data, a 59-egg clutch is a double, or possibly even a triple clutch. We agree with Rivero's assumption that these eggs probably were deposited by different females. Female *E. coqui* were estimated to produce a maximum of six clutches per year or one clutch every two months (Townsend and Stewart 1994). Because these two species are thought to be closely related (Schwartz 1969; Joglar 1989) and on the basis of other reproductive similarities documented in this study, we suggest that females of *E. cooki* could have a similar interclutch interval of approximately 60 days. If the interclutch interval is about 60 days and the developmental period is 22–29 days, finding eggs or hatchlings of a female's first oviposition with eggs or hatchlings of her second oviposition would be impossible. Most observed multiple clutches had developmental differences corresponding to a 7–14-day oviposition interval. These observations may indicate that male *E. cooki* contribute to the genetic makeup of double and triple clutches by mating with different females. Thus, male *E. cooki* are reducing paternal investment and, at the same time, maximizing the number of offspring per guarding effort.

Why "Guajones" Leave Caves at Night
Our limited observations on nocturnal activity suggest two possible explanations, both of which are associated with activities common to most anurans: foraging and rehydrating. At least one additional cave-dwelling species, *Eleutherodactylus gaigeae* in South America, also leaves caves at night to forage (J. D. Lynch, personal communication). *Eleutherodactylus cooki* probably is forced to leave caves at night to forage because of the scarcity of food in the caves.

Terrestrial amphibians must rehydrate in order to compensate for evaporative water loss (Duellman and Trueb 1986 [1985]). Opportunities for *Eleutherodactylus cooki* to rehydrate are probably better outside of the caves at night, given that dew and rain, which provide moisture for rehydration, are not available in the caves.

Is the Species Declining?
Our data suggest that the species is still present in all of its historically known localities. We lack data on the abundance of this species at our study site prior to October 1991, and we do not know the status of other populations in other localities. The population we studied was stable during the study period. Because part of the geographical range is an area being developed (Moreno 1991) and eastern Puerto Rico is one of the fastest growing and developing areas on the island, protecting the species before a detrimental impact is noted would be wise. We consider that the species could be at risk because its distribution is limited to a small geographical area and all known localities are on privately owned land. Moreover, *Eleutherodactylus cooki* is a habitat specialist, as are other Puerto Rican species that are declining or have become extinct. Conservation efforts should focus on research and the acquisition and protection of privately owned land within the range of *E. cooki*. Other populations of *E. cooki* should be studied in order to determine the status of the species throughout its range. Because *E. cooki* is a cave-dweller, with a naturally fragmented habitat and restricted distribution within eastern Puerto Rico, studies of genetic variability within and among populations should be emphasized. Knowledge gained from such study will provide a quantitative measure of its vulnerability to extinction and may suggest management strategies to ensure the survival of the species. For a complete list of recommendations for the protection of *E. cooki*, see Appendix B.

Acknowledgments

Many persons and institutions assisted generously in various aspects of this research. For providing information on museum specimens we are grateful to C. W. Myers, American Museum of Natural History; the late C. J. McCoy, Carnegie Museum of Natural History; W. E. Duellman, Natural History Museum, University of Kansas; J. Rosado, Museum of Comparative Zoology, Harvard University; A. G. Kluge, Museum of Zoology, University of Michigan; J. A. Rivero, Universidad de Puerto Rico, Recinto de Mayagüez; W. R. Heyer, National Museum of Natural History; and M. Vélez, Museo de Biología, Universidad de Puerto Rico, Recinto de Río Piedras and Yale University Peabody Museum of Natural History. We are grateful to José García, Fernando Rivera, Carlos Morales, David Bermúdez, Janice Alemán, Sandra Santiago, María Maymí, Alex Puig, Javier Caldera, José Ruíz, Pedro Ayala, and Ingrid Cruz for assistance in the field. Special thanks go to Pablo Rivera and Marcela Muñoz for allowing us to work on their property. We thank William E. Duellman, Linda Trueb, and two anonymous reviewers for comments on the manuscript, and Helio da Silva for figures. This research was funded by United States Fish and Wildlife Service through a collaborative agreement with La Fundación Puertorriqueña de Conservación (from 1991–1992); Fondos Institucionales Para la Investigación, Decanato de Estudios Graduados e Investigación, Universidad de Puerto Rico, Recinto de Río Piedras (from 1991–1994); and supported by the Departamento de Biología, Universidad de Puerto Rico–Recinto Universitario de Río Piedras, who provided release time and logistics. In addition, Proyecto Coqui, Inc. provided funds for one student's summer salary.

LITERATURE CITED

Clutton-Brock, T. H. 1991. The Evolution of Parental Care. Princeton Univ. Press, Princeton, New Jersey.

Drewry, G. E. 1986. Golden Coqui recovery survey and brief status evaluation of five others Puerto Rican *Eleutherodactylus* species. Rept. Carib. Field Off. United States Fish Wildl. Serv., Boquerón, Puerto Rico.

— and A. S. Rand. 1983. Characteristics of an acoustic community: Puerto Rican frogs of the genus *Eleutherodactylus*. Copeia 1983: 941–953.

Duellman, W. E. and L. Trueb. 1986 (1985). Biology of Amphibians. McGraw Hill, New York.

Grant, C. 1932. A new frog from Puerto Rico. Jour. Dept. Agric. Puerto Rico 16: 145–148.

Hedges, S. B. 1989. Evolution and biogeography of the West Indian frogs of the genus *Eleutherodactylus*: slow-evolving loci and the major groups, p. 305–370. *In* C. A. Woods (ed.), Biogeography of the West Indies: Past, Present and Future. Sandhill Crane Press, Gainesville, Florida.

Joglar, R. L. 1981. Estudio fenético del género *Eleutherodactylus* en Puerto Rico. Unpubl. M. S. Thesis, Univ. Puerto Rico, Recinto de Mayagüez.

—. 1983. Estudio fenético del género *Eleutherodactylus* en Puerto Rico. Carib. Jour. Sci. 19: 33–40.

—. 1989. Phylogenetic relationships of the West Indian frogs of the genus *Eleutherodactylus*: a morphological analysis, p. 371–408. *In* C. A. Woods (ed.), Biogeography of the West Indies: Past, Present, and Future. Sandhill Crane Press, Gainesville, Florida.

Moreno, J. A. (ed). 1991. Status y Distribución de los Reptiles y Anfibios de la Región de Puerto Rico. Publ. Cien. Misc. no. 1. Depto. Rec. Nat. Puerto Rico, San Juan.

Rivero, J. A. 1978. Los Anfibios y Reptiles de Puerto Rico. Ed. Univ., Univ. Puerto Rico, San Juan.

Schwartz, A. 1969. The Antillean *Eleutherodactylus* of the *auriculatus* group. Stud. Fauna Curaçao Carib. Isl. 114: 99–115.

— and R. Thomas. 1975. A Check-list of West Indian Amphibians and Reptiles. Carnegie Mus. Nat. Hist. Spec. Publ. no. 1. Pittsburgh, Pennsylvania.

— and R. W. Henderson. 1985. A Guide to the Identification of the Amphibians and Reptiles of the West Indies Exclusive of Hispaniola. Milwaukee Public Mus., Milwaukee, Wisconsin.

— and —. 1991. Amphibians and Reptiles of the West Indies: Descriptions, Distributions, and Natural History. Univ. Florida Press, Gainesville, Florida.

Stewart, M. and A. S. Rand. 1991. Vocalization and the defense retreat sites by male and females frogs, *Eleutherodactylus coqui*. Copeia 1991: 1013–1024.

Townsend, D. S. and M. M. Stewart. 1985. Direct development in *Eleutherodactylus coqui* (Anura: Leptodactylidae): a staging table. Copeia 1985: 423–436.

— and —. 1994. Reproductive ecology of the Puerto Rican frog *Eleutherodactylus coqui*. Jour. Herpetol. 28: 34–40.

—, —, and F. H. Pough. 1984. Male care and its adaptive significance in a Neotropical frog. Anim. Behav. 32: 421–431.

—, —, —, and P. Brussard. 1981. Internal fertilization in a oviparous frog. Science 212: 469–471.

Woolbright, L. L. 1989. Sexual dimorphism in *Eleutherodactylus coqui*: selection pressures and growth rates. Herpetologica 45: 68–74.

APPENDIX A. Locality records for *Eleutherodactylus cooki*.

SIERRA DE PANDURA REGION: Yabucoa: Pandura Mountains, southeastern Puerto Rico; under vines growing over granite boulders, elev. 305 m; Bo. Quebradillas, 245 m; Bo. Quebradillas, Finca Sr. Carrasquillo; Bo. Quebradillas, at the ground entrance beside the third business along that road; 3.2 km SW of Yabucoa; 7.6 km WSW of Yabucoa; Bo. Quebradillas, Rte. 900, 6.9 km E of Yabucoa, 220 m; Rte. 3, 2.9 km SW of Yabucoa. SAN LORENZO PATILLAS REGION: Rte. 181, km 14.5.

APPENDIX B. Recommendations for the protection of *Eleutherodactylus cooki*.

1. To determine the status of *Eleutherodactylus cooki*, its abundance, genetic variability, and gene flow should be studied.
2. Because the following habitats are privately owned, they should be protected by conservation easements, cooperative agreements with landlords, or acquired: land in the vicinity of Barrio Espino (Rte. 181, from km 14–18) between San Lorenzo and Patillas, and in Barrio Quebradillas (vicinity of Rte. 900) in Yabucoa.
3. Stop all development management practices that could have a negative impact (e.g., logging, construction, etc.) in the habitat of the species.
4. Design and implement an environmental awareness program to incorporate the local community into the conservation effort.
5. Undertake long-term studies in Puerto Rico on acid rain, soil acidity, air and water pollution, and global climatic change, and make efforts to show how these factors might be associated with declining amphibian populations.
6. Coordinate conservation efforts among protection agencies.

Resumen.—Informamos sobre varios aspectos de la biología del Guajón o Demonio de Puerto Rico (*Eleutherodactylus cooki*). Con la excepción de *E. coqui*, es muy poco lo que conocemos acerca de las fluctuaciones poblacionales y biología reproductiva del género *Eleutherodactylus* en Puerto Rico. *Eleutherodactylus cooki* habita un sistema de cuevas en el sureste de la Isla. Por medio de un estudio de 33 meses determinamos que su actividad reproductiva y las densidades poblacionales disminuyen durante los meses de invierno, cuando la temperatura del aire y la precipitación también disminuyen. Todas las camadas de huevos fueron encontradas adheridas en la roca, la mayoría de ellas en depresiones y en menor cantidad, en grietas y superficies lisas. La especie exhibe cuidado parental a cargo de los machos, los cuales cuidan de una a tres camadas que han sido depositadas por distintas hembras. Consideramos que la especie podría estar en riesgo y que es necesario continuar estudios para determinar su estado poblacional. Un paso importante, debe ser la protección de su hábitat el cual al día de hoy, se encuentra exclusivamente en terrenos privados.

Résumé.—À l'exception d'*Eleutherodactylus cooki*, on sait peu de choses sur les fluctuations démographiques et la biologie reproductive d'*Eleutherodactylus*, à Porto Rico. *Eleutherodactylus cooki* colonise un réseau de grottes dans la région sud-est de l'île. Dans le cadre d'une étude de 33 mois, nous avons observé que l'activité reproductive et la densité démographique diminuaient pendant l'hiver lorsque la température de l'air et les précipitations baissaient. La plupart des oeufs étaient pondus dans des dépressions rocheuses; il est rare de trouver des oeufs dans les crevasses ou sur des surfaces planes. La plupart des oeufs sont surveillés par les mâles qui s'occupent de 1 à 3 pontes de différentes femelles. Nous pensons que *E. cooki* est à risque et devrait faire l'objet d'une surveillance plus étroite afin de déterminer son statut. Le premier effort en ce sens vise la protection des habitats qu'elle colonise, lesquels appartiennent tous à des particuliers.

Joglar, R. L., P. A. Burrowes, and N. Rios. 1996. Biology of the Puerto Rican cave-dwelling frog, *Eleutherodactylus cooki*, and some recommendations for its conservation, p. 251–258. *In* R. Powell and R. W. Henderson (eds.), Contributions to West Indian Herpetology: A Tribute to Albert Schwartz. Society for the Study of Amphibians and Reptiles, Ithaca (New York). Contributions to Herpetology, volume 12.

Larval Habitat, Ecology, and Parental Investment of *Osteopilus brunneus* (Hylidae)

Rebecca L. Thompson[1]

Abstract.—The Jamaican brown frog, *Osteopilus brunneus*, provides eggs to its young as a larval food resource. Whereas trophic (inviable) egg provisioning is well described in some dendrobatid frogs and suggested in several other treehole- and bromeliad-breeding species, *O. brunneus* differs by providing fertilized eggs. Females of *O. brunneus* deposit either fertilized or unfertilized eggs depending on larval development. Fertilized eggs are initially deposited into nests while larvae are still too small to consume a full egg complement. Eggs that escape cannibalism develop and join the larval brood. A switch is eventually made to providing unfertilized eggs once larvae are advanced enough to consume an entire egg complement. This strategy can be viewed as an alternative means of dealing with the harsh conditions common to small water catchments. Through continued fertilized egg depositions, *O. brunneus* is able to counter variable, but always high, embryo mortality, by producing additional offspring. This seemingly unique facultative strategy allows females to avoid the irreversible tradeoff of offspring quantity for quality inherent in trophic egg production. This chapter describes the habitat and natural history of these bromeliad-dwelling larvae and how these factors may be related to the evolution of the parental care strategy, both in regard to the use of fertilized eggs and to the timing of the switch to unfertilized egg use.

Introduction

Osteopilus brunneus lives and breeds in arboreal and terrestrial tank bromeliads of *Hohenbergia* spp. and *Aechmea paniculigera* in Jamaica. These frogs range throughout the island wherever these bromeliads abound, i.e., all but the most arid regions and at extreme altitudes, as in the John Crow and Blue mountains (Lynn and Grant 1940). Schwartz and Henderson (1991) noted that these frogs provide their young with unfertilized eggs as a larval food resource (as cited from Lannoo et al. 1987). More recent work (Thompson 1992; unpublished data), however, suggests that the situation surrounding these tadpoles is more complex than had been realized. Life in the bromeliads for a Jamaican tadpole is a particularly harsh existence, with habitat extremes that must have required similarly extreme adaptations for survival.

Osteopilus brunneus is the most common and wide-ranging of the five species of hylid frogs in Jamaica and can be distinguished by its loud, distinctive call and relatively large size (average snout-vent length for males = 45 mm and for females = 65 mm). These treefrogs are adapted to a land in which the water table was periodically inundated by rising Pleistocene sea levels (Pregill and Olson 1981, Liebherr 1988). This, compounded by Jamaica's highly porous limestone base, led to a paucity of available fresh water for more common modes of anuran reproduction. As a result, all of the Jamaican hylids and many other lotic fauna moved their breeding (with aquatic eggs and larvae) to the water-filled leaf-axils of bromeliads. The shift in anuran breeding sites is believed to have led to the wide array of morphological and behavioral modifications now observed in both *O. brunneus* larvae and adults (Dunn 1926; Laessle 1961; Jones 1967; Lannoo et al. 1987).

The environment within a bromeliad reservoir has several potential stressors for anuran larvae. Ac-

[1] Department of Zoology, University of Texas at Austin, Austin, Texas 78712, USA.

cording to previous studies of Jamaican bromeliads (Laessle 1961; Jones 1967), these small stagnant pools can be acidic (pH range = 4.0–6.5), are low in dissolved oxygen (0.03–2.70 ppm), and are almost always extremely food-limited. To deal with the low oxygen levels larvae use aerial breathing through early lung development and greatly reduced gill filaments, and with an increased resistance to hypoxia (Lannoo et al. 1987). Because of the lack of more typical larval food within the reservoirs, *Osteopilus brunneus* larvae feed almost exclusively on conspecific eggs deposited within their reservoir. This sporadic, yet renewable, food supply has led to morphological adaptations in the tadpole mouth, buccal cavity, and digestive system (Dunn 1926; Jones 1967; Lannoo et al. 1987), allowing for more efficient utilization of this atypical energy source. Egg consumption is facilitated by a wide mouth and buccal cavity, and enlarged buccal pump muscles. The liver is enlarged, likely aiding digestion of the fatty food resource, and the stomach is enlarged and expansible to store eggs for later digestion in the intestine. Larval mechanisms for dealing with acidity have not yet been investigated.

The conspecific eggs, which act as the predominant, if not sole, larval food resource, are actively provided to the broods. Females feed their young by depositing additional eggs into the nests every 4.4 ± 1.7 days until all young reach metamorphosis or die (R. L. Thompson 1992, unpublished data). Initially, these eggs are fertilized and any eggs not eaten can hatch and join other tadpoles in the clutch. Thus a range of age classes is found within a given bromeliad. At some point, usually soon after the first larvae have begun metamorphosis and always after at least some larvae have reached the stage of visible limb bud development, deposition of fertilized eggs ceases and the larvae then feed on unfertilized eggs for the remainder of their larval development.

Materials and Methods

I studied *Osteopilus brunneus* during two field seasons (19 May to 10 August 1990 and 15 May to 12 August 1991) on and around the Campbell Farm in Windsor, Trelawney, Jamaica (18°06'N, 77°38'W, elevation 120–350 m). This site lies at the northern periphery of the karst-limestone region known as the Cockpit Country. The typical annual rainfall is between 1500 and 2000 mm, with a bimodal mean monthly rainfall distribution peaking in May and October and an extended dry period from January to April (Asprey and Robbins 1953).

The Cockpit Country is described as a wet limestone forest with a canopy at approximately 20 m (Asprey and Robbins 1953). Bromeliads, orchids, and mosses cover the trees and, in more open-forest areas, bromeliad species may constitute the dominant terrestrial vegetation as well. A tributary to the Martha Brae River emerges in the area, but due to the porous nature of the substrate (bauxitic clay soil atop a limestone base) no standing water exists except for that found in axils of tank-bromeliads. All five Jamaican hylid species live and breed in these bromeliads in and around the study site (R. I. Crombie, personal communication; R. L. Thompson, personal observations).

I conducted the 1990 field research primarily as a pilot study to gather general information on the habitat and natural history of *Osteopilus brunneus*. During the 1991 season, I studied the reproductive behavior and larval natural history of *O. brunneus* in more detail. The data presented are from the 1991 field season unless noted otherwise. Of the two genera of tank-bromeliads available at the study site, I found many more nest sites in *Hohenbergia* (n = 173) than in *Aechmea* (n = 4). All information presented here was collected from nest sites in *Hohenbergia*.

All anuran larval stages presented follow the staging table of Gosner (1960). Due to the continuous nature of larval development and some inaccuracies in field staging abilities, larval development levels may be given as hyphenated stages. I have defined success of a larva by attainment of Stage 41 in development. This stage is characterized as that immediately prior to forelimb emergence and, in the case of *Osteopilus brunneus*, was typically the final observation possible for that particular individual. Once forelimbs emerge, a tadpole immediately leaves the nest reservoir.

Nest Site Monitoring

I searched bromeliads for frogs, tadpoles, and other associated macrofauna by probing plant axils and removing aquatic organisms (pumping) with a 50-cc syringe equipped with a 25-cm section of clear rubber tubing. Due to the possibility of killing embryos and hatchlings, reservoirs with fertilized eggs afloat or larvae less than nine days of age (Stage 25, about 15 mm in length) were not disturbed by pumping and were only visually inspected. I monitored previously-identified nest sites and searched for new *Osteopilus brunneus* nests every day in the field (64 days in 1990 and 77 days in 1991). The number of plants searched and the amount of time spent checking bromeliads each day varied greatly and depended on time and plants available. Plants with *O. brunneus* young, if not collected, were monitored at one- to four-day intervals for the remainder of the season to determine egg deposition frequencies and larval development rates.

Nest site monitoring provided most of the information on tadpole feeding and development. Twelve nest sites were monitored in 1990 and 127 were monitored in 1991. The larvae have translucent venters which allow for direct observation of consumed eggs. During the first two weeks of the 1991 season I monitored clutches at one-day intervals to discern how axil waters and tadpole stomach contents changed relative to the time since the last egg deposition. In this way, I was able to determine when eggs were deposited within a brood axil during monitoring intervals. Stomach distention and the number of visible eggs decreased with time. Larger tadpoles consumed a greater number of eggs and retained eggs in their stomachs longer than smaller tadpoles. Immediately following egg deposition, axil waters became highly viscous, presumably because of the gelatinous nature of associated egg jellies. The viscosity of axil waters decreased with time and provided a valuable clue in determining whether eggs had been deposited between monitoring sessions.

I determined fertilization by egg color and size. *Osteopilus brunneus* ova are black and less than 1 mm in diameter. Unfertilized eggs turn pale brown and swell to 1 mm or more within 12 hrs (R. L. Thompson, personal observation). These eggs could usually be distinguished easily, often even within larval stomachs. Eggs of questionable color and size were sampled and checked for fertilization by inspection with a hand lens and by allowing them the opportunity to develop. However, fertilization was not always determined, particularly if the only observable eggs were those consumed by larvae.

Clutch size estimates proved difficult to obtain and usually only rough estimates of clutch size were determined. New clutches (up to about nine days of age, stages 25 to 26) were prone to failure if disturbed, and by the 20th day of development (about Stage 30), the tadpoles began to leave the central reservoir and move into adjacent axils. These axils usually could not be checked for young without damaging the bromeliads and affecting their abilities to hold water.

In addition to the larval feeding and development information, I also recorded certain plant characteristics of these monitored bromeliads, including plant diameter (the maximum distance between leaf tips), central axil diameter and water depth, plant height (the distance between the ground and the plant's base), sun exposure (the amount of direct sunlight available to the bromeliad, categorized into four ordinal ranks from little to full sun), and central-axil water volume (estimated by calculating the volume as a cone from the measured axil diameter and water depth). The pH of 24 central reservoirs was measured over a five-day period in June with a pocket pH-meter (Fisher; further measurements were not available due to failure of the meter).

Bromeliad Censusing

In order to determine *Osteopilus brunneus* nesting microhabitat requirements, in 1991 I censused 87 tank-bromeliads (22 *Aechmea paniculigera* and 65 *Hohenbergia* spp.) every three days. By determining the initial characteristics of the plants eventually used as nesting sites, as compared to those plants that were unused by *O. brunneus*, I identified bromeliad characteristics associated with nest site choice. Information recorded included plant diameter, central axil diameter, and water depth, plant height, sun exposure, and leaf litter density (the density of forest litter within the central axil of the bromeliad categorized into five ordinal ranks, from no leaf litter to heavy leaf litter). Central axil water volume was estimated as described for monitored nests. I also measured central reservoir pH on three occasions (25 May, 3 June, and 9 June). These characteristics were chosen for analysis due to their obvious variabilities and potential for limiting anuran reproductive success. In addition to the above mentioned plant characteristics, I also recorded all macrofauna encountered within the bromeliads, including *O. brunneus* larvae.

RESULTS

Larval Development

All eggs were deposited at night. On clutch day 0 (the morning following egg deposition), live embryos were afloat within one to three layers of egg jellies across the surface of the reservoir (n = 63). Day 0 embryos ranged in development from stages 3–11, and in diameter from 0.64–0.77 mm (mean ± s.d. = 0.71 ± 0.03 mm, n = 27 eggs from 2 clutches, formalin-preserved eggs measured to the nearest 0.01 mm with a microscope). By day 1, I observed embryos, stages 17–19 (approximately 3–4 mm in total length), at the surface of the reservoir and, by day 2, Stage 24–25 tadpoles (approximately 6 mm) swam just beneath the water's surface with at least one egg in their stomachs (n = 33 and 43 clutches, respectively). Embryos just below those at the surface developed more slowly and did not reach Stage 25 until day 3 (n = 35 clutches).

After reaching Stage 25, larval development slowed relative to the rapid embryonic development and the first larvae did not reach Stage 29–30 (the first field identifiable limb-bud development) until about day 18 (Table 1). During this period, a nest typically received four egg depositions. Approximately 19 days and three to four egg depositions later, the first larvae reached Stage 41, immediately prior to foreleg emergence during metamorphosis. Of 21 clutches reaching metamorphosis and observed to completion, one to 20 larvae reached Stage 41. After the first larva reached Stage 41, the remaining larvae continued to develop to this stage over a period of one to 69 days (mean ± s.d. = 19.9 ± 16.7 days). The interval between the first larva to reach Stage 41 and the last was positively correlated with the number of larvae reaching metamorphosis (r = 0.815, n = 21, P = 0.0003, Spearman's Rank Correlation).

Clutches received egg depositions more frequently during the first half of their larval period than during their last half (mean ± s.d. number of egg depositions/period length in days = 0.21 ± 0.03, n = 24 and 0.19 ± 0.04, n = 18 respectively, U = 114.5, P = 0.008, Mann-Whitney U-test; Table 1).

Small clutches (with fewer than 30 tadpoles by day 9), reduced clutches (with more than 30 tad-

TABLE 1. Feeding and development rates of *Osteopilus brunneus* clutches. "Development period" is based only on the developmental stages of the oldest and most-developed larvae within the clutch (assumed to be from the initiating egg deposition, see text).

Development period	Variable	$\bar{x}$ ± s.d.	Range	Number of clutches
Egg to Stage 41	Duration (days)	36.6 ± 3.8	31–44	15
	Number of feedings	7.3 ± 1.2	6–9	13
	Feedings/duration	0.20 ± 0.03	0.17–0.27	13
Egg to Stage 29–30	Duration (days)	18.0 ± 2.7	12–22	25
	Number of feedings	3.8 ± 0.8	3–6	24
	Feedings/duration	0.21 ± 0.03	0.17–0.32	24
Stage 29–30 to Stage 41	Duration (days)	19.0 ± 4.2	13–28	20
	Number of feedings	3.5 ± 0.9	2–5	18
	Feedings/duration	0.19 ± 0.04	0.13–0.31	18

TABLE 2. Bromeliad characteristics as measured from the bromeliad census in 1991. Leaf litter categories for the central reservoir ranged from 1 to 5 (no litter to heavy leaf litter). Sun exposure categories ranged from 1 to 4 (little to full exposure). Significance was calculated using the Mann-Whitney U-test.

Characteristic	*Aechmea*		*Hohenbergia*	
	$\bar{x} \pm$ s.d.	*n*	$\bar{x} \pm$ s.d.	*n*
Plant width (m)***	1.22 ± 0.32	22	0.87 ± 0.25	65
Axil diameter (cm)**	6.0 ± 1.3	20	5.1 ± 1.7	63
Water depth (cm)	13.6 ± 3.5	20	15.3 ± 4.1	63
Water volume (cm^3)	144.2 ± 88.9	20	130.3 ± 105.4	63
Leaf litter**	3.1 ± 0.4	21	2.3 ± 1.1	40
Sun exposure	1.9 ± 0.3	22	2.2 ± 0.9	65
pH*	6.0 ± 0.3	21	5.7 ± 0.6	64

* P < 0.05; ** P < 0.01; *** P < 0.001.

poles by day 9 and then experimentally reduced to 25 tadpoles), and larger clutches (with more than 30 tadpoles by day 9 and not reduced) differed in the length of their developmental periods. Small clutches developed more quickly than did reduced clutches and larger clutches in the first half of larval development (P = 0.007, d.f. = 2, *n* = 24; Kruskal-Wallis test) and small clutches and reduced clutches developed more quickly than did larger clutches in the second part of development (P = 0.007, d.f. = 2, *n* = 18; Kruskal-Wallis test).

In all nests observed through a complete larval cycle (from clutch-initiating egg deposition until the first metamorphic larvae, Stage 41), females deposited fertilized eggs during the first several weeks of larval development and then, in 29 of 34 nests, switched to unfertilized egg depositions. Once this switch occurred, all subsequent depositions were similarly unfertilized. The earliest observed switch to unfertilized eggs occurred in a 19-day-old brood (when the most developed larvae had reached Stage 29–30); however, the majority of broods (*n* = 19, 65.5% of the switching clutches) did not receive unfertilized eggs until at least some young had reached metamorphosis. Of the five broods which did not receive unfertilized eggs, four had stopped receiving all eggs just prior to metamorphosis. The fifth did not begin to reach metamorphosis until the end of the 1991 field season and could have received unfertilized eggs after this point.

The timing of the switch to unfertilized eggs was positively correlated with the length of the larval development period, particularly the latter half of larval development (P = 0.024 and P = 0.0006, respectively; Kendall Rank correlation). This correlation was not significant, however, for the first half of the development period (P = 0.12). No correlation was observed between the timing of the switch to unfertilized eggs and any of the feeding rates throughout larval development (P = 0.92, P = 0.45, P = 0.27 for the full larval period, the first half of development, the latter half of development, respectively); nor were any correlations for these feeding rates and the lengths of the development periods significant (P = 0.38, P = 0.62, P = 0.36 for the full larval period, the first half of development, the latter half of development, respectively).

Bromeliad Characteristics

The bromeliad censuses provided information on bromeliad characteristics and how apparently appropriate bromeliads compared to those selected as nest sites by *Osteopilus brunneus*. The two tank-bromeliad genera (*Aechmea* and *Hohenbergia*) differed (Mann-Whitney U-test) in plant width (P < 0.001), leaf litter (P < 0.01) and central axil diameter (P < 0.01), yet had similar water volumes (P > 0.05, Table 2). The genera differed in their central axil pH values as well (P < 0.05), which could be due to differing biochemical properties of the genera, the different amounts of central axil leaf litter, or the fact that the central axils of *Hohenbergia* have much more of the plant's surface in contact with the water. I never observed *O. brunneus* larvae within the central reservoir of an *Aechmea*; the four clutches found in *Aechmea* (one in the 1990 census and three from non-censused plants in 1991) were in side axils. In contrast, all clutches found in *Hohenbergia* (six in census 1990, four in census 1991, and more than 160 non-census plants) were all placed in the central reservoir.

By combining the data from censused *Hohenbergia* (*n* = 65), monitored nests (*n* = 119), and additional *Hohenbergia* not used by *Osteopilus brunneus* (n = 19), I found significant differences between those plants chosen as nest sites and those not used (Table 3). *Hohenbergia* used as nest sites were significantly smaller (Mann-Whitney U-test) than those not used by *O. brunneus* in regard to average plant width ($P < 0.05$), central axil diameter ($P < 0.01$), water depth ($P < 0.01$), and water volume ($P < 0.01$). The *Hohenbergia* being used as nest sites also were significantly more acidic than those not used ($P < 0.001$). The plants did not differ in amounts of exposure to sunlight.

The differences observed between nested and non-nested *Hohenbergia* may in large part be related to the high leaf litter density common to many non-nested bromeliads. A nest site must be free of litter. When nests and non-nested *Hohenbergia* with litter-free central axils are compared the differences in pH are lost (4.95 ± 0.19, 4.56 ± 0.74, respectively; $P = 0.41$, $n = 30$).

Mortality and Macrofauna

Of 154 clutch initiations detected, 51% (n = 79) failed by the second observation (one to four days later). Of those that failed, 39% (n = 33) were next seen to be associated with aquatic invertebrates. The other 61% (n = 46) demonstrated no apparent cause of failure; embryos sank into the reservoir and died. If eggs fell below the water surface they failed to hatch, presumably due to the hypoxic and hypercapnic conditions of bromeliad waters. Disturbance to bromeliads was at least partially responsible for clutch failures; of eight known cases of human disturbances of newly deposited eggs, all clutches sank into reservoir waters and failed to produce a single hatchling.

The only macrofauna commonly associated with *Osteopilus brunneus* nests were of three insect families: Dytiscidae (predaceous diving beetles), Chironomidae (midge flies), and Culicidae (mosquitoes). Adult dytiscids were observed in 54 different nests. Six clutches became associated with dytiscids at the end of the season and thus the consequences of these infestations could not be determined. On 25 occasions the beetles became associated with the original egg deposition. All of these clutches failed to have any larvae reach Stage 41. I saw beetles actively feeding on the embryos on five occasions. Although predation on larvae was not observed, beetles were seen feeding on recently dead larvae on four occasions. If these predaceous beetles appeared in a nest before any larvae reached Stage 29–30 the clutch suffered complete failure (n = 7). Of 16 older clutches associated with dytiscids, 14 had some larvae reach metamorphosis. However, tadpoles less developed than Stage 29–30 within these clutches were seriously reduced in number, if not eliminated, by the beetles. In 75% (n = 12) of these infested older clutches, the dytiscids first appeared when a large number of uneaten unfertilized eggs (more than 25) began to spoil. The destruction caused by these beetles could be rapid. In 15 cases, nest sites containing what I considered healthy clutches one day, contained only swarms of dytiscids by the next monitoring session (one to three days later).

Chironomid larvae (Chironomidae) also occasionally were found within *Osteopilus brunneus* nests (n = 26). I observed chironomids feeding on *O. brunneus* eggs and dead tadpoles. These insects did not seem to prey on living tadpoles, but apparently contributed to clutch demise by reducing larval food supplies (as suggested by the emaciated appearance of surviving larvae). Like dytiscids, chironomids were found on the surface of initial egg depositions (n = 10 clutches) or in older clutches with uneaten, unfertilized eggs (n = 11 clutches).

TABLE 3. Characteristics of *Hohenbergia* differentiated by use as nest sites by *Osteopilus brunneus* in 1991. Sun exposure categories ranged from 1 to 4 (little to full exposure). Significance was calculated using the Mann-Whitney U-test.

Characteristic	Non-nest sites		Nest sites	
	$\bar{x}$ ± s.d.	n	$\bar{x}$ ± s.d.	n
Plant width (m)*	0.85 ± 0.25	80	0.77 ± 0.20	123
Axil diameter (cm)**	4.9 ± 1.7	77	4.1 ± 1.2	121
Water depth (cm)*	14.7 ± 4.2	77	13.4 ± 3.3	121
Water volume (cm^3)**	116.4 ± 101.8	77	130.3 ± 105.4	121
Sun exposure	1.9 ± 0.3	80	2.2 ± 0.9	123
pH***	6.0 ± 0.3	64	5.7 ± 0.6	24

* $P < 0.05$, ** $P < 0.01$, *** $P < 0.001$.

These insect larvae fed rapidly on the eggs, consuming entire egg complements before any eggs could hatch ($n = 6$).

I found culicid larvae associated with 11 active nest sites. Each of these clutches appeared to be on the verge of failure by the time the insects were discovered. The clutches were either clutch initiating depositions with few embryos remaining afloat (range = 2–25 embryos, $n = 5$ clutches) or nests that had not received an egg deposition for seven to 20 days and thus contained less viscous waters than the more glycerine-like water of a well fed clutch ($n = 6$). Eight of the 11 *Osteopilus brunneus* clutches soon failed after mosquito larvae appeared, and the other three clutches had tadpoles survive to the end of the study without reaching metamorphosis. No culicids were ever observed feeding on eggs or larvae, and appear to represent a symptom of a failing clutch as opposed to a cause of failure.

In contrast to the low number of nests containing culicids, 71% of the censused *Hohenbergia* without *Osteopilus brunneus* larvae contained mosquito larvae (only plants with particularly low water levels and/or heavy leaf litter in the central axil did not contain culicids). All unused *Hohenbergia* with open central reservoirs ($n = 15$) contained culicid larvae. Other macrofauna associated with open central reservoirs in *Hohenbergia* were oligochaetes ($n = 5$), arachnids ($n = 5$), gastropods ($n = 6$), anisopteran larvae ($n = 4$), zygopteran larvae ($n = 12$), chironomids ($n = 3$), and dytiscids ($n = 3$).

In order to determine if the predaceous insects were differentially assorted between used (containing an *Osteopilus brunneus* nest) and unused *Hohenbergia*, I compared the census macrofauna data to a subset of the nest monitoring data matched by date. By matching these data sets by date, I hopefully accounted for any temporal changes in predator population sizes or movements. As predicted, significant differences existed between the distributions of dytiscid beetles in bromeliads used as nests and those left unused ($P = 0.0001$, $n = 50$, paired t-test); a greater percentage of plants with nests contained dytiscids than did plants without nests (mean ± s.d. = 6.98 ± 15.4%, 0.99 ± 2.8%, respectively; $P = 0.031$, $n = 50$, unpaired t-test). Chironomid larvae also differed in their distributions ($P = 0.0001$, $n = 50$); however, this difference did not translate into an overabundance of these predators in either used or unused reservoirs (2.30 ± 5.7%, 0.98 ± 2.7%, respectively; $P = 0.31$, $n = 50$).

DISCUSSION

Larval Development

Larval development rate is difficult to determine in *Osteopilus brunneus*. Because of the age-class structure within the clutches, only the oldest and fastest-developing individuals could actually be timed. This required the assumption that some larvae hatched from the first egg deposition survived and were the first to reach particular stages of development. However, this assumption seems to be justified because the largest individuals were being monitored regularly at one- to four-day intervals and deviations from advancing development would have been noted.

These first larvae took 31–44 days to develop from fertilized egg to Stage 41. Clutches with a relatively larger number of tadpoles took longer to develop, and with continuous deposition of eggs into a nest, the longer it took to develop, the more egg depositions a clutch received. These data suggest that larval development rate is density-dependent, perhaps with food limitation acting as the primary factor. *Osteopilus brunneus* larvae are dependent on conspecific eggs for growth and survival (R. L. Thompson 1992, unpublished data). Assuming no association between the number of eggs in each deposition and embryonic mortality, more larvae within a clutch would increase competition for food, and would result in fewer eggs being available to each tadpole. The decreased number of eggs available could cause the slower development observed in large clutches.

The effects that larval density can have on anuran growth and development rates are well known (Brockelman 1969; Salthe and Mecham 1974; Wilbur and Collins 1973; Wilbur 1977*a*). Brockelman (1969) clearly showed with *Bufo americanus* that larval densities can have marked effects on developmental rates. The larval periods of individuals in low-density pens were significantly shorter than those in high-density pens. Food availability was considered the underlying reason for this density-dependent effect. Similarly, competitive exclusion at high densities and its effects on food availability have been shown to decrease subsequent growth and development rates in *Rana sylvatica* (Wilbur and Collins 1973; Wilbur 1977*b*).

Thus, as in many species with parental care, a parent-offspring conflict of interest may exist with regard to clutch sizes in *Osteopilus brunneus*. A larger clutch corresponds to fewer eggs per individual. By feeding on fertilized eggs, the larvae remove potential competitors for future food resources, in addition to the obvious and direct nutritional benefits. During early clutch development, the parents are able to swamp their cannibalistic young by depositing more eggs than the larvae can consume. Because young larvae cannot consume all of the eggs deposited, parents have the best chance of producing additional hatchlings during the first few weeks of a clutch. The shorter egg-deposition interval during early clutch development may be a mechanism for swamping the larvae to increase the clutch size.

Shorter egg-deposition intervals at the beginning of clutch development could also be a means of insuring sufficient larval food supplies. The small stomachs of young larvae can only hold a few eggs compared to the large number of eggs that can be consumed by older and larger tadpoles. Young, small larvae are prone to starvation and predation and may benefit from fast growth rates. From the data available, predation risks from dytiscid beetles apparently decrease with the growth and development of the larvae, such that larvae past Stage 29–30 are relatively invulnerable to predation by these beetles. Thus, a high egg-deposition frequency could have a greater impact on larval growth and survivorship during the early part of larval development.

Risks of dytiscid predation may also be affected by the use of fertilized eggs during this critical larval period. The rapid spoiling of uneaten unfertilized eggs appeared to attract these predators to nest sites, as was evidenced by the high proportion of beetle infestation among nests associated with spoiling eggs. With the inability of young broods to consume a full egg complement, spoiling unfertilized eggs would attract these predators when larvae are most vulnerable to attack, as well as result in the fouling of the reservoir environment.

Parental Care and Egg Size

Osteopilus brunneus produces some of the smallest eggs of all anurans (for a review of anuran egg sizes see Duellman and Trueb 1986 [1985]), which is surprising with respect to the large body size of females and the presence of extensive parental care. In many externally-fertilizing ectotherms, egg size and parental care are positively correlated, relative to female size and reproductive mode (McDiarmid 1978; Gross and Shine 1981; Salthe and Mecham 1974). Several investigators have discussed the evolutionary associations between the two characters: increased egg size in response to parental care, the "safe harbor" hypothesis (Shine 1978; Sargent et al. 1987); increased parental care in response to large egg size (Nussbaum 1985; Nussbaum and Schultz 1989); and coevolution of egg size and parental care in response to environmental conditions (Nussbaum and Schultz 1989). At first appearance, *O. brunneus* would seem to be an exception to this trend. However, the care so often correlated with egg size is care for embryos. *Osteopilus brunneus* provides care for larvae. The few known cases of anuran larval care appear as an extension from embryonic care (McDiarmid 1978; Wells 1981). Unless males provide some form of care for embryos, *O. brunneus* may present the only anuran example of a parental care strategy that foregoes any care for embryos. In this light, the lack of positive association between egg size and parental care should be expected.

Reduced egg size may be a response to the high risk of embryonic mortality caused by hypoxia. Small eggs develop and hatch more quickly than do larger eggs (Salthe and Duellman 1973), and thus embryos can pass through this critical period more quickly. The high risk of embryonic death also could select for small egg size as an adaptation for quick egg production. With 51% of initial egg depositions failing to produce a single hatchling and only

10% succeeding in having at least one larva develop to Stage 41 (Thompson 1992), one might expect selection to favor females that can produce a new egg complement as quickly as possible after clutch failure. With a limited amount of maternal investment that can be placed into egg production, a female can produce either a large number of small eggs or fewer larger eggs (Salthe and Duellman 1973; Salthe and Mecham 1974). The evolution of smaller eggs, which presumably can be produced relatively quickly and less expensively, would decrease the period of time required for the production of new clutches.

Evolution of Parental Care

Somewhere within the lineage leading to the reproductive strategy currently used by *Osteopilus brunneus*, breeding moved to bromeliad water catchments. This change in nesting site required a means of countering the hypoxic conditions and the lack of sufficient food within bromeliads to sustain larval development. These frogs have adapted to these conditions by producing small, rapidly developing eggs, and by feeding larvae both fertilized and unfertilized eggs.

Other anuran species (of the genus *Dendrobates*) are known to feed their larvae only unfertilized eggs as food. This nutritive egg production is similar to the strategy employed by *Osteopilus brunneus* during the latter part of its reproductive cycle, but contrasts with the provision of fertilized eggs to young *O. brunneus* clutches. Apart from the deposition of fertilized eggs, major differences in the number and size of eggs produced exist between dendrobatids and *O. brunneus*. Dendrobatids feed each of one to four larvae (Weygoldt 1980, 1987) one to seven large eggs, whereas *O. brunneus* females provide approximately 250–500 small eggs to as many as 170 larvae at each egg deposition (Lannoo et al. 1987; R. L. Thompson, unpublished data). Although these behaviors are similar, what factors should be considered in order to explain the differences observed?

These strategies of bromeliad water use and nutritive egg production have evolved from two different origins, particularly with regard to oviposition sites. *Osteopilus dominicensis* and *O. septentrionalis*, probably the closest non-Jamaican relatives of *O. brunneus* (S. B. Hedges, personal communication) and its only currently-recognized congeners (although see Hedges, this volume), lay small eggs in small ephemeral pools of water and provide no parental care (Schwartz and Henderson 1991), a reproductive strategy one might claim as the ancestral condition for *Osteopilus*. The primitive dendrobatid reproductive strategy includes care for large eggs deposited on an aerial substrate (Weygoldt 1987). After the eggs hatch, all dendrobatids for which the reproductive biology is known transport their well-developed larvae to water. The species that feed their larvae transport the young to bromeliad reservoirs. Thus, dendrobatid nutritive egg provisioning has arisen in response to a change in larval development sites (Weygoldt 1980, 1987), whereas *O. brunneus* has had to respond additionally to a change in the site of embryonic development.

The use of fertilized egg feedings in *Osteopilus brunneus* is most likely tied into the difference in egg sizes. The large dendrobatid eggs hatch into large, relatively advanced hatchlings (Weygoldt 1987). These larvae can easily consume the few large eggs provided as food. The small hatchlings of *O. brunneus* have small stomachs and can only consume a few eggs. Additionally, with the high embryonic mortality in *O. brunneus*, continued fertilized egg production allows for clutch size enhancement. Dendrobatid eggs are adapted for terrestrial development, and thus clutch enhancement in axil waters is not possible. Therefore, little reason exists for these species to fertilize eggs presented to larvae.

Acknowledgements

This research was funded by a National Science Foundation Predoctoral Fellowship and grants from Sigma Xi and the Latin American Studies Department, University of Florida. I would like to thank W. E. Duellman and M. J. Lannoo for their patience and helpful comments on the first draft of this manuscript, and R. Powell and R. W. Henderson for including me in this volume. I would especially like to thank those friends and colleagues in Windsor, Jamaica, who made this research possible: Rudie Diesel, Wolfgang Janetzky, Mother-Miri Campbell, Sugar Belly, Mr. Mike, Pim-Pim, Malibu, Dongo, Bigger, Raquel, Prudence, Ozzie, and Dusky.

Literature Cited

Asprey, G. F. and R. G. Robbins. 1953. The vegetation of Jamaica. Ecol. Monogr. 23: 359–412.

Brockelman, W. Y. 1969. An analysis of density effects and predation in *Bufo americanus* tadpoles. Ecology 50: 632–644.

Duellman, W. E. and L. Trueb. 1986 (1985). Biology of Amphibians. McGraw-Hill Book Co., New York.

Dunn, E. R. 1926. The frogs of Jamaica. Proc. Boston Soc. Nat. Hist. 38: 111–130.

Gosner, K. L. 1960. A simplified table for staging anuran embryos and larvae with notes on identification. Herpetologica 16: 183–190.

Gross, M. R. and R. Shine. 1981. Parental care and mode of fertilization in ectothermic vertebrates. Evolution 35: 775–793.

Jones, D. A. 1967. Green pigmentation in Neotropical frogs. Ph.D. dissertation, Univ. Florida, Gainesville.

Lannoo, M. J., D. S. Townsend, and R. J. Wassersug. 1987. Larval life in the leaves: arboreal tadpole types, with special attention to the morphology, ecology, and behavior of the oophagous *Osteopilus brunneus* (Hylidae) larva. Fieldiana, Zool. 38: 1–31.

Laessle, A. M. 1961. A micro-limnological study of Jamaican bromeliads. Ecology 42: 499–517.

Liebherr, J. K. 1988. The Caribbean: fertile ground for zoogeography, p. 1–14. *In* J. K. Liebherr (ed.), Zoogeography of Caribbean Insects. Cornell Univ. Press, Ithaca, New York.

Lynn, W. G. and C. Grant. 1940. The herpetology of Jamaica. Bull. Inst. Jamaica, Sci. Ser. 1: 1–148.

McDiarmid, R. W. 1978. Evolution of parental care in frogs, p. 127–147. *In* G. W. Burghardt and M. Bekoff (eds.), The Development of Behavior: Comparative and Evolutionary Aspects. Garland Press, New York.

Nussbaum, R. A. 1985. The evolution of parental care in salamanders. Misc. Publ. Mus. Zool. Univ. Michigan 169: 1–50.

— and D. L. Schultz. 1989. Coevolution of parental care and egg size. Amer. Nat. 133: 591–603.

Pregill, G. K. and S. L. Olson. 1981. Zoogeography of West Indian vertebrates in relation to Pleistocene climatic cycles. Ann. Rev. Ecol. Syst. 12: 75–98.

Salthe, S. N. and W. E. Duellman. 1973. Quantitative constraints associated with reproductive mode in anurans, p. 229–249. *In* J. L. Vial (ed.), Evolutionary Biology of the Anurans: Contempory Research on Major Problems. Univ. Missouri Press, Columbia.

Salthe, S. N. and J. S. Mecham. 1974. Reproductive and courtship patterns, p. 309–521. *In* B. Lofts (ed.), Physiology of the Amphibia, Vol. II. Academic Press, New York.

Sargent, R. C., P. D. Taylor, and M. R. Gross. 1987. Parental care and the evolution of egg size in fishes. Amer. Nat. 129: 32–46.

Schwartz, A. and R. W. Henderson. 1991. Amphibians and Reptiles of the West Indies: Descriptions, Distributions, and Natural History. Univ. Florida Press, Gainesville.

Shine, R. 1978. Propagule size and parental care: the "safe harbor" hypothesis. Jour. Theor. Biol. 75: 417–424.

Thompson, R. L. 1992. Reproductive behaviors and larval natural history of the Jamaican brown frog, *Osteopilus brunneus*. M.S. thesis, Univ. Florida, Gainesville.

Wells, K. D. 1981. Parental behavior in male and female frogs, p. 184–197. *In* R. D. Alexander and D. W. Tinkle (eds.), Natural Selection and Social Behavior: Recent Research and New Theory. Chiron Press, Newton, Massachusetts.

Weygoldt, P. 1980. Complex brood care and reproductive behavior in captive poison-arrow frogs, *Dendrobates pumilio* O. Schmidt. Behav. Ecol. Sociobiol. 7: 329–332.

—. 1987. Evolution of parental care in dart poison frogs (Amphibia: Anura: Dendrobatidae). Zeitschr. Zool. Syst. Evolut.–Forsch. 25: 51–67.

Wilbur, H. M. 1977*a*. Density-dependent aspects of growth and metamorphosis in *Bufo americanus*. Ecology 58: 196–200.

—. 1977*b*. Interactions of food level and population density in *Rana sylvatica*. Ecology 58: 206–209.

— and J. P. Collins. 1973. Ecological aspects of amphibian metamorphosis. Science 182: 1305–1314.

Resumen.—El sapo marrón de Jamaica, *Osteopilus brunneus*, le provee huevos a sus crías como un recurso alimenticio para las larvas. Mientras que la provisión trófica de huevos no viables está bien documentada en algunos sapos dendrobátidos y ha sido sugerida para varias otras especies que se reproducen en bromelias o en huecos en los árboles, *O. brunneus* es diferente porque le provee a sus crías huevos fecundados. Las hembras de *O. brunneus* depositan huevos fecundados o no fecundados dependiendo del grado de desarrollo de las larvas. Inicialmente, los huevos fecundados son depositados en nidos donde las larvas son todavía demasiado pequeñas para consumir un complemento completo de huevos. Los huevos que escapan del canibalismo se desarrollan y se unen al grupo de larvas. Eventualmente se proveen huevos no fecundados una vez que las larvas han madurado lo suficiente para consumir un complemento completo de huevos. Esta estrategia puede ser vista como una forma alterna de lidiar con las condiciones adversas comunes a charcos pequeños de agua. Mediante deposiciones continuas de huevos, *O. brunneus* puede contrarrestar la variable, pero siempre alta mortalidad embriónica al producir prole adicional. Esta facultad reproductiva aparentemente única le permite a las hembras evitar el conflicto irreversible entre la cantidad y la calidad de descendientes inherente a la producción trófica de huevos. Este artículo describe el habitáculo y la historia natural de estas larvas que viven en las bromelias y cómo estos factores pueden estar relacionados a la evolución de la estrategia de cuidado parental, tanto en relación al uso de huevos fecundados como en la transición al uso de huevos no fecundados.

Résumé.—La grenouille brune jamaïquaine *Osteopilus brunneus* fournit des oeufs à ses jeunes comme ressource alimentaire larvaire. Alors que l'approvisionnement en oeufs trophiques (non viables) est bien décrite chez certaines grenouilles dendrobates et vraisemblable chez plusieurs autres espèces qui se reproduisent dans le creux des arbres ou dans les broméliacées, *O. brunneus* diffère en ce qu'elle fournit des oeufs fécondés. Les femelles d'*O. brunneus* déposent soit des oeufs fécondés soit des oeufs non fécondés, selon le stade du développement larvaire. Les oeufs fécondés sont d'abord déposés dans les nids alors que les larves sont trop petites pour les manger entièrement. Les oeufs qui échappent au cannibalisme se développent et se joignent aux autres larves. Lorsque les larves sont suffisamment avancées pour manger tout l'oeuf, les femelles leur fournissent des oeufs non fécondés. Cette stratégie peut être perçue comme un moyen de pallier aux conditions très difficiles qui prévalent dans les petits points d'eau. En déposant continuellement des oeufs fécondés, *O. brunneus* est capable de contrer la mortalité embryonnaire variable mais toujours élevée. Cette stratégie unique permet aux femelles d'éviter l'échange irréversible de la quantité pour la qualité, inhérente à la production d'oeufs trophiques. Cet article décrit l'habitat et l'histoire naturelle des larves qui vivent dans des broméliacées et comment ces facteurs peuvent être liés à l'évolution des stratégies qu'adoptent les parents, tant au titre de l'utilisation d'oeufs fécondés que du moment qu'ils choisissent pour passer aux oeufs non fécondés.

Thompson, R. L. 1996. Larval habitat, ecology, and parental investment of *Osteopilus brunneus* (Hylidae), p. 259–269. *In* R. Powell and R. W. Henderson (eds.), Contributions to West Indian Herpetology: A Tribute to Albert Schwartz. Society for the Study of Amphibians and Reptiles, Ithaca (New York). Contributions to Herpetology, volume 12.

Occurrence of the Nematode *Skrjabinoptera scelopori* in the Cuban Treefrog, *Osteopilus septentrionalis*: Mainland and Island Comparisons

Walter E. Meshaka, Jr.[1, 2]

Abstract.—The Cuban treefrog *Osteopilus septentrionalis* (Hylidae) is infected by a parasitic nematode, *Skrjabinoptera scelopori* (Caballero Rodríquez 1971). Species of *Skrjabinoptera* are gastric parasites of reptiles, principally of iguanid, teiid, gekkonid, and chamaeleonid lizards. This new host record represents the first report of *S. scelopori* in an amphibian. The location of the parasite in the host, free in the body cavity and encysted on mesenteries, suggests that *O. septentrionalis* may be a paratenic (transport) host, and, therefore, not required for completion of the parasite life cycle. A survey of museum specimens and recent collections from Florida and the West Indies is presented. Prevalence of infection, highest in specimens collected on the Cayman Islands, is discussed with respect to geographic distribution, gender, and body size of the host.

Introduction

The Cuban treefrog, *Osteopilus septentrionalis* (Hylidae), is a large mesophytic forest-dweller of the West Indian region. The species' geographic distribution is widespread through much of the Greater Antilles and is a recent adventive of man in mainland Florida, Puerto Rico, and isolated islands of the Lesser Antilles (Barbour 1931; Carr 1940; Duellman and Schwartz 1958; Lazell 1989; Schwartz and Henderson 1991; Powell et al. 1992; Kaiser and Henderson 1995). A broad diet, including conspecifics, other frogs, and lizards contributes to the success of *O. septentrionalis* in a variety of habitats and locations (Meshaka 1994).

Euryphagy may predispose *O. septentrionalis* to infection by parasites whose larval stages, in part, take place in insects. Members of the helminth genus *Skrjabinoptera* (Physalopteroidea) have been found in many insectivorous Caribbean lizards (Greve and Powell 1989; Powell et al. 1990*a*, 1990*b*; Fobes et al. 1992; Moster et al. 1992; Lenart et al. 1994 [1995]), and the immature stages of *Skrjabinoptera* sp. have been reported in the stomachs of the Hispaniolan endemic toad, *Bufo guentheri* (Parsons et al. 1995). In this study, I examined specimens of *O. septentrionalis* from across its geographic range, and recorded the presence of *S. scelopori* with consideration to geographic distribution, gender, and body size of the host.

Materials and Methods

Body cavities of *Osteopilus septentrionalis* from museum collections and those collected for this study on buildings on Georgetown (Grand Cayman Island) and Nassau (New Providence Island) and in southern and central Florida (including Long Pine Key and Paradise Key of Everglades National Park, Dade County) were examined for the presence of helminth parasites. Snout-vent lengths (SVL) of infected treefrogs were measured with vernier calipers to the nearest 0.1 mm. Sex and maturity of each treefrog was determined by examination of the gonads. All treefrogs collected for this study are

[1] Archbold Biological Station, P.O. Box 2057, Lake Placid, Florida 33852, USA.

[2] Current address: Museum, Everglades National Park, Homestead, Florida 33034–6733, USA.

deposited in the National Museum of Natural History, Washington, D. C., and voucher specimens of *Skrjabinoptera scelopori*, with exception of those from the Everglades, are deposited in the University of Nebraska State Museum (accession number 38226).

RESULTS AND DISCUSSION

I examined 237 male and 207 female *Osteopilus septentrionalis* and found *Skrjabinoptera scelopori* encysted on the outer lining of the stomach, free in the body cavity, and occasionally entwined in the gonads of the hosts. Larval *Porrocaecum* and larval spirurids also were found in the body cavity.

The role of *Osteopilus septentrionalis* in the life cycle of *Skrjabinoptera scelopori* is depicted in a hypothesized life cycle (Figure 1). Eggs are ingested by one or a series of the intermediate host, presumably an arthropod; intermediate hosts carrying the larval parasites are in turn eaten by the definitive host, such as a lizard; the parasite reproduces in the digestive tract of the definitive host; eggs are passed in the fecal material.

Location of the parasite in the body cavity of *Osteopilus septentrionalis* suggests that the frog is neither an intermediate nor definitive host, but rather a paratenic host; that is, one in which development does not occur, but the parasite continues to be infective to the definitive host. As a paratenic host, *O. septentrionalis* may acquire the parasite by ingesting an intermediate or definitive host. By either acquisition, the parasite would exit the stomach and reside, but not reproduce, in the body cavity as observed in *O. septentrionalis*. This avenue may be the end of the parasite's life cycle, but may also bridge the ecological gap between intermediate host and a definitive host which consumes an infected *O. septentrionalis*.

TABLE 1. Prevalence of *Skrjabinoptera scelopori* in *Osteopilus septentrionalis*: each entry indicates the number of infected individuals followed in parentheses by the size of the sample examined from that location. Locations marked with an asterisk (*) denote samples collected by the author in 1992.

Location	Males	Females
United States		
1. Everglades	6 (14)	1 (14)
Cayman Islands		
2. Cayman Brac	3 (9)	2 (8)
3. Grand Cayman*	0 (50)	3 (104)
4. Grand Cayman	2 (11)	1 (10)
Bahamas		
5. Nassau*	1 (17)	0 (22)
6. Long Island	1 (25)	1 (10)
7. Andros Island	2 (25)	1 (4)
Cuba		
8. Havana	5 (44)	3 (11)
9. Oriente	1 (6)	0 (1)
10. Yumuri	0 (1)	2 (4)
11. Isla de Pinos	2 (6)	0 (2)
12. Cabanas	2 (4)	0 (3)
13. Guantanamo	2 (19)	0 (7)
14. Cabo Cruz	0 (6)	1 (7)
Total	27 (237)	15 (207)
Prevalence	11.4%	7.2%

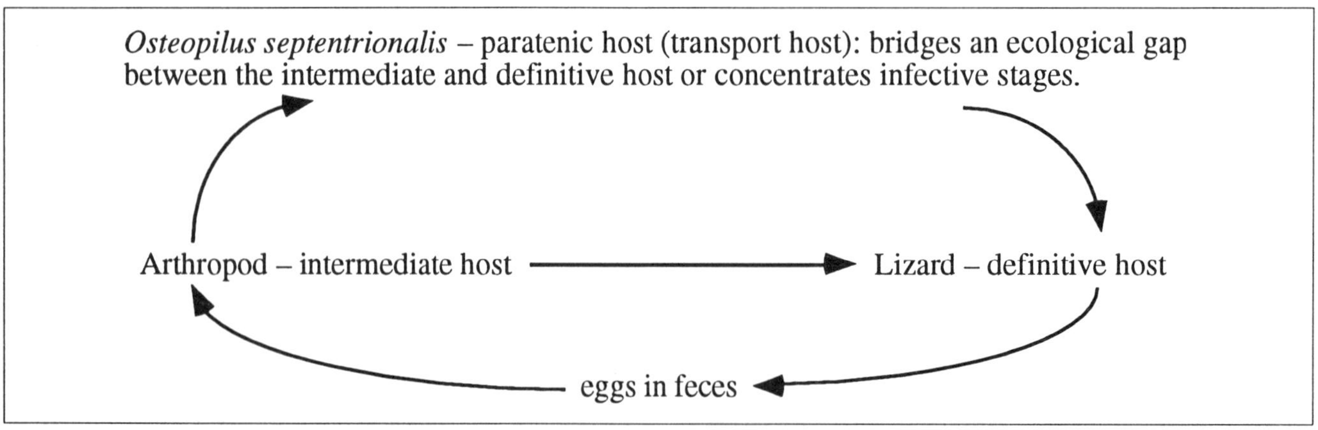

FIGURE 1. Hypothetical life cycle of *Skjrabinoptera scelopori*, including the hypothesized role of *Osteopilus septentrionalis*.

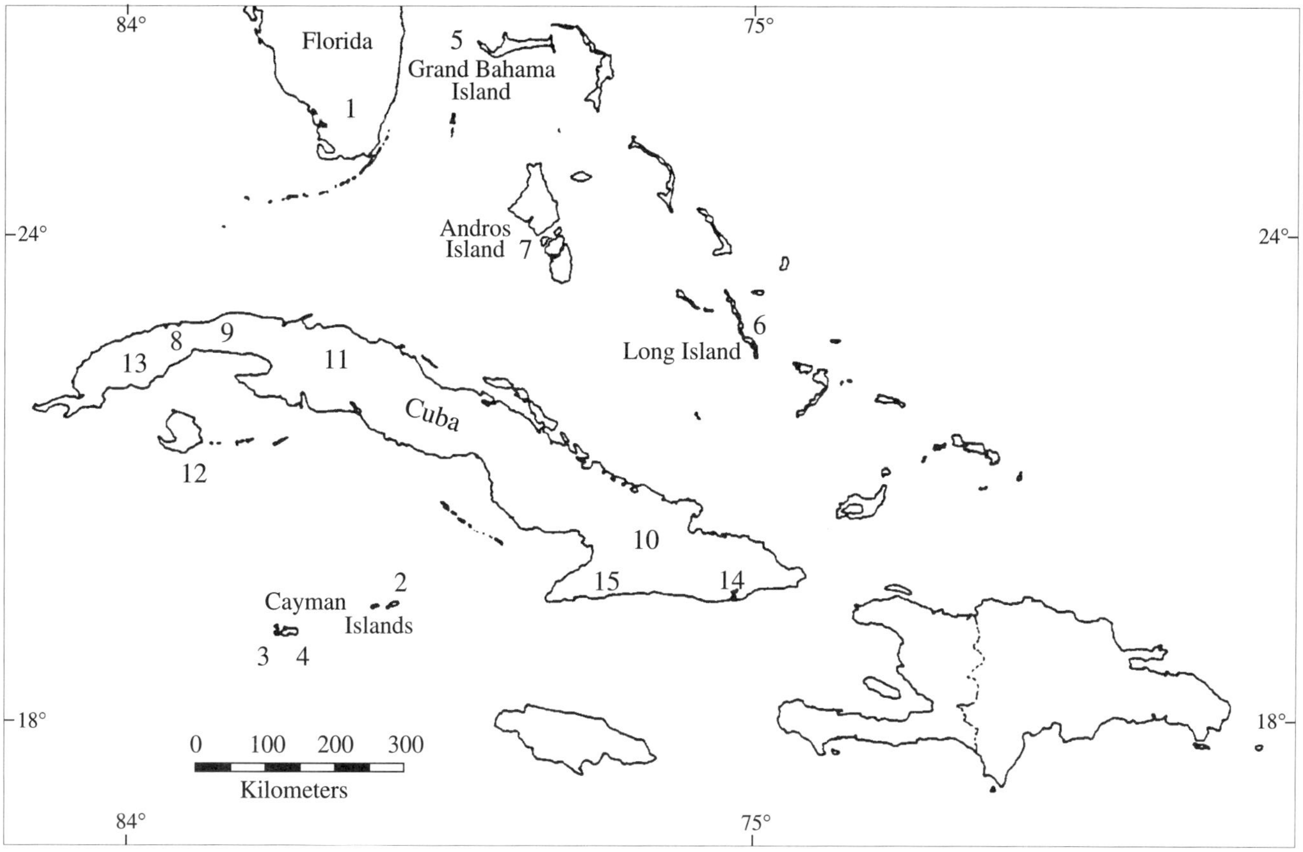

FIGURE 2. Localities where *Skrjabinoptera scelopori* was found in *Osteopilus septentrionalis*. Numbers refer to sites as listed in Table 1. The map was modified from Schwartz and Henderson (1991).

Although not presently known to be hosts of *Skrjabinoptera scelopori*, West Indian reptiles known to eat *Osteopilus septentrionalis* include the Cuban knight anole, *Anolis equestris* (Brach 1976) and members of the colubrid snake genus *Alsophis* (Schwartz and Henderson 1991; Henderson and Sajdak, this volume). In this way the paratenic host could increase the number of pathways for completion of the parasite's life cycle as well as concentrate infective stages of the parasite.

Examination of *Osteopilus septentrionalis* across its geographic range revealed infections by *Skrjabinoptera scelopori* in both native and introduced populations (Figure 2, Table 1). Reasons for the wide range in the prevalence of infections from populations in the West Indies are unclear; however, annual variation in the population sizes of the parasites or site-specific differences in the population sizes of the final host(s) could explain this phenomenon.

Considering all sites, the frequency of infection was greatest in the subtropical regions of southern mainland Florida where the treefrog has been present for more than 40 years (Schwartz 1952). Treefrogs from the Everglades were sampled from sites where their relative abundances were high, sometimes exceeding 100 individuals per 15 minutes of nocturnal searching (Meshaka 1994). High population densities of the treefrog, a paucity of the final host(s), or high densities of an intermediate host or hosts which are exploited by *Osteopilus septentrionalis* could contribute to the high rates of prevalence in a population more proximate to the West Indies than other sampled Florida localities.

Although *Osteopilus septentrionalis* occurs almost continuously through central Florida (Meshaka 1996), *Skrjabinoptera scelopori* does not appear to have expanded with its host into the temperate regions of Florida north of the Everglades. From 1991 to 1995, nearly 500 individuals were collected from

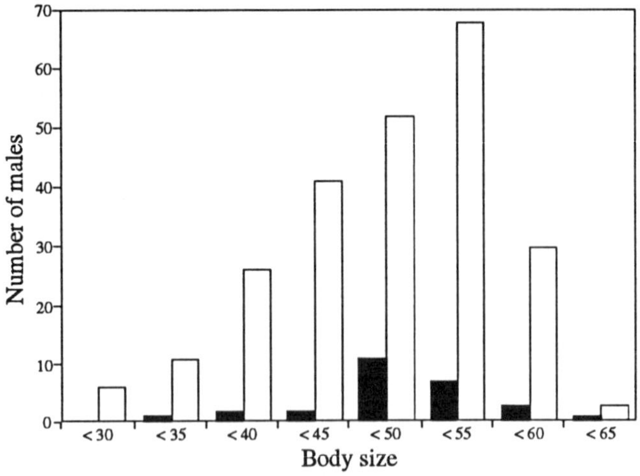

FIGURE 3. Distribution of size classes of male *Osteopilus septentrionalis* infected with *Skrjabinoptera scelopori* (solid bars) and the total number of hosts examined (open bars).

south-central and central Florida and none was found to be infected by *S. scelopori*. Because the sample size was large and collections were made over an extended period of time, I suggest that the parasite was absent or present only at very low frequencies.

If the absence of *Skrjabinoptera scelopori* is not an artifact of sampling, perhaps a critical link in the parasite life cycle is missing. For example, intermediate or definitive hosts may not occur in more northerly areas of Florida. Alternatively, *Osteopilus septentrionalis* may not be exposed to infection because of non-overlap of niches. In this connection, that none of 150 treefrogs examined from Tampa (Hillsborough County) was infected is interesting, especially as *S. scelopori* was found in the stomach of a brown anole (*Anolis sagrei*), an introduced species of West Indian origin. The anole, in turn, was found in the stomach of a female *O. septentrionalis*.

Adult body sizes of *Osteopilus septentrionalis* range widely in both males (27–85 mm SVL) and females (45–165 mm SVL). Mean body size of females is larger than males, and a corresponding disparity exists in the dietary niche breadth in prey size and to some extent in prey taxa (Meshaka 1994). Consequently, an ability to differentially exploit the intermediate host might predispose large females to greater prevalence of infection than males and smaller females.

This prediction was not supported by the data. The prevalence of infection did not differ ($\chi^2 = 2.11$, $P > 0.05$; d.f. = 1) between the sexes (Table 1). Furthermore, no significant difference existed between the distribution of body size of infected and examined males (Figure 3) and females (Figure 4) ($\chi^2 = 4.12$, $P > 0.05$, d.f. = 7 and $\chi^2 = 10.76$, $P > 0.05$, d.f. = 11, respectively). Thus, large body size of the female treefrog did not result in a greater rate of infection.

Perhaps the primary, although not exclusive, means of parasite acquisition was via ingestion of an infected intermediate host. Absence of infections in immature *Osteopilus septentrionalis* (*n* = 47) may reflect a lack of exposure: the older the individual, the more likely it is to have encountered an intermediate or definitive host. Alternatively, the treefrog may need to be large enough to ingest an intermediate host. If either or both hypotheses are true, that parasites were absent in immature *O. septentrionalis* and nearly so in the smallest males is not surprising. This assumption best fits those data that indicate that above a minimum body size or age, no body-size component to infection was present in either sex of the host.

ACKNOWLEDGMENTS

Thanks go to John W. Fitzpatrick, Executive Director of Archbold Biological Station, for support of this project. Michael Kinsella and Mary C. Garvin kindly identified parasites. John H. Greve com-

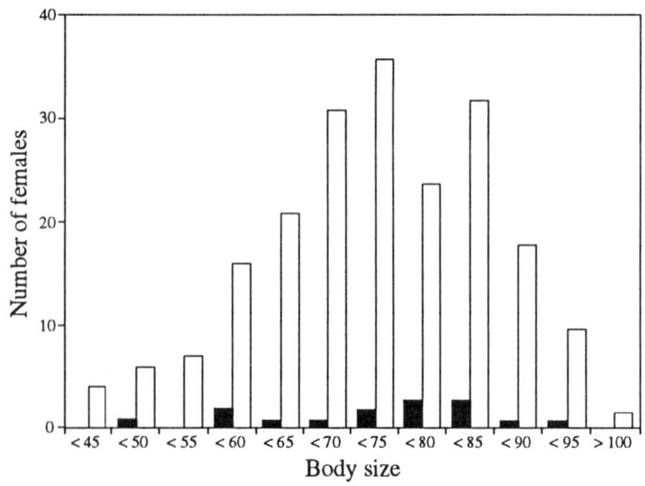

FIGURE 4. Distribution of size classes of female *Osteopilus septentrionalis* infected with *Skrjabinoptera scelopori* (solid bars) and the total number of hosts examined (open bars).

mented on the hypothetical life cycle of the parasite. I am grateful for the loan of specimens by John Simmons (University of Kansas Museum of Natural History), Roy McDiarmid (National Museum of Natural History), Linda Ford (American Museum of Natural History), and Jens Vindum (California Academy of Science). I thank Pedro Fransisco Quintaña-Ascencio for translating the abstract into Spanish. Permission to collect in Everglades National Park was made possible by permit no. 910039 issued by Daniel Foxen. Lastly, I appreciate the good fortune of having had the opportunity to participate in this volume honoring Albert Schwartz.

Literature Cited

Barbour, T. 1931. Another introduced frog in North America. Copeia 1931: 140.

Brach, V. 1976. Habits and food of *Anolis equestris* in Florida. Copeia 1976: 187–189.

Caballero Rodríquez, G. 1971. Contribución al conocimiento de los nematodos que parasitan a los reptiles de México. II. Sobre una nueva fisaloptera parasita de iguanidos (*Sceloporus torquatus torquatus* Wiegmann, 1828). Rev. Biol. Trop. 19: 165–170.

Carr, A. F. 1940. A contribution to the herpetology of Florida. Univ. Florida Publ. Biol. Sci. Serv. 3: 1–118.

Duellman, W. E. and A. Schwartz. 1958. Amphibians and reptiles of southern Florida. Bull. Florida State Mus. 3: 181–324.

Fobes, T. M., R. Powell, J. S. Parmerlee, Jr., A. Lathrop, and D. D. Smith. 1992. Natural history of *Anolis cybotes* (Sauria: Polychridae) from an altered habitat in Barahona, Dominican Republic. Carib. Jour. Sci. 28: 200–207.

Greve, J. H. and R. Powell. 1989. *Skrjabinoptera leiocephalorum* N. Sp. (Nematoda: Physalopteroidea) in *Leiocephalus* spp. (Reptilia: Iguanidae) from Hispaniola. Jour. Parasitol. 75: 677–679.

Kaiser, H. and R. W. Henderson. 1995. The conservation status of Lesser Antillean frogs. Herpetol. Nat. Hist. 2(2): 41–56.

Lazell, J. D., Jr. 1989. Wildlife of the Florida Keys: A Natural History. Island Press, Washington, D. C.

Lenart, L. A., R. Powell, J. S. Parmerlee, Jr., D. D. Smith, and A. Lathrop. 1994 (1995). The diet and a gastric parasite of *Anolis armouri*, a cybotoid anole from montane pine forests in southern Hispaniola. Herpetol. Nat. Hist. 2(2): 97–100.

Meshaka, W. E., Jr. 1994. Ecological correlates of successful colonization in the life history of the Cuban treefrog, *Osteopilus septentrionalis* (Anura: Hylidae). Unpubl. doctoral dissertation. Florida International Univ., Miami, Florida.

—. 1996. Vagility and the Florida distribution of the Cuban treefrog (*Osteopilus septentrionalis*). Herpetol. Rev. 27: in press.

Moster, J. A., R. Powell, J. S. Parmerlee, Jr., D. D. Smith, and A. Lathrop. 1992. Natural history notes on a small population of *Anolis brevirostris* (Sauria: Polychridae) from altered habitat in the Dominican Republic. Bull. Maryland Herpetol. Soc. 28: 150–161.

Parsons, K. J., R. Powell, and J. H. Greve. 1995. Life History Notes: *Peltophryne guentheri*. Parasitism. Herpetol. Rev. 26: 31–32.

Powell, R., P. J. Hall, and J. H. Greve. 1990*a*. Occurrence of *Skrjabinoptera leiocephalorum* (Spirurida: Physalopteridae) in *Leiocephalus* spp. (Sauria: Iguanidae) from Hispaniola. Jour. Helminthol. Soc. Washington 57: 75–77.

—, R. J. Passaro, and R. W. Henderson. 1992. Noteworthy herpetological records from Saint [sic] Maarten, Netherlands Antilles. Carib. Jour. Sci. 28: 234–235.

—, D. J. Pflanz, J. H. Greve, and D. D. Smith. 1990*b*. Life History Notes: *Leiocephalus semilineatus*. Parasitism. Herpetol. Rev. 21: 60–61.

Schwartz, A. 1952. *Hyla septentrionalis* Duméril and Bibron on the Florida mainland. Copeia 1952: 117–118.

— and R. W. Henderson. 1991. Amphibians and Reptiles of the West Indies: Descriptions, Distributions, and Natural History. Univ. Florida Press, Gainesville.

Resumen.—La rana arborícola cubana *Osteopilus septentrionalis* (Hylidae) es parasitada por el nemátado *Skrjabinoptera scelopori* (Caballero Rodríguez 1971). Las especies de *Skrjabinoptera* son parásitos estomacales de reptiles, principalmente de lagartijas en las familias Iguanidae, Teiidae, Gekkonidae, y Chamaeleonidae. Este registro de un nuevo huésped representa el primer reporte de *S. scelopori* en un anfibio. La ubicación del parásito en el huésped paraténico (transporte), y por lo tanto, no indispensable para completar su ciclo biológico. Se presenta un estudio de especímenes de museo y colectas recientes provenientes de Florida y el Caribe. Se discute la prevalencia de la infección, mayor en individuos colectados en las Islas Caimán, con respecto a su distribucion geográfica, entre sexos, y dentro del cuerpo del huésped.

Résumé.—La grenouille arboricole cubaine *Osteopilus septentrionalis* (Hylidés) est parasitée par un nématode du nom de *Skrjabinoptera scelopori* (Caballero Rodríguez 1971). *Skrjabinoptera* est un parasite des reptiles qui infeste principalement le système digestif des Iguanidés, des Téiidés, des Gekkonidés et des Lézards chaméléonidés. C'est la première fois que l'on signale la présence de *S. scelopori* chez un amphibien. La localisation du parasite dans l'hôte, libre dans la cavité corporelle et enkysté dans les mésentères, donne à penser que *O. septentrionalis* est peut-être un hôte paraténique (transport) et qu'il n'est par conséquent pas essentiel au cycle vital du parasite. L'étude de spécimens de musée et de collections récentes provenant de Floride et des Antilles sont présentés. La prévalence de l'infection, très importante dans les spécimens capturés sur les îles Cayman, est analysée en fonction de la répartition géographique, du sexe des spécimens et de leur taille.

Meshaka, W. E., Jr. 1996. Occurrence of the nematode *Skrjabinoptera scelopori* in the Cuban treefrog, *Osteopilus septentrionalis*: mainland and island comparisons, p. 271–276. *In* R. Powell and R. W. Henderson (eds.), Contributions to West Indian Herpetology: A Tribute to Albert Schwartz. Society for the Study of Amphibians and Reptiles, Ithaca (New York). Contributions to Herpetology, volume 12.

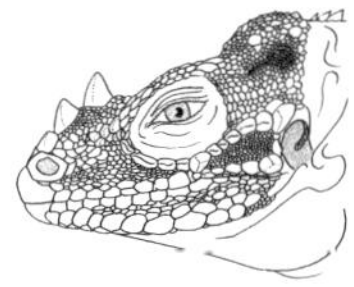

THE EVOLUTION OF SEXUAL SIZE DIMORPHISM IN THE TEIID LIZARD *AMEIVA PLEI*: A TEST OF ALTERNATIVE HYPOTHESES

ELLEN J. CENSKY[1]

Abstract.—The evolution of sexual size dimorphism was studied in an active foraging non-territorial lizard, *Ameiva plei*. Three hypotheses (ecological niche, food limitation, sexual selection) were tested to explain the differences found between a dimorphic population (Anguilla, sexual size dimorphism index: 1.37) and a relatively non-dimorphic population (Dog Island, sexual size dimorphism index: 1.17). Males were significantly larger than females on both islands, and body size in the Anguilla population was significantly larger than that in the Dog Island population. The ecological niche hypothesis was not supported because males and females on Anguilla did not partition prey by size or species, nor were there differences in sizes of prey between the islands. The hypothesis that food limitation contributed to size differences between islands was supported. Insect prey abundances were lower and some food items were missing from Dog Island. In addition, competition for food from hermit crabs was more intense on Dog Island than Anguilla. Lizards on Dog Island grew more slowly and ceased growth at a smaller size than Anguilla lizards. In addition to food limitation, differences in sexual selection may contribute to differences in size dimorphism between islands. Males on Dog Island seem to be unable to guard females through the whole of their fertile periods so large males may have less of a mating advantage on Dog Island than on Anguilla.

INTRODUCTION

MOST STUDIES OF THE EVOLUTION OF SEXUAL SIZE dimorphism have investigated the action of one or more selective agents on size within a population. For example, I have studied sexual selection on size in a population of teiid lizards, *Ameiva plei*, on Anguilla in the Lesser Antilles (Censky 1995*a*). I found that, in this population, males fought over a female at the beginning of the females' receptive period. The male who won the agonistic interactions guarded the female throughout her entire receptive period, and was the only male who mated with the female. Only large males won agonistic encounters and, therefore, only large males were reproductively successful. Large females also had a reproductive advantage because clutch size and number of clutches were positively correlated with body size. The selective advantage of large size was weaker for females because all females large enough to be sexually mature reproduced, although at different rates (i.e., number of clutches in a season was positively correlated with body size). Thus sexual selection can be invoked as the explanation of size dimorphism in *A. plei* on Anguilla. Sexual selection has been implicated in the evolution of size dimorphism in other lizards as well (e.g., Trivers 1976; Ruby 1981; Dugan 1982; Vitt and Cooper 1985; Cooper and Vitt 1987; Stamps 1983; Lewis and Saliva 1987; Anderson and Vitt 1990; Hews 1990).

Another approach to the study of size dimorphism is to compare two or more populations within a species that differ in dimorphism. This approach asks whether explanations for the occurrence of dimorphism within a population can be extended to explain differences in the degree of dimorphism between populations. In this chapter, I apply this comparative approach to size dimorphism in *Ameiva plei* in order to test three hypotheses.

[1] Section of Amphibians and Reptiles, Carnegie Museum of Natural History, Pittsburgh, Pennsylvania 15213, *and* Department of Biological Sciences, University of Pittsburgh, Pittsburgh, Pennsylvania 15260, USA.

The teiid lizard, *Ameiva plei*, is endemic to the islands of the Anguilla Bank, northern Lesser Antilles, where it occurs on three major islands (Anguilla, St. Martin, and St. Barthélémy) and numerous smaller islands. The population on Anguilla, like most of the populations, is sexually dimorphic in size, with males attaining body sizes larger than those of females, and exhibiting a sexual size dimorphism index (SDI) of 1.37 (Censky and Paulson 1992; Censky 1995*a*). The population on Dog Island is smaller in body size and shows less dimorphism (SDI = 1.17) than the Anguilla population. In this chapter I present the results of tests for a number of alternative explanations for the patterns of size dimorphism on the two islands. The successful hypothesis must explain both why *A. plei* is larger and why it is more dimorphic on Anguilla than on Dog Island.

The ecological niche hypothesis (Selander 1966; Schoener 1967, 1969) suggests that males and females evolve sexual size dimorphism to reduce competition for food between the sexes. Larger males feed on larger prey, while the smaller females feed on smaller prey. This hypothesis could explain the pattern of size dimorphism in *Ameiva plei* if males on Anguilla eat larger prey items or different prey items than do females, and if those prey items are absent from Dog Island, thus preventing the evolution of niche differences.

Another hypothesis, the food limitation hypothesis, assumes that, based on independent evidence, large size in males is favored by sexual selection on Dog Island as well as Anguilla, but that food is a limiting factor on Dog Island. This hypothesis predicts that food abundance, or food abundance per lizard, will be less on Dog Island than on Anguilla. In turn, growth rates will be lower and body size smaller on Dog Island. If males are unable to grow to large sizes, then size dimorphism will be less. Many studies have shown that growth rates, and thus body size, are affected by food abundances in lizards (Andrews 1976; Dunham 1978; Lewis 1986). Scarcity of food has also been shown to reduce reproductive output in females (Ballinger 1977). Therefore, clutch size or clutch frequency may also be affected on Dog Island.

Finally, the sexual selection hypothesis assumes that both the extreme size dimorphism on Anguilla and the relative lack of dimorphism on Dog Island can be explained by sexual selection alone. This hypothesis predicts that large males on Anguilla have an advantage in agonistic interactions, access to females (= mate guarding) and mating success, but that no advantage, or less advantage, accrues to larger individuals in Dog Island males. This hypothesis predicts that the mating system is somehow different on Dog Island, so that large size is not advantageous to males in the same way as on Anguilla. Examination of these alternative hypotheses should elucidate important factors influencing body size and sexual size dimorphism in *Ameiva plei*, and should be of general applicability to understanding sexual dimorphism as an evolutionary phenomenon.

Materials and Methods

The study was conducted on Anguilla and Dog Island in the northern Lesser Antilles. Dog Island lies approximately 18 km WNW of Anguilla. The Anguilla study site, described in detail elsewhere (Censky 1995*a*), is primarily a tall manchineel forest (*Hippomane mancinella*) with a salt pond in the center. The study site on Dog Island is similar in that it is primariy manchineel, but with a salt pond adjacent to the site rather than in its center. The study site on Dog Island was smaller than that on Anguilla (0.68 ha versus 1.02 ha, respectively) due to the impenetrable cockspur (*Castella erecta*) which surrounded the Dog Island site. Details of the grid setup and home range analysis are in Censky (1995*a*).

Lizards were marked with colored beads, and weighed (to nearest g) and measured to the nearest 0.5 mm snout-vent length (SVL) (see Censky1995*a* for more details). To test whether a significant difference existed in the degree of size dimorphism between the two populations, I did a two-way analysis of variance (ANOVA) of size with sex and island as main effects.

Lizard activity data were collected using two methods, focal animal observations and instantaneous observations. Methods for gathering those data are described in Censky (1995*a*).

Home range size was analyzed to determine differences between sexes and islands. Lizards on Dog Island were warier than those on Anguilla. As a result, fewer observations were obtained for some analyses. To include as many individuals as possible in the home range analysis, I compared all individuals with 80% (20 observation points) cumulative home range (see Censky 1995*a*) on both islands. Rose (1982) considered 80% an adequate representation of home range when analyzing lizard data. Dunham's Home Range Analysis (5.0) computer program was used to calculate home range size. All data were log transformed to normalize.

Food abundance was sampled by using 30 x 35 cm boards coated with Tanglefoot™. For each sample period, six boards were placed at randomly generated coordinates within the study site. After 24 h the boards were removed and insects were collected from the boards, rinsed in turpentine, and placed in ethanol. Because of logistics, the two study sites could not be sampled at the same period of time. Within each week, both study sites were sampled for 2–3 consecutive days. Samples were taken between 15 June and 13 July 1992. Insect abundances between islands were compared using a Mann-Whitney U-test, and ANOVA was used to compare sizes of insects between the islands.

Insects were identified to order, when possible. To determine whether available prey were of equal size on the two islands, insects from Tanglefoot™ samples were measured (0.1 mm) and compared between the islands using ANOVA. In order to determine whether relatively immobile items (such as snails and *Anolis* eggs) were available for food, litter samples from randomly distributed 1-m^2 plots were sorted by hand at the sites. Eight samples from Anguilla and ten samples from Dog Island were collected and placed in 70% ethanol.

To determine what lizards were eating, lizards were collected and preserved, and their stomach contents were removed and analyzed. Because I did not want to sacrifice study animals in either site, lizards were collected from areas adjacent to the main study sites. Some prey items found in the stomachs may not appear as Tanglefoot™-sampled items because lizards feed on immobile prey and prey items may vary from site to site. Likewise, some trap items may not appear in stomach samples. Nevertheless, prey abundance based on trap samples should reflect the overall abundance of prey on each island. Stomach items were identified to order, where possible. The largest prey item in each stomach was measured. Because prey items were often partially digested, elytra or legs, etc. were compared to whole insects from the Tanglefoot™ samples for size. These whole insects were then measured and comparisons were made using ANOVA. Lizards were also observed feeding and the type of prey item recorded, when possible.

Hermit crabs (*Paguristes* sp.) were very common in the Dog Island study site. To determine whether hermit crabs might have an impact on food resources and thus be in competition with *Ameiva* for food, an exclosure experiment was conducted. A 15 x 15 m exclosure was constructed of PVC pipes and wire mesh in the Dog Island study site. The purpose of the exclosure was to keep hermit crabs out. Twenty different lizards were introduced into the exclosure, two or three at a time (a normal density for lizards). Each was allowed time to adjust. Food was then introduced into the exclosure and each lizard was scored as to whether the food was found and eaten. Additionally, food was placed outside the exclosure at 20 different sites, and a record was made of which individual found the food first and whether it was eaten. Likewise, on Anguilla food was placed in 20 different sites and data recorded as to which individual found it first and ate it. A G-test was employed to compare rates of suc-

TABLE 1. Snout-vent length (± standard deviation) of male and female *Ameiva plei* on Anguilla and Dog Island.

Island	Males	Females
Anguilla	117.86 ± 34.58 (n = 59)	86.25 ± 16.19 (n = 57)
Dog Island	91.79 ± 13.18 (n = 38)	78.15 ± 6.99 (n = 41)

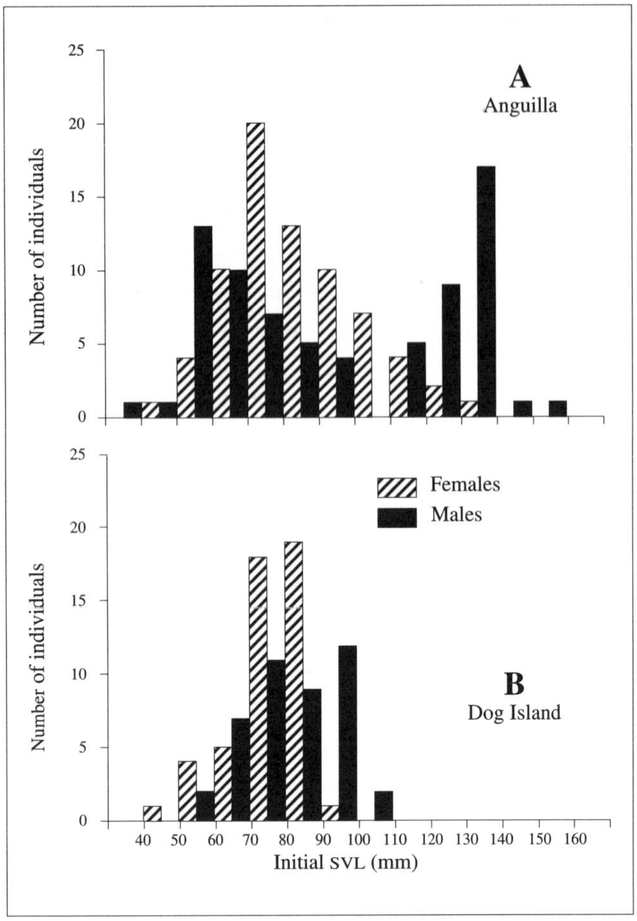

FIGURE 1. Size distribution of male and female *Ameiva plei* from Anguilla (A) and Dog Island (B) study sites. Sizes are those measured after the first capture of an individual.

cess of *Ameiva* between islands. "Natural" food would have been ideal for the experiments, but food was scarce and hard to find on Dog Island. In order to maintain the same conditions for each run of the experiment, bananas and mealworms were used in both study sites. These foods resemble natural alternatives because *Ameiva* are opportunistic, often feeding on insects and whatever fruits are in season (Lewis 1986; Vitt and Colli 1994; E. J. Censky, personal observation).

RESULTS

Size and Sexual Size Dimorphism

Mean SVLs of adult males and females from the Anguilla and Dog Island study sites are given in Table 1. On both islands, males and females mature at approximately 64 mm SVL, as determined by smallest reproductive individual (Censky 1995*b*). This size is attained at one year of age, determined by mark-recapture data. Males were significantly larger than females on both Anguilla and Dog Island ($F_{1, 193}$ = 9.151, $P = 0.003$; Figure 1). The SDI (males/females) for Anguilla was 1.37 and for Dog Island was 1.17.

Home Range and Density

Home range was positively correlated with body size on both Anguilla ($r^2 = 0.36$, $P = 0.04$, $n = 24$) and Dog Island ($r^2 = 0.28$, $P = 0.04$, $n = 14$; Figure 2). Home range size was not significantly different between the sexes on Anguilla (Censky 1995*a*). Too few males and females with home range estimates were available to determine this relationship for Dog Island, but it was assumed to be the same as for Anguilla. To determine whether a difference in home range size existed between the islands, an analysis of covariance (ANCOVA) of log home range with island as a factor and log SVL as a covariate was performed. The effect of body size on home range did not differ between islands ($F_{1, 31} = 0.001$, $P = 0.97$), nor did I find a difference between the islands in home range after controlling for body size effect ($F_{1, 31} = 1.05$, $P = 0.31$). The density of lizards was higher on Dog Island than on Anguilla (91 versus 74 lizards/ha, respectively).

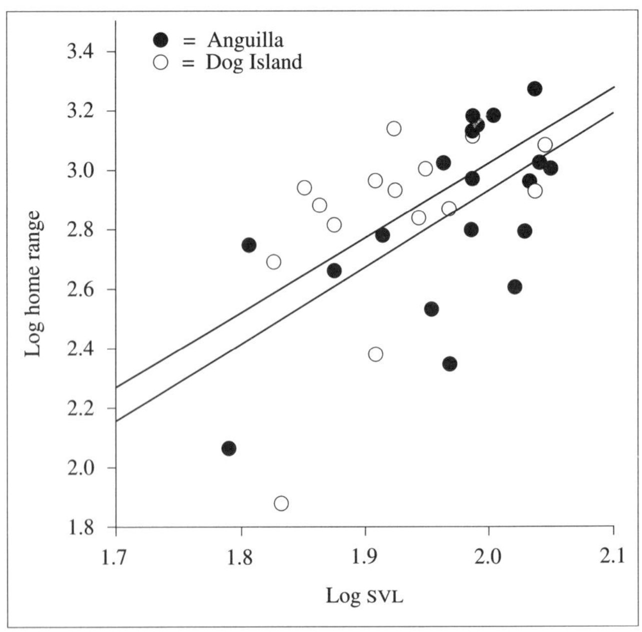

FIGURE 2. Regression of *Ameiva plei* home range (log) on SVL (log) for Anguilla (lower line) and Dog Island (top line).

Growth

Absolute growth rate was calculated by dividing the change in body size by the number of days between captures. Month-month and year-year recaptures were measured, but only the year-year measurements were analyzed. This should alleviate any problems caused by month-month fluctuations in food supply and its reflection on short-term growth, which might have arisen in month-month measurements as they were not all taken within the same two months. The growth rates were plotted against initial SVLs (Figure 3). Male and female growth rates decreased with increasing size on both islands. To determine whether a difference in growth rates existed between the sexes, an ANCOVA with sex as a factor and SVL as a covariate was performed. The effect of body size on growth rate did not differ between the sexes on Dog Island ($F_{1, 18} = 0.54$, $P = 0.47$) or Anguilla ($F_{1, 30} = 3.79$, $P = 0.06$). However, a sex difference in growth rates was seen after controlling for body size effects for both Anguilla ($F_{1, 30} = 12.34$, $P = 0.001$) and Dog Island ($F_{1, 18} = 32.02$, $P < 0.001$). This indicates that the average growth rate of males is higher than female growth rate, and females stop growing at a smaller size than males on both islands. In addition, an ANCOVA revealed that the effect of body size on growth rates did not differ between islands for males ($F_{1, 20} = 3.55$, $P = 0.8$), but a difference was evident after controlling for body size effects ($F_{1, 20} = 4.19$, $P = 0.05$). Likewise, the effect of body size on growth rates was not significantly different between females on Anguilla and Dog Island ($F_{1, 28} = 0.17$, $P = 0.68$), but a difference was evident after controlling for body size effects ($F_{1, 28} = 25.61$, $P < 0.001$). All of these differences indicate that the average growth rates for Dog Island lizards are lower than for Anguilla lizards and that Dog Island lizards stop growing at a smaller size than do Anguilla lizards.

Prey Availability and Size

Abundance of all insects in the study sites was relatively constant. Eleven prey types were identified on the Tanglefoot™ traps. Nine of these were found on both islands (Table 2). Adult Diptera formed a large portion of the insect samples, yet a relatively small portion of stomach contents (see below). Therefore, insect samples were also analyzed excluding Diptera. A Mann-Whitney U-test verified that abundance of insects other than Diptera was lower on Dog Island than on Anguilla ($Z = 2.003$, one-tailed $P = 0.02$; Figure 4). No significant difference existed in the size of prey captured on Tanglefoot™ traps on the two islands ($F_{1, 209} = 2.33$, $P = 0.14$; Figure 5).

The most common prey items found in the stomachs of lizards of both sexes on both islands were Coleoptera (Table 2). However, these were not the most common prey taken on Tanglefoot™ traps on either island. Hymenoptera, Lepidoptera, Gryllidae, and Blattodea occurred in traps at higher percentages.

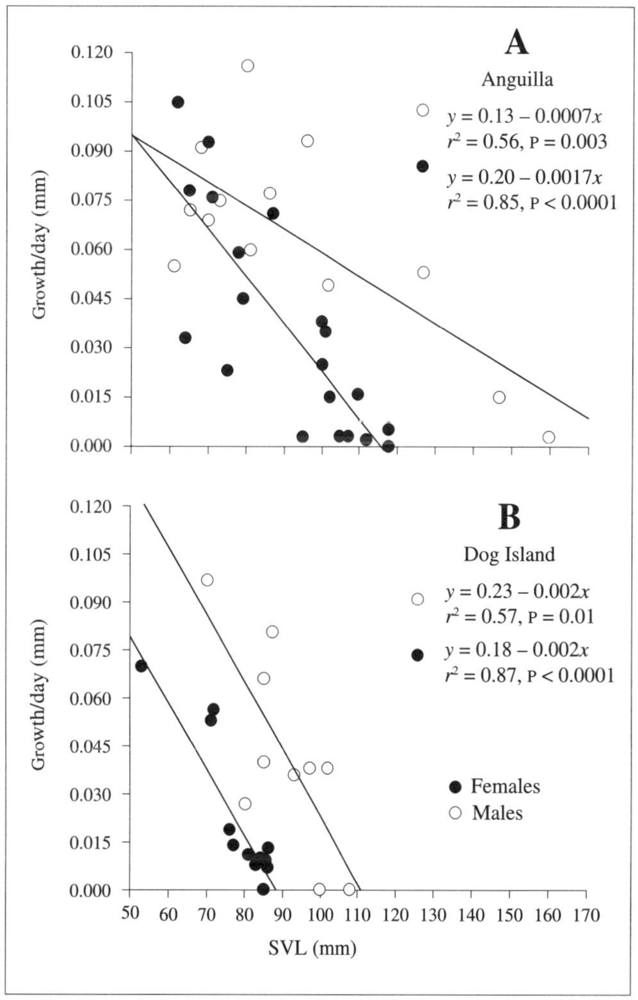

FIGURE 3. Regression of growth rate (mm/day) on SVL for male (upper lines) and female (lower lines) *Ameiva plei* on Anguilla (A) and Dog Island (B).

I measured the largest prey item in the stomachs of each of 26 males and 16 females from Anguilla to determine whether a difference in maximum prey size existed between the sexes. No significant difference was evident in the size of the largest prey taken by males and females ($F_{1,41} = 0.16$, $P = 0.68$; Figure 6). The largest prey item overall (*Blatteria*, 3.4 mm) was taken by a small female (72 mm SVL).

Eight leaf litter samples from Anguilla and 10 leaf litter samples from Dog Island were analyzed for "immobile" prey. The only immobile prey found in the samples were snails. An average of 9.5 (± 6.9) snails were found per sample in the Anguilla leaf litter. No snails were found in the Dog Island samples.

Observations of individuals feeding verified that snails are taken as prey items by both males and females on Anguilla. Prey items were identified in only 23% (59) of the 252 observed feedings. Snails comprised 59% (35) of identified items. Other items observed being eaten were *Blatteria* (8), *Anolis* eggs (1), Gryllidae (4), and fruit (11). These observations were on Anguilla. While foraging behavior (active digging, searching, and tongue flicking) was frequent on Dog Island, only nine actual feedings were observed. In only one case was the food identified (fruit of cockspur).

Competition

In the experimental trials in which food was placed at scattered sites on the grounds, the food items were obtained by *Ameiva plei* in 16 of 17 cases on Anguilla. On Dog Island, *A. plei* obtained the food in 0 of 20 cases in which food was placed outside the exclosures, a significantly lower rate of success ($G = 43.01$, $P << 0.0001$). In all the Dog Island trials, hermit crabs obtained the food instead of

TABLE 2. Insect samples for Anguilla and Dog Island exclusive of adult Diptera. Number in parentheses is total sample.

Food item	Trap samples Number (%)		Stomach contents Number (%)			
	Dog Island (*n* = 66)	Anguilla (*n* = 60)	Dog Island males (*n* = 14)	Dog Island females (*n* = 16)	Anguilla males (*n* = 35)	Anguilla females (*n* = 27)
Hymenoptera	13 (9.9)	32 (19.6)	6 (9.0)	2 (2.1)	23 (12.0)	6 (3.7)
Neuroptera	—	—	—	2 (2.1)	1 (0.5)	1 (0.6)
Psocoptera	1 (0.8)	1 (0.6)	—	—	—	—
Homoptera	15 (11.5)	10 (6.1)	12 (17.9)	10 (10.6)	3 (1.6)	8 (5.0)
Hemiptera (Cydnidae)	—	—	—	1 (1.1)	8 (4.2)	5 (3.1)
Lepidoptera (larvae)	—	—	2 (3.0)	1 (1.1)	4 (2.0)	3 (1.9)
Lepidoptera (adults)	14 (10.7)	49 (30.1)	2 (3.0)	—	2 (1.0)	4 (2.5)
Gryllidae	30 (22.9)	26 (16.0)	—	—	1 (0.5)	1 (0.6)
Blattodea	19 (14.5)	22 (13.5)	5 (7.5)	1 (1.1)	21 (11.0)	6 (3.7)
Coleoptera	22 (16.8)	6 (3.7)	36 (53.7)	62 (65.9)	54 (28.4)	59 (36.9)
Diptera (maggots)	—	—	—	3 (3.2)	20 (10.5)	6 (3.7)
Isopods	2 (1.5)	1 (0.6)	—	6 (6.4)	13 (6.8)	23 (14.3)
Nematodes	—	—	—	1 (1.1)	11 (5.8)	10 (6.2)
Spiders	13 (9.9)	8 (4.9)	1 (1.5)	—	3 (1.6)	—
Solpyugidae	—	—	—	1 (1.1)	—	—
Snail	—	7 (4.3)	—	—	6 (3.2)	3 (1.9)
Crustacean	—	—	—	1 (1.1)	1 (0.5)	—
Anolis eggs	—	—	3 (4.4)	3 (3.2)	—	2 (1.2)
Seeds	—	—	—	—	11 (5.8)	21 (13.0)
Fruit	—	—	—	—	2 (1.0)	—
Other	2 (1.5)	1 (0.6)	—	—	6 (3.2)	2 (1.2)

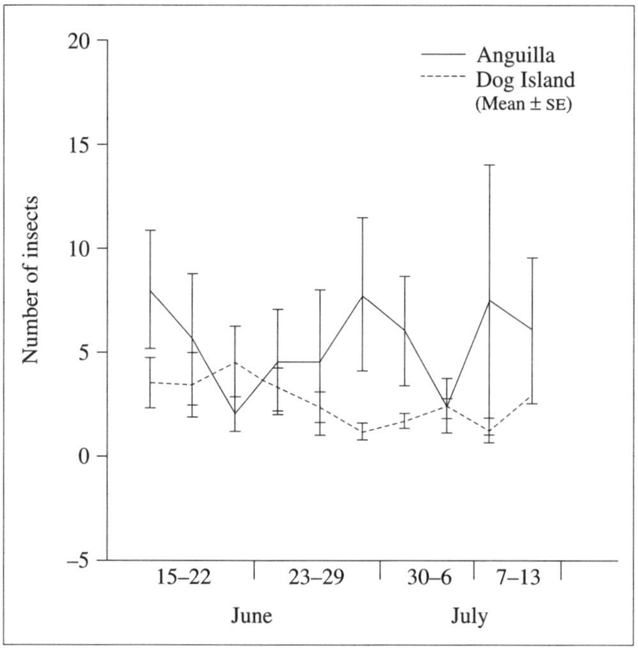

FIGURE 4. Average number of insects per 24-h period collected on Tanglefoot™ traps, excluding adult Diptera.

Ameiva. To test whether the food used was acceptable to *Ameiva* on Dog Island, lizards were tested with the same types of food in exclosures. In 18 of 20 exclosure trials, lizards found and ate the food, although in all cases hermit crabs were trying to enter the exclosure within seconds of the food being placed in the exclosure, and always before lizards had found the food. This verifies that, given the chance on Dog Island, lizards did eat the food presented.

I also compared temporal latency to food discovery between Anguilla lizards and lizards in the Dog Island exclosure to determine whether lizards on Dog Island found food faster, possibly indicating that they foraged more intensely. Individuals in the exclosure on Dog Island found food in significantly less time than did those on Anguilla (7.98 ± 6.69 min versus 43.98 ± 39.35 min, respectively, $F = 15.46$, $P < 0.001$), though it should be noted that Anguilla lizards were not in exclosures.

Agonistic Interactions

On Anguilla, larger males won significantly more intrasexual agonistic encounters than smaller males, and females always won male-female interactions despite the fact that females were always smaller than males (Censky 1995*a*). On Dog Island, I observed a total of 17 agonistic interactions in which both individuals were identified. These interactions ranged from a small lunge of one lizard toward another, to a chase of up to several meters (Censky 1995*a*). Eight of these were male-male interactions. In all but one case, the larger male won the encounter. A sign test confirmed that the larger male in the dyad won significantly more often than the smaller male ($Z = 1.84$, $P = 0.03$). Six agonistic encounters were between males and females. In all cases the male was larger than the female, and in all cases the larger individual in the dyad won ($Z = 2.23$, $P = 0.01$).

Mate Guarding

On Anguilla, large males guard females for their entire receptive period (1–4 days). I observed 16 copulations and all were by guarding males. While females never copulated with more than one male within an individual receptive period, the guarding male did copulate with the guarded female more than once during that period. Seven of 16 observed copulations were repeat within-bout copulations. Only one case occurred where the female was guarded by a different male on a successive receptive period (Censky 1995*a*). Males that guarded females were

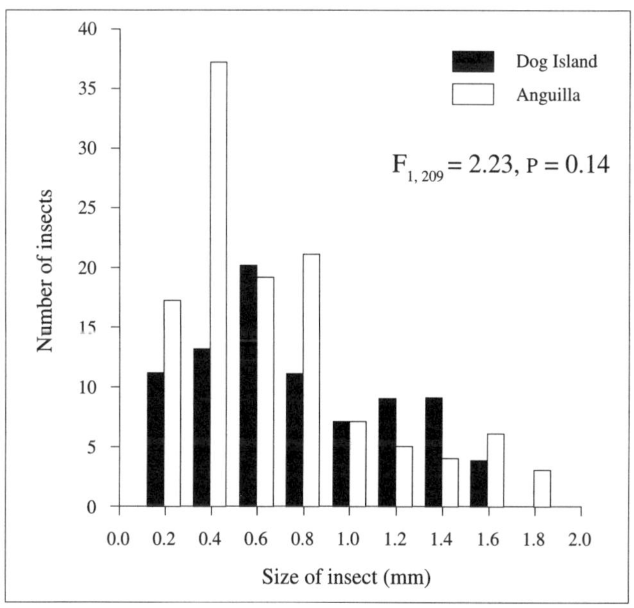

FIGURE 5. Size distribution of insects (mm) collected on Tanglefoot™ traps, separated by island.

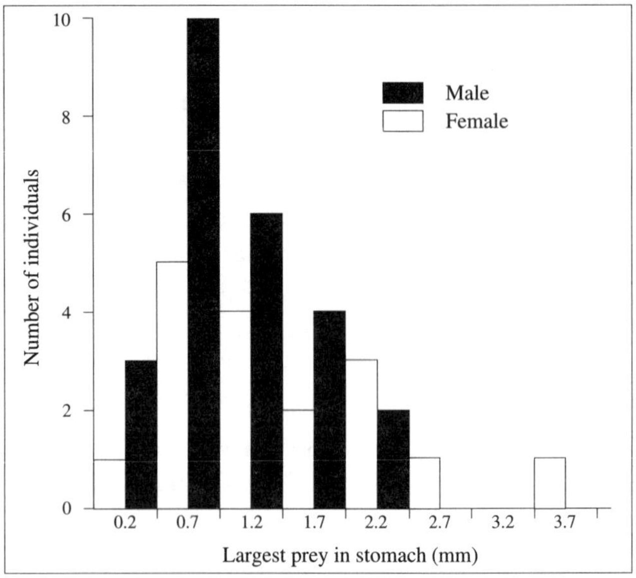

FIGURE 6. Size distribution of largest prey items found in *Ameiva plei* stomachs, separated by sex.

larger than males not seen guarding (Z = 3.15, P = 0.0008).

On Dog Island, males were very rarely seen in attendance of females. Only ten pairs were seen and none of them was seen on consecutive days. Only two pairs were seen together for an entire day, and these pairs were not seen to copulate. Only two copulations were observed on Dog Island, and in both cases the male left the female immediately after mating, and resumed foraging. All males seen with females were large males. Of the 38 adult males in the Dog Island study site, only five were seen in attendance of females. Those five were in the largest size classes (Figure 7). A Mann-Whitney U-test verified that males seen with females were larger than those not seen in attendance of females (Z = 2.72, P = 0.003). Because males did not attend females for extended periods as on Anguilla (Censky 1995*a*), I was less likely to observe matings on Dog Island. This could account for the large number of males in the largest size classes which were not observed mating.

DISCUSSION

Males were significantly larger than females on Anguilla as well as Dog Island (Figure 1). In addition, the lizards on Anguilla were larger than the lizards on Dog Island, with males on Anguilla significantly larger in SVL than males on Dog Island. Likewise, females on Anguilla were significantly larger than on Dog Island. The population on Anguilla is also more sexually dimorphic in size than the Dog Island population (1.37 versus 1.17). Below I summarize the evidence bearing on the hypotheses to explain the difference in degree of sexual size dimorphism between islands.

Ecological Niche Hypothesis

Both Selander (1966) and Schoener (1967) proposed the hypothesis that males and females evolved sexual size dimorphism to reduce competition for food between the sexes. A number of studies have demonstrated that sexual size dimorphism, once evolved, leads to resource partitioning between the sexes (Powell and Russell 1985; Camilleri and Shine 1990; Houston and Shine 1993; Madsen and Shine 1993). For *Ameiva plei*, this hypothesis predicts that larger males on Anguilla should feed on larger prey and smaller females on smaller prey, or that the two sexes should have little overlap in prey consumption. The hypothesis could explain the relative lack of dimorphism on Dog Island if large prey were

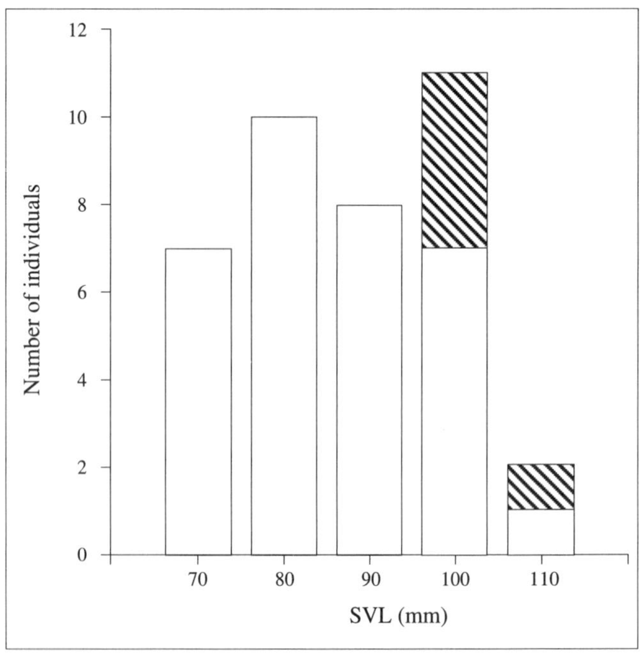

FIGURE 7. Size distribution of male *Ameiva plei* on Dog Island. Cross-hatched bar indicates males seen copulating with a female or in attendance of a female.

absent on Dog Island or if certain sex-specific prey are absent, thus preventing specialization on certain prey by males.

No significant difference existed in the size of insects found on the two islands. In addition, nine of 11 prey types were found on both islands, thus indicating that type and size of resource are not a limiting factor on Dog Island.

Ameiva plei is an opportunistic forager and is extremely catholic in its diet, with both males and females feeding on many types of invertebrates and fruits as well as on *Anolis* eggs (Table 2). Lewis (1989) reported a similar diet for *A. exsul* in Puerto Rico. Coleoptera were the most frequent prey in stomachs of both males and females on both islands (Table 2) indicating overlap in type of prey taken by the sexes. In addition, stomach content analysis revealed no difference in the size of prey taken by males and females on Anguilla, counter to the prediction (Figure 6). In fact, the largest prey item observed overall, *Blatteria* (3.4 mm), was taken by a small female (72 mm SVL). While Coleoptera were found in lizard stomachs at a higher percentage than collected on Tanglefoot™ traps, this does not nessesarily indicate selectivity on the part of *Ameiva*. Roaches, crickets, and ants, which accounted for the bulk of the Tanglefoot™ sample, are far-ranging insects and may wander across traps at a higher rate than beetles which tend to fly short distances, land and forage in smaller areas (R. Davidson, personal communication). These data do not support the ecological niche hypothesis.

Food Limitation Hypothesis

Numerous studies have demonstrated that growth rates, and therefore absolute size, are influenced by food abundance (Andrews 1976; Dunham 1978; Lewis 1986) and prey size (Madsen and Shine 1993; Perez-Mellardo and de la Riva 1993). In all cases, growth rates were lower when food abundance was lower, whether different localities were compared or the same locality in different years. In addition, Madsen and Shine (1993) found that the degree of sexual size dimorphism in *Natrix natrix* was lower on a food-limited island than on the mainland as a direct result of reduced prey availability. In *N. natrix*, however, males are less affected by food scarcity than are females, due to the fact that, in general, males tend to eat less and grow more slowly than females (Madsen 1983). Food limitation could explain the pattern of size dimorphism in *Ameiva plei* if large male size was favored on both Anguilla and Dog Island by some selection pressure, such as sexual selection, and if food was more limited on Dog Island, preventing males from growing to large size on Dog Island. The hypothesis predicts that food should be scarcer overall, and less obtainable by *Ameiva* on Dog Island than on Anguilla or both, that growth rates would be lower on Dog Island, and that growth should stop at a smaller size on Dog Island, particularily among males.

Food was scarcer on Dog Island in a number of aspects. First, if we exclude adult Diptera (found rarely in *Ameiva* stomachs, yet over-represented in traps) from the Tanglefoot™ samples, Dog Island had a lower trap-capture rate per 24-h period (Figure 4). In addition, *Ameiva* are known to feed on fruits (Lewis 1986; E. J. Censky, personal observations) and more species of fruit trees occur on Anguilla (*Hippomane mancinella*, *Canella winterana*, *Coccoloba uvifera*, *Tamarindus indica*, *Malpighia emarginata*, *Meliococcus bijugatus*) than on Dog Island (*Hippomane mancinella*, *Castella erecta*). These fruits ripen at different times of the year, allowing for a more steady food source on Anguilla. All of these fruits were readily eaten by *A. plei* (E. J. Censky, personal observations; analysis of stomach contents). Other sources of food on Anguilla, but absent from Dog Island, were land snails and possibly scarab beetles. Land snails were very abundant in the Anguilla study site and have been recorded as lizard food. No obvious additional food source occurred on Dog Island which might compensate for the absence of snails and many fruit trees.

Food abundance for *Ameiva* may also be affected by competition from the hermit crab (*Paguristes* sp.), which occurs on both islands. Hermit crabs are considered scavengers (Barnes 1974), foraging in the leaf litter, yet capable of climbing and digging for food. They are known to feed on carrion, insects, fruit, and table scraps (E. J. Censky, personal observations). Likewise, much of the foraging activity of *Ameiva* is in the leaf litter, along with digging for food. Absolute densities of hermit crabs are not known; however anecdotal evidence indicates that hermit crab numbers are very high on Dog Island. For example, in one instance I inadvertently dropped food on Dog Island, and within three minutes counted a total of 302 hermit crabs in a 2-m^2 area surrounding the item. Food experiments demonstrated that hermit crabs were more efficient foragers on Dog Island, reaching and consuming food placed on the ground more quickly than *Ameiva*. The exclosure experiments showed that lizards would consume the food, given a hermit crab-free environment. On Anguilla, however, lizards reached and consumed the food in the majority of cases.

Reduced food abundances on Dog Island may be manifested in lower growth rates and ultimately in the smaller size of individuals. Males on Anguilla have higher growth rates and grow to larger sizes than females, and this has been attributed to the investment of more energy into reproduction than growth in females and to a mating advantage of large size in males (Censky 1995*a*). Likewise, males grow faster and larger than females on Dog Island. The explanation may be the same. However, males and females on Anguilla have higher growth rates than males and females on Dog Island and lizards on Dog Island cease growing at smaller sizes.

Case (1978) predicted that, in the absence of predators and competitors, island population densities should rise. Interestingly, the density of lizards on Dog Island is higher than on Anguilla, yet the home ranges are the same sizes (Figure 2). These lizards may be foraging at a maximum home range size imposed by energy constraints. Many predators occur on Anguilla (*Alsophis rijersmai*, *Falco sparvarius*, *Nyctanassa violacea*, *Margarops fuscatus*, rats, cats, dogs). Rats are the only predators known to occur on Dog Island. However, as indicated above, a potentially major competitor exists in the hermit crab. This should reduce the density of Dog Island lizards, yet it is higher than the density of lizards on Anguilla.

Case (1978) also predicted that, in the absence of predators, a potential exists for over-exploitation of a food resource, and that these populations should exist at lower levels of food, and therefore smaller sizes. Prey availability may be a major determinant of body size, as suggested by Madsen and Stille (1988). Over-exploitation of the food resource on Dog Island may occur because of increased competition and increased density of lizards, with the result being smaller size of lizards. These data strongly suggest that growth rates of Dog Island lizards may be reduced because of low food availability, but whether lower growth rates and thus smaller size on Dog Island are an evolved response to lower food supply or an environmental effect is unknown. Provisioning or transplant experiments would be needed to test these alternatives.

Why might the sexual dimorphism index respond to food shortage? One possible explanation is that the poor nutritional status due to food shortage manifests itself differently in the two sexes. This has been shown to be the case in snakes (Madsen and Shine 1993) and red deer (Clutton-Brock et al. 1982). Though both males and females are smaller on Dog Island than on Anguilla, males average 26 mm smaller on Dog Island whereas females average only 8 mm smaller. Females on both islands grow more slowly than do males, presumably because they invest energy into reproduction as opposed to growth (Censky 1995*a*, 1995*b*). On Anguilla, size of clutch, as well as number of clutches, is significantly correlated with SVL (Censky 1995*b*). On Dog Island, clutch size is positively correlated with SVL, but females do not produce a second clutch in a season, or if they do, a long period of time occurs between clutches (Censky 1995*b*). Reduced food supply may affect reproduction more than growth of females on Dog Island, whereas males are affected with drastically reduced growth rates. The evidence strongly supports food limitation as at least a partial explanation of reduced size and size dimorphism in the Dog Island population.

Sexual Selection Hypothesis

Many studies have shown that large size in male reptiles confers a selective advantage in mating success (e.g., Trivers 1972; Shine 1978; Vitt and Cooper 1985; Cooper and Vitt 1987). This is the case for *Ameiva plei* on Anguilla (Censky 1995*a*). Only large males guard females and only males that guard females mate. Sexual selection alone could explain the pattern of dimorphism on the two islands if large males have less of a mating advantage on Dog Island than on Anguilla.

A selective advantage was evident for large size in males on Dog Island. Large individuals won agonistic encounters, and only large males were seen attending females or in copulation. However, a limitation to the advantage may exist. On Anguilla, an individual male guarded a female throughout her receptive period, copulating with her several times during this period (Censky 1995*a*). Males spent significantly less time foraging while guarding females, which is undoubtedly energetically expensive. Most of the time was spent in active attendence of the female, either watching and guarding, or trailing. Males on Dog Island were never seen in attendance of females for more than one day. In addition, the two observed copulations involved no mate guarding. The male pursued the female, almost at a chase, until the female allowed the male to copulate. In both cases, immediately after copulation, the male left the female to forage. Food limitation may be what is driving the differences in mating systems between Dog Island and Anguilla. Males on Dog Island may not be able to gain enough energy to forego foraging for an extended period of time, and therefore do not guard females. Although large individuals on Dog Island win agonistic encounters, males may not be assured of paternity because they do not guard females. Therefore, selection for large size in males may be less intense on Dog Island than on Anguilla because of the overriding food limitation on Dog Island. However, an advantage of large size in mating is apparent on Dog Island with males seen in attendance of females that are larger, on average, than males not seen with females; data are insufficient to test whether the advantage is weaker than on Anguilla.

Conclusion

These data do not support the ecological niche hypothesis as an explanation of the difference in sexual size dimorphism between Anguilla and Dog Island. However, the food limitation and sexual selection hypotheses, taken together, do explain much of the variation found on the two islands. Whereas sexual selection is a strong determinant of large size in *Ameiva plei* on Anguilla, it appears to be mediated by a reduced food supply on Dog Island. Food was less abundant on Dog Island than on Anguilla, and as a result, growth rates were reduced. Apparently, ecological factors have an overriding influence, not only on phenotypic expression, but on behavioral expression in the form of mating strategy, with a shift from mate guarding where food is adequate, to strict promiscuity where food supply is low. The mating strategy for *A. plei* may be extremely plastic, shifting from promiscuity to mate guarding, depending on the availability of food.

Acknowledgments

I thank the late C. J. McCoy, Curator of Amphibians and Reptiles at the Carnegie Museum of Natural History, who was my mentor and friend, and without whom I would not have ventured down the path to a Ph.D. My special thanks are extended to my committee, William Searcy, Steven Gaulin, Joseph Merritt, and Robert Raikow, for direction and support throughout this study. My field work could not have been accomplished without the help of many Anguillans. I extend a deep-felt thank you to David Carty (former Parliamentary Secretary–Environment) for logistical help and for the many interesting conversations about Anguilla's environment and history. Special thanks go to Edward Carty and Roland Hodge for much appreciated boat transport to Dog Island, and to Rev. John Gumbs for the use of his land for a study site. I would also like to thank David Duvall for catching that one crippled lizard, and for all the help on Dog Island. Finally, I would like to thank Robert Davidson for his help with invertebrate identification, and Laurie Vitt and anonymous reviewers for their helpful comments. This study was supported by a grant from the National Geographic Society (4497–91) and the Carnegie Museum of Natural History (Edward O'Neil Research Fund).

Literature Cited

Anderson, R. A. and L. J. Vitt. 1990. Sexual selection versus alternative causes of sexual dimorphism in teiid lizards. Oecologia 84: 145–157.

Andrews, R. M. 1976. Growth rate in island and mainland anoline lizards. Copeia 1976: 477–482.

Ballinger, R. E. 1977. Reproductive strategies: food availability as a source of proximal variation in a lizard. Ecology 58: 628–635.

Barnes, R. D. 1974. Invertebrate Zoology. W. B. Saunders Co., Philadelphia, Pennsylvania.

Camilleri, C. and R. Shine. 1990. Sexual dimorphism and dietary divergence: differences in trophic morphology between male and female snakes. Copeia 1990: 649–658.

Case, T. J. 1978. A general explanation for insular body size trends in terrestrial vertebrates. Ecology 59: 1–18.

Censky, E. J. 1995*a*. Mating strategy and reproductive success in a teiid lizard, *Ameiva plei*. Behaviour 132: 529–557.

—. 1995*b*. Reproduction in two Lesser Antillean populations of *Ameiva plei* (Teiidae). Jour. Herpetol. 29: 553–560.

— and D. R. Paulson. 1992. Revision of the *Ameiva* (Reptilia: Teiidae) of the Anguilla Bank, West Indies. Ann. Carnegie Mus. 61: 177–195.

Clutton-Brock, T. H., F. E. Guinness, and S. D. Albon. 1982. Red Deer. Behavior and Ecology of Two Sexes. Univ. Chicago Press, Chicago, Illinois.

Cooper, W. E. and L. J. Vitt. 1987. Deferred agonistic behavior in a long-lived scincid lizard *Eumeces laticeps*: field and laboratory data on the roles of body size and residence in agonistic strategy. Oecologia 73: 321–326.

Dugan, B. 1982. The mating behavior of the green iguana, *Iguana iguana*, p. 320–341. *In* G. M. Burghardt and A. S. Rand (eds.), Iguanas of the World: Their Behavior, Ecology and Conservation. Noyes, Park Ridge, New Jersey.

Dunham, A. E. 1978. Food availability as a proximate factor influencing individual growth rates in the iguanid lizard *Sceloporus merriami*. Ecology 59: 770–778.

Hews, D. 1990. Examining hypotheses generated by field measures of sexual selection of male lizards, *Uta palmeri*. Evolution 44: 1956–1966.

Houston, D. and R. Shine. 1993. Sexual dimorphism and niche divergence: feeding habits of the Arafura filesnake. Jour. Anim. Ecol. 62: 737–748.

King, R. B. 1989. Body size variation among island and mainland snake populations. Herpetologica 45: 84–88.

Lewis, A. R. 1986. Body size and growth in two populations of the Puerto Rican ground lizard (Teiidae). Jour. Herpetol. 20: 190–195.

—. 1989. Diet selection and depression of prey abundance by an intensively foraging lizard. Jour. Herpetol. 23: 164–170.

— and J. E. Saliva. 1987. Effects of sex and size on home range, dominance, and activity budgets in *Ameiva exsul* (Lacertilia: Teiidae). Herpetologica 43: 374–383.

Madsen, T. 1983. Growth rates, maturation and sexual size dimorphism in a population of grass snakes, *Natrix natrix*, in southern Sweden. Oikos 49: 129–132.

— and R. Shine. 1993. Phenotypic plasticity in body sizes and sexual size dimorphism in European grass snakes. Evolution 47: 321–325.

— and B. Stille. 1988. The effect of size dependent mortality on colour morphs in male adders, *Vipera berus*. Oikos 52: 73–78.

Perez-Mellardo, V. and I. de la Riva. 1993. Sexual size dimorphism and ecology: the case of a tropical lizard, *Tropidurus melanopleurus* (Sauria: Tropiduridae). Copeia 1993: 969–976.

Powell, G. L. and A. P. Russell. 1985. Growth and sexual size dimorphism in Alberta populations of the eastern short-horned lizard, *Phrynosoma douglassi brevirostre*. Canadian Jour. Zool. 63: 139–154.

Rose, B. 1982. Lizard home ranges: methodology and functions. Jour. Herpetol. 16: 253–269.

Ruby, D. E. 1981. Phenotypic correlates of male reproductive success in the lizard, *Sceloporus jarrovi*, p. 96–107. *In* R. D. Alexander and D. W. Tinkle (eds.), Natural Selection and Social Behavior. Chiron Press, New York.

Schoener, T. W. 1967. The ecological significance of sexual dimorphism in size in the lizard *Anolis conspersus*. Science 155: 474–477.

—. 1969. Models of optimal size for solitary predators. Amer. Nat. 103: 277–313.

Selander, R. K. 1966. Sexual dimorphism and differential niche utilization in birds. Condor 68: 113–151.

Shine, R. 1978. Sexual size dimorphism and male combat in snakes. Oecologia 33: 269–277.

Stamps, J. 1983. Sexual selection, sexual dimorphism, and territoriality, p. 169–204. *In* R. B. Huey, E. R. Pianka, and T. W. Schoener (eds.), Lizard Ecology: Studies of a Model Organism. Harvard Univ. Press, Cambridge, Massachusetts.

Trivers, R. L. 1972. Parental investment and sexual selection, p. 136–179. *In* B. Campbell (ed.), Sexual Selection and the Descent of Man 1871–1971. Aldine Publ. Co., Chicago, Illinois.

—. 1976. Sexual selection and resource-accruing abilities in *Anolis garmani*. Evolution 30: 253–269.

Vitt, L. J. and G. R. Colli. 1994. Geographical ecology of a Neotropical lizard: *Ameiva ameiva* (Teiidae) in Brazil. Canadian Jour. Zool. 72: 1986–2008.

— and W. E. Cooper. 1985. The evolution of sexual dimorphism in the skink *Eumeces laticeps*: an example of sexual selection. Canadian Jour. Zool. 63: 995–1002.

Resumen.—Se estudió la evolución de dimorfismo sexual en tamaño de *Ameiva plei*, una lagartija forrajera activa no territorialista. Tres hipótesis (nicho ecológico, limitación de alimento, selección sexual) fueron examinadas para explicar las diferencias encontradas entre una población dimórfica (Anguila, índice de dimorfismo sexual en tamaño 1.37) y una población relativamente no dimórfica (Isla Dog, índice de dimorfismo sexual en tamaño 1.17). Los machos fueron significativamente más grandes que las hembras en ambas islas, y el tamaño del cuerpo en la población de Anguila fue significativamente mayor que en la población de la Isla Dog. La hipótesis del nicho ecológico no fue apoyada porque los machos y las hembras en Anguila no dividieron las presas a partir de la especie o el tamaño de las mismas; tampoco hubo diferencias en los tamaños de las presas entre las islas. Sin embargo, se sostuvo la hipótesis de que la limitación de alimento contribuyó a las diferencias en tamaño entre las islas. La abundancia en las especies de insectos que fueron presas fue menor en la Isla Dog, y algunos tipos de presas estuvieron ausentes en esa isla. Además, la competencia por el alimento con cangrejos ermitaños fue más intensa en la Isla Dog que en Anguila. Las lagartijas en la Isla Dog crecieron más lentamente y dejaron de crecer cuando habían alcanzado un tamaño más pequeño que las lagartijas de Anguila. En adición a la limitación de alimento, diferencias en selección sexual pueden contribuir a diferencias en dimorfismo en tamaño entre las islas. Los machos en la Isla Dog parecen ser incapaces de proteger a las hembras por la totalidad de su período fértil, de manera que los machos grandes pueden tener una menor ventaja de apareo en la Isla Dog que en Anguila.

Résumé.—L'évolution du dimorphisme de la taille sexuelle a été étudiée chez *Ameiva plei*, un lézard non territorial actif. Trois hypothèses (niche écologique, restriction alimentaire, sélection sexuelle) ont été vérifiées pour expliquer les différences observées entre une population dimorphique (Anguilla, SDI:1.37) et une population relativement non dimorphique (Dog Island, SDI:1.17). Les mâles sont significativement plus grands que les femelles sur les deux îles et la taille de la population d'Anguilla est significativement plus grosse que celle de la population de Dog Island. L'hypothèse de la niche écologique n'a pu être vérifiée car les mâles et les femelles d'Anguilla ne différencient pas les proies en fonction de la taille ou de l'espèce; nous n'avons pas observé de différences de taille dans les proies entre les îles. L'hypothèse voulant que les restrictions alimentaires contribuent aux différences de taille entre îles a été vérifiée. Il y a moins d'insectes à Dog Island et certains aliments ne s'y retrouvent pas. Les Bernard-l'ermite font une plus grande concurrence aux *A. plei* sur Dog Island qu'à Anguilla. Les lézards de Dog Island grandissent moins vite et sont moins gros que ceux d'Anguilla. En marge des restrictions alimentaires, les différences au titre de la sélection sexuelle peuvent contribuer aux différences de dimorphisme entre les îles. Les mâles de Dog Island semblent être incapables de garder les femelles sous surveillance pendant toute la période de fécondité ce qui explique peut-être pourquoi les gros mâles ont moins d'avantages en matière d'accouplement sur Dog Island que sur Anguilla.

Censky, E. J. 1996. The evolution of sexual size dimorphism in the teiid lizard *Ameiva plei*: a test of alternative hypotheses, p. 277–289. *In* R. Powell and R. W. Henderson (eds.), Contributions to West Indian Herpetology: A Tribute to Albert Schwartz. Society for the Study of Amphibians and Reptiles, Ithaca (New York). Contributions to Herpetology, volume 12.

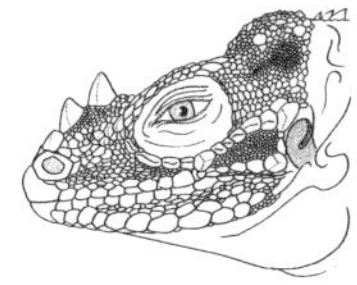

Morphology, Ecology, and Behavior of the Twig Anole, *Anolis angusticeps*

Duncan J. Irschick[1] and Jonathan B. Losos[1]

Abstract.—The polychrotid lizard *Anolis angusticeps* is superficially similar to twig anoles, an anole ecomorph that has arisen several times independently in the Greater Antilles. We examined the habitat use, locomotor behavior, and morphological characteristics of *A. angusticeps* to determine whether it exhibits characteristics similar to other twig anoles. Our results indicate that in all respects, *A. angusticeps* is a typical twig anole. As predicted, *A. angusticeps* is typically found on supports of narrow dimensions, walks slowly, displays infrequently, and is similar morphologically to other twig anoles. The cryptic coloration, slow movements, and infrequent display behavior of twig anoles support the hypothesis that these lizards have evolved to be inconspicuous. The crypticity of twig anoles may have arisen because they use a habitat (narrow supports) in which rapid escape may be difficult and therefore have evolved to hide from potential predators.

Introduction

Within the Greater Antillean islands of the West Indies, an amazing radiation of *Anolis* lizards has occurred, producing a set of morphologically distinct forms, termed "ecomorphs" (Rand and Williams 1969; Williams 1972, 1983). Previous work has shown that these ecomorphs differ in habitat use, morphology, and behavior (Moermond 1979*a*, 1979*b*; Mayer 1989; Losos 1990*a*, 1990*b*, 1990*c*). Remarkably, these forms have, with some exceptions, convergently evolved on each of the islands, producing largely similar within-island sets of ecomorphs.

Although many Greater Antillean *Anolis* can be placed into a particular ecomorph category (Losos 1992), the classification of some species is still uncertain. *Anolis angusticeps*, believed to have originated on Cuba (Williams 1969), but now found on several Bahamian islands, is such a species. Superficially, *A. angusticeps* resembles twig anoles by its possession of a slender body, short tail, and short legs (although these characteristics have not been quantified to date). However, observations on habitat use in *A. angusticeps* are conflicting; in the Bahamas, Schoener (1968) reported that this species uses the typical habitat of twig anoles (narrow supports on the periphery of trees and bushes). In contrast, others have described the structural habitat of this species in Cuba as typical of trunk-ground or trunk anoles. For example, Collette (1961) observed *A. angusticeps* primarily on tree trunks, fence posts, rocks and the ground, whereas Schwartz and Thomas (1969) described finding this species on rocks, shrubs, fence posts, trees, and the ground. Observations by Ruibal (1964) and Hardy (1967) also describe finding this species in habitats other than narrow branches or twigs. Clearly, more data are needed to clarify the ecomorph status of *A. angusticeps*.

We examined whether *Anolis angusticeps* exhibits morphological, ecological, and behavioral characteristics of other twig anoles. These characteristics are as follows: (1) twig anoles are slender and possess a short tail and short legs (Williams 1983; Hedges and Thomas 1989; Losos 1990*a*); (2) twig anoles use narrow branches, move by walking slowly, and are highly cryptic (Williams and Rand 1969; Webster 1969; Hicks and Trivers 1983; Losos

[1] Department of Biology, Washington University, St. Louis, Missouri 63130–4899, USA.

1990*a*, 1990*b*). Furthermore, several authors (Hicks and Trivers 1983; Losos 1990*b*) have demonstrated that the twig anole *A. valencienni* displays infrequently relative to other anoles. One possible reason for the infrequent displays of *A. valencienni* is that this species, like other twig anoles, relies on crypticity to avoid detection. Therefore, we predicted that *A. angusticeps* should also display infrequently so as to avoid bringing attention to itself. Finally, in addition to testing the above predictions, we also examine intersexual differences in habitat use and locomotor behavior in *A. angusticeps*.

MATERIALS AND METHODS

Morphometrics

To examine whether *Anolis angusticeps* is similar morphologically to other twig anoles, we compared external morphological characteristics of *A. angusticeps* from Cuba ($n = 13$) and South Bimini Island ($n = 11$) to 31 other Greater Antillean *Anolis* whose ecomorph status is known. Appendix 1 provides museum numbers for Cuban specimens. The ecomorph status for most of these 31 anoles has been established based on previous morphological, ecological, and behavioral analyses (see Losos 1990*a*, 1990*b*, 1990*c*; Losos 1992 for details on how anole ecomorphs are characterized), although values for a few species mentioned here have not been published. We took the following measurements on preserved adult male (data from other anoles are from adult males) *A. angusticeps* (following Losos 1990*a*): snout-vent length (SVL), number of subdigital lamellae under the fourth phalanx of the hindfoot, and length of the forelimb, hindlimb, and tail. Similar data are available for 31 other Greater Antillean species (Losos 1990*a*, 1992, and unpublished data; see Appendix 2 for species used in morphometric comparisons).

Ecological and Behavioral Data

To examine whether *Anolis angusticeps* exhibits the predicted ecological and behavioral characteristics, we collected field data on 16 adult male and 8 adult female *A. angusticeps* on South Bimini, Bahamas from 13–20 April 1992. South Bimini is a small (approximately 8 km^2), flat island primarily covered by bushes and trees. The weather during the week of observation was sunny and warm with few clouds and relatively little wind. Given the difficulty of distinguishing adult females and subadult males by sight, we only included animals in our analyses that were caught after the observational period and sexed. During normal activity hours (0900–1700 h), we examined available habitats (e.g., high in trees, rocks on the ground) on relatively undisturbed parts of the island. Upon sighting a lizard, we observed the lizard from a distance of 2–7 m for 30 min, or until it disappeared (range = 11–30 min, mean [SE] = 25.4 ± 1.2 min). All lizards that were obviously disturbed or observed for less than 10 min were not included in the analysis. All observational data were recorded into a microcassette recorder and later transcribed.

Each movement was categorized as a walk, run, or jump. Walks (equivalent to "crawls"; e.g., Moermond 1979*a*) were slow movements where all four limbs remained in close contact with the substrate during movement. Runs were quick movements whereby one or more limbs were cycled quickly above the substrate. Jumps were quick movements where the lizard intentionally jumped either between or on substrates. The percentage of time that each lizard spent displaying was measured by recording all display events (i.e., head-bobs, dewlap extensions) and summing the amount of time that lizards engaged in these activities. The following ecological and behavioral measurements were quantified for each lizard: (1) perch type, height, and diameter; (2) percent time spent displaying; and (3) percent of total movements that were walks, runs, and jumps. Perch type was the substrate on which the lizard was initially observed (e.g., branch, tree trunk). Perch height was the height above the ground at which the lizard was first observed. Perch diameter was the diameter of the substrate on which the lizard was first observed.

Statistical Analyses

All statistical analyses were performed on SYSTAT (Wilkinson 1990). All morphological and habitat variables were $\log_{10}$ transformed prior to statistical analyses. All morphological variables increase with

TABLE 1. Morphological, ecological, and behavioral characteristics of Greater Antillean *Anolis* ecomorphs. Data were gathered from Losos (1990*a*) and Losos (unpublished data). Ecomorphs are indicated as follows: trunk-ground (T-G), trunk-crown (T-C), grass-bush (G-B), crown-giant (C-G), twig (Tw), and trunk (Tr). The numbers in parentheses in the ecomorph columns for morphology, behavior, and ecology are the number of species representing that ecomorph type for that data base. See Appendix 2 for the list of species used in these comparisons. For the morphological variables, numbers not enclosed in brackets are mean (± 1 SE) non-size removed values for each ecomorph type, whereas numbers in brackets are mean (± 1 SE) size-removed (residual) values for each variable. All measurements are in mm unless otherwise indicated. The behavioral and ecological variables are not size-removed values. Percent (%) display is the mean (± 1 SE) amount of time that ecomorphs spend displaying (per observation period). Percent (%) walks, runs, or jumps are the mean (± 1 SE) percentages of the total number of movements (per observation period) that ecomorphs perform these locomotor movements.

Characteristic	T-G	T-C	G-B	C-G	Tw	Tr
Morphology	9	9	6	3	4	1
SVL	62.3 ± 2.6	59.5 ± 4.0	44.7 ± 1.0	132.6 ± 14.0	48.8 ± 2.0	51.3
Hindlimb length	52.0 ± 3.0 (0.07 ± 0.01)	39.9 ± 2.6 (–0.02 ± 0.02)	34.8 ± 1.1 (0.06 ± 0.01)	94.9 ± 8.2 (–0.03 ± 0.02)	25.2 ± 4.7 (–0.13 ± 0.02)	38.3 (0.03)
Forelimb length	30.8 ± 1.8 (0.06 ± 0.01)	26.0 ± 1.6 (0.01 ± 0.02)	17.7 ± 0.7 (–0.01 ± 0.01)	58.7 ± 6.0 (–0.05 ± 0.01)	15.9 ± 3.4 (–0.11 ± 0.01)	25.9 (0.09)
Lamellae number	18.9 ± 0.2 (–0.06 ± 0.01)	24.7 ± 1.2 (0.07 ± 0.01)	18.9 ± 0.5 (0.0 ± 0.01)	29.8 ± 0.8 (0.01 ± 0.01)	18.8 ± 1.6 (–0.02 ± 0.02)	18.0 (–0.04)
Tail length	120.6 ± 5.3 (0.02 ± 0.01)	107.4 ± 10.5 (–0.02 ± 0.01)	122.6 ± 5.0 (0.17 ± 0.02)	258.8 ± 21.2 (0.01 ± 0.03)	60.6 ± 10.3 (–0.17 ± 0.03)	71.5 (–0.12)
Behavior	5	6	4	1*	3	3
% walks	24.4 ± 3.3	49.3 ± 10.3	44.1 ± 8.1	45.0	80.1 ± 4.5	15.2 ± 7.3
% runs	54.2 ± 2.9	39.8 ± 9.0	30.6 ± 4.9	45.0	8.6 ± 4.1	83.1 ± 8.0
% jumps	21.4 ± 2.5	10.9 ± 4.5	25.3 ± 4.3	10.0	11.3 ± 0.8	1.7 ± 0.8
% display	5.7 ± 1.2	6.0 ± 1.4	3.4 ± 0.9	—	1.3 ± 0.002	2.0 ± 0.8
Ecology	5	6	4	1	3	3
Perch height (m)	1.1 ± 0.2	3.7 ± 0.8	0.6 ± 0.1	3.5	2.7 ± 0.5	1.9 ± 0.5
Perch diameter (cm)	22.5 ± 3.7	19.3 ± 3.9	2.9 ± 1.7	33.2	2.9 ± 1.7	22.2 ± 2.6

* No data available for display behavior in *Anolis garmani*.

size (SVL) (one-tailed least-squares regresssions, P < 0.001); to remove the effects of size we calculated residuals from regressions (ordinary least squares) between each morphological variable (dependent variable) and our measure of size (SVL) (independent variable). These residuals represent SVL-allometry free shape variables (Bookstein 1989). Mean residual values for each species were subsequently used in statistical analyses. We examined whether *Anolis angusticeps* exhibits morphological characteristics similar to other twig anoles by performing a discriminant function analysis (DFA) with both Cuban and Bahamian *A. angusticeps* and 31 other Greater Antillean *Anolis*. To test with which of the ecomorphs the Cuban and Bahamian populations of *A. angusticeps* would group, both populations were classified *a posteriori* based on the analysis of the 31 other species. To determine how strongly each of the morphometric variables contributed to each of the discriminant axes, we performed multiple regressions using as dependent variables the discriminant function scores and as independent variables the size-removed variables. Percentage data were arcsin square-root transformed prior to statistical analyses. All statistical analyses were two-tailed unless otherwise indicated.

For comparative purposes, we present data on habitat use and behavior for a subset of the 31 anoles used in morphometric comparisons (data from Losos 1990*a*, unpublished data; see Appendix 2 for species used in ecological and behavioral comparisons). However, some anoles used in ecological/behavioral comparisons were not used in morphological comparisons and vice versa. Ecological and behavioral data were collected in a manner similar to that described for *A. angusticeps* (see Losos 1990*a* for exact protocol). As in morphometric comparisons, because previously collected data were gathered on adult males, only values for adult male *A. angusticeps* are included in these comparisons.

Results

Morphometric Analyses

Table 1 provides mean (± 1 SE) values for ecomorphs used in morphometric analyses and Table 2 provides mean (± 1 SE) morphological values for Cuban and Bahamian *Anolis angusticeps*. The discriminant function analysis based on morphology of 31 Greater Antillean *Anolis* species was significant (Figure 1; Wilk's lambda = 0.003, $F_{16,76} = 16.76$, $P < 0.001$) and classified all 31 species to predesignated ecomorph categories. *Anolis angusticeps* from both Cuba and the Bahamas were classified *a posteriori*

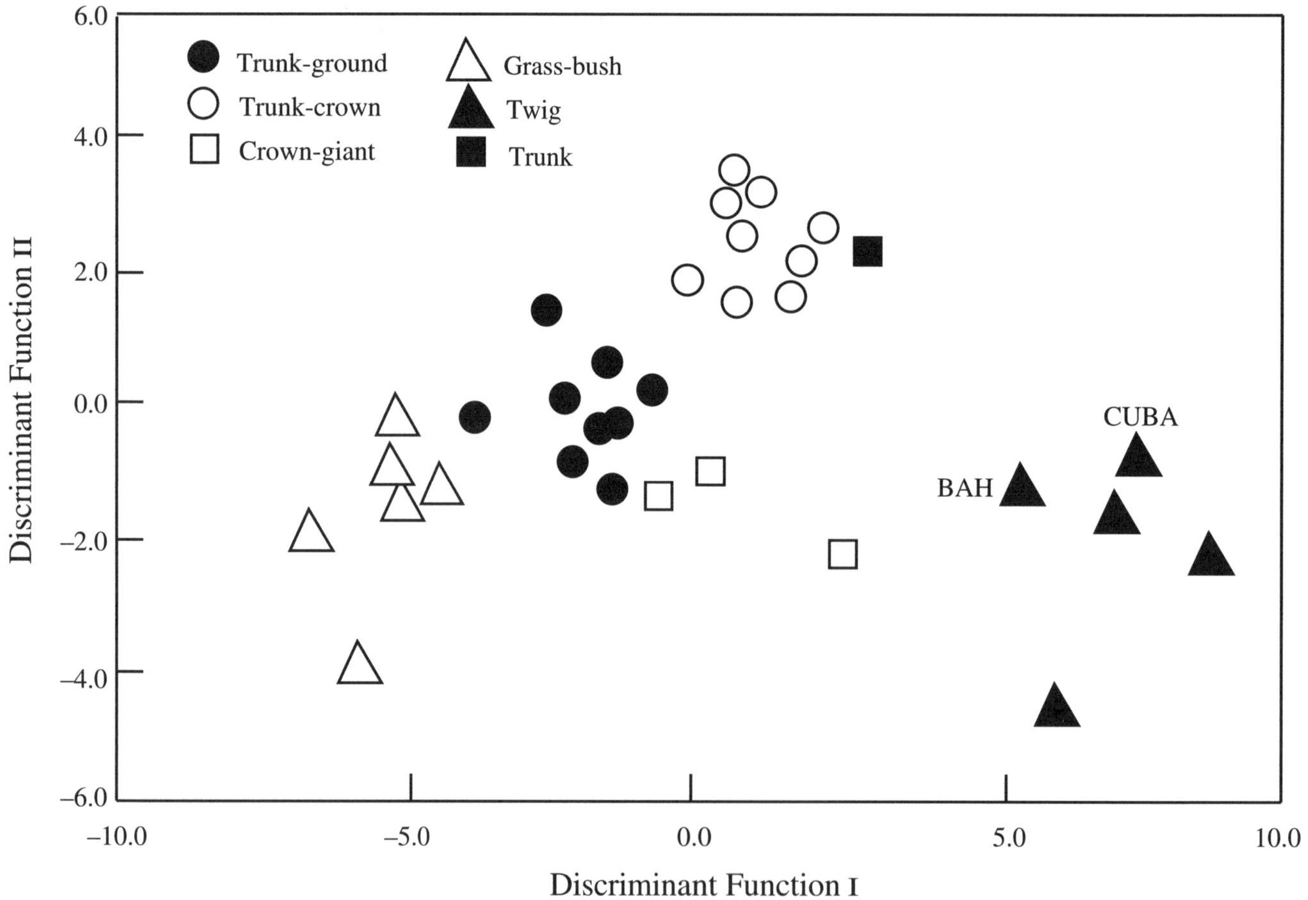

FIGURE 1. A plot of discriminant function II (y-axis) against discriminant function I (x-axis). Discriminant functions were generated from a discriminant function analysis of 31 *Anolis* species belonging to six ecomorph types (both Cuban and Bahamian *A. angusticeps* populations were included *a posteriori* in the DFA, and are presented here for purposes of comparison to other *Anolis*). Each point represents a different species, each symbol an ecomorph type. Symbols with BAH and CUBA adjacent are Bahamian and Cuban samples of *A. angusticeps*. See text for details of the analysis.

TABLE 2. Morphological, ecological, and behavioral characteristics of *Anolis angusticeps*. Values not in parentheses for morphological variables are mean (± 1 SE) non size-removed values for Bahamian and Cuban samples. Values in brackets are size-removed (residual) values generated from regressions with 31 other anole species. Because mean values for *A. angusticeps* were used to remove size effects, no standard error bars are given for size-removed values. All measurements are in mm unless otherwise indicated. The ecological/behavioral variables also represent mean (± 1 SE) values, but for Bahamian males and females. Percent (%) walks, runs, or jumps is the mean percentage of locomotor movements that males and females used during observational periods. Percent (%) display is the mean amount of time that males and females spent displaying during observational periods.

Variable	Sample	
Morphology	Bahamian (n = 11)	Cuban (n = 13)
SVL	47.7 ± 1.1	37.6 ± 0.8
Forelimb length	16.6 ± 0.3 (–0.07)	11.3 ± 0.4 (–0.11)
Hindlimb length	26.0 ± 0.5 (–0.11)	17.7 ± 0.5 (–0.15)
Lamellae number	19.6 ± 0.4 (0.00)	19.4 ± 0.3 (0.05)
Tail length	64.9 ± 2.0 (–0.14)	47.0 ± 1.2 (–0.17)
Ecology / Behavior	Males (n = 16)	Females (n = 8)
Perch height (m)	2.2 ± 0.3	1.8 ± 0.4
Perch diameter (cm)	2.3 ± 0.5	0.7 ± 0.2
% walks	89.3 ± 2.0	92.4 ± 3.7
% runs	1.9 ± 0.9	0.5 ± 0.4
% jumps	8.8 ± 1.7	7.1 ± 3.6

as twig anoles (probability = 1.0 for both populations). Standardized partial regression coefficients for the multiple regressions between discriminant functions and the morphometric variables are in Table 3. The first three discriminant functions exhibited high canonical correlations (the higher the correlation the greater the amount of variation explained along that axis) but the fourth function had a markedly lower canonical correlation (Table 3). Therefore, the fourth function explains little of the variance among ecomorphs.

With the exception of hindlimb length and DFA I, all morphometric variables are significantly correlated with each of the discriminant functions ($P < 0.001$). Tail length exhibits a high negative correlation with DFA I, indicating the utility of this character in segregating ecomorphs. In general, twig anoles (including Bahamian and Cuban *Anolis angusticeps*) exhibit very high values of DFA I (short tails) whereas grass-bush anoles exhibit very low values of DFA I (long tails) (Figure 1). Forelimb length exhibits a high positive correlation with DFA II indicating that anoles with high values for this variable possess long forelimbs. Trunk-crown anoles and the one trunk anole (*A. distichus*) tend to have high positive values for DFA II, indicating long forelimbs for these anoles. Twig anoles and grass-bush anoles tend to exhibit intermediate or low values of DFA II, indicating moderately short forelimbs for both ecomorphs. Hindlimb length loaded highly only on DFA IV, but inspection of Tables 1 and 2 shows that Bahamian and Cuban *A. angusticeps* and twig anoles in general possess high negative residual values for hindlimb length, indicating short hindlimbs for these anoles. Overall, Bahamian and Cuban populations of *A. angusticeps* differ only slightly from one another in size-removed morphological characteristics (Figure 1, Table 2).

Ecology and Behavior of *Anolis angusticeps*

Both male and female *Anolis angusticeps* perched on branches or twigs in bushes, crowns of trees, or on branches jutting from fallen trees. None of the

TABLE 3. Standardized partial regression coefficients from multiple regressions between each of the four discriminant functions (dependent variable) and four size-removed morphometric variables (independent variables). Discriminant function scores were generated from a discriminant function analysis of 31 *Anolis* species. Canonical correlations are also provided (higher correlations represent a greater proportion of variance explained by that axis). See text for details.

	Function			
Variable	I	II	III	IV
Hindlimb length	0.02	–1.07	0.15	–4.13
Forelimb length	–0.27	1.58	0.15	2.40
Lamellae number	0.12	0.53	–0.65	–1.11
Tail length	–0.88	0.14	–0.50	2.11
Canonical correlations	0.97	0.89	0.82	0.05

24 *A. angusticeps* were found on trunks or on the ground and the majority of individuals were observed oriented parallel with the perch occupied. Both sexes of *A. angusticeps* were found relatively high in the canopy and did not differ significantly in this measure (Table 2; ANOVA, $F_{1,22} = 0.22$, $P > 0.50$). Both sexes were found on relatively narrow perches, with males tending to perch on broader supports than females (Table 2; ANOVA, $F_{1,22} = 5.40$, $P < 0.05$). Male and female *A. angusticeps* both walked as their primary means of locomotion and did not differ significantly in this behavior (Table 2; ANOVA, $F_{1,22} = 1.06$, $P > 0.25$). Male *A. angusticeps* spent more time displaying than females (ANOVA, $F_{1,22} = 3.09$, $P < 0.05$, one-tailed test due to the *a priori* expectation that males would display more than females).

The habitat, locomotor, and display characteristics of *Anolis angusticeps* are similar to other twig anoles and differed strikingly from most other anole ecomorphs (Tables 1 and 2, Figure 2). The other twig anoles for which locomotor and display behavior data are available (*A. valencienni* and *A. insolitus*) also use narrow supports, walk frequently, and display infrequently relative to other ecomorphs (Table 1, Figure 2). The ecomorph most distinctive from the twig anoles are the trunk-ground anoles. Trunk-ground anoles use broad supports, move primarily by running, and spend about four times as much time displaying as twig anoles. In contrast, grass-bush anoles are much more similar to twig anoles. Grass-bush anoles also use narrow supports, move about 44% of the time by walking, and display only about twice as much as twig anoles. The remaining ecomorphs (trunk-crown, crown-giant, trunk) exhibit habitat, locomotor, and display characteristics intermediate between these extremes.

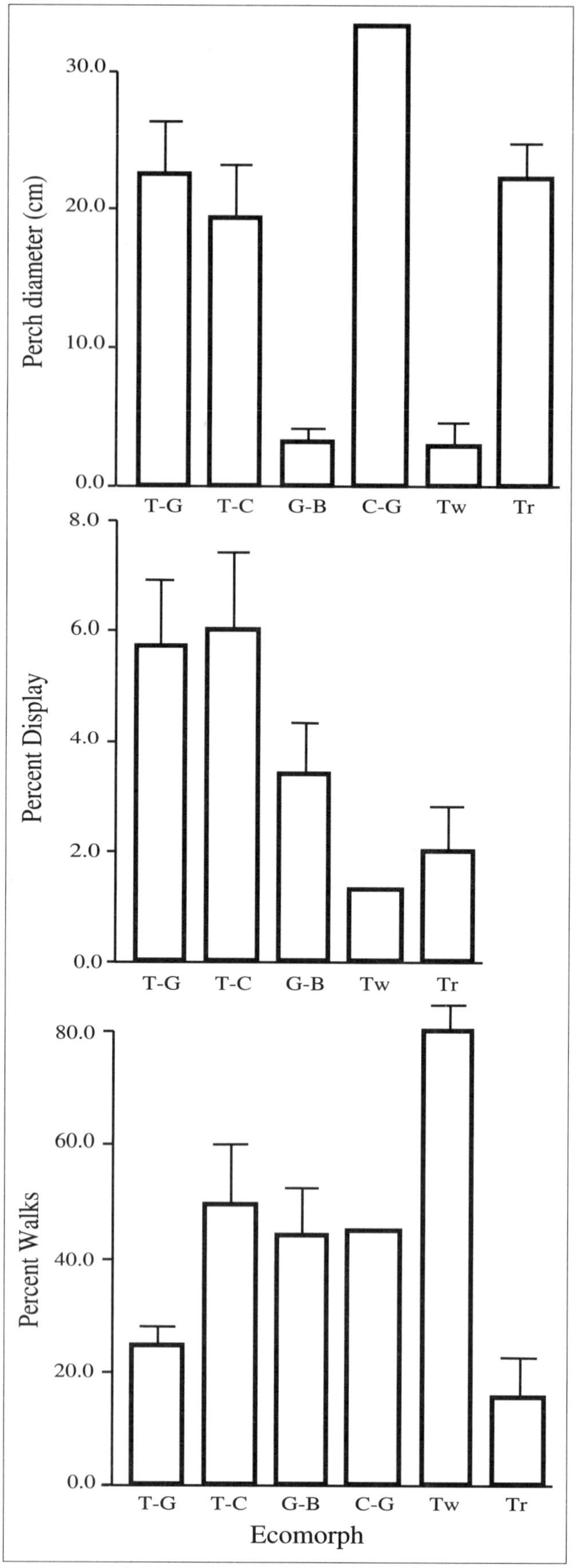

FIGURE 2. Mean (± 1 SE) perch diameter, percentage of time displaying, and percentage of walks for West Indian *Anolis* (see Appendix 2 for list of species included in ecological and behavioral comparisons). Ecomorphs are indicated as in Table 1. No standard error bars are provided for the crown-giant ecomorph type because only one species is included. No standard error bars are indicated on the twig ecomorph type for the percent display comparison because the error was too small to be seen on the plot.

Values for ecological and behavioral characteristics of some ecomorphs should be given greater consideration than others due to uneven sample sizes (e.g., six trunk-crown anoles and one crown-giant anole).

Discussion

In all respects, *Anolis angusticeps* appears to be a twig anole; it is similar morphologically to other twig anoles, uses narrow perches, tends to move by walking slowly, and displays infrequently. These data show that, at least on South Bimini, this species uses a habitat and behaves like a twig anole. Although no quantitative habitat or behavioral data are available for *A. angusticeps* from Cuba, Cuban individuals are morphologically similar to Bahamian *A. angusticeps*.

How, then, can we explain the reported differences in habitat use of *Anolis angusticeps* on Cuba and the Bahamas (Collette 1961; Ruibal 1964; Hardy 1967; Schoener 1968; Schwartz and Thomas 1969)? One possibility is that, for some reason, *A. angusticeps* evolved a twig morphology on Cuba even though it does not use a typical twig habitat. A more likely possibility is that workers on Cuba (Collette 1961; Ruibal 1964; Hardy 1967; Schwartz and Thomas 1969) did not detect *A. angusticeps* in its twig habitat. This is not unlikely, as *A. angusticeps* is likely to be much more obvious when they use substrates such as tree trunks or rocks rather than twigs. Indeed, several authors have noted that *A. angusticeps* is a difficult species to find (Barbour 1937; Oliver 1948; Schoener 1968). For example, Oliver (1948) did not discover *A. angusticeps* on South Bimini until the end of a month long trip, missing this species in its natural habitat despite the fact that this species is not uncommon on the island (Schoener 1968; D. J. Irschick and J. B. Losos, personal observation). Therefore, detailed searching on available habitat in Cuba may reveal whether *A. angusticeps* uses a habitat similar to that used by the species on South Bimini Island.

Ecology-morphology Correlations

Correlations between ecology and morphology have long been considered as evidence that species have evolved adaptations for their microhabitats (Karr and James 1975; Ricklefs and Travis 1980, but see Wiens and Rotenberry 1980; Wiens 1984). Previous studies of *Anolis* lizards from Costa Rica, Hispaniola, Puerto Rico, and Jamaica have documented strong correlations between habitat use and morphology (Moermond 1979*a*; Pounds 1988; Losos 1990*a*). The most striking of these correlations exists between the preferred perch diameter and limb dimensions of *Anolis* species; species that use broad surfaces possess long hindlimbs whereas species using narrow perches possess shorter hindlimbs. The data presented here are consistent with this correlation; *A. angusticeps* is typically found on narrow perches and possesses short hindlimbs.

This correlation between limb length and perch diameter in *Anolis* lizards is consistent with biomechanical models (Cartmill 1985). On broad surfaces, where there is little risk of losing balance, *Anolis* species have maximized their ability to run quickly and jump farther by evolving long legs which enhance maximal running and jumping ability (Emerson 1985; Hildebrand 1985; Losos 1990*a*, 1990*c*). In contrast, on narrow surfaces anoles have evolved short legs to prevent toppling while moving. Although a correlation between limb length and perch diameter has been documented in three twig anoles (*A. valencienni*, *A. angusticeps*, and *A. insolitus*), quantitative ecological and morphological data are unavailable for others considered to be twig anoles (e.g., *A. placidus*, *A. sheplani*, *A. darlingtoni*, and *A. fowleri* of Hispaniola). We recommend quantitative assessment of microhabitat and morphology of these species to determine whether they also use narrow supports and possess short limbs.

Locomotor Behavior and Performance

The habitat matrix model predicts that species will evolve locomotor behavior appropriate for their microhabitats (Moermond 1979*a*, 1979*b*; Fleagle and Mittermeier 1980; Robinson and Holmes 1982; Pounds 1988; Losos 1990*b*). Previous studies have interpreted the behavior of slow walking as a means to avoid losing balance while moving on narrow surfaces (Pounds 1988; Losos 1990*b*). The use of narrow supports and slow movements of *Anolis angusticeps* supports this generalization. In contrast, anoles that use supports in which loss of balance is less of a concern (e.g., trunk-ground and trunk-crown anoles that use broad supports) move primarily by rapid movements that might be more difficult on narrow supports.

The microhabitat of twig anoles may also exert a strong influence on their locomotor capabilities. Laboratory tests of locomotor capabilities indicate that whereas the twig anole *Anolis valencienni* moves with little difficulty on narrow surfaces, it exhibits poor sprinting and jumping capabilities on broad surfaces relative to other, similarly-sized *Anolis* (Losos and Sinervo 1989; Losos 1990*a*, 1990*c*). These limited maximal locomotor capabilities of *A. valencienni* result, in part, from the short limbs of this anole which, in turn, may have evolved to allow slow, steady movements on narrow supports (Losos and Sinervo 1989; Losos 1990*a*). These findings demonstrate that the evolution of morphological characteristics appropriate for movement on specific surfaces involves trade-offs. Further tests of the locomotor capabilities of *A. angusticeps* may reveal if this species, by evolving a morphology and locomotor behavior appropriate for walking on narrow surfaces, exhibits reduced maximal sprinting and jumping capabilities on broad supports.

Display Behavior

The microhabitat of twig anoles may exert a strong influence on their escape behavior and display behavior. Because lizards run quickly on narrow supports only with difficulty (D. J. Irschick, personal observation), the ability of anoles using such surfaces to escape predators (e.g., snakes) rapidly may be compromised. Furthermore, twig anoles possess limited maximal running and jumping capabilities, making running and jumping quickly an especially inefficient mode of escape (Losos and Sinervo 1989; Losos 1990*a*, 1990*c*). An alternative method of avoiding predators is to be highly inconspicuous and, when detected, to elude capture by hiding. Possibly as a consequence of the above factors, twig anoles have opted for this second approach. All twig anoles known to date are highly cryptic and, when disturbed, slowly move out of sight (Schoener 1968; Williams and Rand 1969; Hicks and Trivers 1983; D. J. Irschick and J. B. Losos, personal observation). If twig anoles have evolved to be highly inconspicuous, then frequent displays might render them highly visible to predators. The need to avoid predation may thus explain, in part, why twig anoles examined to date display infrequently relative to other anole ecomorphs. Because anoles display to defend territories, infrequent displays might thus make active defense of a territory unfeasible. Although few data are available on territoriality in twig anoles, that at least one twig anole (*Anolis valencienni*) appears to be much less territorial than other anoles (Hicks and Trivers 1983) is in support of this hypothesis. One means of testing if display behavior is directly correlated with behavioral characteristics such as territorial behavior would be to examine whether anole ecomorphs which display frequently (e.g., trunk-ground anoles) are more territorial than ecomorphs which display less frequently (e.g., twig anoles).

Intersexual Differences in *Anolis angusticeps*

Intersexual differences in microhabitat use have been previously documented in *Anolis* lizards (Andrews 1971; Lister 1976). We observed a significant propensity for males to perch on broader supports than females. In a much more extensive study at the same locality, Schoener (1968) observed no significant intersexual differences in perch diameter in *A. angusticeps*. Due to his much larger data set, Schoener's data probably best reflects intersexual ecological relationships. We also observed that male *A. angusticeps* spent more time displaying than females. This difference in display behavior is consistent with data which indicate that male lizards are more likely to actively defend territories or attempt to attract mates than females (Stamps 1983).

Summary

Few quantitative data exist on the behavior and ecology of twig anoles. Their cryptic habits make such studies especially difficult. Here we examined whether an anole superficially similar to twig anoles, *Anolis angusticeps*, exhibits morphological, ecological, and behavioral characteristics similar to other twig anoles for which data are available. In particular, we predicted that, if *A. angusticeps* is a twig anole, it should be similar morphologically to other twig anoles, use narrow perches, move by walking slowly, and display infrequently. These predictions were confirmed. We recommend further studies investigating the morphological, ecological and behavioral characteristics of twig anoles to verify whether the above-mentioned characteristics are more broadly applicable.

Acknowledgments

We thank R. Powell and R. Henderson for allowing us to participate in the SSAR symposium, "West Indian Herpetology: A Tribute to Albert Schwartz." We would like to thank the Government of the Bahamas for permission to conduct this research. For assistance and advice, we thank M. Isaacs, T. Schoener, C. Whaling, and J. Cheverud. We thank M. Hedin, K. de Querioz, T. Hrbek, and two anonymous reviewers for comments on previous drafts of the manuscript. Finally, we thank the Smithsonian Institution for access to their specimens. This work was supported by the National Geographic Society and the National Science Foundation (DEB-9318642 to J. B. Losos and BSR-9020052 to T. W. Schoener). This research was performed when the senior author was an undergraduate student at the University of California at Davis.

Literature Cited

Andrews, R. M. 1971. Structural habitat and time budget of a tropical *Anolis* lizard. Ecology 52: 262–270.

Barbour, T. 1937. Third list of Antillean reptiles and amphibians. Bull. Mus. Comp. Zool., Harvard Univ. 82: 77–166.

Bookstein, F. 1989. "Size and shape": a comment on semantics. Syst. Zool. 38: 173–180.

Cartmill, M. 1985. Climbing, p. 73–88. *In* M. Hildebrand, D. M. Bramble, K. F. Liem, and D. B. Wake (eds.), Functional Vertebrate Morphology. Belknap Press, Cambridge, Massachusetts.

Collette, B. B. 1961. Correlations between ecology and morphology in anoline lizards from Havana, Cuba and Southern Florida. Bull. Mus. Comp. Zool., Harvard Univ. 125: 137–162.

Crome, F. H. and G. C. Richards. 1988. Bats and gaps: microchiropteran community structure in a Queensland rain forest. Ecology 69: 1960–1969.

Emerson, S. B. 1985. Jumping and leaping, p. 58–72. *In* M. Hildebrand, D. M. Bramble, K. F. Liem, and D. B. Wake (eds.), Functional Vertebrate Morphology. Belknap Press, Cambridge, Massachusetts.

Fleagle, J. G. and R. A. Mittermeier. 1980. Locomotor behavior, body size and comparative ecology of seven Surinam monkeys. Amer. Jour. Phys. Anthropol. 20: 440–453.

Hardy, J. D., Jr. 1967. Geographic variation in the West Indian *Anolis* lizard, *Anolis angusticeps*, with the description of a new form, *Anolis angusticeps paternus*, subsp. nov., from the Isle of Pines, Cuba (Reptilia: Iguanidae). Carib. Jour. Sci. 6: 23–31.

Hedges, S. B. and R. Thomas. 1989. A new species of *Anolis* (Sauria: Iguanidae) from the Sierra de Neiba, Hispaniola. Herpetologica 45: 330–336.

Hicks, R. A. and R. L. Trivers. 1983. The social behavior of *Anolis valencienni*, p. 570–595. *In* A. G. J. Rhodin and K. Miyata (eds.), Advances in Herpetology and Evolutionary Biology: Essays in Honor of Ernest E. Williams. Mus. Comp. Zool., Cambridge, Massachusetts.

Hildebrand, M. 1985. Walking and running, p. 38–57. *In* M. Hildebrand, D. M. Bramble, K. F. Liem, and D. B. Wake (eds.), Functional Vertebrate Morphology. Belknap Press, Cambridge, Massachusetts.

Karr, J. R. and F. C. James. 1975. Ecomorphological configurations and convergent evolution, p. 258–291. *In* M. L. Cody and J. M. Diamond (eds.), Ecomorphological Configurations and Convergent Evolution in Species and Communities. Belknap Press, Cambridge, Massachusetts.

Lister, B. C. 1976. The nature of niche expansion in West Indian *Anolis* lizards. I: Ecological consequences of reduced competition. Evolution 30: 659–676.

Losos, J. B. 1990*a*. Ecomorphology, performance capability, and scaling of West Indian *Anolis* lizards: an evolutionary analysis. Ecol. Monogr. 60: 369–388.

—. 1990*b*. Concordant evolution of locomotor behavior, display rate and morphology in *Anolis* lizards. Anim. Behav. 39: 879–890.

—. 1990*c*. The evolution of form and function: morphology and locomotor performance in West Indian *Anolis* lizards. Evolution 44: 1189–1203.

—. 1992. The evolution of convergent structure in Caribbean *Anolis* communities. Syst. Biol. 41: 403–420.

— and B. Sinervo. 1989. The effects of morphology and perch diameter on sprint performance of *Anolis* lizards. Jour. Exper. Biol. 145: 23–30.

Mayer, G. C. 1989. Deterministic patterns of community structure in West Indian reptiles and amphibians. Ph.D. dissertation. Harvard Univ., Cambridge, Massachusetts.

Moermond, T. C. 1979*a*. Habitat constraints on the behavior, morphology and community structure of *Anolis* lizards. Ecology 60: 152–164.

—. 1979*b*. The influence of habitat structure on *Anolis* foraging behavior. Behaviour 70: 147–166.

Oliver, J. A. 1948. The anoline lizards of Bimini, Bahamas. Amer. Mus. Novit. 1383: 1–36.

Pounds, A. J. 1988. Ecomorphology, locomotion, and microhabitat structure: patterns in a tropical mainland *Anolis* community. Ecol. Monogr. 58: 299–320.

Rand, A. S. and E. E. Williams. 1969. The anoles of La Palma: aspects of their ecological relationships. Breviora 327: 1–19.

Ricklefs, R. E. and J. Travis. 1980. A morphological approach to the study of avian community organization. Auk 97: 321–338.

Robinson, S. K., and R. T. Holmes. 1982. Foraging behavior of forest birds: the relationship among search tactics, diet, and habitat structure. Ecology 63: 1918–1931.

Ruibal, R. 1964. An annotated checklist and key to the anoline lizards of Cuba. Bull. Mus. Comp. Zool., Harvard Univ. 130: 475–520.

Schoener, T. W. 1968. The *Anolis* lizards of Bimini: resource partitioning in a complex fauna. Ecology 49: 704–726.

Schwartz, A. and R. Thomas. 1968. A review of *Anolis angusticeps* in the West Indies. Quart. Jour. Florida Acad. Sci. 31: 51–69.

Stamps, J. A. 1983. Sexual selection, sexual dimorphism, and territoriality, p. 169–204. *In* R. B. Huey, E. R. Pianka, and T. W. Schoener (eds.), Lizard Ecology: Studies on a Model Organism. Harvard Univ. Press, Cambridge, Massachusetts.

Webster, T. P. 1969. Ecological observations on *Anolis occultus* William and Rivero (Sauria, Iguanidae). Breviora 312: 1–5.

Wiens, J. A. 1984. On understanding a non-equilibrium world: myth and reality in community patterns and processes, p. 439–477. *In* D. R. Strong, Jr., D. Simberloff, L. G. Abele, and A. B. Thistle (eds.), Ecological Communities: Conceptual Issues and the Evidence. Princeton Univ. Press, Princeton, New Jersey.

— and J. T. Rotenberry. 1980. Patterns of morphology and ecology in grassland shrubsteppe bird populations. Ecol. Monogr. 50: 287–308.

Wilkinson, L. 1990. SYSTAT: The System for Statistics. SYSTAT, Inc., Evanston, Illinois.

Williams, E. E. 1969. The ecology of colonization as seen in the zoogeography of anoline lizards on small islands. Quart. Rev. Biol. 44: 345–389.

—. 1972. The origin of faunas. Evolution of lizard congeners in a complex island fauna: a trial analysis. Evol. Biol. 6: 47–89.

—. 1983. Ecomorphs, faunas, island size, and diverse end points in island radiations of *Anolis*, p. 326–370. *In* R. B. Huey, E. R. Pianka, and T. W. Schoener (eds.), Lizard Ecology: Studies on a Model Organism. Harvard Univ. Press, Cambridge, Massachusetts.

— and A. S. Rand. 1969. *Anolis insolitus*, a new dwarf anole of zoogeographic importance from the mountains of the Dominican Republic. Breviora 326: 1–21.

APPENDIX 1. National Museum of Natural History (USNM) specimen numbers for Cuban samples of *Anolis angusticeps* examined in this study. Data for Bahamian specimens of *A. angusticeps* are available upon request from the authors.

USNM 51847, Cape San Antonio, Cuba; USNM 54416, Santa Cruz River, Cuba; USNM 160915, Habana, Cuba; USNM 160920, Habana, Cuba; USNM 160921, Habana, Cuba; USNM 160922, Habana, Cuba; USNM 160924, Bosque de Habana, Cuba; USNM 160928, La Habana, Cuba; USNM 160937, La Habana, Cuba; USNM 337640, Matanazas, Playa Larga, Cuba; USNM 337641, Matanazas, Playa Larga, Cuba; USNM 337642, Matanazas, Playa Larga, Cuba; USNM 337644, Matanazas, Playa Larga, Cuba.

APPENDIX 2. *Anolis* species included in morphological (designated by an M), ecological (E), and behavioral (B) analyses, followed by the ecomorph designation and island of origin: HISP = Hispaniola, JA = Jamaica, PR = Puerto Rico. Sample sizes for ecological analyses are noted in parentheses (sample sizes for morphological and behavioral analyses are not provided).

Anolis sagrei — Trunk-ground, JA (18) [M, E, B]; *A. lineatopus* — Trunk-ground, JA (18) [M, E, B]; *A. cooki* — Trunk-ground, PR [M]; *A. cristatellus* — Trunk-ground, PR (28) [M, E, B]; *A. gundlachi* — Trunk-ground, PR (28) [M, E, B]; *A. cybotes* — Trunk-ground, HISP (71) [M, E, B]; *A. longitibialis* — Trunk-ground, HISP [M]; *A. marcanoi* — Trunk-ground, HISP [M]; *A. strahmi* — Trunk-ground, HISP [M]; *A. grahami* — Trunk-crown, JA (18) [M, E, B]; *A. opalinus* — Trunk-crown, JA (18) [M, E, B]; *A. evermanni* — Trunk-crown, PR (28) [M, E, B]; *A. stratulus* — Trunk-crown, PR (28) [M, E, B]; *A. chlorocyanus* — Trunk-crown, HISP (86) [M, E, B]; *A. aliniger* — Trunk-crown, HISP (36) [M, E, B]; *A. coelestinus* — Trunk-crown, HISP (40) [M]; *A. singularis* — Trunk-crown, HISP [M]; *A. porcatus* — Trunk-crown, HISP [M]; *A. pulchellus* — Grass-bush, PR (28) [M, E, B]; *A. poncensis* — Grass-bush, PR (18) [M, E, B]; *A. krugi* — Grass-Bush, PR (28) [M, E, B]; *A. olssoni* — Grass-Bush, HISP (46) [M, E, B]; *A. semilineatus* — Grass-Bush, HISP [M]; *A. bahorucoensis* — Grass-bush, HISP [M]; *A. distichus* — Trunk, HISP (44) [M, E, B]; *A. christophei* — Trunk, HISP [E, B]; *A. brevirostris* — Trunk, HISP [E, B]; *A. valencienni* — Twig, JA (15) [M, E, B]; *A. occultus* — Twig, PR [M]; *A. insolitus* — Twig, HISP (4) [M, E, B]; *A. cuvieri* — Crown-giant, PR (5) [M]; *A. garmani* — Crown-giant, JA (18) [M, E, B]*; *A. barahonae* — Crown-giant, HISP (5) [M].

* No data for display behavior are available.

Resumen.—La lagartija policrótida *Anolis angusticeps* es superficialmente similar a las lagartijas anolinas enanas que viven en ramitas, un ecomorfo que ha surgido independientemente varias veces en las Antillas. Examinamos la utilización del habitáculo, conducta de locomoción y características morfológicas de *A. angusticeps* para determinar si éste exhibe características similares a las de otras lagartijas anolinas enanas que viven en ramitas. Nuestros resultados indican que en todos los aspectos *A. angusticeps* es una típica lagartija anolina enana que vive en ramitas. Como fue predicho, *A. angusticeps* se encuentra característicamente en perchas de dimensiones estrechas, camina lentamente, raras veces muestra alguna conducta, y es morfológicamente similar a otras lagartijas anolinas que viven en ramitas. La coloración críptica, movimientos lentos, y exhibición infrecuente de conductas de las lagartijas anolinas enanas que viven en ramitas apoyan la hipótesis de que estas lagartijas han evolucionado para ser inconspicuas. Las características crípticas de las lagartijas anolinas enanas que viven en ramitas pudieron haber surgido porque estas lagartijas utilizan un habitáculo (perchas estrechas) en el cual escapar rápidamente puede ser difícil y por lo tanto han evolucionado para esconderse de depredadores potenciales.

Résumé.—Nous avons étudié la répartition de la faune herpétologique dans 31 îles de Exuma Cays Land and Sea Park (eclsp) aux Bahamas dans le cadre de quatre visites intensives effectuées entre 1990 et 1993. La faune herpétologique se compose de 15 espèces dont deux (peut-être quatre) ont été introduites. Trois modèles zoogéographiques fondamentaux sont mis en évidence par la faune herpétologique, c'est-à-dire par les espèces dont la répartition est soit ubiquiste, soit réservée aux îles centrales, soit répandue. On ne trouve qu'une seule espèce sur presque toutes les îles, trois sur les îles les plus grandes et les plus complexes sur le plan structurel (îles centrales) et quatre sur une vaste gamme d'îles dont la taille varie de grande à petite. Trois espèces sont si rares que nous n'avons pas pu les classifier. Les deux autres espèces devaient probablement être plus répandues avant mais semblent aujourd'hui confinées à certains territoires. La taille de l'île explique à elle seule 42% de la variation au titre de la richesse des espèces. La taille de l'île centrale la plus proche et la distance entre l'île centrale la plus proche et les îles satellites n'influent nullement sur la diversité faunique. La répartition de la faune herpétologique de eclsp résulte vraisemblablement d'une série d'événements géologiques et historiques associés à des manifestations écologiques dynamiques permanentes. L'impact potentiel des prédateurs exotiques et des établissements humains ne saurait être sous-estimé pour expliquer le modèle zoogéographique actuel.

Irschick, D. J. and J. B. Losos. 1996. Morphology, ecology, and behavior of the twig anole, *Anolis angusticeps*, p. 291–301. *In* R. Powell and R. W. Henderson (eds.), Contributions to West Indian Herpetology: A Tribute to Albert Schwartz. Society for the Study of Amphibians and Reptiles, Ithaca (New York). Contributions to Herpetology, volume 12.

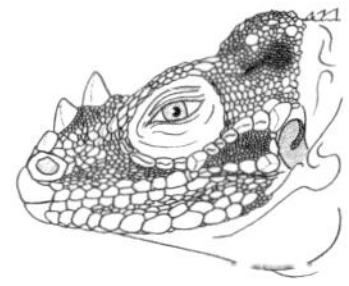

A TEST OF ASSORTATIVE MATING BETWEEN SIBLING LIZARD SPECIES, *ANOLIS WEBSTERI* AND *A. CAUDALIS*, IN HAITI

THOMAS A. JENSSEN[1]

Abstract.—Field experiments were conducted (June-July) at a Haitian contact zone between the sibling species, *Anolis websteri* and *A. caudalis*. Using a tree which was isolated from adjacent habitat as a "natural enclosure," daylong (0630–1600 h) observations were made over a period of 12 days on each of three combinations of subjects: (I) a male and 3 females from each species; (II) a male of each species and 6 female *A. caudalis*; and (III) a male *A. caudalis* with a *websteri*-colored dewlap and 3 females of each species. In Experiment I, female *A. websteri* copulated with similar frequency with both conspecific and heterospecific males, while female *A. caudalis* were not seen to copulate. Conclusion: unequivocally, male and female *A. websteri* and male *A. caudalis* did not assortatively mate, while female *A. caudalis* remained an unknown. In Experiment II, female *A. caudalis* mated only with conspecific males, but few copulations were recorded. Conclusion: tentatively, *A. caudalis* females appeared to discriminate against congeners. In Experiment III, the abnormal dewlap color of the male *A. caudalis* did not interrupt conspecific breeding. Conclusion: dewlap color may not be an important criterion for mate discrimination by female *A. caudalis*.

Heterospecific differences in coitus duration and frequency were marked. Copulatory durations for male *Anolis websteri* were ≤1 s. By contrast, mean conspecific intromission for male *A. caudalis* was 13.4 min, decreasing to 3.5 min when with *A. websteri* females. In Experiment I, individual female *A. websteri* averaged 0.64 copulations/day as contrasted with 0.04 for female *A. caudalis* in Experiment II. Most *A. websteri* females engaged in multiple copulations during their observation periods.

Despite the sibling species having distinctly different courtship display traits, little evidence was found for an effective pre-mating behavioral reproductive isolation mechanism (RIM). At the least, the potential for gene flow from *Anolis caudalis* to *A. websteri* populations exists. However, no significant hyrid zone was apparent, suggesting a post-mating RIM such as hybrid sterility.

INTRODUCTION

VERRELL (1988) CONSIDERED A CENTRAL PROBLEM in evolutionary biology to be the continued confusion regarding the processes of speciation. The slowness of the speciation phenomenon, except for cases of polyploidy and some chromosomal mechanisms, is not conducive to direct observation. Thus, the reconstruction of events leading to speciation are largely inferential, relying on deductive generalizations (Mayr 1982). Testing these generalizations has drawn upon laboratory settings using microbes and invertebrates with rapid generation rates or has employed examples of artificial selection. Therefore, data from natural populations which deal with natural selection and incipient speciation continue to be of great interest (Endler 1986). A likely place to search for microevolutionary processes and the specific mechanisms of genetic isolation would be at the point of contact between populations in their initial phase of genetic divergence (e.g., Barton and Hewitt 1985; Hewitt 1988, 1989).

Anoline lizards, with their notable radiation within tropical America and the Caribbean islands, provide a spectrum of speciation events. One example, the *Anolis brevirostris* complex in Haiti, forms a parapatric arrangement of sibling species

[1] Department of Biology, Virginia Polytechnic Institute and State University, Blacksburg, Virginia 24061, USA.

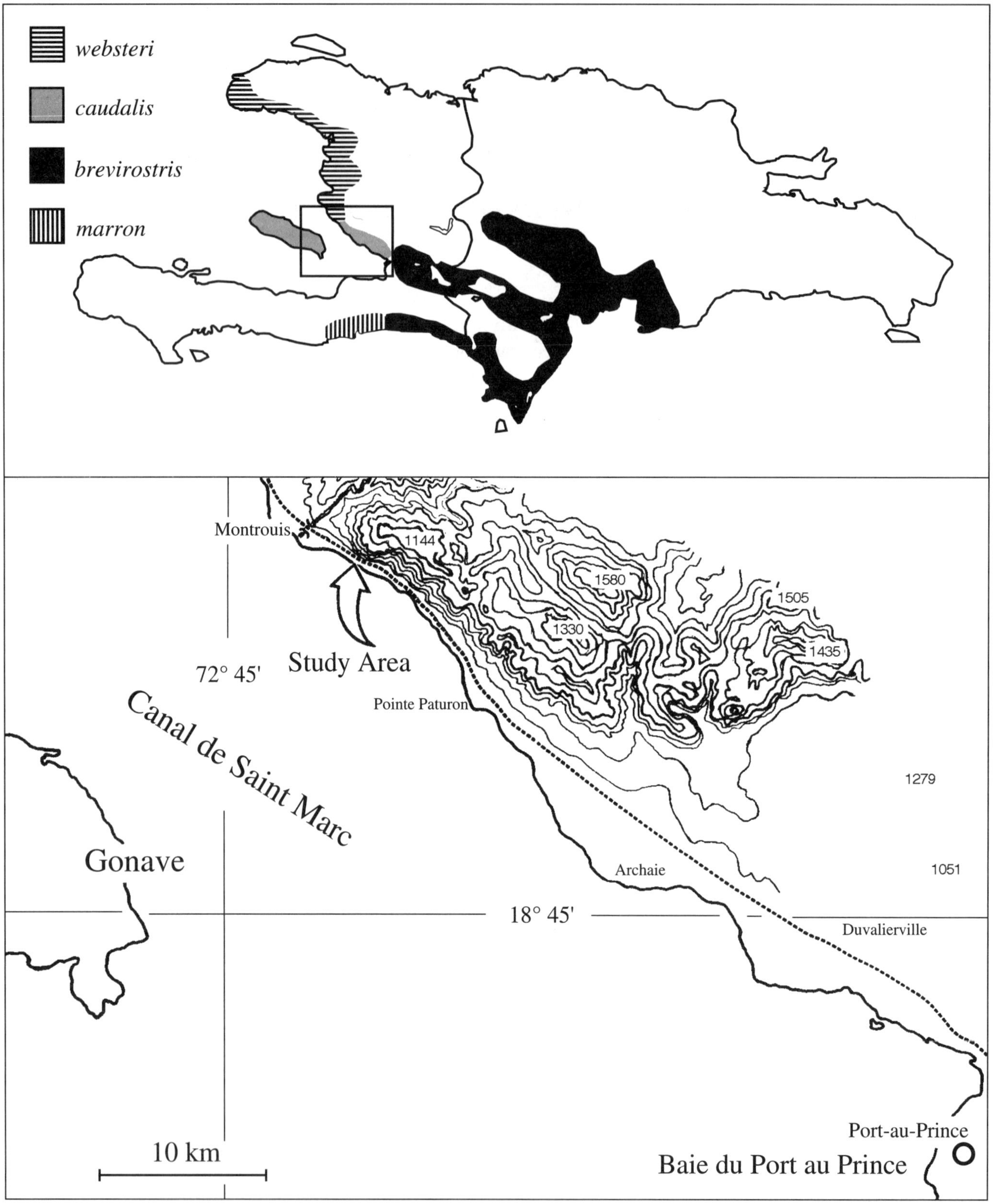

FIGURE 1. The distribution of the four species in the *Anolis brevirostris* complex on Hispaniola (top). The insert box is detailed below, showing the road (dashed line) out of Port-au-Prince to Montrouis and the topography in the vicinity of the study area (numbers give altitude in meters). The site of the study area (indicated by the arrow) is at the interface of the ranges of *A. websteri* and *A. caudalis*.

(Figure 1) of undoubtedly recent origin which may still reflect the microevolutionary processes of their speciation. Of the four species in the *brevirostris* complex (Arnold 1980), three are distributed contiguously along the west coast of Haiti, and were first separated on the basis of electrophoretic evidence (Webster and Burns 1973). One primary morphological means of distinguishing among the species has focused on the color of the dewlap, a characteristic anoline structure (Font and Rome 1990) used in visual displays (Jenssen 1977). More recently, Jenssen and Gladson (1984) examined the visual displays of the three parapatric species in the *brevirostris* complex (*A. websteri*, *A. caudalis*, and *A. brevirostris*) and found their headbob patterns to be stereotyped and distinctly species-specific. For both dewlap color and display patterns, the greatest interspecific differences occur in areas where the species have their closest contact, suggesting that the phenomenon of reproductive character displacement (Brown and Wilson 1957) is operative. Because the dewlap and complex patterns of headbobbing are dedicated to communication, a premating ethological isolating mechanism is very likely operating within the courtship context where sibling species are in contact.

My field experiments were devised to look for assortative mating between the parapatric sibling species, *Anolis websteri* and *A. caudalis*. The study area and particular populations for observation were chosen to maximize the intimacy of heterospecific contact and the degree of divergence between the communication traits of the sibling species. The study's objectives were to: (1) use free-ranging subjects, natural environment, and non-disruptive protocol, where possible, for manipulating mating opportunities; (2) determine if assortative mating occurs; (3) if heterospecific matings are avoided, determine which species and sex most likely practice discrimination; and (4) if discrimination occurs, examine the relative effects of individual-unique display aberrations on mating success.

Materials and Methods

Species/Study Area

Arnold (1980), Jenssen and Gladson (1984), and Webster and Burns (1973) indicated the greatest degree of interspecific differences in display pattern and dewlap color of the *brevirostris* complex occurs at the interface of *Anolis websteri* and *A. caudalis*. The interspecific border is approximately

TABLE 1. Collecting areas south of Montrouis (by kilometers from the town's bridge on Route 100 over the Riviere la Couline), with *Anolis websteri* and *A. caudalis* habitat and dewlap color. Color code: 1 = slight pigmentation (grayish) to pale yellow, 2 = light yellow, 3 = yellow, 4 = orange-yellow, 5 = orange.

Landmarks	km	Habitat	Dewlap		
			Background	Border	Scales
St. Paul's Church	0.7	Large trees	5	3 only at anterior half	3
Dirt road to beach	1.6	Last groves of large trees	5	3 along entire margin	3
Pasture	1.7–2.7	Closely cropped without lizards	—	—	—
8 fenced properties *	2.7–3.2	Scrub / scattered moderate trees	4, with some 3 at mid-margin	2	2
Study site	3.2	Scrub / trees	1 to 2	2	2
Southern sites	3.2–8.2	Scrub / trees	Much variation (all 1 to almost all 4)	2 or 3	2 or 3

* Area of suspected hybridization; some dewlaps of intermediate species characteristics.

A
B

72 km north of Port-au-Prince on Route 100, 2.4 km south of Montrouis (latitude 18°56', longitude 72°40'; Figure 1). At this point, the road passes along a severely constricted strip of *websteri*/*caudalis* habitat, with the ocean on the west and a 900-m escarpment on the east (Figures 1 and 2). Censuses in the xeric, scrub habitat up the mountain side at the contact zone produced no specimens of either species. Therefore, a 0.3–1 km wide north-south bottleneck existed in the region of the species' contact. For this reason, the road (Route 100) was an effective census transect for estimating the location and overlap of the *A. websteri* and *A. caudalis* distributions (Table 1).

Both species are small and of similar size, with male and female snout-vent lengths (SVL) of ≤51 mm and ≤47 mm, respectively. The two species inhabit an arboreal/trunk habitat niche (Arnold 1980; T. A. Jenssen, personal observation), and their body coloration (grays, browns, and black) is cryptically adapted for the bark substrate on which they perch. Body patterns of both species (and the species complex) is characterized by a black spot, or ocellus, on each shoulder. Superficially, *Anolis websteri* and *A. caudalis* appear indistinguishable. Species assignments of collected lizards, however, could be made with the following criteria.

Anolis websteri males have orange dewlaps which are extended during the species' distinctive Type A and B headbobbing displays (see Jenssen and Gladson 1984). Both males and females have a light yellow cast on their white venters. *Anolis caudalis* male dewlap color varies along a 5-km north-south transect from the contact zone (Table 1). Dewlaps are white at the northern edge of the *A. caudalis* distribution (i.e., 3.2 km south of Montrouis, Table 1). Beginning less than 0.2 km from the northern border, however, *A. caudalis* dewlaps may contain a medial yellow spot (of varying size) on a white or pale yellow background. The variability increases as one collects to the south. A rare specimen may have a dark yellow dewlap with only a thin border of white. In contrast, display behavior is uniform, regardless of collection area. *Anolis caudalis* males have a distinctive, species-specific set of Types A and B displays. An easily recognized *A. caudalis* display trait is the "tail lift," which concludes most (85%) of the males' advertisement (Type A) displays (see Jenssen and Gladson 1984). Venters of both males and females are white, lacking the prominent yellow wash found on the undersides of *A. websteri*.

FIGURE 2 (opposite page). (A) View to the north showing Montrouis (area of tree cover on peninsula and along road). Trees, including those south of Montrouis, are allopatric *Anolis websteri* habitat. (B) View of area just south of that depicted in (A), showing: (1) the habitat constriction between the ocean and the escarpment, and (2) the region of *A. websteri* and *A. caudalis* overlap.

Experimental Design

During June-July 1978, well within the breeding season of other Caribbean anoles (reviewed in Jenssen and Nunez 1994), private property and housing were rented at the *websteri*/*caudalis* contact zone to conduct an undisturbed series of extended social manipulations. Different combinations of mating assemblages were brought together in a "natural enclosure," which consisted of an isolated, moderately sized *Acacia* tree (approximately 5.6 m high, with a 7.7-m canopy diameter, Figure 3) known to be a species used by both lizard species (Arnold 1980; T. A. Jenssen, personal observation). The tree was chosen because: (1) its compound leaves and physiognomy minimized observational blind spots; and (2) it was located on a beach; the lack of ground cover effectively discouraged lizards from leaving the tree.

Subjects were taken within 0.5–4.0 km of the contact zone. Before noosing, lizards were watched to monitor their display characteristics (if a male), then captured and morphological traits also checked for species identification. Each subject was given an individual-specific toe clip and a corresponding acrylic paint mark code on the tail tip in yellow for *Anolis websteri* and white for *A. caudalis*. Lizards were then released onto the observation tree. Availability of food was facilitated by placing a number of crushed mangos in the forks of the tree branches to attract insects. Water occurred naturally as rain and dew. With the conclusion of an observation period, lizards were returned to the site of their original capture.

FIGURE 3. The observation tree used as a "natural enclosure" for *Anolis websteri* and *A. caudalis* observations.

During the study, no more than eight subjects (four of each species) were on the observation tree at a time. This is a high density condition, but not unrealistic. From a census study of syntopic *Anolis caudalis* and *A. cybotes* (Jenssen et al. 1988), as many as two male and four female *A. caudalis* and a pair of *A. cybotes* were found to share a tree. Data collection began the day after a replicate of lizards was placed on the observation tree. Two observers with binoculars, situated several meters on either side of the observation tree (Figure 3), followed male-female interactions between approximately 0630 to 1630 h. Behavioral variables were dictated into tape recorders, with significant behavioral events recorded on film (Nizo S800 camera). Courtship variables transcribed from tape and film were: (1) species and sex; (2) time of day; (3) display pattern and frequency of occurrence; (4) individual that initiated courtship; (5) courtship disposition (i.e., female lacked interest and moved away, male lacked interest and moved away, or both interested and copulation resulted); and (5) copulatory duration. Focal animal sampling (*sensu* Altmann 1974) was used, with each observer following one male. Copulatory events were recorded over the entire observation period, whereas other variables were sampled in random blocks. Because the objectives of each experiment were conditional (i.e., a subsequent experiment was contingent upon the results of its predecessor), the protocol used for the various experiments is addressed chronologically below.

RESULTS

Experiment I

Question: Individuals of which species and sex may be rejecting interspecific breeding opportunities? This experiment was replicated three times using fresh subjects. Each replicate (consisting of 1 male and 3 females of each species) was placed on the observation tree. Each replicate then was watched for four days.

Over 12 days, 23 copulations were observed; 13 involved *Anolis websteri* males, 10 *A. caudalis* males, and all mated females were *A. websteri* (Table 2). The conclusions were: (1) males of both species mate indiscriminately; (2) female *A. websteri* exhibited no apparent mate preference, and provided a potential avenue for hybridization; and, oddly, (3) female *A. caudalis* appeared sexually unavailable, rejecting all conspecific and heterospecific courtship invitations (see Courtship Behavior below).

Experiment II

Question: Males of which species will mate successfully if only *Anolis caudalis* females are present? This experiment was replicated three times using fresh subjects. Each replicate (consisting of 1 male of each species and 6 female *A. caudalis*) was placed on the observation tree. Each replicate then was watched for four days.

Over 12 days of observation and with the same ratio of males to females as in Experiment I (2:6), three copulations were seen; all by male *Anolis caudalis* (Table 2). The conclusions were: (1) female *A. caudalis* appear to have a low copulatory frequency (approximately 13% of that of female *A. websteri* in Experiment I); and (2) female *A. caudalis* may be practicing assortative mating.

TABLE 2. Frequency of hetero- and conspecific copulations within groups of *Anolis caudalis* and *A. websteri*: (I) 1 male and 3 females of each species; (II) 1 male of each species and 6 female *caudalis*; and (III) a *websteri*-dewlap male *caudalis* and 3 females of each species. Each experiment ran for 12 days in natural surroundings, using three replicates.

Subject	Male *Anolis caudalis*	Male *Anolis websteri*
Experiment I		
Female *Anolis websteri*	10	13
Female *Anolis caudalis*	0	0
Experiment II		
Female *Anolis caudalis*	3	0
Experiment III		
Female *Anolis websteri*	2	—
Female *Anolis caudalis*	4	—

Experiment III

Question: Will an *Anolis caudalis* male with a species-specific display pattern and an *A. websteri* dewlap color be discriminated against by female *A. caudalis*? While collecting 2.4 km south of the contact zone, a male *A. caudalis* was observed performing its species-specific headbob displays (verified by subsequently filmed and analyzed displays), including the distinctive "tail lift." However, the dewlap was orange-yellow with a thin pale-yellow border, very similar to the *A. websteri* condition. The male was collected, confirmed to be *A. caudalis* from agreement with other characters, and used as a mixed stimulus of *caudalis* bob pattern and *websteri* dewlap color. This experiment was replicated three times using fresh female subjects. Each replicate (the orange-dewlap *A. caudalis* male and 3 females of each species) was placed on the observation tree. Each replicate then was watched for four days.

During 12 days of observation, the male was seen copulating four times with *Anolis caudalis* and twice with *A. websteri* females (Table 2). The conclusions were: (1) as in Experiment I, the male showed little or no discrimination between hetero- and conspecific females; (2) female *A. caudalis*, as in Experiment II, exhibited a similar low frequency of copulation; (3) *A. caudalis* females accepted the orange-dewlap male with a copulatory frequency comparable to that for conspecific males with white dewlaps in Experiment II; and (4) based on these preliminary data, dewlap color may not be of singular importance in *A. caudalis* mate choice when the species-specific headbob pattern is still present.

Courtship Behavior

Display behavior, courtship frequency, and courtship disposition were quantified for 23 h of recorded observations from Experiment I, during which time a balanced complement of sexes represented the two species. For the purposes of this analysis, the data were assumed to be sufficient to permit the comparison of behavioral variables across species.

Display rates were not significantly different between *Anolis websteri* and *A. caudalis* males, at 52 (SE 10.6) and 45 (SE 5.8) displays/h, respectively. The composition of these displays consisted of 22% dewlap pulsing, 54% Type A displays, and 24% Type B displays (see Jenssen and Gladson 1984 for display descriptions and use).

Initiation of courtship, defined as a male approaching and displaying to a female, had a similar frequency among male *Anolis websteri* and *A. caudalis*, averaging approximately one every 13 min. In outcomes of the male-female contacts, conspecific females avoided *A. websteri* and *A. caudalis* males 43% and 50% of the time, respectively, while males moved away from conspecific females in 28% of *A. websteri* contacts and 11% of *A. caudalis* contacts. Male *A. caudalis* engaged in heterospecific contacts 39% of the time as compared with 29% of heterospecific contacts by *A. websteri* males. Because female *A. caudalis* were not seen to copulate during Experiment I, their reluctance to mate may have shifted the attention of male *A. caudalis* more toward *A. websteri* females.

Coitus Duration

Coitus duration was discovered to be dramatically different between *Anolis websteri* and *A. caudalis*. Mean duration for 13 copulations involving *A. websteri* males was 1.2 s (SE 0.2 s). Male *A. websteri* copulations were quite unusual for anolines, being analogous to a bird's "cloacal kiss." Cinema records (18 frames/s) of four instances of male *A. websteri* copulations revealed the hemipenis in the female for 4–8 frames (0.3–0.5 s). Timings from field observations were only accurate to about a second as start/stop times were based on relative cloacal juxtaposition and observer reaction time. The briefness of male *A. websteri* copulations might have caused some of their couplings to go unnoticed if it were not for the unusual behavior preceding intromission. After taking the usual "male-over-female" position, a male would engage in what appeared to be tactile stimulation by rubbing his body across and along the female's back, interspersed with headbob displays and some dismounting-remounting. Duration of this physical contact ranged 23–131 s. Then abruptly, and sometimes without a mouth hold, the male would quickly wrap his pelvic region under the female's, give one to several rapid thrusts, and dismount. Frame-by-frame analysis of films documented hemipenis penetration, and in one sequence, fluid was recorded on film leaving the female's cloaca shortly after copulation.

Male *Anolis caudalis* copulation durations may be conservative because the initiation of coitus was not always seen. Nevertheless, these durations were influenced by the species of the partner. In seven conspecific matings, mean coital duration was 13.4 min (SE 1.34 min), and in 12 matings with female *A. websteri* it was 3.5 min (SE 0.33 min), a significant difference from conspecific matings (Wilcoxon Rank Sums Test, $Z = 2.75$, $P = 0.006$). As with most anolines, male *A. caudalis* would initiate coitus by using his mouth to take a pinch of nuchal skin on the female's neck, then swinging his cloaca around and under the base of the female's tail to effect intromission. During coitus, female *A. caudalis* were passive, only headbobbing occasionally. In contrast, when *A. caudalis* males mated with *A. websteri*, the females soon became agitated, frequently headbobbing, lashing their tails, body squirming, and attempting to move away. This restlessness is believed to have resulted in the much shorter copulatory durations than when *A. caudalis* males were with conspecific females. The agitation exhibited by a female *A. websteri* during extended copulation with a male *A. caudalis* probably reflects her expectations of a 1 s-conspecific intromission.

Coitus Frequency

The frequency of copulations by species and sex was probably a reliable indicator of natural propensities. Density of lizards on the observation tree was high, but not abnormal. No conspecific or heterospecific couplings appeared to be the result of forced copulations because no behavioral sequences of unwillingness were observed. A forced copulation would be identified if a female ran from a male, was chased down, restrained, and then mated. On the contrary, male *Anolis websteri*, for example, preceded their 1-s copulations with a long episode of mounting and dismounting during which the female was free of a neck hold (see Coitus Duration).

Comparison of the sexual activity profiles between heterospecific males was quite similar. During Experiment I, *Anolis websteri* and *A. caudalis* males averaged approximately four female contacts/h, and copulated about once/day, with a similar copulatory success (13 and 10, respectively).

Comparison of the heterospecific female profiles, however, was strikingly different. For the nine *Anolis websteri* females involved in Experiment I, eight were seen to copulate. Of the eight who mated, two females mated only once; one female with an *A. websteri* male and the other with an *A. caudalis* male. The remaining six females mated either twice or three times during their respective 4-day observation periods. In all cases of multiple copulations, each female *A. websteri* mated with both conspecific and heterospecific males. Durations between an individual female's copulations ranged from 1.5 h to two days. The frequency of a female's sexual acceptance (copulations/female/day) averaged 0.64 (23 copulations/9 females/4 days); this was unexpectedly high.

For the nine *Anolis caudalis* females of Experiment I, the frequency of sexual acceptance was 0.0 (0 copulations/9 females/4 days); this was unexpectedly low. Even if *A. caudalis* females were practicing assortative mating, one would expect at least one receptive female. A very speculative hypothesis is drawn from the time of day when the seven *A. caudalis* copulations were recorded during Experiments II and III. All occurred early in the morning (0615–0830 h). An anecdotal observance during a collecting trip also noted several pairs of *A. caudalis* in copula at about 1900 h. Perhaps *A. caudalis* females tend to be sexually active in the early morning and then again in the evening. If this is true, their copulations may have been under-represented during the experimental observation periods (0630–1600 h).

The *Anolis websteri* copulations were unusual because of their brief duration and their high frequency of occurrence by individual females. This raises a host of proximate and ultimate questions concerning such phenomona as sperm transfer, sperm storage, and the relaxed association of female receptivity from the estrus cycle. However, besides their acknowledgment, a discussion of these implications is beyond the scope of this chapter.

DISCUSSION

Theoretical Predictions

The evolution of reproductive isolating mechanisms (RIM) is believed to proceed in two ways, both of which can involve hybrid zones where diverging subspecies or semispecies contact each other (Grant 1991). In the first mode ("incidental"), a RIM develops as an incidental byproduct during genetic divergence while populations are in allopatry. Secondary contact reinforces the RIM and perfects any premating RIM in instances of negative heterosis in hybrid individuals (Ayala 1981).

In the second mode ("reinforcement"), referred to as the Wallace effect or character displacement (Grant 1991), selection is for reproductive isolation *per se*. It occurs when sympatric contact between species produces adaptively inferior hybrids and one or both species are polymorphic with respect to characteristics affecting reproductive isolation. An ethological premating RIM, such as an interspecific divergence in species-specific courtship behavior, is more likely to be a selected result of this second mode of isolation than it is to be as a byproduct of the first mode (but see Paterson 1985 as well as Verrell 1988).

Hybrid zones, though not always leading to isolating mechanisms and speciation (e.g., Barton and Hewitt 1985; Moore 1987), would theoretically develop linkage disequilibria of assortative mating genes with nonallelic genes, leading to reduced hybridization (Endler 1977; Felsenstein 1981). Case and Williams (1984) and Webster (1974, 1978) reported subspecific and specific hybridization within and between species in the two closely allied anoline *distichus* and *brevirostris* complexes, known as "disticoids." Where studied (e.g., *Anolis brevirostris* x *A. distichus ravitergum*), meiosis is disturbed by hybridization; males have active testes, but are sterile (Webster 1974, 1978). One would predict that premating RIMs should be in evidence for disticoids because of mismatings documented in many geographic areas of heterospecific contact (Webster 1978).

Theoretically, the pair of parapatric disticoid species most likely to demonstrate a premating RIM and character displacement would be the sibling species, *Anolis websteri* and *A. caudalis*, at their Montrouis contact zone. This expectation is based on five lines of evidence. First, the habitat of both species is similar throughout the contact zone (Arnold 1980; T. A. Jenssen, personal observation), with no environmental cline between the species which might be selecting for an ecological effect on dewlap color (Fitch and Hillis 1984). Second, the *A. websteri* and *A. caudalis* courtship displays are markedly divergent (Jenssen and Gladson 1984), providing a behavioral mechanism for species recognition and mate choice. Third, the color of the male dewlap (i.e., the throat fan extended during displays) is also distinctly different between the sibling species at the Montrouis contact zone (white in *A. caudalis* and orange in *A. websteri*), providing still another cue for species recognition and mate choice. Fourth, the *A. caudalis* dewlap color is variable and forms a cline over the species' north/south distribution, shifting away from yellow/orange to white at the northern border (Webster and Burns 1973). This shift is postulated to be in response to the orange-dewlap *A. websteri*, and suggests evidence for an incipient ethological barrier as a direct result of heterospecific contact (e.g., Coyne and Orr 1989). Fifth, the B displays of the two species (those used in courtship) incorporate much dewlap extension and the flashing of species-distinct dewlap color. The relationship between social context and the species-differentiated display traits implies the display's importance to mate choice. Thus, it is not surprising that West-Eberhard (1983) found the only good prospect for documenting reproductive character displacement in reptiles to be the cline in *A. caudalis* dewlap colors focused at the *websteri/caudalis* contact.

I expected my study at the *websteri/caudalis* interface, to document assortative mating, and provide field evidence to support the species recognition theory proposed for anoline dewlap/headbobbing signals (e.g., Crews and Williams 1977; Jenssen 1977; Losos 1985; Macedonia et al. 1994; Rand and Williams 1970; Williams and Rand 1977). The results were quite different from initial expectations.

Predictions Examined

Female *Anolis websteri*, in spite of the considerable divergence between heterospecific male headbob displays, unequivocally showed no evidence of assortative mating. *Anolis websteri* females bred almost at random with conspecific and heterospecific males. My results argue that display traits, even though species-specific, are not actively serving as species-recognition signals nor being used for female choice. The almost similar preference that female *A. websteri* showed toward conspecific and heterospecific displaying males is a rarely-reported phenomenon, but one which Ryan and Rand (1993) observed for female túngara frog preference toward conspecific and heterospecific vocalizations. The evolutionary interpretation which best fits the *A. websteri* female behavior of my study is that the display differences between the sibling species occurred during allopatry (the incidental mode; *sensu* Mayr 1988), and not as a premating RIM during sympatry (the reinforcement mode; *sensu* Grant 1991).

The female *Anolis caudalis* responses must be interpreted much more cautiously due to low numbers of observed copulations. However, my *a priori* expectation that *A. caudalis* females would avoid heterospecific matings was not refuted because females were seen copulating only with conspecific males. Observed matings also were limited to the morning and evening. If female *A. caudalis* receptivity is crepuscular (though highly speculative from current data), there would be a decreased likelihood of receptive *A. caudalis* females encountering courting *A. websteri* males whose females copulate throughout the day. The suspected temporal separation in circadian breeding periods of the sibling species could be an ecological rather than a social response. If both species experience significant predation from diurnal predators, the long copulatory durations of *A. caudalis* ($\bar{x}$ = 13 min) would be safer during low light conditions, while *A. websteri* may be compensating for diurnally exposed copulations with extremely brief coital durations ($\leq$1 s).

With the assumption that female *Anolis caudalis* assortatively mate, conspecific male display signals should provide cues by which females discriminate against heterospecific males. The species-typical displays are composed of stereotyped headbobbing cadences and patches of color when the dewlap is extended. Taking these two components separately, the headbob cadence for *A. caudalis* displays is very uniform across the species' range, including samples from the island of Gonave (Jenssen and Gladson 1984). Because of their lack of variance between widely separated populations, the A and B headbob patterns appear to be under stable selection; this suggests a species recognition function.

The dewlap color, on the other hand, reflects a geographic cline, contrasting optimally with dewlap colors of parapatric sibling species; this suggests the result of sexual selection. However, evidence contrary to a sexual selection interpretation comes from Experiment III. Though only suggestive because of low sample sizes, the results indicated that the orange-dewlap *Anolis caudalis* male was not handicapped in his acceptance by conspecific females. His mating success was 0.11 copulations/female/day (4 copulations/9 conspecific females/4 days) as compared with the white-dewlap *A. caudalis* males in Experiment II which registered 0.04 copulations/female/day (3 copulations/18 conspecific females/4 days). As further evidence, Case (1990) found that within and among populations of *A. distichus* and *A. brevirostris*, variability in dewlap color was unrelated to electrophoretic variability or level of genetic divergence among populations. Her data argue against dewlap color as a trait for assortative mating and genetic isolation in disticoids. Therefore, current observations encourage alternative hypotheses, such as social selection (West-Eberhard 1983), to explain causation behind the display characteristics. Dewlap color may have little to do with mate choice in these anoles.

Summary

My current assessment of the *websteri*/*caudalis* contact zone and its evolutionary interpretation is as follows. The depth of the contact zone appeared limited (<0.5 km; Table 1, Figure 2B). This is probably the result of: (1) the topographically restricted contact zone due to the proximity of sea and mountains, (2) the low vagility of anoles (e.g., Andrews and Rand 1983; Rand 1967), and (3) either a recent contact between the species and/or hybrid sterility as found at other disticoid contact sites (Webster 1978). As female *Anolis websteri* were seen to outbreed and female *A. caudalis* were not, suspected gene flow between the species is probably asymmetric, moving from *A. caudalis* to *A. websteri* populations. Any disadvantageous heterosis resulting from outbreeding should select for a premating RIM; however, the proposed unilateral nature of the assortative mating may reflect a recent secondary contact between the species, with insufficient time to have fully developed a mutual premating RIM.

The behavioral interactions between parapatric *Anolis websteri* and *A. caudalis* present a natural experiment for the opportunistic examination of a host of theoretical issues relating to the evolution of communication systems, mate choice, and species. However, contrary to my expectation, available evidence appears to widen the realm of possibilities, rather than confirming pre-existing theoretical explanations for the *A. websteri* and *A. caudalis* display/dewlap traits. Hopefully, sufficient behavioral and genetic data will be gathered in the future to clarify this microevolutionary event.

Acknowledgments

This study is dedicated to the memory of Albert Schwartz, whose diligent descriptions of interpopulational diversity have provided theorists with their raison d'être. I am indebted to L. Jenssen, C. Pague, and M. Selover for their help in the field, and R. M. Andrews, M. Lovern, A. S. Rand, and E. E. Williams for comments on the manuscript. Financial assistance was provided by a National Science Foundation grant (DEB 74–2143).

Literature Cited

Altmann, J. 1974. Observational study of behavior: sampling methods. Behaviour 49: 227–267.

Andrews, R. M. and A. S. Rand. 1983. Limited dispersal of

juvenile *Anolis limifrons*. Copeia 1983: 429–434.

Arnold, D. L. 1980. Geographic variation in *Anolis brevirostris* (Sauria: Iguanidae) in Hispaniola. Breviora 461: 1–31.

Ayala, F. J. 1981. Gradualism versus punctualism in speciation: reproductive isolation, morphology, genetics, p. 51–66. *In* C. Barigozzi (ed.), Mechanisms of Speciation. Alan Liss, Inc., New York, New York.

Barton, N. H. and G. M. Hewitt. 1985. Analysis of hybrid zones. Ann. Rev. Ecol. Syst. 16: 113–148.

Brown, W. L. and E. O. Wilson. 1957. Character displacement. Syst. Zool. 5: 49–64.

Case, S. M. 1990. Dewlap and other variation in the lizards *Anolis distichus* and *A. brevirostris* (Reptilia: Iguanidae). Biol. Jour. Linnean Soc. 40: 373–393.

— and E. E. Williams. 1984. Study of a contact zone in the *Anolis distichus* complex in the central Dominican Republic. Herpetologica 40: 118–137.

Coyne, J. A. and H. A. Orr. 1989. Patterns of speciation in *Drosophila*. Evolution 43: 362–381.

Crews, D. and E. E. Williams. 1977. Hormones, behavior, and speciation. Amer. Zool. 17: 271–286.

Endler, J. A. 1977. Geographic Variation, Speciation, and Clines. Princeton Univ. Press, Princeton, New Jersey.

—. 1986. Natural Selection in the Wild. Princeton Univ. Press, Princeton, New Jersey.

Felsenstein, J. 1981. Skepticism towards Santa Rosalia, or why are there so few kinds of animals? Evolution 35: 124–138.

Fitch, H. S. and D. M. Hillis. 1984. The *Anolis* dewlap: interspecific variability and morphological associations with habitat. Copeia 1984: 315–323.

Font, E. and L. Rome. 1990. Functional morphology of dewlap extension in the lizard *Anolis equestris* (Iguanidae). Jour. Morphol. 206: 245–258.

Grant, V. 1991. The Evolutionary Process: A Critical Study of Evolutionary Theory. 2nd ed. Columbia Univ. Press, New York, New York.

Hewitt, G. H. 1988. Hybrid zones — natural laboratories for evolutionary studies. Trends Ecol. Evol. 3: 158–167.

—. 1989. The subdivision of species by hybrid zones, p. 85–110. *In* D. Otte and J. A. Endler (eds.), Speciation and Its Consequences. Sinauer and Assoc., Sunderland, Massachusetts.

Jenssen, T. A. 1977. Evolution of anoline lizard display behavior. Amer. Zool. 17: 203–215.

— and N. L. Gladson. 1984. A comparative display analysis of the *Anolis brevirostris* complex in Haiti. Jour. Herpetol. 18: 217–230.

—, D. L. Marcellini, and E. P. Smith. 1988. Seasonal microdistribution of sympatric *Anolis* lizards in Haiti. Jour. Herpetol. 22: 266–274.

— and S. C. Nunez. 1994. Male and female reproductive cycles of the Jamaican lizard, *Anolis opalinus*. Copeia 1994: 767–780.

Losos, J. B. 1985. An experimental demonstration of the species-recognition role of *Anolis* dewlap color. Copeia 1985: 905–910.

Macedonia, J. M., C. S. Evans, and J. B. Losos. 1994. Male *Anolis* lizards discriminate video-recorded conspecific and heterospecific displays. Anim. Behav. 47: 1220–1223.

Mayr, E. 1982. Process of speciation in animals, p. 1–19. *In* C. Barigozzi (ed.), Mechanisms of Speciation. Alan Liss, Inc., New York, New York.

—. 1988. The why and how of species. Biol. Philos. 3: 431–441.

Moore, W. S. 1987. Random mating in the northern flicker hybrid zone: implications for the evolution of bright and contrasting plumage patterns in birds. Evolution 41: 539–546.

Paterson, H. E. H. 1985. The recognition concept of species, p. 21–29. *In* E. Vrba (ed.), Species and Speciation. Transvaal Mus. Monogr. no. 4, Pretoria, South Africa.

Rand, A. S. 1967. Ecology and social organization in the iguanid *Anolis lineatopus*. Proc. United States Natl. Mus. 122: 1–79.

— and E. E. Williams. 1970. An estimation of redundancy and information content of anole displays. Amer. Nat. 104: 99–103.

Ryan, M. J. and A. S. Rand. 1993. Species recognition and sexual selection as a unitary problem in animal communication. Evolution 47: 647–657.

Verrell, P. A. 1988. Stabilizing selection, sexual selection and speciation: a view of specific-mate recognition systems. Syst. Zool. 37: 209–215.

Webster, T. P. 1974. Report, p. 57–60. *In* E. E. Williams (ed.), Second *Anolis* Newsletter. Mus. Comp. Zool., Harvard Univ., Cambridge, Massachusetts.

—. 1978. Hybridization of Hispaniolan lizards in the *Anolis distichus* species group, p. 166–173. *In* E. E. Williams (ed.), Third *Anolis* Newsletter. Mus. Comp. Zool., Harvard Univ., Cambridge, Massachusetts.

— and J. M. Burns. 1973. Dewlap color variation and electrophoretically detected sibling species in a Haitian lizard, *Anolis brevirostris*. Evolution 27: 368–377.

West-Eberhard, M. J. 1983. Sexual selection, social competition, and speciation. Quart. Rev. Biol. 58: 155–183.

Williams, E. E. and A. S. Rand. 1977. Species recognition, dewlap function and faunal size. Amer. Zool. 17: 261–270.

Resumen.—Se efectuaron experimentos de campo entre las especies hermanas *Anolis websteri* y *A. caudalis* en una zona de contacto haitiana. Utilizando un árbol que estaba apartado de un habitáculo adyacente como un "cercado natural," se observó durante todo el día (de las 0630 a las 1600 horas) por un período de 12 días cada una de tres combinaciones de lagartijas: (I) un macho y tres hembras de cada especie; (II) un macho de cada especie y seis hembras de *A. caudalis*, y (III) un macho de *A. caudalis* con una gaita pintada del color de la gaita de *A. websteri* y tres hembras de cada especie. En el Experimento I, la hembras de *A. websteri* copularon con machos conespecíficos y heteroespecíficos con una frecuencia similar, mientras que no se observó a las hembras de *A. caudalis* copulando. Conclusión: inequívocamente, machos y hembras de *A. websteri* y machos de *A. caudalis* no se aparean exclusivamente; se desconoce si las hembras de *A. caudalis* lo hacen. En el Experimento II, las hembras de *A. caudalis* se aparearon sólo con machos conespecíficos, pero se observaron pocas copulaciones. Conclusión: tentativamente, las hembras de *A. caudalis* discriminan en contra de congéneres. En el Experimento III, el color anormal de la gaita de los machos de *A. caudalis* no interrumpió el apareamiento conespecífico. Conclusión: es posible que el color de la gaita no sea un criterio importante para la discriminación de la pareja por las hembras de *A. caudalis*.

Las diferencias heteroespecíficas en la duración y la frecuencia del coito fueron notables. La duración de las copulaciones para los machos de *Anolis websteri* fue ≤1 s. En comparación, la duración promedio del coito con hembras conespecíficas fue 13.4 min para los machos de *A. caudalis*, disminuyendo a 3.5 min con hembras de *A. websteri*. En el Experimento I, cada hembra de *A. websteri* promedió 0.64 copulaciones/día, en contraste con 0.04 copulaciones/día para cada hembra de *A. caudalis* en el Experimento II. La mayoría de las hembras de *A. websteri* participó en múltiples copulaciones durante los períodos de observación.

A pesar del hecho de que las especies hermanas exhiben conductas de cortejo marcadamente diferentes, se encontró poca evidencia para la existencia de mecanismos efectivos de aislamiento reproductivo por comportamiento antes del apareamiento (MAR). Cuando menos, el potencial existe para el flujo genético de poblaciones de *Anolis caudalis* a poblaciones de *A. websteri*. Sin embargo, ninguna zona híbrida significativa es aparente, lo que sugiere la existencia de mar postapareamiento, tales como esterilidad del híbrido.

Résumé.—Des expériences sur le terrain ont été entreprises (juin-juillet) dans une zone de contact haïtienne entre les espèces jumelles *Anolis websteri* et *A. candalis*. Un arbre isolé de l'habitat environnant a été utilisé comme "enceinte naturelle," des observations diurnes (06 h 30–16 h) ont été effectuées pendant 12 jours pour chacune des trois combinaisons de sujets suivantes: (I) un mâle et trois femelles de chaque espèce; (II) un mâle de chaque espèce et six femelle *A. caudalis*; et (III) un mâle *A. caudalis* avec un fanon de la même couleur que celui de *A. websteri* et trois femelles de chaque espèce. Dans le cadre de l'expérience I, la femelle *A. websteri* a copulé avec la même fréquence avec les deux mâles con-spécifique et hétérospécifique alors que les observations n'ont pas permis de voir la femelle *A. caudalis* copuler. Conclusion: il n'y a incontestablement pas d'homogamie entre les mâles et femelles *A. websteri* et le mâle *A. caudalis* alors que l'inconnu subsiste pour la femelle *A. caudalis*. Dans le cadre de l'expérience II, la femelle *A. caudalis* s'est accouplée seulement avec les mâles con-spécifiques mais peu de copulations ont été observées. Conclusion: il semble que les femelles *A. caudalis* font des distinctions entre leurs congénères. Dans l'expérience III, la couleur anormale du fanon du mâle *A. caudalis* n'a pas empêché l'accouplement con-spécifique. Conclusion: la couleur du fanon ne semble pas être un critère important dans le choix des partenaires pour les femelles *A. caudalis*.

Des différences hétérospécifiques dans la durée du coït et sa fréquence ont été remarquées. La durée de la copulation pour les mâles *Anolis websteri* est de 1 seconde. Par contre, l'intromission con-spécifique moyenne pour le mâle *A. caudalis* est de 13,4 minutes, et de 3,5 minutes s'il s'agit d'une femelle *A. websteri*. Dans l'expérience I, la femelle *A. websteri* a copulé en moyenne 0,64 fois par jour contre 0,04 pour la femelle *A. caudalis* dans l'expérience II. La plupart des femelles *A. websteri* ont copulé plusieurs fois pendant les périodes d'observation.

Même si ces espèces jumelles se distinguent au niveau de la parade nuptiale, rien ne permet d'affirmer qu'il y a effectivement intervention d'un mécanisme d'isolement reproductif avant l'accouplement. Il existe tout au moins la possibilité d'un flux de gènes entre *Anolis caudalis* et *A. websteri*. Toutefois, aucune zone hybride significative n'est apparente ce qui donne à penser qu'il y a intervention d'un mécanisme d'isolement reproductif après l'accouplement, comme la stérilité des hybrides.

Jenssen, T. A. 1996. A test of assortative mating between sibling lizard species, *Anolis websteri* and *A. caudalis*, in Haiti, p. 303–315. *In* R. Powell and R. W. Henderson (eds.), Contributions to West Indian Herpetology: A Tribute to Albert Schwartz. Society for the Study of Amphibians and Reptiles, Ithaca (New York). Contributions to Herpetology, volume 12.

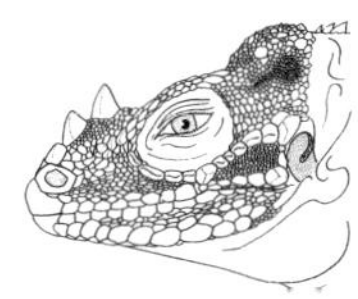

EVIDENCE OF SPATIAL NICHE PARTITIONING BY A HISPANIOLAN LIZARD COMMUNITY IN A XERIC HABITAT

ROBERT POWELL,[1] JOHN S. PARMERLEE, JR.,[1] AND DONALD D. SMITH [1, 2]

Abstract.—We investigated aspects of niche partitioning by lizards at a xeric study site near Baní, Provincia de Peravia, República Dominicana, elevation ~130 m. The site is characterized by a generally dry river bed extending through a predominately *Acacia* scrub/woodland. Six distinctly different microhabitats were identified and distinguished primarily on the basis of vegetation patterns, the associated availability of shade, and substrate. During dry seasons in 1986 through 1993, 394 lizards (6 species in 3 genera and 3 families) were taken and/or observed at the site. Preliminary investigations of food habits indicate that all of these species are opportunistic feeders, leading to the hypothesis that resources are partitioned spatially. Lizard community composition in the six microhabitats was significantly different. Spatial relationships were vertical where possible, or horizontal (by area) between forms with similar vertical niches. Where two species' habitat selections overlapped, size or foraging strategy may serve to effectively partition resources. Nearly half (185) of the lizards were recorded during one visit in June 1991 when conditions were characterized by unusually moderate temperatures and abundant moisture. Observed patterns of habitat use were very similar to those seen under the harsher, more commonly encountered conditions except for *Ameiva chrysolaema*, which was found in greater numbers and in more microhabitats than during other visits.

INTRODUCTION

BIOLOGISTS WORKING IN TROPICAL AREAS FREquently are attracted to the admittedly fascinating challenges posed by studies in rainforests; and concerns regarding the conservation of Antillean biodiversity often are centered on the remaining tracts of moist tropical forest. While these areas support a tremendously complex flora and fauna (Organization of American States 1969), focusing attention on them often leads to the neglect of less spectacular, but equally threatened communities of other types. Here we present some preliminary data on one aspect of community structure in just such a neglected area, specifically the spatial partitioning of resources by a lizard community in a xeric region of the República Dominicana.

Anecdotal observations in January 1986 during a visit to the Llanos de Azua led to an informal association of lizard species with specific microhabitats, some only a few meters apart. Structural habitats have been recognized and formally defined (Rand 1964) for anoline lizards in vertically complex habitats, and information about the natural history of West Indian lizards summarized by Schwartz and Henderson (1991) could be compiled to provide a larger picture of multi-family lizard communities, but no formal studies of spatial niches of Hispaniolan lizards occupying other than forested areas have been published. In March 1986, we initiated a series of visits to a site in the Llanos de Azua, Provincia de Peravia, characterized by a dry riverbed, in order to address the question of whether in fact lizards partition resources in a consistent and identifiable pattern.

In recent years, investigations of lizard communities have become quite common in resource-partitioning studies. This research attempts to determine how species differ ecologically and to impli-

[1] Department of Natural Sciences, Avila College, Kansas City, Missouri 64145, USA.

[2] Division of Allergy & Rheumatology, University of Kansas Medical Center, Kansas City, Kansas 66160, USA.

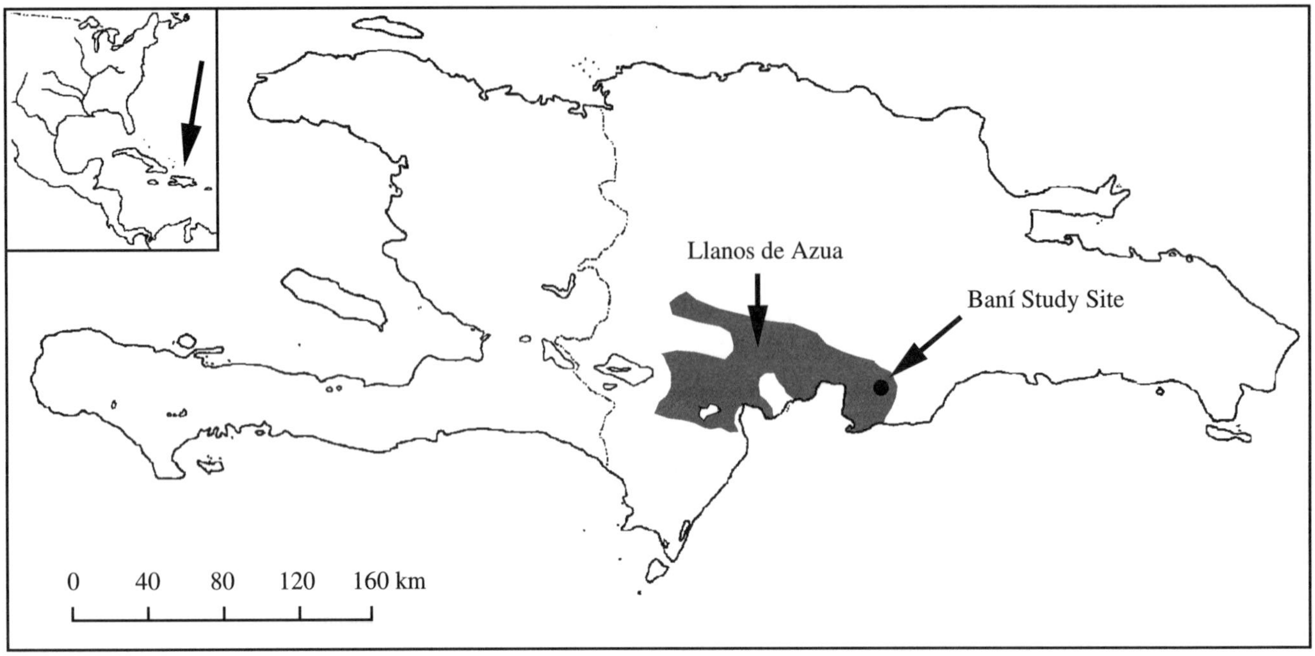

FIGURE 1. Location of the Baní study site on Hispaniola.

cate specific causal factors. Motivating most such studies is the hypothesis that food competition is important (Schoener 1983). The relevant literature has been reviewed recently by Pianka (1986) and Scheibe (1987). Sympatry demands that one or more mechanisms function to reduce possible competition, especially if essential resources are limited. Lack of strife may result from variations in food preferences, microdistribution, optimal body temperatures, and/or activity periods. Recent saurian studies (e.g., Smith et al. 1987, Smith 1989) examined these variables in efforts to evaluate their relative degrees of importance.

Spatial niches may be evaluated by correlating frequencies of encounters to microhabitat. The often subtle differences in place niches of terrestrial and arboreal lizards were reviewed by Pianka (1986) and Schoener (1968), respectively. Temperatures of active individuals often reflect times of activity (Pianka 1969). Interdependence of thermal regimes and space niches was demonstrated by Gorman and Hillman (1977) and Pacala and Roughgarden (1985). In this study, data collected in order to evaluate food habits and temperature relationships of sympatric lizards provided further insights in recognition of distinct spatial patterns of microhabitat utilization.

THE STUDY SITE AND LIZARD COMMUNITY

The study site is situated in the easternmost extension of the Llanos de Azua, about 8 km WNW of Baní, Provincia de Peravia (Figure 1), at an elevation of approximately 130 m (18°18'65" N, 70°23'62" W). The site is characterized by a (generally) dry river bed extending through an *Acacia* scrub/woodland. The area is quite xeric. Mean annual rainfall in Baní is 931 mm, but that in Azua (although more distant from the study site, this city actually is more similar in climate) is only 673 mm (Salcedo et al. 1983). Mean annual temperature in Azua is 27°C (Salcedo et al. 1983). The site (Figure 2) parallels highway 2 and is approximately 180 m x 600 m (ca. 0.1 km^2) in size. The river bed enters the northwestern corner of the area, shifts westerly until it essentially parallels the highway (and long axis of the area), and creates in a small and easily accessible area a heterogeneous complex of distinctly different microhabitats (Table 1).

During 1986–1993 an assemblage of three morphologically dissimilar groups of lizards (6 species in 3 genera and 3 families) (Table 2) was sampled and/or observed at the site. Differences in morphology reflect microhabitat preferences: *Ameiva* and *Leiocephalus* are ground-dwellers, whereas anoles

TABLE 1. Characteristics of habitats at the Baní study site. Names and numbers (in parentheses) of each habitat are used throughout this chapter.

Habitat	Characteristics
Acacia woods (1)	Most common plant: *Acacia macrocantha* (to 4 m) Other common plants: *Lemairocereus* sp., grasses Cover: dense (> 75% shaded) Substrate: mixed sand and small gravel, occasional boulders
Acacia scrub (2)	Most common plant: *Acacia macrocantha* (to 2 m) Other common plants: *Lantana* sp., scattered grasses Cover: sparse (< 50% shaded) Substrate: sand
River bed (3)	Most common plant: *Lantana* sp. (to 0.5 m) Other plants: none of significance Cover: very sparse (< 10% shaded) Substrate: gravel and boulders on sand
Islands (4)	Most common plant: *Lantana* sp. (to 1 m) Other common plants: *Acacia macrocantha*, *Cylindropuntia* sp., widely scattered clumps of grasses Cover: sparse (about 25% shaded) Substrate: sand with some gravel and a few boulders
Mixed forest (5)	Most common plant: *Acacia macrocantha* (to 4 m) Other common plants: *Lemairocereus* sp., *Casearia* sp., *Mimosa* sp., *Cylindropuntia* sp., scattered grasses Cover: moderate (about 60% shaded) Substrate: gravel and occasional boulders on sand
Edge (river bank) (6)	Most common plant: *Acacia macrocantha* Other common plants: *Cylindropuntia* sp. Cover: variable, often sparse Substrate: steep compressed sand, few to numerous rocks

(*Anolis*) are arboreal to various degrees. Field observations indicated further differences in foraging tactics displayed by ground-dwelling forms. Unless discovered under cover, *Ameiva* spp. were usually found while actively foraging, whereas *Leiocephalus* nearly always was encountered while stationary. These observations are consistent with the dichotomy often displayed between teiid and iguanian lizards (Pianka 1973): *Ameiva* spp. are intensive foragers (Regal 1983), but *Leiocephalus* spp. prefer to employ a sit-and-wait strategy (Schoener 1970). In discussing fundamental niche dimensions, Pianka (1973), Huey and Pianka (1983), and Schoener (1983) stated that foraging modes can influence all dimensions of resource partitioning, reducing interspecific competition by influencing the types of prey encountered and the composition of lizards' diets. A similar dichotomy has been described in *Anolis* species (Moermond 1979), which, in addition, often display significant degrees of vertical stratification in their microdistribution.

Anolis brevirostris is a xeric-adapted member of the *A. distichus* species group, and its range encompasses a wide variety of habitats (Arnold 1980, Schwartz and Henderson 1991). *Anolis whitemani* is the most xeric-adapted of Hispaniolan anoles, and its range is restricted to very dry regions of the island (Powell and Carr 1990). The study site is situated at the easternmost extension of the known ranges of both *A. brevirostris* and *A. whitemani*. *Anolis olssoni* is a grass anole with a very broad distribution on Hispaniola (Schwartz and Henderson 1991) and an apparent tolerance to a variety of conditions. However, this species was the least frequently encountered of those considered in this study. *Ameiva chrysolaema* is a widely-distributed generalist (Schell et al. 1993*b*) and lives in conditions ranging from mesic to semi-arid (Schwartz and Henderson 1991). Both *Ameiva lineolata* and *Leiocephalus semilineatus* are exceedingly xerophilic and, like *A. brevirostris* and *A. whitemani*, have ranges apparently limited by the more mesic conditions immediately east of the study site (Powell et al. 1989, Schwartz and Henderson 1991).

METHODS

Eight visits to the study site occurred in January (1991), March (1986, 1988, 1990), May (1989), June (1991, 1993), and August (1987), all outside the area's pronounced (albeit very moderate) rainy season, which extends from late September through

TABLE 2. The Baní study site lizard community: species (family) and a summary statement of some niche dimensions.

Species (Family)	Characteristics
Anolis brevirostris (Polychrotidae)	Arboreal (trunk) Sit-and-wait foraging strategy
Anolis olssoni (Polychrotidae)	Grass/shrub Sit-and-wait foraging strategy
Anolis whitemani (Polychrotidae)	Arboreal (ground/trunk) Sit-and-wait foraging strategy
Ameiva chrysolaema (Teiidae)	Terrestrial Active foraging strategy Body size large
Ameiva lineolata (Teiidae)	Terrestrial Active foraging strategy Body size small
Leiocephalus semilineatus (Tropiduridae)	Terrestrial Sit-and-wait foraging strategy

mid-December (Salcedo et al. 1983). Investigations in the area were between 0900 and 1400 hours, periods during which ambient temperatures were rising from 26–27°C to highs of 33–38°C (except during June 1991, when temperatures rose from 23–28°C and August 1987 when the highest temperature recorded was 43°C). These temporal restrictions do allow limited comparisons, especially of distributional patterns under prevailing conditions, but preclude the use of these data to draw conclusions regarding absolute or even relative abundances of the species observed. The number of persons involved in any visit varied from three (August 1987 and January 1991) to 14 (March 1986 and 1988) ($\bar{x} = 9.5 \pm 5.2$) and person-hours from nine (January 1991) to 65 (June 1991 and 1993) ($\bar{x} = 39.1 \pm 23.6$). Total person-hours were 413.

Specific microhabitats in which lizards were collected or observed were noted. During most visits, temperatures of the substrate and of the air (at 2 cm and 1 m) were recorded with quick-reading thermometers (Miller & Weber, Queens, New York) at each site where a lizard was initially observed. Air temperatures were taken in the shade and sheltered from any prevailing wind. To evaluate the temperatures in the microhabitats of lizards, the substrate and air (at 2 cm) temperatures were averaged, and only those for which all data were available were used in our analysis. Cloacal temperatures were taken from lizards captured immediately after initial observation. Except for temperatures, environmental conditions were evaluated qualitatively.

Only data for collected specimens are available for 1986–1989; in 1990 few specimens were taken, but transects perpendicular to the river bed (and highway) were walked and locations were noted for all observed lizards. In 1991 and 1993 data were recorded for both observations and collected specimens. To prevent repeated observations of the same animals, when data from observations only were included we walked transects only once. If an area within the study site was visited more than once during any trip, only data from collected animals were considered. Collected specimens were accessioned into the Bobby Witcher Memorial Collection at Avila College, Kansas City, Missouri, USA. All actions complied with "Guidelines for Use of Live Amphibians and Reptiles in Field Research" (American Society of Ichthyologists and Herpetologists, Herpetologists' League, and Society for the Study of Amphibians and Reptiles 1987).

Those data treated statistically failed tests for normality. As a result, only non-parametric tests were used. For all tests, $\alpha \leq 0.05$. Means are presented ± one standard deviation.

RESULTS AND DISCUSSION

In contrast to the sunny, hot, and dry conditions that prevailed during all other visits, in June 1991 we experienced mild temperatures (from 23°C at 0900 h to only 28°C at 1400 h) and mostly cloudy skies with scattered light showers. Also, the area obviously had been subjected to considerably more rainfall than we encountered at other times (evidenced by the verdant landscape, large number of forbs in bloom, and considerable moisture in subsurface soil). Because these conditions were notably different, we have discussed data collected during this visit separately from those of all other visits.

From all visits except that in June 1991, precise locality data were recorded for a total of 209 lizards. *Anolis brevirostris* (n = 32) was restricted to *Acacia* woods (n = 29) and mixed forest (n = 3), both areas where shade was most available and insolation less intense than in other microhabitats. Every individual was on the main trunk or a larger branch, usually of *Acacia* (n = 26). Whether this species' absence from other areas of the site was due to a lack of vertical structure or shelter from intense insolation is unknown and beyond the scope of this study. *Anolis olssoni* (n = 10) was found only where grass and shade were readily available, all but two were in *Acacia* woods, and the others were in *Acacia* scrub. All were on stems of tall (to 50 cm) clumps of grasses or small shrubs. *Anolis whitemani* (n = 29), as expected, was far more tolerant of exposed areas and was found largely where suitable perches were available in areas apparently too exposed for *A. brevirostris*. All but three were found in scattered shrubs along the riverbank, on islands, or in *Acacia* scrub; the other individuals were in *Acacia* woods (n = 2) and under trash in the river bed, but within 3 m of the edge. Except for the last-mentioned specimen, all were on main trunks or large branches of trees or shrubs, or on roots exposed by erosion along the river bank. During the periods surveyed in this study, every individual was in the shade. *Ameiva chrysolaema* (n = 38), thought to be the least xeric-adapted of the terrestrial species, was largely restricted to *Acacia* forest (n = 16), *Acacia* scrub (n = 6), and islands (n = 10), and was nearly always observed in the shade. Six additional individuals were foraging among trash deposited in the river bed during the later years of the study. All but two juveniles (found under cover) were first seen actively foraging for food. The true xerophiles, *A. lineolata* (n = 42) and *Leiocephalus semilineatus* (n = 58), were the species most commonly encountered. Although individuals were found in all habitats, the vast majority of both species was in the river bed (n = 73 of both species)

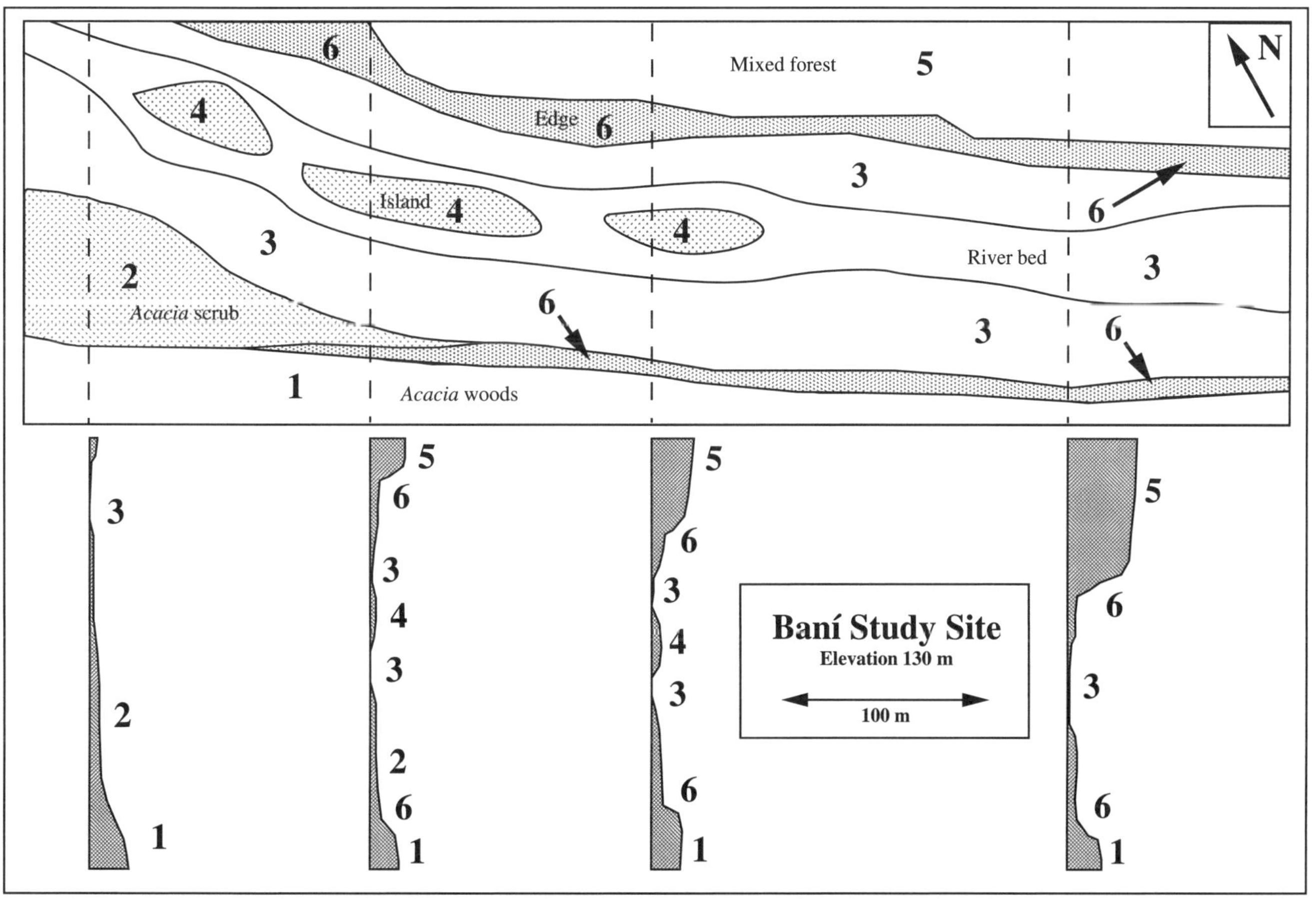

FIGURE 2. Diagrammatic representation of the Baní study site: overview and cross-sections. Numbers represent habitats named and characterized in Table 1.

and along the edges of islands ($n = 17$ of both species). Those found in *Acacia* woods, scrub, and mixed forest were usually along the edges of the road bed, in a clearing associated with a powerline right of way, or in breaks in the vegetation formed by grazing goats or trucks driving into the river bed to mine gravel or dump trash. Fewer than half of both species (slightly higher in *L. semilineatus*) were found under cover (rocks and logs that had been washed down the stream bed during intermittent flooding, cracks in the dry mud of the river bed, or trash dumped in increasing quantities over the period of the study). All of the remaining *A. lineolata* were first seen actively foraging. All but 14 of the remaining *L. semilineatus* were observed initially in stationary positions alongside cover, the other 14 were perched, presumably basking and/or searching for prey, on rocks or, in two instances, on masses of logs and branches accumulated during rare periods of apparently intense flooding.

Temperatures of microhabitat used by each of the six species differed significantly (Friedman Test, d.f. = 5, $\chi^2_{\text{corrected for ties}} = 88.75$, P = 0.0001). The lowest mean microhabitat temperatures were those of *Anolis brevirostris* ($n = 7$, $\bar{x} = 29.9 \pm 2.7$°C, range = 27.0–35.4°C) and *A. olssoni* ($n = 22$, $\bar{x} = 31.6 \pm 2.7$°C, range = 27.0–35.4°C). Microhabitats used by *A. whitemani* were warmer ($n = 10$, $\bar{x} = 31.6 \pm 3.6$°C, range = 32.5–43.0°C), approaching those of *Leiocephalus semilineatus* ($n = 25$, $\bar{x} = 35.5 \pm 3.0$°C, range = 29.2–40.4°C) and *Ameiva lineolata* ($n = 20$, $\bar{x} = 34.5 \pm 3.2$°C, range = 29.8–43.0°C), although the highest temperatures were recorded in association with individuals found in the river bed. Surprisingly, the highest mean microhabitat temperatures were those of *A. chrysolaema* (N = 8, $\bar{x} = 38.9 \pm 3.8$°C, range = 35.4–43.0°C), although these were predominantly associated with lizards foraging among debris in the river bed. Had more data been available from the more commonly used areas, the figure undoubtedly would have been lower. Still, these temperatures largely support the qualitative evaluations of the areas most often frequented by the various species.

In June 1991, all six species reported from the area were either captured or observed, and for all but one species the distributional data reinforced those previously discussed. All *Anolis brevirostris* ($n = 13$) were in *Acacia* woods. *Anolis whitemani* ($n = 23$) also was found in locations similar to those described above: three were in *Acacia* scrub, six on the north edge of a central island, eight on the north

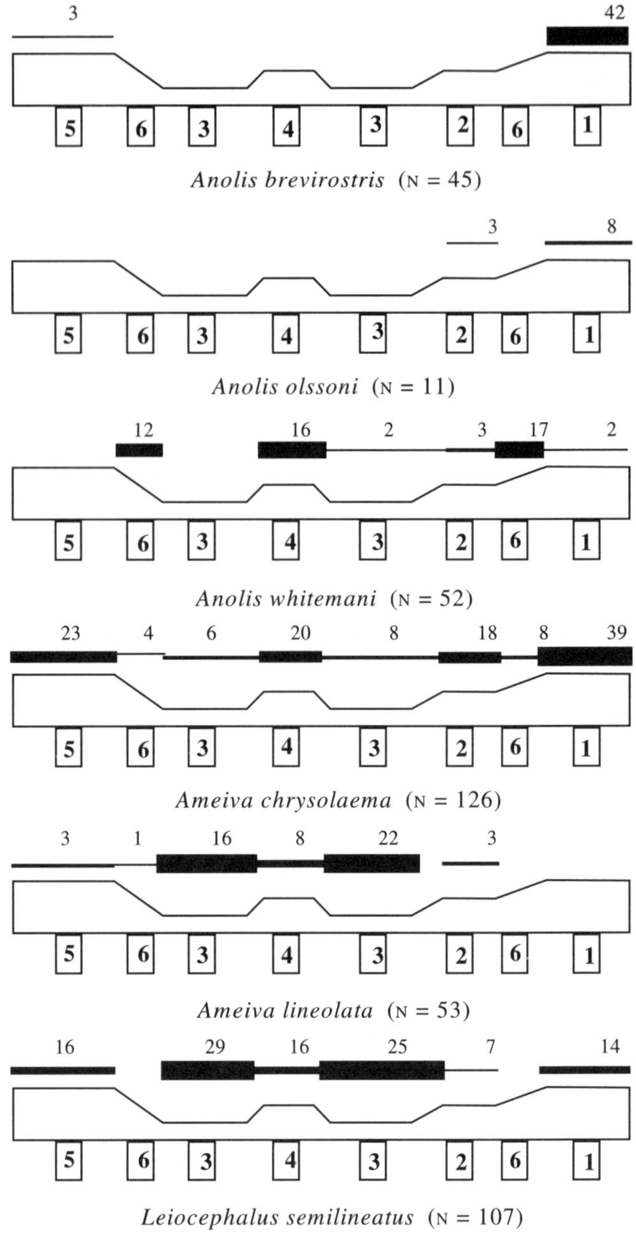

FIGURE 3. Distribution of lizard species at the Baní study site. Numbers below composite cross-sections of the site identify habitats named and characterized in Table 1. The thicknesses of lines above cross-sections are proportional to the frequencies in which each species was encountered in each habitat (see text).

river bank, five on the south river bank, and one juvenile on a rock in the river bed (but close to the south bank). The lone *A. olssoni* was on grass in *Acacia* scrub. *Leiocephalus semilineatus* (n = 49, nearly as many as had been seen on all other visits) was found throughout the area: 26 were either in *Acacia* woods or mixed forest, the majority in open areas along the road or in a cleared area under a power line, 11 of these were under cover (either rocks or logs), the remainder were in the open, five (based on posture) were basking; five were in *Acacia* scrub; 18 were either in the river bed or on islands. One individual, when disturbed, retreated into a burrow occupied by a large theraphosid spider, but no interaction of any kind was observed. All *Ameiva lineolata* (n = 11) were observed in the river bed or in open areas in mixed forest or *Acacia* scrub. Only observations of *Ameiva chrysolaema* (n = 88) differed substantially from those reported above. More individuals were observed than had been seen on all other visits, suggesting that this species may be the most common lizard in the area, and present in numbers even greater than those of *L. semilineatus*. Individuals were observed most frequently at shaded sites, but often were seen foraging in more exposed areas: 23 in *Acacia* woods, 23 in mixed forest, 12 in *Acacia* scrub, 10 on islands, and 20 in the river bed — 12 along the banks and 8 actively foraging amidst trash dumped at the site.

Preliminary investigations of diets (Carr et al. 1989, Burns et al. 1992, Moster et al. 1992, Schell et al. 1993*a*, Smith et al. 1994, and R. Powell, unpublished data) indicate that all of these species to varying degrees are opportunistic feeders (presumably advantageous, if not essential, in xeric habitats) and consume many of the same arthropodan prey. As a result, if these data are interpreted to imply potential competition for food resources, these resources apparently need to be partitioned by some mechanism (e.g., size).

The most obvious pattern that emerged from our data was one of spatial partitioning (Figure 3). The community composition of the six identified microhabitats (Table 1, Figure 3) were significantly different (Friedman Test, $\chi^2_{\text{corrected for ties}} = 13.05$, $P = 0.023$). This presumed division of resources was horizontal (by area) between *Anolis brevirostris* (largely restricted to shady woodlands) and *A. whitemani* (characterized by the data as an inhabitant of the more exposed edges of the river bed or islands). These anoles have comparable vertical niches (at least in this area). Similarly, horizontal partitioning apparently played a role in separating the niches of *Ameiva chrysolaema* (almost exclusively in shaded areas) and the other ground-dwelling taxa (found largely in the open river bed or clearings in other habitats). However, some partitioning of space was obviously vertical (when permitted by vegetation patterns). The separation between the tree- and shrub-dwelling anoles, the grass anoles, and the terrestrial species fall into this category. The consistency of these observations over eight years and a number of visits (albeit under relatively similar weather conditions of clear days, high temperatures, and dry substrate) speaks strongly in favor of this partitioning mode working at least under the presumably harsh conditions that apparently prevail during the dry season.

To suggest, however, that only spatial partitioning is operative would be overly simplistic. In areas where two species' microhabitats overlap at least to some degree, body size (between *Ameiva chrysolaema* and *A. lineolata*) or foraging strategy (between *Ameiva* spp. and *Leiocephalus semilineatus*) may function in partitioning essential resources of food and shelter. Differing thermal regimes may also play a role, but our data to date are insufficient to evaluate this dimension.

Conclusions

Resources seem to be partitioned largely by space among the lizards in this dry riverbed community, at least under the frequently encountered conditions of intense insolation, high temperatures, and dry substrate. However, some operative constraints (especially those limiting the distribution of *Ameiva chrysolaema*) apparently break down under the unusual or only seasonally prevalent milder conditions of moderate temperatures and readily available moisture. The high frequency of encounters with *A. chrysolaema* and its broad distribution throughout the site in June 1991 was surprising in light of the data from our other visits. This species is obviously more tolerant of the prevailing xeric conditions than we had surmised, but apparently emerged to forage only during periods (daily, seasonal, or adventitious) when conditions are less extreme than those we encountered at other times. This complicates the somewhat simplistic patterns presented above by superimposing considerations of temperature, obviously related to temporal (and seasonal?) factors, on the spatial partitioning observed during most of our visits. Reasonable hypotheses include: (1) lizards less capable of utilizing resources under harsh conditions take full advantage of opportunities provided by changing weather patterns to exploit areas not normally accessible; or (2) only during moderate conditions are resources sufficiently abundant in exposed portions of the study site to justify exploitation by lizards operating at the very limits of their tolerance ranges. In order to address these possibilities, an extended visit (or a series of visits) during the rainy season is necessary.

Recommendation

This small and severely disturbed area, subjected frequently to harsh environmental conditions, is nevertheless capable of supporting a rather high β-diversity (*sensu* Whittaker 1965) of lizard species. Unfortunately, like so many other areas throughout the Dominican Republic and the Antilles in general (Schwartz and Henderson 1991), this site is being severely altered. The causal factors at this location include overgrazing by goats, uncontrolled mining of streambed gravel, and dumping of refuse. In other areas, agricultural activities and expanding towns or urban sprawl might be responsible. As none of these conditions are likely to improve (in light of the growing human population of the region) and the biodiversity of this and similar sites has not been adequately evaluated, we recommend that investigators working in the West Indies be encouraged to pursue studies in these less-than-pristine sites — while not neglecting efforts in threatened forested areas, currently (and appropriately) receiving the bulk of researchers' attention.

Acknowledgments

Assistance in the field was provided by Martin L. Jolley, Jaime D. Villa, Sascha Oerter, Ellen J. Censky, and students in Avila College field biology classes and "Research Experiences for Undergraduates" programs. Thomas A. Zanoni, then of the Jardin Botánico, Santo Domingo, confirmed the identification of plants. Sixto J. and Yvonne A. Incháustegui, Grupo Jaragua, José A. Ottenwalder, United Nations Development Programme, and Andreas Schubert, Servicio Alemán and Departamento de Vida Silvestre, often have facilitated our opportunities for field research in the Dominican Republic. Permits were provided by Emílio A. Bautista M., Director, Departamento de Vida Silvestre, República Dominicana. Some support for field work in 1991 and 1993 was provided by National Science Foundation grants (BBS-9100410, BIR-9224486) awarded to the senior author.

Literature Cited

American Society of Ichthyologists and Herpetologists, Herpetologists' League, and the Society for the Study of Amphibians and Reptiles. 1987. Guidelines for use of live amphibians and reptiles in field research. Copeia 1987 (suppl.): 1–14.

Arnold, D. L. 1980. Geographic variation in *Anolis brevirostris* (Sauria: Iguanidae) in Hispaniola. Breviora 461: 1–31.

Burns, J. K., C. A. Cunningham, R. A. Dupuis, M. N. Trask, J. S. Tulloch, R. Powell, J. S. Parmerlee, Jr., K. L. Kopecky, and M. L. Jolley. 1992. Lizards of the Cayos Siete Hermanos, Dominican Republic, Hispaniola. Bull. Chicago Herpetol. Soc. 27: 225–232.

Carr, D. K., R. Powell, J. S. Parmerlee, Jr., and D. D. Smith. 1989. Life history notes (food habits): *Anolis whitemani whitemani*. Herpetol. Rev. 20: 49, 52.

Gorman, G. C. and S. Hillman. 1977. Physiological basis for climatic niche partitioning in two species of Puerto Rican *Anolis* (Reptilia, Lacertilia, Iguanidae). Jour. Herpetol. 11: 337–340.

Huey, R. B. and E. R. Pianka. 1983. Temporal separation of activity and interspecific dietary overlap, p. 281–290. *In* R. B. Huey, E. R. Pianka, and T. W. Schoener (eds.), Lizard Ecology: Studies of a Model Organism. Harvard Univ. Press, Cambridge, Massachusetts.

Moermond, T. C. 1979. Habitat constraints on the behavior, morphology, and community structure of *Anolis* lizards.

Ecology 60: 152–164.

Moster, J. A., R. Powell, J. S. Parmerlee, Jr., D. D. Smith, and A. Lathrop. 1992. Natural history notes on a small population of *Anolis brevirostris* (Sauria: Polychridae) from altered habitat in the Dominican Republic. Bull. Maryland Herpetol. Soc. 28: 150–161.

Organization of American States. 1969. Survey of the natural resources of the Dominican Republic. Study on the development and planning of natural resources. Report of the Technical Assistance Mission to the Dominican Republic carried out by the Natural Resources Unit of the Department of Economic Affairs in 1965–1966. Gen. Secretariat of the Organization of American States, Washington, D.C.

Pacala, S. W. and J. Roughgarden. 1985. Population experiments with the *Anolis* lizards of St. Maarten and St. Eustatius. Ecology 66: 129–141.

Pianka, E. R. 1969. Sympatry of desert lizards (*Ctenotus*) in western Australia. Ecology 50: 1012–1030.

—. 1973. The structure of lizard communities. Ann. Rev. Ecol. Syst. 4: 53–74.

—. 1986. Ecology and Natural History of Desert Lizards. Princeton Univ. Press, Princeton, New Jersey.

Powell, R. and D. K. Carr. 1990. *Anolis whitemani*. Cat. Amer. Amph. Rept.: 462.1–462.2.

—, J. S. Parmerlee, Jr., P. Ward, and D. D. Smith. 1989. Geographic distribution: *Anolis whitemani whitemani*. Herpetol. Rev. 20: 12.

Rand, A. S. 1964. Ecological distribution in anoline lizards of Puerto Rico. Ecology 45: 745–752.

Regal, P. J. 1983. The adaptive zone and behavior of lizards, p. 105–118. *In* R. B. Huey, E. R. Pianka, and T. W. Schoener (eds.), Lizard ecology: studies of a model organnism. Harvard Univ. Press, Cambridge, Massachusetts.

Salcedo, R. L., J. Czerwenka, and E. Bolay. 1983. Atlas de Diagramas Climáticos de la República Dominicana. Secretaría de Estado de Agricultura, Santo Domingo.

Scheibe, J. S. 1987. Climate, competition, and the structure of temperate zone lizard communities. Ecology 68: 1424–1436.

Schell, P. T., J. S. Parmerlee, Jr., and R. Powell. 1993*a*. *Ameiva chrysolaema*. Cat. Amer. Amph. Rept.: 563.1–563.6.

—, R. Powell, J. S. Parmerlee, Jr., A. Lathrop, and D. D. Smith. 1993*b*. Notes on the natural history of *Ameiva chrysolaema* (Sauria: Teiidae) from Barahona, Dominican Republic. Copeia 1993: 859–862.

Schoener, T. W. 1968. The *Anolis* lizards of Bimini: resource partitioning in a complex fauna. Ecology 49: 704–726.

—. 1970. Nonsynchronous spatial overlap of lizards in patchy habitats. Ecology 51: 408–418.

—. 1983. Overview of part III, p. 233–239. *In* R. B. Huey, E. R. Pianka, and T. W. Schoener (eds.), Lizard Ecology: Studies of a Model Organism. Harvard Univ. Press, Cambridge, Massachusetts.

Schwartz, A. and R. W. Henderson. 1991. The Amphibians and Reptiles of the West Indies: Descriptions, Distributions, and Natural History. Univ. Florida Press, Gainesville.

Smith, D. D. 1989. A comparison of food habits of sympatric *Cnemidophorus exanguis* and *Cnemidophorus gularis* (Lacertilia: Teiidae). Southwest. Nat. 34: 232–234.

—, P. A. Medica, and S. R. Sanborn. 1987. Ecological comparison of sympatric populations of sand lizards (*Cophosaurus texanus* and *Callisaurus draconoides*). Great Basin Nat. 47: 175–185.

Smith, J. W., R. Powell, J. S. Parmerlee, Jr., D. D. Smith, and A. Lathrop. 1994. Natural history notes on a population of grass anoles, *Anolis olssoni* (Sauria: Polychrotidae), from the Dominican Republic. Bull. Maryland Herpetol. Soc. 30: 67–75.

Whittaker, R. H. 1965. Dominance and diversity in land plant communities. Science 147: 250–260.

Williams, E. E. 1983. Ecomorphs, faunas, island size, and diverse end points in island radiations of *Anolis*, p. 326–370. *In* R. B. Huey, E. R. Pianka, and T. W. Schoener (eds.), Lizard Ecology: Studies of a Model Organism. Harvard Univ. Press, Cambridge, Massachusetts.

Resumen.—Investigamos aspectos de la división del nicho por lagartijas en un lugar xérico cerca de Baní, Provincia de Peravia, República Dominicana, elevación ~130 m. El lugar está caracterizado por un cauce generalmente seco que se extiende por una región de malezas y bosque dominada por *Acacia*. Se identificaron seis microhabitáculos claramente diferentes que fueron distinguidos principalmente a base de patrones de vegetación, la disponibilidad asociada de sombra, y el sustrato. Durante las temporadas secas de 1986 a 1993, 394 lagartijas (seis especies en tres géneros y tres familias) fueron capturadas y/o observadas en el lugar. Investigaciones preliminares de los hábitos alimentarios indicaron que todas estas especies son depredadoras oportunistas, lo que condujo a la hipótesis que los recursos son divididos espacialmente. La composición de la comunidad de lagartijas en los seis microhabitáculos fue significativamente diferente. Las relaciones espaciales fueron verticales donde era posible y horizontales (por área) entre formas con nichos verticales similares. Cuando dos especies coinciden en la selección del habitáculo, el tamaño o la estrategia de forrajeo pueden servir para dividir los recursos efectivamente. Casi la mitad (185) de las lagartijas fue observada durante una visita en junio de 1991 cuando las condiciones se caracterizaron por temperaturas inusitadamente moderadas y humedad abundante. Los patrones observados de uso del habitáculo fueron muy similares a aquellos observados bajo las condiciones más severas que son más comunes en el área, excepto para *Ameiva chrysolaema*, que fue encontrada en números mayores y en más microhabitáculos que durante otras visitas.

Résumé.—Nous avons étudié les aspects du cloisonnement des niches chez des lézards d'un habitat sec près de Baní, Province de Peravia en République Dominicaine, à une altitude d'environ 130 m. Le site se caractérise par la présence d'un lit de rivière sec en général qui traverse un bois où dominent les *Acacias*. Six micro-habitats différents ont été identifiés et distingués essentiellement en fonction de la végétation qui y croissait, des zones d'ombre et du substrat. Pendant les saisons sèches de 1986 à 1993, 394 lézards (six espèces de trois genres et trois familles) ont été capturés et(ou) observés sur place. Les études préliminaires des habitudes alimentaires révèlent que toutes ces espèces sont opportunistes, ce qui a conduit à la formulation de l'hypothèse voulant que les ressources soient cloisonnées sur le plan spatial. La composition de la communauté de lézards dans les six micro-habitats est très différentes. Les rapports spatiaux sont verticaux dans la mesure du possible ou horizontaux (par section) entre les formes ayant des niches verticales similaires. Lorsque deux espèces se disputent le même habitat, la taille ou la stratégie d'alimentation intervient dans le cloisonnement efficace des ressources. Près de la moitié des lézards (185) ont été répertoriés dans le cadre d'une visite en juin 1991 où les conditions se caractérisaient par des températures anormalement modérées et une humidité abondante. Les habitudes observées au titre de la colonisation de l'habitat sont comparables à celles observées dans le cadre de conditions plus dures et plus courantes à l'exception d'*Ameiva chrysolaema* que l'on a retrouvé en plus grand nombre et dans plus de micro-habitats qu'à l'occasion d'autres visites.

Powell, R., J. S. Parmerlee, Jr., and D. D. Smith. 1996. Evidence of spatial niche partitioning by a Hispaniolan lizard community in a xeric habitat, p. 317–326. *In* R. Powell and R. W. Henderson (eds.), Contributions to West Indian Herpetology: A Tribute to Albert Schwartz. Society for the Study of Amphibians and Reptiles, Ithaca (New York). Contributions to Herpetology, volume 12.

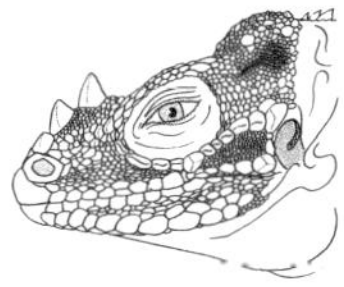

Diets of West Indian Racers (Colubridae: *Alsophis*): Composition and Biogeographic Implications

Robert W. Henderson[1] and Richard A. Sajdak[2]

Abstract.—The xenodontine colubrid snake genus *Alsophis* is represented by 11 species in the West Indies and no other colubrid genus is more widely distributed (Bahamas to Dominica) or occurs on more islands (about 100) in the West Indies. Species of *Alsophis* eat invertebrates, fish, frogs, turtles, lizards (sphaerodactylines to iguanines), snakes, birds, and mammals (bats and rodents), but frogs (*Eleutherodactylus*, *Osteopilus*) and lizards (primarily *Anolis*) are the most frequently consumed prey. Our data indicate that species of *Alsophis* are opportunistic, indiscriminate foragers. Inter- and intraspecific diet diversity is a reflection of the diversity of the available prey on a particular island. Larger snakes within and between species exploit a more diversified (taxonomically and by size) prey fauna than do smaller snakes. With increasing size, species of *Alsophis* take, on the average, larger prey items, but the size range of items also increases, as does the minimum prey size. Larger species exhibit a more marked ontogenetic shift in diet.

Introduction

The xenodontine colubrid snake genus *Alsophis* is represented by 11 species in the West Indies, and no other colubrid genus is more widely distributed (Bahamas to Dominica) or occurs on more islands (about 100, ranging in size from 0.2 km^2 to over 100,000 km^2) in the West Indies. Hispaniola is the only island that harbors more than one species. Ironically, no other West Indian snakes have had their numbers reduced, been extirpated, or become extinct as often as members of this enigmatic genus (Henderson and Sajdak 1986; Sajdak and Henderson 1991; Henderson 1992; Dodd 1993). The probable agent for the population declines and extirpations is the introduced Indian mongoose (*Herpestes auropunctatus*). On small islands where the mongoose has not been introduced (Bahama Islands, some of the Virgin Islands, some of the Lesser Antilles), *Alsophis* populations appear healthy and densities are frequently high, whereas on small islands where the mongoose has been introduced, populations are depressed or have been extirpated (see Table 2 in Sajdak and Henderson 1991). Three species are likely extinct (*A. ater* on Jamaica, *A. melanichnus* on Hispaniola, *A. sanctaecrucis* on St. Croix), several populations have been extirpated (*A. antiguae* on Antigua; *A. antillensis* on Marie Galante; *A. rufiventris* on St. Christopher and Nevis; *A. portoricensis* on Vieques and St. Thomas), and other populations have been reduced to small, localized areas (*A. antillensis* on Guadeloupe; *A. rijersmai* on St.-Martin) (Henderson 1992). The mongoose has been introduced to all of the Greater Antilles, but racer populations remain relatively healthy on Cuba and Puerto Rico. This is likely due to the larger size of these islands, where the deleterious effects of alien introductions are not as rapidly apparent (although indications exist that *Herpestes* also may be taking its toll on these large islands; R. Thomas, personal communication).

Despite the wide distribution, sometimes high densities, and conservation needs of *Alsophis*, little

[1] Section of Vertebrate Zoology, Milwaukee Public Museum, Milwaukee, Wisconsin 53233–1478, USA.

[2] Milwaukee County Zoo, Milwaukee, Wisconsin 53226, USA.

is known about their ecology. We do know that they are oviparous, diurnal, largely ground-dwelling (but sometimes ascending into low shrubs and even the crowns of rainforest trees), visually-oriented active foragers that occur in a wide variety of macrohabitats. They are opisthomegadonts and usually kill their prey with a venom secreted by Duvernoy's gland (Thomas and Prieto-Hernández 1985; Rodríguez-Robles 1992). At least *A. portoricensis* (Rodríguez-Robles 1992) and *A. cantherigerus* (Novo Rodríguez and Arazoza Rodríguez 1986; R. W. Henderson, personal observations) are capable of subduing prey by constriction, and small (e.g., *Sphaerodactylus* geckos) or harmless prey items may be swallowed without first killing (or at least immobilizing) them (Rodríguez-Robles and Leal 1993*b*; J. A. Rodríguez-Robles, personal communication; R. W. Henderson, personal observations). Although the venom is not lethal to humans, *A. cantherigerus* (Jaume and Garrido 1980) and *A. portoricensis* bites (Heatwole and Banuchi 1966; Thomas and Prieto-Hernández, 1985) have caused severe reactions.

In order to gain further insights into *Alsophis* ecology, we examined the stomachs of over 800 preserved specimens of *Alsophis* in order to ascertain their diets and determine the influence of geography and prey species diversity on diet breadth. Based on *a priori* knowledge regarding an active foraging mode and taxonomically generalized diet of *Alsophis*, we expected our data to further indicate indiscriminate foraging. What was unknown was the role of biogeography in defining trophic breadth in this genus. We predicted that, as active foraging, opportunistic predators, those species or populations that have the potential for encountering more prey taxa would have broader trophic niches than those species or populations lacking similar opportunities.

METHODS

Preserved specimens of *Alsophis* (representing all species except Hispaniolan *A. melanichnus*) were examined from the following collections: American Museum of Natural History (AMNH), Florida Museum of Natural History at the University of Florida (UF), Milwaukee Public Museum (MPM), Museum of Comparative Zoology at Harvard University (MCZ), National Museum of Natural History (USNM), and the University of Kansas (KU, which includes the Albert Schwartz collection). Each snake that contained prey had the following measurements taken: snout-vent length (SVL), head width (HW), and midbody circumference (MBC). Prey remains were removed, identified to species if possible, and volume was determined whenever possible by water displacement in a graduated cylinder. An intact specimen of the same species and size was sometimes used to obtain an accurate volume if some digestion had occurred. Prey volumes were not determined for prey items in which considerable digestion had occurred and identification to species was not possible.

Data from two or more islands were sometimes pooled to increase sample sizes and also to examine broad geographic patterns of prey exploitation. Basic geographic units were: Bahama Islands, Cuba (including Isla de la Juventud), Cayman Islands, Puerto Rico (including Culebra and Isla Piñero), Virgin Islands, and the Lesser Antilles. Sample sizes from Hispaniola (*A. anomalus* and *A. melanichnus*) and Jamaica (*A. ater*) were so small (0–2) that they were not included in analyses of islands or island groups. Sample sizes of prey for six species (*A. anomalus*, *A. antiguae*, *A. ater*, *A. melanichnus*, *A. rijersmai*, *A. sanctaecrucis*) were either zero or too small to do a species analysis.

We used prey genera rather than species as our index of diet breadth, based on the rationale of Henderson et al. (1988), although we do advocate identifying prey taxa to the most refined level possible. In determining the number of potential prey genera that were available to a particular population of *Alsophis*, we used the genera documented as prey in *Alsophis* diets as our criterion (with the exception of fish and turtles). That is, if a snake belonging to the genus *Arrhyton* was eaten by *Alsophis portoricensis* on Puerto Rico, snakes of the genus *Arrhyton* were considered potential prey for any species or population of *Alsophis* if *Arrhyton* occurred on the island occupied by that species or population (even if that prey item had not been documented in the diet of that species or population). Birds, bats, and rodents were each considered a single prey class.

Statistical analyses were performed using SAS. Data were analyzed using one-way analysis of variance (ANOVA), Pearson and Spearman rank correlation, and linear regression. Differences among group means were determined with a Tukey studentized range test (TSRT). Alpha level for statistical significance ≤ 0.05.

RESULTS

Prey Diversity

Species of *Alsophis* eat invertebrates, fish, frogs, hatchling sea turtles, lizards, snakes (including conspecifics), birds (wild and domestic), and mammals (rodents and bats). Following is a list of the species of *Alsophis* in the West Indies, islands included in our sample for each species, and the prey genera/species found in each species on each island. Information on the distribution of each species is available in Table 1, and Schwartz and Henderson provided additional details of distribution (1988) and distribution maps (1991).

TABLE 1. Species of *Alsophis* in the West Indies: maximum snout-vent length (SVL) and distributions.

Species of *Alsophis*	Maximum SVL (mm)	Distribution
anomalus	1770	Hispaniola; Île de la Tortue; Isla Beata
antiguae	660	Antigua; Great Bird Island
antillensis	930	Montserrat; Guadeloupe Archipelago; Dominica
ater	850	Jamaica
cantherigerus	1220	Cuba and associated islands; Cayman Islands; Little Swan Island
melanichnus	630	Hispaniola
portoricensis	923	Puerto Rico and satellites; Isla Mona; Isla Desecheo; Virgin Islands
rijersmai	790	Anguila; St.-Martin; St. Barthélémy
rufiventris	920	Saba; St. Eustatius; St. Christopher; Nevis
sanctaecrucis	1025	St. Croix
vudii	855	Bahama Islands

Alsophis anomalus.—HISPANIOLA: lizards: 1 *Anolis* sp. ÎLE DE LA TORTUE: mammals: 1 *Mus musculus*.

Alsophis antiguae.—ANTIGUA: lizards: 1 *Anolis* sp.; mammals: 1 *Mus musculus*. GREAT BIRD ISLAND: lizards: 1 *Anolis* sp.

Alsophis antillensis.—MONTSERRAT: frogs: 6 *Eleutherodactylus johnstonei*; lizards: 2 *Anolis lividus*; *Eleutherodactylus* and *Anolis* occurred in the same stomach. GUADELOUPE: frogs: 2 *Eleutherodactylus* sp.; lizards: 1 *Anolis marmoratus*. TERRE DE BAS (LES SAINTES): frogs: 1 *Eleutherodactylus martinicensis*; lizards: 1 *Anolis marmoratus*; *Eleutherodactylus* and *Anolis* occurred in the same stomach. DOMINICA: frogs: 4 *Eleutherodactylus martinicensis*; lizards: 1 *Anolis oculatus*, 1 *Ameiva fuscata*.

Alsophis ater.—JAMAICA: lizards: 1 *Celestus* sp.

Alsophis cantherigerus.—CUBA: frogs: 2 *Eleutherodactylus* sp., 3 *Osteopilus septentrionalis*; turtles: 1 *Eretmochelys imbricata*; lizards: 1 *Sphaerodactylus* sp., 9 *Anolis* sp., 1 *A. angusticeps*, 4 *A. ophiolepis*, 1 *A. porcatus*, 1 *A. quadriocellifer*, 1 *A. sagrei*, 1 *Cyclura nubila*, 1 *Leiocephalus cubensis*, 2 *Diploglossus delasagra*; snakes: 2 *Antillophis andreae*; birds: 2 unidentified; mammals: 3 *Mus musculus*, 2 unidentified rodents, 1 unidentified bat; an unidentified bird and an anole occurred in the same stomach, and an unidentified bird and a *Cyclura nubila* occurred in the same stomach. CAYMAN ISLANDS: GRAND CAYMAN: 5 *Eleutherodactylus planirostris*; lizards: 10 *Anolis* sp., 6 *A. conspersus*. CAYMAN BRAC: lizards: 1 *Sphaerodactylus argivus*, 3 *Anolis sagrei*; mammals: 1 *Mus musculus*.

Hardy (1957) reported two snakes from Cuba containing 1 and 2 bats (*Phyllonycteris poeyi*), respectively. Novo Rodriguez and Arazoza Rodríguez (1986) reported that stomach content analyses yielded remains of the hylid frog *Osteopilus septentrionalis*, the lizards *Anolis* sp., and *Leiocephalus* sp., the snakes *Tropidophis* sp. and *Antillophis andreae*; in addition, their field observations indicated that *A. cantherigerus* preyed on the domestic fowl *Gallus gallus*. On Cayman Brac, Grant (1940) reported finding *Anolis sagrei* and a honey creeper in racer stomachs.

Alsophis portoricensis.—PUERTO RICO: frogs: 3 *Eleutherodactylus* sp., 2 *E. antillensis*; lizards: 10 *Anolis* sp., 10 *A. cristatellus*, 1 *A. pulchellus*, 1 *Ameiva* sp., 3 *A. exsul*; snakes: 1 *Typhlops platycephalus*; ISLA PIÑERO: lizards: 2 *Anolis* sp. CULEBRA: frogs: 1 *Eleutherodactylus antillensis*; lizards: 3 *Sphaerodactylus macrolepis*, 4 *Anolis* sp., 3 *A. pulchellus*, 2 *A. stratulus*, 1 *Ameiva* sp., 1 *A. exsul*. CAJA DE MUERTES: lizards: 4 *Anolis* sp., 2 *A. cristatellus*, 2 *Ameiva exsul*. UNITED STATES VIRGIN ISLANDS: ST. THOMAS: lizards: 1 *Anolis* sp., 1 *A. pulchellus*, 1 *Ameiva exsul*. HANS LOLLICK: lizards: 2 *Sphaerodactylus macrolepis*. WATER ISLAND: lizards: 1 *Iguana iguana*. BRITISH VIRGIN ISLANDS: GUANA: lizards: 1 *Sphaerodactylus macrolepis*. PETER ISLAND: frogs: 1 *Eleutherodactylus* sp. VIRGIN GORDA: lizards: 1 *Sphaerodactylus* sp., 2 *Anolis* sp., 1 *A. stratulus*. ANEGADA: lizards: 1 *Anolis* sp.

In addition, Rodríguez-Robles and Leal (1993*b*) report *Eleutherodactylus coqui*, *Sphaerodactylus townsendi*, *Anolis evermanni*, *Arrhyton exiguum*, and *Rattus rattus* from the stomachs of *Alsophis portoricensis* on Puerto Rico. Manuel Leal (unpublished data) recorded *Ameiva wetmorei* from Puerto Rico, and *Hemidactylus mabouia* and *S. monensis* from Mona Island. Grant (1932) reported that one snake from Peter Island regurgitated a *H. mabouia*, and that one from Water Island contained three "young" *Iguana iguana*. Norton (1993) observed a snake on Congo Cay "masticating" dried fish (*Herangula* sp.) lost from a brown pelican (*Pelecanus occidentalis*) nest. On Little Saba Cay, Norton (1993) observed an *A. p. richardi* enter a nest containing two eggs of the ground-nesting Zenaida dove

TABLE 2. Some factors that may be critical in influencing diet breadth in West Indian *Alsophis*: maximum snout-vent length (SVL), island area, minimum number of prey genera expoited, potential number of prey genera available, percent of available prey genera that were exploited, and the number of prey items in the sample for each population.

Species of *Alsophis*	Maximum SVL (mm)	Island area (km^2)	Minimum number of prey genera exploited	Potential prey genera	Percent genera exploited	Number of prey items
antiguae	660					
Antigua and Great Bird Island		282	3	11	27.3	3
antillensis	930					
Montserrat		101	2	11	18.2	6
Dominica		790	3	11	27.3	6
cantherigerus	1220					
Cuba		111,463	14	17	82.4	38
Grand Cayman		197	6	12	50.0	21
Cayman Brac		36	4	13	30.8	5
portoricensis	923					
Puerto Rico		8,860	7	14	50.0	31
Culebra		28	4	10	40.0	15
rufiventris	920					
St. Eustatius		20	1	10	10.0	5
Saba		13	2	9	22.2	8
vudii	855					
Cat Island		388	7	11	63.6	27
Eleuthera		518	6	11	54.5	14

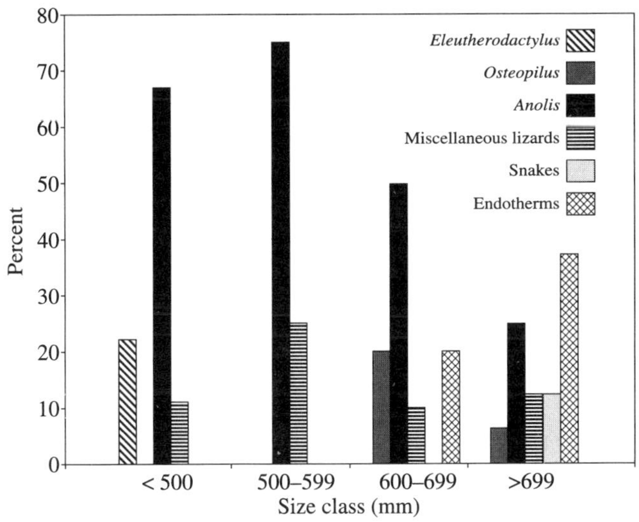

FIGURE 1. Prey genera or groups exploited by different size classes (SVL) of *Alsophis cantherigerus* on Cuba. Miscellaneous lizards include individuals from the genera *Sphaerodactylus*, *Leiocephalus*, *Cyclura*, and *Diploglossus*.

(*Zenaida aurita*); the snake positioned its head over the eggs while the adult dove stood nearby in a combative posture; the snake fled when disturbed.

Alsophis rijersmai.—ANGUILLA: lizards: 1 *Anolis gingivinus*. ST. BARTHÉLÉMY: lizards: 2 *Anolis gingivinus*.

Alsophis rufiventris.—SABA: frogs: 5 *Eleutherodactylus johnstonei*; lizards: 3 *Anolis sabanus*. ST. EUSTATIUS: lizards: 2 *Anolis* sp., 3 *A. wattsi*. ST. CHRISTOPHER: lizards: 1 *Anolis* sp., 1 *Ameiva erythrocephala*. NEVIS: lizards: 1 *Anolis* sp. One snake from Saba (MPM 23117) had four *Eleutherodactylus johnstonei* froglets in its stomach.

Alsophis vudii.—GRAND BAHAMA: frogs: 1 *Eleutherodactylus planirostris*; lizards: 1 *Anolis* sp., 4 *A. sagrei*. SOUTH BIMINI ISLAND: insects: 1 unidentified Phymatidae. BERRY ISLANDS: GREAT HARBOR CAY: lizards: 1 *Anolis* sp. NEW PROVIDENCE: frogs: 1 *Osteopilus septentrionalis*; lizards: 1 *Sphaerodactylus* sp., 3 *Anolis* sp., 1 *Ameiva auberi*. ELEUTHERA: frogs: 2 *Eleutherodactylus planirostris*, 1 *Osteopilus septentrionalis*; lizards: *Sphaerodactylus* sp., 6 *Anolis* sp., 2 *Ameiva auberi*; snakes: 2 *Alsophis vudii*, one *A. vudii* (KU 267379) was collected while in the process of swallowing another. ANDROS: frogs: 1 *Eleutherodactylus* sp., 1 *E. planirostris*, 2 *Osteopilus septentrionalis*; lizards: 1 *Sphaerodactylus* sp., 3 *Anolis* sp., 2 *A. sagrei*; snakes: 1 *Alsophis vudii*. CAT ISLAND: frogs: 2 *Eleutherodactylus planirostris*, 6 *Osteopilus septentrionalis*; lizards: 4 *Sphaerodactylus* sp., 8 *Anolis* sp., 4 *A. sagrei*; snakes: 1 *Alsophis vudii*; birds: 1 unidentified; mammals: 1 *Mus musculus*. GREAT EXUMA: lizards: 1 *Sphaerodactylus notatus*, 2 *Anolis* sp. LONG ISLAND: lizards: 1 *Anolis* sp. CROOKED ISLAND: frogs: 2 *Osteopilus septentrionalis*; lizards: 3 *Anolis brunneus*. ACKLINS ISLAND: lizards: 1 *Anolis brunneus*, 1 *Leiocephalus punctatus*. GREAT INAGUA: lizards: *Sphaerodactylus* sp., 5 *Ameiva maynardi* (the gecko occurred in the same stomach as *Ameiva*). INAGUA: SHEEP CAY: lizards: 3 *Anolis scriptus*, 1 *Ameiva maynardi*; *A. scriptus* and *A. maynardi* shared the same stomach.

On Cat Island, nine *Osteopilus* froglets occurred in the same stomach as an *Anolis* (MCZ 39514); and *Osteopilus* froglets occurred in multiples of 5 (MCZ 39534), 18 (MCZ 39504), and 26 (MCZ 39531), respectively, in three different snakes. Each instance of multiple froglets was counted as one prey item.

For the entire sample, we encountered 258 prey items of which 72.9% were lizards, 17.8% frogs, 3.9% mammals, 3.1% snakes, 1.2% birds, and the remainder invertebrates (0.4%), turtles (0.4%), and an unidentified vertebrate (0.4%). Prey taxa by genus included 56.3% *Anolis*, 11.8% *Eleutherodactylus*, 7.5% *Ameiva*, 6.9% *Sphaerodactylus*, 5.9% *Osteopilus*, 3.5% *Mus*, 2.0% other *Alsophis*, 1.2% unidentified birds, and one or two records each of phymatid insects, the lizards *Leiocephalus*, *Iguana*, *Cyclura*, *Celestus*, and *Diploglossus*, the sea turtle *Eretmochelys*, the snakes *Typhlops* and *Antillophis*, and an unidentified bat.

Values for population variables tested appear in Table 2, and we chose those populations with the largest sample sizes of prey items. Island area was correlated with the number of potentially exploitable prey genera (= more species) (Spearman, $r = 0.73$, $P < 0.01$) and the number of exploited prey genera ($r = 0.73$, $P < 0.01$). Snake SVL was not correlated with exploited prey genera (Spearman, $r = 0.22$, $P > 0.05$), but a weak (but not significant) correlation existed between SVL and potentially exploitable genera ($r = 0.58$, $P < 0.10$).

Four of the five species (*cantherigerus*, *portoricensis*, *rufiventris*, *vudii*) for which samples sizes are adequate to do a species analysis, had *Anolis* representing 49–77% of all prey items. The diet of *Alsophis antillensis* included only 29.4% *Anolis* (*Eleutherodactylus* accounted for 64.7% of its diet).

Ontogenetic Shifts in Diets

Overall, except for snakes less than 300 mm SVL and for those greater than 800 mm SVL, anoles comprise the most frequently eaten prey items in West Indian *Alsophis*. In the entire sample, small racers (<300 mm SVL) took a higher percentage of *Eleutherodactylus* frogs and the diminutive gekkonid *Sphaerodactylus*. Large racers (>800 mm SVL) took a variety of prey items (frogs, snakes, birds, mammals), and *Mus* was the most common. *Alsophis cantherigerus* on Cuba is the largest species of *Alsophis* for which an adequate prey sample is available. Snake size and the diversity of potential prey species on Cuba gave *A. cantherigerus* the most diverse diet and the most dramatic diet ontogeny (Figure 1). From a typical diet of *Eleutherodactylus* and *Anolis*, at about 700 mm SVL *A. cantherigerus* begins preying on small introduced rodents (primarily or exclusively *Mus*, but probably *Rattus* as well). The Bahamian racer, *Alsophis vudii*, exhibits an ontogenetic shift in diet from *Eleutherodactylus*, small *Osteopilus*, *Sphaerodactylus*, and *Anolis* to predominantly medium-large *Osteopilus* and *Anolis* (Figure 2).

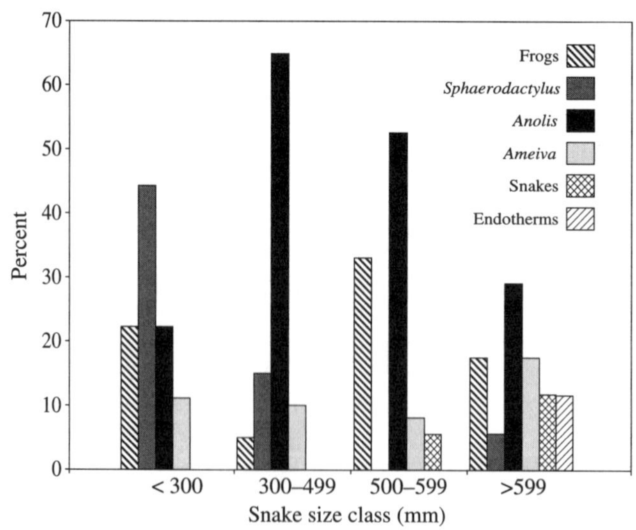

FIGURE 2. Prey genera or groups exploited by different size classes of *Alsophis vudii* (from many islands). Frogs include individuals in the genera *Eleutherodactylus* and *Osteopilus*.

Prey Size

The entire sample of prey volumes ($n = 141$) ranged from 0.1–50.0 cm^3 ($\bar{x}$ = 2.8 cm^3). Prey size (volume) for the entire sample was correlated (Pearson) with three morphological variables: SVL ($n = 140$, P < 0.0001), MBC ($n = 123$, P < 0.0001), and HW ($n = 120$, P < 0.0001). Table 3 illustrates the relationship between snake size, prey size, and prey diversity. With increasing size, species of *Alsophis* take, on the average, larger prey items, but the size range of prey items also increases (Figure 3), as does the minimum prey size. Also, larger snakes have more taxonomically diverse diets: e.g., snakes in the 500–599 mm SVL size class contained 91 prey items of nine prey genera; snakes in the greater than 800 mm SVL size class contained only eight prey items (8.8% of the larger sample), yet seven prey genera were represented (87.5% of the larger sample).

The single largest prey item ingested was an *Alsophis vudii* by another *A. vudii*. ANOVA indicated significant differences (P < 0.0001, F = 32.03, d.f. = 140) in mean size (volume) of prey items taken in each genus, and TSRT indicated that mean volume of the prey items in the Group 1 genera were significantly different from the mean volume of prey items in the Group 2 genera (P < 0.05, d.f. = 129): Group 1 (mean volume in cm^3 is given in parentheses) — *Cyclura* (23.0), *Antillophis* (20.0), *Alsophis* (19.7), *Iguana* (17.0), *Eretmochelys* (15.0), *Mus* (11.8); Group 2 — *Ameiva* (3.2), *Osteopilus* (2.6), *Leiocephalus* (2.1), *Anolis* (1.4), *Eleutherodactylus* (0.7), and *Sphaerodactylus* (0.3). Of the genera in group one, only *Mus* is preyed on with any frequency; the rest occurred in the sample infrequently ($n = 1$–3). The frogs and lizards in the genera of the second group were, with the exception of *Leiocephalus*, preyed upon much more often (n = 10–78) than the members of the first group.

An individual *Alsophis antillensis* that weighed 45.0 g was collected on Terre-de-Bas (Guadeloupe Archipelago) and regurgitated a recently ingested male *Anolis marmoratus* that weighed 7.2 g; mass ratio (prey mass/predator mass) was 0.160. Crother

(1986) collected an *A. cantherigerus* (830 mm SVL, 242 g) that had eaten a hatchling sea turtle (*Eretmochelys imbricata*); the turtle weighed 15 g (B. I. Crother, personal communication), giving a mass ratio of 0.062. Prey mass ratios were calculated for four *A. rufiventris* and they ranged from 0.004–0.027 with a mean of 0.016. The smallest mass ratio was for a snake that weighed 132 g and had eaten an *E. johnstonei* that weighed, at most, 0.5 g. On Waderick Wells Cay, Franz and Dodd (1994) collected a 196 g *A. vudii* that had just eaten a 34.5 g *Ameiva auberi*; the mass ratio was 0.176. Mass ratios are, in all cases, quite small in this small sample and no snake consumed a prey item that weighed more than 18% of its own mass. Although our small sample of prey-predator mass ratios precludes comparisons with other snake species, our general impression based on studies of other colubrids (e.g., Seib 1984, 1985) is that species of *Alsophis* take, on the average, relatively small prey items. The distribution of prey sizes in Figure 3 also indicates that small prey items are eaten by racers with SVLs up to about 800 mm.

Four *Alsophis vudii* on Cat Island were collected with from 5–26 *Osteopilus septentrionalis* froglets in their stomachs. Live, recently metamorphosed *O. septentrionalis* froglets weigh about 0.15–0.41 g (W. Meshaka, in litt.); 0.3 g is about average. The *A. vudii* that consumed 26 froglets was 583 mm SVL, and weighed about 85 g (based on data from R. Franz, in litt.); therefore, each froglet had a mass ratio of 0.0035. In four of five instances when froglets occurred in *A. vudii* stomachs, they occurred in multiples. Furthermore, all of the froglets were at a similar stage of digestion, indicating that they had all been ingested over a fairly short period of time (probably < 1 h). If the 26 frogs are viewed as a single prey item, the mass ratio escalates to 0.092. Ingestion of multiple *Osteopilus* (*dominicensis*) froglets also has been observed in the Hispaniolan colubrid *Uromacer catesbyi*. Recently metamorphosed froglets frequently occur in aggregations (R. W. Henderson, R. Powell, personal observations). A solitary *Osteopilus* froglet may be ignored as potential food by a foraging adult *A. vudii*, being regarded as unprofitable because of an extremely low net energy gain (low energy prey of Arnold 1993), but when several froglets can be ingested quickly with little additional search effort, the net energy gain may be profitable. An alternative hypothesis is that single froglets are so inconspicuous that they are overlooked by a foraging racer, whereas an aggregation will attract the attention of the same snake.

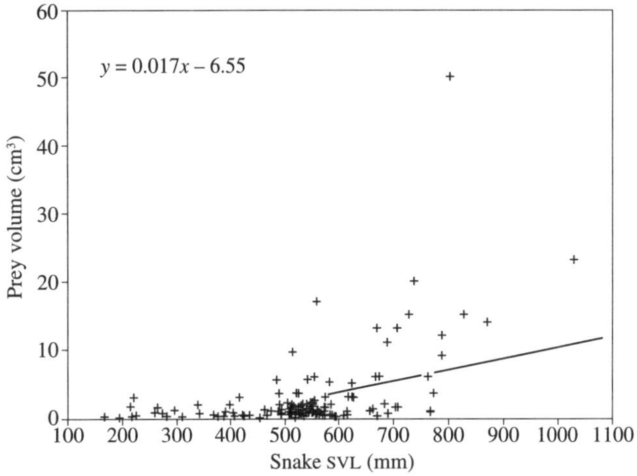

FIGURE 3. The relationship between prey volume and snake snout-vent length for all *Alsophis*.

Island and Island Group Comparisons

Island group comparisons were made between the Bahamas, the Greater Antilles, and the Lesser Antilles. Only *Anolis* figured prominently in the diets of *Alsophis* on all three island groups, representing more than 50% of the total diet for all size classes combined (Figure 4). Within our sample (i.e., excluding published records), the number of prey genera exploited by *Alsophis* in the Greater Antilles was 16, in the Virgin Islands eight, in the Bahamas seven, and in the Lesser Antilles four. On islands where potential prey diversity is greater, species of *Alsophis* exploit a greater diversity of prey genera (Spearman, $r = 0.73$, $P < 0.01$; Table 1).

In adult snakes, frogs of the genus *Eleutherodactylus* were important prey items in the Lesser Antilles, but less so in the Bahamas and Greater Antilles. The hylid frog *Osteopilus septentrionalis* is an important prey species in the Bahamas, but much less so in the Greater Antilles (although not native to Greater Puerto Rico or the Lesser Antilles, it has become locally established in both areas; Schwartz and Henderson 1991). ANOVA indicated no significant differences between broad geographic areas (Bahamas, Greater Antilles, Lesser Antilles) in size of prey eaten ($P > 0.05$, $F = 1.52$). However, refinement of the analysis (Bahamas, Cuba, Cayman Islands, Puerto Rico, Virgin Islands, Lesser Antilles) did indicate geographic differences in prey volume (ANOVA; $P < 0.002$, $F = 3.99$), but only mean prey size on Cuba was significantly greater than those from the other geographic samples (TSRT, $P < 0.05$).

DISCUSSION

West Indian *Alsophis* have been described as trophic generalists and opportunists (Henderson et al. 1988; Henderson and Crother 1989; Rodríguez-Robles and Leal 1993) and our data support that observation. Species of *Alsophis* are opportunistic predators on a broad taxonomic array of vertebrate prey, representing a wide range of sizes but, in general, favoring small prey items. As in West Indian snakes in general (Henderson and Crother 1989), lizards comprise the single most important component of *Alsophis* diets. Conversely, birds and mammals are rare in *Alsophis* diets, but the only West Indian endemics that prey routinely on endotherms are several boids (*Epicrates* spp.) and two viperids (*Bothrops* spp.). *Alsophis* is believed to be among the most primitive of colubrids, but the phylogenetic relationships of *Alsophis* are unknown, and many conflicting opinions have been published (Maglio 1970; Rabb and Marx 1973; Thomas 1976; Cadle 1984; Crother 1989). Xenodontine snakes that may share an evolutionary history with *Alsophis* are likely to prey predominantly on frogs and lizards also, as bird and mammal predation is rare in xenodontine snakes on the Neotropical mainland (Henderson and Crother 1989; Cadle and Greene 1993).

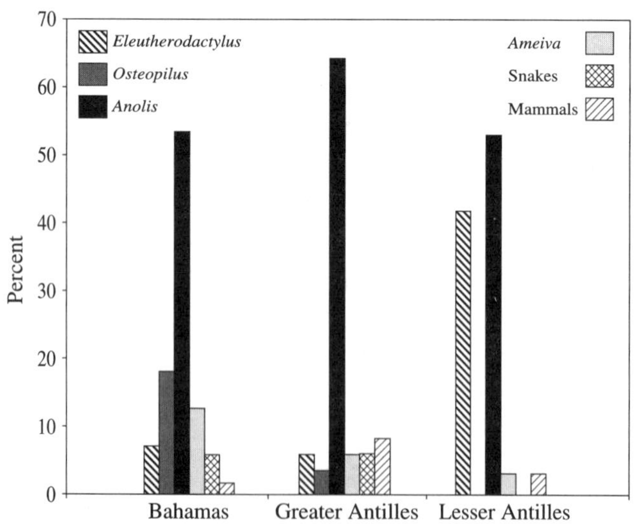

FIGURE 4. Percent of prey genera exploitation by *Alsophis* from three broadly-defined geographic regions of the West Indies.

As has been noted elsewhere (Henderson et al. 1988; Henderson and Crother 1989), the incidence of anoles in the diets of West Indian snakes is remarkable, and it is likely that nowhere else in the world is a snake fauna so dependent on a single group of prey species belonging to the same genus. In Australia, the rich elapid fauna preys heavily on scincid lizards, but on representatives of many genera (Shine 1991*a*). Although some species of *Alsophis* (*cantherigerus*, *portoricensis*, *vudii*) are remarkably euryphagous, *Anolis* makes an important contribution to the diets of all species for which we have an adequate sample.

We have documented that, on Cuba, *Alsophis cantherigerus* exploits at least 14 of 17 potential vertebrate prey genera (82.4%; Table 2). On Puerto Rico, the largest island (next to Cuba) for which we have an adequate sample, *A. portoricensis* preys on members of at least seven of 14 vertebrate genera. This number is matched in the Bahamas, but racers in the Lesser Antilles exploit members of only one to three genera per island.

Examination of the information presented in Table 2 indicates an obvious bias in these data, but one that can be used in an enlightening manner. Populations of *Alsophis* for which we have larger sample sizes exploit, in general, a higher percentage of the potential prey genera (Spearman, $r = 0.79$, $P < 0.005$). That is, as prey sample size increases, so does the number of prey genera exploited. The number of exploited prey genera is also correlated with the number of prey items in the sample ($r = 0.89$, $P < 0.01$; Figure 5). So, although depressed

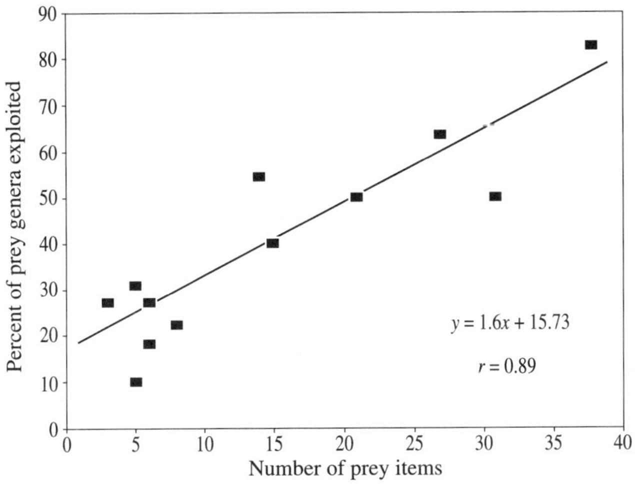

FIGURE 5. The relationship between prey sample size and the number of prey genera exploited.

sample sizes may present a biased picture, the relationship between sample size and actual or potential prey suggests trophic opportunism and extreme euryphagy. Given a large enough sample size, most, if not all, available prey genera eventually may be exploited (as on Cuba). This implies that *Alsophis* is indiscriminate as far as vertebrates go, and that the prey most often consumed is the prey most often encountered. On Hispaniola, a diurnal, ground-dwelling, active forager, *Antillophis parvifrons*, is *Alsophis*-like in behavior, but smaller (maximum SVL = 530 mm) than any species of *Alsophis*. Despite the smaller size of *A. parvifrons*, its diet exhibits characteristics similar to *Alsophis* — at least eight vertebrate genera and at least 25 species that were predominantly lizards (*Anolis* accounted for 70.9% of 199 prey items) and, as the prey sample increased, so too did the taxonomic breadth of its diet (Henderson et al. 1987).

Inter- and intraspecific diet diversity is a reflection of the diversity of available prey on a particular island as the number of genera actually exploited is correlated with the number of potential prey genera on a particular island, and island species diversity is the result of a variety of factors (area, physiography, proximity to other land masses, etc.). *Alsophis cantherigerus* exploits amphibians, reptiles (turtles, lizards, and snakes), birds, and mammals (rodents, bats), and demonstrates euryphagy that is rivaled by few other snake species, including other West Indian taxa (Henderson et al. 1988). Snake species that do exhibit comparable euryphagy are also active foragers (e.g., *Coluber constrictor*; Fitch 1963), but not all are ground-dwelling (e.g., *Boiga dendrophila* and *B. irregularis*; Greene 1989). In general, little is known about diets of colubrid snakes on the Neotropical mainland (Cadle and Greene 1993), but Seib (1984, 1985) examined the diets of four ground-dwelling, active-foraging, racer-like snakes (three colubrines and one xenodontine) that occur in Central America. As in *Alsophis*, frogs and lizards were the most common prey items, and individual prey item size tended to be relatively small.

TABLE 3. The relationship between snake size classes, prey size, and prey diversity for all species of West Indian *Alsophis* combined.

Size class (mm):	<300	300–499	500–599	600–699	700–799	>800
Number with prey	12	26	63	16	11	4
Mean prey size ± SE (cm^3)	0.9 ± 0.24	1.2 ± 0.25	1.7 ± 0.31	3.6 ± 0.95	7.6 ± 2.00	25.5 ± 8.41
Range	0.1–3.0	0.1–5.5	0.1–17.0	0.3–13.0	1.5–20.0	14.0–50.0
Prey maximum – prey minimum (cm^3)	2.9	5.4	16.9	12.7	18.5	36.0
Number of prey items	14	38	91	44	25	8
Number of prey genera in the sample	5	5	8	9	6	7

We conclude, then, that trophic breadth in species of *Alsophis* is limited by the available prey fauna which is, in turn, geographically constrained. A secondary factor may be that *Alsophis* populations with a smaller mean SVL have a more limited ability to subdue and swallow larger available prey and, therefore, the taxonomic diversity of the diet will be limited. That is, frogs, lizards, snakes, birds, bats, and rodents (pre- and post-Columbus; Pregill et al. 1994) occur on many *Alsophis*-inhabited islands but, in the Lesser Antilles for example, only rarely are prey items other than *Eleutherodactylus* and *Anolis* taken. Although racers in the Lesser Antilles encounter fewer potential prey taxa, they do not necessarily encounter fewer potential prey items (Henderson and Bourgeois 1993). Since snakes are gape-limited, larger snakes are capable of eating both large and small prey items, whereas small snakes obviously are restricted to small prey items (Shine 1991*b*, 1994). As has been suggested for other West Indian snake populations (Henderson and Bourgeois 1993), and with snake sizes being equal, the species or population of *Alsophis* that has access to more kinds of vertebrate prey will have a broader trophic niche than the species or population with accessibility to a prey fauna depressed by geographic constraints.

ACKNOWLEDGEMENTS

Personnel responsible for several collections were extremely generous in loaning large numbers of specimens that sometimes necessitated multiple loans, and for granting permission to examine them for prey remains: C. J. Cole and P. Damiani (AMNH), D. Auth (UF), J. P. Rosado (MCZ), E. J. Beldon, W. R. Heyer, and G. R. Zug (USNM), and W. E. Duellman, A. Nieto, J. E. Simmons, and E. Wild (KU). We are grateful to R. Franz, J. C. Gillingham, M. Leal, and J. A. Rodríguez-Robles for sharing data on *Alsophis* diets, and W. E. Meshaka, Jr. for providing data regarding *Osteopilus*. We received valuable criticisms from two anonymous reviewers and especially from Javier A. Rodríguez-Robles. R. W. Bourgeois was, as usual, indispensable in collecting data. Kevin Lyman was instrumental in producing the graphs. Field work with *Alsophis* has been generously funded by the Institute for Museum Services (Conservation Grant no. IC-60084-86), the Milwaukee Public Museum, the Milwaukee County Zoo, and the late Albert Schwartz. We have been ably assisted in the field by R. M. Henderson, T. J. McCarthy, and S. L. Sajdak.

LITERATURE CITED

Arnold, S. J. 1993. Foraging theory and prey-size – predator-size relations in snakes, p. 87–115. *In* R. A. Seigel and J. T. Collins (eds.), Snakes: Ecology and Behavior. McGraw-Hill, New York.

Cadle, J. E. 1984. Molecular systematics of neotropical xenodontine snakes: I. South American xenodontines. Herpetologica 40: 8–20.

— and H. W. Greene. 1993. Phylogenetic patterns, biogeography, and the ecological structure of neotropical snake assemblages, p. 281–293. *In* R. E. Ricklefs and D. Schluter (eds.), Species Diversity in Ecological Communities. Univ. Chicago Press, Chicago, Illinois.

Crother, B. I. 1986. *Alsophis cantherigerus* (Cuban Racer): diet. Herpetol. Rev. 17: 47.

—. 1989. Phylogeny and biogeography of the West Indian xenodontine snakes: a biochemical perspective. Unpubl. Ph.D. Dissertation, University of Miami, Coral Gables, Florida.

Dodd, C. K., Jr. 1993. Strategies for snake conservation, p. 363–393. *In* R. A. Seigel and J. T. Collins (eds.), Snakes: Ecology and Behavior. McGraw-Hill, New York, New York.

Fitch, H. S. 1963. Natural history of the racer *Coluber constrictor*. Univ. Kansas Publ., Mus. Nat. Hist. 15: 351–468.

Franz, R. and C. K. Dodd, Jr. 1994. *Alsophis vudii vudii* (Brown Runner): diet and growth. Herpetol. Rev. 25: 28.

Grant, C. 1932. Herpetology of Tortola; notes on Anegada and Virgin Gorda, British Virgin Islands. Jour. Dept. Agr. Porto Rico 16: 339–346.

—. 1940. The herpetology of the Cayman Islands. Inst. Jamaica, Sci. Ser. 2: iv, 1–56.

Greene, H. W. 1989. Ecological, evolutionary, and conservation implications of feeding biology in Old World cat snakes, genus *Boiga* (Colubridae). Proc. California Acad. Sci. 46: 193–207.

Hardy, J. D., Jr. 1957. A note on the feeding habits of the Cuban racer, *Alsophis angulifer* (Bibron). Copeia 1957: 49–50.

Heatwole, H. and I. B. Banuchi. 1966. Envenomation by the colubrid snake *Alsophis portoricensis*. Herpetologica 22: 132–134.

Henderson, R. W. 1992. Consequences of predator introductions and habitat destruction on amphibians and reptiles in the post-Columbus West Indies. Carib. Jour. Sci. 28: 1–10.

— and R. W. Bourgeois. 1993. Notes on the diets of West Indian *Liophis* (Serpentes: Colubridae). Carib. Jour. Sci. 29: 253–254.

— and B. I. Crother. 1989. Biogeographic patterns of predation in West Indian colubrid snakes, p. 479–518. *In* C. A. Woods (ed.), Biogeography of the West Indies: Past, Present, and Future. Sandhill Crane Press, Gainesville, Florida.

—, B. I. Crother, T. A. Noeske-Hallin, A. Schwartz, and C. R. Dethloff. 1987. The diet of the Hispaniolan snake *Antillophis parvifrons* (Colubridae). Jour. Herpetol. 21: 330–334.

—, T. A. Noeske-Hallin, B. I. Crother., and A. Schwartz. 1988. The diets of Hispaniolan colubrid snakes II. Prey species, prey size, and phylogeny. Herpetologica 44: 55–70.

— and R. A. Sajdak. 1986. West Indian racers: a disappearing act or a second chance? Lore (Milwaukee Public Mus.) 36: 13–18.

Jaume, M. L. and O. H. Garrido. 1980. Notas sobre mordidas de jubo *Alsophis* (Serpentes: Colubridae). Misc. Zool. Acad. Cien. Cuba 11: 2–3.

Maglio, V. J. 1970. West Indian xenodontine colubrid snakes: their probable origin, phylogeny, and zoogeography. Bull. Mus. Comp. Zool., Harvard Univ. 141: 1–54.

Norton, R. L. 1993. *Alsophis portoricensis richardi* (Ground Snake): feeding. Herpetol. Rev. 24: 34.

Novo Rodríguez, J. and F. de Arazoza Rodríguez. 1986. Cría en cautiverio de *Alsophis cantherigerus* (Ophidia: Colubridae). Rep. Invest. Inst. Zool., Acad. Cienc. Cuba 30: 1–15.

Pregill, G. K., D. W. Steadman, and D. R. Watters. 1994. Late Quaternary vertebrate faunas of the Lesser Antilles: historical components of Caribbean biogeography. Bull. Carnegie Mus. Nat. Hist. 30: iii, 1–51.

Rabb, G. B. and H. Marx. 1973. Major ecological and geographic patterns in the evolution of colubroid snakes. Evolution 27: 69–83.

Rodríguez-Robles, J. A. 1992. Notes on the feeding behavior of the Puerto Rican racer, *Alsophis portoricensis* (Serpentes: Colubridae). Jour. Herpetol. 26: 100–102.

— and M. Leal. 1993*a*. Effects of prey type on the feeding behavior of *Alsophis portoricensis* (Serpentes: Colubridae). Jour. Herpetol. 27: 163–168.

— and —. 1993*b*. *Alsophis portoricensis* (Puerto Rican Racer): diet. Herpetol. Rev. 24: 150–151.

Sajdak, R. A. and R. W. Henderson. 1991. Status of West Indian racers in the Lesser Antilles. Oryx 25: 33–38.

Schwartz, A. and R. W. Henderson. 1988. West Indian amphibians and reptiles: a check-list. Milwaukee Public Mus. Contrib. Biol. Geol. 74: 1–264.

— and —. 1991. Amphibians and Reptiles of the West Indies: Descriptions, Distributions, and Natural History. Univ. Florida Press, Gainesville.

Seib, R. L. 1984. Prey use in three syntopic Neotropical racers. Jour. Herpetol. 18: 412–420.

—. 1985. Euryphagy in a tropical snake, *Coniophanes fissidens*. Biotropica 17: 57–64.

Shine, R. 1991*a*. Australian Snakes: A Natural History. Cornell Univ. Press, Ithaca, New York.

—. 1991*b*. Why do larger snakes eat larger prey items? Func. Ecol. 5: 493–502.

—. 1994. Allometric patterns in the ecology of Australian snakes. Copeia 1994: 851–867.

Thomas, R. and J. A. Prieto-Hernández. 1985. The use of venom by the Puerto Rican snake, *Alsophis portoricensis*. Décimo Simposio de Recursos Naturales 1983: 13–22a.

Thomas, R. A. 1976. A revision of the South American colubrid snake genus *Philodryas* Wagler, 1930. Unpubl. Ph.D. Dissertation, Texas A & M Univ., College Station.

Resumen.—El género de serpientes colúbridas xenodontinas *Alsophis* está representado por 11 especies en las Antillas; ningún otro género de colúbridos está más ampliamente distribuido (desde las Islas Bahamas hasta Dominica) o habita en más islas (aproximadamente 100) de las Antillas. Las especies de *Alsophis* comen invertebrados, peces, sapos, lagartijas (desde esferodactilinos hasta iguáninos), serpientes, aves, y mamíferos (murciélagos y roedores), pero los sapos (*Eleutherodactylus*, *Osteopilus*) y las lagartijas (principalmente *Anolis*) son las presas consumidas más frecuentemente. Nuestros datos indican que las especies de *Alsophis* son forrajeras oportunistas e indiscriminadas. La diversidad inter e intraespecífica de la dieta de las especies de *Alsophis* es un reflejo de la diversidad de presas disponibles en una isla en particular. Las serpientes más grandes de una misma especie y de especies diferentes explotan una mayor diversidad (taxonómica y de tamaño) de presas que las culebras más pequeñas. A medida que las especies de *Alsophis* aumentan en tamaño, se alimentan, en promedio, de presas más grandes, pero la variación en el tamaño de las presas y el tamaño mínimo de éstas también aumenta. Las especies más grandes exhiben un cambio ontogenético en su dieta más notable que el que exhiben las especies más pequeñas.

Résumé.—Le colubridé xénodontine du genre *Alsophis* est représenté par 11 espèces dans les Antilles. C'est le genre le plus largement distribué (des Bahamas à la Dominique) et que l'on retrouve sur le plus grand nombre d'îles (environ 100) dans les Antilles. Les *Alsophis* se nourrissent d'invertébrés, de poissons, de grenouilles, de tortues, de lézards (Sphaerodactylidés aux Iguanidés), de serpents, d'oiseaux et de mammifères (chauves-souris et rongeurs). Les grenouilles (*Eleutherodactylus*, *Osteopilus*) et lézards (essentiellement *Anolis*) sont toutefois leurs proies les plus fréquentes. Nos données indiquent que les *Alsophis* sont des espèces opportunistes qui consomment leurs proies sans discrimination. La diversité alimentaire inter- et intraspécifique témoigne de la diversité des proies disponibles sur chaque île. Les gros serpents inter- et intraespèces exploitent une faune beaucoup plus diversifiée (tant sur le plan taxonomique que sur le plan de la taille) que les serpents plus petits. Plus leur taille est grande plus les *Alsophis* capturent en moyenne des proies plus grosses; on observe également une augmentation du spectre des dimensions et de la taille minimum des proies. L'alimentation des plus grosses espèces affiche un glissement ontogénique beaucoup plus marqué.

Henderson, R. W. and R. A. Sajdak. 1996. Diets of West Indian racers (Colubridae: *Alsophis*): composition and biogeographic implications, p. 327–338. *In* R. Powell and R. W. Henderson (eds.), Contributions to West Indian Herpetology: A Tribute to Albert Schwartz. Society for the Study of Amphibians and Reptiles, Ithaca (New York). Contributions to Herpetology, volume 12.

ECOLOGICAL PATTERNS IN GREATER ANTILLEAN MACROSTOMATAN SNAKE ASSEMBLAGES, WITH COMMENTS ON BODY-SIZE EVOLUTION IN *EPICRATES* (BOIDAE)

JAVIER A. RODRÍGUEZ-ROBLES[1] AND HARRY W. GREENE[1]

Abstract.—Fifty-one species of macrostomatan snakes inhabit the Greater Antilles; 32 species are small, 11 are of medium size, and eight are large. Small- and medium-sized Greater Antillean macrostomatans prey mainly or entirely on lizards and frogs, whereas six of the seven large species for which dietary data are available primarily eat mammals. Several colubrid clades have diversified on the four largest Greater Antillean islands, and macrostomatan species richness is positively correlated with island area. Only Cuba supports more than one species of *Tropidophis* and on that island one species eats birds and mammals with some frequency, whereas the other nine prey mostly on lizards and frogs; perhaps they have diversified mainly in terms of habitats. No island supports more than one large specialist on mammals (i.e., *Epicrates*); on islands with more than one species of *Epicrates,* the additional species is (are) smaller and feeds predominantly on lizards when adult. The genus *Epicrates* includes six large, robust species and four smaller, more slender species. Phylogenetic analyses suggest that size reduction has occurred in the mainland *E. cenchria* and twice within the Antillean clade, whereas size increase has occurred in *E. angulifer*. The hypothesis that a shift to an adult diet of lizard prey was a likely factor in the evolution of smaller body length in Caribbean *Epicrates* is supported by a phylogenetic analysis of the evolution of this character. Antillean macrostomatan snakes offer rich opportunities for studying historical and contemporary ecological phenomena, but to realize fully that potential we need much additional information on their natural history and phylogenetic relationships.

INTRODUCTION

ECOLOGISTS HAVE BUILT A STRONG CASE FOR THE importance of food as the primary biotic resource for many species, and one worker even suggested that the niche be operationally defined as "the nutritional role of the animal in its ecosystem, that is, its relations to all the foods available to it" (Weatherly 1963: 15). Numerous life history and ecological studies of snake assemblages have emphasized the dietary habits of the serpents (e.g., McCosker 1975; Shine 1977; Brown and Parker 1982; Fitch 1982; Mushinsky et al. 1982; Reynolds and Scott 1982; Vitt and Vangilder 1983; Voris and Voris 1983; Henderson et al. 1988; Cadle and Greene 1993). All these works have shown that at least a partial understanding of the trophic relationships of sympatric snakes can shed light on ways in which these communities are organized (Vitt 1987).

Based on energetic and physical considerations, the body size of snakes and other solitary predators should be related to the body size of their prey. Three lines of evidence support this deduction. First, optimal foraging theory (e.g., Stephens and Krebs 1986; Krebs and Kacelnik 1991) predicts that predators try to maximize their net energy gain while foraging. Consequently, smaller animals may eat smaller prey than larger animals because handling time and energy costs increase with prey body size (e.g., Pough and Andrews 1985; Pulliam 1985; Calver et al. 1988; Shine 1991; Arnold 1993; Rodríguez-Robles and Leal 1993*b*). Second, for

[1] Museum of Vertebrate Zoology and Department of Integrative Biology, University of California, Berkeley, California 94720–3160, USA.

gape-limited predators such as snakes that swallow their prey whole and do not experience a size-specific trophic specialization or change in shape in their feeding apparatus, the maximum prey size that can be ingested is determined by the physical limitations of the predator's trophic morphology (Emerson et al. 1994); that is, larger animals can eat prey of certain sizes that smaller predators are physically incapable of swallowing. Third, the microhabitat of the prey also imposes constraints on a predator's body size. For example, for a predator that gets inside the prey's refuge (e.g., rock crevices, burrows, trunk holes) to capture it, the diameter of the predator's body cannot exceed that of the cavity. All of these factors suggest that a correlation should exist between predator and prey body size and, if the body size of the prey changes, the predator's body size should change accordingly. Several authors have indeed considered the potential effects of diet on body size (e.g., Schoener 1969, 1971; Case 1978, 1979; Gittleman 1985; Case and Schwaner 1993; Madsen and Shine 1993; Emerson et al. 1994), but this issue rarely has been addressed within a phylogenetic context (e.g., Richman and Price 1992).

Extant snakes are divided into two major clades (Kluge 1991; Cundall et al. 1993): Scolecophidia (blind snakes of three families) and Alethinophidia (all other snakes, 15 families). Macrostomata is a

TABLE 1. Greater Antillean boid, tropidophiid, and colubrid snake species and their maximum snout-vent length (SVL). Data were taken from Schwartz and Henderson (1991), Hedges and Garrido (1992*a*, 1992*b*), and Henderson and Sajdak (this volume).

Species	Distribution	Maximum SVL (mm)
Boidae		
Epicrates angulifer	Cuba	4000
Epicrates chrysogaster	Caicos	1310
Epicrates exsul	Little Bahama	810
Epicrates fordii	Hispaniola	860
Epicrates gracilis	Hispaniola	905
Epicrates inornatus	Puerto Rico	1860
Epicrates monensis	Puerto Rico	1010
Epicrates striatus	Hispaniola	2330
Epicrates subflavus	Jamaica	2000
Colubridae		
Alsophis anomalus	Hispaniola	1770
Alsophis ater	Jamaica	850
Alsophis cantherigerus	Cuba	1220
Alsophis melanichnus	Hispaniola	630
Alsophis portoricensis	Puerto Rico	923
Alsophis vudii	Great Inagua and Little Bahama	855
Antillophis andreai	Cuba	650
Antillophis parvifrons	Hispaniola	530
Arrhyton ainictum	Cuba	363
Arrhyton callilaemum	Jamaica	465
Arrhyton dolichura	Cuba	249
Arrhyton exiguum	Puerto Rico	438
Arrhyton funereum	Jamaica	479
Arrhyton landoi	Cuba	250
Arrhyton polylepis	Jamaica	455
Arrhyton procerum	Cuba	207
Arrhyton supernum	Cuba	255
Arrhyton taeniatum	Cuba	448
Arrhyton tanyplectum	Cuba	306
Arrhyton vittatum	Cuba	207
Darlingtonia haetiana	Hispaniola	305
Hypsirhynchus ferox	Hispaniola	773
Ialtris agyrtes	Hispaniola	576
Ialtris dorsalis	Hispaniola	990
Ialtris parishi	Hispaniola	745
Nerodia clarkii	Cuba	933
Tretanorhinus variabilis	Cuba	800
Uromacer catesbyi	Hispaniola	830
Uromacer frenatus	Hispaniola	930
Uromacer oxyrhynchus	Hispaniola	1500
Tropidophiidae		
Tropidophis canus	Great Inagua	373
Tropidophis feicki	Cuba	460
Tropidophis fuscus	Cuba	304
Tropidophis greenwayi	Caicos	337
Tropidophis haetianus	Cuba, Hispaniola, and Jamaica	636
Tropidophis maculatus	Cuba	370
Tropidophis melanurus	Cuba	950
Tropidophis nigriventris	Cuba	355
Tropidophis pardalis	Cuba	312
Tropidophis pilsbryi	Cuba	418
Tropidophis semicinctus	Cuba	419
Tropidophis wrighti	Cuba	400

monophyletic group that accounts for more than 95% of Alethinophidia and comprises the Boidae, Pythonidae, Bolyeriidae, Tropidophiidae, Acrochordidae, Viperidae, Atractaspididae, Colubridae, and Elapidae. The clade is characterized, among other traits, by morphological synapomorphies correlated with increased mobility of the jaws and increased gape (Rieppel 1988), hence the name Macrostomata ("big mouth") (except where otherwise noted, "snakes" and "serpents" in this chapter hereafter refer exclusively to macrostomatans). In the Greater Antilles, macrostomatans are represented by one boid genus (*Epicrates*), one tropidophiid genus (*Tropidophis*), and nine colubrid genera (*Alsophis*, *Antillophis*, *Arrhyton*, *Darlingtonia*, *Hypsirhynchus*, *Ialtris*, *Nerodia*, *Tretanorhinus*, and *Uromacer*; Schwartz and Henderson 1991); six of the latter are endemic to the West Indies. Herein we will examine the universe of prey potentially available to Greater Antillean macrostomatans and review the dietary habits of these serpents, searching for patterns that suggest ways in which the use of trophic resources may influence the structure of these snake communities. We will emphasize the general relationships between macrostomatan diets and body sizes; we then will use an explicitly historical perspective to explore the hypothesis that smaller adult body size in Greater Antillean *Epicrates* is related to a shift in diet from endothermic to ectothermic (i.e., lizard) prey (Nellis et al. 1983).

Our goal is to offer plausible (although not exhaustive) explanations for the patterns we observe. These hypotheses can be tested by appropriate field and laboratory experiments, by collecting additional natural history information, and/or by analyzing available data in novel ways (e.g., in a historical context when phylogenies for different lineages of Greater Antillean macrostomatans are available).

METHODS

Size Patterns and Species Richness

We compiled a list of Greater Antillean macrostomatan snakes and their maximum body length (snout-vent length, SVL), regardless of sex (Table 1). We have no assurance, of course, that maximum length individuals have been found for all species, but general trends can be noted (Cadle and Greene 1993). We divided macrostomatans into three size categories: "small" (SVL ≤ 800 mm), "medium" (800 < SVL ≤ 1200 mm), and "large" (SVL > 1200 mm). We defined the Greater Antilles as the large islands of Cuba (excluding Isla de la Juventud and the Cayman Islands), Jamaica, Hispaniola (excluding Île de la Gonâve and Navassa Island), Puerto Rico (excluding Desecheo, Mona, Monito, and the Virgin Islands), and three Bahamian island banks: Great Inagua, Caicos (Providenciales, North, South, East, West, Grand [=Middle] Caicos), and Little Bahama (Grand Bahama, Great Abaco, Little Abaco). We excluded several smaller Greater Antillean islands and Bahamian island banks because data on their fossil and/or extant amphibian, lizard, and mammalian fauna are not available.

TABLE 2. Island area (taken from Morgan 1989; Woods 1990) and number of species of macrostomatan snakes, amphibians and lizards, and terrestrial mammals on the Greater Antilles (as defined in the text). Data are taken from Estrada and Hedges (1991), Schwartz and Henderson (1991), Garrido and Hedges (1992), Hedges (1992), Hedges and Garrido (1992*a*, 1992*b*, 1993), Hedges and Thomas (1992*a*, 1992*b*), Hedges et al. (1992), Thomas and Hedges (1992, 1993), and Thomas et al. (1992). Abbreviations: macrostomatan snakes (MS), amphibians and lizards (AL), mammals (MA).

Island or island bank	Area (km^2)	Number of species		
		MS	AL	MA
Cuba	111,463	23	136	76
Hispaniola	76,193	15	179	47
Jamaica	11,424	6	51	28
Puerto Rico	8,865	4	46	22
Little Bahama	2,794	2	5	5
Great Inagua	1,627	2	7	8
Caicos	800	2	7	5

We assembled a list of terrestrial, vertebrate ectothermic and mammalian prey potentially available to macrostomatans (see Table 2 for a summary of number of snakes and prey species per island), assuming that all prey on a given island or island bank are available to all snakes on that island, an assumption which undoubtedly inflated prey species richness. We did not consider invertebrates because Greater Antillean macrostomatans nearly exclusively eat vertebrates throughout their lives (Henderson and Crother 1989; Schwartz and Henderson 1991). We also did not include crocodiles or turtles because no records of snake predation upon the former group exist and only one case of the Cuban colubrid *Alsophis cantherigerus* eating hatchling sea turtles is known (Crother 1986). Although some snakes (i.e., *Alsophis*, *Antillophis*, *Arrhyton*, *Tropidophis*, *Typhlops*) are eaten by macrostomatans (Schwartz and Henderson 1991; Henderson and Sajdak, this volume), this occurs so infrequently that we omitted them from our list of possible prey. We did not include birds because they are only rarely eaten by Greater Antillean serpents and several migratory species are not available throughout the year. We also excluded prey species that have been introduced by human activities into some of the islands.

The list of possible mammalian prey includes extant and extinct species (Woods 1990), but the catalog of potential amphibian and lizard prey is restricted to living forms because we only have information on a few fossil amphibians and lizards for Hispaniola (Rieppel 1980; Böhme 1984; Poinar and Cannatella 1987) and Puerto Rico (Pregill 1981; MacPhee and Wyss 1990). By using extant prey communities to study how feeding patterns may influence the structure of Greater Antillean macrostomatan snake assemblages, we assume that modern (or in the case of mammals, Pleistocene and Holocene [1.5 million years ago, mya, to Recent]) prey communities reflect ancient ones that were present when macrostomatans were diverging. We have no way to assess whether our assumption is conservative or liberal, so if new evidence shows that ancient Greater Antillean prey communities were significantly more (or less) species-rich than those of the present day, a reconsideration of our conclusions will be necessary. Data on diets of Greater Antillean macrostomatans were taken from Greene (1980, 1983, unpublished data), Henderson and Crother (1989), Schwartz and Henderson (1991), Rodríguez-Robles and Leal (1993*a*), Tolson and Henderson (1993), and Henderson and Sajdak (this volume).

Body-Size Evolution in *Epicrates*

Nine of the ten species of *Epicrates* are found exclusively on the Bahamas and the Greater Antilles (Tolson and Henderson 1993); only *E. cenchria* occurs on the mainland, from Costa Rica to Argentina (Tolson 1987). The ten species of *Epicrates* can be divided into large and robust snakes (*E. angulifer*, *E. cenchria*, *E. chrysogaster*, *E. inornatus*, *E. striatus*, *E. subflavus*), and small to medium-sized, more slender snakes (*E. exsul*, *E. fordii*, *E. gracilis*, *E. monensis*). Juveniles of large *Epicrates* and juveniles and adults of small *Epicrates* mainly eat lizards and, occasionally, frogs, rodents, and birds (Henderson and Crother 1989; Tolson and

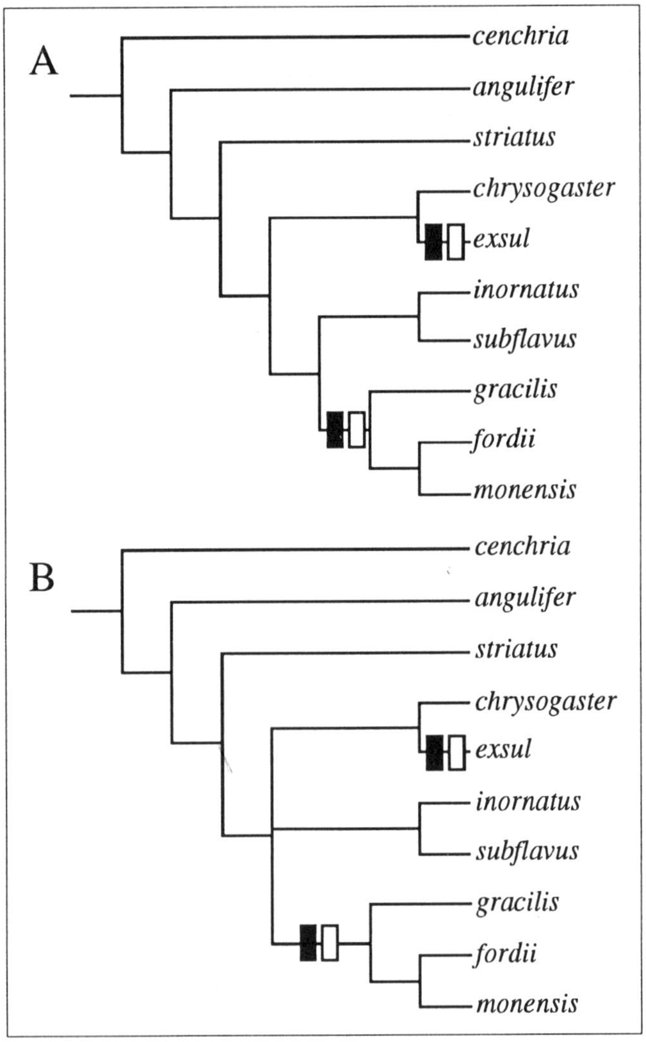

FIGURE 1. Two hypotheses of phylogenetic relationships within the boid snake genus *Epicrates*: (A) after Tolson (1987); (B) after Kluge (1989). The black bar represents an adult diet of predominantly *Anolis* lizards and the white bar indicates the evolution of smaller, adult body size in *Epicrates*.

Henderson 1993; P. J. Tolson, in litt.), whereas adults of large *Epicrates* feed mostly on mammals and birds and only rarely on lizards and snakes (Serié 1920; Hardy 1957; Duellman 1978; Rodríguez and Reagan 1984; Henderson et al. 1987*b*).

Phylogenetic hypotheses provide the essential framework for addressing patterns of ecological, ethological, morphological, and physiological character evolution (e.g., Brooks and McLennan 1991; Harvey and Pagel 1991; Garland and Carter 1994; Losos and Miles 1994). Tolson (1987) used 24 biochemical and nine morphological characters to infer interspecific relationships of *Epicrates* (Figure 1A). Kluge (1989) added 44 osteological characters to Tolson's data set, selected a different outgroup (*Eunectes*, rather than *Corallus*; see Kluge [1991] for details), and produced an alternative hypothesis of phylogenetic relationships within the genus (Figure 1B). The most parsimonious estimate of the evolution of smaller SVL in *Epicrates* is identical for Tolson's and Kluge's cladograms. Because the clades (*E. exsul*, *E. chrysogaster*) and (*E. gracilis* (*E. fordii*, *E. monensis*)) in Figures 1A and 1B are supported by several synapomorphies in addition to adult SVL (Tolson 1987; Kluge 1989), we have no reservations in using Kluge's phylogeny to analyze the evolution of SVL in *Epicrates* (see Donoghue and Sanderson 1992; Kluge and Wolf 1993).

We gathered information on maximum SVL for seven outgroup taxa of *Epicrates* (*sensu* Kluge 1991): *Eunectes deschauenseei*, *E. murinus*, *E. notaeus*, *Boa constrictor*, *B. dumerili*, *B. madagascariensis*, and *B. mandrita* (Table 3). These two genera include relatively uncommon or very large species and SVL data are either scarce or questionable (particularly for *B. constrictor* and *Eunectes*); therefore, we only used apparently reliable reports of their maximum SVL. For heuristic purposes, we inferred ancestral states of SVL for all nodes on the *Epicrates* cladogram using the "minimum evolution" algorithms of "Wagner" (Farris 1970; Swofford and Maddison 1987) and "squared-change" parsimony (Maddison 1991; see Martins and Garland 1991 for a description of minimum evolution methods) as implemented in the computer software MacCLADE (Maddison and Maddison 1992). Wagner parsimony minimizes the sum of the absolute value of the changes on the branches of a phylogenetic tree. This often yields a range of equally parsimonious values at each branch, and tends to concentrate evolutionary changes on relatively few branches of the tree. Squared-change parsimony minimizes the sum of the squared changes on the branches of the tree. Because these methods can only infer ancestral states of continuous characters in fully dichotomous phylogenies, we performed the analyses on each of the three possible resolutions to the trichotomy in Kluge's (1989) cladogram of *Epicrates* (Figures 2A, 2B, and 2C).

TABLE 3. Maximum snout-vent length (SVL) of the continental *Epicrates cenchria* and seven outgroup taxa of *Epicrates* (*sensu* Kluge 1991). Data taken from: (1) Cope (1862), (2) Dunn and Conant (1936), (3) Roze (1966), (4) Pérez-Santos and Moreno (1988), (5) Cei (1993), (6) Jorge da Silva (1993), (7) Strimple (1993), (8) V. Wallach (in litt. to H. W. Greene), and (9) W. R. Branch (in litt. to H. W. Greene).

Species	Maximum SVL (mm)	References
Epicrates cenchria	1719	6
Outgroup taxa		
Eunectes deschauenseei	2261	2
Eunectes murinus	9000	4, 7
Eunectes notaeus	2438	1
Boa constrictor	4000	3, 5
Boa dumerili	1971	9
Boa madagascariensis	2684	9
Boa mandrita	1820	8

We calculated standardized, phylogenetically independent contrasts (Felsenstein 1985; Garland et al. 1992) using the computer program PDTREE (Version 2.0; Garland et al. 1993) to remove the confounding effects of phylogeny from our analysis of correlation between SVL and adult diet in *Epicrates*. Because information on branch lengths was not available, we assumed that all branches were equal, which is tantamount to assuming that amount of evolutionary change in SVL has been speciational (cf. Rohlf et al. 1990; Garland et al. 1993). We used the nonparametric Sign test (T. Garland, Jr., personal communication) to determine if an association exists between changes in SVL and dietary shifts from endothermic to ectothermic prey in *Epicrates*. The null hypothesis is one-tailed, because we are testing the hypothesis that a dietary switch is related to a decrease in SVL. As with any other comparative study, this analysis should be redone if a completely resolved phylogeny for the genus, information on branch lengths, and/or more precise dietary data for each species of *Epicrates* (e.g., knowledge of the percentage of mammalian and lizard prey in their adult diet, prey/predator mass relationships) become available.

RESULTS

Size Patterns and Species Richness

Fifty-one species (Table 1) of macrostomatan snakes occur on the Greater Antilles: 23 on Cuba, 15 on Hispaniola, six on Jamaica, four on Puerto Rico, and two each on Little Bahama, Great Inagua, and Caicos (*Alsophis vudii* occurs on Great Inagua and Little Bahama and *Tropidophis haetianus* is found on Cuba, Hispaniola, and Jamaica). Number of macrostomatan species is positively correlated to island area (both axes ln-transformed; $r^2 = 0.94$, $t = 9.71$, d.f. = 5, two-tailed P = 0.0002), as is number of terrestrial, vertebrate ectotherm and mammalian prey species potentially available to them (both axes l_n-transformed; $r^2 = 0.89$, $t = 7.12$, d.f. = 5, two-tailed P = 0.0008; Figure 3). Number of snake species is also positively correlated to number of possible prey species (both axes l_n-transformed; $r^2 = 0.90$, $t = 7.32$, d.f. = 5, two-tailed P = 0.0007). The regression of number of macrostomatan species (l_n-transformed) against the residuals from the regression of prey species on island area (both axes l_n-transformed) was not significant ($r^2 = -0.19$, $t = 0.20$, d.f. = 5, two-tailed P = 0.85), suggesting that the relationship between macrostomatan species richness and prey species richness is spurious because it results from the relationship between each variable and island area.

Most Greater Antillean macrostomatans are small ($n = 31$), followed by medium-sized ($n = 12$), and large snakes ($n = 8$; Figure 4). We tested for differences in SVL between genera with more than three species (i.e., *Alsophis*, *Arrhyton*, *Epicrates*,

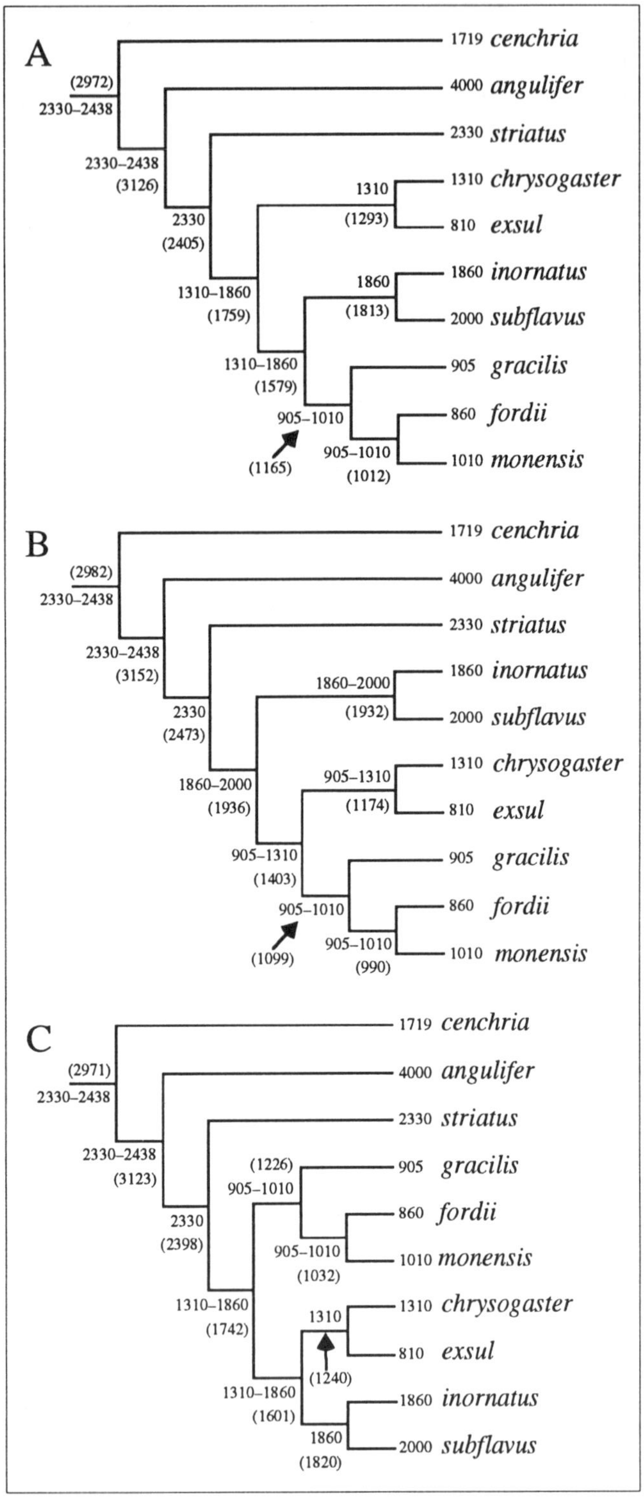

FIGURE 2. The three possible resolutions to the trichotomy in Kluge's (1989) cladogram of *Epicrates*. Numbers at nodes indicate ranges of equally parsimonious, estimated ancestral values of maximum snout-vent length (SVL) using Wagner parsimony. Figures in parentheses are inferred values of the same nodes using square-change parsimony. Numbers at the end of the branches represent the maximum SVL of that species (data are from Schwartz and Henderson 1991; Jorge da Silva 1993).

Tropidophis) with the nonparametric Kruskal-Wallis one-way ANOVA by ranks. Because the results of the test were significant (Kruskal-Wallis $H = 25.86$, $n = 39$, two-tailed $P < 0.0001$), we performed a nonparametric multiple comparison test to evaluate all pairwise comparisons (Siegel and Castellan 1988; Table 4). Species of *Alsophis* and *Epicrates* are not significantly different in SVL, nor are species of *Alsophis* and *Tropidophis* or those of *Arrhyton* and *Tropidophis*. Species of *Epicrates* are larger than those of *Arrhyton* and *Tropidophis*, whereas species of *Alsophis* are larger than species of *Arrhyton*.

Dietary information exists for only 33 of the 51 species of Greater Antillean macrostomatans. Fourteen of 16 small-bodied snakes are not known to eat mammals, whereas among medium-sized serpents, species that occasionally feed on mammals (6) occur with almost the same frequency as species that do not (4); no small- or medium-sized snake is a mammalian specialist. Six of seven large macrostomatans eat mammalian prey, the exception being the Hispaniolan *Uromacer oxyrhynchus*, that preys exclusively on lizards (Henderson et al. 1987*b*). Distribution of maximum SVL of serpents that occasionally or commonly prey on mammals exhibits a greater skewness to the right (coefficient of skewness = 1.88) than those snakes that do not eat mammalian prey (coefficient of skewness = 1.49; Figure 5).

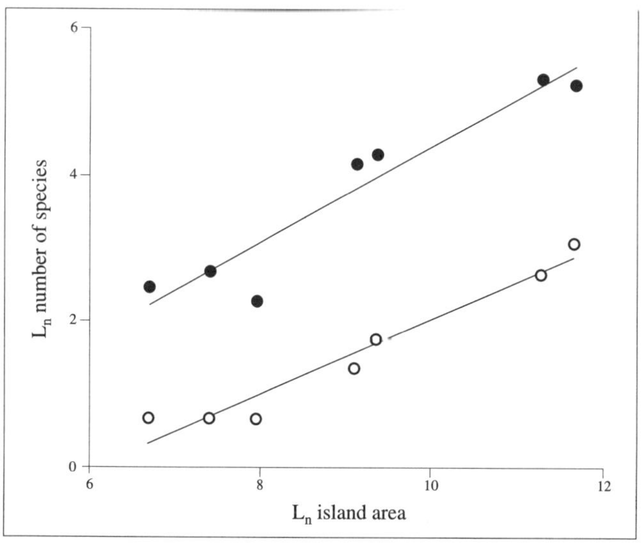

FIGURE 3. Number (l_n-transformed) of species of Greater Antillean macrostomatan snakes (circles) and of their potential amphibian, lizard, and mammalian prey (dots) as a function of (l_n-transformed) island area.

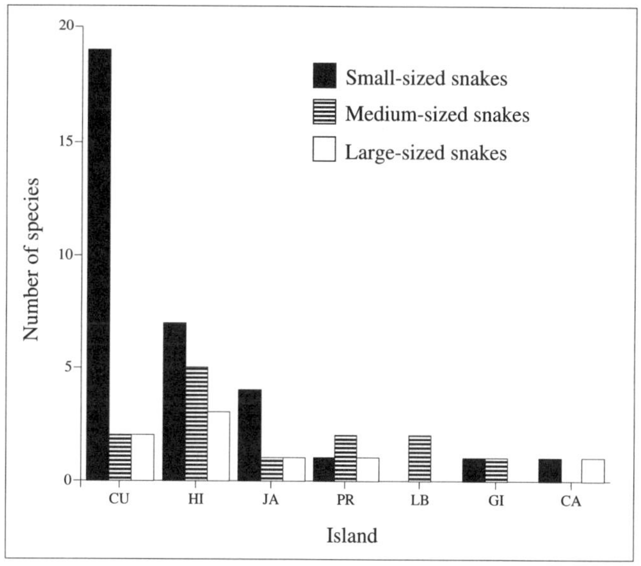

FIGURE 4. Frequency distribution of small- (solid bars), medium- (hatched bars), and large-sized (open bars) macrostomatan snakes on islands of the Greater Antilles. Abbreviations: Cuba (CU); Hispaniola (HI); Jamaica (JA); Puerto Rico (PR); Little Bahama (LB); Great Inagua (GI); Caicos (CA).

The maximum number of large snakes on the four largest Greater Antillean islands is three (Figure 4), and in each case the largest serpent is an *Epicrates* and the second largest is an *Alsophis*. Nonetheless, the maximum number of large species that eat mammals is two per island. Because *Alsophis* has a catholic diet (Rodríguez-Robles and Leal 1993*a*; Henderson and Sajdak, this volume), only large *Epicrates* feed almost exclusively on endotherms as adults. The three species of *Epicrates* that occur on the Bahama Islands have allopatric distributions. *Epicrates exsul* inhabits Little Bahama (northern Bahamas), *E. chrysogaster* is found on Turks and Caicos Islands and Great Inagua (southern Bahamas), and *E. striatus* occurs on islands in between (Great Bahama Bank). Islands of the Greater Antilles apparently can support only one mammalian specialist, a large species of *Epicrates*. Hispaniola and Puerto Rico are the only islands of the Greater Antilles that have more than one species of *Epicrates* (three and two, respectively), and on both islands the additional species is (are) always smaller and mostly or exclusively feeds on

Anolis lizards throughout their lives (Henderson et al. 1987*b*; Tolson and Henderson 1993). In the case of Hispaniola, the two smaller species (*E. fordii* and *E. gracilis*) evolved after the larger one (*E. striatus*), and the same evolutionary sequence could have taken place on Puerto Rico with *E. inornatus* and *E. monensis* (Figures 2A and 2B).

Hispaniola has eight genera of macrostomatans, six of which are colubrids; of the latter, four are endemic to the island (*Darlingtonia*, *Hypsirhynchus*, *Ialtris*, *Uromacer*). Hispaniolan snakes include lizard specialists (*Epicrates gracilis*, *Hypsirhynchus ferox*, *Uromacer frenatus*; Henderson 1984; Henderson et al. 1987*a*, 1987*b*) and frog specialists (*Darlingtonia*; Henderson and Schwartz 1986) (in preliminary phylogenetic analyses, *Darlingtonia* is nested within *Arrhyton* [Hedges and Garrido 1992*b*], but we recognize the former genus here with the caveat that it may not be a valid taxon). As with all *Arrhyton*, most species of *Tropidophis* feed almost exclusively on lizards and frogs (Greene 1980, 1983, unpublished data; Henderson and Crother 1989; Schwartz and Henderson 1991). These observations suggest that the resource spectrum of lizard and amphibian prey can support many species of macrostomatans and that, at least on Hispaniola, these prey resources are easily divisible among snakes.

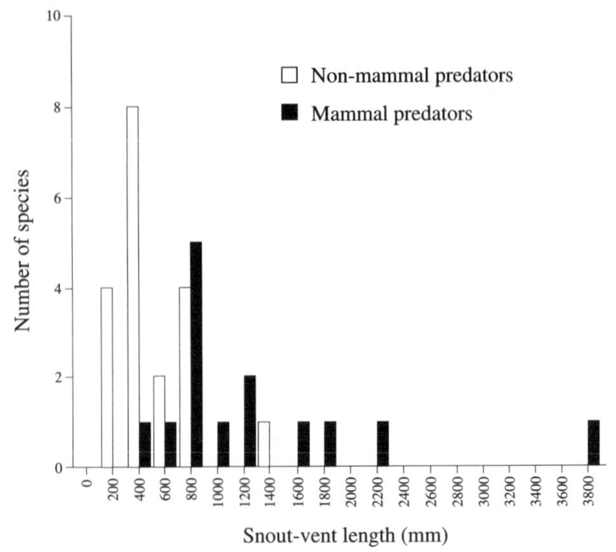

FIGURE 5. Frequency distribution of body size of Greater Antillean macrostomatan snakes that do not prey on mammals (open bars) and of those which do (solid bars). The three size categories herein recognized are "small" (SVL ≤ 800 mm), "medium" (800 < SVL ≤ 1200 mm), and "large" (SVL > 1200 mm).

Finally, Bogert (1969; see also Greene 1980) suggested that advanced snakes (colubrids on the Greater Antilles) have competitively excluded boids (*sensu lato*, including tropidophiids) from many modern snake faunas. If that were so, species richness of the latter should be negatively correlated with that of colubrids. We performed a regression of (l_n-transformed) number of species of boids plus tropidophiids and of (l_n-transformed [X + 1]) number of colubrid species on (l_n-transformed) island area to remove the confounding effects of area on these two variables. We then plotted the residuals from the first regression against those from the second. The relationship is in the predicted direction but not significant ($r^2 = -0.045$, $t = -0.86$, d.f. = 5, two-tailed P = 0.43), thus not supporting the hypothesis of competitive exclusion between these two groups of snakes on the Greater Antilles.

TABLE 4. Results of the nonparametric multiple comparison test for differences in snout-vent length among Greater Antillean species of *Alsophis*, *Arrhyton*, *Epicrates*, and *Tropidophis*. Abbreviations: absolute value of the difference between mean ranks (AV), critical value (CV), significant (S), nonsignificant (NS).

Comparison	AV	CV	
Alsophis – Arrhyton	18.2	15.04	S
Alsophis – Epicrates	4.1	15.85	NS
Alsophis – Tropidophis	14.2	15.04	NS
Epicrates – Arrhyton	22.3	13.26	S
Arrhyton – Tropidophis	4.0	12.28	NS
Epicrates – Tropidophis	18.3	13.26	S

Body Size Evolution in *Epicrates*

Decreased adult SVL in *Epicrates* has evolved three times within the genus (Tolson 1987; Kluge 1989): in the continental *E. cenchria* relative to its first and second outgroup taxa, *Eunectes* and *Boa*, respectively; in *E. exsul*; and in the most recent common ancestor of *E. gracilis*, *E. fordii*, and *E. monensis* (Figures 1A, 1B). On the other hand, a considerable increase in SVL occurred in the Cuban *E. angulifer* relative to that of *E. cenchria* (Figures 2A, 2B, 2C). Perhaps interspecific competition from small-bodied Cuban snakes (i.e., *Antillophis*, *Arrhyton*, *Tropidophis*) may have forced *E. angulifer* to eat larger prey and therefore to evolve a larger SVL. Recent experimental evidence shows that interspecific competition for resources can promote morphological divergence (Schluter 1994).

The association between phylogenetically independent contrasts for adult diet and SVL in *Epicrates* was statistically significant (Sign test, $n = 9$, one-tailed $P = 0.02$, for Figures 2A, 2B, 2C). This result supports the hypothesis that a dietary shift to lizard prey by adult *Epicrates* was a likely factor in the evolution of decreased SVL in these snakes. Of course, a change in behavior *per se* could not cause the morphological transition from larger to smaller SVL, it could only influence the maintenance (and subsequent propagation) or suppression of variants already existing in a population.

DISCUSSION

Size Patterns and Species Richness

Some authors (e.g., Arnold 1972; Rogers 1976; Murphy and Ottley 1984; Vitt 1987) have suggested that snake species richness at a given locality is best explained on the basis of the number of potential prey species present at that site. We also found a strong relationship between potential prey species richness and the number of macrostomatan species on islands of the Greater Antilles, but the correlation of these two variables is more parsimoniously explained by their common relationship to island area. Hence, ascertaining if the relationship that has been found in other studies between species richness of snakes and their potential prey remains after removing the effects of area from those analyses would be interesting.

The Greater Antillean macrostomatan fauna is dominated by small serpents. Nineteen of the 23 Cuban snakes are of small size (Figure 4), and all but two belong to the genera *Arrhyton* or *Tropidophis* (Table 1). Hispaniolan and Jamaican ophidian faunas also have a majority of small species, although not as overwhelmingly so as in Cuba. All small Jamaican snakes are either *Arrhyton* or *Tropidophis* species; the only small Puerto Rican macrostomatan belongs to the genus *Arrhyton*, and both small Bahamian serpents are species of *Tropidophis*. Only on Hispaniola do most small snakes belong to different genera. Overall, 24 of the 31 Greater Antillean small macrostomatans are species of *Arrhyton* or *Tropidophis* (12 species each), which suggests phylogenetic constraints (*sensu* Losos and Miles 1994) in body size in these two genera, and may explain why most Cuban snakes are of small size — namely because the serpent fauna of that island is dominated by species of *Arrhyton* and *Tropidophis*. This, however, raises another intriguing question: why, compared to other Greater Antillean islands, do so many species of *Arrhyton* and *Tropidophis* occur on Cuba? Smaller snakes usually take smaller prey than larger serpents (e.g., Voris and Moffett 1981; Seib 1985; Jayne et al. 1987; Slip and Shine 1988; Greene 1989; Shine and Webb 1990; Arnold 1993; Houston and Shine 1993). The preponderance of small species in the Greater Antillean macrostomatan fauna may account for the fact that most of them eat relatively small prey (i.e., lizards and frogs). In fact, all species of *Arrhyton* and all but two of *Tropidophis* do not prey on mammals, probably be-

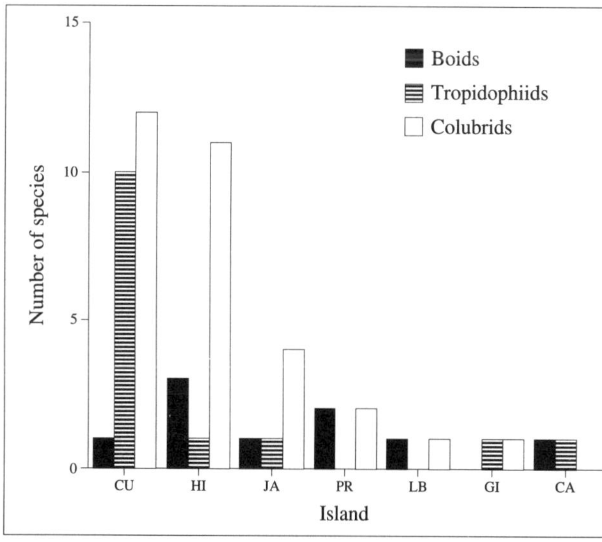

FIGURE 6. Frequency distribution of boid (solid bars), tropidophiid (hatched bars), and colubrid (open bars) species on islands of the Greater Antilles. Abbreviations: Cuba (CU); Hispaniola (HI); Jamaica (JA); Puerto Rico (PR); Little Bahama (LB); Great Inagua (GI); Caicos (CA).

cause they do not get as big as do species of *Alsophis*, *Epicrates*, and *Tropidophis* that take mammalian prey (Figure 7).

The Neotropical family Tropidophiidae includes the genera *Exiliboa*, *Trachyboa*, *Tropidophis*, and *Ungaliophis* (Kluge 1991; but see Zaher 1994). Only *Tropidophis* (16 species) has apparently radiated; the remaining genera contain either one (*Exiliboa*) or two (*Trachyboa* and *Ungaliophis*) species (*Trachyboa* is the sister taxon of *Tropidophis*; *sensu* Kluge 1991). Thirteen species of *Tropidophis* occur on the West Indies and 10 of these are native to Cuba (Figure 6). On Cuba, all but two *Tropidophis* eat mainly *Anolis* lizards and occasionally frogs, the exceptions being *T. haetianus* and *T. melanurus*, that also feed on rodents (*T. haetianus*, but only rarely) and birds and rodents (*T. melanurus*; Greene 1980, 1983, unpublished data; Schwartz and Henderson 1991). *Tropidophis haetianus* is the second largest *Tropidophis*, whereas *T. melanurus*, the largest *Tropidophis* (Table 1), is the only one that takes endotherms with some frequency. Outside of Cuba, no more than one species of *Tropidophis* inhabits an island (e.g., *T. canus* and *T. greenwayi* are allopatric in the Bahamas, although each occurs on several islands), and no intrafamilial sympatry occurs on the mainland. Cuban *Tropidophis* (except *T. haetianus* and *T. melanurus*) have similar diets and perhaps partition habitat interspecifically; they differ among species in color patterns and morphology (e.g., the presence or absence of "vestigial" hind limbs, smooth or keeled scales), traits that are suggestive of locomotor and microhabitat differences (cf. Jackson et al. 1976; Marx and Rabb 1972; see Tolson and Henderson 1993 for photographs of Antillean *Tropidophis*). Detailed characterizations of habitats and locomotor behavior of Cuban *Tropidophis* are necessary to test this hypothesis.

On the Greater Antilles, only Hispaniola and Puerto Rico have more than one species of *Epicrates*, and on both islands only one of the species is large. If this biogeographic pattern differs significantly from a null model of species distributions (e.g., Simberloff 1978; Connor and Simberloff 1979, 1984), it may indicate: (1) differential rates of colonization of those islands, combined with evolutionary diversification, (2) higher predation levels on large snakes, (3) differential susceptibility to parasites of large, sympatric species of *Epicrates* (e.g., Jeffries and Lawton 1984; Settle and Wilson 1990; Grosholz 1992), or (4) competitive exclusion of species with similar adult diets (see Brown 1995; Kelt et al. 1995; Winston 1995).

The Hispaniolan *Uromacer oxyrhynchus* is the only Greater Antillean large snake that does not feed on mammals. *Uromacer oxyrhynchus* is a slender, largely arboreal snake capable of exploiting very thin branches. As most West Indian mammals (excluding bats) are terrestrial, the fact that *U. oxyrhynchus* feeds almost exclusively on scansorial prey is not surprising. By contrast, the continental, arboreal colubrids *Leptophis ahaetulla* and *Oxybelis fulgidus* occasionally eat birds, as well as frogs and lizards (Oliver 1948; Henderson and Binder 1980; H. W. Greene, unpublished data). *Leptophis ahaetulla* and *O. fulgidus* attain a maximum SVL similar to that of *U. oxyrhynchus* (1500 mm), but *Leptophis* and *Oxybelis* have much larger body masses than does *Uromacer*. Similarly, although Antillean species of *Epicrates* and *Alsophis* do not significantly differ in SVL, large species of *Epicrates* have larger body masses than do species of *Alsophis*, a difference that may be reflected in the average

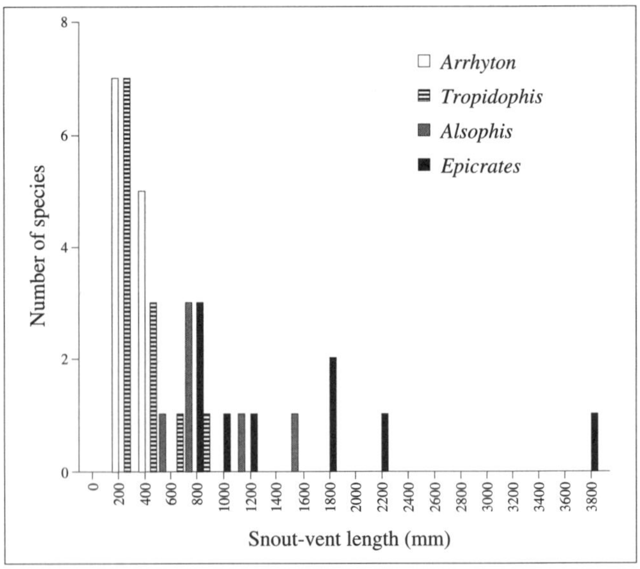

FIGURE 7. Frequency distribution of body size of Greater Antillean species of *Arrhyton* (open bars), *Tropidophis* (hatched bars), *Alsophis* (shaded bars), and *Epicrates* (solid bars).

relative prey mass ingested by adults of large species of *Epicrates*, compared to that consumed by adults of all species of *Alsophis*.

Inclusion of phylogenetic information into evolutionary ecological studies can provide new, sometimes more parsimonious explanations that are not apparent in the absence of a historical perspective (Cadle and Greene 1993; Losos 1994). For example, on the Greater Antilles, only on Cuba do snakes occur which specialize on eating fish (*Nerodia clarkii* and *Tretanorhinus variabilis*). We could postulate that for some unknown reason(s), freshwater fish on the remaining Greater Antilles represent such an unpredictable food resource that no fish specialist has evolved on those islands, or the specialists on Cuba represent a recent invasion from the mainland. Similarly, no Greater Antillean snakes eat leptotyphlopid serpents, and only one (*Arrhyton taeniatum*; Schwartz and Henderson 1991) is known to take amphisbaenians, despite the fact that, at least on Hispaniola, the latter are sympatric with macrostomatans and are common at some localities (R. Thomas, personal communication). Typhlopid snakes also are infrequently taken as prey by Antillean serpents (Schwartz and Henderson 1991). Again, we could argue that the secretive habits of amphisbaenians, leptotyphlopids, and typhlopids make the evolution of snakes that prey upon them unlikely. Although those proximate explications are plausible, we also should consider an alternative explanation involving historical contingencies.

Tretanorhinus is a member of the Central American lineage of xenodontine colubrids (Cadle 1985), whereas *Nerodia* is a North American natricine. In other words, the only Greater Antillean snakes that eat fish most likely dispersed to Cuba from mainland sites (Hedges, this volume) and *N. clarkii* belongs to a clade of fish specialists. Greater Antillean colubrids (except for *Nerodia* and *Tretanorhinus*) are phylogenetically related to members of the South American xenodontine colubrid clade (Cadle 1985). Among South American xenodontines, fish-eating colubrid taxa are relatively common (e.g., *Helicops*, *Hydrops*, *Hydrodynastes*, *Liophis*, *Pseudoeryx*), and at least three genera (*Apostolepis*, *Elapomorphus*, *Erythrolamprus*; Cadle and Greene 1993) include species that eat snakes and/or amphisbaenians with some regularity. Invertebrates represent yet another prey group nearly completely absent from the diet of Greater Antillean snakes (one record exists of a phymatid hemipteran insect eaten by an *Alsophis vudii* on South Bimini Island; Henderson and Sajdak, this volume). The only West Indian macrostomatans that eat invertebrates with any regularity at some time during their lives are Lesser Antillean species of *Liophis* (Henderson and Bourgeois 1993), a South American xenodontine genus (Cadle 1984) of about 40 species. Because West Indian serpent communities are depauperate relative to the mainland (Williams 1989; Hedges, this volume), that Greater Antillean snake assemblages lack representatives of any of the aforementioned genera is not surprising. Hence the absence or near absence of some taxa from the diets of western Caribbean serpents may simply reflect the lack of colonization of taxa capable of predation on them.

Although we emphasize the importance of historical events, we do not deny the contribution of contemporary phenomena in structuring Greater Antillean macrostomatan snake assemblages. In fact, combining both approaches should provide a more complete understanding of the factors influencing the character of ecological communities (Vitt 1987; Cadle and Greene 1993). Thus the effort is worthwhile to ask why serpents that eat invertebrates, fish, or leptotyphlopids have not evolved *in situ* on the Greater Antilles, given the case that, for example, most snakes that prey on invertebrates are small, as are most Greater Antillean macrostomatans; furthermore, Hispaniola has as many species of freshwater fish as Cuba (Burgess and Franz 1989), and northern South American localities similar in area to Hispaniola have at least one species of fish-eating snake (Cadle and Greene 1993). Do, then, any physiological, morphological, and/or behavioral constraints preclude the evolution of invertebrate-, fish-, or leptotyphlopid-eaters in lineages of Greater Antillean macrostomatans? Have Antillean snakes been on those islands long enough to evolve new dietary habits? Or, perhaps, does the diversity, ecological ubiquity, and frequently high population densities of *Anolis* make the exploitation of other prey types unnecessary (Henderson and Crother 1989)?

Body Size Evolution in *Epicrates*

Using independent contrasts to remove the effects of "phylogenetic inertia" from our analysis of SVL and adult diet in *Epicrates*, we found evidence for the correlated evolution of both traits in these snakes. Although a correlation between a change in diet to ectothermic prey and smaller SVL does not establish that those attributes co-evolved, we favor that view for the following reasons.

When prey handling ability is closely related to prey size (as in snakes), predator body size may change in response to a shift in the size spectrum of its food. In fact, one disadvantage for large serpents that continue to feed regularly on small prey is the likelihood of trying to swallow prey in the vicinity of their lower "breaking point." The lower breaking point is defined as the smallest size of prey that a snake is incapable of ingesting (Arnold 1993). Swallowing performance in the elapid ophidian *Pseudechis porphyriacus* suggests that swallowing time, and presumably energy spent in feeding, increase disproportionately as the breaking point is approached, hence decreasing the net energy gain per feeding event (Shine 1991). In snakes, having a body size that is closely related to the size of their most common prey is advantageous in order to maximize the energy acquired per feeding event.

Juveniles of both large and small *Epicrates* eat mostly *Anolis*, with the adults of larger species switching ontogenetically to endothermic prey (Henderson et al. 1987*a*; Tolson and Henderson 1993). *Anolis* lizards are the most conspicuous and abundant land vertebrates on islands of the West Indies (Williams 1983). On the Bahamas and the Greater Antilles, species of *Anolis* achieve densities of up to 0.97 (*A. sagrei* on Abaco; Schoener and Schoener 1980) and greater than two lizards per square meter (*A. stratulus* on Puerto Rico; Reagan 1992). If modern anole densities are similar to those between 10 and 15 mya (middle Miocene, when *Epicrates* and *Anolis* were already present on the Greater Antilles, see below), and if *Anolis* was then a common prey for *Epicrates*, that some *Epicrates* continued to feed predominantly or exclusively on anoles throughout their lives is not surprising.

Although mathematical models predict that optimal body size relates directly to food availability (e.g., Schoener 1969; Case 1978), biotic or physical factors can counteract or reverse this prediction. Most West Indian *Anolis* are diurnally active and sleep at night on leaves or very thin branches (Goto and Osborne 1989; Chandler and Tolson 1990; Clark and Gillingham 1990), habitats that are physically inaccessible to heavy-bodied snakes. Of the four small *Epicrates*, *E. monensis* and *E. gracilis* are exclusively or largely arboreal (they occur mostly on slender branches, bushes, vines, and tangled vegetation), whereas *E. fordii* is usually terrestrial, but also may be found in arboreal situations (Tolson and Henderson 1993). Nevertheless, adults of *E. monensis*, *E. gracilis*, and *E. fordii* forage at night for their principal prey items: sleeping anoles (Henderson et al. 1987*a*; Chandler and Tolson 1990). Structural habitat constraints thus may have favored an evolutionary decrease in SVL in large *Epicrates* that as adults encountered mainly *Anolis* as prey. Kluge (1989) suggested that small *Epicrates* are morphologically paedomorphic, and we propose that they are ecologically paedomorphic as well. Our scenario for the evolution of smaller SVL in *Epicrates* requires that *Anolis* and *Epicrates* were simultaneously present on the Greater Antilles when the latter speciated. Estimates from the albumin molecular clock for *Epicrates* (Dessauer et al. 1987) and the fossil record for *Anolis* (Rieppel 1980) indicate that *Anolis* and the common ancestors of large and small *Epicrates* were in fact simultaneously present on the Greater Antilles at least 10–15 mya (middle Miocene).

Changes in body size in response to diet apparently also have occurred in the elapid snake *Notechis ater* on many islands off the southern coast of Australia. Maximum body size on any given island is related to prey size and availability, and these size differences seem to be at least partly genetic (Barnett and Schwaner 1984; Schwaner 1985). On mainland southern Australia, populations of *N. ater* are among those with the smallest adult sizes, and in all of those sites, small lizards and frogs are the most frequently eaten prey (Shine 1987). On the Japanese Izu islands, the colubrid snake *Elaphe quadrivirgata* becomes smaller on islands where the snake feeds mainly on small lizards, and grows larger on islands where it eats larger sea birds (Hasegawa and Moriguchi 1989). We conclude that a shift to an adult diet of predominantly lizard prey was also a likely factor in the evolution of smaller SVL in West Indian *Epicrates*. Behavioral changes commonly precede external, morphological evolution (e.g., Plotkin 1988; Wcislo 1989; West-Eberhard 1989; Aboitiz 1990; see also Arnold 1992), and although multiple selective pressures always are acting on the size of an individual (e.g., predation, competition, abiotic factors [McKinney 1990]), prey size is one of the key determinants of predator body size (Peters 1983; Mori 1994).

Concluding Remarks

With at least 51 species in 11 genera and three families, the Greater Antillean macrostomatan snake fauna holds special promise for investigating patterns of evolutionary and ecological divergence in gape-limited, vertebrate predators. Our review and the extensive autecological studies of Henderson and colleagues (summarized in Henderson et al. 1988; Henderson and Crother 1989; Schwartz and Henderson 1991; Tolson and Henderson 1993; Henderson and Sajdak, this volume) reveal interesting intra- and inter-island patterns of species richness, size distributions, and the presence or absence of sympatry in closely related and morphologically similar species. The potential of these assemblages for shedding light on historical determinants of contemporary distributional and ecological patterns is heightened by the fact that at least three divergence events in snake evolutionary history are represented (Kluge 1991; Cundall et al. 1993).

A deeper understanding of evolutionary and ecological patterns in West Indian snake assemblages (like most of those elsewhere) is hampered by several problems with existing knowledge. First, we have used an inadequate measure of body size (snout-vent length), only because it is relatively easily derived from the literature and preserved specimens. Body mass is perhaps the best overall parameter of body size (e.g., Greene 1983; Pough 1983; Hedges 1985), and we urge herpetologists to follow the lead of other vertebrate field biologists and record fresh weights in their notes and on collecting tags — after all, three portable spring balances spanning 0.01 to 1000 g weigh no more than a couple of sandwiches. Second, we lack even fragmentary data on the diets of more than a third of the species of Antillean snakes, and lack also detailed characterizations of habitats for more than half. Third, *Epicrates* is the only clade of Antillean snakes for which a robust phylogeny is available, and thus the rich possibilities of a phylogenetic perspective are largely denied for the time being. Fourth, an inadequate fossil record for both macrostomatans and their potential prey precludes a detailed analysis of the evolutionary and ecological history of these communities. Finally, as others have emphasized (e.g., Steadman et al. 1984; Wing 1989; Henderson 1992; Pregill et al. 1994), the Antilles have been subject to long-standing and profound human impact. Anthropogenic landscape changes pose interesting dilemmas for interpreting past processes in the light of existing ecological conditions (e.g., Pregill 1986); they also place a premium on rapidly learning more about Antillean faunas, in part so that knowledge can inform conservation (Greene 1994).

Acknowledgments

We thank J. B. Losos, C. Guyer, R. W. Henderson, M. Leal, C. J. Bell, T. Garland, Jr., and S. B. Ashe for helpfully criticizing earlier versions of this manuscript; W. R. Branch, S. B. Hedges, and V. Wallach for providing valuable information; T. Garland, Jr. and K. Autumn for patiently explaining to us the intricacies of comparative phylogenetic analyses; and J. R. Mendelson III for examining stomachs of *Epicrates exsul* for us. We have benefited greatly from discussions with K. Autumn, T. Garland, Jr., A. Graybeal, C. Guyer, and J. B. Losos about phylogenetic systematics and the analysis of character evolution. The senior author thanks R. Thomas for introducing him to the study of West Indian amphibians and reptiles and for his continuous encouragement. Preparation of the manuscript was supported by the Patricia Roberts Harris Fellowship from the United States Department of Education and by a grant from the Carl B. Koford Memorial Fund from the Museum of Vertebrate Zoology, University of California at Berkeley, to the senior author.

LITERATURE CITED

Aboitiz, F. 1990. Behavior, body types and the irreversibility of evolution. Acta Biotheoret. 38: 91–101.

Arnold, S. J. 1972. Species densities of predators and their prey. Amer. Nat. 106: 220–236.

—. 1992. Constraints on phenotypic evolution. Amer. Nat. 140: S85–S107.

—. 1993. Foraging theory and prey-size — predator-size relations in snakes, p. 87–115. *In* R. A. Seigel and J. T. Collins (eds.), Snakes — Ecology and Behavior. McGraw-Hill, New York.

Barnett, B. and T. D. Schwaner. 1984. Growth in captive born tiger snakes (*Notechis ater serventyi*) from Chappell Island: implications for field and laboratory studies. Trans. Roy. Soc. South Australia 109: 31–36.

Bogert, C. M. 1969. Boas — a paradoxical family. Anim. Kingdom 72: 18–25.

Böhme, W. 1984. Erstfund eines fossilen Kugelfingergeckos (Sauria: Gekkonidae: Sphaerodactylinae) aus dominikanischem Bernstein (Oligozän von Hispaniola, Antillen). Salamandra 20: 212–220.

Brooks, D. R. and D. H. McLennan. 1991. Phylogeny, Ecology, and Behavior: A Research Program in Comparative Biology. Univ. Chicago Press, Chicago, Illinois.

Brown, J. H. 1995. Macroecology. Univ. Chicago Press, Chicago, Illinois.

Brown, W. S. and W. S. Parker. 1982. Niche dimensions and resource partitioning in a Great Basin desert snake community, p. 59–81. *In* N. J. Scott (ed.), Herpetological Communities. United States Fish Wildl. Serv. Wildl. Res. Rept. 13.

Burgess, G. H. and R. Franz. 1989. Zoogeography of the Antillean freshwater fish fauna, p. 263–304. *In* C. A. Woods (ed.), Biogeography of the West Indies: Past, Present, and Future. Sandhill Crane Press, Gainesville, Florida.

Cadle, J. E. 1984. Molecular systematics of Neotropical xenodontine snakes: I. South American xenodontines. Herpetologica 40: 8–20.

—. 1985. The Neotropical colubrid snake fauna (Serpentes: Colubridae): lineage components and biogeography. Syst. Zool. 34: 1–20.

— and H. W. Greene. 1993. Phylogenetic patterns, biogeography, and the ecological structure of Neotropical snake assemblages, p. 281–293. *In* R. E. Ricklefs and D. Schluter (eds.), Species Diversity in Ecological Communities: Historical and Geographical Perspectives. Univ. Chicago Press, Chicago, Illinois.

Calver, M. C., J. S. Bradley, and D. R. King. 1988. The relationship between prey size and handling time and prey size and capture success in three sympatric species of dasyurid marsupials. Australian Wildl. Res. 15: 615–623.

Case, T. J. 1978. A general explanation for insular body size trends in terrestrial vertebrates. Ecology 59: 1–18.

—. 1979. Optimal body size and an animal's diet. Acta Biotheoret. 28: 54–69.

— and T. D. Schwaner. 1993. Island/mainland body size differences in Australian varanid lizards. Oecologia (Berlin) 94: 102–109.

Cei, J. M. 1993. Reptiles del Noroeste, Nordeste y Este de la Argentina. Herpetofauna de las Selvas Subtropicales, Puna y Pampas. Monogr. XIV. Mus. Reg. Sci. Nat., Torino.

Chandler, C. R. and P. J. Tolson. 1990. Habitat use by a boid snake, *Epicrates monensis*, and its anoline prey, *Anolis cristatellus*. Jour. Herpetol. 24: 151–157.

Clark, D. L. and J. C. Gillingham. 1990. Sleep-site fidelity in two Puerto Rican lizards. Anim. Behav. 39: 1138–1148.

Connor, E. F. and D. Simberloff. 1979. The assembly of species communities: chance or competition? Ecology 60: 1132–1140.

— and —. 1984. Neutral models of species' co-occurrence patterns, p. 316–331. *In* D. R. Strong, Jr., D. Simberloff, L. G. Abele, and A. B. Thistle (eds.), Ecological Communities: Conceptual Issues and the Evidence. Princeton Univ. Press, Princeton, New Jersey.

Cope, E. D. 1862. Synopsis of the species of *Holcosus* and *Ameiva*, with diagnoses of new West Indian and South American Colubridae. Proc. Acad. Nat. Sci. Philadelphia 14: 60–82.

Crother, B. I. 1986. Natural History Notes: *Alsophis cantherigerus* (Cuban Racer). Diet. Herpetol. Rev. 17: 47.

Cundall, D., V. Wallach, and D. S. Rossman. 1993. The systematic relationships of the snake genus *Anomochilus*. Zool. Jour. Linnean Soc. 109: 275–299.

Dessauer, H. C., J. E. Cadle, and R. Lawson. 1987. Patterns of snake evolution as suggested by their proteins. Fieldiana, Zool. n. s. 34: 1–34.

Donoghue, M. J. and M. J. Sanderson. 1992. The suitability of molecular and morphological evidence in reconstructing plant phylogeny, p. 340–368. *In* P. S. Soltis, D. E. Soltis, and J. J. Doyle (eds.), Molecular Systematics of Plants. Chapman and Hall, New York.

Duellman, W. E. 1978. The biology of an equatorial herpetofauna in Amazonian Ecuador. Misc. Publ. Univ. Kansas Mus. Nat. Hist. 65: 1–352.

Dunn, E. R. and R. Conant. 1936. Notes on anacondas, with descriptions of two new species. Proc. Acad. Nat. Sci. Philadelphia 88: 503–506.

Emerson, S. B., H. W. Greene, and E. L. Charnov. 1994. Allometric aspects of predator-prey interactions, p. 123–139.

In P. C. Wainwright and S. M. Reilly (eds.), Ecological Morphology: Integrative Organismal Biology. Univ. Chicago Press, Chicago, Illinois.

Estrada, A. R. and S. B. Hedges. 1991. Nueva especie de *Eleutherodactylus* de la región oriental de Cuba. Carib. Jour. Sci. 27: 139–145.

Farris, J. S. 1970. Methods for computing Wagner trees. Syst. Zool. 19: 83–92.

Felsenstein, J. 1985. Phylogenies and the comparative method. Amer. Nat. 125: 1–15.

Fitch, H. S. 1982. Resources of a snake community in prairie-woodland habitat of northeastern Kansas, p. 83–97. *In* N. J. Scott (ed.), Herpetological Communities. United States Fish Wildl. Serv. Wildl. Res. Rept. 13.

Garland, T., Jr. and P. A. Carter. 1994. Evolutionary physiology. Ann. Rev. Physiol. 56: 579–621.

—, A. W. Dickerman, C. M. Janis, and J. A. Jones. 1993. Phylogenetic analyses of covariance by computer simulation. Syst. Biol. 42: 265–292.

—, P. H. Harvey, and A. R. Ives. 1992. Procedures for the analysis of comparative data using phylogenetically independent contrasts. Syst. Biol. 41: 18–32.

Garrido, O. H. and S. B. Hedges. 1992. Three new grass anoles from Cuba (Squamata: Iguanidae). Carib. Jour. Sci. 28: 21–29.

Gittleman, J. L. 1985. Carnivore body size: ecological and taxonomic correlates. Oecologia (Berlin) 67: 540–554.

Goto, M. M. and M. A. Osborne. 1989. Nocturnal microhabitats of two Puerto Rican grass lizards, *Anolis pulchellus* and *Anolis krugi*. Jour. Herpetol. 23: 79–81.

Greene, H. W. 1980. Evolutionary biology of the dwarf boas (Serpentes: Tropidophiidae). Yearb. Amer. Philos. Soc. 1979: 206–207.

—. 1983. Dietary correlates of the origin and radiation of snakes. Amer. Zool. 23: 431–441.

—. 1989. Ecological, evolutionary, and conservation implications of feeding biology in Old World cat snakes, genus *Boiga* (Colubridae). Proc. California Acad. Sci. 46: 193–207.

—. 1994. Systematics and natural history, foundations for understanding and conserving biodiversity. Amer. Zool. 34: 48–56.

Grosholz, E. D. 1992. Interactions of intraspecific, interspecific, and apparent competition with host-pathogen population dynamics. Ecology 73: 507–514.

Hardy, J. D., Jr. 1957. Bat predation by the Cuban boa, *Epicrates angulifer* Bibron. Copeia 1957: 151–152.

Harvey, P. H. and M. D. Pagel. 1991. The Comparative Method in Evolutionary Biology. Oxford Univ. Press, Oxford.

Hasegawa M. and H. Moriguchi. 1989. Geographic variation in food habits, body size and life history traits of the snakes on the Izu islands, p. 414–432. *In* M. Matsui, T. Hikida, and R. C. Goris (eds.), Current Herpetology in East Asia. Herpetol. Soc. Japan, Kyoto.

Hedges, S. B. 1985. The influence of size and phylogeny on life history variation in reptiles: a response to Stearns. Amer. Nat. 126: 258–260.

—. 1992. A reconsideration of two montane species of *Eleutherodactylus* in Hispaniola (Anura: Leptodactylidae). Carib. Jour. Sci. 28: 11–16.

—, A. R. Estrada, and R. Thomas. 1992. Three new species of *Eleutherodactylus* from eastern Cuba, with notes on vocalizations of other species (Anura: Leptodactylidae). Herpetol. Monogr. 6: 68–83.

— and O. H. Garrido. 1992*a*. A new species of *Tropidophis* from Cuba (Serpentes: Tropidophiidae). Copeia 1992: 820–825.

— and —. 1992*b*. Cuban snakes of the genus *Arrhyton*: two new species and a reconsideration of *A. redimitum* Cope. Herpetologica 48: 168–177.

— and —. 1993. A new species of gecko (*Sphaerodactylus*) from central Cuba. Jour. Herpetol. 27: 300–306.

— and R. Thomas. 1992*a*. A new marsh-dwelling species of *Eleutherodactylus* from Haiti (Anura: Leptodactylidae). Jour. Herpetol. 26: 191–195.

— and —. 1992*b*. Two new species of *Eleutherodactylus* from remnant cloud forest in Haiti (Anura: Leptodactylidae). Herpetologica 48: 351–358.

Henderson, R. W. 1982. Trophic relationships and foraging strategies of some New World tree snakes (*Leptophis*, *Oxybelis*, *Uromacer*). Amphibia-Reptilia 3: 71–80.

—. 1984. The diet of the Hispaniolan snake *Hypsirhynchus ferox* (Colubridae). Amphibia-Reptilia 5: 367–371.

—. 1992. Consequences of predator introductions and habitat destruction on amphibians and reptiles in the post–Columbus West Indies. Carib. Jour. Sci. 28: 1–10.

— and M. H. Binder. 1980. The ecology and behavior of vine snakes (*Ahaetulla*, *Oxybelis*, *Thelotornis*, and *Uromacer*). Milwaukee Public Mus. Contrib. Biol. Geol. 37: 1–38.

— and R. W. Bourgeois. 1993. Notes on the diets of West Indian *Liophis* (Serpentes: Colubridae). Carib. Jour. Sci. 29: 253–254.

— and B. I. Crother. 1989. Biogeographic patterns of predation in West Indian colubrid snakes, p. 479–517. *In* C. A. Woods (ed.), Biogeography of the West Indies: Past, Present, and Future. Sandhill Crane Press, Gainesville, Florida.

—, T. A. Noeske–Hallin, B. I. Crother, and A. Schwartz. 1988. The diets of Hispaniolan colubrid snakes II. Prey species, prey size, and phylogeny. Herpetologica 44: 55–70.

—, —, J. A. Ottenwalder, and A. Schwartz. 1987*a*. On the diet of *Epicrates striatus* on Hispaniola, with notes on *E. fordi* and *E. gracilis*. Amphibia-Reptilia 8: 251–258.

—, A. Schwartz, and T. A. Noeske–Hallin. 1987*b*. Food habits of three colubrid tree snakes (genus *Uromacer*) on Hispaniola. Herpetologica 43: 241–248.

— and A. Schwartz. 1986. The diet of the Hispaniolan colubrid snake, *Darlingtonia haetiana*. Copeia 1986: 529–531.

Houston, D. and R. Shine. 1993. Sexual dimorphism and niche divergence: feeding habits of the Arafura filesnake. Jour. Anim. Ecol. 62: 737–748.

Jackson, J. F., W. Ingram III, and H. W. Campbell. 1976. The dorsal pigmentation pattern of snakes as an antipredator strategy: a multivariate approach. Amer. Nat. 110: 1029–1053.

Jayne, B. C., H. K. Voris, and Kiew Bong Heang. 1987. Diet, feeding, behavior, growth, and numbers of a population of *Cerberus rynchops* (Serpentes: Homalopsinae) in Malaysia. Fieldiana, Zool. n. s. 50: 1–15.

Jeffries, M. J. and J. H. Lawton. 1984. Enemy free space and the structure of ecological communities. Biol. Jour. Linnean Soc. 23: 269–286.

Jorge da Silva, N. 1993. The snakes from Samuel hydroelectric power plant and vicinity, Rondônia, Brazil. Herpetol. Nat. Hist. 1: 37–86.

Kelt, D. A., M. L. Taper, and P. L. Meserve. 1995. Assessing the impact of competition on community assembly: a case study using small mammals. Ecology 76: 1283–1296.

Kluge, A. G. 1989. A concern for evidence and a phylogenetic hypothesis of relationships among *Epicrates* (Boidae, Serpentes). Syst. Zool. 38: 7–25.

—. 1991. Boine snake phylogeny and research cycles. Misc. Publ. Mus. Zool. Univ. Michigan 178: 1–58.

— and A. J. Wolf. 1993. Cladistics: what's in a word? Cladistics 9: 183–199.

Krebs, J. R. and A. Kacelnik. 1991. Decision-making, p. 105–136. *In* J. R. Krebs and N. B. Davies (eds.), Behavioural Ecology: An Evolutionary Approach. 3rd ed. Blackwell Scientific Publ., Oxford.

Losos, J. B. 1994. Historical contingency and lizard community ecology, p. 319–333. *In* L. J. Vitt and E. R. Pianka (eds.), Lizard Ecology: Historical and Experimental Perspectives. Princeton Univ. Press, Princeton, New Jersey.

— and D. B. Miles. 1994. Adaptation, constraint, and the comparative method: phylogenetic issues and methods, p. 60–98. *In* P. C. Wainwright and S. M. Reilly (eds.), Ecological Morphology: Integrative Organismal Biology. Univ. Chicago Press, Chicago, Illinois.

MacPhee, R. D. E. and A. R. Wyss. 1990. Oligo-Miocene vertebrates from Puerto Rico, with a catalog of localities. Amer. Mus. Novit. 2965: 1–45.

Maddison, W. P. 1991. Squared-change parsimony reconstructions of ancestral states for continuous-valued characters on a phylogenetic tree. Syst. Zool. 40: 304–314.

— and D. R. Maddison. 1992. MacClade Version 3 — Analysis of Phylogeny and Character Evolution. Sinauer Assoc., Sunderland, Massachusetts.

Madsen, T. and R. Shine. 1993. Phenotypic plasticity in body sizes and sexual size dimorphism in European grass snakes. Evolution 47: 321–325.

Martins, E. P. and T. Garland, Jr. 1991. Phylogenetic analyses of the correlated evolution of continuous characters: a simulation study. Evolution 45: 534–547.

Marx, H. and G. B. Rabb. 1972. Phyletic analysis of fifty characters of advanced snakes. Fieldiana, Zool. 63: 1–321.

McCosker, J. E. 1975. Feeding behavior of Indo-Australian Hydrophiidae, p. 217–232. *In* W. E. Dunson (ed.), The Biology of Sea Snakes. Univ. Park Press, Baltimore, Maryland.

McKinney, M. L. 1990. Trends in body-size evolution, p. 75–118. *In* K. J. McNamara (ed.), Evolutionary Trends. Univ. Arizona Press, Tucson.

Morgan, G. S. 1989. Fossil Chiroptera and Rodentia from the Bahamas, and the historical biogeography of the Bahamian mammal fauna, p. 685–740. *In* C. A. Woods (ed.), Biogeography of the West Indies: Past, Present, and Future. Sandhill Crane Press, Gainesville, Florida.

Mori, A. 1994. Ecological and morphological characteristics of the Japanese rat snake, *Elaphe climacophora*, on Kammuri-jima island: a possible case of insular gigantism. Snake 26: 11–18.

Murphy, R. W. and J. R. Ottley. 1984. Distribution of amphibians and reptiles on islands in the Gulf of California. Ann. Carnegie Mus. 53: 207–230.

Mushinsky, H. R., J. J. Hebrard, and D. S. Vodopich. 1982. Ontogeny of water snake foraging. Ecology 63: 1624–1629.

Nellis, D. W., R. L. Norton, and W. P. Maclean. 1983. On the biogeography of the Virgin Islands boa, *Epicrates monensis granti*. Jour. Herpetol. 17: 413–417.

Oliver, J. A. 1948. The relationships and zoogeography of the genus *Thalerophis* Oliver. Bull. Amer. Mus. Nat. Hist. 92: 157–280.

Pérez-Santos, C. and A. G. Moreno. 1988. Ofidios de Colombia. Monogr. VI. Mus. Reg. Sci. Nat., Torino.

Peters, R. H. 1983. The Ecological Implications of Body Size. Cambridge Univ. Press, Cambridge.

Plotkin, H. C. (ed.). 1988. The Role of Behavior in Evolution. MIT Press, Cambridge, Massachusetts.

Poinar, G. O., Jr. and D. C. Cannatella. 1987. An upper Eocene frog from the Dominican Republic and its implication for Caribbean biogeography. Science 237: 1215–1216.

Pough, F. H. 1983. Amphibians and reptiles as low-energy systems, p. 141–188. *In* W. P. Aspey and S. I. Lustick (eds.), Behavioral Energetics. Ohio State Univ. Press, Columbus.

— and R. M. Andrews. 1985. Energetic costs of subduing and swallowing prey for a lizard. Ecology 66: 1525–1533.

Pregill, G. K. 1981. Late Pleistocene herpetofaunas from Puerto Rico. Misc. Publ. Univ. Kansas Mus. Nat. Hist. 71: 1–72.

—. 1986. Body size of insular lizards: a pattern of Holocene dwarfism. Evolution 40: 997–1008.

—, D. W. Steadman, and D. R. Watters. 1994. Late Quaternary vertebrate faunas of the Lesser Antilles: historical components of Caribbean biogeography. Bull. Carnegie Mus. Nat. Hist. 30: 1–51.

Pulliam, H. R. 1985. Foraging efficiency, resource partitioning, and the coexistence of sparrow species. Ecology 66: 1829–1836.

Reagan, D. P. 1992. Congeneric species distribution and abundance in a three-dimensional habitat: the rain forest anoles of Puerto Rico. Copeia 1992: 392–403.

Reynolds, R. P. and N. J. Scott, Jr. 1982. Use of a mammalian resource by a Chihuahuan snake community, p. 99–118. *In* N. J. Scott (ed.), Herpetological Communities. United States Fish Wildl. Serv. Wildl. Res. Rept. 13.

Richman, A. D. and T. Price. 1992. Evolution of ecological differences in the Old World leaf warblers. Nature 355: 817–821.

Rieppel, O. 1980. Green anole in Dominican amber. Nature 286: 486–487.

—. 1988. A review of the origin of snakes. Evol. Biol. 22: 37–130.

Rodríguez, G. A. and D. P. Reagan. 1984. Bat predation by the Puerto Rican boa, *Epicrates inornatus*. Copeia 1984: 219–220.

Rodríguez-Robles, J. A. and M. Leal. 1993*a*. Natural History Notes: *Alsophis portoricensis* (Puerto Rican Racer). Diet. Herpetol. Rev. 24: 150–151.

— and —. 1993*b*. Effects of prey type on the feeding behavior of *Alsophis portoricensis* (Serpentes: Colubridae). Jour. Herpetol. 26: 163–168.

Rogers, J. S. 1976. Species density and taxonomic diversity of Texas amphibians and reptiles. Syst. Zool. 25: 26–40.

Rohlf, F. J., W. S. Chang, R. R. Sokal, and J. Kim. 1990. Accuracy of estimated phylogenies: effects of tree topology and evolutionary model. Evolution 44: 1671–1684.

Roze, J. A. 1966. La Taxonomía y Zoogeografía de los Ofidios de Venezuela. Univ. Cent. Venezuela. Ed. Bibl., Caracas.

Schluter, D. 1994. Experimental evidence that competition promotes divergence in adaptive radiation. Science 266: 798–801.

Schoener, T. W. 1969. Models of optimal size for solitary predators. Amer. Nat. 103: 277–313.

—. 1971. Theory of feeding strategies. Ann. Rev. Ecol. Syst. 2: 369–404.

— and A. Schoener. 1980. Densities, sex ratios, and population structure of four species of Bahamian *Anolis* lizards. Jour. Anim. Ecol. 49: 19–53.

Schwaner, T. D. 1985. Population structure of black tiger snakes, *Notechis ater niger*, on offshore islands of South Australia, p. 35–46. *In* G. Grigg, R. Shine, and H. Ehmann (eds.), Biology of Australasian Frogs and Reptiles. Surrey Beatty and Sons, Sydney.

Schwartz, A. and R. W. Henderson. 1991. Amphibians and Reptiles of the West Indies: Descriptions, Distributions, and Natural History. Univ. Florida Press, Gainesville.

Seib, R. L. 1985. Euryphagy in a tropical snake, *Coniophanes fissidens*. Biotropica 17: 57–64.

Serié, P. 1920. Notas sobre la alimentación de algunos ofidios. Rev. Jardín Zool. (Buenos Aires, Diciembre 1919): 3–24.

Settle, W. H. and L. T. Wilson. 1990. Invasion by the variegated leafhopper and biotic interactions: parasitism, competition, and apparent competition. Ecology 71: 1461–1470.

Shine, R. 1977. Habitats, diets, and sympatry in snakes: a study from Australia. Canadian Jour. Zool. 55: 1118–1128.

—. 1987. Ecological comparisons of island and mainland populations of Australian tigersnakes (*Notechis*: Elapidae). Herpetologica 43: 233–240.

—. 1991. Why do larger snakes eat larger prey items? Funct. Ecol. 5: 493–502.

— and J. K. Webb. 1990. Natural history of Australian typhlopid snakes. Jour. Herpetol. 24: 357–363.

Siegel, S. and N. J. Castellan, Jr. 1988. Nonparametric Statistics for the Behavioral Sciences. 2nd ed. McGraw-Hill, New York.

Simberloff, D. 1978. Using island biogeographic distributions to determine if colonization is stochastic. Amer. Nat. 112: 713–726.

Slip, D. J. and R. Shine. 1988. Feeding habits of the diamond python, *Morelia s. spilota*: ambush predation by a boid snake. Jour. Herpetol. 22: 323–330.

Steadman, D. W., G. K. Pregill, and S. L. Olson. 1984. Fossil vertebrates from Antigua, Lesser Antilles: evidence for late Holocene human-caused extinctions in the West Indies. Proc. Natl. Acad. Sci. USA 81: 4448–4451.

Stephens, D. W. and J. R. Krebs. 1986. Foraging Theory. Princeton Univ. Press, Princeton, New Jersey.

Strimple, P. D. 1993. Overview of the natural history of the green anaconda (*Eunectes murinus*). Herpetol. Nat. Hist. 1: 25–35.

Swofford, D. L. and W. P. Maddison. 1987. Reconstructing ancestral character states under Wagner parsimony. Math. Biosci. 87: 199–229.

Thomas, R. and S. B. Hedges. 1992. An unusual new *Sphaerodactylus* from Hispaniola (Squamata: Gekkonidae). Jour. Herpetol. 26: 289–292.

— and —. 1993. A new banded *Sphaerodactylus* from eastern Hispaniola (Squamata: Gekkonidae). Herpetologica 49: 350–354.

—, —, and O. H. Garrido. 1992. Two new species of *Sphaerodactylus* from eastern Cuba (Squamata: Gekkonidae). Herpetologica 48: 358–367.

Tolson, P. J. 1987. Phylogenetics of the boid snake genus *Epicrates* and Caribbean vicariance theory. Occas. Pap. Mus. Zool. Univ. Michigan 715: 1–68.

— and R. W. Henderson. 1993. The Natural History of West Indian Boas. R & A Publ. Ltd., Taunton, Somerset.

Vitt, L. J. 1987. Communities, p. 335–365. *In* R. A. Seigel, J. T. Collins, and S. S. Novak (eds.), Snakes: Ecology and Evolutionary Biology. McGraw-Hill, New York.

— and L. D. Vangilder. 1983. Ecology of a snake community in northeastern Brazil. Amphibia-Reptilia 4: 273–296.

Voris, H. K. and M. W. Moffett. 1981. Size and proportion relationship between the beaked sea snake and its prey. Biotropica 13: 15–19.

— and H. H. Voris. 1983. Feeding strategies in marine snakes: an analysis of evolutionary, morphological, behavioral, and ecological relationships. Amer. Zool. 23: 411–425.

Wcislo, W. T. 1989. Behavioral environments and evolutionary change. Ann. Rev. Ecol. Syst. 20: 137–169.

Weatherly, A. H. 1963. Notions of niche competition among animals, with special reference to freshwater fish. Nature 197: 14–17.

West-Eberhard, M. J. 1989. Phenotypic plasticity and the origins of diversity. Ann. Rev. Ecol. Syst. 20: 249–278.

Williams, E. E. 1983. Ecomorphs, faunas, island size, and diverse end points in island radiations of *Anolis*, p. 326–370. *In* R. B. Huey, E. R. Pianka, and T. W. Schoener (eds.), Lizard Ecology: Studies of a Model Organism. Harvard Univ. Press, Cambridge, Massachusetts.

—. 1989. Old problems and new opportunities in West Indian biogeography, p. 1–46. *In* C. A. Woods (ed.), Biogeography of the West Indies: Past, Present, and Future. Sandhill Crane Press, Gainesville, Florida.

Wing, E. S. 1989. Human exploitation of animal resources in the Caribbean, p. 137–152. *In* C. A. Woods (ed.), Biogeography of the West Indies: Past, Present, and Future. Sandhill Crane Press, Gainesville, Florida.

Winston, M. R. 1995. Co-occurrence of morphologically similar species of stream fishes. Amer. Nat. 145: 527–545.

Woods, C. A. 1990. The fossil and recent land mammals of the West Indies: an analysis of the origin, evolution, and extinction of an insular fauna, p. 641–680. *In* Biogeographical Aspects of Insularity. Accad. Naz. Lincei, Rome.

Zaher, H. 1994. Les Tropidopheoidea (Serpentes; Alethinophidia) sont-ils réellement monophylétiques? Arguments en faveur de leur polyphylétisme. Comp. Rend. l'Acad. Sci. Paris, ser. III — Sci. Vie 317: 471–478.

Resumen.—Cincuenta y una especies de serpientes macrostomatas habitan en las Antillas Mayores; 32 son de tamaño pequeño, 11 son medianas, y 8 son grandes. Las serpientes macrostomatas pequeñas y medianas se alimentan principal o exclusivamente de lagartijas y sapos, mientras que seis de las siete especies de tamaño grande para las cuales hay datos sobre sus hábitos alimenticios consumen mayormente mamíferos. Algunos linajes de colúbridos se han diversificado en las cuatro islas más grandes de las Antillas Mayores, y la riqueza de especies de serpientes macrostomatas está positivamente relacionada con el área de la isla. Solamente en Cuba habita más de una especie de *Tropidophis*; una de estas especies se alimenta de aves y mamíferos con alguna frecuencia, mientras que las restantes nueve especies comen principalmente lagartijas y sapos. Quizás estas nueve especies se han diversificado mayormente en términos de los hábitats que ocupan en esa isla. Ninguna isla tiene más de una serpiente que se especialice en comer mamíferos (i.e., *Epicrates*), y en islas donde hay más de una especie de *Epicrates* la especie o especies adicionales tienen un tamaño menor y se alimentan principalmente de lagartijas cuando son adultas. El género *Epicrates* incluye seis especies grandes y robustas y cuatro especies de menor tamaño y más delgadas. Los análisis filogenéticos sugieren que ha ocurrido una reducción en el tamaño de las *Epicrates* en tres ocasiones: en la especie continental *E. cenchria* y dos veces en las especies de *Epicrates* de las Antillas. La hipótesis de que un cambio en la dieta de los adultos a una presa de lagartijas fue probablemente un factor importante en la reducción en tamaño en las *Epicrates* del Caribe es apoyada por un análisis filogenético de la evolución de esta característica. Las serpientes macrostomatas de las Antillas ofrecen excelentes oportunidades para estudiar fenómenos históricos y ecológicos, pero para alcanzar por completo ese potencial es necesario obtener mucha información adicional sobre su historia natural y relaciones filogenéticas.

Résumé.—Cinquante et une espèces de serpents macrostomates peuplent les Grandes Antilles; 32 d'entre elles sont des petites espèces, 11 sont de taille moyenne et 8 sont grosses. Les petits serpents et ceux de taille moyenne se nourrissent essentiellement ou exclusivement de lézards et de grenouilles alors que six des sept grosses espèces pour lesquelles nous possédons des données alimentaires se nourrissent essentiellement de mammifères. Plusieurs clades de Colubridés se sont diversifiés sur les quatre plus grandes îles des Grandes Antilles et il existe une corrélation positive entre la diversité des espèces et l'île. Seule Cuba abrite plus d'une espèce de *Tropidophis* et seule une espèce se nourrit d'oiseaux et de mammifères assez régulièrement alors que les neuf autres espèces se contentent principalement de lézards et de grenouilles; leur diversification est peut-être uniquement liée à l'habitat. Aucune île n'abrite plus d'une grosse espèce se nourrissant de mammifères (comme *Epicrates*) et sur les îles où vivent plus d'une espèce d'*Epicrates*, les autres espèces sont généralement plus petites et se nourrissent essentiellement de lézards lorsqu'elles ont atteint leur taille adulte. Le genre *Epicrates* comporte six grosses espèces robustes et quatre espèces plus petites et plus fines. Les analyses phylogénétiques donnent à penser que la diminution de la taille est intervenue à trois reprises chez *Epicrates*: une fois chez *E. cenchria* (continent) et deux fois au sein du clade antillais. L'hypothèse selon laquelle le glissement vers la taille adulte propice à la prédation des lézards soit un facteur vraisemblable de l'évolution vers un corps plus petit chez *Epicrates* des Antilles est confirmée par l'analyse phylogénétique de l'évolution de cette caractéristique. Les serpents macrostomates des Antilles offrent de multiples occasions d'étude des phénomènes écologiques contemporains et historiques. Pour pouvoir tirer pleinement parti de ce potentiel, il faut toutefois augmenter les données sur leur histoire naturelle et leurs relations phylogénétiques.

Rodríguez-Robles, J. A. and H. W. Greene. 1996. Ecological patterns in Greater Antillean macrostomatan snake assemblages, with comments on body-size evolution in *Epicrates* (Boidae), p. 339–357. *In* R. Powell and R. W. Henderson (eds.), Contributions to West Indian Herpetology: A Tribute to Albert Schwartz. Society for the Study of Amphibians and Reptiles, Ithaca (New York). Contributions to Herpetology, volume 12.

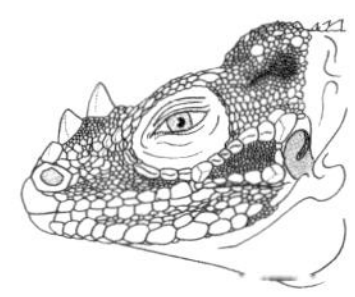

Species Richness and Biogeography of the Herpetofauna in the Exuma Cays Land and Sea Park, Bahamas

C. Kenneth Dodd, Jr.[1] and Richard Franz[2]

Abstract.—We examined herpetofaunal distribution patterns on 31 islands in the Exuma Cays Land and Sea Park (ECLSP), Bahamas, as a result of four intensive site visits between 1990 and 1993. The herpetofauna consists of 15 species, two (possibly four) of which are introduced. Three basic zoogeographic patterns are shown by the herpetofauna, i.e., species with ubiquitous, core island, or widespread distribution patterns. A single species was found on nearly all islands surveyed, three species were found only on the largest and most structurally complex (=core) islands, and four species were found on a wide variety of small to large sized islands. Three species were so rare that they could not be classified. The remaining two species were probably more widespread at one time, but appear to be much more restricted now. Island size alone explains 42% of the variation in island species richness. The size of the nearest core island and the distance of the nearest core island from its satellite islands did not influence species richness. The distribution of the herpetofauna of the ECLSP probably results from a combination of historical and geological events coupled with ongoing dynamic ecological processes. The potential impact of both exotic predators and human settlement cannot be overlooked when attempting to explain present zoogeographical patterns.

Introduction

Islands have always intrigued biologists and geographers, and the study of island biotas has led to important insights concerning ecological and evolutionary processes (e.g., MacArthur and Wilson 1967; Case and Bolger 1991; Paulay 1994). Islands are uniquely suited to the study of the dynamic nature of speciation, competition, dispersal, colonization, trophic interactions, and the effects of patch size on species richness and diversity. Island-based studies on such topics not only have resulted in a better understanding of biological concepts, but also in recognition of the practical value of such knowledge, particularly for conservation biology and natural resource management (see references in Meffe and Carroll 1994). Given the proximity of the Bahamas to continental North America, that so little is known about the biology of the herpetofauna of these ephemeral islands is surprising. The taxonomic and distributional work by Thomas Barbour and Albert Schwartz and the later ecological investigations by Thomas Schoener and colleagues illustrate the potential that these islands offer for productive research.

During the last major low Ice Age sea level (approximately 15,000–16,000 years BP), the Great Bahama Bank was emergent as one vast island (Great Bahama Island) (Sealey 1985). Terrestrial herpetofauna from the Greater Antilles colonized Great Bahama Island across what were then narrow channels. Most of the indigenous amphibians and reptiles in the Bahamas originated from Hispaniola, Puerto Rico, or, more importantly, Cuba (Schwartz 1967, 1968; Schwartz and Crombie 1975; Buden 1981). Rising ocean levels during the Holocene, coupled with erosion and subsidence, reduced Great Bahama Island to its present configuration of numerous scattered small islands, and resulted at least

[1] National Biological Service, Gainesville, Florida 32653, USA.

[2] Florida Museum of Natural History, University of Florida, Gainesville, Florida 32611, USA.

partially in the current distribution of the native Exuma Archipelago herpetofauna.

The Exuma Islands are located on the Great Bahama Bank. They form a linear archipelago stretching from the southeast to the northwest, and are isolated to the east by Exuma Sound and to the west by the Tongue of the Ocean. The Great Bahama Bank extends northward to New Providence and Eleuthera. The largest part of the bank extends south of the Exuma chain and west of Andros Island. Islands on the bank are separated by shallow seas.

In 1990, we began an intensive inventory of the amphibians and reptiles of the Exuma Archipelago (Franz et al. 1993). The purpose of the present survey was to collect basic distributional and biological information on the herpetofauna of the Exuma Cays Land and Sea Park (ECLSP) which could be used to assist park personnel with management of the islands' resources. The ECLSP is located between Wax Cay Cut (24°34'07"N 76°47'27"W at Bush Hill Cay) in the north and Conch Cut (24°16'39"N 76°32'30"W at Rocky Dundas Cay) in the south. The Park is a rectangle 14.8 km x 40.8 km (455.8 km^2). The islands are low in elevation (<26 m) and are composed of oolitic limestone. They range in size from small jagged rocks barely above water to islands of nearly 1900 ha in area. Vegetation on the islands include coastal rock, sand strand, and coastal coppice communities as listed in Correll and Correll (1982). Soils in the ECLSP are mostly sedimentary sands, although organic soils (*sensu* Sealey 1985) commonly are found in depressions between rock outcrops, in valleys, on the lower slopes of ridges, and in caves and pits. Extensive areas of organic soil form only on the larger islands.

During the course of our surveys, certain patterns and seeming anomalies in the distribution of ECLSP herpetofauna became apparent. In this chapter, we describe these biogeographic patterns and examine some of the factors that might explain present herpetofaunal distribution.

METHODS

Collecting was conducted during four visits to the Exuma Cays Land and Sea Park between 1990 and 1993 (August 1990; May 1991; March 1992; July 1993). Each visit, defined as one sampling period, lasted approximately two weeks. Specimens were collected by hand, by stunning with rubber bands, by noosing, and by wire mesh funnel traps (only on Waderick Wells Cay). Voucher specimens are deposited in the Florida Museum of Natural History. We also examined museum records to determine the presence of species overlooked during our surveys (see Franz et al. 1993).

Ground litter, tree trunks and branches, and caves were searched thoroughly. Because islands ranged in size from 5 to nearly 1900 ha, the amount of time spent sampling varied roughly in proportion to island size (1–2 person-hours on the smaller islands to more than 20 person-hours on the larger islands). Most large islands were visited more than once, usually in different sampling periods. Additional details concerning methods, the geographic coordinates of the islands visited within the park, and a checklist of the amphibians and reptiles known from each island are provided elsewhere (Franz et al. 1993). Additional reptilian records for the ECLSP are in Franz and Dodd (1994*a*, 1994*b*, 1994*c*, 1994*d*, 1994*e*).

We determined the area of the islands surveyed ($n = 31$) using a planimeter on a 1:25,000 scale map of the ECLSP. Areas were measured three times for each island, and the average was used in statistical analyses. The park was partitioned into five geographic areas based on proximity to nearest "core" island. A core island was defined as a large island containing sufficient topographic relief and area to sustain a complex vegetative structure and a variety of floral communities. Core groups (large islands plus their satellite islands, Table 1) generally were separated from one another by channel cuts with strong east-west currents. The five core islands were Shroud, Hawksbill, Waderick Wells, Halls Pond, and Bells cays (Figure 1). Several large islands (Little Bells, O'Briens) were not considered core islands because of low topographic relief which resulted in rather homogenous vegetative structure.

The relationship between reptilian species richness and island size was examined using linear regression on $\log_e$ transformed data. We visually examined data (plots of reptilian species richness versus core island area and species richness of satellite islands versus distance to nearest core island) to look for trends. These data were further examined using linear regression (area data transformed to $\log_e$). We excluded the two introduced land iguanas (*Cyclura*

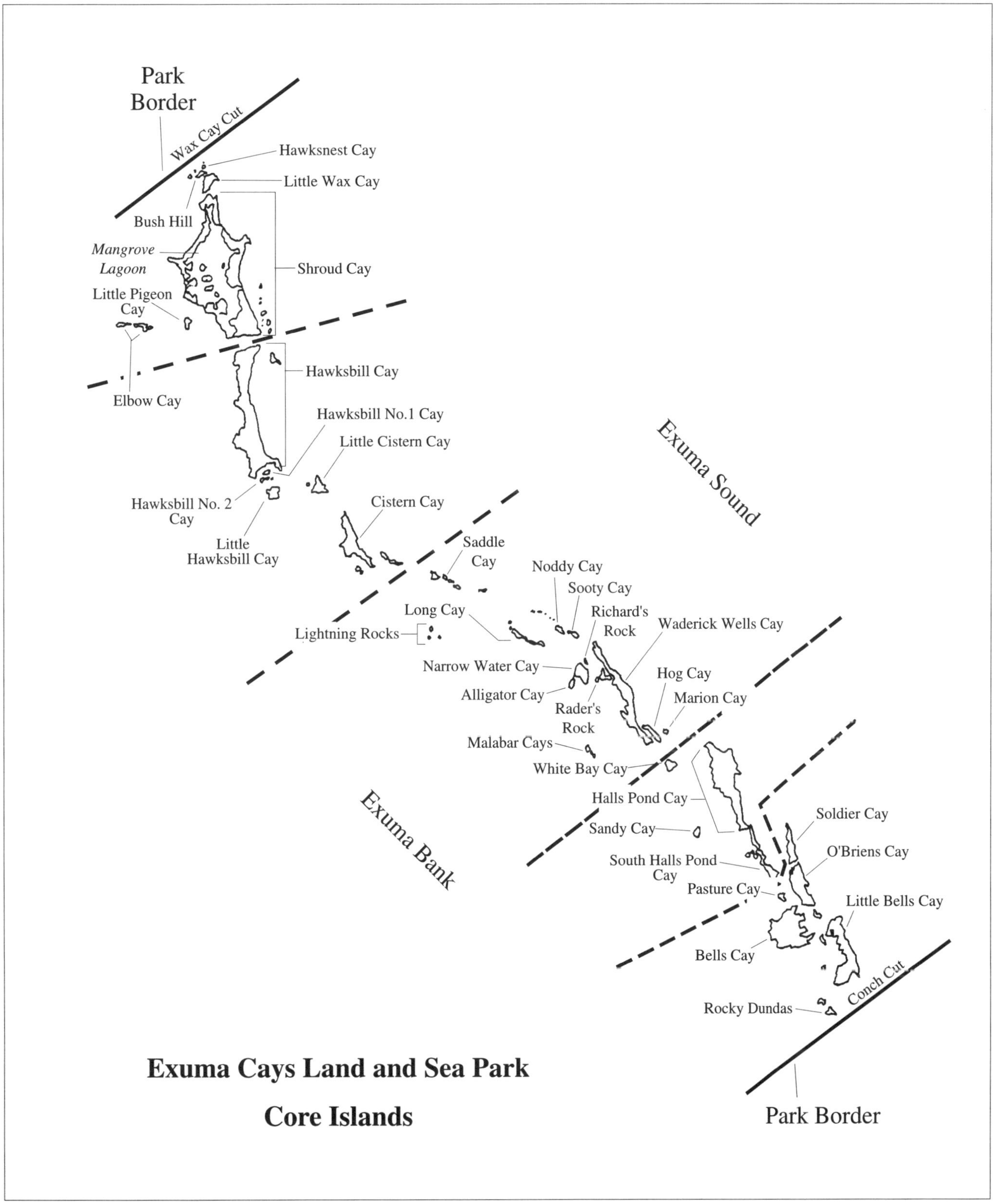

FIGURE 1. Map of the Exuma Cays Land and Sea Park showing the distribution of core island groups.

cychlura and *C. rileyi*) from the analysis of geographic distribution patterns.

We obtained information on exotic rat (*Rattus rattus*) and cat (*Felis catus*) introductions from park wardens. Data on human settlement patterns were obtained from records of the ECLSP and the Bahamas government. We examined the association between the presence of humans and/or exotic predators and the presence or absence of the ground-dwelling diurnal lizards *Ameiva* and *Leiocephalus* using χ^2 contingency tables with Fisher's exact test. All analyses were carried out using the SAS program for microcomputers (SAS Institute Inc. 1988).

RESULTS

Three species of anoles (*Anolis distichus*, *A. sagrei*, *A. smaragdinus*) were found in the ECLSP. All are diurnal and were easily observed during field surveys. The Bahamian brown anole (*A. sagrei*) was found on all of the islands we surveyed (Figure 2B), with the exception of Lightning Rocks, which supported no reptiles. The Bahamian brown anole inhabited both the smallest vegetated rocks and the largest islands. *Anolis distichus* and *A. smaragdinus*

TABLE 1. Islands surveyed for herpetofauna in the Exuma Cays Land and Sea Park, Bahamas, 1990 to 1993. See text for definition of core island. A core island group consists of a core island plus its satellites.

Core island	Satellite islands
Shroud Cay	Bush Hill, Elbow, Little Wax
Hawksbill Cay	Cistern, Hawksbill No. 1, Hawksbill No. 2, Little Cistern, Little Hawksbill
Waderick Wells Cay	Alligator, Hog, Lightning Rocks,* Long, Malabar, Marion, Narrow Water, Noddy, Rader's Rock, Richard's Rock, Sooty
Halls Pond Cay	Sandy, South Halls Pond, White Bay
Bells Cay	Little Bells, O'Briens, Pasture, Rocky Dundas, Soldier

* No herpetofauna; excluded from all analyses.

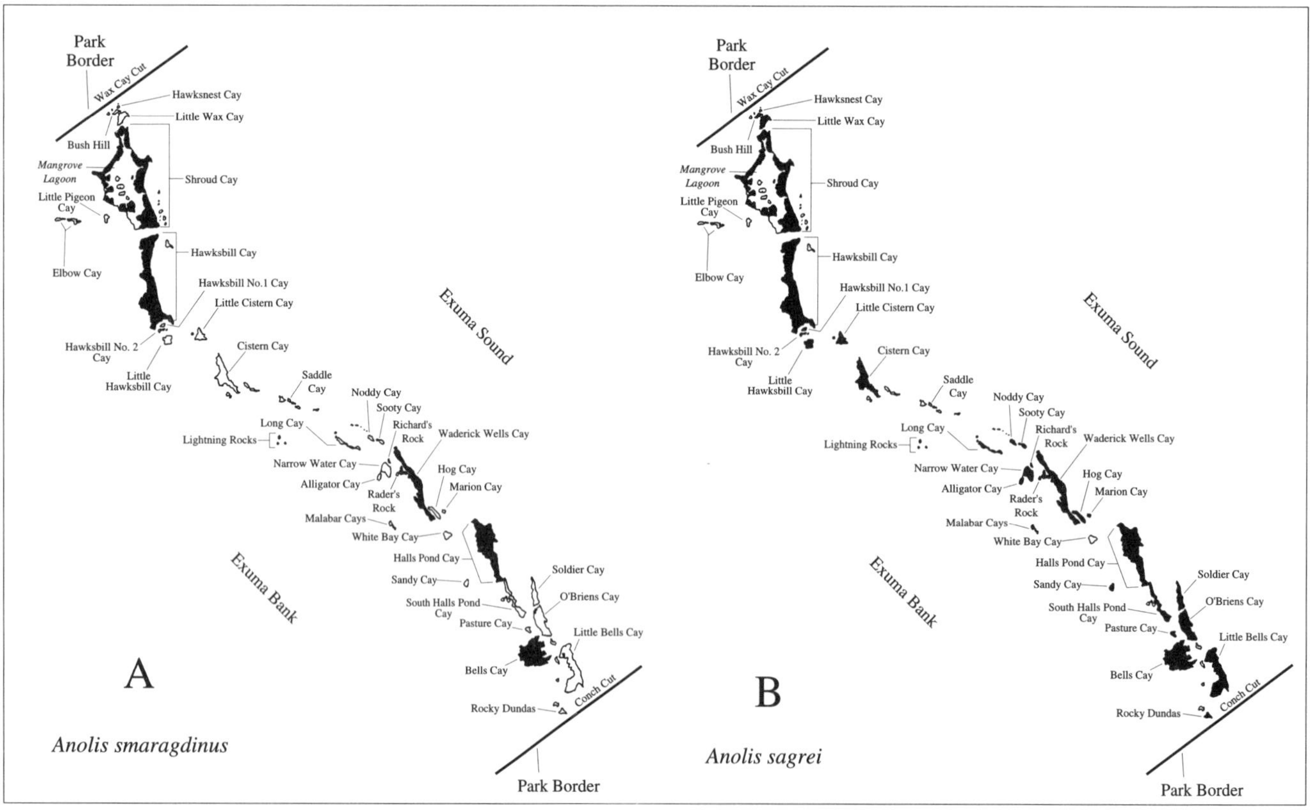

FIGURE 2. Distribution maps showing two patterns of herpetofaunal distribution within the Exuma Cays Land and Sea Park. A: Distribution restricted to core islands. B: Ubiquitous distribution throughout the archipelago.

(Figure 2A) were observed only on the core islands in forests of complex vegetation structure and composition. The frog *Eleutherodactylus planirostris* also was found only on core islands, although an unconfirmed sight record of *E. planirostris* exists for White Bay Cay. The frog was found in caves or was heard calling from under rocks at night. Neither *A. distichus* nor *E. planirostris*, however, have been collected from all core islands (Franz et al. 1993).

One snake and three lizards (*Alsophis vudii*, *Sphaerodactylus nigropunctatus*, *S. notatus*, *Tarentola americana*) were found both on core and satellite islands, including rather small islands (Richard's Rock, Long Rock Cay, Marion Cay). The Bahamian brown runner (*A. vudii*) was encountered infrequently and only during diurnal surveys; the geckos *S. nigropunctatus* and *S. notatus* were collected from under thick leaf litter, usually after intensive effort; and the gecko *T. americana* was collected in caves or at night on buildings. Because of the secretive habits or specialized habitats of these species, they may be more widespread than our samples indicate.

TABLE 2. Relationship between exotic predators, past and present human settlement, and presence and absence of the diurnal lizards *Ameiva auberi* and *Leiocephalus carinatus*.

Island	Exotic predators	Human settlement	*A. auberi*	*L. carinatus*
Little Wax	+	—	—	—
Shroud	+	—	+[1]	+[1]
Hawksbill	—	+[2]	+	+
Cistern	+	+	—	—
Waderick Wells		+[3]	+	+
Halls Pond	+	+	+	—
Soldier	+	+	—	—
Bells	+	+	+	—
Little Bells	+	—	+	—

[1] Although present, these species are very scarce.
[2] This island has not been inhabited in recent times.
[3] This island has not been settled in recent times; present habitation consists only of park personnel.

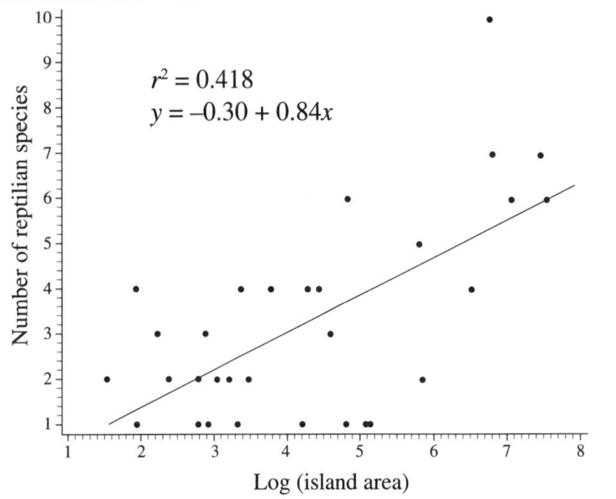

FIGURE 3. Relationship between reptilian species richness and island area in the Exuma Cays Land and Sea Park. Areas are in hectares.

The lizards *Ameiva auberi* and *Leiocephalus carinatus* are diurnal, easily observed, terrestrial lizards. Where present, they or their trails were conspicuous, particularly along beaches and sandy dunes behind beaches. *Ameiva auberi* was present on all core islands as well as the larger uninhabited satellite islands (Little Hawksbill, Little Cistern, Narrow Water, O'Briens). This species was absent from Soldier and Cistern cays. *Leiocephalus carinatus* was present on most of the islands, including several rather small islands (e.g., Narrow Water, Alligator, White Bay cays), but was absent from several core (Halls Pond, Bells) and larger satellite (Soldier, Cistern, Little Bells) islands.

The boa *Epicrates striatus* is known from sightings by visitors on Bells Cay (a core island) and by residents of Cistern Cay (a large satellite island of Hawksbill Cay). A blind snake (*Typhlops* sp.) was sighted on Waderick Wells Cay during building construction (Franz et al. 1993). Neither species was seen during our surveys. The treefrog *Osteopilus septentrionalis* is known only from Bells Cay (Franz et al. 1993), although unconfirmed reports exist from Hawksbill. This species may be native to the ECLSP, or it might have been introduced when the islands were inhabited.

A significant ($P < 0.0001$) positive correlation existed between island area and reptilian species richness (Figure 3), with the regression line explaining approximately 42% of the variation. For lizards, no trend towards increasing species richness with core island size (P = 0.2751; Figure 4) was seen. On islands greater than 1,000 ha, lizard species richness was similar among islands. No relationship was noted between reptilian species richness on satellite islands and distance from nearest core island (P = 0.8809). However, considerable inter-island variation in species richness existed (Figure 5). The largest core island (Shroud) had the greatest percentage of species found only on core islands, whereas the smallest core island (Waderick Wells) had the lowest percentage of species found only on a core island (Figure 6).

Our collections suggest a negative association between the distribution of the diurnal terrestrial lizard *Leiocephalus carinatus* and the presence of exotic predators (Table 2). Similar but less obvious trends are apparent for *Ameiva auberi* and for the effects of human settlement on *Leiocephalus* (Table 2). We could not demonstrate statistical significance except for the negative effect of exotic predators on *L. carinatus* (P = 0.008). Although the arboreal and semi-arboreal anoles do not seem to be affected, we were unable to gauge impacts of human settlement or exotic predators on other herpetofaunal species.

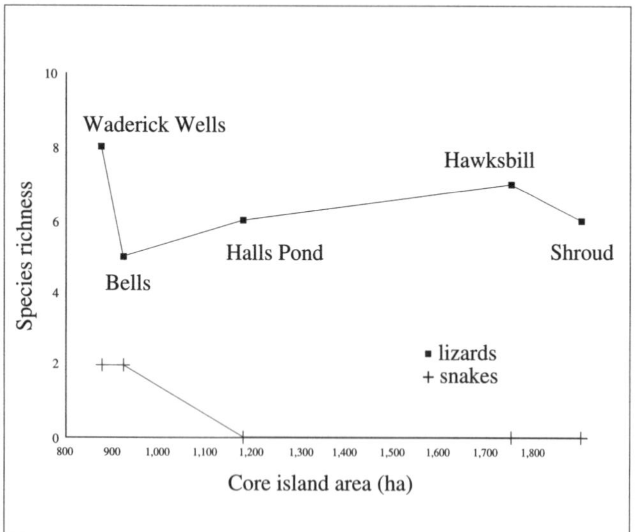

FIGURE 4. Relationship between lizard and snake species richness and core island area, Exuma Cays Land and Sea Park. The absence of snakes recorded from large islands undoubtedly stems from collecting bias (see text).

Discussion

Zoogeographic patterns of the herpetofauna of the ECLSP could be the result of isolation as Holocene sea levels rose from 16,000 to 6,000 years BP, stranding species on higher elevations that became present-day islands. Present distribution patterns also could have resulted from a dynamic process of extinction and colonization, or from a combination of the above factors. In addition, island size, isolation, and habitat structure likely contribute to present herpetofaunal distribution patterns. Current patterns, although shared by more than one species, may have resulted from different processes.

Typically, both island size and structural habitat diversity are positively correlated with species richness in diurnal lizards (e.g., Schoener and Schoener 1983*b*). That herpetofaunal species richness was positively correlated with island size in the ECLSP is therefore not surprising. If collecting bias is taken into consideration, more than 42% of the variance likely could be explained by island size alone. In addition, large islands had more structurally diverse vegetation, allowing potential for resource partitioning among the anoles.

Both predation (Schoener and Schoener 1978, 1982) and isolation affect population densities (Schoener and Schoener 1983*c*). Previous studies have examined the effects of isolation on lizard species richness when the distances between islands were large (Schoener and Schoener 1983*c*). Isolation may be an important predictor of species richness if the distances are great between land masses. However, none of the islands within the Exuma Archipelago are more than a few kilometers from one another, and most islands are separated by only a few hundred meters or less of relatively shallow water. As such, neither geologic nor geographic isolation is likely an important predictor of reptilian species' richness.

The Exuma Cays Land and Sea Park contains two amphibians and 13 reptiles, including two introduced species of *Cyclura* (Franz et al. 1993). Considering the relatively small size and lack of topographic elevation of these islands, species richness on some of the larger islands is high compared

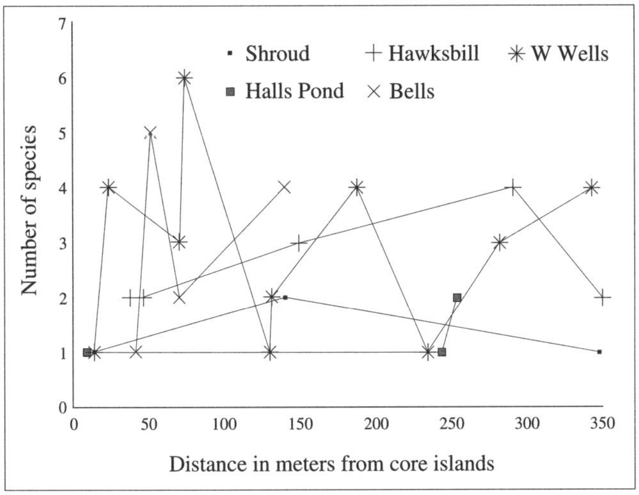

FIGURE 5. Relationship between species richness and distance to nearest core island among satellite islands within core island groups, Exuma Cays Land and Sea Park.

to other, much larger topographically diverse oceanic islands (e.g., Richman et al. 1988; Bullock and Evans 1990; Corke 1992). Relatively high species richness in the Exuma islands may be a product of the recent origin of the herpetofauna and the short distance between the Bahamas and the Greater Antilles.

Records for two species (*Osteopilus septentrionalis*, *Epicrates striatus*) are scarce. Their association with islands that have a history of post-European colonization suggests that they may have been introduced deliberately or accidentally by humans. The small fossorial snake *Typhlops* is likely present on all large islands because of substantial amounts of available habitat, although its habits make the possibility of collection during relatively brief surveys improbable. *Typhlops lumbricalis* has been collected elsewhere in the Exuma Archipelago, and occupies a fairly wide range of habitats (Schwartz and Henderson 1991).

Excluding the above-named species, three biogeographical patterns are shown by Exuma Island herpetofauna: ubiquitous species (species found on every island surveyed with at least some vegetation), core island species (species found only on large, structurally diverse islands), and widespread species (species found on most islands except the smallest, i.e., less than approximately 5 ha, vegetated islands).

Not surprisingly, the Bahamian brown anole, *Anolis sagrei*, is the only ubiquitous species. This lizard is known to be an aggressive over-water colonizer (Williams 1969) with excellent rafting abilities and inclinations (Schoener and Schoener 1983*a*; Schoener and Schoener 1984). It can reach extremely high population densities on islands (Schoener and Schoener 1980) and is found on the ground and rocks in close proximity to water. Schoener and Schoener (1983*d*) successfully introduced *A. sagrei* onto previously uninhabited islands using as few as five lizards as propagules, and we observed *A. sagrei* on White Bay Cay more than 15 years after lizards were introduced by the Schoeners. They speculated that the distribution of Bahamian brown anoles on small, vegetated islands was the product of immigration and extinction during and immediately after hurricanes.

The core island species include two arboreal *Anolis* lizards and one frog. The extensively developed organic soils and cave systems on larger islands likely would provide more moist and sheltered habitats than the sandy soils and exposed habitats on small islands. These conditions would result in more favorable retreats on large islands for the direct-developing *Eleutherodactylus*. Both anoles occupy dense vegetation and, unlike *A. sagrei*, are not typically terrestrial. *Anolis smaragdinus* is a trunk-crown species whereas *A. distichus* is a trunk species (Schoener and Schoener 1982). Both are found on larger trees and in the higher canopy forest habitats found only on the larger islands. As noted by Schoener and Schoener (1983*b*), structural habitat diversity, in addition to island area, is particularly important for predicting the occurrence of diurnal lizards on Bahamian islands. Although their rafting potential is unknown, densities of both species are much less than *A. sagrei* (Schoener and Schoener 1980). Structural habitat diversity, in addition to island size, may account for the present distribution of *Anolis* in the Exumas.

Several reptilian species (*Sphaerodactylus* spp., *Tarentola*, *Alsophis*) are found on a wide variety of different sized islands, although none of these species was the sole reptile found on any island. The inability to find these species on certain medium-

and larger-sized islands is probably an indication of secretiveness or difficulty of capture (i.e., collecting bias) rather than absence. For example, no *Alsophis* or *Tarentola* are known yet from Shroud Cay (1894 ha), although *Alsophis* has been recorded from Long Rock Cay (74 ha) and *Tarentola* has been captured on Richard's Rock (9.4 ha). No *a priori* reason why these species should not be present on Shroud Cay is apparent.

At least one of the two Exuma species of *Sphaerodactylus* is known from most islands larger than 7 ha, although no pattern in their distribution is obvious. Both species can occur on small islands (e.g., 30-ha Rocky Dundas). Inasmuch as both species can be found sympatrically, their present distributions may be the result of chance isolation as sea levels rose. Both ubiquitous and widespread island species appear to be generalists able to take advantage of ecological opportunities on a variety of islands of different sizes, assuming some minimum requirements of habitat and prey base are met.

The Role of Exotic Predators

Black rats and/or feral cats presently are found on Cistern, Soldier, Halls Pond, Little Bells, and Bells cays, although cats may have died out within the last few years on Halls Pond. No population estimates are available. Human settlements were located during the Loyalist Era (late 1700s to the early 1830s) on Hawksbill, Waderick Wells, and Bells cays. Waderick Wells Cay was settled as late as 1887 (Bahamas Government Land and Surveys Department 1929) when all lands were returned to the Crown. Waderick Wells now is inhabited by park personnel (generally 1 or 2 people). A casino/resort was operated on Halls Pond Cay from the 1960s to the late 1970s (Pavlidis and Darville 1994), but the island is presently unoccupied. Cistern and Soldier cays are privately-owned, inhabited islands within ECLSP and have been extensively modified by residents.

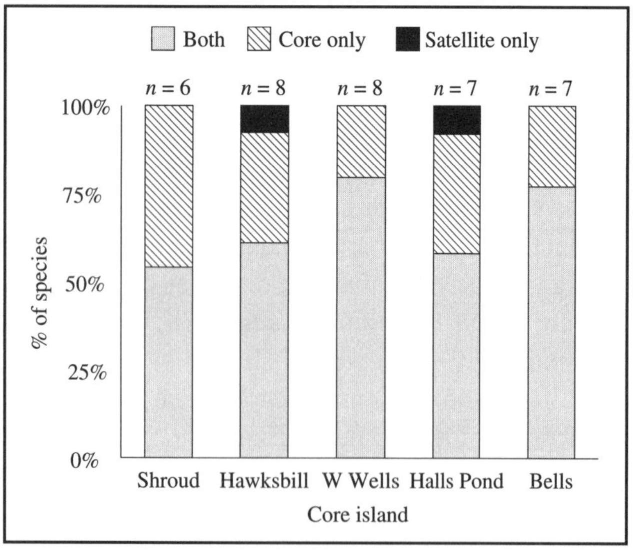

FIGURE 6. Percentage of species found only on core islands, only on satellite islands, or in both categories of islands, Exuma Cays Land and Sea Park.

Nothing is known concerning the natural predators of the Exuma herpetofauna nor what effect predation by native predators has had on species distribution. However, the negative effect of exotic predators on island populations of reptiles is well-documented (Iverson 1978; Case and Bolger 1991; Henderson 1992; Arnaud et al. 1993). Cats, in particular, seem very adept at decimating small vertebrate faunas (see Potter 1991 and references therein). In the ECLSP, cats and rats are present on several large islands, and may have been present on them and smaller islands in the past.

We recognize that our suggestion that cats have affected reptilian distribution is based on circumstantial evidence. Cat predation has been implicated in increased wariness among island populations of lizards (Stone et al. 1994). Cats are present on Cistern and Soldier cays, and both lack the conspicuous diurnal lizards *Leiocephalus* and *Ameiva* despite an abundance of available habitat. On large nearby islands such as Hawksbill and Waderick Wells, which do not have exotic predators, both lizards are abundant. At least two cats were present on Halls Pond Cay until recently. We are not aware of cats on Shroud, Bells, Little Wax, and Little Bells cays at present, although rats do occur.

Although *Ameiva* is present on Bells, Little Bells, and Halls Pond cays, its survival is probably a function of its quick and alert movements as it forages, its nighttime use of burrows, and the lack of, or decreased densities of, cats. *Leiocephalus*, on the other hand, is not very wary, is easily approached, and shelters at night under surface leaf litter and rocks. These habits would make it particularly vulnerable to both rats and cats. On the one island (Shroud) where both lizards occur despite the presence of exotics, lizard abundance is very low.

We believe the statistical failure to demonstrate adverse impacts from exotic predators on *Ameiva* and from human settlement on both species results from the small sample size ($n = 9$ islands) available for analysis. Presence/absence data do not take into account the abundance of the affected species, the abundance of the exotic predator, the temporal aspects of human settlement, and exotic predator colonization and extinction. On islands presently undergoing substantial impacts from both exotic predators and human activities (Cistern, Soldier), both *Ameiva* and *Leiocephalus* have been extirpated. Populations of *Ameiva* and *Leiocephalus* on Waderick Wells and Hawksbill may have survived exotic predators or recolonized after predator extinction.

Interpreting biogeographic patterns of the islands within the ECLSP becomes difficult when considering all possible scenarios of human and exotic predator effects over a 200-year period. On islands inhabited during the Loyalist Era, humans probably introduced many domesticated animals in addition to cats (dogs, pigs, chickens). Such introductions may have had adverse effects on island herpetofaunas. The effects of clearing and planting the land are unknown. Humans have had direct detrimental impacts on island faunas throughout the Caribbean (e.g., Steadman et al. 1984) and past and present human habitation quite possibly has affected island species richness in the ECLSP. Human disturbance exacerbates extinction processes that occur even without widespread interference (Richman et al. 1988).

A great deal of both observational and experimental data on what affects species richness of Exuma and other Bahamian island herpetofaunas has been generated (see works of Schoener and colleagues). Despite widespread knowledge of the effects of exotic predators on other island herpetofaunal species throughout the world, the potential impact of exotic species and human settlement has never been examined or even discussed for Exuma species. These questions deserve more consideration. For example, patchiness in the distribution of *Leiocephalus* (Schoener and Schoener 1983*b*) may have resulted from human perturbations instead of competitive interactions among vertebrate groups. We suggest that the absence of *Leiocephalus* and the decreased abundance of *Ameiva* may be indicators of human impact. In order to understand present herpetofaunal zoogeography in the Bahamas, the history of human activities as well as geologic history and dynamic ecological processes must be understood.

Acknowledgments

Permits to collect in the Exuma Cays Land and Sea Park were obtained from the Bahamas National Trust and from the Department of Agriculture, Nassau, Bahamas. We thank Gary Larson (Bahamas National Trust) and Maurice Isaacs (Bahamas Department of Agriculture) for their cooperation and interest in the project; Lester Albury, Miranda Albury, Ray Darville, Jaci Darville, and Peggy Hall (Park Wardens at ECLSP) for their hospitality and support during our visits; Sandy Buckner, Peggy Hall, Mary Kay Clark, David Lee, Don Buden, Shelley Franz, Kevin Jordan, Gary Morgan, and Laurie Wilkins for assistance in collecting material and reporting observations; Sandy Buckner for data on human settlement in the ECLSP; Lora Smith for computing island areas; Jennifer Shoemaker for drawing Figures 1 and 2; and Robert Henderson, John Iverson, Max Nickerson, and two anonymous reviewers for comments on the manuscript.

Literature Cited

Arnaud, G., A. Rodriguez, A. Ortega-Rubio, and S. Alvarez-Cardenas. 1993. Predation by cats on the unique endemic lizard of Socorro Island (*Urosaurus auriculatus*), Revillagigedo, Mexico. Ohio Jour. Sci. 93: 101–104.

Bahamas Government Land and Surveys Department. 1929. Aranha's Report on the Exuma Cays. Nassau, Bahamas.

Buden, D. W. 1981. Endemism and patterns of distribution among terrestrial vertebrates in the Bahamas. Bahamas Nat. 5(2): 2–18.

Bullock, D. J. and P. G. H. Evans. 1990. The distribution,

density and biomass of terrestrial reptiles in Dominica, West Indies. Jour. Zool., London 222: 421–443.

Case, T. J. and D. T. Bolger. 1991. The role of introduced species in shaping the distribution and abundance of island reptiles. Evol. Ecol. 5: 272–290.

Corke, D. 1992. The status and conservation needs of the terrestrial herpetofauna of the Windward Islands (West Indies). Biol. Conser. 62: 47–58.

Correll, D. S. and H. B. Correll. 1982. Flora of the Bahama Archipelago, including the Turks and Caicos Islands. J. Cramer, Vaduz, Liechtenstein.

Franz, R. and C. K. Dodd, Jr. 1994*a*. Geographic distribution: *Anolis sagrei ordinatus* (Bahamian Brown Anole). Herpetol. Rev. 25: 164.

— and —. 1994*b*. Geographic distribution: *Ameiva auberi obsoleta* (Exuma Islands Ameiva). Herpetol. Rev. 25: 163–164.

— and —. 1994*c*. Geographic distribution: *Sphaerodactylus notatus amaurus* (Great Bahama Bank Ground Gecko). Herpetol. Rev. 25: 166.

— and —. 1994*d*. Geographic distribution: *Sphaerodactylus nigropunctatus nigropunctatus* (Exuma Islands Black-spotted Ground Gecko). Herpetol. Rev. 25: 166.

— and —. 1994*e*. Geographic distribution: *Leiocephalus carinatus virescens* (Exuma Islands Curlytail Lizard). Herpetol. Rev. 25: 165.

—, —, and D. W. Buden. 1993. Distributional records of amphibians and reptiles from the Exuma Islands, Bahamas, including the first reports of a freshwater turtle and an introduced gecko. Carib. Jour. Sci. 29: 165–173.

Henderson, R. W. 1992. Consequences of predator introductions and habitat destruction on amphibians and reptiles in the Post-Columbus West Indies. Carib. Jour. Sci. 28: 1–10.

Iverson, J. B. 1978. Impact of feral cats and dogs on populations of the West Indian rock iguana, *Cyclura carinata*. Biol. Conser. 14: 63–73.

MacArthur, R. H. and E. O. Wilson. 1967. The Theory of Island Biogeography. Princeton Univ. Press, Princeton, New Jersey.

Meffe, G. K. and C. R. Carroll. 1994. Principles of Conservation Biology. Sinauer Associates, Inc., Sunderland, Massachusetts.

Paulay, G. 1994. Biodiversity on oceanic islands: its origin and extinction. Amer. Zool. 34: 134–144.

Pavlidis, S. J. and R. Darville. 1994. A cruising guide to the Exuma Cays Land and Sea Park. Night Flyer Enterprises, Covington, Georgia.

Potter, C. (ed.). 1991. The impact of cats on native wildlife. Proceedings of a workshop held on 8–9 May 1991. Endangered Species Unit, Australian National Parks and Wildlife Service, Canberra, ACT, Australia.

Richman, A. D., T. J. Case, and T. D. Schwaner. 1988. Natural and unnatural extinction rates of reptiles on islands. Amer. Nat. 131: 611–630.

SAS Institute, Inc. 1988. SAS/STAT User's Guide. Release 6.03 Edition. SAS Institute, Inc., Cary, North Carolina.

Schoener, A. and T. W. Schoener. 1984. Experiments on dispersal: short-term floatation of insular anoles, with a review of similar abilities in other terrestrial animals. Oecologia 63: 289–294.

Schoener, T. W. and A. Schoener. 1978. Inverse relation of survival of lizards with island size and avifaunal richness. Nature 274: 685–687.

— and —. 1980. Densities, sex ratios, and population structure in four species of Bahamian *Anolis* lizards. Jour. Anim. Ecol. 49: 19–53.

— and —. 1982. The ecological correlates of survival in some Bahamian *Anolis* lizards. Oikos 39: 1–16.

— and —. 1983*a*. On the voluntary departure of lizards from very small islands, p. 491–498. *In* A. G. J. Rhodin and K. Miyata (eds.), Advances in Herpetology and Evolutionary Biology. Mus. Comp. Zool., Harvard Univ., Cambridge, Massachusetts.

— and —. 1983*b*. Distribution of vertebrates on some very small islands. I. Occurrence sequences of individual species. Jour. Anim. Ecol. 52: 209–235.

— and —. 1983*c*. Distribution of vertebrates on some very small islands. II. Patterns in species number. Jour. Anim. Ecol. 52: 237–262.

— and —. 1983*d*. The time to extinction of a colonizing propagule of lizards increases with island area. Nature 302: 332–334.

Schwartz, A. 1967. The *Leiocephalus* of the southern Bahama islands. Ann. Carnegie Mus. 39: 153–185.

—. 1968. The geckos (*Sphaerodactylus*) of the southern Bahama Islands. Ann. Carnegie Mus. 39: 227–271.

— and R. I. Crombie. 1975. A new species of the genus *Aristelliger* from the Caicos Islands. Proc. Biol. Soc. Washington 88: 305–313.

— and R. W. Henderson. 1991. Amphibians and Reptiles of the West Indies: Descriptions, Distributions, and Natural History. Univ. Florida Press, Gainesville.

Sealey, N. E. 1985. Bahamian Landscapes. An Introduction to the Geography of the Bahamas. Collins Caribbean, London.

Steadman, D. W., G. K. Pregill, and S. L. Olsen. 1984. Fossil vertebrates from Antigua, Lesser Antilles: evidence for late Holocene human-caused extinctions in the West Indies. Proc. Natl. Acad. Sci., USA 81: 4448–4451.

Stone, P. A., H. L. Snell, and H. M. Snell. 1994. Behavioral diversity as biological diversity: introduced cats and lava lizard wariness. Conser. Biol. 8: 569–573.

Williams, E. E. 1969. The ecology of colonization as seen in the zoogeography of anoline lizards on small islands. Quart. Rev. Biol. 44: 345–389.

Resumen.—Examinamos patrones de la distribución de la herpetofauna en 31 islas de los Cayos Exuma y Sea Park (ECLSP) en las Islas Bahamas como resultado de cuatro visitas intensas a las islas entre 1990 y 1993. La herpetofauna consiste de 15 especies, dos (posiblemente cuatro) de las cuales son introducidas por el hombre. La herpetofauna muestra tres patrones zoogeográficos básicos: especies con patrones de distribución ubicuos, de centro de isla, o amplios. Se halló una sola especie presente en casi todas las islas inspecciondas, tres especies fueron encontradas solamente en las islas más grandes y estructuralmente más complejas (islas núcleo), y se encontraron cuatro especies en una amplia variedad de islas grandes y pequeñas. Tres especies fueron tan raras que no pudieron ser clasificadas. Las otras dos especies probablemente estuvieron más ampliamente distribuidas en el pasado, pero parecen estar mucho más restringidas ahora. El tamaño de la isla explica de por sí el 42% de la variación en la riqueza de especies en la isla. El tamaño de la isla núcleo más cercana y su distancia a las islas satélites no influyeron en la riqueza de especies. La distribución de la herpetofauna de ECLSP es probablemente consecuencia de una combinación de eventos históricos y geológicos, en conjunto con procesos ecológicos dinámicos actuales. El impacto potencial de depredadores exóticos y la colonización por el hombre no pueden ser ignorados cuando se tratan de explicar los patrones zoogeográficos presentes.

Résumé.—Nous avons étudié la répartition de la faune herpétologique dans 31 îles de Exuma Cays Land and Sea Park (ECLSP) aux Bahamas dans le cadre de quatre visites intensives effectuées entre 1990 et 1993. La faune herpétologique se compose de 15 espèces dont deux (peut-être quatre) ont été introduites. Trois modèles zoogéographiques fondamentaux sont mis en évidence par la faune herpétologique, c'est-à-dire par les espèces dont la répartition est soit ubiquiste, soit réservée aux îles centrales, soit répandue. On ne trouve qu'une seule espèce sur presque toutes les îles, trois sur les îles les plus grandes et les plus complexes sur le plan structurel (îles centrales) et quatre sur une vaste gamme d'îles dont la taille varie de grande à petite. Trois espèces sont si rares que nous n'avons pas pu les classifier. Les deux autres espèces devaient probablement être plus répandues avant mais semblent aujourd'hui confinées à certains territoires. La taille de l'île explique à elle seule 42% de la variation au titre de la richesse des espèces. La taille de l'île centrale la plus proche et la distance entre l'île centrale la plus proche et les îles satellites n'influent nullement sur la diversité faunique. La répartition de la faune herpétologique de ECLSP résulte vraisemblablement d'une série d'événements géologiques et historiques associés à des manifestations écologiques dynamiques permanentes. L'impact potentiel des prédateurs exotiques et des établissements humains ne saurait être sous-estimé pour expliquer le modèle zoogéographique actuel.

Dodd, C. K., Jr. and R. Franz. 1996. Species richness and biogeography of the herpetofauna in the Exuma Cays Land and Sea Park, Bahamas, p. 359–369. *In* R. Powell and R. W. Henderson (eds.), Contributions to West Indian Herpetology: A Tribute to Albert Schwartz. Society for the Study of Amphibians and Reptiles, Ithaca (New York). Contributions to Herpetology, volume 12.

Declining Amphibian Populations in Puerto Rico

Rafael L. Joglar[1,2] and Patricia A. Burrowes[2]

Abstract.—In the last 20 years, several species of Puerto Rican amphibians have experienced drastic population declines. The results of eight years of field work (1986–1994), which included long-term population studies, are presented in order to document the current population status of all species of Puerto Rican frogs in the genus *Eleutherodactylus*. *Eleutherodactylus karlschmidti*, *E. jasperi*, and *E. eneidae* have not been seen or heard since 1974, 1981, and 1990, respectively, and could be extinct. Three of seven populations of amphibians that were studied on a long-term basis were found to be declining. Five other populations have become extinct in two of our study sites in the last eight years, and no recolonization has occurred. Ten (55.6%) of the 18 endemic species of Puerto Rican amphibians are declining. Some trends have been observed. The declining and disappearing species are highly specialized (morphologically and/or ecologically) and occur at high elevations. Paradoxically, El Yunque, in eastern Puerto Rico, the best-conserved forest on the island, seems to be more affected than other localities.

Introduction

The widespread decline of amphibian populations has become apparent to many herpetologists (Barinaga 1990; Wyman 1990; Wake 1991; Wake and Morowitz 1991; Tyler 1991; Burrowes and Joglar 1991; Joglar 1992; Carey 1993; Vial and Saylor 1993; Blaustein et al. 1994*b*; Pounds and Crump 1994). That the decline of amphibians is an artifact of sampling and that it could be explained by natural population fluctuations has been suggested (Wake 1991; Pechmann et al. 1991). Others (Wake 1991; Pechmann and Wilbur 1994) have suggested that the disappearance of some species of amphibians corresponds to periods of inactivity of the species. Herpetologists concerned with these issues have recommended that long-term studies be initiated of amphibian population patterns with consideration of metapopulation dynamics (Pechmann and Wilbur 1994). Additional compilation of information on habitat type, quality, and degree of fragmentation, and the use of historical data to understand past versus present situations also has been recommended (Pechmann et al. 1991; Wake 1991; Blaustein et al. 1994*b*; McCoy 1994; Pechmann and Wilbur 1994).

Puerto Rico is no exception to the global trend in amphibian disappearance. In the last 20 years, some species of amphibians in Puerto Rico have experienced drastic population declines, and some of these species could be extinct, endangered, or threatened (Burrowes and Joglar 1991; Joglar 1992; Vial and Saylor 1993). Some of the declining species have not been seen in recent years despite the fact that scientists have searched sites where they occurred in the past (Drewry 1986; Moreno 1987; Burrowes and Joglar 1991; Hedges 1993).

The golden coqui, *Eleutherodactylus jasperi*, has been listed as a threatened species since 11 November 1977 under the United States Endangered Species Act. This species, as well as *E. karlschmidti* and *E. eneidae*, was declared threatened by the Department of Natural Resources of Puerto Rico in 1985. In 1986, one of us (RLJ), while searching for these species, noted that other species that were abundant in the 1970s and early 1980s, were scarce

[1] Departamento de Biología, Universidad de Puerto Rico, Recinto de Río Piedras, Apartado 23360, San Juan, Puerto Rico 00931–3360, USA.

[2] Natural History Museum, The University of Kansas, Lawrence, Kansas 66045–2454, USA.

or had virtually disappeared from certain areas. In 1987, we started an active search for all of the potentially declining species. By 1989, it was apparent that the status of most species of amphibians in Puerto Rico was unknown. In an effort to provide some baseline information on the patterns of activity and population fluctuations of *Eleutherodactylus*, we began monitoring several populations at different localities. The purposes of this study were: (1) to survey species of Puerto Rican *Eleutherodactylus* that appeared to be declining, with an emphasis on *E. jasperi*, *E. karlschmidti*, and *E. eneidae*; (2) to determine if any of these species were extant and, if so, to survey their distribution and abundance; and (3) to conduct long-term studies of population fluctuations of five species of *Eleutherodactylus* at three different localities.

Methods and Materials

Seven museum collections and the literature were consulted in order to determine the type and historical localities for *Eleutherodactylus jasperi*, *E. karlschmidti*, and *E. eneidae*. In addition to these localities, others that provided potential habitat for these species were visited by us or other investigators during different times of the year, by day or night, depending on the biology of the species involved. Surveys of *E. karlschmidti* and *E. eneidae* were conducted during hours of reported maximum activity for each species (Drewry and Rand 1983). During diurnal and nocturnal surveys, both acoustic and visual searches for individuals were performed. In the search for *E. jasperi*, bromeliads were examined by introducing a probe between axils. Trees were climbed with Klein Tree and Pole Climbers in order to reach bromeliads above 6 m, and with an Aluminum Extension Ladder to reach those below 6 m. At each locality visited, species that were heard or seen were noted in order of abundance. Museum research and several personal interviews provided information on collection sites and dates, number of collected specimens, collectors' names, and species habitat. Data were obtained and analyzed from the following museums and institutions: American Museum of Natural History; Carnegie Museum of Natural History; Natural History Museum, The University of Kansas; Museum of Comparative Zoology, Harvard University; Museum of Zoology, University of Michigan; Universidad de Puerto Rico, Recinto de Mayaguez; National Museum of Natural History; and Museo de Biología, Universidad de Puerto Rico, Recinto de Río Piedras.

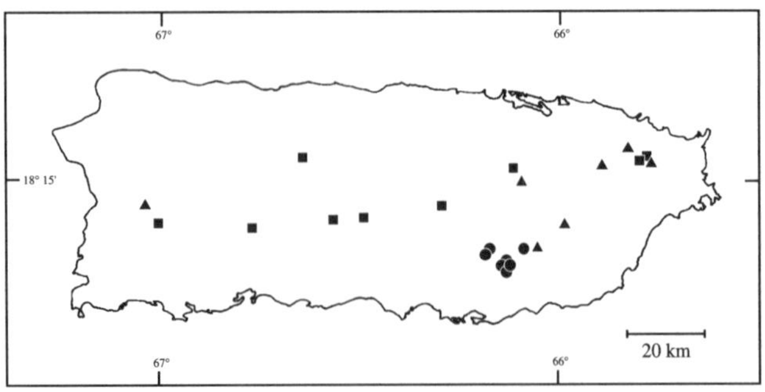

FIGURE 1. Map of locality records in Puerto Rico for *Eleutherodactylus karlschmidti* (▲), *E. jasperi* (●), and *E. eneidae* (■).

In order to determine whether some extant populations of amphibians were declining in Puerto Rico and to acquire a realistic notion of activity patterns and population fluctuations, long-term field studies on seven populations of five species of *Eleutherodactylus* were conducted at three localities. Results on one of these species, *E. cooki*, are presented in Joglar et al. (this volume) and will not be discussed in this paper. Two 50 x 3-m transects were established near the University of Puerto Rico Field Station, at an elevation of 661 m in Holdridge's Subtropical Wet Life Zone (Ewel and Whitmore 1973) at the Luquillo Experimental Forest (hereafter referred to as El Yunque) in eastern Puerto Rico. Relative abundance of *E. coqui* and *E. portoricensis* on these transects was studied for five years. Relative abundance was calculated as a ratio of the total number of individuals seen active over the area of the transect. In each transect survey, amphibians were counted, identified to species, and categorized by sex and age. These two populations were surveyed 100 times between February 1989 and June 1994. The surveys were conducted from 2000–2200 h, twice a month from 1989 to 1991, and once a month from 1992 to 1994. These populations were not surveyed in July 1989, June 1991, December 1992, and April–May 1993.

In June 1990, three 25 x 6-m transects were established near the Mt. Britton Trail at an elevation of 850 m in a cloud forest, within Holdridge's Subtropical Lower Montane Wet Life Zone (Ewel and Whitmore 1973) at El Yunque in eastern Puerto Rico. On these transects, the relative abundances of four species (*Eleutherodactylus coqui*, *E. portoricensis*, *E. gryllus*, and *E. unicolor*) were studied for four years. Relative abundances of *E. gryllus* and *E. unicolor* were calculated as a ratio of the total number of individuals heard, rather than seen, in the transect. These four populations were surveyed 71 times from June 1990 until June 1994. Surveys were conducted from 2200–2400 h, two or three times a month from June 1990 to March 1992, and once each month from April 1992 to June 1994. These populations were not surveyed in July and December 1992 and April and May 1993.

To determine the relationship between relative abundance and time, we performed regression analyses for each population. Populations were considered to be declining when probabilities for negative slopes were significant. Information on past species composition of El Yunque was obtained from tape recordings and field notes taken by RLJ in 1977, 1980, and 1982. These data provided a means of comparison with the results generated by the present study.

EXTINCTION OF THREE PUERTO RICAN *ELEUTHERODACTYLUS*?

Eleutherodactylus karlschmidti

The following distribution is known for *Eleutherodactylus karlschmidti*: Puerto Rico, primarily eastern mountains and in the west, at Las Vegas near Maricao. The altitudinal distribution for the species is 45–630 m above sea level (Figure 1). No *E. karlschmidti* were seen or heard in our surveys (see Appendix A for a complete list of localities visited). *Eleutherodactylus karlschmidti* was collected for the first time in 1931 and for the last time in 1974. In the 1930s and 1960s, the species was present in the eastern and western extremes of Puerto Rico (Joglar 1992). This bipolar distribution may be indicative of an island-wide distribution prior to 1930. Museum data suggest that by the 1960s the species was declining and disappearing from some of its historical localities, probably because by then it was limited to more protected areas such as El Yunque and El Verde in northeastern Puerto Rico (Joglar 1992).

More than 20 herpetologists have searched streams in El Yunque for several years without finding *Eleutherodactylus karlschmidti* (Burrowes and Joglar 1991; Hedges and Thomas 1991). We hypothesize that the extinction of *E. karlschmidti* may be related to particular aspects of its biology, e.g., limited distribution, highly specialized morphology and ecology, and its large body size. *Eleutherodactylus karlschmidti* has extensive toe webbing, a characteristic that rarely occurs in the genus, but when present, indicates a close association with streams (Lynch 1971). Only four species of West Indian *Eleutherodactylus* have toe webbing (Joglar 1989) and three of these seem to be declining or are extinct (Hedges 1993).

Eleutherodactylus jasperi

This species is known from Puerto Rico, Sierra de Cayey, including the Carite Forest (Figure 1). The species was collected for the first time on 22 May 1973 and for the last time in 1977. Most specimens reported were collected between 1973 and 1975. At that time, Drewry estimated the population of *E. jasperi* to be between 1000 and 2000 individuals for the entire Sierra de Cayey (Diaz 1984). The last time the species was seen or heard was in 1981 (R. Thomas, personal communication). No *E. jasperi* were seen or heard in our surveys (see Appendix A for a complete list of localities visited).

Several intrinsic factors may have contributed to the disappearance of *Eleutherodactylus jasperi*, including small population size, restricted distribution, highly specialized reproductive mode (ovoviviparous), specific microhabitat (water-filled bromeliads), and limited dispersal capacity owing to small size and habitat requirements. Burrowes and Joglar (1991) and Joglar (1992) discussed extrinsic factors that also might have contributed to the decline of this species; among these are: (1) acidic water in bromeliads (pH < 4); (2) hurricanes; and (3) recurrent fires.

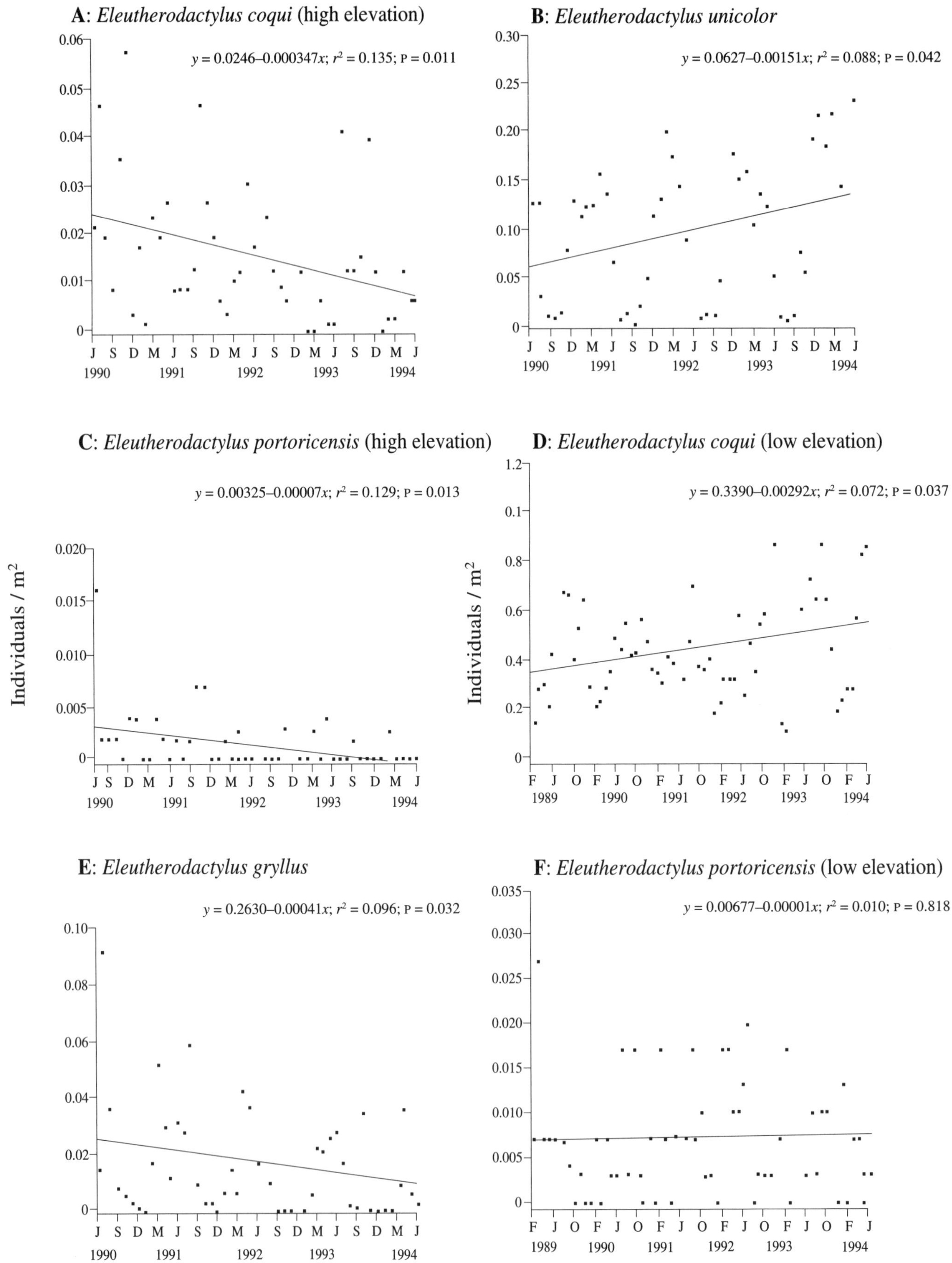
A: Eleutherodactylus coqui (high elevation)
y = 0.0246–0.000347x; r² = 0.135; P = 0.011
J S D M J S D M J S D M J S D M J
1990 1991 1992 1993 1994
B: Eleutherodactylus unicolor
y = 0.0627–0.00151x; r² = 0.088; P = 0.042
J S D M J S D M J S D M J S D M J
1990 1991 1992 1993 1994
C: Eleutherodactylus portoricensis (high elevation)
y = 0.00325–0.00007x; r² = 0.129; P = 0.013
Individuals / m²
J S D M J S D M J S D M J S D M J
1990 1991 1992 1993 1994
D: Eleutherodactylus coqui (low elevation)
y = 0.3390–0.00292x; r² = 0.072; P = 0.037
Individuals / m²
F J O F J O F J O F J O F J O F J
1989 1990 1991 1992 1993 1994
E: Eleutherodactylus gryllus
y = 0.2630–0.00041x; r² = 0.096; P = 0.032
J S D M J S D M J S D M J S D M J
1990 1991 1992 1993 1994
F: Eleutherodactylus portoricensis (low elevation)
y = 0.00677–0.00001x; r² = 0.010; P = 0.818
F J O F J O F J O F J O F J O F J
1989 1990 1991 1992 1993 1994

Eleutherodactylus eneidae

The distribution of this species is the interior uplands of Puerto Rico, 303–1152 m above sea level (Figure 1). The species was collected for the first time on 11 April 1947, and the last collecting record is on 4 May 1984. Although two acoustic records are known from 1990 at El Yunque and the Cordillera Central (Hedges 1993; J. Rivero, personal communication), we have neither seen nor heard the species during our 1986–1994 surveys (see Appendix A for a complete list of localities visited). Because the historical distribution for *Eleutherodactylus eneidae* was more extensive than those of *E. jasperi* and *E. karlschmidti* and little is known about the species' biology, its disappearance is difficult to explain.

Because, despite search efforts, *Eleutherodactylus karlschmidti*, *E. jasperi*, and *E. eneidae* have not been seen or heard since 1974, 1981, and 1990, respectively, we believe that they are extinct. If any of these species is still extant, it is extremely rare and has experienced dramatic population declines.

Other Declining Amphibians: What Are the Long-Term Studies Telling Us?

Three of the four populations monitored at the cloud forest locality were found to be declining, *Eleutherodactylus coqui*, *E. portoricensis*, and *E. gryllus* (Figure 2). The following populations were not declining: *E. unicolor* at the cloud forest locality, and *E. coqui*, and *E. portoricensis*, both of which occur at the University of Puerto Rico Field (UPR) Station locality (Figure 2).

We cannot explain the causes of these declines. The fact that the only species that is not declining at the high-elevation locality (*Eleutherodactylus unicolor*) occurs under mosses, rocks, and roots suggests that factors which are causing declines in the cloud forest at El Yunque might affect mostly amphibians which are exposed above ground. Because fog, which is common in this type of forest, effectively concentrates pesticides and other contaminants (Pounds and Crump 1994), the possibility of toxic contaminants of air should be examined. In addition, scientists should be encouraged to search for other causes, such as an increased exposure to ultraviolet radiation (Blaustein et al. 1994*a*).

FIGURE 2 (opposite page). Relative abundance over time of six populations of four species of *Eleutherodactylus* at two localities in El Yunque, Puerto Rico.

Other Declining Amphibians: The Case of Extinct Populations of Extant Species

Analysis of tape recordings and field notes taken by one of us (RLJ) in 1977, 1980, and 1982 provided data to document the extinction of five populations of three species in our study areas. In the UPR Field Station (661 m), *Eleutherodactylus wightmanae*, common during the 1980s, was last heard during our surveys in November 1991. In addition, a population of *E. richmondi* observed and recorded in 1980 and 1982, never was found during our surveys. Thus, we assume it became extinct sometime before 1986. At the Mt. Britton locality, three populations became extinct: *E. wightmanae* after February 1992, *E. richmondi* after July 1980, and *E. locustus* after July 1990.

None of the sites where these local extinctions occurred has been recolonized by the disappearing species. Although some of these extinctions are relatively recent, occurring only 28 months ago, some occurred eight years ago, and none are likely to represent natural population fluctuations. Long-term population studies of some species of Puerto Rican *Eleutherodactylus* have demonstrated that activity fluctuates with changes in air temperature and precipitation (R. L. Joglar and P. A. Burrowes, unpublished data). Rather than undergoing long periods of inactivity, the species remain active all year. The fact that reproduction also fluctuates with temperature and precipitation suggests that population size could be affected by seasonal patterns.

Travis (1994) presented a model ($P = 0.25\ [1 - \exp(-\pi p d_c^2)]$) to determine the probability (P) of recolonization of a locality from which a population has become extinct. In this model, the probability that recolonization will occur in a single generation depends on two factors: (1) the probability that migrants will move from a nearby population (set to 25%), and (2) the probability that the site is attainable, which in turn, depends on the density of populations in the area (p) and the critical movement distance (d_c) of the organism. We believe that the d_c is particularly small for the declining anurans in Puerto Rico, not only because of low vagility of amphibians in general (Blaustein et al. 1994*b*), but because most of them have specific habitat requirements and may encounter intolerable gaps in their displacements. In addition, if the populations of these species are declining at many other sites in El Yunque, the p-value in Travis' equation would be smaller, and thus, the probability of recolonization of a site would be low.

What is happening at El Yunque?

Whereas El Yunque is considered the best-conserved forest reserve in Puerto Rico, the extinctions of *Eleutherodactylus karlschmidti* and *E. eneidae* and the declines of *E. richmondi*, *E. locustus*, *E. portoricensis*, *E. gryllus*, and *E. wightmanae* are difficult to explain. Even more difficult is an explanation of why *E. richmondi* and *E. wightmanae* are almost extinct in El Yunque, but are still present in the Sierra de Cayey and Cordillera Central (Joglar 1992). Ariel Lugo (personal communication) has suggested that perhaps some of these declining species are not capable of coping with mature forest conditions. The evolution of these species on a Caribbean island certainly would suggest that some endemic species may be better adapted to naturally disturbed forests (e.g., hurricane effects) than to mature stable forests. If this is the case, we might expect a population recuperation sometime after a forest disturbance, such as a hurricane. However, five years after Hurricane Hugo caused considerable damage in many areas of El Yunque, only one population (*E. portoricensis* at the UPR Field Station) has shown indications of recovery.

Biologists working in Puerto Rico with other vertebrates are reporting similar patterns. Several species of birds are declining at El Yunque while their populations in other parts of the island are stable or declining at lower rates (Carlos Delannoy, personal communication). A possible explanation of this phenomenon could be the geographical isolation of El Yunque from other forested areas on the Island (Francisco Vilella, personal communication). Because of the intensive urban and industrial development around El Yunque and the lack of natural corridors to other forested mountains, the possibilities of recolonization after local extinctions are limited. In the case of amphibians, for reasons explained previously, the possibility of recolonization from other parts of the island is probably nonexistent.

Pechmann and Wilbur (1994) suggested that scientists investigating amphibian declines should test one of two hypotheses. If the declines were occurring in pristine habitat the null hypothesis should be that they resulted from natural fluctuations. If the declines could be associated with human impact, the null hypothesis should be that amphibians were not affected more than other organisms. Here we report on a situation that is not that simple. El Yunque represents a human view of pristine forest. However, amphibian populations are declining at El Yunque and not recovering as we predicted they might from natural fluctuations. Perhaps our concept of pristine forest is flawed biologically; i.e., some of these areas may be affected in ways that are not obvious to humans. At least one other group of vertebrates (birds) seems to be declining at El Yunque. Among the possible causes of these declines may be contaminants, and natural and human-induced habitat fragmentation. We suggest that surveys be conducted on other organisms to determine if similar declines are occurring on a broader biological spectrum.

Trends and Recommendations

Some common denominators exist for all amphibian declines in Puerto Rico. All extinct and declining species of *Eleutherodactylus* are highly specialized (morphologically and/or ecologically) and occur at high elevations (Figure 3). El Yunque, in northeastern Puerto Rico, seems to be the most seriously affected geographic area on the island.

Our data document the decline of eight species of *Eleutherodactylus* in Puerto Rico. Several of these are already extinct. Other investigators have reported declines in additional species. The Department of Natural Resources has reported that *E. cooki* and *E. hedricki* are declining in some areas and has listed *Bufo lemur* as endangered (Moreno 1991). No quantitative data exist for five species: *E. antillensis*, *E. brittoni*, *E. cochranae*, *E. monensis*, and *Leptodactylus albilabris*. Except for *E. monensis*, these species are abundant ecological generalists with lowland-wide distributions on Puerto Rico, and thus, probably are not declining. *Eleutherodactylus monensis* is endemic to Mona Island, a 56.1-km^2 xerophitic island west of Puerto Rico. We confirmed the presence of this species on Mona from 1986 to 1992, but we have no data on its current status. Although the presence of a species in a locality has been used as evidence that the species is not declining (Hedges 1993), we think that this approach is inappropriate because it could lead to erroneous conclusions, particularly when dealing with island populations. Because Mona is a small island and *E. monensis* is restricted to a small area of appropriate

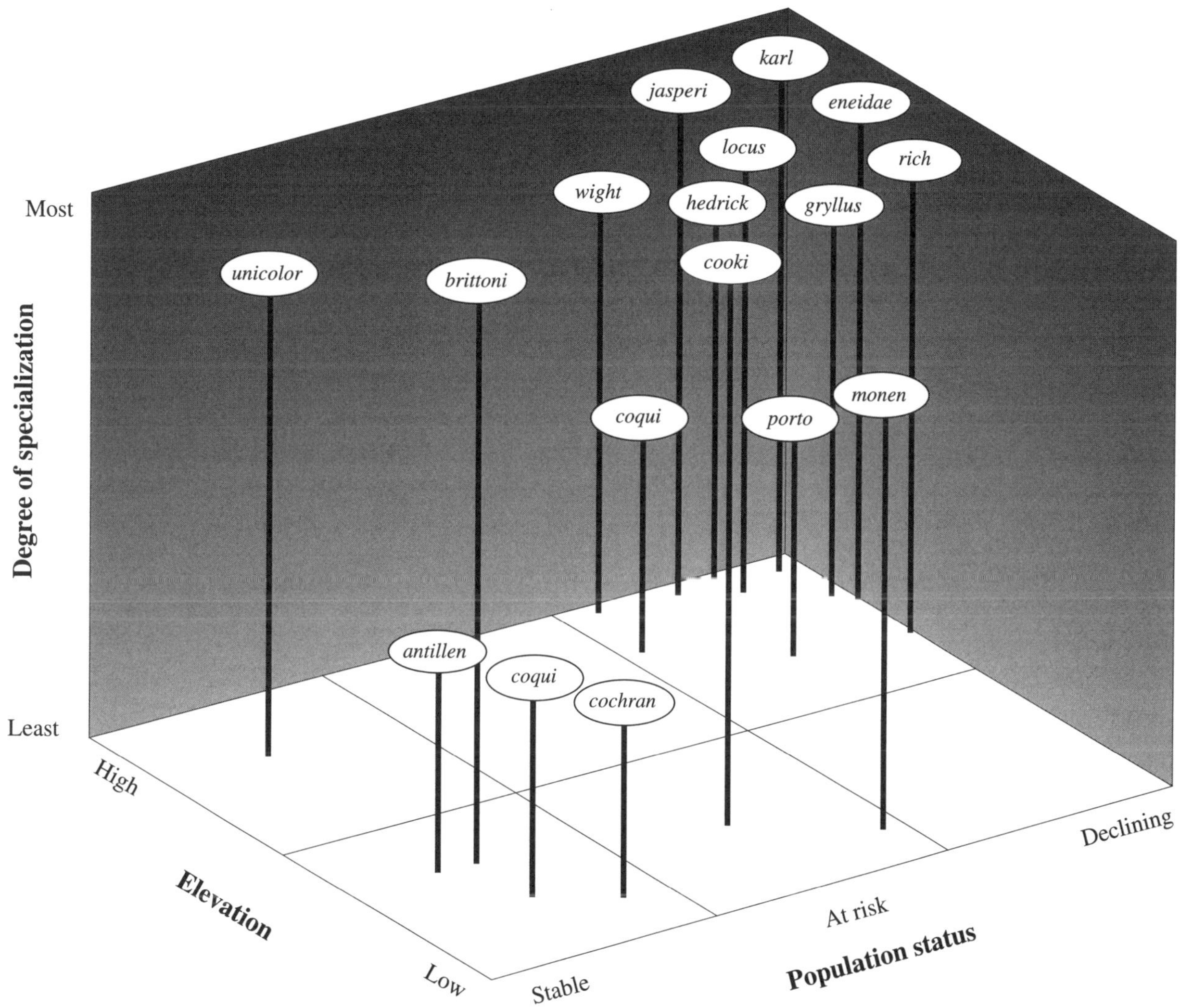

FIGURE 3. Elevation and degree of specialization trends among declining amphibian populations of *Eleutherodactylus* in Puerto Rico. Populations of *E. coqui* from high and low elevations are indicated separately. Abbreviations: *E. antillensis* (*antillen*), *E. cochranae* (*cochran*), *E. hedricki* (*hedrick*), *E. karlschmidti* (*karl*), *E. locustus* (*locus*), *E. monensis* (*monen*), *E. portoricensis* (*porto*), *E. richmondi* (*rich*), and *E. wightmanae* (*wight*).

habitat, we consider the species to be at risk and in need of monitoring.

CONCLUSIONS

Among 18 endemic species of amphibians in Puerto Rico, three (16.7%) are extinct, seven (38.9%) are declining, two (11.1%) are at risk, and six (33.3%) probably are stable. All species found to be declining in our surveys, except for *Eleutherodactylus coqui*, also are declining in other parts of the island (Drewry 1986; Moreno 1987; Burrowes and Joglar 1991; Joglar and Burrowes 1991; Moreno 1991; Woolbright 1991; Joglar 1992). Because of its economic and political stability, Puerto Rico is one of the most forested and better-conserved of the Greater Antillean islands. If 55.6% of the Puerto Rican species of amphibians are declining, we hypothesize that in Cuba, Jamaica, and Hispaniola, a similar or probably greater percentage of species may be declining. In general, extinctions are more common on islands than on continents (Pregill 1981; Williamson 1983; Olson 1989); thus, declining amphibian populations in Puerto Rico or any other island should not be underestimated. Our responsibility as scientists who are aware of these declines is to: (1) notify government agencies and demand conservation actions, research funding, and education; and (2) collaborate with other scientists within the region to try to identify patterns that may lead to a knowledge of cause and effect.

ACKNOWLEDGMENTS

Numerous people and institutions assisted generously in various aspects of this research. For providing information on specimens collected, dates, and localities we are grateful to the following persons and their institutions: Charles W. Myers (American Museum of Natural History), the late C. J. McCoy (Carnegie Museum of Natural History), William E. Duellman (Natural History Museum, The University of Kansas), José Rosado (Museum of Comparative Zoology, Harvard University), Arnold G. Kluge (Museum of Zoology, University of Michigan), Juan A. Rivero (Universidad de Puerto Rico, Recinto de Mayaguez), W. Ronald Heyer (National Museum of Natural History, Washington, D. C.), and Manuel Vélez (Museo de Biología, Universidad de Puerto Rico, Recinto de Río Piedras), and Yale University Peabody Museum of Natural History. Juan A. Rivero (UPR, Mayaguez), Hector Quintero (Universidad Interamericana at San Germán), and Peter Ortiz (Department of Natural Resources) provided interesting discussions as well as information on specimens collected and localities. We are grateful to all our research students at UPR-Recinto de Río Piedras for their assistance in the field. In particular we thank José García, Ricardo Meléndez, Juan Morales, Carlos Morales, David Bermúdez, Janice Alemán, Neftalí Rios, Alex Puig, and Ingrid Cruz for their dedication to this project. We also thank Angel Luis Umpierre for granting permission to visit Cerro El Gato in Cayey. We are especially grateful to William E. Duellman, Linda Trueb, and two anonymous reviewers for their comments on the manuscript, and to Linda Trueb, Helio da Silva, and Errol Hooper for figures. This research was funded by: (1) Fondos Institucionales para la Investigación, Decanato de Estudios Graduados e Investigación de la Universidad de Puerto Rico–Recinto de Río Piedras (from 1987–1989 and 1991–1994); (2) United States Fish and Wildlife Service through a collaborative agreement with La Fundación Puertorriqueña de Conservación (from 1991–1992); (3) División de Ecología Terrestre (UPR) through its LTER program (from 1989–1991); (4) Proyecto Coqui, which provided funds for one student's summer salary and materials and equipment. This project was supported by the Departamento de Biología, Universidad de Puerto Rico–Recinto Universitario de Río Piedras by providing release time and logistics (from 1987–1994).

LITERATURE CITED

Barinaga, M. 1990. Where have all the froggies gone? Science 247: 1033–1034.

Blaustein, A. R., P. D. Hoffman, D. G. Hokit, J. M. Kiesecker, S. C. Walls, and J. B. Hays. 1994. UV repair and resistance to solar UV-B in amphibian eggs: a link to population declines? Proc. Natl. Acad. Sci. (USA) 91: 1791–1795.

—, D. B. Wake, and W. P. Sousa. 1994. Amphibian declines: judging stability, persistence, and susceptibility of populations to local and global extinctions. Conserv. Biol. 8: 60–71.

Burrowes, P. A. and R. L. Joglar. 1991. A survey of the population status and an ecological evaluation of three Puerto Rican frogs, p. 42–46. *In* J. A. Moreno (ed.), Status y Distribución de los Anfibios y Reptiles de Puerto Rico. Publ. Cien. Misc. no. 1. Depto. Rec. Nat. Puerto Rico, San Juan, Puerto Rico.

Carey, C. 1993. Hypothesis concerning the causes of disappearance of the boreal toads from the mountains of Colorado. Conserv. Biol. 7: 355–362.

Diaz, C. 1984. Recovery Plan for the Golden Coqui (*Eleutherodactylus jasperi*). United States Fish Wildl. Serv., Atlanta, Georgia.

Drewry, G. E. 1986. Golden Coqui Recovery Survey and Brief Status Evaluation of Five Other Puerto Rican *Eleutherodactylus* Species. Trip Rept., Caribbean Field Off. United States Fish Wildl. Serv., Boquerón, Puerto Rico.

— and A. S. Rand. 1983. Characteristics of an acoustic community: Puerto Rican frogs of the genus *Eleutherodactylus*. Copeia 1983: 941–953.

Ewel, J. J. and J. L. Whitmore. 1973. The Ecological Life Zones of Puerto Rico and the Virgin Islands. United States For. Serv. Res. Pap. ITF-18. Inst. Trop. Forestry, Río Piedras, Puerto Rico.

Hedges, S. B. 1993. Global amphibian declines: a perspective from the Caribbean. Biodiv. Conserv. 2: 290–303.

— and R. Thomas. 1991. The importance of systematic research in the conservation of amphibian and reptile populations, p. 56–61. *In* J. A. Moreno (ed.), Status y Distribución de los Anfibios y Reptiles de Puerto Rico. Publ. Cien. Misc. no. 1. Depto. Rec. Nat. Puerto Rico, San Juan, Puerto Rico.

Joglar, R. L. 1989. Phylogenetic relationships of the West Indian frogs of the genus *Eleutherodactylus*: a morphological analysis, p. 371–408. *In* C. A. Woods (ed.), Biogeography of the West Indies: Past, Present, and Future. Sandhill Crane Press, Gainesville, Florida.

—. 1992. Status Survey of Four Species of *Eleutherodactylus*: Final Report. Rept. Carib. Field Off. United States Fish Wildl. Serv., Boquerón, Puerto Rico.

— and P. A. Burrowes. 1991. El efecto del Huracán Hugo sobre una comunidad de anfibios en El Yunque, Puerto Rico y algunas recomendaciones para la protección de las especies del género *Eleutherodactylus*, p. 47–53. *In* J. A. Moreno (ed.), Status y Distribución de los Anfibios y Reptiles de Puerto Rico. Publ. Cien. Misc. no. 1. Depto. Rec. Nat. Puerto Rico, San Juan, Puerto Rico.

Lynch, J. D. 1971. Evolutionary relationships, osteology, and zoogeography of leptodactyloid frogs. Univ. Kansas Mus. Nat. Hist., Misc. Publ. 53: 1–238.

McCoy, E. D. 1994. "Amphibian Decline": A scientific dilemma in more ways than one. Herpetologica 50: 98–103.

Moreno, J. A. 1987. Final Report: Golden Coqui (*Eleutherodactylus jasperi*) Status Survey. Puerto Rico Dept. Nat. Res. Rept., San Juan, Puerto Rico.

— (ed.). 1991. Status y Distribución de los Anfibios y Reptiles de Puerto Rico. Publ. Cien. Misc. no. 1. Depto. Rec. Nat. Puerto Rico, San Juan, Puerto Rico.

Olson, S. L. 1989. Extinctions on islands: man as a catastrophe, p. 50–53. *In* D. Western and M. C. Pearl (eds.), Conservation for the Twenty-first Century. Oxford Univ. Press, Oxford.

Pechmann, J. H., D. S. Scott, R. D. Semlitsch, J. P. Caldwell, L. J. Vitt, and J. W. Gibbons. 1991. Declining amphibian populations: the problem of separating human impacts from natural fluctuations. Science 253: 892–895.

— and H. M. Wilbur. 1994. Putting amphibian populations in perspective: natural fluctuations and human impacts. Herpetologica 50: 65–84.

Pounds, J. A. and M. L. Crump. 1994. Amphibian declines and climate disturbance: the case of the golden toad and the harlequin frog. Conserv. Biol. 8: 72–85.

Pregill, G. 1981. Late Pleistocene herpetofaunas from Puerto Rico. Misc. Publ. Univ. Kansas Nat. Hist. Mus. 71: 1–72.

Travis, J. 1994. Calibrating our expectations in studying amphibian populations. Herpetologica 50: 104–108.

Tyler, M. J. 1991. Declining amphibian populations — a global phenomenon? An Australian perspective. Alytes 9: 43–50.

Vial, J. L. and L. Saylor. 1993. The Status of Amphibian Populations: A Compilation and Analysis. IUCN/SSC DAPTF Working Doc. no.1. Corvallis, Oregon.

Wake, D. 1991. Declining amphibian populations. Science 253: 860.

— and H. J. Morowitz. 1991. Declining amphibian populations — a global phenomenon? Findings and recommendations. Alytes 9: 33–42.

Williamson, M. 1983. Island Populations. Oxford Univ. Press, Oxford.

Woolbright, L. 1991. The impact of Hurricane Hugo on forest frogs in Puerto Rico. Biotropica 23: 462–467.

Wyman, R. 1990. What's happening to amphibians? Conserv. Biol. 4: 350–352.

APPENDIX A. A complete list of the localities that were visited in the search for *Eleutherodactylus jasperi*, *E. karlschmidti*, and *E. eneidae* since 1986. Legend: T = type locality; H = historical locality (where the species has been found in the past); PH = potential habitat (appropriate habitat for the species is present); number of visits; "a" indicates surveys conducted by Joglar and Burrowes in 1986–1994, "b" marks information collected by Drewry in 1986, and "c" denotes data from a survey by Moreno in 1987. In all surveys the number of observed or heard frogs from each of these species was zero.

Eleutherodactylus karlschmidti.—Sierra de Luquillo: El Yunque Forest (T); Río Icacos (PH, 3a); Rte. 191, km 19.2 landslide to dam (PH, 2a); Naguabo, Rte. 191 kms 22.0–28.2 (PH, 1a); Naguabo, Río Blanco (H, 2a); Naguabo, Río Prieto (PH, 3a); Naguabo, Río Cubuy (PH, 3a); Naguabo, Río Sabana (H, 1a); Rte. 186, Qbda. Sonadora (H, 1a); Rte. 186, Río Espiritú Santo (H, 1a); Rte. 191, Qbda. Juan Diego (H, 13a); Rte. 191, Qbda. La Coca (PH, 12a); Rte. 191, Qbda. La Mina, Big Tree Trail (H, 15a). Toro Negro Forest, Villalba–Jayuya: Rte. 143, kms 18.8–21.2 (PH, 5ab); Lago Guineo (PH, 3a); Recreational area (PH, 2a). Maricao Forest, Maricao: Rtes. 105, 106, 120, 357 (T, 2ab).

Eleutherodactylus jasperi.—Cayey: Rte. 7738 km 2.5, Cerro El Gato (T, 36abc); Rte. 15, km 15, Cerro La Tabla (T, 5a); Rte. 15, kms 16.1, 16.4, and 16.9, Jajome Alto (T, 33abc); Rte. 15, km 17.4, power transformers (H, 6a); Rte. 7741, km 5.9, 6 km SSE of Cayey (T, 11ac); Rte. 7741, kms 5.6, 5.7, 6.3, and 7.0 (TH, 29abc); Rte. 715, Cerro Avispa (H, 14abc); Rte. 741, km 1.2 (H, 18abc); Rte. 742, Lago Carite (PH, 13abc); Rte. 738, km 1.2, end of road (PH, 24abc); Rte. 184, kms 14–27, Carite Forest, Charco Azul, Cerro La Santa (PH, 29abc); Rte. 7740, km 6.9 (PH, 18a). Guayama: Rte. 179, kms 14.0–14.7, 18.6 (H, 24abc).

Eleutherodactylus eneidae.—Sierra de Luquillo, El Yunque Forest (H): Rte. 988, Bisley (PH, 4a); Rte. 191, kms 13.1–19.3 (H, 32a); Rte. 930, UPR Field Station (PH, 100a); Road to communication towers, Palma de Sierra Association (PH, 43a); Mt. Britton Area, Elfin Forest (H, 71a); El Yunque Rock (PH, 10a); El Yunque Towers (H, 4a); East Peak Road (PH, 8a); Rte. 9966 to El Verde (PH, 1a). Toro Negro Forest, Villalba, Jayuya (T): Rte. 143, kms 18.8–21.2 (H, 5ab); Lago Guineo (PH, 3ab); Rte. 577, Cerro Maravilla (PH, 3a); Rte. 143, Cerro La Puntita (PH, 2a); Recreational area (PH, 2a). Maricao Forest, Maricao (H): Rte. 120, km 14.0 (PH, 2a); Rte. 120, Casa de Piedra (PH, 2a); Rte. 120, km 11.8 (PH, 2a).

Resumen.—En los últimos 20 años varias especies de anfibios puertorriqueños han experimentado disminuciones poblacionales drásticas. Se presentan los resultados de ocho años de trabajo de campo (1986–1994) incluyendo estudios poblacionales a largo plazo, para documentar el estado actual de las especies puertorriqueñas del género *Eleutherodactylus*. *Eleutherodactylus karlschmidti*, *E. jasperi*, y *E. eneidae* no han sido observados ni escuchados desde 1974, 1981, y 1990, respectivamente, y podrían estar extintos. Tres de las siete poblaciones estudiadas a largo plazo están disminuyendo. En los últimos ocho años, otras cinco poblaciones se han extinguido en dos de nuestras áreas de estudio y ninguna de ellas ha recolonizado. Diez (55.6%) de las 18 especies de anfibios endémicos de Puerto Rico están disminuyendo. Hemos observado algunas tendencias o denominadores comunes. Las especies que están disminuyendo y las que han desaparecido, son sumamente especializadas (en morfología y en ecología) y son mayormente de montaña alta. Paradogicamente, El Yunque, la reserva forestal mejor conservada de la Isla, parece estar más afectada que otras localidades.

Résumé.—Au cours des 20 dernières années, la population de plusieurs espèces d'amphibiens portoricains a enregistré une baisse spectaculaire. Les résultats d'une étude sur le terrain de huit ans (1986 à 1994) qui englobe des études démographiques à long terme, sont présentés de manière à faire le bilan de la situation des populations actuelles de toutes les espèces de grenouilles portoricaines du genre *Eleutherodactylus*. *Eleutherodactylus karlschmidti*, *E. jasperi* et *E. eneidae* n'ont pas été vues ou entendues respectivement depuis 1974, 1981 et 1990. Il se pourrait fort bien que ces espèces aient disparu. Trois des sept populations d'amphibiens étudiées à long terme sont en voie de disparition. Cinq autres populations ont disparu de deux de nos sites d'observation au cours des huit dernières années et aucune recolonisation n'a été observée. Dix (55,6%) des 18 espèces endémiques d'amphibiens portoricains ont vu leur nombre décliner. Certaines tendances ont été observées. Les espèces dont le nombre diminue et qui disparaissent sont hautement spécialisées (morphologiquement et(ou) écologiquement) et vivent en altitude. Paradoxalement, El Yunque, à l'est de Porto Rico, qui est la forêt la mieux protégée de l'île semble être plus affectée que les autres localités.

Joglar, R. L. and P. A. Burrowes. 1996. Declining amphibian populations in Puerto Rico, p. 371–380. *In* R. Powell and R. W. Henderson (eds.), Contributions to West Indian Herpetology: A Tribute to Albert Schwartz. Society for the Study of Amphibians and Reptiles, Ithaca (New York). Contributions to Herpetology, volume 12.

The Current Status of Sea Turtles in Haiti

José A. Ottenwalder[1]

Abstract.—Aerial surveys and interviews were conducted in May and November 1982 and in May 1983 to investigate nesting activity and status of sea turtles in Haiti. Green, hawksbill, and leatherback turtles still nest in Haiti. Results indicate that nesting intensity was low and that sea turtle abundance in the coastal waters has declined noticeably during historical times. In view of the current socio-economic and political trends in Haiti, little hope of survival exists for the remaining nesting populations unless a drastic change in local attitudes toward their conservation is achieved.

Introduction

ACCORDING TO EARLY CHRONICLERS, SEA TURTLES once abounded on the coasts of Haiti. During Columbus' first historic voyage, he named Île de la Tortue off of the Hispaniolan coast to commemorate the taking of a giant turtle (Parsons 1962). Detailed accounts of marine turtles in Hispaniola and some of its offshore islands, particularly Tortue and Saona, were included by A. O. Exquemeling (or Oexmelin, often Anglicized as John Esquemeling) in his memoirs of buccaneering in the Caribbean in the second half of the 17th century (Esquemeling 1967). His descriptions included aspects of the anatomy, harvest methods, and exploitation of turtles by the European settlers, particularly the French and Spaniards. The eggs of sea turtles were harvested from the beach and the adults were taken there and at sea. In fact, Esquemeling (1967) gave excellent descriptions of the different turtles found on this island, shared by the Dominican Republic and Haiti, including remarks about their eggs and the names given to them by the Spaniards and French. His accounts suggest that hawksbill, green, loggerhead, and leatherback turtles nested on Haiti in the past. He referred to greens and hawksbills as the most common species. However, except for Île de la Tortue and Île Grand Cayemite, little historical information is available about nesting localities along the Haitian coast.

Virtually no information exists regarding the present status of sea turtles in Haiti. Haiti was not covered during the surveys conducted from 1978 to 1981 to assess populations and critical habitats of sea turtles in the Western Atlantic (Carr et al. 1980, 1982), and according to that study, information on marine turtles from the area is among the poorest for the whole Western Atlantic. The only mention of Haiti in the regional assessment of Caribbean sea turtle stocks by the Western Central Atlantic Fishery Commission (WECAF) (Bacon 1981) was the sighting of adult and juvenile *Eretmochelys* foraging at Île de la Gonâve. This chapter provides a full report on the results of surveys conducted in Haiti on behalf of the Western Atlantic Turtle Symposium (Ottenwalder 1983 provided a preliminary report, which was presented in part by Kavanaght 1984).

Methods

Aerial surveys were conducted late 1982 and 1983 to assess sea turtle nesting activity and critical habitat. Observations of the distribution and habitats of sea turtles were also recorded between 10–13 May 1982, during aerial surveys to investigate the status of West Indian manatees, *Trichechus manatus*

[1] Conservation and Management of Biodiversity Project, United Nations Development Program (UNDP/GEF), P. O. Box 1424, Santo Domingo, República Dominicana.

(Rathbun et al. 1985), and American flamingos, *Phoenicopterus ruber* (Ottenwalder et al. 1990) in Haiti. Survey duration and coverage was as follows: 10–13 May 1982 (12.5 h), entire coast; 14 November 1982 (3 h), south-western Tiburon Peninsula, from Côtes de Fer to Jérémie; 5–7 May 1983 (13.5 h), entire coastline with the exception of the north-eastern portion between Cap-Haïtian and the mouth of the Massacre River. All surveys were conducted between 0600 and 1000. Sea turtle surveys were flown at speeds of 130–170 km/h, altitudes of 50–125 m, and distances of 20–50 m offshore. During manatee surveys, the aircraft was flown at speeds of 120–200 km/h, altitudes of 150–300 m, and distances of about 500 m offshore. All flights were flown in a clockwise direction around the coast, following the contour of the shoreline, including bays, the mouths of rivers, inlets, and offshore islands. At least two observers were present on all flights, and three were present on most flights. Single engine, four-seat aircraft with the wing above the cockpit were used on all flights. Numbers and location of sea turtles, nesting crawls on beaches, and observations on the human use of coastal habitats were recorded on 1:250,000-scale topographic charts.

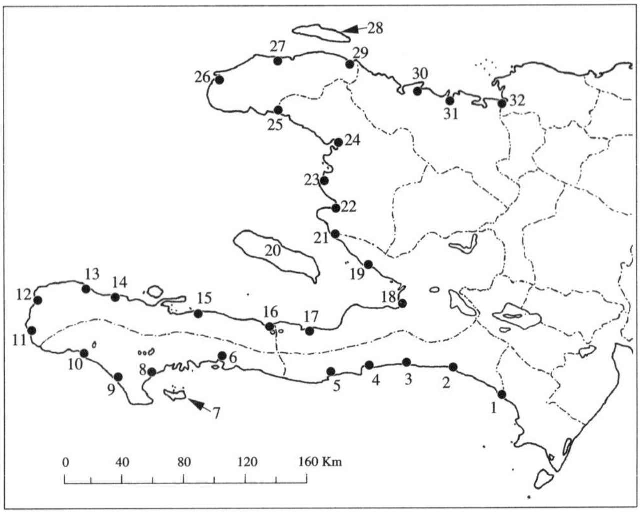

FIGURE 1. Haiti, with the coastline divided into numbered sections (see text).

To facilitate data recording and analysis, the Haitian coastline was divided into sections (Figure 1), using major towns and relevant shoreline contours as landmarks. The selected physiographic locations and section numbers used were: Anse à Pitres (1), Belle-Anse (2), Marigot (3), Jacmel (4), Bainet (5), Aquin (6), Île-à-Vache (7), Les Cayes (8), Roche à Bateau (9), Chardonnières (10), Les Irois (11),

TABLE 1. Sea turtle nesting crawls recorded during aerial surveys in May 1983. Species: *Eretmochelys imbricata* (Ei), *Chelonia mydas* (Cm), *Dermochelys coriacea* (De). See also Figure 2.

Coastal section	Number of Nests: Fresh	Number of Nests: Old	Species	Beach location and description
5	1	0	Ei	Between Laborieux and Pointe des Lataniers; some parts dune backed with open surf or fringe reef
5	0	3	Cm	Between Laborieux and Pointe des Lataniers; some parts dune backed with open surf or fringe reef
7	0	1	Ei	Pointe Est, Île-à-Vache; fringed by reef, backed by mangrove
7	1	0	Cm	Pointe Diamant, Île-à-Vache; open surf, backed by mangrove and lagoon
10	1	0	De	Near Tiburon; high energy, gray sand, backed by dunes and scrubby, dry vegetation
12	0	1	Ei	Near Anse d'Azur; pocket beach
13	0	1	?	East of Jérémie; long stretch gray sand, good slope but tide somewhat high
21	0	1	Cm	Near Freycinau west of St. Marc; small pocket beach
Total	3	7		

Dame Marie (12), Jérémie (13), Roseaux (14), Petit Trou de Nippes (15), Miragoâne (16), Petit Goâve (17), Port-au-Prince (18), Arcahaie (19), Île de la Gonâve (20), Montrouis (21), St. Marc (22), Grand Saline (23), Gonaïves (24), Anse Rouge (25), Môle St. Nicholas (26), Baie du Port à l'Ecu (27), Île de la Tortue (28), Anse à Foleur (29), Cap-Haïtien (30), Caracol (31), and Manzanillo (32). Data on nesting season, distribution, relative abundance, and exploitation of sea turtles were recorded at every opportunity during several additional visits to Haiti between 1982 and 1991.

Results

Populations, Distribution, and Nesting

The information gathered during the surveys indicated that *Chelonia* and *Eretmochelys* were the most abundant species of marine turtles in Haitian waters. We found few medium to large *Caretta* carapaces. *Dermochelys* apparently is poorly known and we were unsuccessful in obtaining relevant information about this species. Except for a single *Lepidochelys* carapace for sale in the local market, further information about their status in Haiti is lacking.

During the aerial survey of 5–7 May 1983, we recorded only 10 nesting crawls (Table 1, Figure 2). All observed tracks but one were recorded on beaches of the Tiburon Peninsula, Département du Sud. Reports also pointed consistently to the south coast as the most important nesting area of Haiti. Furthermore, 90% of the turtles observed at the surface (Table 3, Figure 3) during all three pelagic surveys were recorded in the same coastal area (Table 2). No nesting crawls were found during our aerial surveys of May and November 1982. So far no evidence supports the possibility that concentrated nesting might still occur at any Haitian locality. Available data suggest that three (maybe four, if *Caretta* is included) species of sea turtles might still nest in Haiti in very reduced numbers. Greens and hawksbills were the most consistently reported nesting turtles, but little information exists concerning nesting numbers, whereas nesting by leatherbacks is seemingly occasional. Only fragmented and uncon-

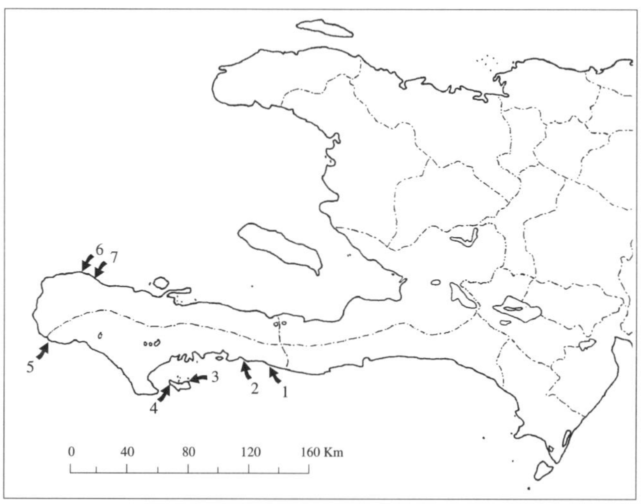

FIGURE 2. Locations of sea turtle nesting crawls recorded during the aerial survey of May 1983 (see Table 1).

TABLE 2. Dominant marine habitats in Haiti by coastal section.

Coastal Section	Description
Anse-à-Pitres to Côtes de Fer	Only isolated patches of reefs and grass flats
Côtes de Fer to Pointe Morne Rouge	Excellent feeding and nursery habitat
Pointe Morne Rouge to Pointel'Abacou, including Baie des Cayes	Extensive grass flats, reef communities, and shallow waters
Tiburon to Jérémie	Extensive offshore shelf containing large areas of reefs and grass beds
Pointe de la Patte Large to Petit Trou de Nippes	Abundant patches of shallow reefs and vast seagrass beds through the Channel of Grand Cayemite Island and inside Baie de Baraderes
Baie de Miragoâne	Scattered patches of suitable habitat
Île de la Gonâve fringing reefs and fairly	Surrounded by shallow extensive grass beds
St. Marc to Gonaïves	Good foraging areas
Canal de la Tortue to the mouth of the Massacre River, particularly between Cap-Haïtien and Bahía de Manzanillo	Good foraging areas

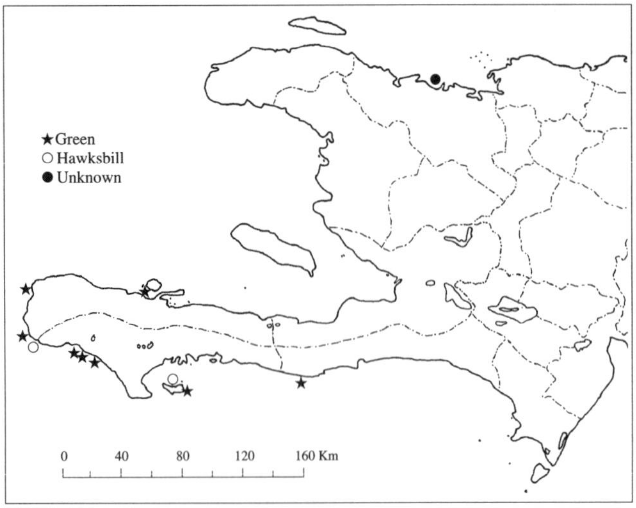

FIGURE 3. Observations of sea turtles on the surface recorded during aerial surveys in Haiti (see also Table 3).

firmed data concern the nesting of loggerheads, and their nesting appears to be only sporadic. Lack of adequate information precluded assessment of turtle harvest and the size of nesting populations. Such quantification is further complicated by the fact that all turtle species are locally referred to as "caret" and "tortue." However, an apparently safe assumption is that nesting activity is either low in frequency (hawksbill, green) or occurs as isolated events (leatherback, loggerhead).

Eretmochelys imbricata

Hawksbill nesting sites (Figure 4) were identified between Anse-à-Pitres to Belle-Anse, Laborieux and Pointe de Trois Lataniers, Île-à-Vache at Pointe de l'Est, and near Anse d'Azur (Table 1). Nesting was also reported by Kavanaght (1984) at beaches within the coastal sections of Anse-à-Pitres to Belle-Anse, Mayette to Côtes de Fer, Les Cayes to St. Jean, Pointe à Gravois to Port Salut, and Anse à Chou Chou. Nesting season is from May to August, perhaps extending to November, with a peak in June and July. Bacon (1981) reported adults and juveniles in foraging habitats around Île de la Gonâve. The accounts by Esquemeling (1967) suggested that hawksbill turtles were probably the second most abundant species in Haitian waters during the late 1700s. Today, populations seem severely reduced and, although nesting numbers are unknown, these are probably low.

Chelonia mydas

Nesting crawls of green turtles (Table 1, Figure 2) were identified at beaches between Laborieux and Pointe des Trois Lataniers, Île-à-Vache, at Pointe Diamant, east of Jérémie, and Freycinau, west of St. Marc. Kavanaght (1984) reported nesting (Figure 4) at unspecified beaches within the following coastal sections: Anse-à-Pitre to Belle-Anse, Cayes Jacmel to Raymond, Mayette to Mouillage, Baie de Caracol, Petit Anse, Anse à Chou Chou, and Fond Larange. The nesting season apparently extends from May to October, but most nesting was reported to occur from May to August. The size of the nesting population is unknown, and no evidence of important populations resulted from any of our surveys. The accounts by Esquemeling (1967) suggested that green turtles were probably the most abundant species in the coastal waters of Haiti during the late 1700s. Today, nesting populations appear to be seriously depleted and probably are still declining.

Caretta caretta

No confirmed nesting records or evidence of any kind were obtained during our surveys, but according to Kavanaght (1984), nesting sites were presumably reported by fishermen from the following coastal areas: Anse-à-Pitre to Belle-Anse, Cayes-Jacmel to Raymond, Côtes de Fer to Mouillage, Les Cayes to St. Jean, Anse du Diable, Pointe à Gravois

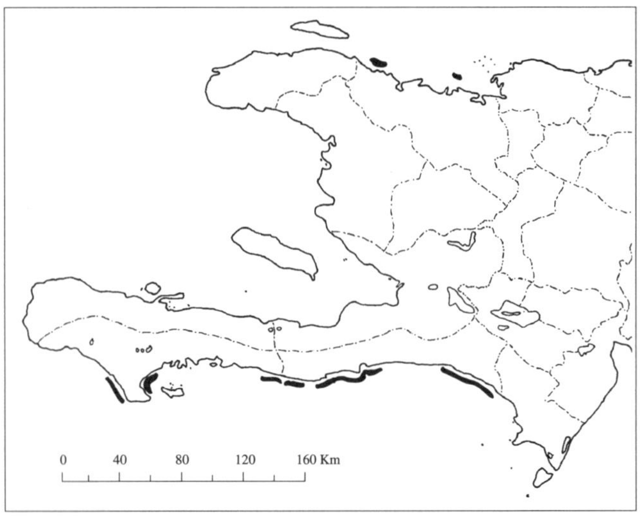

FIGURE 4. Sea turtle nesting areas (most were confirmed during ground-truth surveys between 1982 and 1991).

to Port Salut, Petite Anse and Fond Larange. Nesting presumably occurs from May to September. Nesting numbers are unknown. Trends in numbers can only be estimated from historical information. According to Esquemeling (1967), loggerheads once abounded in the waters around Haiti in the late 1700s. Today, adult nesting populations seem to be depleted and presumably are locally extirpated from some coastal areas. Loggerheads probably were and might still be more common in Haiti than in the Dominican Republic. Surveys from these two countries (Ottenwalder 1981, 1983, 1987*a*, 1987*b*) suggest that, in Hispaniola, *Caretta* is less common than the other three species and that the abundance of this species in the West Indian region of the Caribbean decreases east of Cuba.

Dermochelys coriacea

The only nesting site (Figure 4) was recorded on a high-energy beach (one open to heavy surf, with substantial portions above high tide and a moderately steep profile) near Tiburon, on the southwestern corner of the Tiburon Peninsula (aerial survey of May 1983). The nesting season probably extends from late March to May as reported from the Dominican Republic (Ottenwalder 1981). Nesting numbers are unknown. Data from our surveys and interviews suggest that nesting is only sporadic. Assuming that sea turtles, including leatherbacks, were once abundant in the waters around Haiti, current trends in nesting numbers suggest that today these populations are most likely depleted.

Lepidochelys olivacea

A single carapace of an Olive Ridley was found in a market gift shop in Port-au-Prince. We were unable to determine the origin of the specimen. Ridleys in Haitian waters are probably migrating.

Coastal and Pelagic Habitats

Haiti is located on the western portion of the island of Hispaniola and has a 1,535-km coastline of highly variable relief. The country is mountainous and steep cliffs rise abruptly, exposing a rocky coast. Such a coastal topography can be found throughout the Caribbean, Golfe de la Gonâve, and Atlantic coasts of the country. This coastal structure is interrupted in places by extensive mangrove communities, coastal lagoons, and sandy stretches. The shelf is about 5000 km² in area, but is very narrow, such that water depth frequently reaches 100 fathoms near the coastal cliffs.

A summary of observations recorded during the surveys on coastal characteristics (Table 2), particularly on nesting, foraging, and developmental habitats, are summarized below.

South Coast

From Anse-à-Pitres to Côtes de Fer, the shoreline is predominantly rocky with intermittent pockets of beach. Gravel beaches are the most common, but small sandy beaches can be found dispersed throughout. Only isolated patches of reefs and grass flats were observed. From Côtes de Fer to Pointe l'Abacou, extensive stretches of sandy beaches appear to be suitable for nesting, and are adjacent to marine habitats that are apparently suitable for feeding and the growth of young. Large mangrove communities and tidal swamps are on the mainland shore, and extensive grass flats characterize the bottoms of tidal creeks and estuaries between Pointe Morne Rouge and Pointe l'Abacou. Patch reefs are common and the one around Île-à-Vache, particularly the formation in the channel, may be the richest along the southern coast. The morphology from Pointe l'Abacou to Chardonnières is mostly rocky.

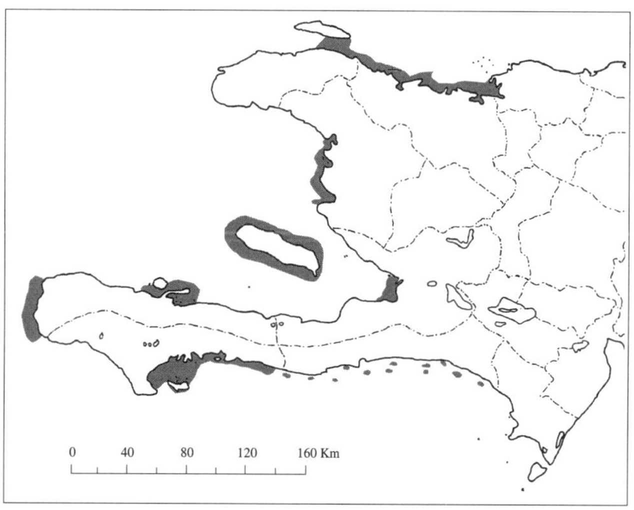

FIGURE 5. Important sea turtle foraging areas identified during aerial surveys.

Sandy beaches are located between Port Salut and Coteaux, interrupted occasionally by a few gravel beaches. Unfortunately, the sandy beaches are invariably associated with large towns, tourist developments, or small coastal villages (Figure 6). Approximately 12 km of gray-to-brown, fine sand extends between La Cahoune and Tiburon. Although the sand appears to be compact along some stretches, much of it has a fairly high profile, is backed by dunes and dry forest, and is exposed to strong surf. The area is remote and human presence is relatively innocuous.

On the westernmost tip of the Tiburon Peninsula, between the settlements of Tiburon and Jérémie, the coast is mostly rocky and dotted with pockets of sandy beaches. A large offshore shelf at 20 fathoms extends to the west for several kilometers. The area seems to be the scenario of an active fishery, and the many villages have numerous boats (of which approximately 30% are dugout canoes). Grass flats and reef patches might be present, but we were unable to survey the area in detail. To the east of Jérémie, some stretches of gray sand beach are as long as several kilometers and are presumably suitable for nesting. Unfortunately, most of these high-energy beaches are densely settled by people (Figure 6).

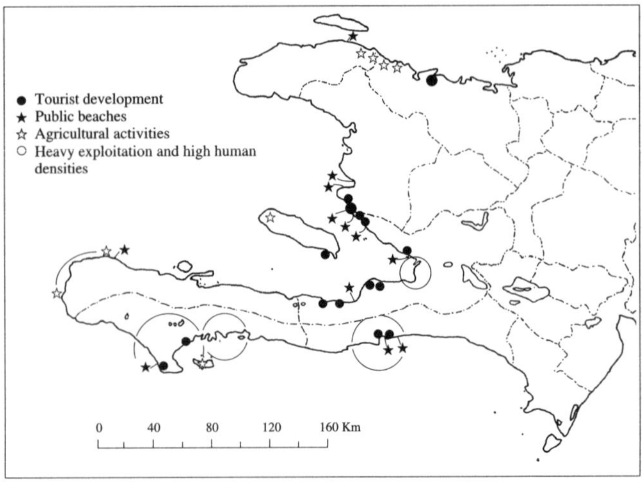

FIGURE 6. Human impact on coastal habitats and potential sea turtle nesting sites in Haiti.

West Coast (Golfe de la Gonâve)

From the area just past Pointe de la Patte to Petit Trou de Nippes, large foraging (Figure 5) and nursery habitats may be found along the coast. Here, abundant patches of shallow reefs and vast communities of grass beds and mangroves extend through the channel of Grand Cayeite Island, and inside Baie des Baraderes (Table 2). Long tidal creeks are also found here. From Petit Trou de Nippes east to Port-au-Prince, a marked alternation occurs in the topography of the coast between low, steep cliffs and the more common gravel beaches. Occasional sandy beaches and mangroves occur as well. Reefs and grass beds are less impressive in Baie de Miragoâne. Here, we examined polished carapaces of three large green turtles and of one loggerhead caught in the waters off-shore from Miragoâne a few years be-

TABLE 3. Observations of sea turtles at the water surface during aerial (pelagic) surveys along the coasts of Haiti. Species: *Eretmochelys imbricata* (Ei), *Chelonia mydas* (Cm); size class: small (S), large (L); habitat type: patch reef (PR), grass flat (GF), shallow reef (SR), deep water reef (DWR).

Coastal section	Survey date	Turtle species	Turtle size	Habitat type	Location
7	7 May 1983	Ei	S	PR	North Île-à-Vache
7	7 May 1983	Cm	S	GF	East Île-à-Vache
9	14 November 1982	Cm	L	GF	Near Port-à-Piment
9	14 November 1982	Cm	L	DWR	0.7 km south of Chardonnières
10	14 November 1982	Cm	L	GF	Between Chardonnières and Les Anglais
10	14 November 1982	Ei	S	PR	1 km southwest of Anse du Milieu
11	14 November 1982	Cm	L	GF	Baie de Irois
12	7 May 1983	Cm	S	GF	Baie de Dame Marie
14	7 May 1983	Cm	S	SR	Baie de Grande Cayemite
30	12 May 1982	?	S	PR	0.5 km north of Bord de Mer

fore 1982. Human densities increase steadily from Miragoâne to Port-au-Prince, particularly after Baie de Petit Goâve. Most beaches along the stretch show tourist development, beach houses, fishing villages, or some major towns (Figure 6). Most suitable nesting habitats from Port-au-Prince to St. Marc are also invaded by presently inactive resorts and private beach houses. Other human land uses include plantations, factories, free-roaming livestock, rock fences, and numerous private docks. Vessel and boat traffic is heavy on this coast. The areas left undeveloped represent mostly mangroves and rocky cliffs.

The southwestern shore of Île de la Gonâve is characterized by a low-profile coast with some long stretches of sandy beach that might be used by turtles for nesting. These low-energy beaches are backed by mangrove swamps, small lagoons, or even some dry forests. They invariably have small villages or dispersed houses. The northeastern end of the island is rocky, and the northern coast is predominantly covered by mangrove swamp, with the few sandy stretches used for human habitation. Île de la Gonâve is populous, surrounded by a shallow fringing reef and vast grass flats.

From St. Marc to Gonaïves, the coast is dominated by mangrove habitats with numerous tidal flats and long tidal creeks (Table 2). A fringing reef extends along the shore from St. Marc to Grand Saline. Sandy barrier beaches flank mangroves in some places. From Gonaïves to Môle St.-Nicolas the coastline is mainly rocky with steep cliffs of various heights. Pockets of beaches are occasional.

North Coast

On Île de la Tortue, suitable beaches for nesting are found only on the south coast, facing the mainland. An excellent beach of white sand with a wide, steep profile is situated on the western tip near Gros Raisinier. Much of the southern shore consists of sandy but populous beaches, and most human traffic from one end of the island to the other is on the beach (Figure 6). A fringing reef extends all along the south shore.

Some sandy beaches occur from Môle St.-Nicolas to Port-de-Paix, often just pockets, alternating with the dominantly rocky coast (Table 2). Beaches along the coast from Port-de-Paix to Cap-Haïtien are mostly sandy river banks, and appear appropiate for nesting. Several pocket beaches (small arcuate sand beaches between cliff promontories) occur between Le Borgne and Baie l'Acul. Patches of reef rise from the bottom outside the bay. A small but suitable isolated and sandy beach surrounds Rat Island, a tiny cay off l'Acul Bay. The bays of Cap-Haïtien and Caracol are largely mangroves, with some sandy stretches suitable for nesting, but most are too narrow to provide suitable nesting sites. An almost continuous fringing reef extends from Cap-Haïtien Bay to the mouth of the Massacre River on the border of the Dominican Republic. The bottom is extensively vegetated in this area.

Foraging Habitat

Observations on suitable foraging and nursery habitats for sea turtles recorded during the aerial surveys along the Haitian coast are summarized in Tables 2 and 3. The most significant foraging areas (Figure 5) were located between Côtes de Fer and Pointe l'Abacou (on the south coast) and from Tiburon to Jérémie (on the westernmost tip of the Tiburon Peninsula).

Exploitation and Management: Socio-economics and Fisheries

Sea turtle meat and eggs are highly appreciated by Haitians. When available, these can be obtained at seafood shops in Port-au-Prince. Data from interviews indicated that most of the captures are juveniles caught by divers. Official statistics on the quantity of turtles and turtle products harvested for human consumption and other uses do not exist. Small to large carapaces of hawksbill and green turtles are readily available in markets and tourist shops (Table 4). Reportedly, a large proportion of the large carapaces were on display and for sale in these establishments for several years. Many of the small and medium-sized carapaces in stock are stored in crates, and only a small fraction are actually on display in shops. Most carapaces offered by fishermen in recent years are of small or medium size. Small and medium-sized carapaces are also the most frequently sold because of their lower prices, greater availability, and convenience for air transportation.

In a visit made on November 1982 to the warehouse of a major local dealer in Port-au-Prince, we found one crate containing 20 unpolished hawksbill carapaces ranging from dinnerplate to medium size (at US $25–40 each) and 18 large bags of tortoiseshell with an approximate total weight of 600 kg, valued at US $180/kg. All carapaces were reported to have come from the southern Tiburon Peninsula. Most hawksbills and tortoiseshell were said to come from Belle-Anse and Les Cayes. I also have evidence that Haitian fishermen fish extensively in Dominican waters from the political boundary towns of Anse-à-Pitre on the south coast and Fort Libertéé and Dérac on the north coast. Hunting methods include the use of ordinary trammel nets, "folles" or turtle decoys, harpoons, and spearguns.

No information about the intensity of the turtle fisheries is available from Haitian sources. However, between 1959 and 1991, Japan imported at least 48,432 kg of tortoiseshell from Haiti (Table 5), hence at least 42,860 hawksbills were presumably taken from Haitian waters for trade with Japan alone (estimates are based on conversion factors originally developed by TRAFFIC: a single turtle yields between 0.75 and 1.5 kg of tortoiseshell in the form of 13 carapacial scutes, with an average of about 1 kg — the 27 smaller marginal scutes are less in demand and are usually not included in the average weight [see also Groombridge and Luxmoore 1989]). Kavanaght (1984) reported that the localities of Côtes de Fer, Raymond-les-Bains, Port Salut, and Coridon served as landing sites for turtles and the availability of turtle products.

TABLE 4. Sea turtle carapaces recorded in our surveys and by others (T. Tucker, E. Roet, J. Thorbjarnarson, personal communications) in markets, hotels, and tourist shops of Port-au-Prince, Petionville, and Cap-Haïtien during October and November 1983. Size class criteria from carapace length measured over the curve (in cm). Species: *Eretmochelys imbricata* (Ei), *Chelonia mydas* (Cm), and *Caretta caretta* (Cc). All large hawksbill shells had scutes removed.

Size class (cm)	Number of individuals			Total	Percent
	Ei	Cm	Cc		
20	0	0	0	0	0
20–30	40	0	0	40	12.9
30–50	81	17	2	100	31.5
50–70	14	54	0	68	21.9
70–90	31	71	0	102	32.8
90	1	1	0	2	0.6
Total	167	143	2	312	100.0
Percent	(53.5)	(45.8)	(0.6)	(100.0)	

Based on estimates from Weil et al. (1973) and the Centre d'Etudes de Geographie Tropicale (CEGET-CNRS) (1985), between 11,000 and 15,000 Haitians, largely working independently, are engaged in fishing. Only about one-third are full-time fishermen, fishing an average of 150 days per year. Most fishermen remain close to shore, with only a few venturing into deeper waters. Fishing practices are archaic. Fishermen are essentially confined to a level of economical subsistence due to, among other reasons, the limited productivity of the waters, the low efficiency of traditional fishing techniques, the lack of equipment, and the weakness of the island market.

Part-time fishermen, actually agriculturalists from the coastal zone, make up nearly two-thirds of the total. Both professional and part-time fishermen are very poor. Consequently, the purchase of tools and equipment is not feasible. They commonly form small groups to pool their resources to rent a boat, surrendering part of their catch as payment. Many full-time fishermen belong to guilds sponsored by fishing patrons. These groups disperse widely and lead a semi-nomadic life on various beaches.

Skilled fisheries activities are greatly influenced by easterly tradewinds, which are almost constant along the Haitian coasts, and by the nature of the fishing fleet. Few fishermen have motorized boats. The traditional boats of the Haitian coast have sails, oars, or both; and whose maximum fishing radius under optimal conditions is about 5 km. Canoes with oars are typically 2–10 m long, and about 80% are between 3.5 and 5 m in length. Three other types of boats are used: (1) "bois fouille," which is a dugout trunk of a tree; (2) "pris-pris," 2.5 to 3 m long, made of assembled bamboo, and maneuvered with a till or sail (these are particularly common along the north coast); and (3) "corallaire," a small boat less than 3 m in length in the shape of a canoe with a flat bottom and a square stern, and often used on coral reefs. Motorized boats are owned by fishing patrons. These are rare and usually only have 6 to 25 horsepower motors. Current estimates of the total number of fishing boats in Haiti are not available. A 1956 estimate listed 1,744 boats as follows: 1,454 canoes, 50 bois fouille, 200 pris-pris, and 40 corallaires. More recent estimates (Food and Agriculture Organization of the United Nations [FAO] 1983; CEGET-CNRS 1985) tallied 3,000 boats, of which 2,000 were canoes and 1,000 of the other three types.

Legislation

Through the early 1990s, the government agency responsible for the enforcement of fisheries regulations was the Service des Pêches, Direction des Ressources Naturelles, Ministère de l'Agriculture, des Resources Naturelles et du Dévelopment Rural (DARNDR), Port-au-Prince, Haiti. Reportedly, 17 fisheries inspectors were assigned throughout the Haitian coasts for this task: (Aquin 1, Les Cayes 1, from Tiburon to Jérémie 3, Miragoâne 1, Arcahaie 1, St. Marc 1, and from Gonaïves to Port-de-Paix 9). The only known regulations concerning conservation and management of sea turtles in Haiti are quoted (in translation) below from the Fisheries Law of 27 October 1978. As far as is known, these regulations have been totally ignored.

Article 97. Formally prohibited are: paragraph (b) to fish "the tortue," "the caret" during the months of May to October (laying season); (c) to collect the eggs of turtles of all species in the territorial waters, especially those of "caret" and of "tortue;" and (d) to capture the "sea turtles," "the carets" on the beach.

Article 122. Prohibited is: (2) the exploitation of "caret" and "turtle" meat, and their shells without an authorization from the Service of Fisheries.

Trade

Together with Cuba and the Cayman Islands, Haiti was among the main exporters of tortoiseshell in the Caribbean region (Barber 1980). Between 1959 and 1990, Haiti maintained an average supply of 1.5 tons/year (Table 5), and exports represented around 10% of all the bekko (Japanese trade name

TABLE 5. Haitian imports and exports (in kg) of bekko and other tortoiseshell between 1959 and 1991.

Year	Exports		Imports
	Japan[a]	Other countries	Cayman Islands
1959	311		
1960	0		
1961	827		
1962	2,090		
1963	899		
1964	944		
1965	687		
1966	820		
1967	1,016		
1968	1,468		
1969	1,490		
1970	1,497		
1971	1,983		
1972	1,954		
1973	2,390		
1974	678		
1975	831		
1976	1,094	90[b]	
1977	1,173		
1978	959		
1979	1,689		
1980	1,020		8,404.6[c]
1981	892	8,413[d]	
1982	1,188		
1983	1,788		
1984	1,988		
1985	2,203		
1986	2,767		
1987	2,867[e]		
1988	4,940[e]		
1989	2,427[e]		
1990	1,374[e]		
1991	178[e]		
Totals	48,432	8,503	8404.6

[a] Japanese customs statistics; Groombridge and Luxmore (1989)
[b] To Italy (Mack et al. 1979)
[c] Luxmore and Canin (1985)
[d] Originated in the Cayman Islands, re-exported from Haiti to Germany (Luxmore and Canin 1985)
[e] TRAFFIC (Japan); WCMC/CITES

for tortoiseshell derived from *Eretmochelys*) and other tortoiseshell imported by Japan from the West Indies (Table 5), and almost 4% of Japan's total bekko imports. During that period, amounts fluctuated noticeably, apparently influenced both by internal and external market demands. From 1959 to 1966, Haiti supplied Japan with an average of 800 kg/year. Exports between 1967 and 1984 increased to an average of 1.4 tons/year and, at a time during which CITES controls gradually reduced trade from other traditional sources of bekko in the region, imports from Haiti (a non-CITES nation) increased further to an average of 2.7 tons/year from 1985 to 1990. As a direct result of the socio-political crisis caused by the coup d'etat, Haitian exports to Japan dropped substantially (to 178 kg) in 1991, and halted completely in 1992. No imports have been reported by Japanese customs since (A. Ishiara, TRAFFIC Japan, personal communication). That Haitian exports were not renewed thereafter probably is due to the fact that Japan set a zero quota on the import of unworked bekko in January 1993 and finally withdrew its reservation on *Eretmochelys* in July 1994.

DISCUSSION AND CONCLUSIONS

The nesting season of sea turtles in Haiti is probably similar to that reported for the Dominican Republic (Ottenwalder 1981, 1987*a*), namely from late March to November (the peak varying by species). Local fishermen report that May to August are the most important nesting months in Haiti, with peaks in June and July. Little nesting activity was apparent during the surveys carried out in early May. However, more comprehensive data are needed on the seasonality of nesting before accurate predictions can be made about the size of nesting populations based on number of nests and the extent of the nesting season. Although the number of turtle tracks recorded during our aerial surveys in May 1983 may reflect trends in nesting activity for the past few years, meaningful estimates on relative abundance are difficult when based on a single survey.

With numerous steep cliffs and many kilometers of abrupt slopes, Haiti has a more mountainous topography than does the Dominican Republic; as a result, long stretches of high-energy, sandy beaches are not as frequent in Haiti as in the Dominican Republic. Most current habitat suitable for nesting is scattered along the coast in numerous pocket beaches. Consequently, current nesting activity probably is dispersed. This factor hampers the estimation of population size and status of sea turtle populations.

Human activity has had a very negative effect on sea turtle populations and their nesting habitat in Haiti. High human population densities probably represent the major threat. People are found even in the most isolated and remote coastal areas. Furthermore, areas that are isolated from suitable transportation, supplies, and other service infrastructures suffer from a scarcity of food, and consequently increase their dependence on the gathering of marine wildlife, including turtles. Chances that nesting females or nests will be overlooked in inhabited areas are slight. However, the fishing pressure on sea turtles probably has decreased in Haitian waters (by divers and nets) due to the loss of boats with people who left the country during the summer of 1994 as the socio-political and economic crisis caused by the military coup and embargo worsened. Tourism was an important source of income to the weak local economy until the mid 1980s. Resorts, beach houses, and other recreation developments have usurped many kilometers of excellent nesting habitat. Other development activities, such as sand removal and agriculture, also have had negative impacts on nesting habitats. Before the embargo, sand removal was a common practice. Subsistence agriculture and livestock farming also extend to beaches, because of the limited availability of fertile lowlands and valleys inland.

The contradiction between the apparent scarcity of turtles in Haitian waters and the high volume of tortoiseshell exported every year from Haiti to Japan have raised considerable skepticism concerning the origin of the Haitian tortoiseshell. Either tortoiseshell originating from other Caribbean countries was being channeled through Haiti for the trade with Japan, as was generally suspected, or foraging hawksbills are still common in Haitian waters. CITES annual reports from 1977 to 1985 also indicated exports of sea turtle products from Haiti to France (65 kg) and the United States (19 shells and 100 carvings). The imports to Haiti from the Cayman Islands of 8,405 kg of shell, later re-exported from Haiti to Germany was a reflection of the unquantified interisland trade of tortoiseshell suggested for the Caribbean by Meylan and Mack (1984). Carr et al. (1982) considered it unlikely that any but the most vestigial numbers of sea turtles would be found in the country today. Furthermore, the quantities of tortoiseshell imported to Japan from Haiti are not only large, but increased steadily from 1981 to a high of 4,940 kg in 1988. Groombridge and Luxmoore (1989) and Milliken and Tokunaga (1987) independently suggested that Haiti either provided a route through which Caribbean turtle products were exported to the rest of the world, or that Haiti increased exploitation of her own hawksbill resources.

In reference to Haiti, Carr et al. (1982), concluded that "it seems unlikely that more than the most vestigial stocks of turtles of any age group remain anywhere about the island." In the course of our surveys, we found no evidence to justify anything but complete agreement with these remarks.

Recommendations

Available information indicates that the numbers and nesting sites of sea turtles in Haiti have been seriously reduced in historical times. The status of the remaining nesting populations is, based on all evidence, critical; and the prospects for sea turtle conservation in the country are not encouraging. The following recommendations for further research and management are made: (1) Aerial and ground-truth surveys should be conducted to compile data on nesting distribution, numbers, and activity, particularly along selected coastal areas where nesting sites have been recorded in recent years, and especially along the southeastern coast of the Tiburon Peninsula between Côtes de Fer and Pointe Morne Rouge. (2) A total ban on all forms of exploitation of sea turtles in Haiti is badly needed. However, such a ban would be difficult to reconcile with the socio-economic problems currently faced by Haitians. An alternative would be to ban all commercialization of sea turtle products, and to allow only subsistence hunting outside of the nesting season. Current regulations must be reviewed, updated, and enforced. The survival of remnant populations of sea turtles will depend on the enforcement of such regulations. Steps to promote the accession of Haiti to CITES should be re-taken as soon as possible, taking advantage of international efforts to establish an organized governmental infrastructure, which presumably will include wildlife management agencies and authorities. (3) Marine parks and reserves must be created and effectively managed along the Haitian coast. Areas of particular importance for sea turtles and other critical marine resources which should receive priority are: Bahie des Cayes, including Île-à-Vache (proposed); the Presqu'ile des Baradères-Les Cayemites; Les Arcadines–Île de la Gonâve; and Bahie des Labadie to Bahie de Fort Liberté.

Acknowledgements

Financial support for the aerial surveys was provided by the United States National Marine Fisheries Service (NMFS/NOAA), the Sea Grant Program, and the Caribbean Conservation Corporation for sea turtle surveys, and by the United States Fish and Wildlife Service for manatee surveys. Logistical support and materials were provided by the Florida Museum of Natural History, University of Florida, the Haitian Institut de Sauvegarde du Patrimonie National (ISPAN), and the Département de l'Agriculture, Resources Naturelles et du Development Rural. I express my appreciation to Charles A. Woods, who accompanied me in some of the aerial surveys and provided invaluable logistical support and local contacts; and to Paul Parisky for assistance with local permits and plane arrangements. Thanks also to Akiko Ishihara, TRAFFIC (Japan), and John Caldwell, Wildlife Trade Monitoring Unit/WCMC, for providing the last recorded trade data on Japanese imports of bekko from Haiti. This chapter was drafted while the author was being supported by grants from the Wildlife Conservation Society/NYZS. Special thanks also go to the late Archie Carr for his guidance and support, and for introducing me to the conservation of sea turtles. This paper is dedicated to my old Caribbean friend and colleague, Albert Schwartz. When I was a student, he exposed me to the broad scope of zoological research, hard field work, and in 1975 he reviewed my first paper. Yo te saludo, Al.

Literature Cited

Bacon, P. 1981. The status of sea turtles stocks management in the Western Central Atlantic. UNDP/FAO, Interregional Fisheries Development and Management Programme (WECAF Component), Western Central Atlantic Fishery Commission, WECAF Studies no. 7.

Barber, J. 1980. Survey of Caribbean wildlife products. Traffic Bull. 2(2): 13–14.

Carr, A., K. Bjorndal, D. Carr, P. Carr, T. Carr, A. Meylan, P. Meylan, J. Mortimer, G. Cuellar, A. Ruiz, W. Rainey, and others. 1980. Survey and preliminary census of marine turtle populations in the Western Atlantic. Final Report to NMFS, Caribbean Conservation Corporation. San Juan, Puerto Rico.

—, A. Meylan, J. Mortimer, K. Bjorndal, and T. Carr. 1982. Surveys of sea turtle populations and habitats in the Western Atlantic. NOAA Tech. Memorandum NMFS-SEFC-91.

Centre d'Etudes de Geographie Tropicale (CEGET-CNRS). 1985. Université de Bordeaux 3. Atlas d'Haiti.

Esquemeling, J. 1967. The buccaneers of America. Dover Publ., Inc., New York.

Food and Agriculture Organization of the United Nations (FAO). 1983. Rapport de la Mission d'Identification dans les Secteurs de la Pêche Maritime et la Pisciculture en Haiti. Programme de Coopération FAO/FIDA, Rapport no. 72/83 IF-HAI 11.

Groombridge, B. and R. Luxmore. 1989. The Green Turtle and Hawksbill (Reptilia: Cheloniidae): World Status, Exploitation and Trade. CITES Secretariat, Lausanne.

Kavanaght, R. 1984. National Report for Haiti. Proceedings of the Western Atlantic Turtle Symposium (WATS), July 1983, Costa Rica. RSMAS Printing, Miami, Florida.

Luxmore, R. and J. Canin. 1985. International trade in raw sea turtle shell. Traffic Bull. 7(2): 30–39.

Mack, D., N. Duplaix, and S. Wells. 1979. The Sea Turtle: An Animal of Divisible Parts. International Trade in Sea Turtle Products. Traffic (USA) Special Report I. Washington, D. C.

Meylan, A. and D. Mack. 1984. A study of the Caribbean hawksbill turtle. WWF Monthly Rept. (October): 269–275.

Milliken, T. and H. Tokunaga. 1987. The Japanese Sea Turtle Trade 1970–1986. Traffic (Japan) Spec. Rept. Washington, D. C.

Ottenwalder, J. A. 1981. Situación, Distribución, y Biología Reproductiva de las Tortugas Marinas en la República Dominicana. Tesis Lic. Biol., Univ. Auton. Santo Domingo, Santo Domingo, República Dominicana.

—. 1983. Status of Sea Turtles Nesting in Haiti: Preliminary Results of the 1982–1983 Aerial Surveys. Rept. Western Atlantic Turtle Symposium (WATS) and the National Marine Fisheries Service (NMFS/NOAA) (unpublished).

—. 1987*a*. Conservation Status, Exploitation, and Management of Sea Turtles in the Dominican Republic. National Report for the Country of the Dominican Republic. Technical Report prepared for the Western Atlantic Turtle Symposium (WATS II) under contract with the Sea Grant Program, University of Puerto Rico (unpublished).

—. 1987*b*. The Status of Sea Turtles in Haiti: Ad-hoc National Report for the Country of Haiti. Prepared for the Western Atlantic Turtle Symposium (WATS II) under contract with the National Marine Fisheries Service, NOAA.

—, C. A. Woods, G. B. Rathbun, and J. B. Thorbjarnarson. 1990. Status of the greater flamingo in Haiti. Colon. Waterbirds 13: 115–123.

Parsons, J. J. 1962. The Green Turtle and Man. Univ. Florida Press, Gainesville.

Rathbun, G. B., C. A. Woods, and J. A. Ottenwalder. 1985. The manatee in Haiti. Oryx 19: 234–236.

Weil, T. E., J. K. Black, H. I. Blutstein, K. T. Johnston, D. S. McNorrris, and F. P. Munson. 1973. Area Handbook for Haiti. United States Govt. Print. Off., Washington, D. C.

Resumen.—La situación y actividad de anidamiento de las tortugas marinas en Haití, fueron investigadas por medio de reconocimientos aéreos y encuestas realizadas en mayo y noviembre de 1982, y en mayo de 1983. Las observaciones obtenidas indican que la tortuga verde, el carey, y el tinglar aún anidan en Haití, pero la información disponible sugiere que la intensidad de anidamiento es muy baja para las dos primeras especies, y particularmente reducida para la tercera. La abundancia de tortugas marinas en las aguas costeras ha disminuido considerablemente en tiempos históricos. En vista de las tendencias socio-económicas y políticas dominantes en Haití durante las últimas dos décadas, parece existir poca esperanza para un cambio local de actitudes hacia su protección efectiva en el futuro cercano, u por lo tanto, para la supervivencia de las poblaciones remanentes que aún anidan en las playas del país.

Résumé.—Des études aériennes et des entrevues ont été menées en mai et novembre 1982 puis en mai 1983 pour étudier les activités de nidification et la situation des tortues de mer en Haïti. Des tortues vertes, des tortues imbriquées et des tortues luth nichent toujours en Haïti. Les résultats indiquent que l'intensité de nidification est faible et que l'abondance des tortues de mer dans les eaux côtières a décliné de façon marquée à différents moments historiques. Au vu des tendances socio-économiques actuelles en Haïti, la survie des populations restantes est très compromise.

Ottenwalder, J. A. 1996. The current status of sea turtles in Haiti, p. 381–393. *In* R. Powell and R. W. Henderson (eds.), Contributions to West Indian Herpetology: A Tribute to Albert Schwartz. Society for the Study of Amphibians and Reptiles, Ithaca (New York). Contributions to Herpetology, volume 12.

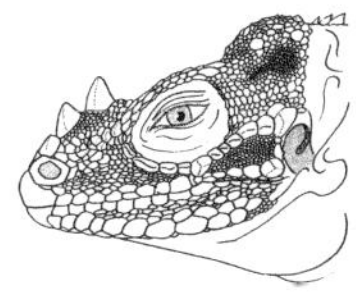

CONSERVATION STRATEGY FOR THE JAMAICAN IGUANA, *CYCLURA COLLEI*

PETER VOGEL,[1] RICHARD NELSON,[1,2] AND RHEMA KERR[3]

Abstract.—The Jamaican Iguana (*Cyclura collei*) seems to have declined dramatically during the second half of the last century, perhaps because of the introduction of the mongoose (*Herpestes auropunctatus*), changing land use patterns, and human population growth. At the beginning of the present century, the species was thought to survive only on the Goat Islands, two small keys off the coast of the Hellshire Hills. Many biologists believed the iguana to be extinct after the population on the Goat Islands disappeared in the 1940s. However, the species' survival was confirmed in 1970 and again in 1990 in the Hellshire Hills. A preliminary survey in 1990 revealed a small population in the least disturbed central and western sections of the Hills, as well as two nesting sites. The iguanas disappeared from the northern and eastern sections because of extensive charcoal production, pig hunters using dogs, and human encroachment. Iguanas rapidly vanish from the forested areas. Exotic predators, including mongooses, cats, and stray dogs, prey on iguanas. Development interests propose large-scale limestone mining and human settlements in the entire eastern half of the area in which the iguanas live. Current field efforts concentrate on emergency conservation measures. Traditional forest users are directed away from the iguana population, stray dogs are removed, and both nesting females and emerging hatchlings are protected from predation. A comprehensive program to conserve *C. collei* and its habitat is unfolding. More than 100 hatchlings have been caught in the field from 1991 to 1994. They will be used for captive propagation, and for repatriation into the wild. Mongooses will be controlled in juvenile habitats. The Jamaican Conservation Authorities are preparing a submission to the Jamaican Government recommending the declaration of the Hellshire Hills as a protected area. A management scheme is being developed which foresees sustainable forest use in buffer zones implemented through close collaboration among conservationists, traditional forest users, and other interest groups along the boundaries of the Hellshire Hills.

INTRODUCTION

THE ENDEMIC JAMAICAN IGUANA (*CYCLURA COLLEI*) is one of eight species of *Cyclura*, a genus restricted to the northern Caribbean archipelago (Schwartz and Carey 1977). The Jamaican species was thought extinct by many (e.g., Haynes et al. 1989) until recently, when its survival in the Hellshire Hills on Jamaica's south coast was confirmed. Rediscovery of this rare species has sparked many initiatives, both locally and internationally, to conserve the Jamaican Iguana and its habitat. In this chapter, we describe the decline of the species, its rediscovery in the Hellshire Hills, major threats to the surviving population, ongoing conservation efforts, and plans for future management.

We drafted a first conservation strategy in 1992, a little over one year after the start of the project (Jamaican Iguana Research and Conservation Group 1992). A Population and Habitat Viability Analysis (PHVA) further shaped our ideas. Completed in Kingston in 1993, the PHVA brought together iguana specialists from throughout the world, and established an important level of cooperation between local and international efforts to save the Jamaican Iguana and its habitat (Alberts 1993). A final report is in preparation.

[1] Department of Zoology, University of the West Indies, Kingston 7, Jamaica.

[2] Natural Resources Conservation Authority, Kingston 10, Jamaica.

[3] Hope Zoological Gardens, Kingston 6, Jamaica.

The Decline of the Jamaican Iguana

According to Sloane (1725), who visited the island in 1688, the iguana was once common in Jamaica, although its historic distribution seems to have been restricted to the drier sections near the southern coast. Three Amerindian middens on the south coast (White Marl, Bellevue, and Rodney House) contained *Cyclura* remains, whereas three on the north coast did not (Wing 1972, 1989, personal communication; Scudder 1991). Johnson (1976) tentatively identified bones from a midden on the north side of the island as belonging to *Cyclura*; however, reexamination proved the identification to be wrong (E. S. Wing, personal communication). MacPhee (1984; see also Morgan 1993) reported *Cyclura* fossils from Long Mile Cave in the Cockpit Country of northwestern Jamaica. Based on the results of more extensive excavations, however, he now suspects that the material (which is lost) was misidentified (R. D. E. MacPhee, personal communication). The southeastern range of the iguana has given rise to the name of the coastal land on which the Capital of Kingston is now located: the Liguanea Plain (MacMillan 1957).

Amerindians in Jamaica hunted iguanas for food, as indicated by the presence of iguana bones in middens (Wing 1989). This custom also seemed to have been popular among the early European colonists and enslaved Africans. Iguanas were "commonly sold for half a crown a piece in the public markets" (Sloane 1725) and "frequently served up in fricassees" (Browne 1756). Exploitation for food eventually subsided as the species declined. Today, most Jamaicans are disgusted by the thought of eating iguanas.

In 1846, Richard Hill of Spanish Town wrote about the Jamaican Iguana to Gosse (Gosse 1848): the "low limestone chain of hills, along the shore from Kingston Harbour and Goat Island, onto its continuation in Vere, is its ordinary haunt; and it is not infrequently taken in the plains between those sea-coast hills and the more inland mountains, being found in hollow trees in the pastures, where they congregate, several of them together." Hill's report suggests that by the middle of the last century the species was still relatively common and had survived not only in forested hills, but also in human-derived habitats such as pastures.

The situation appears to have changed dramatically in the following 60 years. Barbour (1910) was the first author since Gosse to write about the iguana. He stated that the species was reported to have survived in the Hellshire Hills, "but to be very rare," and in a small population on Goat Island off the Hellshire Hills coast. Barbour was not able to confirm survival in the Hellshire Hills, and eventually concluded that "the species is probably entirely extinct on the mainland" (Barbour 1922).

Thus the iguana apparently declined dramatically in the late 19th century, coincident with the introduction of the mongoose (*Herpestes auropunctatus*) in 1872 and its subsequent spread (Lewis 1949; Hoagland et al. 1989). Many authors suggested that the mongoose fed on iguana eggs and young, and so caused its extinction or near-extinction on the mainland (e.g., Barbour and Noble 1916; Lewis 1944); yet other factors might have contributed to the species' decline. The country underwent great social change after the abolition of slavery in 1837 (Beckford and Witter 1980). Many former slaves left the plantations and established small farms on marginal lands. The human population approximately doubled from 1850 to 1900, with the greatest increase occurring in the Liguanea Plain, the center of the iguana's range.

By 1940, the general belief was that the iguana had survived only on the two Goat Islands off the west coast of the Hellshire Hills peninsula (Grant 1940; Lewis 1944). The two islands are connected by a dense mangrove wetland that fringes much of the coastline. The mongoose was introduced there much later than on the mainland, probably between 1920 and 1925 by fishermen and woodcutters "fearing the very numerous yellow snakes (*Epicrates subflavus*) and disliking the iguanas" (Lewis 1944). The Natural History Division of the Institute of Jamaica made arrangements to exterminate the mongooses and establish an iguana sanctuary on the Goat Islands, but abandoned the plans upon learning that the islands were to be included in a United States Base Area. Eventually, only Little Goat Island became part of the base, whereas Great Goat Island retained its native forest.

In a frantic effort to save the lizard, 22 iguanas were caught over a six-month period in 1940 and kept in a pen in the garden of the Institute of Jamaica in Kingston. Unfortunately, the iguanas never bred in captivity, and by the end of the decade, all had died. The population on the Goat Islands also seemed to have disappeared at around the same time (Lewis 1946; T. Farr, personal communication). Many then believed that the Jamaican Iguana had become extinct.

THE REDISCOVERY OF THE JAMAICAN IGUANA

The Hellshire Hills form a rugged limestone area with fringing wetlands and beaches. They are located about 20 km to the west of Kingston and form a peninsula extending over approximately 114 km^2. Despite their closeness to Jamaica's capital and most densely populated area, the Hills have persisted as a wilderness because of their ruggedness and the lack of surface water, which made them unsuitable for both agriculture and settlement.

The vegetation consists of varying formations of tropical dry forest, the most endangered major tropical ecosystem (Janzen 1988). The Hellshire Hills forest has so far remained in an almost pristine condition in the central and western sections. About 300 species of higher plants have been identified in the area, including 53 Jamaican endemics; some of these endemics occur exclusively or nearly so in the Hellshire area. An exquisite assemblage of endemic vertebrate species also is found, and is represented by typical Caribbean taxa. Species include the Jamaican Iguana and the Jamaican Hutia (*Geocapromys brownii*). The area appears to be the last refugium of the skink *Mabuya bistriata* in Jamaica. North American migrant birds add to the diversity of native birds from autumn to spring; numerous terrestrial and marine species find refuge in either the dry forest or the fringing wetlands. The Hellshire Hills represent an outstanding example of the Caribbean natural heritage, and probably contain the most important remaining dry forest in the Caribbean archipelago.

During a biological survey of the Hellshire Hills conducted by the University of the West Indies and the Institute of Jamaica (Woodley 1971), researchers encountered feral pig hunters. Several hunters claimed that their dogs occasionally attacked and killed iguanas. Subsequently, one hunter hung an iguana killed by his dogs in a tree, and the carcass was later retrieved by researchers (Woodley 1980). This was the first confirmation of a Jamaican Iguana from the main island in the 20th century.

Not until 20 years later did the next iguana appear, but this time a live specimen was retrieved by Edwin Duffus. He had hunted pigs for many years in the westernmost section of the Hellshire Hills, inheriting the tradition from his father. Never before had he met an iguana, nor does he recall his father telling him about such encounters. However, in June 1990, Duffus decided to shift his hunting grounds to the center of the Hellshire Hills; there he saw two iguanas on his first trip. His dogs attacked one of the lizards, which fled into a log where it was partially hidden. He saved the iguana from his dogs, stuck it into a bag, carried it on his shoulders about two hours to the edge of the forest, and then rode another two hours home on his bicycle. He eventually alerted the Hope Zoo of his discovery.

Finding a live specimen led to the formation of a local Jamaican Iguana Research and Conservation Group composed of representatives from the Department of Zoology at the University of the West Indies, the Hope Zoo, the Natural Resources Conservation Authority, and the Natural History Division of the Institute of Jamaica. The group conducted an initial field survey for five weeks in 1990 (Vogel and Kerr 1992 and in press). During this survey, 23 sightings representing at least 15 individuals were made. In addition, a number of traces were found, including fecal pellets, tail impressions, rock holes used by iguanas, and pieces of shed skin. The survey discovered two nesting sites about 110 m apart (Vogel 1994). Since then, we have continuously monitored these sites and witnessed four nesting seasons from 1991 to 1994; each year, about eight females deposited eggs.

Most sightings and traces were concentrated in two areas (Figure 1). The first area in the central part of the Hills extended over about 4 km in a north-south direction. This area yielded 21 of the 23 sightings and the highest number of iguana traces, as well as the two nesting sites. The second area was in the western section and was much smaller. Within a range of about 800 m, two active holes, two tail impressions, and several fecal pellets were discovered, and one iguana was observed. Only one iguana and a few isolated droppings were found outside these two sites. Extensive field work since 1990 covering the entire Hellshire Hills has not revealed major additional iguana areas or nesting sites.

An estimation of population size has proven to be extremely difficult because of the secrecy of the iguanas' habits. For example, we appear to cross the home range of one individual along the trail to the major iguana site. We have often walked this trail several times a week during the last four years. However, only once did we see the iguana, and only a few times did we find fecal pellets. Another individual, characterized by conspicuous scars, once was seen along the same trail. A male, which we captured and tagged within the major iguana area, was only resighted after one year. One can obviously walk through the home range of an individual dozens of times without detecting its presence. We hope that future tagging efforts will lead to reliable estimates of numbers for the major sites, but the process of tagging is slow, given the difficulties of sighting and capturing iguanas. The total population size must be very low, perhaps less than one hundred adults.

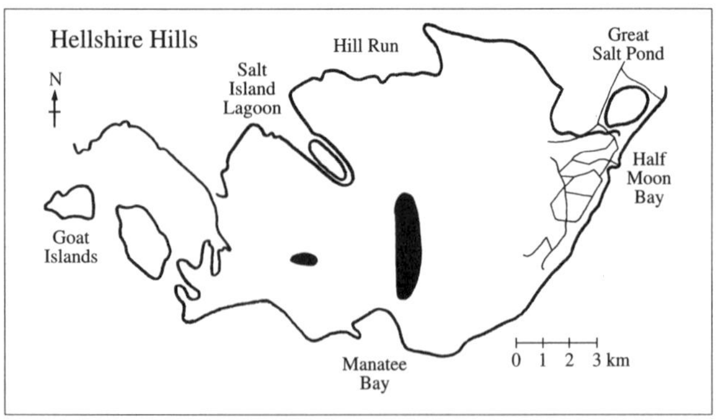

FIGURE 1. The Hellshire Hills region of Jamaica showing major iguana sites (solid) in the central and western parts, and the present road system in the east.

THREATS TO THE SURVIVING POPULATION

The surviving iguana population is threatened by exotic predators, habitat destruction by traditional forest users, and proposed development projects.

Exotic Predators

Exotic predators — especially mongooses, cats, and dogs — are probably the most important factor responsible for the low density or absence of iguanas in areas with well-preserved forest in the Hellshire Hills. The introduction of cats and dogs during a resort development on Pine Cay in the Turks and Caicos Islands extirpated a thriving population of *Cyclura carinata* within five years (Iverson 1978, 1979). Feral pigs also are able to prey on iguana nests, as reported for *C. cornuta stejnegeri* on Mona Island, Puerto Rico (Wiewandt 1977). However, we have not yet observed any such attempts by pigs in the Hellshire Hills.

Mongooses are very common throughout the Hellshire Hills, including the major iguana areas and nesting sites. Our observations suggest that, at least occasionally, mongooses prey upon eggs. During the nesting season they show vivid interest in iguana trial diggings, which they inspect frequently. In 1990, we found iguana egg shells scattered across one of the nesting sites during the incubation period. In 1994, a mongoose challenged a burrowing female, but she chased it away. After the female had left the nesting site, the same or another mongoose returned and started to dig repeatedly in the vicinity of the nest; we eventually removed the animal and protected the site with wire mesh.

Like previous authors, we suspect that mongooses prey heavily on hatchlings and young juveniles, but we have few data so far to prove it. We presently are monitoring the fate of hatchlings using telemetry. Seven hatchlings have been equipped with transmitters. One radio was retrieved in a hole with iguana skin attached to it. The hole appeared to be a mongoose retreat and contained bird bones and feathers; the iguana most likely was eaten by the mongoose.

We have observed feral cats at various places in the Hellshire Hills, including the nesting sites. Although the density of cats appears to be low, their secretive habits might confer a biased impression. Similar to mongooses, they probably feed on young iguanas.

Hunters use dogs, up to six or more per party, to chase feral pigs. The dogs roam far off the trails and are able to kill even large iguanas (Woodley 1980). As the hunters only leave the trails when they hear a pig being attacked by the dogs, iguanas may be killed without notice by the hunters.

The problem is aggravated by charcoal burners, who often bring dogs into the forest. Some of these dogs are not fed, and must find their own food. They penetrate deep into the forest, including areas populated by iguanas. Dogs also might intrude into the forest from settlements in the eastern part of the Hellshire Hills.

Dogs are able to survive in the forest for several weeks, although no feral, self-sustaining population exists. Over the last four years, we have witnessed dogs in the vicinity of the nesting sites on more than a dozen occasions. Since females congregate for nesting and guard their clutches for up to 16 days (Vogel 1994), a pack of dogs could wipe out a substantial proportion of the adult female population within a few days.

Predation pressure increases with the advance of charcoal burners into deeper sections of the forest. As feral pigs disappear from areas where char coal burners are active, pig hunters are forced into more remote sections of the forest. One pig hunter told us that he shifted his hunting range about 2 km deeper into the forest as a result of charcoal burning in the 1980s, and that he witnessed his dogs killing at least six iguanas in the new area within a few years. This area is only about 1 km away from the two known nesting sites, and today contains only one or a very few isolated iguanas.

Habitat Destruction by Traditional Forest Users

The production of charcoal is a major economic activity in Jamaica, and provides at least partial income for about 10,000 persons (Eyre 1988). Poor people in Kingston form the largest customer base for commercially produced charcoal. Rapid population growth, migration from rural areas to Kingston, and the erosion of real income over the last 20 years has led to a steady increase in charcoal demand. Given its proximity to the capital, the Hellshire Hills region is one of thc primc production centers for charcoal. The charcoal burners encroach into the Hills from the nearest roads using donkeys for transportation. Most burners chop the trees with a cutlass or an axe, but an alarming trend exists towards the use of chainsaws.

The northeastern section of the Hills has been heavily denuded. Along the north-central border, the charcoal burners have intruded 2–3 km into the forest, and are just about 1.5 km away from the major iguana concentration, including the two known nesting sites. Along the south coast, charcoal burners land from boats to cut down the coastal forest; they have not yet expanded their activities northward into the limestone hills. The central and most of the western sections of the Hellshire Hills still are covered with little-disturbed, primary tropical dry forest.

The iguanas disappear from the intact forest in front of the charcoal burners. To what degree the cutting of trees affects the iguanas is not known. In moderately exploited sections of the forest, many trees survive, and regrowth of shrubbery and small trees is rapid. Such areas might provide habitat for iguanas if the exotic predators could be controlled. Severe degradation of the forest destroys the iguanas' food resources.

Except for charcoal burners, very few other traditional forest users are degrading the forest. Along the south coast, fishermen occasionally cut sticks to build fish traps, and a few herbalists collect medicinal plants. Their impact has remained minimal.

Development Projects

During the 1960s, plans were made to develop much of the eastern half of the Hellshire Hills as a new settlement area for several hundred thousand people. Subsequently, an access road was built into the easternmost section of the Hills, and more and more houses were built. Simultaneously, large new settlements were built immediately east of the Hills. The present extent of the Hellshire Hills settlement is reflected by the road system illustrated in Figure 1.

If the development plans are carried out as planned, the new town would border on and overlap the major area presently used by the iguana population (Figure 2). Human-induced disturbances spreading from such an urban area would endanger critically the entire Hellshire Hills ecosystem. The rapid degradation of northern, eastern, and coastal sections of the forest, all of which are accessible from the nearest roads or by sea, provides a preview. Neither existing legislation nor verbal commitments by the responsible development agency have been successful in protecting these areas.

About 75% of Jamaica's surface is covered with limestone. Limestone provides the raw material for high quality tiles and other products sold both nationally and internationally. Many limestone quarries exist along the roads bordering or leading into the Hills. Plans to extend the mining activities across all of the eastern half of Hellshire exist, probably as a prelude to the expansion of the settlements. A small number of local quarries in the interior might not have much direct influence on the natural habitat and on the iguanas. However, such activities would demand new roads, which would allow charcoal burners, pig hunters, and other forest users to encroach further to the interior.

Conservation of Biodiversity in Jamaica

Until recently, remoteness, inaccessibility, and the lack of surface water have been the only protection

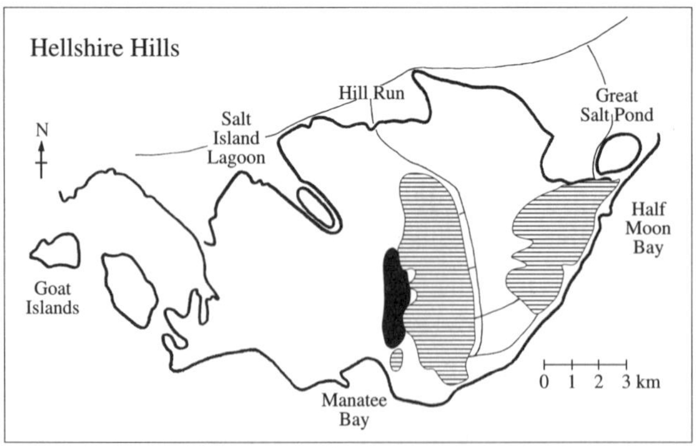

FIGURE 2. Planned housing developments in the Hellshire Hills: two large cities (hatched areas) connected by new roads. The western city would overlap with the present major iguana site (solid area).

for Jamaican natural habitats despite seemingly endless numbers of reports, recommendations, conservation plans, and even laws (International Institute for Environment and Development 1987). Ironically, most of the remaining ecologically important forests are owned by the government and protected by law as forest reserves under the Forestry Act of 1937. This act even forbids carrying a cutlass. However, the act has not been enforced. Charcoal burning, slash-and-burn agriculture, squatting, and other destructive forest uses have and are still perpetrated with little control, mainly carried out by poor people desperately seeking a livelihood. Commercial forestry and the illegal planting of marijuana add to the problem. Jamaica appears to have one of the highest rates of deforestation in the world, although considerable disagreement and confusion prevail over the widely varying estimates (Eyre 1987). World Resources Institute (1994) listed the annual rate of deforestation in Jamaica as 5.3% — the highest among all countries included in that study. Unprecedented rates of inflation and currency devaluation over the last few years may have aggravated destructive forest use.

Hope for conserving Jamaican ecosystems comes from a recent environmental awakening in Jamaican society. Public interest in environmental matters has increased greatly. New environmentally concerned organizations have been founded, and older ones have started to flourish again. The environment has suddenly become a matter of interest to the media. The business community and politicians have started to take notice. For the first time in Jamaica's history, authorities are promising the establishment of national parks, the Protected Areas Resource Conservation (PARC) project (Lee 1992). Despite these encouraging trends, however, many concerns remain.

Ongoing Conservation Activities

Emergency Conservation Measures

Since 1990, we have pursued a number of emergency conservation measures in the field. One of our major problems is the charcoal burners moving south, deeper into the forest, and approaching the major iguana site. We do not foresee eliminating this problem in the foreseeable future. Thus, our short-term policy is to establish good personal relationships with the burners and to try to convince them to move east or west rather than south. We are defending a line running east to west about 2.5 km from the northern border of the Hills. Beyond that line, we will not allow destructive forest use. This approach has been fairly successful, as many other areas are available and suitable for charcoal burning. We also monitor activities of pig hunters, and we direct hunting parties into areas without iguanas. These measures demand extensive presence in the field, from 100 to 200 days each year since 1991.

Other emergency measures include the protection of the iguanas from exotic predators. We patrol the major site at least twice a week and search for stray dogs. If any are present, we remove them. During the four-to-six-week nesting and hatching season, we monitor the nesting sites continuously.

Iguana Field Study

A field study on the biology of the iguana is being undertaken by Richard Nelson for a postgraduate thesis at the University of the West Indies. The work focuses on aspects relevant to the species' conservation. A systematic assessment of the habitat quality across the entire Hellshire Hills and the Goat Islands has been completed. Iguanas are captured for measurement and individual marking. Sightings are recorded using a Global Positioning System. The reproductive success of females is monitored by observing clutch deposition. After the emergence of hatchlings, the clutch is located by digging along the emergence tunnel. This permits us to determine clutch size and the number of successful and unsuccessful eggs. All fecal pellets are collected and their contents will be analyzed. Since the last hatching season in September 1994, hatchlings are being tracked using radio telemetry. Nesting females will be tracked in 1995. The results of these studies will be available within the next two years.

Captive Breeding Program

A captive breeding program is being developed in cooperation with zoological parks in the United States, including zoos in Fort Worth, San Diego, and Indianapolis. The program aims to safeguard against a catastrophic loss in the wild population, and to serve as a genetic reservoir. Since 1991, we have captured approximately 50% of the hatchlings as they emerged from their nest burrows (Vogel 1994). They are being reared at the Hope Zoological Garden in Kingston. Twelve individuals have been sent to cooperating zoos in 1994, and a further contingent will be sent in 1995. Following protocols developed by the Captive Breeding Specialist Group of the International Union for the Conservation of Nature (IUCN) and the Lizard Advisory Group of the American Zoo and Aquarium Association (AZA), we plan to sustain a captive breeding population of 50 males and 50 females (Hudson et al. 1994). This group should reach carrying capacity within 10 years of the first captive reproduction expected in 1995 or 1996.

The captive breeding program is augmented by a genetic study to measure levels of heterozygosity and relatedness among individual clutches. Blood samples have been collected from over 80 individuals kept in captivity.

Release of Headstarted Iguanas

Because we suspect that the mongoose is a major predator on juveniles, we have begun to headstart hatchlings. They are reared at the Hope Zoo in a newly constructed hatchery to a size at which we hope they will be safe from mongoose predation. A release pilot project is planned for early 1996. Iguanas hatched in 1991 will be selected based on a screening procedure; special care will be taken to maintain the genetic diversity of the wild population and to prevent the transmission of parasites acquired in the zoo. The individuals will be closely monitored using telemetry. Subsequent releases will depend on the success of the pilot project.

Mongoose Control

Given the size of the Hellshire Hills and their broad connection with the rest of Jamaica, eradication of the mongoose is impossible. However, a permanent catching program to reduce local densities within selected areas is feasible. Target areas should include habitats of hatchlings and juveniles. Our preliminary data generated by radio-tracking hatchlings suggest that such areas surround the nesting sites. All of seven hatchlings followed over four months so far have remained within 200 m of their nests.

Establishment of New Populations

As a further safeguard against extinction, additional populations should be established at suitable sites by releasing iguanas bred and/or reared in captivity. The most promising site is Great Goat Island, where a population survived until at least the late 1940s. However, the numerous goats on this island would have to be removed and the ground vegetation given a chance to recover before release. At the same time, mongooses should be eliminated from the island. Further sites could be Little Goat Island and the Portland Ridge area. The latter is similar to the Hellshire Hills and has retained extensive areas of relatively undisturbed dry forest. Experiences gained from translocating and reestablishing iguanas in Guantanomo, Cuba (Alberts 1994) and Guana Island, British Virgin Islands (Goodyear and Lazell 1994) will assist our efforts.

Hellshire Hills Protected Area

The Government of Jamaica owns most of the Hellshire Hills, which have been declared a forestry reserve under the Forestry Act of 1937. More recently, large sections of the Hills were assigned to a quasi-governmental organization with a mandate for urban development. The Jamaican Iguana Research and Conservation Group together with supporting Jamaican non-governmental organizations therefore has submitted a request to the Natural Resources Conservation Authority (NRCA): (1) to recommend to the Minister for the Environment the declaration of a Hellshire Hills Protected Area, including all areas to the west of the present road system, all fringing wetlands, the entire coastline, and the two Goat Islands; (2) to prevent any expansion of the present road system into the proposed protected area; (3) to prevent any encroachment of charcoal burning upon undisturbed sections of the forest; and (4) to allow only such use of the proposed protected area as sustains the forest and its native wildlife. This request was endorsed by many organizations and individuals both from Jamaica and elsewhere. The Hellshire Hills also have been identified as a priority area for protection in a recent island-wide assessment of ecologically important sites in Jamaica; the assessment is part of a national plan to develop a system of protected areas (Jamaica Conservation and Development Trust 1992).

The NRCA is preparing a submission to the Government of Jamaica recommending the declaration of the Hellshire Hills as a "Protected Area" under the Natural Resources Conservation Authority Act of 1991. Such a declaration would give the NRCA a mandate to preserve the unique biological features of the forest.

A protected area would represent a promising legal instrument to prevent the expansion of large-scale development projects into the forest of the Hellshire Hills. Although we recognize the need for limestone mining and urban development, we think that alternative sites exist. These sites already are ecologically degraded but economically underutilized. Destruction of the few remaining primary forests of Jamaica runs counter to the development and well-being of the nation.

New environmental legislation will hardly be effective with regard to small-scale forest users. The NRCA has only two full-time field officers for the entire island. The security forces are burdened with so many other tasks that the patrolling of remote forest areas is assigned a very low priority. That financial resources will be forthcoming in the near future to employ forest rangers in sufficient numbers to control the many dozens of charcoal burners appears unlikely.

Management of Buffer Zones

In this section, we describe our vision of a management system for buffer zones which integrates conservationists, traditional forest users, and other groups and individuals with interests along the periphery of the Hellshire Hills. A very similar concept has been successfully applied in a project to conserve a mangrove habitat in St. Lucia (Smith and Berkes 1993). To illustrate why we believe that interactive conservation can work, we present two experiences with charcoal burners.

A married couple (Mas Sony and his wife) were among those we found furthest in the forest at the beginning of the project in 1990. They had been burning coal in the Hellshire Hills for more than ten years. For most of the week they would live in the forest with their two donkeys, and leave only over the weekend. Mas Sony had traversed the Hills repeatedly from north to south to collect crabs at Manatee Bay. On a few of these trips he saw iguanas. We explained the purpose of our visits to the forest, that is, to save the iguanas from extinction. The Sonys probably found this a bit odd, but they accepted it as a valid goal. They agreed not to burn coal to the south of a specified line. Whenever we met them in the forest they greeted us warmly. Eventually, without being asked to do so, they retreated from their advanced position and continued to burn coal closer to the edge of the forest. They said that their former area should belong to us, and that they do not want to interfere with our conservation work. Over the last year, they stayed home more and more because of Sony's poor health.

Our second story concerns four charcoal burners searching for a new area in which to work. In addition to the usual machetes, they carried a chainsaw. They had approached the major iguana site when they were met by Edwin Duffus, the pig hunter who rediscovered the iguana in 1990 and who has since worked with us as a full-time conservation officer. As they talked, Duffus divulged some of his hard-to-believe pig hunting stories. He eventually told them that, unfortunately, the area they were in was not really a good place to burn coal. He claimed that it was under strict government control, and that the army often would use it for dangerous maneuvers. A small airplane then passed fairly low over them. Duffus suggested that this must be the government looking for intruders. Soon after that experience, he led the four out of the forest and we never saw them again.

Based on our experiences in the forest, we suggest that present forest users should be licensed and given restricted exploitation rights over specific areas within a designated buffer zone. We propose that the charcoal burners should be required to: (1) use assigned areas only, (2) employ sustainable harvesting methods, (3) discourage new users from entering the assigned areas, and (4) leave dogs at home. If a licensee disobeys, his license should be revoked. In addition, the areas in which charcoal production is allowed should be gradually reduced in size when present producers retire. A similar scheme should be applied for pig hunters; in addition, they should be trained in alternative hunting methods (i.e., without dogs).

Licensed forest users should be employed as much as possible for any work created by the conservation project and related research activities. For instance, we regularly employ charcoal burners to carry water to our field station. As a result of income-earning opportunities, most charcoal burners become more inclined to listen to our arguments.

A number of persons safeguard and maintain resources along the periphery of the Hills. Fish farmers employ security personnel to prevent theft. Sugar cane farmers employ workers throughout the year to maintain the fields and roads. Gun clubs employ rangers to prevent charcoal burners from cutting down the trees in areas used for shooting of game birds. Inhabitants of settlements employ guards for protection against criminals. The Coast Guard regularly patrols the coast looking for smugglers. These persons should be co-opted by the forest management team. Their prime conservation task would be to monitor forest users entering and leaving the Hills. If they notice unlicensed forest users, they should either intercept them or alert a full-time conservation officer. The task could be simplified by reducing the number of entry points into the forest. The persons maintaining resources along the periphery have already constructed extensive physical security systems, such as fences and roadblocks, which could be expanded to the mutual benefit of both the users of the periphery and the forest conservationists.

In addition to co-opted personnel, the recruitment of full-time field officers will be necessary, whenever possible from among the present forest users. These workers will be responsible to coordinate and assist the access control, and to monitor the exploitation activities inside the forest. They also could assist in research and species management projects.

SUMMARY

The Jamaican Iguana can only be conserved if present efforts are increased and sustained over an extended period. Several full-time jobs eventually will need to be funded on a permanent basis in order to manage the buffer zones and the threatened species. So far, finding salaries for just one or two men in the field has been an arduous task. We have not yet found a funding agency willing to make a commitment for more than one year, and the rule has been a few weeks or months. We realize that no donor will pay salaries on an indeterminate basis. Ultimately, the project has to generate its own income. However, just like the iguana hatchlings, we hope for a headstart.

AKNOWLEDGEMENTS

Our sincerest thanks go to Edwin Duffus, the former pig hunter, now conservation officer, who started the present project by saving an iguana from his dogs in 1990. He has since worked with us for the conservation of the iguanas, and has become a role model of a forest-user turned conservationist. Lisa Henlin (Institute of Jamaica) and Yvette Strong (Natural Resources Conservation Authority) have contributed much to the work of the Jamaican Research and Iguana Group. We have received assistance in the field and in the care of the captive iguanas from many persons too numerous to mention here; we are very grateful to them. Our international cooperation is based on the invaluable efforts of Rick Hudson, Allison Alberts, Scott Davis, Simon Hicks, Quentin Bloxam, Peter Tolson, and others. Many institutions supported us financially and through the donation of equipment: Audubon Society, Environmental Foundation of Jamaica, Gladys Porter Zoo, Jersey Wildlife Preservation Trust, Natural Resources Conservation Authority, The Nature Conservancy, World Wildlife Fund (USA), and the zoological gardens of Fort Worth, San Diego, and Indianapolis. Stafford Henri and the National Wildlife Foundation of Jamaica assisted our project by administrating a one-year grant received from the Environmental Foundation of Jamaica. Catherine Levy, Jeremy Woodley, and two anonymous reviewers made valuable comments on our manuscript.

LITERATURE CITED

Alberts, A. C. 1993. The most endangered lizard in the world: the Jamaican iguana, *Cyclura collei*. Vivarium 5: 12–14.

—. 1994. A reintroduction program for the iguanas of Guantanamo. Iguana Times 3: 10–12.

Barbour, T. 1910. Notes on the herpetology of Jamaica. Bull. Mus. Comp. Zoöl., Harvard Univ. 52: 273–301.

—. 1922. Annotated list of the Amphibia and Reptilia of Jamaica. Handb. Jamaica 1922: 666–669.

— and G. K. Noble. 1916. A revision of the lizards of the genus *Cyclura*. Bull. Mus. Comp. Zoöl., Harvard Univ. 60: 158–160.

Beckford, G. and M. Witter. 1980. Small Garden . . . Bitter Weed. Maroon Publ. House, Morant Bay, St. Thomas, Jamaica.

Browne, P. 1756. The Civil and Natural History of Jamaica. Osborne and Shipton, London.

Eyre, L. A. 1987. Jamaica: a test case for tropical deforestation? Ambio 16: 333–343.

—. 1988. Charcoal burning in Jamaica and its impact on the environment, p. 74–121. *In* UNDP/World Bank Energy Sector Management Assistance Program Report. Washington, D. C.

Goodyear, N. C. and J. Lazell. 1994. Status of a relocated population of endangered *Iguana pinguis* on Guana Island, British Virgin Islands. Restoration Ecol. 2: 43–50.

Gosse, P. 1848. On the habits of *Cyclura lophoma*. Proc. Zool. Soc. London 1848: 99–104.

Grant, C. 1940. The herpetology of Jamaica. II. The reptiles. Bull. Inst. Jamaica Sci. Ser. 1: 61–148.

Haynes, A. M., R. L. Sutton, and K. D. Harvey. 1989. Conservation trends, and the threats to endemic birds in Jamaica, p. 827–838. *In* C. A. Woods (ed.), Biogeography of the West Indies: Past, Present, and Future. Sandhill Crane Press, Gainesville, Florida.

Hoagland, D. B., G. R. Horst, and C. W. Kilpatrick. 1989. Biogeography and population biology of the mongoose in the West Indies, p. 611–634. *In* C. A. Woods (ed.), Biogeography of the West Indies: Past, Present, and Future. Sandhill Crane Press, Gainesville, Florida.

Hudson, R., A. Alberts, S. Ellis, and O. Byers. 1994. Conservation Assessment and Management Plan for Iguanidae and Varanidae. IUCN/SSC Conservation Breeding Specialist Group, Apple Valley, Minnesota.

International Institute for Environment and Development.

1987. Jamaica Country Environmental Profile. Kingston.

Iverson, J. B. 1978. The impact of feral cats and dogs on populations of the West Indian rock iguana, *Cyclura carinata*. Biol. Conserv. 14: 63–73.

—. 1979. Behavior and ecology of the rock iguana *Cyclura carinata*. Bull. Florida State Mus. Biol. Sci. 24: 175–358.

Jamaican Conservation and Development Trust. 1992. A Plan for a System of Protected Areas in Jamaica. Unpubl. Rept., Kingston.

Jamaican Iguana Research and Conservation Group. 1992. Conservation Strategy for the Jamaican Iguana and its Habitat. Univ. West Indies, Kingston, Jamaica.

Janzen, D. H. 1988. Tropical dry forest: the most endangered major tropical ecosystem, p. 130–137. *In* E. O. Wilson (ed.), Biodiversity. National Academy of Sciences Press, Washington, D. C.

Johnson, K. F. 1976. A preliminary faunal analysis of Cinnamon Hill, Jamaica. Archaeol.-Jamaica 76-1: 9–18.

Lee, D. 1992. The PARC project: a new beginning. Jamaica Nat. 2: 17–21.

Lewis, C. B. 1944. Notes on *Cyclura*. Herpetologica 2: 93–98.

—. 1946. Jamaican Lizards, p. 34–36. *In* C. B. Lewis (ed.), Glimpses of Jamaican Natural History, Vol. II. Nat. Hist. Soc. Jamaica and Inst. Jamaica, Kingston.

—. 1949. Rats and the Mongoose, p. 16–18. *In* C. B. Lewis (ed), Glimpses of Jamaican Natural History, Vol. I. 2nd ed. Nat. Hist. Soc. Jamaica and Inst. Jamaica, Kingston.

MacMillan, M. 1957. The Land of Look Behind: A Study of Jamaica. Faber and Faber, London.

MacPhee, R. D. E. 1984. Quaternary mammal localities and heptaxodontid rodents of Jamaica. Amer. Mus. Novit. 2803: 1–34.

Morgan, G. S. 1993. Quaternary land vertebrates of Jamaica, p. 417–442. *In* R. M. Wright and E. Robinson (eds.), Biostratigraphy of Jamaica. Geol. Soc. Amer. Mem. 182.

Schwartz, A. and W. M. Carey. 1977. Systematics and evolution in the West Indian iguanid genus *Cyclura*. Studies Fauna Curaçao Carib. Isl. 53: 15–97.

Scudder, S. 1991. Early Arawak subsistence strategies on the south coast of Jamaica. Rept. Archaeol.-Anthropol. Inst. Netherlands Antilles 9: 297–312.

Sloane, H. 1725. A Voyage to the Islands Madera, Barbados, Nieves, S. Christophers and Jamaica with the Natural History of the Herbs and Trees, Four-Footed Beasts, Insects, Reptiles etc. of the Last of Those Islands. Vol. II. Published by the author, London.

Smith, A. H. and F. Berkes. 1993. Community-based use of mangrove resources in St. Lucia. Internatl. Jour. Environ. Studies 43: 123–131.

Vogel, P. 1994. Evidence of reproduction in a remnant population of the endangered Jamaican Iguana, *Cyclura collei* (Lacertilia, Iguanidae). Carib. Jour. Sci. 30: 234–241.

— and R. Kerr. 1992. The conservation status of the Jamaican Iguana. Jamaica Nat. 2: 11–16.

— and —. In press. The Jamaican Iguana: survival in a threatened habitat. Proceedings of the Fourth Annual Conference on Science and Technology, 1990. Sci. Res. Council, Kingston.

Wiewandt, T. A. 1977. Ecology, Behavior, and Management of the Mona Island Ground Iguana, *Cyclura stejnegeri*. Ph. D. Dissertation, Cornell Univ., Ithaca, New York.

Wing, E. S. 1972. Identification and interpretation of faunal remains, p. 18–35. *In* J. Silverberg (ed.), The White Marl Site in Jamaica: Report of the 1964 Robert R. Howard Excavation. Dept. Anthropol., Univ. Wisconsin-Milwaukee.

—. 1989. Human exploitation of animal resources in the Caribbean, p. 137–152. *In* C. A. Woods (ed.), Biogeography of the West Indies: Past, Present, and Future. Sandhill Crane Press, Gainesville, Florida.

Woodley, J. D. (ed). 1971. Hellshire Hills Scientific Survey, 1970. Univ. West Indies and Inst. Jamaica, Kingston.

—. 1980. Survival of the Jamaican Iguana, *Cyclura collei*. Jour. Herpetol. 14: 45–49.

Resumen.—El número de iguanas de Jamaica (*Cyclura collei*) parece haber disminuido dramáticamente durante la segunda mitad del siglo pasado, quizás debido a la introducción de la mangosta (*Herpestes auropunctatus*), cambios en las formas de utilización de las tierras, y al crecimiento de la población humana. Al comienzo de este siglo, se creyó que la especie sobrevivía solamente en la Islas Goat, dos cayos pequeños en las afueras de la costa de las Colinas Hellshire. Muchos biólogos creyeron que la iguana se había extinguido luego de que la población de las Islas Goat desapareció en los años 1940. Sin embargo, la supervivencia de la especie en las Colinas Hellshire fue confirmada en 1970 y nuevamente en 1990. Una inspección preliminar en 1990 reveló que una población pequeña aún existía en las regiones centrales y occidentales (las menos perturbadas) de las Colinas. También se encontraron ahí dos lugares de anidaje. Las iguanas desaparecieron de las regiones septentrionales y orientales de las Colinas debido a la excesiva producción de carbón, a cazadores de cerdos que usaban perros, y a la intrusión humana. Las iguanas desaparecen rápidamente de áreas deforestadas. Los depredadores exóticos, incluyendo mangostas, gatos, y perros realengos, se alimentan de iguanas. Los intereses de desarrollo proponen que se mine a gran escala los depósitos de piedra caliza y que se establezcan personas en toda la mitad oriental del área donde viven las iguanas.

Los esfuerzos actuales se concentran en medidas de conservación de emergencia en el campo. Estas medidas incluyen alejar a los usuarios tradicionales del bosque de la población de iguanas, remover a los perros realengos, y proteger de depredación a las hembras que están anidando y a las iguanas recién nacidas. Se está desarrollando un amplio programa para conservar a *C. collei* y a su habitáculo. Más de 100 iguanas recién nacidas han sido capturadas en el campo desde 1991 hasta 1994. Las mismas serán usadas en un programa de reproducción en cautiverio y serán repatriadas al campo. Así mismo, se controlará a las mangostas en los habitáculos de las iguanas juveniles. Las Autoridades Jamaiquinas de Conservación están preparando una declaración al Gobierno Jamaiquino recomendando designar las Colinas Hellshire como un área protegida. Se está desarrollando un plan de manejo que prevee el uso sostenible del bosque en zonas de amortiguación a lo largo de los límites de las Colinas Hellshire que será implementado por medio de una colaboración íntima entre ambientalistas, usuarios tradicionales del bosque, y otros grupos de interés.

Résumé.—Les populations d'iguanes de la Jamaïque (*Cyclura collei*) semblent avoir considérablement décliné pendant la deuxième moitié du siècle dernier, phénomène qui est peut-être attribuable à l'introduction de la mangouste (*Herpestes auropunctatus*), à de nouveaux modes d'utilisation des sols et à la croissance démographique. Au début du XXe siècle, on pensait que ces espèces n'avaient survécu que sur les Goat Islands, deux petites îles côtières au large des Hellshire Hills. De nombreux biologistes pensaient que l'iguane avait complètement disparu après la disparition de la population des Goat Islands dans les années 1940. Toutefois, la survie de l'espèce a été confirmée en 1970 puis en 1990 dans les Hellshire Hills. Une enquête préliminaire menée en 1990 a révélé l'existence d'une petite population d'iguanes dans les parties occidentales et centrales les moins perturbées des Hills ainsi que deux sites de nidification. Les iguanes ont disparu des parties nord et orientales en raison d'une production extensive de charbon de bois, de la chasse aux cochons sauvages avec des meutes de chiens et des établissements humains. Les iguanes ont rapidement disparu des zones forestières. Les prédateurs exotiques comme les mangoustes, les chats et les chiens errants sont leurs principaux ennemis. L'exploitation du calcaire à grande échelle et plusieurs établissements humains sont prévus dans le cadre de projets de développement dans la moitié orientale de la zone où vivent les iguanes. Les efforts déployés actuellement sur le terrain sont principalement axés sur les mesures de conservation d'urgence. Les utilisateurs traditionnels de la forêt sont éloignés des populations d'iguanes, les chiens errants sont capturés et les femelles qui couvent ainsi que les portées sont protégées des prédateurs. Un programme complet de protection de *C. collei* et de son habitat est sur le point d'être déployé. Plus de 100 jeunes ont été capturés sur le terrain entre 1991 et 1994. Ils seront utilisés pour la reproduction en captivité et la réinsertion à l'état sauvage. Les mangoustes feront l'objet d'une surveillance plus étroite dans les habitats où vivent les jeunes iguanes. Les autorités jamaïquaines de conservation préparent une soumission qu'ils présenteront aux pouvoirs publics et dans laquelle elles recommandent que la région des Hellshire Hills devienne une zone protégée. Un plan de gestion est en cours d'élaboration qui prévoit l'utilisation durable de la forêt dans les zones tampons par le biais de l'étroite collaboration des naturalistes, des utilisateurs traditionnels de la forêt et d'autres groupes d'intérêt vivant le long des Hellshire Hills.

Vogel, P., R. Nelson, and R. Kerr. 1996. Conservation strategy for the Jamaican Iguana, *Cyclura collei*, p. 395–406. *In* R. Powell and R. W. Henderson (eds.), Contributions to West Indian Herpetology: A Tribute to Albert Schwartz. Society for the Study of Amphibians and Reptiles, Ithaca (New York). Contributions to Herpetology, volume 12.

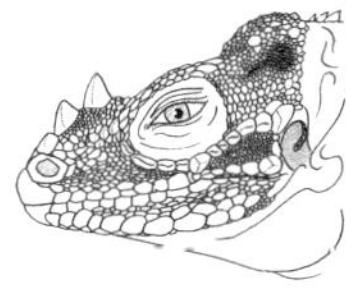

CONSERVATION OF *EPICRATES MONENSIS* ON THE SATELLITE ISLANDS OF PUERTO RICO

PETER J. TOLSON[1]

Abstract.—*Epicrates monensis* is a small (usually < 1 m snout-vent length) endangered boid endemic to Isla Mona and the Puerto Rico Bank. Although extant on Puerto Rico and a few of the islands to its east, scattered localities in the United States and British Virgin Islands, and Isla Mona, the boa has been extirpated from a number of other islands. Threats to the snake include predation from exotic mammals, habitat destruction, and stochastic processes resulting from population fragmentation and isolation. In cooperation with the U. S. Fish and Wildlife Service, the Departamento de Recursos Naturales de Puerto Rico, and the Division of Wildlife, U. S. Virgin Islands, the Toledo Zoological Gardens has been implementing recovery efforts for this snake. Toledo Zoo Conservation staff surveyed more than 20 islands on the Bank from 1985–1989, implemented studies of habitat, initiated a program of ecological research and feral mammal eradication on several islands, and initiated a cooperative breeding program under the Species Survival Plan© of the American Zoo and Aquarium Association. The last step resulted in the reintroduction of the boa to Cayo Ratones, Puerto Rico, in August 1993.

INTRODUCTION

EPICRATES MONENSIS IS A SMALL BOID SNAKE (USUally < 1 m snout-vent length) which exhibits an extremely disjunct distribution on the islands of the Puerto Rico Bank. The bank includes Puerto Rico, Culebra, Vieques, the La Cordillera Cays, the United States and British Virgin Islands, and Isla Mona, a relatively large island in the Mona Passage between Hispaniola and Puerto Rico (Figure 1). The nominate subspecies, the Mona boa (*E. m. monensis*), is endemic to Isla Mona (Zenneck 1898; Schmidt 1926), whereas the Virgin Islands boa (*E. m. granti*) is found on Puerto Rico and scattered islands and cays from Cayo Diablo eastward through the U. S. and British Virgin Islands. Included in the species' range are St. Thomas, Tortola, and Virgin Gorda (Grant 1932; Stull 1933; Nellis et al. 1983; Mayer and Lazell 1988). Judging from the present distribution, the historical range of *E. monensis* probably encompassed low-elevation coastal habitats over virtually the entire Puerto Rico Bank.

The boa is a nocturnal, semi-arboreal opportunistic predator, feeding primarily on *Anolis cristatellus* (by *E. m. granti*) or *A. monensis* (by *E. m. monensis*). Snakes also will prey on *Ameiva exsul*, rodents, and small birds (Sheplan and Schwartz 1974; Tolson 1988). The species seems to be most abundant in closed canopy littoral forest composed of *Coccoloba uvifera*, *Conocarpus erectus*, and other salt-tolerant species, but may be found in upland plant communities as well. Snakes have been captured in disturbed vegetation, and in trees and shrubs of the following genera: *Acacia*, *Ficus*, and *Croton*. By day individuals take refuge in the termitaria of *Nasutitermes* sp., in tree hollows, and beneath rocks and debris.

The Virgin Islands boa is listed as Endangered (Federal Register, 13 October 1970, 35: 16047), and is absent from far more islands on the Puerto Rico Bank than on which it is resident. The Mona boa is listed as Threatened (Federal Register, 3 February 1978, 43: 4618). The extremely disjunct distribu-

[1] Department of Conservation, Toledo Zoological Gardens, Toledo, Ohio 43609, USA.

tion of this species as an Isla Mona/Puerto Rico Bank endemic (Nellis et al. 1983; Mayer and Lazell 1988) provides evidence for extirpation and decline on numerous islands and cays of the Bank since the Pleistocene (Sheplan and Schwartz 1974). Apparently, judging from museum collection holdings, neither the Mona boa nor the Virgin Islands boa have ever been common in this century. That this decline resulted from a single causitive factor is unlikely. Instead, the disappearence of the snake from several localities, such as St. John in the U. S. Virgin Islands, appears to be due to a complex series of biotic, environmental, and stochastic interactions. Habitat destruction was among the earliest threats, as widespread areas of Isla Culebra, Isla Mona (primarily the coastal terrace from Playa Sardinera to Playa Uvero), and theVirgin Islands were denuded for charcoal production and agricultural uses, primarily cultivation of sugar cane and cattle ranching. The introduction of exotic predators also had serious adverse effects on boa populations. The black rat (*Rattus rattus*) likely was introduced to Puerto Rico shortly after the beginning of Spanish colonization (Atkinson 1985), followed by the Indian mongoose (*Herpestes auropunctatus*) in 1887 (Nellis and Everard 1983). The house cat (*Felis catus*) was present on Isla Mona as early as 1898 (Hübener 1898) and presumably became feral on

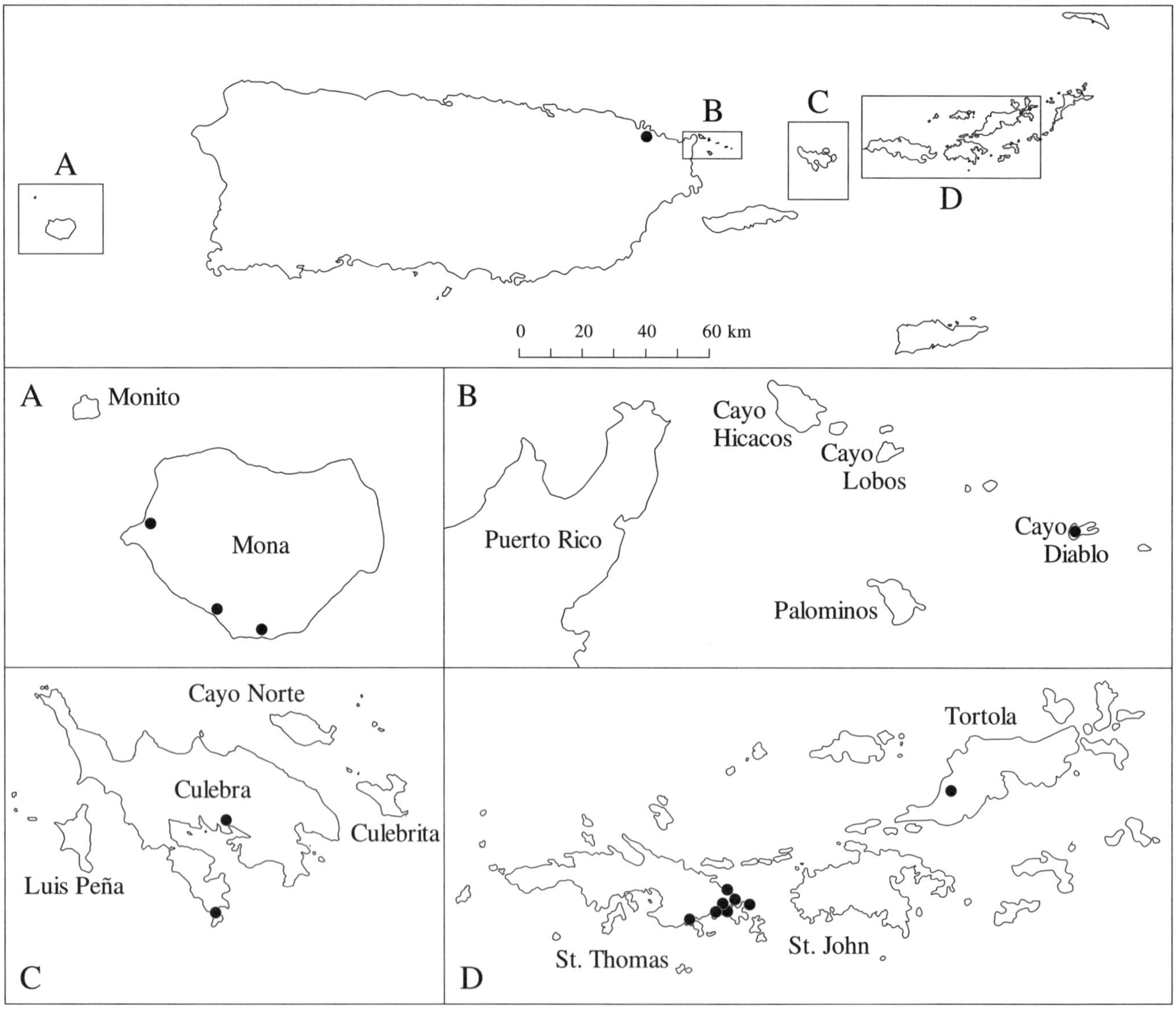

FIGURE 1. Isla Mona and the Puerto Rico Bank (enlarged areas are not to the same scale). Dots indicate capture localities for *Epicrates monensis* (see text).

Puerto Rico much earlier. Climatic changes during the Pleistocene (Pregill 1982), eustatic events, and the subsequent genetic isolation of several populations, and the stochastic processes associated with the fragmentation of a once-contiguous range have also doubtless contributed to the decline of this taxon (United States Fish and Wildlife Service [USFWS] 1986; Tolson 1988).

Survey Work and Status Determination

One of the first steps in implementing USFWS-mandated recovery activities for this species was intensive survey work at localities likely to harbor previously unknown populations of the boa (USFWS 1984, 1986). Most of the efforts were directed toward studies on the offshore cays of Puerto Rico and the U. S. Virgin Islands, as the majority of these islands are free of cats and mongooses, and have larger tracts of littoral forest than Cayo Diablo. This species can exist in high densities on small islands. Prior to Hurricane Hugo, capturing 8–10 snakes in three hours of hunting was not unusual in one *Coccoloba* grove on Cayo Diablo (P. J. Tolson, personal observations). The following Islands were surveyed: Buck Island, Cas Cay, Isla Culebra, Isla Culebrita, Cayo Diablo, Cayo Icacos, Cayo Lobos, Cayo Ratones, Congo Cay, Great St. James Cay, Isla Magueyes, Isla Mona, Isla Monito, Outer Brass Cay, Saba Cay, Salt Cay, Savanna Cay, and Steven Cay. Coastal areas near Río Grande and Guanica, Puerto Rico were also searched.

The presence or absence of the boa at a particular locality was assessed by searching all habitat types. During the period of April 1983 to September 1987, surveys were carried out independently of habitat analysis. Typically, two or more weeks were spent searching larger islands and three to five days were spent searching smaller cays. Within each 24-h period, two persons spent four hours per night searching likely foraging sites (e.g., vine tangles, terminal branches of trees, palm crowns, and littoral vegetation). Only one night was spent on Cayo Lobos, as the native vegetation was all but completely destroyed by human activity and all densely vegetated areas could be searched repeatedly in a single night. During the daylight hours, refugia such as debris piles, termitaria, and palm axils were examined. No new populations of the boa were discovered on any of the small islands, but fortuitously, a crab hunter brought in a dead boa to the USFWS office on Isla Culebra which I identified as an *Epicrates monensis granti*. A population of boas was subsequently found on Isla Culebra by M. A. García and P. J. Tolson (unpublished data). Despite repeated searching in the coastal forests of Río Grande and Guanica, no additional populations of the boa have been discovered on Puerto Rico. I am optimistic, however, that additional populations will be found in Guanica or perhaps in the area surrounding the Reserva Natural de Las Cabezas de San Juan on the northeastern coast.

Feral Mammal Control

I consider the house cat and the black rat to be two of the most serious threats to the continued existence of *Epicrates monensis*. Cats hunt and kill adult boas on both Isla Mona and St. Thomas. A 40-mm head of a boa was dropped by a cat when it was surprised by a hunter on the El Faro Road on Isla Mona in March 1993, and three boas imported by the Toledo Zoo (TZG) from St. Thomas in 1991 were obtained after they were rescued from cats by residents of that island. On Isla Mona, cats are shot by rangers of the Departamento de Recursos Naturales de Puerto Rico (DRN), and sport hunters are urged to eliminate them when possible, but the cats rapidly learn to avoid humans. Although some cat trapping, using box traps, is carried out by TZG and DRN personnel, box traps are generally ineffective for cats (Veitch 1985), and the cats rapidly become trap-shy. These methods cannot adequately control the population given the high fecundity and long life span of feral cats.

The evidence against rats is more circumstantial. In fact, direct demonstrations of rat predation on any reptile or amphibian is difficult as rats macerate their food so thoroughly that identification of stomach or gut contents is almost impossible. Whittaker (1973, 1978) demonstrated that lizard diversity was greater on small islands free of rats than on larger islands which had rats. Boas clearly coexist with rats on the large islands of the Puerto Rico Bank (e.g., Isla Culebra, Puerto Rico, and St. Thomas) and Isla Mona. However, TZG surveys have never detected boas on any of the smaller islands where rats are present. This may be due to the fact that rats are far more abundant on most of the smaller cays than in equivalent habitat on Isla Mona, Puerto Rico, or the Virgin Islands. For example, on Isla Mona, two 100-m traplines with traps set at 10-m intervals gave capture rates of 0.016 and 0.015 rats/trap/h, whereas equivalent traplines on Buck Island, Congo Cay, and Outer Brass Island gave capture rates of 0.043–0.092 rats/trap/h. Capture rates are based on three sequential nights of trapping from 1900–2100 h, and were calculated using the methods of Strecker et al. (1962).

The higher densities of rats on the smaller islands may be due either to the absence of introduced predators such as *Felis catus* and *Herpestes auropunctatus*, and/or the characteristics of small island rodent populations (e.g., low migration rates and higher longevity), which allow greater rodent carrying capacities (Gliwicz 1984). The absence of boas on most of the smaller islands of the Puerto Rico Bank does not seem to be due to small island size. The land areas of Buck Island (16.61 ha), Congo Cay (10.60 ha) and Outer Brass Island (43.69 ha) are all considerably larger than that of Cayo Diablo, Reserva Natural de La Cordillera, (5.70 ha), which has an extant population of the Virgin Island boa.

Fortunately, rats can be eliminated on small islands fairly easily with rodenticides. The most effective rat poisons in general use today are the second-generation anticoagulant compounds (Lund 1985). Brodifacoum, currently available as Talon WeatherBlok®, and bromadiolone, currently available as Contrac® and Maki® paraffin blocks, are two of the most effective and easily-used rodenticides. Anticoagulant compounds kill rodents by reducing the clotting ability of the blood and causing severe internal hemorrhaging. The active ingredients are imbedded in a paraffin matrix which is relatively impervious to weather, allowing for greater long-term bait exposure to the rat population. The 10-g blocks of poison are easily carried and distributed, and can be placed in runs, burrows, nests, and tree holes without difficulty. The bait remains palatable for a minimum of three months in the field, does not cause bait shyness, and rodents may consume a lethal dose in one feeding. Talon® holds up much better than Maki® in exposed locations, because Maki® has ground corn imbedded in the matrix, and this makes the block friable and less water resistant.

On Cayo Ratones, Puerto Rico, rats were eliminated by placing 8–10 blocks of anticoagulant poison at each interstice of a 10-m^2 grid which covered the entire island. Baits were replenished as they were consumed for a period of three days. This regimen was repeated on two successive visits spaced six months apart. In addition to grid poisoning, bait blocks also were placed at sites likely to be visited by rats, such as in hollows of trees, under rocks, in debris piles, and along the shoreline. This poisoning strategy was previously tested by the TZG and the Division of Wildlife, U. S. Virgin Islands, and resulted in successful rat eradication on Dog Cay, Kalkun Cay, and Steven Cay, with no harm to the native biota (Nellis 1984). Pre- and post-poisoning activity by rats on Cayo Ratones was monitored by removal trapping along three 100-m trapline transects as described above for Isla Mona. After poisoning, the pre-poisoning rat index of 0.160 rats/trap/h on Cayo Ratones, calculated from trapping on 1–3 September 1991, dropped to an activity level of 0 rats/trap/h on the next two visits to Cayo Ratones on 1–3 March 1993 and 5–7 June 1993. To detect low levels of rat activity vegetable oil-soaked chew sticks were placed for one week in each habitat type on the island and were checked for rat chew marks (Moors 1985). No rats were detected using this method.

HABITAT UTILIZATION AND PROTECTION

Although the population of Virgin Islands boas at Río Grande, Puerto Rico, in the northern foothills of the Sierra de Luquillo, occurs in subtropical moist forest, subtropical dry forest is the habitat where *Epicrates monensis* is most commonly observed — particularly on Isla Culebra, Isla Mona, and St. Thomas. This forest is characterized by small (< 5 m) deciduous trees with small, coriaceous or succulent leaves and thorns, spines, and secondary defensive compounds (Ewel and Whitmore 1973). Examination of the present range of the boa indicates that the species is generally syntopic with dry subtropical forest on the Puerto Rico Bank (Ewel and Whitmore 1973). This distribution is most apparent on St. Thomas, where *E. monensis* is restricted to the dry eastern end of the island despite the presence of more mesic habitat elsewhere (Nellis et al. 1983). Common tree species on these islands include *Bursera simaruba*, *Cephalocereus royenii*, *Pictetia aculeata*, *Bucida buceras*, *Guaiacum officinale*, *Leucaena glauca*, *Tamarindis indica*, *Melicoccus bijugatus*, *Acacia* spp., and *Capparis cynophallophora* (Little and Wadsworth 1964). More xeric areas on Isla Mona contain many *Byrsonima lucida*, *Euphorbia petiolaris*, and *Metopium toxiferum*. The trees in most dry forest habitats on the Puerto Rico Bank are deciduous, and have spreading crowns which which often interlock with those of neighboring trees. In coastal areas, the littoral forest may have only two to five dominant tree species, many with compound trunks, primarily *Coccoloba uvifera*, *Conocarpus erectus*, *Hippomane mancinella*, *Pisonia subcordata*, and *Thespesia populnea*. Foraging boas are not restricted to trees, however, and also use salt-tolerant shrublands just above the high tide line. This vegetation is composed primarily of *Suriana* and *Tournefortia*.

Captures and sightings of the Mona boa have been limited to three distinct localities: dry plateau forest above Playa Sardinera, Playa Uvero, and Playa Pájaros; *Coccoloba uvifera* groves of Playa Pájaros; and cliffside vegetation adjacent to Playa Sardinera (Campbell and Thompson 1978; Rivero et al. 1982; P. J. Tolson, unpublished data). In Puerto Rico, the Virgin Islands boa has been encountered repeatedly only in the area of Río Grande, on the northeastern coast of the island, and on two offshore islands: Isla Culebra and Cayo Diablo. On Cayo Diablo, *Coccoloba uvifera* is the habitat where foraging *Epicrates monensis* are most commonly observed (Tolson 1988). On Isla Culebra, boas are usually found in roadside *Acacia* and *Leucaena*.

One of the most difficult problems the boa faces on the smaller islands is the loss of littoral forest habitat, primarily by tree cutting for firewood and by clearing for the construction of squatters' houses. Puerto Rican law clearly prohibits these activities, and on Culebra, the USFWS is attempting to remove squatter housing from refuge lands. The problem is particularly acute in the Punta Soldado area. Because of damage by campers and fishermen, the DRN is seeking to prohibit visitation to the smaller uninhabited islands of the Reserva Natural de La Cordillera (between latitude18°23'N, longitude 65°30'W on a northwest to southeast trending axis to latitude18°21'N, longitude 65°25'W). This chain of cemented dune islands are part of a submarine ridge, and includes Cayo Diablo, Cayo Icacos, Cayo Lobos, and Cayo Ratones. Fortunately, prohibitions against destruction of vegetation are rigorously enforced on Isla Mona.

Captive Breeding

In cooperation with the USFWS, the DRN, and the Division of Fish and Wildlife, U. S. Virgin Islands, the TZG initiated a captive-breeding program for this species in 1985. This program resulted in the first successful captive breeding of the species in 1986 (Tolson 1989), publication of an American Zoo and Aquarium Association-sanctioned Regional Studbook in 1987, and development of a Species Survival Plan© in 1990. Species survival plans are cooperative management programs in which several zoos pool resources to develop conservation strategies for endangered wildlife (Wiese and Hutchins 1994). The focus of the TZG program is the production of snakes for reintroduction programs. The basis for this management strategy was developed under the USFWS Recovery Plan for the Virgin Islands boa (USFWS 1986), and included two years of field research on the ecology and demography of the snake on Isla Culebra, Isla Mona, Puerto Rico, and the U. S. Virgin Islands. Data continue to accumulate from wild populations through an ongoing mark-recapture study of over 250 marked individuals on Cayo Diablo. These efforts were coupled with a comprehensive reproductive research program using the Cuban boa, *Epicrates angulifer*, and the Haitian boa, *E. striatus*, as models to establish the proximate environmental and social factors critical for reproduction in this species. These factors include sufficient stimulation during courtship by the male and a night-time temperature drop of 5–8°C during the winter (Tolson and Teubner 1987; Tolson 1994). The reproductive program has been very successful. More than 100 offspring have been produced by ten founders at the TZG. Representatives of two distinct populations, one from Cayo Diablo, Puerto Rico and one from St. Thomas, U. S. Virgin Islands, are now maintained in captivity. As each population exhibits a distinct coloration, the decision was made to maintain intra-population integrity in the captive snakes and to manage them as two separate genetic units; that is, individuals captured on Cayo Diablo are not crossbred with stock originating on St. Thomas. These two populations lately have been shown to exhibit differences in mtDNA restriction enzyme cleavage sites (M. Gach and P. J. Tolson, unpublished data).

The environmental and social parameters necessary to induce reproduction can be duplicated fairly easily in captivity, and the husbandry of the species is not difficult (Tolson 1989, 1994). Space requirements are minimal — adult snakes can be housed in 18.9-l (five-gallon) aquaria and bred in 110-l (29-gallon) aquaria. Neonates are usually housed in plastic shoeboxes. Survivorship of captive neonates is extremely high (exceeding 83% at the TZG). Although *Epicrates monensis* reproduces biennially, longevity may exceed 20 years, and a female has the potential to produce 50–75 offspring during her lifetime (Tolson 1994). This longevity allows the generation time in captivity to be extended, which preserves genetic diversity in the captive population. The probability of inbreeding is high in several wild populations due to isolation and/or the founder effect, but because no indications of genetic anomalies or inbreeding depression have been observed, the genetic load probably is not excessive. However, the degree of heterozygosity within populations and the degree of relatedness between isolated populations are unknown at this time.

Reintroduction

Reintroduction of boas to a second site on La Cordillera was considered to be one of the critical elements of a strategy for recovery of this species (USFWS 1986). The almost total destruction of nearby Cayo Diablo by Hurricane Hugo in September 1989, and the subsequent three-year reproductive failure of the boa population there (P. J. Tolson, unpublished data) have placed what was once the densest population of this snake in peril. Reintroduction is a reasonable strategy to assist in the recovery of *Epicrates monensis*, because the elements deemed to be important in the success of a captive-release program have generally been fulfilled (Kleiman 1990). Research in the genetics and reproductive biology of the captive population has continued, the basic natural history of the species has been documented, habitat preservation and management is continuing, and adequate preparation for the reintroduction in terms of choice of candidates (e.g., genetic background, sex ratio, age class, behavioral attributes) has been completed. Reintroductions to the wild only involved those individuals originating from a source population closest to the release site. Government-owned, relatively undisturbed cays which exhibit attributes identified as typical habitat where this species currently occurs naturally, (e.g., high densities of shrubs, palms, trees with compound trunks, and continuity of vegetation; Tolson 1988).

Site Selection and Preparation

Three sites were originally selected for reintroduction attempts: Cayo Icacos, Reserva Natural de La Cordillera, Puerto Rico; Congo Cay; and Steven Cay. These islands are government-owned, have extensive stands of closed-canopy littoral forest, have densities of *Anolis* comparable to Cayo Diablo, and are small enough that rodent eradication could be accomplished if attempted. A site on Cayo Icacos initially identified (Tolson 1988) as the primary candidate for reintroduction was severely damaged by Hurricane Hugo in September 1989, and an alternate site on Cayo Ratones was chosen instead. Major hurricanes are a frequent feature of annual weather patterns in the Antilles (Walker et al. 1991). As hurricane storm surges can inundate low-lying cays, topography was an important consideration in the selection of release sites. Cayo Ratones is situated on the same submarine ridge as Cayo Diablo, the source of the founder stock of the snakes used in the release program. A lowering of sea level of only 5 fathoms would unite all of the cays of La Cordillera, and in the Pleistocene Cayo Diablo and Cayo Ratones likely were once part of the same land mass situated on a broad submarine bank (Kaye 1959).

The only management activity needed on Cayo Ratones prior to the reintroduction was the eradication of rodents, which was carried out as described above. Eradication of exotic mammals prior to reintroduction is probably the greatest problem that needs to be overcome in any management program for this species.

Preparation of Snakes for Reintroduction

Captive-born adult and subadult snakes destined for release were quarantined at the TZG for 30 days. During that period, they were examined by staff veterinarians and screened for disease and parasites. They were fed live *Anolis cristatellus* to ensure that they found natural food to be palatable. None of the snakes destined for reintroduction ever refused to eat these lizards. After feeding successfully on natural prey, the snakes were tested for hunting ability in a 2 x 2 x 1.5-m mesh enclosure containing live *A. cristatellus*. The anoles were able to escape if the snake was unsuccessful in its predatory attempts. Of the 23 snakes tested for the August 1993 release, only two were unsuccessful in capturing *Anolis* the first night. These two snakes were not released. Neonates were released within one week of birth. No attempts were made to feed these individuals prior to their release.

Evaluation of Reintroduction Strategies

During August 1993, 28 captive-born boas from seven different zoos (Buffalo, Denver, Fort Worth, Milwaukee, San Antonio, Virginia, and Toledo) were released on Cayo Ratones. Completion of rodent eradication and monitoring of *Anolis cristatellus* densities on Cayo Ratones in June 1993 indicated that the site was adequate for a reintroduction attempt. The release sites on Cayo Ratones (a small depression forest dominated by *Pisonia subcordata*, and a littoral forest dominated by *Coccoloba uvifera* and *Conocarpus erectus*) were representative of habitat elsewhere on the Puerto Rico Bank where high densities of boas were present (Tolson 1988). The snakes released were offspring of adults captured on nearby (ca. 6 km) Cayo Diablo. All released snakes were adults and subadults. Each individual was identified using passive integrated transponders (InfoPet Identification Systems, Inc., Burnsville, Minnesota); neonates were identified by photographs of the dorsal body pattern. Four adult animals were implanted with SM-1 radiotransmitters (AVM Instrument Co., Ltd., Livermore, California) for monitoring.

Data on activity patterns and foraging behavior of released snakes are being collected presently, and will be analyzed using the methods of Chandler and Tolson (1990). The initial releases included radiotracking individuals implanted with transmitters to assess to what degree captive-raised individuals duplicated behavior of wild snakes. Transmitter-implanted snakes were intermittently monitored by TZG and DRN personnel from August to October 1993, until battery failure. Minimum estimates of survival of all released snakes, based on periodic recaptures, are 73.9% through February 1994, and 56.5% through August 1994. Six of seven neonates released in August 1993 survived at least one year. To obtain a truly representative analysis of behavior and survivorship, quarterly monitoring was extended through 1994 and semi-annual monitoring through 1995. Annual monitoring will continue through 2005.

Reproduction has been documented on Cayo Ratones within the first year following release. In August 1994, a gravid female was recaptured, and she gave birth to five healthy young two days later. The young were released in the area where the female was encountered — a grove of *Cordia rickseckerii*. Two of these five neonates were recaptured in December 1994, four months after their birth.

CONCLUSIONS

The best strategies to elect for a conservation program are highly dependent on the natural history of the organism and the degree and types of threats faced by the wild population (Conant 1988; Kleiman 1990). Given the considerable dissent on the efficacy of reintroduction as a conservation strategy (Povolitis 1990; Burke 1991; Dodd and Seigel 1991), a conservative approach for recovery efforts for a species would be one which attempts to mitigate as many threats as possible. We have attempted a multifaceted approach for the Mona/Virgin Islands boa conservation program. Although claiming success for the reintroduction phase of the project using the criteria of Dodd and Seigel (1991) is premature, we have established that captive-born *Epicrates monensis*: (1) can learn to hunt and capture their natural prey after release; (2) will reproduce readily after release; and (3) have a first-year survival rate when released on Cayo Ratones which surpasses that of the wild population on Cayo Diablo (P. J. Tolson, unpublished data).

Federal, Departamento de Recursos Naturales de Puerto Rico (DRN), and TZG management personnel have a seven-year history of successful cooperation in recovery efforts for *Epicrates monensis*. With continued commitment from these agencies, future management actions should continue to be successful.

ACKNOWLEDGEMENTS

I acknowledge Conservation Project Support from the Institute of Museum Services from 1987–present (grants IC-70095-87 and IC-90453-89), which have made field research on *Epicrates monensis* possible, and support from the AZA Conservation Endowment Fund and the USFWS (grant 1448-0004-94-967) for reintroduction and associated follow-up activities. In addition, the following persons were indispensible for their help in the field: Edwin Carrasquillo, Carlos Diéz, Miguel A. García, Valerie Hornyak, Pablo M. Reyes, Joanne M. Roehrs, and José Sustache. Luis O. Nieves and Raúl Pérez of the University of Puerto Rico provided facilities for hunting trials of the boas scheduled for reintroduction. George Proctor and José Sustache identified many of the plants on Cayo Ratones.

LITERATURE CITED

Atkinson, I. A. E. 1985. The spread of commensal species of *Rattus* to oceanic islands and their effects on island avifaunas, p. 35–81. *In* P. J. Moors (ed.), Conservation of Island Birds. Internat. Council Bird Preserv. Tech. Bull. no. 3.

Burke, R. L. 1991. Relocations, repatriations, and translocations of amphibians and reptiles: taking a broader view. Herpetologica 47: 350–357.

Campbell, H. W. and F. G. Thompson. 1978. Observations of a captive Mona Island boa, *Epicrates monensis monensis* Zenneck. Bull. Maryland Herpetol. Soc. 14: 98–99.

Chandler, C. R. and P. J. Tolson. 1990. Habitat use by a boid snake, *Epicrates monensis*, and its anoline prey, *Anolis cristatellus*. Jour. Herpetol. 24: 151–157.

Conant, S. 1988. Saving endangered species by translocation. Bioscience 38: 254–257.

Dodd, C. K. and R. A. Siegel. 1991. Relocation, repatriation, and translocation of amphibians and reptiles: are they conservation strategies that work? Herpetologica 47: 336–350.

Ewel, J. J. and J. L. Whitmore. 1973. The Ecological Life Zones of Puerto Rico and the U. S. Virgin Islands. United States Forest Serv. Res. Pap. ITF-18: 1–72.

Gliwicz, J. 1984. Rodents on islands: living in a crowd. Acta. Zool. Fennica 172: 95–98.

Grant, C. 1932. Herpetology of Tortola; notes on Anegada and Virgin Gorda, British Virgin Islands. Jour. Dept. Agric. Puerto Rico 16: 339–346.

Hübener, T. 1898. Die Inseln Mona und Monito. Globus 74: 368–372.

Kaye, C. A. 1959. Shoreline features and Quaternary shore line changes, Puerto Rico. United States Geol. Surv. Prof. Pap. 317-B: 1–140.

Kleiman, D. G. 1990. Decision-making about a reintroduction: do appropriate conditions exist? Endangered Species Update 8: 18–19.

Little, E. L., Jr. and F. H. Wadsworth. 1964. Common Trees of Puerto Rico and the Virgin Islands. Agric. Handb. no. 249. United States Dept. Agric. Forest Serv., Washington, D. C.

Lund, M. 1985. The "second generation" anticoagulants: a review. Acta Zool. Fennica 173: 149–153.

Mayer, G. C. and J. D. Lazell, Jr. 1988. Distributional records for reptiles and amphibians from the Puerto Rico Bank. Herpetol. Rev. 19: 23–24.

Moors, P. J. 1985. Eradication campaigns against *Rattus norvegicus* on the Noises Islands, New Zealand, using brodifacoum and 1080, p. 143–155. *In* P. J. Moors (ed.),

Conservation of Island Birds. Internat. Council Bird Preserv. Tech. Bull. no. 3.

Nellis, D. W. 1984. Rat control on small tropical cays. Unpubl. Rept. to the Division of Wildlife, U. S. Virgin Islands.

— and C. O. R. Everard. 1983. The biology of the mongoose in the Caribbean. Studies Fauna Curaçao Carib. Isl. 195: 1–162.

—, R. L. Norton, and W. P. MacLean. 1983. On the biogeography of the Virgin Islands boa, *Epicrates monensis granti*. Jour. Herpetol. 17: 413–417.

Povolitis, T. 1990. Is captive breeding an appropriate strategy for endangered species conservation? Endangerered Species Update 8: 20–23.

Pregill, G. K. 1982. Late Pleistocene herpetofaunas from Puerto Rico. Univ. Kansas Mus. Nat. Hist. Misc. Publ. 71: 1–72.

Rivero, J. A., R. Joglar, and I. Vazquez. 1982. Cinco nueves ejemplares del culebrón de la Mona *Epicrates monensis monensis* (Ophidia: Boidae). Carib. J. Sci. 17: 1–4.

Schmidt, K. P. 1926. The amphibians and reptiles of Mona Island, West Indies. Field Mus. Nat. Hist., Zool. Ser. 12: 149–163.

Sheplan, B. R. and A. Schwartz. 1974. Hispaniolan boas of the genus *Epicrates* (Serpentes, Boidae) and their Antillean relationships. Ann. Carnegie Mus. 45: 57–143.

Strecker, R. L., J. T. Marshall, Jr., W. B. Jackson, K. R. Babehenn, and D. L. Johnson. 1962. Pacific island rat ecology. Bernice P. Bishop Mus. Bull. 225: 1–274.

Stull, O. G. 1933. Two new subspecies of the family Boidae. Occas. Pap. Mus. Zool. Univ. Michigan 267: 1–4.

Tolson, P. J. 1988. Critical habitat, predator pressures, and management of *Epicrates monensis* on the Puerto Rico Bank: a multivariate analysis, p. 228–238. *In* R. C. Szaro, K. E. Severson, and D. R. Patton (eds.), Management of Amphibians, Reptiles, and Small Mammals in North America. United States Dept. Agric. Forest Serv. Gen. Tech. Rept., RM-166. Washington, D. C.

—. 1989. Breeding the Virgin Islands boa, *Epicrates monensis granti*, at the Toledo Zoological Gardens. Internatl. Zoo Yearb. 28: 163–167.

—. 1994. The reproductive management of insular species of *Epicrates* (Serpentes:Boidae) in captivity, p. 353–359. *In* J. B. Murphy, K. Adler, and J. T. Collins (eds.), Captive Management and Conservation of Amphibians and Reptiles. Soc. Study Amphib. Rept. Contrib. Herpetol., Ithaca, New York, vol. 11.

— and V. A. Teubner. 1987. The role of social manipulation and environmental cycling in propagation of the boid genus *Epicrates*: lessons from the field and laboratory, p. 606–613. *In* Amer. Assoc. Zool. Parks Aquar. Reg. Conf. Proc., 1987. Amer. Zoo Aquar. Assoc., Wheeling, West Virginia.

USFWS (United States Fish and Wildlife Service). 1984. Mona Boa Recovery Plan. USFWS, Atlanta, Georgia.

—. 1986. Virgin Islands Tree Boa Recovery Plan. United States Fish Wildl. Serv., Atlanta, Georgia.

Veitch, C. R. 1985. Methods of eradicating feral cats from offshore islands in New Zealand, p. 125–141. *In* P. J. Moors (ed.), Conservation of Island Birds. Internat. Council Bird Preserv. Tech. Bull. no. 3.

Walker, L. R., D. J. Lodge, N. V. L. Brokaw, and R. B. Walde. 1991. An introduction to hurricanes in the Caribbean. Biotropica 23: 313–316.

Weise, R. J. and M. Hutchins. 1994. Species Survival Plans — Strategies for Wildlife Conservation. Amer. Zoo Aquar. Assoc., Bethesda, Maryland.

Whitaker, A. H. 1973. Lizard populations with and without Polynesian rats, *Rattus exulans*. Proc. New Zealand Ecol. Soc. 20: 121–130.

—. 1978. The effects of rodents on reptiles and amphibians, p. 75–86. *In* P. R. Dingwall, I. A. E. Atkinson, and C. Hay (eds.), The Ecology and Control of Rodents in New Zealand Nature Reserves. New Zealand Dept. Lands Surv. Infor. Serv. no. 4. Wellington, New Zealand.

Zenneck, J. 1898. Die Zeichnung der Boiden. Zeitschr. Wiss. Zool. 64: 1–384.

Resumen.—*Epicrates monensis* es un bóido pequeño (usualmente < 1 m de longitud de la punta del hocico hasta la cloaca) en peligro de extinción endémico de la Isla de Mona y del Banco de Puerto Rico. Aunque esta boa aún habita en Puerto Rico, en algunas de las islas al este de Puerto Rico, en localidades dispersas en las Islas Vírgenes Británicas y de los Estados Unidos, y en la Isla de Mona, la especie se ha extinguido en varias otras islas. Los peligros para esta culebra incluyen depredación por mamíferos exóticos, destrucción del habitáculo, y procesos estocásticos resultantes de la fragmentación y aislamiento de poblaciones. En cooperación con el Departamento de Pesca y Vida Silvestre de los Estados Unidos de Norteamérica, el Departamento de Recursos Naturales de Puerto Rico y la División de Vida Silvestre de las Islas Vírgenes de los Estados Unidos, el Jardín Zoológico de Toledo ha estado implementando un plan de manejo de esta culebra. El personal de conservación del Zoológico de Toledo inspeccionó más de 20 islas del Banco de Puerto Rico de 1985 a 1989, efectuó estudios del habitáculo, e inició un programa de investigación ecológica y erradicación de mamíferos ferales en varias islas. También comenzó un programa cooperativo de reproducción bajo el Plan de Supervivencia de Especies© del Zoológico Americano y la Asociación de Acuarios. La última gestión resultó en la reintroducción de la boa a Cayo Ratones, Puerto Rico, en agosto de 1993.

Résumé.—*Epicrates monensis* est un petit Boïdé en voie de disparition (dont la longueur ne dépasse généralement pas une mètre) dont les populations sont endémiques sur l'île Mona et dans le Puerto Rico Bank. Même si on le trouve à Porto Rico et dans quelques îles à l'est, dans certaines localités des îles Vierges américaines et britanniques ainsi que sur l'île Mona, ce boa a disparu d'un certain nombre d'autres îles. Le serpent est menacé par les prédateurs mammifères exotiques, la destruction des habitats et les procédés stochastiques résultant de la fragmentation et de l'isolation des populations. En collaboration avec le United States Fish and Wildlife Service (USFWS), le Departamento de Recursos Naturales de Puerto Rico (DRN) et la Division of Wildlife des îles Vierges américaines, le Toledo Zoological Garden (TZG) a lancé un programme de sauvetage. Le personnel du Toledo Zoo Conservation a arpenté plus de 20 îles entre 1985 et 1989, entrepris des études d'habitat, lancé un programme de recherche écologique et d'éradication des mammifères sauvages sur plusieurs îles et mis au point un programme d'élevage coopératif sous les auspices du Species Survival Plan© de l'American Zoo and Aquarium Association. Le dernier effort déployé en ce sens a abouti à la réintroduction du boa à Cayo Ratones (Porto Rico), en août 1993.

Tolson, P. J. 1996. Conservation of *Epicrates monensis* on the satellite islands of Puerto Rico, p. 407–416. *In* R. Powell and R. W. Henderson (eds.), Contributions to West Indian Herpetology: A Tribute to Albert Schwartz. Society for the Study of Amphibians and Reptiles, Ithaca (New York). Contributions to Herpetology, volume 12.

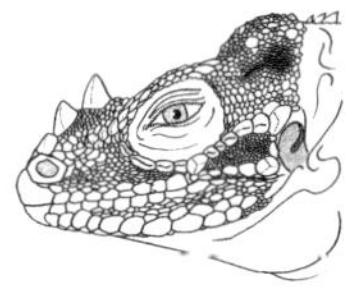

CORALLUS HORTULANUS (SERPENTES: BOIDAE) IN THE POST-COLUMBIAN WEST INDIES: NEW HABITATS, NEW PREY SPECIES, AND NEW PREDATORS

ROBERT W. HENDERSON,[1] R. ALLAN WINSTEL,[2] AND JOEL FRIESCH[3]

Abstract.—The arboreal boid snake *Corallus hortulanus* has a wide geographic distribution on the Neotropical mainland and a limited distribution in the West Indies (St. Vincent and the Grenada Bank). Due to radical alterations in its environment resulting from agricultural practices in the West Indies, *C. hortulanus* now utilizes cultivated tree species that were not previously available, preys on rodents that did not occur on the islands 500 years ago, and may fall prey to vertebrate species that did not live on the islands 100–300 years ago. Because of agricultural practices utilized during the plantation era, *C. hortulanus* is probably now more common than it was in the pre-Columbian (and pre-cultivation) West Indies.

INTRODUCTION

STUDY OF ORGANISMS IN HUMAN-MODIFIED HABITATS, especially those documenting how they respond to the alterations of those habitats, can provide insights into the influence of human history on specific species or an entire community. As the study of animals in undisturbed situations becomes increasingly difficult, attempts must be made to determine the ability of a species to adapt to ecological alterations, be they habitat modification, the introduction of alien predators, or elimination of a food source. Alteration of habitats and the introduction of alien predators have been implicated in extinctions, extirpations, and declining populations, and island faunas may be especially sensitive. Not surprisingly, the ecology of many plant and animal species is closely tied to human history (e.g., Case et al. 1992; Gasc and Lescure 1981; Kaiser and Henderson 1994; Lee 1990).

Corallus hortulanus (formerly referred to as *C. enydris*; McDiarmid et al. 1996) (Figure 1) is a geographically widespread boa that occurs from southern Costa Rica to southern Brazil and on several continental and oceanic islands (i.e., St. Vincent and the Grenada Bank) (Henderson 1993*a*; Henderson et al. 1995). The species is highly arboreal, has a laterally compressed body, a prehensile tail, and attains a maximum snout-vent length (SVL) in the West Indies of 1625 mm. Two subspecies are currently recognized, and the more northern subspecies (*C. h. cooki*) occurs in a wider range of habitats and climatic regimes than does the more southern nominate subspecies (Henderson 1993*a*). *Corallus hortulanus* almost surely colonized the West Indies by overwater dispersal, and molecular evidence indicates a close affinity with populations in northeastern South America (e.g., the Guianas) rather than more geographically proximal Trinidad or Venezuela (Henderson and Hedges 1995).

A series of projects that examined aspects of the ecology of *Corallus hortulanus* in the West Indies was initiated in 1987. Basic to these projects was the determination of where the species did and did not occur, including macro- and microdistribution. Although *C. hortulanus* occurred at a particular locality, it became increasingly apparent that both its macro- and microdistribution were closely associ-

[1] Section of Vertebrate Zoology, Milwaukee Public Museum, Milwaukee, Wisconsin 53233–1478, USA.

[2] 2651 Cornwall Drive, Cincinnati, Ohio 45231, USA.

[3] 165 Alpine Terrace, San Francisco, California 94117, USA.

ated with current patterns of agricultural activity on the islands, and, by inference, the agricultural history of the islands. Similarly, the arrival of Europeans had a significant effect on the diet of *C. hortulanus*. In this chapter, we describe modifications to the biotic environment of an arboreal snake species and how that species has responded.

METHODS

Between 1987 and 1994 we made 10 trips to St. Vin-cent and/or the Grenada Bank, with the most intensive field work concentrated on Grenada. Initial efforts focused on determining the West Indian distribution of *Corallus hortulanus*, and subsequent efforts focused on more specific aspects of the species' natural history. Details of methods employed in various facets of the research can be found elsewhere (distribution: Henderson 1993*a*; diet: Henderson 1993*b*, 1993*c*; habitat: Henderson and Winstel 1995, Henderson and Henderson 1995). Statistical procedures included Spearman Rank Correlations and *chi*-square tests utilizing SAS ($\alpha \leq 0.05$).

RESULTS

Distribution

Corallus hortulanus occurs on St. Vincent and on at least ten islands of the Grenada Bank (Bequia, Île Quarte, Baliceaux, Mustique, Canouan, Mayreau,

FIGURE 1. *Corallus hortulanus* from Upper Pearls, St. Andrew Parish, Grenada.

Union, Carriacou, Petit Martinique, and Grenada). To date, no specimens have been taken on an island with an area of less than 1.0 km^2. The species occurs at elevations that range from sea level to about 520 m, but reaches higher densities at lower elevations (< 200 m).

Habitat

Because arboreal snake species "are highly and convergently specialized in ways that promote functional performance in trees and shrubs," they may be especially sensitive to habitat disturbance (Lillywhite and Henderson 1993). *Corallus hortulanus* occurs in edge habitat ranging from dense *Acacia* scrub to rainforest and human disturbed areas (fruit groves, gardens), and in rainfall regimes that range from less than 1200 mm to over 4000 mm annually.

Habitat selection in tree boas was studied along a 610-m transect (divided into 10-m segments) at a site in St. David Parish, Grenada. This area was generally open, largely without native trees, and characterized by mixed agriculture (sugarcane, pineapples, cacao, bananas, mangos, breadfruit, papaya, coconut, citrus, nutmeg; Figure 2) interspersed with patches of "dry scrub woodland" (Beard 1949). Vegetation in the uncultivated sections of the transect included *Bursera simaruba*, *Pisonia fragrans*, *Tabebuia pallida*, *Ficus* sp., *Leucaena leucocephala*, *Lonchocarpus* sp., *Acacia* spp., *Amyris* sp., *Pithecellobium* sp., *Randia* sp., and occasionally *Bambusa* sp.). The site has been under cultivation for about 200 years (Brizan 1984). With the economic decline of the international sugar market, the amount of land devoted to sugarcane cultivation has been reduced, and these areas are now devoted to producing various kinds of fruit. *Corallus* was not uniformly distributed through the transect, and some segments of the transect received a disproportionate amount of use by the snakes. The contiguous segments of the transect that produced the most observations were 60–120 m in length and had contiguous vegetation, either 100% uncultivated or with a high incidence (75–100%) of mango and breadfruit trees. Those segments that produced few or no observations were under sugarcane or pineapple cultivation or were open with vegetation cropped by grazing livestock. These areas sometimes contained mango or other fruit trees, but they were invariably solitary, lacking crown vegetation that was contiguous with other trees. The total number of observations of *Corallus hortulanus* from throughout the transect was positively correlated with the occurrence of mango trees ($P < 0.0001$; Spearman Rank Correlation) and uncultivated woodlands ($P < 0.0002$). No correlation was detected between the number of observations and the occurrence of cultivated fruit trees exclusive of mangos ($P > 0.05$). The total number of observations was negatively correlated with open (virtually treeless) transect segments on east and west sides of the transect ($P < 0.0001$), and isolated trees or bushes did not harbor snakes. For example, 200 of 201 (99.5%) observations were made while snakes were in trees or bushes that were contiguous with other trees or bushes.

Three size classes (small = snakes < 500 mm SVL; medium = 500–1000 mm SVL; large >1000 mm SVL) of *Corallus hortulanus* exhibited differential use of three habitat categories: mango trees, cultivated trees other than mango, and uncultivated woodlands (Figure 3). Small snakes were encountered mostly in uncultivated woods (78.1%), medium-sized snakes occurred in uncultivated woods (64.3%) more often than in fruit trees (35.7%), but used the latter more often than small snakes (21.9%). Large snakes were most often encountered in mango trees (48.2%) and other cultivated trees (22.2%; e.g., breadfruit, cacao, coconut palm, and banana), and less often in uncultivated woods (29.6%) (3 x 3 contingency, $\chi^2 = 17.17$, d.f. = 4, $P < 0.002$). No sexual differences in habitat selection ($P > 0.05$) were found within each of the three size categories. *Corallus hortulanus* required only a narrow corridor of trees in which to move and forage. Concentrations of tree boas were observed in tree lines only one tree wide, usually with a high incidence of mango and breadfruit trees, contiguous with other trees (either native or introduced), and bordering small plots of sugarcane.

FIGURE 2. A portion of the study site in St. David Parish, Grenada, showing an area of mixed agriculture in the transect. Coconut, mango, cacao, and banana trees are visible in the photograph (by R. A. Winstel).

Diet

Corallus hortulanus is nocturnal and utilizes both active and ambush foraging modes (Henderson 1993*b*, 1993*c*). Throughout its vast range on the Neotropical mainland, this species preys on a variety of vertebrates (frogs, lizards, birds, and mammals), but endotherms are consumed much more frequently than ectotherms. Especially important are small birds (e.g., flycatchers), bats, and rodents (Henderson 1993*c*). In the West Indies, however, *C. hortulanus* preys heavily on lizards (primarily of the polychrotid genus *Anolis*) and rodents (exclusively of the murid genera *Mus* and *Rattus*), but rarely exploits birds and bats (Henderson 1993*b*). Prior to the arrival of Europeans and the introduction of *Mus* and *Rattus*, however, West Indian populations of *C. hortulanus* most likely fed on murid rodents of the genus *Oryzomys* (Lippold 1991;

Pregill et al. 1994), now extinct within the present West Indian range of *C. hortulanus*.

Predators

The only known predator of *Corallus hortulanus* in the West Indies is the introduced marsupial *Didelphis marsupialis* (Allen 1911; Morgan and Woods 1986); a small *C. hortulanus* was found in the stomach of an opossum on Grenada (R. Frederick and A. Joseph, Department of Forestry, personal communication). The African primate *Cercopithecus mona* (Cercopithecidae) probably was introduced to Grenada during slave trade (Groome 1970) and has been observed molesting *C. hortulanus* at Grand Etang National Park (M. Glenn, personal communiation). Of course, tree boas likely fall prey to a variety of other vertebrates (e.g., raptors), including the mongoose (*Herpestes auropunctatus*), which was introduced to Grenada in 1876–1879 and to St. Vincent by 1900 (Hoagland et al. 1989), and is known to occasionally ascend into trees (R. W. Henderson, personal observations).

Discussion

Corallus hortulanus cooki, as currently defined (Henderson 1993*a*), has demonstrated ecological plasticity in habitat, climatic regimes (Henderson and Henderson 1995), and diet (Henderson 1993*a*, 1993*c*). Successful colonization of oceanic islands and persistence in human-disturbed habitats is testimony to the species' adaptability. During the past 500 years on Grenada, *C. hortulanus* has had to: (1) change its diet, as the probable pre-Columbian rodent prey species (*Oryzomys* sp.) is now extinct; (2) exploit trees and habitats that were not on the island previously; and (3) avoid potential predators that were introduced from the Old World or from the Neotropical mainland (Table 1).

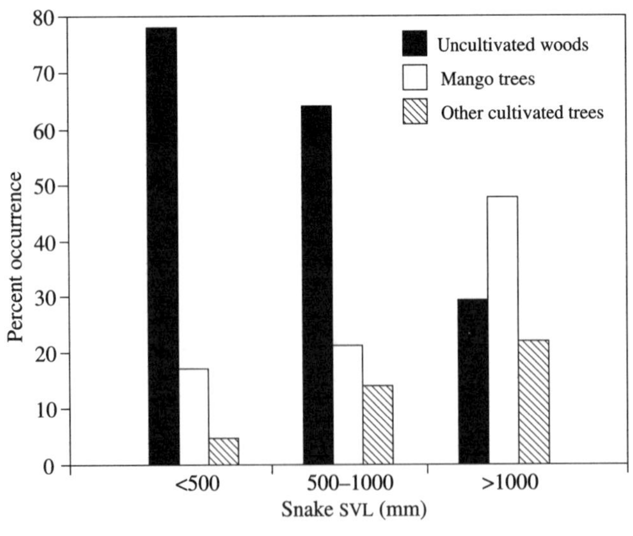

FIGURE 3. Percent occurrence of three size classes of *Corallus hortulanus* in three habitat types in the St. David Parish transect (modified from Henderson and Winstel 1995).

Peoples of the Ortoiroid series of peoples and cultures expanded from South America into the Lesser Antilles about 5000 YBP (Rouse 1989), and habitat alteration undoutedly occurred with their arrival in the region. More specifically, the history of agricultural exploitation on Grenada goes back at least 2000 YBP (Bullen 1964), and a similar chronology probably exists for St. Vincent (Bullen and Bullen 1972). The first attempt by Europeans to establish a settlement on Grenada, by the British in 1609, was a failure. The first successful attempt was by the French in 1650 (Brizan 1984), and the plantation system ushered in the era of mass clearing of forests for sugarcane. Both St. Vincent and Grenada had most cultivatable areas cleared of forests (Beard 1949), and Beard (1949) determined that "In the interior [of Grenada], practically all of the land right to the mountain tops was originally sold to estates, and cultivations were pushed to the highest practicable limit in most cases." Between 1700 and 1750, sugar, indigo, cotton, cacao, and coffee were cultivated, with the latter crop the most important. In 1700, only three sugar plantations existed on Grenada, but 81 were in operation in addition by 1763 to 208 coffee plantations (Brizan 1984). In 1772, 334 estates had been established on Grenada, and 125 of them devoted 12,955 ha to sugarcane (Brizan 1984). Grenada, however, did not rely on a monoculture economy, and in the last half of the 18th century, it was the leading producer of coffee, cotton, sugar, and cacao in the Windward Islands. Therefore, although considerable land was cleared of trees for sugarcane, other land was planted in trees, albeit for other crops. By 1852, only 2,581 ha were planted in sugarcane and, in 1982, 200 years after sugar had reached its economic apex, only 534

TABLE 1. Summary of critical aspects of the ecology of *Corallus hortulanus* in the pre- and post-Columbian West Indies.

Parameter	Pre-Columbian	Post-Columbian
Habitat	Natural vegetation, native to the Neotropics in general and the West Indies in particular	Natural vegetation, but with huge tracts of introduced cultivated orchards (bananas, cacao, citrus, coffee, mango, breadfruit, coconuts, nutmeg)
Diet	Presumably predominantly native *Anolis* lizards and rodents of the murid genus *Oryzomys*; occasionally birds and bats	Predominantly native *Anolis* lizards and introduced rodents of the murid genera *Mus* and *Rattus*; occasionally birds and bats
Potential predators	Presumably native raptors	Presumably native raptors; an introduced marsupial (*Didelphis marsupialis*); possibly an introduced carnivore (*Herpestes auropunctatus*); potentially an introduced primate (*Cercopithecus mona*)

ha were devoted to sugar (Anonymous 1982). In the meantime, cacao, bananas, and nutmeg became economically important (2,821 ha were planted in bananas and bananas mixed with cacao in 1982), and all provided potential arboreal habitat for *C. hortulanus* (although the species rarely occurs in bananas; R. W. Henderson, personal observations).

Along the St. David transect, *Corallus hortulanus* used a high percentage of tree species not native to Grenada, the West Indies, or even the Neotropics (see Table 2 in Henderson and Winstel 1995 for a list of introduced tree species used by *C. hortulanus* and the approximate year of introduction to Grenada or the Caribbean area). Mango (*Mangifera indica*) was introduced into the West Indies (on Barbados) in 1742 (Little and Wadsworth 1964), and has become by far the most frequently used tree species by large *C. hortulanus* in the transect. Although largely circumstantial, and based only on contemporary evidence, the agricultural practices on Grenada may have increased the amount of exploitable edge habitat and thereby facilitated range expansion by *C. hortulanus*, which is now ubiquitous from sea level to an elevation of about 300 m on Grenada. Because *C. hortulanus* requires only a narrow corridor of trees for movements and foraging, expanses of open habitat do not preclude *C. hortulanus* as long as at least a single row of trees connects stands of vegetation.

Habitat choice by *Corallus hortulanus* is undoubtedly influenced by a suite of factors (e.g., tree morphology, prey density), but the single most important criterion related to vegetation is tree crown continuity. Similar relationships between vegetation continuity and arboreal snake occurrence have been observed in a temperate colubrid (*Opheodrys aestivus*, Plummer 1981) and another West Indian boid (*Epicrates monensis*, Tolson 1988). Fieldwork with *C. hortulanus* in habitats that present a mosaic of conditions has indicated that some habitat alteration has had positive ramifications (e.g., creation of more edge situations, introduction and cultivation of some tree species), and the encounter rate (snakes/hour) is higher in agriculturally disturbed areas (at lower elevations; 3.0–4.5 snakes/h) than in less disturbed situations (rainforest at higher elevations; 0.2 snakes/h) (Henderson and Henderson 1995). A critical point, however, is that clear-cutting or lack of contiguous crown vegetation precluded *C. hortulanus* from a given area. Thus, despite potentially catastrophic alterations to its environment in the West Indies, *C. hortulanus* has adjusted to conditions that include new habitat, new prey species, and new predators. In all likelihood, *C. hortulanus* is now more abundant and has a wider distribution on Grenada and St. Vincent than it did in the pre-Columbian, pre-cultivation West Indies.

ACKNOWLEDGMENTS

Funding for fieldwork by R. W. Henderson on Grenada was generously provided by the Milwaukee Public Museum, Inc., the late Albert Schwartz, Jack Puelicher, and the Central Florida Herpetological Society. Permission to conduct fieldwork in Grenada was graciously granted by Alan Joseph and Rolax Frederick of the Department of Forestry (Ministry of Agriculture) and George Vincent (formerly Director of National Parks). Likewise, fieldwork in St. Vincent and the Grenadines was granted by Brian Johnson, Calvin Nicholls, and Cornelius Richards of the Forestry Division. For kindnesses shown to R. W. Henderson while in the field, he is grateful to Brian Alexander of the Mustique Company, Jacques Daudin (Union Island), the staff of the Maffiken Apartments (Grenada), the Westerhall Rum Co. (Grenada), and the United States Peace Corps (St. Vincent and Grenada). Kevin Lyman generously produced Figure 3. For enthusiastic support in the field, R. W. Henderson is grateful to Gary Haas, Ky Henderson, Rose Henderson, and Richard Sajdak.

LITERATURE CITED

Allen, G. M. 1911. Mammals of the West Indies. Bull. Mus. Comp. Zool., Harvard Univ. 54: 173–263.

Anonymous. 1982. Land Use Map, 1982. Ministry of Agriculture, Office of Land Use, St. George's, Grenada.

Beard, J. S. 1949. The Natural Vegetation of the Windward & Leeward Islands. Oxford, Clarendon Press, Oxford, England.

Brizan, G. 1984. Grenada: Island of Conflict. Zed Books Ltd., London.

Bullen, R. P. 1964. The archaeology of Grenada, West Indies. Contrib. Florida State Mus., Soc. Sci. 11: 1–67.

— and A. K. Bullen. 1972. Archaeological investigations on St. Vincent and the Grenadines, West Indies. William L. Bryant Found., Amer. Studies 8: vi, 1–170.

Case, T. J., D. T. Bolger, and A. D. Richman. 1992. Reptilian extinctions: the last ten thousand years, p. 91–125. *In* P. L. Fiedler and S. K. Jain (eds.), Conservation Biology. Chapman and Hall, New York and London.

Gasc, J. P. and J. Lescure. 1981. Effets de l'homme sur la repartition geographique des amphibiens et des lezards dans l'ensemble Guyano-Amazonien. Comp. Rend. Soc. Biogeogr. 57: 33–49.

Groome, J. R. 1970. A Natural History of the Island of Grenada. Caribbean Printers Ltd., Arima, Trinidad.

Henderson, R. W. 1993*a*. *Corallus enydris*. Cat. Amer. Amphib. Rept.: 576.1–576.6.

—. 1993*b*. Foraging and diet in West Indian *Corallus enydris* (Serpentes: Boidae). Jour. Herpetol. 27: 24–28.

—. 1993*c*. On the diets of some arboreal boids. Herpetol. Nat. Hist. 1: 91–96.

— and S. B. Hedges. 1995. Origin of West Indian populations of the geographically widespread boa *Corallus enydris* inferred from mitochondrial DNA sequences. Molec. Phylogen. Evol. 4: 88–92.

— and K. F. Henderson. 1995. Altitudinal variation in body temperature in foraging tree boas (*Corallus enydris*) on Grenada. Carib. Jour. Sci. 31: 73–76.

—, T. Waller, P. Micucci, G. Puorto, and R. W. Bourgeois. 1995. Ecological correlates and patterns in the distribution of Neotropical boines (Serpentes: Boidae): a preliminary assessment. Herpetol. Nat. Hist. 3: 15–27.

— and R. A. Winstel. 1995. Aspects of habitat selection by an arboreal boa (*Corallus enydris*) in an area of mixed agriculture on Grenada. Jour. Herpetol. 29: 272–275.

Hoagland, D. B., G. R. Horst, and C. W. Kilpatrick. 1989. Biogeography and population biology of the mongoose in the West Indies, p. 611–633. *In* C. A. Woods (ed.), Biogeography of the West Indies: Past, Present, and Future. Sandhill Crane Press, Gainesville, Florida.

Kaiser, H. and R. W. Henderson. 1994. The conservation status of Lesser Antillean frogs. Herpetol. Nat. Hist. 2: 41–56.

Lee, J. C. 1990. Creatures of the Maya. Nat. Hist. (January 1990): 45–51.

Lillywhite, H. B. and R. W. Henderson. 1993. Behavioral and functional ecology of arboreal snakes, p. 1–48. *In* R. A. Seigel and J. T. Collins (eds.), Snakes: Ecology and Behavior. McGraw Hill, New York.

Lippold, L. K. 1991. Animal resource utilization by Saladoid peoples at Pearls, Grenada, West Indies, p. 264–268. *In* E. N. Ayubi and J. B. Haviser (eds.), Proceedings of the 13th International Congress for Caribbean Archeology. Rept. Archeol.-Anthropol. Inst. Netherlands Antilles no. 9.

Little, E. L., Jr. and F. H. Wadsworth. 1964. Common Trees of Puerto Rico and the Virgin Islands, Vol. 1. United States Dept. Agric., Agric. Handb. no. 249. Washington, D. C.

McDiarmid, R. W. , T'S. Touré, and J. M. Savage. 1996. The proper name of the tropical American tree boa often referred to as *Corallus enydris* (Serpentes: Boidae). Jour. Herpetol. 30: in press.

Morgan, G. S. and C. A. Woods. 1986. Extinction and zoogeography of West Indian land mammals. Biol. Jour. Linnean Soc. 28: 167–203.

Plummer, M. V. 1981. Habitat utilization, diet and movements of a temperate arboreal snake (*Opheodrys aestivus*). Jour. Herpetol. 15: 425–432.

Pregill, G. K., D. W. Steadman, and D. R. Watters. 1994. Late Quaternary vertebrate faunas of the Lesser Antilles: historical components of Caribbean biogeography. Bull. Carnegie Mus. Nat. Hist. 30: iii, 1–51.

Rouse, I. 1989. Peopling and repeopling of the West Indies,

p. 119–135. *In* C. A. Woods (ed.), Biogeography of the West Indies: Past, present, and Future. Sandhill Crane Press, Gainesville, Florida.

Tolson, P. J. 1988. Critical habitat, predator pressures, and the management of *Epicrates monensis* (Serpentes: Boidae) on the Puerto Rico Bank: a multivariate analysis, p. 228–238. *In* R. C. Szaro, K. E. Severson, and D. R. Patton (eds.), Management of Amphibians, Reptiles, and Small Mammals in North America. USDA Forest Serv., Gen. Tech. Rept. RM-166. Ft. Collins, Colorado.

Resumen.—La serpiente arbórea bóida *Corallus hortulanus* tiene una amplia distribución geográfica en el continente neotropical y una distribución limitada en las Antillas (habita en el Banco de San Vicente y en el de Granada). Debido a alteraciones radicales en su ambiente a causa de prácticas agrícolas en las Antillas, *C. hortulanus* utiliza ahora especies de árboles cultivados que no estaban disponibles previamente, se alimenta de roedores que no habitaban esas islas hace 500 años, y puede ser presa de especies de vertebrados que no vivían en las islas hace unos 100 a 300 años. Debido a prácticas agrarias usadas durante la época de la plantación, *C. hortulanus* es probablemente más abundante ahora que en las Antillas precolombinas y en las Antillas antes de que se practicara el cultivo agrícola.

Résumé.—La répartition géographique du Boïdé arboricole *Corallus hortulanus* est très vaste sur le continent néotropical et relativement limitée dans les Antilles (St.-Vincent et Grenada Bank). Les modifications radicales que l'exploitation agricole a apportées à l'environnement dans les Antilles ont conduit *C. hortulanus* à coloniser les arbres cultivés qui n'existaient pas avant, à chasser les rongeurs qui n'existaient pas sur ces îles il y a 500 ans et à être possiblement prédatés par des espèces vertébrées qui ne vivaient pas sur les îles 100 à 300 ans auparavant. Si l'on tient compte des pratiques agricoles qui existaient à l'époque des plantations, *C. hortulanus* est aujourd'hui probablement plus répandu qu'il ne l'était à l'époque précolombienne (et préagricole).

Henderson, R. W., R. A. Winstel, and J. Friesch. 1996. *Corallus hortulanus* (Serpentes: Boidae) in the post-Columbian West Indies: new habitats, new prey species, and new predators, p. 417–423. *In* R. Powell and R. W. Henderson (eds.), Contributions to West Indian Herpetology: A Tribute to Albert Schwartz. Society for the Study of Amphibians and Reptiles, Ithaca (New York). Contributions to Herpetology, volume 12.

Conservation of the American Crocodile (*Crocodylus acutus*) in the Dominican Republic

Andreas Schubert[1] and Gloria Santana[1]

Abstract.—The American Crocodile (*Crocodylus acutus*), once widespread on Hispaniola, now survives only in Lago Enriquillo, a hyper-saline inland lake in the Dominican Republic, and Etang Saumatre in Haiti. In the late 1980s and early 1990s, the population size and reproduction rate of the Dominican population declined by about two thirds compared to former levels, mainly due to human impact. As a result of a study carried out by the Dominican Wildlife Department, an Action Plan for the conservation of the species was established and presently is implemented. The results to date include: (1) frequent patrols of the lake shore by park rangers and wildlife inspectors have minimized human disturbances; (2) since 1992, the adult/subadult population seems to have increased slightly and today numbers about 200 individuals; (3) since May 1993, a "genetic reserve" of some 130 juvenile crocodiles is maintained in the Santo Domingo Zoo (headstart program); and (4) Lago Enriquillo soon will have protected status as a National Park.

Introduction

The American crocodile (*Crocodylus acutus*) ranges from coastal regions of central Mexico south through Central America to northern Peru on the Pacific and to Venezuela on the Caribbean coast. The species is also present in extreme southern Florida and in the Greater Antilles (Cuba, Jamaica, and Hispaniola) (Groombridge 1987). Once widespread and very abundant, most populations today are severely depleted, mainly due to hunting for skins and habitat destruction (Thorbjarnarson 1992).

On Hispaniola, American crocodiles historically were very abundant along much of the coastline (Descourtilz 1809; Moreau de St. Mery 1797–1798). More recently, in Haiti, small populations of *Crocodylus acutus* could be found in four areas along the coast (Thorbjarnarson 1988). However, subsequently in the Río Massacre area, on the border between the Dominican Republic and Haiti in northern Hispaniola, no crocodiles or signs of their presence were found, despite intensive searches (SEA/DVS 1993). Most of the mangrove habitat in this area is heavily degraded.

Aside from the small populations around the coast of Haiti, the largest viable crocodile populations on Hispaniola live in the inland lakes Etang Saumatre in Haiti and Lago Enriquillo in the Dominican Republic (Figure 1). The Etang Saumatre population was studied in 1983 by Thorbjarnarson (1988). This brackish lake with a salinity of 8–10 ppt has a size of 113 km^2 and is situated at 15 m above sea level. The total crocodile population was estimated at 450 animals, including all size classes; the adult population was estimated to number about 70 individuals. More recently, efforts have been focused on the conservation of the populations in Lago Enriquillo.

Characteristics of the Lago Enriquillo Area

Physical Conditions

Lago Enriquillo is situated in the Neiba Valley in the southwestern part of the Dominican Republic.

[1] Departamento de Vida Silvestre and Servicio Alemán, Apartado 761–2, Santo Domingo, República Dominicana.

The lake is a remnant of a marine channel that once separated the two Hispaniolan paleoislands (Mann et al. 1984). The valley is bordered by two mountain ranges with peaks exceeding 2000 m (Araguás et al. 1993), the Sierra de Neiba to the north and the Sierra de Bahoruco to the south. The climate is arid to semi-arid, with annual precipitation ranging between 470 and 780 mm. Evaporation is estimated to exceed 2000 mm. Generally, two rainy seasons can be distinguished: a moderate April-May period and a usually more prominent September–October season. The lake lies at an elevation of 40 m below sea level, has a water surface of about 200 km^2, a length of 35 km, a width of 11 km, a maximum depth of 22 m, and a salinity of about 70 ppt. Mean air and lake water temperatures are about 30°C. Heavy easterly winds cause considerable wave action, especially in the afternoon. Lago Enriquillo contains one large island (Isla Cabritos) and two small islands (La Islita and La Barbarita). Due to the low lake level at the present time, the latter constitute a peninsular extension from the lake shore.

Water level, lake surface, and salinity vary substantially from year to year. In early 1979, after a prolonged drought, the lake was divided almost in half by a peninsula formed by the emergent islands. However, later that year, Hurricane David initiated a three-month period of heavy rainfall; the water level rose in excess of 5 m and the salinity dropped from more than 90 ppt to about 35 ppt (Margalef 1985).

Vegetation and Fauna

Extensive freshwater wetlands in the eastern and western ends of Lago Enriquillo create conditions that result in a mosaic of different vegetation types: cattail stands (*Typha domingensis*), mangrove forests (*Conocarpus erectus*), and halophyte marshes (*Batis maritima*). Other areas are totally devoid of any vegetation. The typical vegetation around Lago Enriquillo and on the islands is dry forest, composed of spinous hard woods and cacti. This dry forest harbors many species of birds and reptiles, among them the two threatened species of iguana, *Cyclura cornuta* and *C. ricordii*.

Along most of the northern lakeshore, irrigation schemes allow agriculture and pasture usage down to the shoreline. Many canals and creeks enter the lake, most of the latter arising close to the shore in so-called "borbollones." No fresh water is found along most of the southern shore or on the islands. Two endemic species of the genus *Cyprinodon* occur in the lake. In the 1950s, *Tilapia mossambica* was introduced into Lago Enriquillo and today is the most abundant species of fish (Incháustegui et al. 1978).

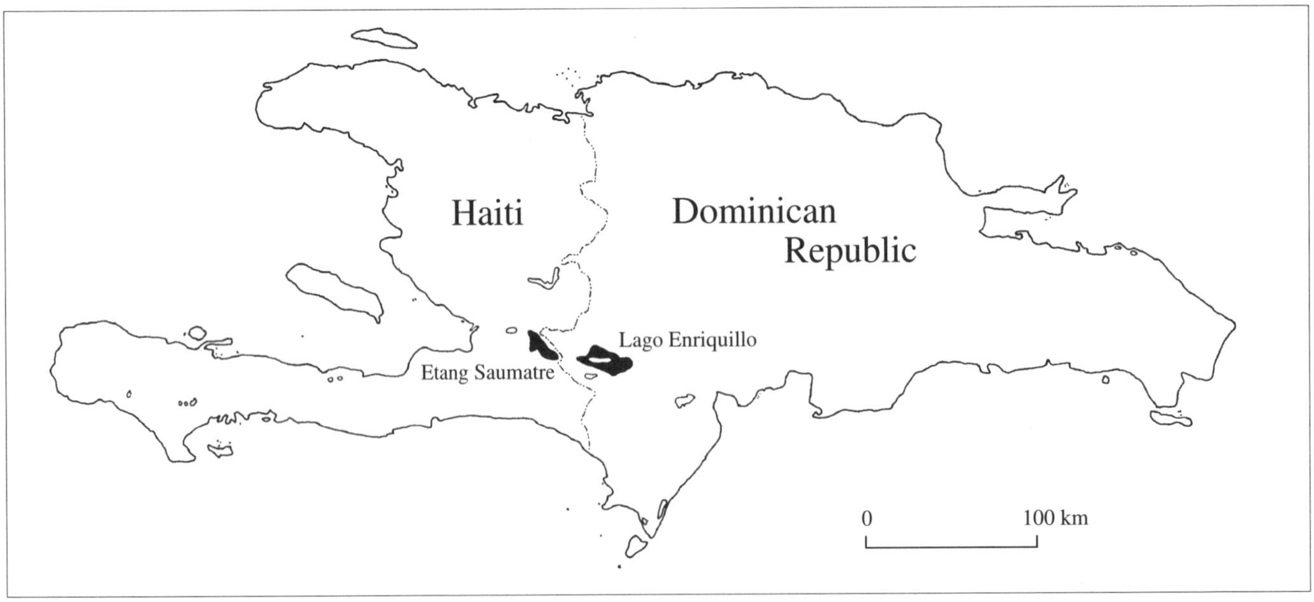

FIGURE 1. Map of Hispaniola illustrating the location of the two inland lakes supporting viable American crocodile populations.

FIGURE 2. Carcass of a crocodile killed in 1991.

HUMAN IMPACT ON THE CROCODILE POPULATION

Before the Isla Cabritos National Park was established in 1974, the impact of human activities on the crocodile population was very severe (SEA/DVS 1993). Sometimes up to 20 crocodiles were killed in a single event. Generally the motive behind this killing was to obtain meat for food, fat for the local pharmaceutical industry, and penises (which are assumed to have aphrodisiac properties), rather than hides.

However, as recently as the early 1980s, the crocodile population of Lago Enriquillo was considered to be the largest and densest throughout the entire range of the species. During the late 1970s and early 1980s different Dominican institutions (CIBIMA, Museo de Historia Natural, and Departamento de Vida Silvestre) studied this population. They estimated an adult crocodile population between 300 and 600 individuals (S. J. Incháustegui and E. A. Bautista, personal communication). Between 70 and 112 nests were found in the years 1977 through 1984 (estimates of the total number of nests vary between 100 and 150 for these years; Incháustegui, *in* SEA/DVS 1993).

In 1990 and 1991 the Dominican Wildlife Department surveyed nesting activities and found alarmingly low numbers of nests, about 30 in 1990 and only 10 in 1991. Very few crocodiles of any age class were seen during these surveys. In one nocturnal count at the end of May 1992, for example, only six crocodiles were seen — and most of them were very thin, as is evident in a series of video recordings from 1992.

At the same time, reports of crocodile killings became more and more frequent (Figure 2). In June 1991, six crocodiles were killed by local inhabitants at one time — apparently just for fun. In February 1992, a crocodile was killed by four fishermen and the meat was sold in a local discotheque.

The Wildlife Department purchased a boat with an outboard motor and a canoe, and initiated field studies in March 1992. The surveys during the first months showed that the status of the crocodile population was even worse than expected. Few adults were seen along the coast and, despite intense searching, only three nests were located. Crocodile bones, including smashed skulls, were found in most of the important basking and nesting areas. Long wooden poles used to capture crocodiles were found and destroyed. The next morning a new pole was seen at the same site. Human tracks were found even on the most remote beaches. At the end of May 1992 we even found bones with flesh and blood still on them. Evidence was common of another way to kill crocodiles, traps built out of sticks in the ground with a snare and bait at the end.

The evaluation of 1992 crocodile monitoring led to an estimate of 160 adult and subadult animals (larger than 1.5 m) for the entire lake (SEA/DVS 1993), indicating that the population had dropped to half or even one-third of its size in the early 1980s. A similar decline was observed for nesting success. In 1993 some 36 nests were found, while in the 1980s between 70 and 112 nests were counted. The cause of the decline is still not totally clear. The Departamento de Vida Silvestre (SEA/DVS 1993) suggested that the main reason for the decline was that crocodiles were killed for their hides. However, no evidence exists to support this supposition.

A more probable reason why male crocodiles were killed was to obtain their penises, which are believed to have aphrodisiac powers. One dried penis 28 cm in length was purchased in a Santo Domingo market, where it was sold secretely and only on request. However, many crocodiles were killed for their meat and fat. The fat was used in the local pharmaceutical industries to produce an anti-rheumatic treatment, and meat was sold in area markets. Lamentably, many crocodiles also were killed indiscriminately, and the corpses were left

FIGURE 3. A local fisherman paddling his small raft in the western part of the lake in 1991.

without being used for any purpose. Mainly young men, but occasionally even children, walked the lakeshore and engaged in these killings. In interviews, several of these young men stated that, once they found a crocodile, some instinct to kill it arose, and they started to beat it or stab it with their machetes. Most of the crocodiles killed in this fashion were smaller individuals.

Another factor affecting the crocodile population was poaching of nests. That only three nests were found in 1992 can be explained by the assumption that the majority of the nests that year were poached. Egg robbers use a narrow stick to probe into the sand. A nest site is identified when the stick comes out with yolk on it. Even with good surveillance installed at all nesting beaches, three of 36 nests were robbed in 1993.

Fishermen in Lago Enriquillo (Figure 3) consider the crocodiles to be potential competitors and even a threat. Fishing in the lake does not have a long tradition. In the early 1960s people used only lines and hooks for fishing. Then, in 1980, after Hurricane David raised the lake level and dropped salinities, populations of *Tilapia* increased, and fishermen began using nets. Crocodiles, apparently attracted by the confined fishes, entangled themselves in the nets, and drowned or were killed by the fishermen. The use of nets had a dramatically negative impact on fish populations, which led, in 1992, to a cessation of all fishing activities in the lake.

Initial Steps to Protect the Crocodile Population

Awareness of the severity of the problems facing the crocodiles caused the Dominican Wildlife Department, which is charged with the protection of wild plants and animals as well as their habitats, to initiate the project entitled "Estudio y Protección del Cocodrilo Americano (*Crocodylus acutus*) en la República Dominicana." Additional technical and financial support was provided by the German Service for Social and Technical Cooperation (DED). Several important steps were taken to save the population from extirpation. In July 1992, a workshop was held in the Santo Domingo Zoo (ZOODOM). A number of different government agencies and nongovernmental organizations participated. As an immediate result of this workshop, a "Surveillance Plan" was established for the whole lake, utilizing park rangers under the National Park Directorate and wildlife inspectors from the Departamento de Vida Silvestre. This continues, and currently, regular patrols are scheduled by some 20 men along the lake shore on foot or by boat. Every month a meeting is held to evaluate the patrols and to plan new activities. While patrolling the lake, rangers and inspectors also collect data on crocodiles and other important species. Another important step was the provision of additional technical support by the United States Peace Corps. Since June 1992, a Peace Corps volunteer has been working on the project with a local counterpart from La Descubierta, one of the villages in the vicinity of the lake.

Action Plan for Conservation

In addition to immediate conservation activities, studies on the crocodile population, socio-economic conditions in the lake area, and impact on the crocodiles were initiated. The results of these studies were published in June 1993 (SEA/DVS 1993) with an "Action Plan for the Conservation of the American Crocodile." The action plan provides for a multi-institutional "Executive Council" to coordinate and supervise the activities of the plan. Activities are carried out by five working teams, each responsible for one facet of the five-part program: (1) surveillance, (2) investigation, (3) reproduction, (4) public relations, and (5) resource management. All teams to date have been quite successful in addressing their goals.

Surveillance

The impact of human disturbances on the crocodiles and on other species has been reduced substantially. Very few incidents (e.g., construction of traps or killing of crocodiles) have been reported since 1992. In 1994, apparently no nests were robbed. Positive side effects of the surveillance are reductions in the frequency of illegal hunting of birds (especially migratory waterfowl and doves) and the capture and the illegal trade of Hispaniolan parrots from the nearby mountains.

Investigation

The year-round monitoring of the lake has yielded important data on abundance and distribution of the crocodiles. According to nocturnal counts, a maximum of 166 adults and subadults was counted. Based on these numbers, we, along with D. Birdsall and H. Mendez, estimated that in 1995 at least 200 individuals larger than 1.5 m live in the lake. The principal habitats for neonates and yearlings have been identified, and more than 20 yearlings were found in April 1995 among cattail stands in the northwestern part of the lake.

Reproduction

In April 1993, a total of 178 eggs from eight nests was taken to ZOODOM to be artificially incubated in a headstart program. From these eggs, 92 neonates hatched (Figure 4). In May another 53 hatchlings were taken to the zoo, for a total of 145 small crocodiles in captivity. More than 130 are still alive and await their liberation in Lago Enriquillo within the next two years.

As 36 nests were identified and monitored in 1993, no eggs or hatchlings were taken to the zoo in 1994. A total of 48 nests was registered in 1994, half of them located on the islands where no freshwater habitat is available (SEA/DVS 1994). Because neonates must move from nesting beaches to freshwater, the decision was made to catch many of the neonates and transport them to recently-identified areas along the northern and northwestern shores.

Before release, each hatchling was measured, weighed, sexed, and individually marked by means of notches in their tail scutes. Two hundred and thirty neonates marked in this manner will enable us to monitor growth, migration patterns, and survival rates of these small crocodiles.

Public Relations

An educational television show was prepared and is being transmitted by different channels throughout the nation. Slide shows on the crocodiles and on Lago Enriquillo have been created and are pre-

FIGURE 4. A newly hatched *Crocodylus acutus* from the headstart program at ZOODOM.

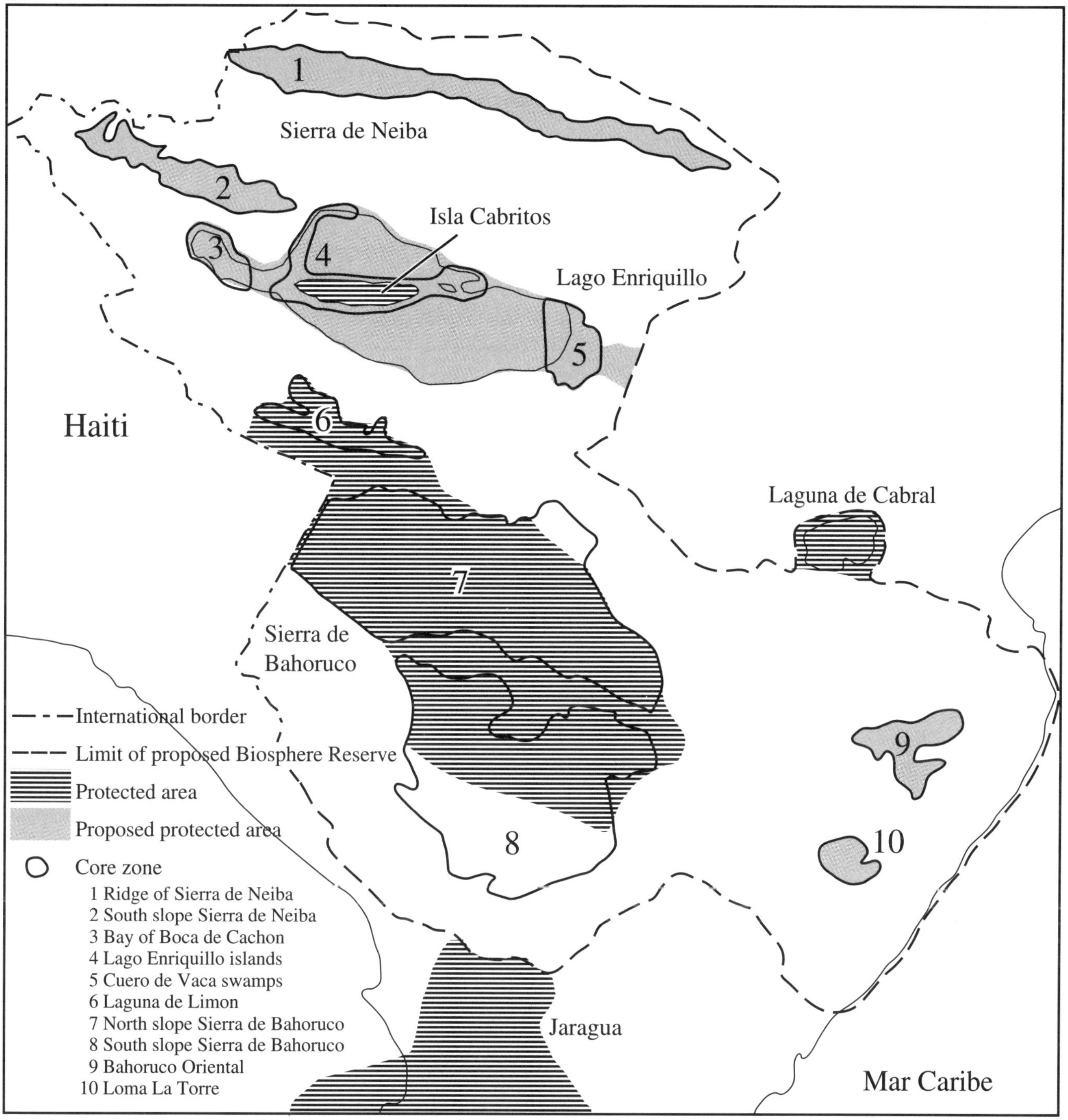

FIGURE 5. The proposed Enriquillo Biosphere Reserve, core zones, and legally protected areas.

sented in communities around the lake. Plans exist to prepare in the near future brochures and posters focusing on the lake and its species; these will be widely distributed in the lake region and throughout the county. To increase the likelihood of success of the conservation plan in the Lago Enriquillo area, the American crocodile is being used as a focal species for all public education efforts.

Resource Management

In 1993, a draft for a presidential decree establishing a new Lago Enriquillo National Park (Figure 5) was written by the staff of the Departamento de Vida Silvestre and forwarded to the president of the Dominican Republic. Undoubtedly due to the 1994 elections, the decree has yet to be signed. The proposed national park includes the entire lake, its shores, and the freshwater swamps to the east and the west. Six areas of substantial ecological importance are scheduled to receive stricter protection. Isla Cabritos is excluded, because it already has protected status as a national park.

Additionally, Vida Silvestre is developing strategies to establish and manage the biosphere reserve "Enriquillo" (Figure 5), which will include Lago Enriquillo, the Sierra de Neiba to the north, and the Sierra de Bahoruco to the south. A proposal for the establishment of this reserve has been drafted and will be channeled to the UNESCO Man and the Biosphere (MaB) headquarters in Paris via the Dominican national MaB committee. The proposed biosphere reserve will include 10 core areas, three of them in Lago Enriquillo and its adjacent wetlands. The proposal also recognizes the lake as an important wetland and seeks to list it as such under the IUCN Ramsar Convention on wetlands.

National and international tourists visit the lake and especially Isla Cabritos, mainly because they wish to see crocodiles and flamingos. To minimize the impact of this growing concern, visitor-guiding schemes still need to be developed, especially in La Azufrada, the point of departure for most tourists visiting the island.

Conclusions

In 1992, the population of *Crocodylus acutus* at Lago Enriquillo was in immediate danger of becoming extirpated. Two years later we can confidently say that this will not happen in the near future. The population size has stabilized and appears to be increasing slowly. The number of nests produced has increased from three in 1992 to 36 in 1993 and to 48 in 1994. More than 500 newly hatched crocodiles were recorded in 1994. If crocodile reproduction continues to increase as it has in the past two years, we may have a population recovery to pre-1980 levels within the next five to ten years. However, conservation efforts must not be relaxed. The establishment of a regional national park, that of the even larger biosphere reserve, and the elaboration and maintenance of appropriate management plans are of the greatest importance.

Acknowledgements

This paper describes the crocodile conservation activities of the Departamento de Vida Silvestre, Secretaría de Estado de Agricultura, mainly carried out by David Birdsall, Wendy James, Hermógenes Méndez, and the authors. Project activities were financed by the German Service for Social and Technical Cooperation (DED) and to the Swiss Association for Development and Cooperation (HELVETAS). We also wish to thank the Sub-secretaría de Recursos Naturales (SURENA), the Dirección Nacional de Parques, the Santo Domingo Zoo (ZOODOM), and the United States Peace Corps, as well as all participating park rangers and wildlife inspectors, for their cooperation.

Literature Cited

Araguás, L., C. Michelen, and J. Febrillet. 1993. Estudio de la dinamica del Lago Enriquillo, informe de avance para el Organismo Internacional de Energía Atomica (OIEA).

Descourtilz, M. E. 1809. Voyages d'un naturaliste, et ses observations. Histoire naturelle du crocodile de Saint Domingue. Dufort pere Libr., Paris 3: 11–108.

Groombridge, B. 1987. The distribution and status of world crocodilians, p. 9–21. *In* G. Webb, C. Manolis, and P. Whitehead (eds.), Wildlife Management: Crocodiles and Alligators. Surrey Beatty & Sons, Chipping Norton (Australia).

Incháustegui, S., W. Gutierrez, V. Rivas, V. Álvarez, N. Nuñez, and I. Bonnelly. 1978. Notas sobre la ecología del Lago Enriquillo, p. 307–342. *In* Centro de Investigaciones de Biología Marina (CIBIMA), Conservación y Ecodesarrollo. Univ. Autónoma Santo Domingo (Dominican Republic).

Mann, P., F. W. Taylor, K. Burke, and R. Kulstad. 1984. Subaerially exposed Holocene coral reef, Enriquillo Valley, Dominican Republic. Bull. Geol. Soc. Amer. 95: 1084–1092.

Margalef, R. 1985. Limnología del Lago Enriquillo. Oecología Acuatica, República Dominicana 8: 1–9.

Moreau de St. Mery, M. L. E. 1797–1798. Description topographic, physique, civile, politique et historique de la parte Francaise de l'Isle de Saint Domingue. 2 vols. Soc. l'Hist. Col. Francaises, Libr. Larose, Paris.

SEA/DVS (Secretaría de Estado de Agricultura/Departamento de Vida Silvestre). 1993. Estudio y Protección del Crocodilo Americano (*Crocodylus acutus*) en la República Dominicana. Secr. Est. Agric., Depto. Vida Silvestre, Santo Domingo.

—. 1994. Mejoramiento de la Situación Ambiental en la Propuesta Reserva de Biosfera "Enriquillo." Secr. Est. Agric., Depto. Vida Silvestre, Santo Domingo.

Thorbjarnarson, J. B. 1988. The status and ecology of the American crocodile in Haiti. Bull. Florida State Mus., Biol. Sci. 33: 1–86.

—. 1992. Crocodiles. An Action Plan for Their Conservation. Internatl. Union Conserv. Nature, Gland (Switzerland).

Resumen.—El cocodrilo americano (*Crocodylus acutus*), una vez ampliamente distribuido en la Española, solamente sobrevive en Lago Enriquillo en la República Dominicana y en Etang Saumatre en Haiti. A finales de los años 1980 y al principio de los 1990 el tamaño de la población y la tasa de reproducción bajaron a un tercio, principalmente causado por impactos humanos. Como resultado de un estudio, ejecutado por el Departamento de Vida Silvestre, un Plan de Accion para la conservación de la especie fue elaborado y actualmente está en implementación. Los resultados principales hasta ahora: (1) el frequente patroullaje de guardaparques e inspectores de vida silvestre en las orillas del lago logró casi por completo parar las perturbaciones; (2) desde el 1992 la población de adultos y subadultos sigue aumentando lentemente y hoy en día se hay alrededor de 200 individuos; (3) desde mayo 1993 una "reserva genetica" de unos 130 neonatos esta en el parque zoológico de Santo Domingo (headstart program); y (4) en un próximo futuro el Lago Enriquillo va a tener protección legal como parque nacional.

Resumé.—Le crocodile americain (*Crocodylus acutus*) qui se distribuait largement en Hispaniola, survit maintenant, seulement dans le lac Enriquillo de la Republique Dominicaine et dans l'Etang Saumatre en Haiti. A la fin des années 1980 et au debut des années 1990, la dimension de la population et le taux de reproduction ont baissés á un tiers, causés principalement par l'impact humain. Comme resultat d'une étude executée par le Departement de la Vie Sauvage un plan d'action pour la conservation de l'espece a été elaboré et actuellement se trouve en execution. Les resultats principaux jusqu'a maintenant: (1) le patrouillage frequent des gardes de parc et des inspecteurs de la vie sauvage au bord du lac ont arretées completement les perturbations humaines; (2) depuis 1992 la population d'adultes et sous-adultes a augmentée et aujourd'hui il existent environ 200 individus; (3) depuis Mai 1993 une "reserve genetique" de quelques 130 jeunes crocodiles grandissent dans le parc zoologique de Santo Domingo (headstart program); et (4) dans un futur proche le lac Enriquillo aura une protection legale comme parc national.

Schubert, A. and G. Santana. 1996. Conservation of the American Crocodile (*Crocodylus acutus*) in the Dominican Republic, p. 425–433. *In* R. Powell and R. W. Henderson (eds.), Contributions to West Indian Herpetology: A Tribute to Albert Schwartz. Society for the Study of Amphibians and Reptiles, Ithaca (New York). Contributions to Herpetology, volume 12.

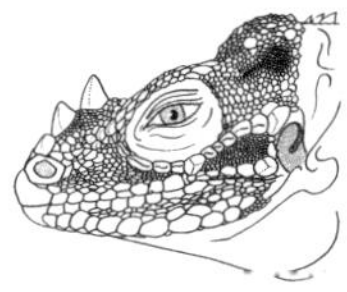

Abstracts of Additional Papers Presented at the Symposium in Honor of Albert Schwartz

Pre-Columbian Human Influences on the Biogeography of Amphibians and Reptiles in the Yucatan Peninsula

Julian C. Lee[1]

Abstract.—To judge by their representation in virtually all aspects of Mayan artistic expression, amphibians and reptiles featured prominently in the thought, mythology, and religion of the ancient Maya. Serpents are ubiquitous, but turtles, crocodiles, lizards, frogs, and toads appear frequently and in various contexts. Apart from their symbolic importance, amphibians and reptiles were also of practical significance, serving as food, providing raw materials for the manufacture of various items, and, perhaps, serving as a source of psychoactive chemicals ingested to induce altered states of consciousness. The zooarcheological record documents the occurrence of several species of Yucatecan reptiles outside of their present geographical ranges, suggesting habitat alteration or transport through human agencies. Moreover, the disjunct distributions of some forest-avoiding reptiles in the Yucatan Peninsula may be attributed to deforestation by the Maya, followed by forest regeneration subsequent to the Mayan collapse around 900 AD. Some paleobotanical data are consistent with this interpretation, but the timing of forest regeneration is uncertain, and may have coincided with the arrival of Europeans in the early 16th century.

Resumen.—Los anfibios y reptiles ocuparon un lugar prominente en el pensamiento, la mitología y la religión de los antiguos mayas, como indican la expresiones artísticas representadas en casi todas las manifestaciones del arte maya. La serpientes son ubicuas, pero las tortugas, los cocodrilos, las lagartijas, las ranas y los sapos aparecen frecuentemente y en contextos diversos. Además de su importancia simbólica, los anfibios y reptiles también tenían una importancia práctica y servían de alimento, de materia prima para la elaboración de diversos artículos y quizás constituían una fuente de sustancias químicas psicoactivas ingeridas con el fin de provocar transformaciones psíquicomentales. El registro zooarqueológico documenta que varias especies de reptiles de Yucatán ocurren fuera de sus ámbitos geográficos actuales, lo cual sugiere que hubo una modificación de su hábitat o que fueron transportadas por conductos humanos. Además, se pueden atribuir las distribuciones alopátricas de algunos reptiles que evitan los bosques en la Península de Yucatán a la deforestación por parte de los maya, sequida de la regeneración de los bosques posterior a la ruina la civilización maya alrededor de 900 AD. Algunos datos paleobotánicos son consistentes con esta interpretación, pero se cesconoce la fecha exacta de la regeneración de los bosques y puede haber coincidido con la llegada de los europeos hacia comienzos del siglo XVI.

Résumé.—Si l'on en juge par leur présence très marquée dans presque toutes les manifestations de l'art maya, les amphibiens et reptiles occupaient une place prédominante dans la pensée, la mythologie et la religion des Mayas. Si les serpents sont ubiquitaires, les tortues, les crocodiles, les lézards, les grenouilles et les crapauds apparaissent fréquemment et dans différents contextes. En marge de leur importance symbolique, les amphibiens et

[1] Department of Biology, University of Miami, Coral Gables, Florida 33124, USA.

reptiles avaient également une importance pratique puisque les Mayas s'en nourrissaient, s'en servaient dans la fabrication de différents articles, voire dans la préparation de substances psychoactives qui permettaient d'induire différentes transformations psychiques. Les registres zooarchéologiques signalent la présence de plusieurs espèces de reptiles du Yucatán en dehors de leurs secteurs géographiques actuels, ce qui suppose peut-être une altération des habitats ou leur transport par l'être humain. Par ailleurs, la répartition disjointe de certains reptiles de la péninsule du Yucatán qui ne vivent pas en forêt est peut-être attribuable aux activités de déforestation des Mayas, suivies par le reboisement qui a fait suite au déclin de l'empire maya vers 900 av. J.-C. Certaines données paléobotaniques confirment cette interprétation mais la date du reboisement reste incertaine et pourrait coïncider avec l'arrivée des Européens, au début du XVIe siècle.

•

Population Differentiation of *Iguana delicatissima* and *I. iguana* in the Lesser Antilles

Mark L. Day[1] and Roger S. Thorpe[2]

Abstract.—Multivariate analysis of body dimensions, scalation, and color patterns, in combination with DNA methods, were used to investigate geographic variation in nine populations of the Lesser Antillean Iguana (*Iguana delicatissima*) and seven populations of the common iguana (*I. iguana*) from the Eastern Caribbean. Multivariate analysis revealed less morphological variation in *I. delicatissima* than in *I. iguana*, but matrix correlation (Mantel tests) indicated a significant association between facets of this variation and climatic variation in *I. delicatissima*. Direct sequencing of a section of the mtDNA cytochrome b gene revealed no variation among *I. delicatissima* populations, but analysis of the region is continuing. The section of cytochrome b investigated revealed 13% bp differences between *I. delicatissima* and *I. iguana*. Hybrids between these two species on Les Iles des Saintes, Guadeloupe were detected by multivariate and molecular methods. This hybridization may be occurring rapidly and could result in local extinctions of *I. delicatissima*.

Resumen.—El análisis multivariado de dimensiones corporales, de patrones de escamas, y de coloración, en combinación con métodos de estudio de ADN, fueron usados para investigar la variación geográfica en nueve poblaciones de la iguana de las Antillas Menores (*Iguana delicatissima*) y en siete poblaciones de la iguana común (*I. iguana*) del Caribe Oriental. El análisis multivariado reveló menos variación morfológica en *I. delicatissima* que en *I. iguana*, pero una matriz de correlación (pruebas Mantel) indicó una asociación significativa en *I. delicatissima* entre aspectos de esta variación y variación climática. La secuenciación directa de un segmento de ADN del gene mitocondrial citocromo b no reveló ninguna variación entre las poblaciones de *I. delicatissima*, pero el análisis de esta región continúa. El segmento estudiado del citocromo b mostró un 13% de diferencia en pares de bases entre *I. delicatissima* e *I. iguana*. Se detectaron híbridos entre estas dos especies en Îles des Saintes, Guadalupe, usando métodos multivariados y moleculares. Esta hibridización puede estar ocurriendo rápidamente y podría causar extinciones locales de *I. delicatissima*.

Résumé.—Les analyses multifactorielles des mesures anthropométriques, des échelles et des couleurs ainsi que des méthodes d'ADN ont été utilisées pour étudier les variations géographiques de neuf populations d'iguanes des Petites Antilles (*Iguana delicatissima*) et de sept populations d'iguanes communs (*I. iguana*) de l'Est des Antilles. L'analyse multifactorielle révèle moins de variations morphologiques dans *I. delicatissima* que dans *I. iguana*. Les tests de Mantel (tests logarithmiques) indiquent un rapport significatif entre les facettes de cette variation et la variation climatique chez *I. delicatissima*. Le séquençage direct d'une section

[1] Department of Zoology, University of Aberdeen, Tillydrone Avenue, Aberdeen AB9 2TN, United Kingdom.

[2] School of Biological Sciences, University College of North Wales, Bangor, Gwynedd LL57 2UW, United Kingdom.

de l'ADNmt du gène cytochrome b n'a révélé aucune variation parmi les populations d'*I. delicatissima* et l'analyse de la région se poursuit. La section du cytochrome b étudiée révèle des différences de 13% bp entre *I. delicatissima* et *I. iguana*. Des hybrides de ces deux espèces ont été décelés sur l'île des Saintes (Guadeloupe), par des méthodes multifactorielles et moléculaires. Cette hybridation peut survenir rapidement et pourrait entraîner la disparition locale d'*I. delicatissima*.

•

Long-Term Study of Activity Patterns of Puerto Rican Rain Forest Frogs[1]

Margaret M. Stewart[2]

Abstract.—A deme of *Eleutherodactylus coqui* was followed from 1979 to 1993 to determine seasonal and annual variation in numbers and activity patterns. The study site was adjacent to the El Verde Field Station at 350 m in the Luquillo Experimental Division, Caribbean National Forest. All visible frogs and predatory spiders in a 50 x 2-m transect in the forest were counted for three consecutive nights semi-monthly for two years, then annually or biannually thereafter for a total of 255 evening counts. Ten all-night counts were made at five different times of the year to determine the time of maximal activity during the night. The maximum numbers were seen at the 2000-h count. The number of visible adults was constant throughout the night, but the number of visible juveniles declined steadily during the night. All frogs were in their diurnal retreats by 0600 h.

Calling and reproduction occurred throughout the year, but slowed during cooler months, October to March. When temperatures fell below 20°C, almost all calling ceased, especially when conditions were also dry. After prolonged periods without rain, activity declined, and few frogs, mostly adults in water-conserving postures, were visible. Population size varied seasonally, with numbers increasing from June until December, followed by a gradual decline until May. The number of adults varied from 1–29/100 m^2, whereas the number of juveniles varied from 0–221/100 m^2. The maximum single count of all frogs was 244/100 m^2. Counts of more than 100 juveniles occurred during October through January in the years 1979–1982, and in 1989. A marked drop in the numbers of frogs occurred in 1983; from 1979–1983, 3–50% of the counts yielded 15 or more adults, whereas the maximum count from 1984–1988 was 11 adults. The drop in numbers was correlated with an increased number of periods of days with 3 mm or less of rain. Neither monthly nor annual rainfall differed significantly during the study, but periods with little or no rain increased in number and length after 1982. The years 1983, 1984, and 1985 had the longest dry periods (22, 28, and 25 days, respectively) during the pre-Hurricane Hugo period. After the decrease in 1983, populations never regained prior levels, although numbers were increasing early in 1989 before Hurricane Hugo (September 1989).

Over the time period 1979–1989, the number of frogs observed was negatively correlated with the longest dry period during the previous year. Juveniles apparently cannot survive extensive drought, and extended dry periods may be lethal to adults, which are inhibited from feeding and breeding because of potential desiccation. A widespread generalist such as *Eleutherodactylus coqui* can recover rapidly from periodic stresses, but the less abundant congeneric species, *E. wightmanae* and *E. portoricensis*, disappeared from the site during the latter part of the study. Small species, or those preferring moist sites, are more vulnerable to long periods without rain.

Predatory spider populations crashed two years following the decline of the frog populations, then disappeared following the hurricane. A 30-day dry period after the hurricane could have been serious for small frogs and other moisture-dependent species. Juvenile frogs declined dramatically after Hurricane Hugo, but adults increased. Deep litter provided ample retreat and nest sites.

[1] Full paper published in the *Journal of Herpetology* (1995, volume 29, number 3, pages 437–446).

[2] Department of Biological Sciences, State University of New York at Albany, Albany, New York 12222, USA.

The most abundant predators in the transect were sparassid crab spiders (*Olios antiguensis*, *Stasina portoricensis*). Other predators seen were the tarantula *Avicularia laeta*, the amblypygid *Phrynus longipes*, the forest crab *Epilobocera situatifrons*, the cricket *Amphiacusta caraibea*, the Puerto Rican screech owl *Otus nudipes*, and the black rat *Rattus rattus*. The arachnid predators were rarely seen following the hurricane.

Rather than total monthly or annual rainfall, the distribution of the rain was important to these rain forest species. Erratic rainfall patterns may have resulted from the 1982–1983 El Niño Southern Oscillation event.

Resumen.—Un deme de *Eleutherodactylus coqui* fue estudiado para determinar la variación temporal y anual en números y patrones de actividad desde 1979 a 1993. El lugar de estudio quedaba adyacente a la Estación de El Verde (elevación 350 m), División Experimental de Luquillo, Bosque Nacional del Caribe. Todas las ranas y arañas depredadoras visibles en un transecto de 50 x 2 m en el bosque fueron contadas por tres noches consecutivas dos veces al mes por dos años, y después una vez al año o cada dos años, para un total de 255 conteos vespertinos. Se efectuaron diez conteos durante toda la noche en cinco ocasiones diferentes al año para determinar el período de mayor actividad durante la noche. Los números máximos fueron vistos en el conteo de las 2000 horas. El número de ranas adultas visibles fue constante durante la noche, pero el número de ranas juveniles visibles disminuyó constantemente durante la noche. Todas las ranas estaban en sus refugios a las 0600 horas.

El canto y la reproducción ocurrieron durante todo el año, pero disminuyeron durante los meses más frescos, octubre a marzo. Cuando las temperaturas descendieron bajo los 20°C, casi todo el canto cesó, especialmente cuando las condiciones también eran secas. Luego de períodos prolongados sin lluvia, la actividad disminuyó y pocas ranas, la mayoría en posturas para conservar agua, fueron visibles. El tamaño de la población varió por temporadas, con los números aumentando de junio a diciembre, seguido por un descenso gradual hasta mayo. El número de ranas adultas varió de 1 a 29/100 m^2, mientras que el número de ranas juveniles varió de 0 a 221/100 m^2. Los conteos de más de 100 ranas juveniles ocurrieron durante el período de octubre hasta enero en los años de 1979 a 1982 y en 1989. Un descenso en el número de ranas ocurrió en 1983; de 1979 a 1983, de 3 a 50% de los conteos totalizó 15 o más ranas adultas, mientras que el conteo máximo de 1984 a 1988 fue de 11 ranas adultas. Esta disminución estuvo correlacionada con un aumento en el número de períodos de días con 3 mm o menos de lluvia. La precipitación pluvial mensual o anual no varió significativamente durante el estudio, pero períodos con poca o ninguna lluvia aumentaron en frecuencia y duración después de 1982. Los años 1983, 1984, y 1985 tuvieron los períodos de sequía más largos (22, 28, y 25 días, respectivamente) antes del Huracán Hugo. Luego del descenso de 1983, las poblaciones nunca regresaron a los niveles anteriores, aunque los números fueron aumentando gradualmente a inicios del 1989, antes del Huracán Hugo (en septiembre de 1989).

Durante el período de 1979 a 1989, el número de las ranas observadas estuvo correlacionado negativamente con el período seco más largo del año anterior. Aparentemente, las ranas juveniles no pudieron sobrevivir la sequía extendida, y extensos períodos secos pueden ser letales para las ranas adultas porque no pueden alimentarse o reproducirse debido al peligro de desecación. Un generalista que ocurre en muchas partes como *Eleutherodactylus coqui* puede recobrarse rápidamente de períodos de estrés, pero las especies congenéricas menos abundantes, *E. wightmanae* y *E. portoricensis*, desaparecieron del lugar durante la última parte del estudio. Las especies pequeñas, o aquellas que preferían sitios más húmedos, fueron más vulnerables a largos períodos sin lluvia.

La población de arañas depredadoras disminuyó estrepitosamente dos años después de la declinación de las poblaciones de ranas y desapareció luego del huracán. Un período seco de 30 días después del huracán pudo haber sido severo para ranas pequeñas y otras especies dependientes de humedad. Las ranas juveniles declinaron dramáticamente luego del

Huracán Hugo, pero las ranas adultas aumentaron. La profundidad de la hojarasca y las ramas caídas proveyeron suficientes lugares de refugio y anidaje.

Los depredadores más abundantes en el transecto fueron arañas cangrejos esparásidas (*Olios antiguensis*, *Stasina portoricensis*). Otros depredadores vistos fueron la tarántula *Avicularia laeta*, el amblipígido *Phrynus longipes*, el cangrejo de bosque *Epilobocera sinuatifrons*, el grillo *Amphicausta caraibea*, el múcaro puertorriqueño *Otus nudipes*, y la rata negra *Rattus rattus*. Los arácnidos depredadores fueron vistos en pocas ocasiones después del huracán.

En lugar de la cantidad total de lluvia mensual o anual, la distribución de la precipitación pluvial fue más importante para estas especies del bosque lluvioso. Los patrones erráticos de precipitación pluvial pueden haber sido causados por el evento de Oscilación Meridional El Niño de 1982 a 1983.

Résumé.—Un dème d'*Eleutherodactylus coqui* a été suivi entre 1979 et 1993 pour déterminer les variations annuelles et saisonnières dans le nombre de grenouilles et leurs activités. Le site étudié jouxtait la station El Verde à 350 m de la Luquillo Experimental Division de la Caribbean National Forest. Toutes les grenouilles visibles et araignées prédatrices d'un transect de 50 x 2 m dans la forêt ont été dénombrées pendant trois nuits consécutives, deux fois par mois pendant deux ans, puis chaque année ou tous les deux ans par la suite, pour un total de 255 décomptes nocturnes. Dix décomptes nocturnes ont été effectués à cinq moments différents de l'année pour déterminer le moment d'activité maximale pendant la nuit. C'est au décompte de 20 h 00 que l'on a observé le plus grand nombre de sujets. Le nombre d'adultes visibles est resté constant pendant la nuit mais le nombre de jeunes visibles a décliné de façon marquée pendant la nuit. Toutes les grenouilles étaient dans leur retraite diurne à 6 h 00.

Les coassements et activités de reproduction surviennent toute l'année mais ralentissent pendant les mois les plus frais (octobre à mars). Lorsque les températures chutent en-deçà de 20°C, presque tous les coassements s'interrompent. Le même phénomène a été observé en cas de sécheresse. Pendant les périodes prolongées sans précipitation, les activités des grenouilles ralentissent et peu de grenouilles sont visibles (la plupart étant des adultes en position de conservation d'eau). La taille des populations varie chaque saison, augmentant entre juin et décembre puis diminuant graduellement jusqu'en mai. Le nombre d'adultes varie entre 1–29/100 m^2 alors que le nombre de jeunes varie entre 0–221/100 m^2. Le décompte maximal de toutes les grenouilles s'est établi à 244/100 m^2. Les décomptes qui permettent de recenser plus de 100 jeunes se font d'octobre à janvier, pour les années 1979–1982 puis en 1989. Un déclin marqué dans le nombre de grenouilles s'observe en 1983; de 1979 à 1983, 3 à 50 % des décomptes ont permis de recenser 15 adultes ou plus alors que le décompte maximal entre 1984 et 1988 a permis de recenser 11 adultes. Ce recul coïncide avec un plus grand nombre de journées avec des précipitations inférieures à 3 mm. Ni les précipitations mensuelles ni les précipitations annuelles n'ont différé significativement pendant l'étude mais les périodes de sécheresse ou de faibles précipitations ont augmenté en nombre et en durée après 1989. Les années 1983, 1984 et 1985 furent les années les plus sèches (sécheresse respectivement de 22, 28 et 25 jours) et précèdaient l'ouragan Hugo. Après le recul de 1983, les populations de grenouilles n'ont jamais récupéré leurs niveaux antérieurs même si leur nombre a augmenté au début de 1989 avant l'ouragan Hugo (septembre 1989).

Entre 1979 et 1989, nous avons pu établir une corrélation négative entre le nombre de grenouilles observées et la sécheresse la plus longue pendant l'année antérieure. Les jeunes ne peuvent apparemment pas survivre aux sécheresses prolongées et celles-ci peuvent être également mortelles pour les adultes qui ne peuvent ni s'alimenter ni se reproduire en raison des risques de déssiccation. *Eleutherodactylus coqui* se remet plus rapidement de ces stress périodiques mais ses congénères moins abondants comme *E. wightmanae* et *E. portoricensis* ont disparu du site pendant la dernière partie de l'étude. Les petites espèces ou celles qui préfèrent des lieux plus humides, sont plus vulnérables aux longues périodes de sécheresse.

Les populations d'araignées prédatrices ont fortement diminué, deux ans après le déclin des populations de grenouilles puis ont disparu après l'ouragan. Une période de sécheresse de 30 jours après l'ouragan aurait pu être très grave pour les petites grenouilles et autres espèces vivant en milieu humide. Les populations de jeunes grenouilles ont diminué de façon spectaculaire après l'ouragan Hugo mais les populations adultes ont augmenté. La litière profonde offrait d'abondants sites de retraite et de reproduction.

Les prédateurs les plus abondants dans le transect étaient les araignées-crabes (*Olios antiguensis*, *Stasina portoricensis*). Parmi les autres prédateurs figurait la tarentule *Avicularia laeta*, l'amblypyge *Phrynus longipes*, le crabe *Epilobocera situatifrons*, le criquet *Amphiacusta caraibea*, le petit-duc maculé portoricain *Otus nudipes* et le rat noir *Rattus rattus*. Les prédateurs arachnides se sont fait plus rares après l'ouragan.

La répartition des pluies est importante pour ces espèces des forêts tropicales, plus que ne le sont les précipitations mensuelles ou annuelles totales. L'irrégularité des précipitations est peut-être due aux variations provoquées par El Niño en 1982–1983.

•

Developmental Migrations and Habitat Use by Green Turtles in the Bahamas

Karen A. Bjorndal[1] and Alan B. Bolten[1]

Abstract.—Tidal bays or "creeks" in the Bahamas are important habitat for immature green turtles, *Chelonia mydas*. A long-term tagging study in Union Creek, Great Inagua, Bahamas, has provided information on the movements of immature green turtles in the Greater Caribbean. Tags placed on green turtles in Union Creek have been returned from the northern Bahamas, Colombia, Cuba, Dominican Republic, Haiti, Nicaragua, and Venezuela. Mitochondrial DNA sequence analyses (in collaboration with P. Lahanas, S. Encalada, and M. Miyamoto) have demonstrated that green turtles in Union Creek arrive from nesting beaches throughout the Atlantic. Conception Island Creek in the central Bahamas supports a population of immature green turtles for which we have estimated residency times. Conception Island is not inhabited by humans and is a Wildlife Sanctuary of the Bahamas National Trust; these factors afford the green turtle population some protection. Tidal bays in the Bahamas, and probably throughout the West Indies, once served as refugia for small green turtles. Today, these areas serve as natural funnel traps. Turtle hunters visit these areas every few weeks or months and easily capture turtles that have accumulated during the interval.

[1] Center for Sea Turtle Research, University of Florida, Gainesville, Florida 32611, USA.

Resumen.—Los estanques de agua o "ensenadas" que se forman en las costas rocosas al bajar la marea en las Islas Bahamas son habitáculos importantes para tortugas verdes sexualmente inmaduras *Chelonia mydas*. Un estudio de marcaje de larga duración en Union Creek, Gran Inagua, Islas Bahamas, ha proporcionado información sobre los movimientos de las tortugas verdes sexualmente inmaduras en el Caribe. Se han devuelto etiquetas colocadas en las tortugas verdes en Union Creek desde Colombia, Cuba, República Dominicana, Haití, Nicaragua, Venezuela, y las Islas Bahamas septentrionales. Los análisis de secuencia del adn mitocondrial (en colaboración con P. Lahanas, S. Encalada y M. Miyamoto) han demostrado que las tortugas verdes en Union Creek llegan desde playas de anidaje de todo el Atlántico. Hemos estimado el tiempo de residencia de la población de tortugas verdes sexualmente inmaduras que se encuentra en la ensenada de la Isla de Concepción en las Islas Bahamas centrales. La Isla de Concepción no está habitada por el hombre y es un Santuario de Vida Silvestre del Fideicomiso Nacional de las Islas Bahamas; estos factores le proveen alguna protección a la población de tortugas verdes. Los estanques de agua que se forman en las costas rocosas al bajar la marea en las Islas Bahamas, y probablemente en todas las Antillas, sirvieron alguna vez de refugio a tortugas verdes pequeñas. En el presente estas áreas sirven como una trampa de embudo natural. Los cazadores de tortugas visitan estos lugares con cierta frecuencia y capturan fácilmente las tortugas que se han acumulado ahí.

Résumé.—Les baies ou criques des Bahamas constituent des habitats privilégiés pour les tortues vertes immatures *Chelonia mydas*. Une étude à long terme par étiquetage menée à Union Creek, Great Inagua (Bahamas), a permis de recueillir des renseignements sur les déplacements des tortues vertes immatures dans les Grandes Antilles. Les étiquettes placées sur les tortues vertes à Union Creek nous ont été retournées en provenance du nord des Bahamas, de Colombie, de Cuba, de la République dominicaine, d'Haïti, du Nicaragua et du Venezuela. Les analyses de séquence d'ADN des mitochondries (en collaboration avec P. Lahanas, S. Encalada et M. Miyamoto) ont démontré que les tortues vertes d'Union Creek proviennent de plages dans différents endroits de l'Atlantique. Conception Island Creek dans le centre des Bahamas accueille une population de tortues vertes immatures dont nous avons évalué le temps de séjour. Conception Island n'est pas habitée par l'homme et il s'agit d'un sanctuaire du Bahamas National Trust; ces facteurs contribuent à la protection des populations de tortues vertes. Les criques des Bahamas et probablement de toutes les Antilles, ont déjà servi de refuge aux petites tortues vertes. Aujourd'hui, ces secteurs constituent de véritables pièges naturels. Les chasseurs de tortues se rendent régulièrement dans ces criques où ils capturent facilement les tortues qui y ont trouvé refuge.

•

Harmful Prey Aversion in Hatchling *Anolis lineatopus*: Learning the Good Things First

Peter Vogel[1]

Abstract.—Some prey-naive hatchlings of the lizard, *Anolis lineatopus*, reject ants and various other prey. The percentage of prey-naive hatchlings rejecting ants decreased with hatchling size, and increased with ant size, and with the abundance of stinging ants, *Solenopsis geminata*, in their habitat. In a growth experiment, prey-naive hatchlings were first tested with a harmless ant and then released to a site where *S. geminata* was abundant. Over a two-week period, small hatchlings which initially had accepted ants grew more slowly than similarly sized rejectors; the results were reversed for large hatchlings. When offered harmless prey together with *S. geminata*, prey-naive acceptors developed an aversion to most prey types. Prey-naive rejectors at first rejected many other prey items besides ants, but then diversified their diets to include all non-ant prey. Although they also attacked *S. geminata* eventually, this did not result in aversion to most other prey. Overall, initial rejectors ate more prey items than did prey-naive acceptors. In conclusion, the early rejection of potentially harmful prey allowed the predator to become familiar with harmless prey: when harmful prey eventually was attacked, the predator was sufficiently experienced to separate the good from the bad.

[1] Department of Zoology, University of the West Indies, Kingston 7, Jamaica.

Resumen.—Algunos especímenes recién nacidos de la lagartija *Anolis lineatopus* que no tienen experiencia capturando su alimento rechazan hormigas y varias otras presas. El porcentaje de lagartijas recién nacidas inexpertas que rechaza hormigas disminuyó con el tamaño de la lagartija recién nacida y aumentó con el tamaño de la hormiga y con la abundancia de las hormigas aguijoneadoras *Solenopsis geminata* en el habitáculo. En un experimento de crecimiento, las lagartijas recién nacidas inexpertas fueron observadas con una hormiga inofensiva y luego liberadas en un lugar donde *S. geminata* era abundante. Durante un período de dos semanas, las lagartijas recién nacidas más pequeñas, que inicialmente habían aceptado hormigas, crecieron más lentamente que las lagartijas del mismo tamaño que habían rechazado las hormigas; las lagartijas recién nacidas de mayor tamaño mostraron resultados contrarios. Las lagartijas inexpertas desarrollaron aversión por la mayoría de los tipos de presas cuando se les ofrecieron hormigas inofensivas simultáneamente con *S. geminata*. Al principio, las lagartijas inexpertas rechazaron muchos otros tipos de presas además de hormigas, pero entonces diversificaron sus dietas para incluir todo tipo de presa que no fueran hormigas. Aunque eventualmente ellas también atacaron a *S. geminata*, esto no resultó en aversión a la mayoría de las demás presas. En general, las lagartijas que inicialmente rechazaron las presas luego comieron más tipos de presas que las lagartijas inexpertas que al principio aceptaron las presas. Por último, el rechazo temprano de presas potencialmente nocivas le permitió al depredador familiarizarse con presas inofensivas; cuando las presas dañinas fueron atacadas posteriormente, el depredador tenía suficiente experiencia como para separar lo bueno de lo malo.

Résumé.—Certains jeunes lézards *Anolis lineatopus* ne connaissant rien des proies rejettent spontanément les fourmis et diverses autres proies. Le pourcentage de jeunes ne connaissant rien des proies qui rejettent les fourmis diminue avec la taille du lézard et augmente avec la taille des fourmis et l'abondance des fourmis piqueuses, *Solenopsis geminata*, dans leur habitat. Dans le cadre d'une expérience de croissance, les jeunes ignorant tout de leurs proies ont d'abord été testés avec une fourmi inoffensive puis relâchés sur un site où *S. geminata* abondait. Après deux semaines, les petits lézards qui acceptaient au début les fourmis grandissaient moins vite que ceux de même taille qui les rejetaient; les résultats sont inversement proportionnels pour les jeunes dont la taille est supérieure. Les lézards qui ignorent tout de leurs proies à qui l'on a offert des proies inoffensives ainsi que *S. geminata* ont développé une aversion contre presque tous les types de proies. Ceux qui rejetaient les fourmis rejetaient au départ de nombreuses autres proies en plus des fourmis, puis ont diversifié leur alimentation pour y incorporer toutes les proies sauf les fourmis. Même s'ils s'attaquaient parfois à *S. geminata*, les lézards n'ont pas développé d'aversion contre la plupart des autres proies. Globalement, ceux qui rejetaient les fourmis au départ mangeaient plus de proies que ceux qui les acceptaient. En conclusion, le rejet précoce des proies potentiellement nocives permet au prédateur de se familiariser avec les proies inoffensives: lorsqu'ils finissent par attaquer des proies nocives, ils sont suffisamment expérimentés pour séparer le bon du mauvais.

•

Anoline Diversity in Three Differentially Altered Habitats in the Sierra de Baoruco, Dominican Republic, Hispaniola[1]

Lori A. Lenart[2] and Scott P. Sowell[3]

Abstract.—We examined three disturbed, but distinctly different sites on the historically forested eastern slope of the Sierra de Baoruco, Provincia de Barahona, Dominican Republic, in order to evaluate the impact of habitat alteration on the composition of anoline lizard communities. Five species of *Anolis* (*A. bahorucoensis*, *A. barahonae*, *A. coelestinus*, *A. cybotes*, *A. distichus*) are known to occur in the immediate area of our study sites. Only three were found in the most substantially altered clear-cut site, whereas all five species occupied the most heterogeneous habitat, a coffee plantation characterized by a three-tiered canopy. Similarly, numbers of individuals of all but one species were greatest in the plot that most closely resembled the natural forests in this area. Calculated niche breadths, based on analysis of stomach contents, of the three species found in all sites increased with the degree of habitat alteration. The higher number of *A. coelestinus* in more heavily altered sites may be attributed to density compensation, the result of this anole taking advantage of microhabitats apparently unavailable to the more specialized species.

Resumen.—Examinamos tres lugares perturbados por el hombre, pero claramente diferentes, en la históricamente forestada ladera oriental de la Sierra de Bahoruco, Provincia de Barahona, República

---

[1] Accepted for publication in *Biotropica*.

[2] Department of Biology, Indiana University, Bloomington, Indiana 47406, USA.

[3] Department of Biology, University of North Florida, Jacksonville, Florida 32224, USA.

Dominicana, para evaluar el impacto de la alteración del habitáculo en la composición de comunidades de lagartijas anolinas. Se sabe que cinco especies de *Anolis* (*A. bahorucoensis*, *A. barahonae*, *A. coelestinus*, *A. cybotes*, *A. distichus*) habitan en el área immediata a nuestros lugares de estudio. Solamente tres especies fueron halladas en el lugar más intensamente deforestado, mientras que las cinco especies viven en el habitáculo más heterógeneo: una plantación de café caracterizada por un dosel de tres niveles. De la misma manera, el número de individuos de todas, excepto una de las especies, fue mayor en la parcela que más se asemejaba a los bosques naturales en esta área. Las amplitudes del nicho, calculadas a base de análisis de contenidos estomacales de las tres especies presentes en todas las áreas, aumentaron con el grado de alteración del habitáculo. El número mayor de *A. coelestinus* en lugares perturbados más intensamente puede atribuirse a compensación de densidad, es decir, al hecho de que esta especie está aprovechándose de microhabitáculos aparentemente no disponibles para las especies más especializadas.

Résumé.—Nous avons étudié trois sites perturbés très différents sur le versant oriental qui était historiquement boisé de la Sierra de Baoruco dans la province de Barahona en République Dominicaine, pour évaluer l'impact des perturbations de l'habitat sur la composition des communautés d'anoles. Cinq espèces d'*Anolis* (*A. bahorucoensis*, *A. barahonae*, *A. coelestinus*, *A. cybotes* et *A. distichus*) vivent à proximité des secteurs que nous avons choisis d'étudier. Seules trois espèces ont été observées dans le site le plus perturbé par les coupes forestières alors que les cinq espèces occupaient l'habitat le plus hétérogène, une plantation de café caractérisée par un couvert forestier à trois niveaux. Parallèlement, le nombre d'animaux de toutes les espèces, sauf une, était supérieur dans la zone qui ressemblait le plus étroitement aux forêts naturelles de la région. D'après l'analyse du contenu des estomacs des trois espèces trouvées sur les sites, plus l'habitat était perturbé, plus la niche écologique était étendue. L'abondance d'*A. coelestinus* sur les sites les plus perturbés était peut-être une réaction à la densité des autres sites; *A. coelestinus* profitait peut-être de micro-habitats apparemment inaccessibles aux espèces plus spécialisées.

•

LEARNING AND MEMORY IN *ANOLIS* LIZARDS

SHARONI SHAFIR[1]
AND JONATHAN ROUGHGARDEN[1]

Abstract.—A model for optimal foraging by a sit-and-wait *Anolis* lizard is presented. The optimal cutoff distance up to which prey should be pursued, and beyond which prey should be ignored, that maximizes the mean rate of net energy intake, depends on the size and abundance of prey. The analytic solution is computationally intensive, and seems to be well beyond a lizard's cognitive abilities. A simple decision-rule is proposed. Simulations reveal that a lizard that follows this rule can learn to adjust its cutoff distance to converge on the optimal cutoff distance. Initial small increases in the memory length of simulated lizards greatly improved the lizards' performances. The reproductive fitness of each simulated lizard was determined based on its energy intake. Lizards that could remember only the last two prey encounters laid almost twice as many eggs per day as lizards that foraged with no memory. Field data agree qualitatively with the model's predictions. Instrumental conditioning experiments show that anoles readily can learn to modify their foraging behavior. In experiments in which lizards were presented with insects at varying distances, cutoff distances were shorter when insect abundance was higher, and cutoff distances were shorter for smaller prey. Lizards responded quickly to changes in insect abundance, suggesting that recent foraging events are weighted heavily in determining cutoff distances.

Resumen.—Se presenta un modelo de conducta de forrajeo óptima de una lagartija *Anolis* que localiza a su presa desde un mismo lugar. La distancia límite

[1] Department of Biological Sciences, Stanford University, Stanford, California 94305, USA.

óptima hasta la cual la presa debe ser perseguida, y luego de la cual debe ser ignorada, y que maximiza la razón promedio de obtención de energía neta depende del tamaño y de la abundancia de la presa. La solución analítica requiere de computación intensa y parece sobrepasar por mucho las habilidades cognoscitivas de una lagartija. Se propone una regla de decisión sencilla. Las simulaciones revelan que una lagartija que sigue esta regla puede aprender a ajustar su distancia límite para que converja con la distancia límite óptima. Los pequeños aumentos iniciales en la duración de la memoria de las lagartijas simuladas aumentaron grandemente el desempeño de las lagartijas. La aptitud reproductiva de cada lagartija simulada fue determinada basándose en su obtención de energía. Las lagartijas que apenas pudieron recordar sus dos últimos encuentros con presas pusieron casi el doble de huevos por día que las lagartijas que forrajearon sin memoria. Los datos de campo concuerdan cualitativamente con las predicciones del modelo. Los experimentos de acondicionamiento instrumental muestran que las lagartijas anolinas pueden aprender fácilmente a modificar su conducta de forrajeo. En experimentos en los cuales se colocaron insectos a varias distancias de las lagartijas, las distancias límites fueron menores cuando la abundancia de los insectos fue mayor y fueron menores para presas más pequeñas.

Résumé.—Nous présentons ici un modèle de recherche de nourriture optimal pour un lézard *Anolis*, caractérisé par l'inaction et l'attente. La distance limite optimale sur laquelle les proies peuvent être poursuivies et au-delà de laquelle il convient de les ignorer, ce qui permet de maximiser le taux moyen d'apport énergétique net, dépend de la taille et de l'abondance des proies. La solution analytique est sur la plan computationnel très intensive et dépasse largement les facultés cognitives du lézard. Nous proposons donc une règle de décision simple. Les simulations révèlent qu'un lézard qui suit cette règle peut apprendre à ajuster sa distance limite et converger vers la distance limite optimale. De petites augmentations initiales dans la longueur mnémonique de lézards simulés améliorent considérablement les performances des lézards. Les facultés reproductives de chaque lézard simulé ont été déterminées sur la base de leurs apports énergétiques. Les lézards qui ne se souviennent que des deux dernières proies qu'ils ont chassées pondent presque deux fois plus d'oeufs par jour que les lézards qui recherchent leur nourriture sans se souvenir des proies qu'ils ont chassées. Les données recueillies sur le terrain coïncident qualitativement avec les prédictions du modèle. Les expériences de conditionnement instrumental montrent que les anoles apprennent facilement à modifier leur comportement de recherche de nourriture. Les expériences dans le cadre desquelles des insectes ont été présentés aux lézards à différentes distances montrent que les distances limites sont plus courtes lorsqu'il y a beaucoup d'insectes et lorsque les proies sont plus petites.

•

The Relationship between Habitat Use and Sexual Dimorphism among Caribbean Anoles

Marguerite A. Butler[1]
and Jonathan B. Losos[1]

Abstract.—We investigated the pattern of adult sexual size dimorphism among *Anolis* lizards of the Greater Antilles (Cuba, Hispaniola, Jamaica, and Puerto Rico). The *Anolis* species on these islands have convergently evolved into one of six habitat types, called "ecomorphs" (*sensu* Williams 1972). Each island contains a representative set of ecomorphs, which are thought to have evolved via independent radiations within each island. This replication of ecomorphs across islands gives statistical power to our analysis. Using adult snout-vent lengths as our measure of body size, we tested for correlated evolution of sexual dimorphism and ecomorph (habitat) type. A phylogeny was not available for the complete data set, so the statistical nonindependence of species was minimized by randomly choosing only one member from each clus-

[1] Department of Biology, Washington University, St. Louis, Missouri 63130–4899, USA.

ter of closely related species. Results from a Kruskal-Wallis and one-way ANOVA were compared. In the Puerto Rican and Jamaican data set, for which a phylogeny was available, we compared the results from a Kruskal-Wallis test (not accounting for phylogeny), a randomization test using simulated data (Martins and Garland 1991), a phylogenetic simulation program to generate a distribution of dimorphism values, and parsimony methods for the analysis of discrete characters. The results generally support a significant effect of habitat on the degree of sexual dimorphism.

Resumen.—Investigamos los patrones de dimorfismo sexual de los adultos de las lagartijas *Anolis* de las Antillas Mayores (Cuba, Jamaica, La Española, y Puerto Rico). Las especies de *Anolis* en estas islas han evolucionado convergentemente en uno de seis tipos de lagartijas características de ciertos habitáculos. Estos tipos de lagartijas se denominan "ecomorfos" (*sensu* Williams 1972). Cada isla posee un grupo representativo de ecomorfos, los cuales se cree que han evolucionado vía radiaciones independientes en cada isla. Esta replicación de ecomorfos a través de las islas le da confiabilidad a nuestro análisis estadístico. Utilizando como nuestra medida del tamaño del cuerpo de lagartijas adultas la longitud desde la punta del hocico hasta la cloaca, investigamos si había una correlación evolutiva entre dimorfismo sexual y tipo de ecomorfo (habitáculo). No había disponible una filogenia para el conjunto completo de datos, de manera que la falta de independencia estadística de las especies fue minimizada escogiendo al azar sólo un miembro de cada grupo de especies relacionadas íntimamente. Comparamos los resultados de un análisis de variancia Kruskal-Wallis en una sola dirección. En el conjunto de datos de Jamaica y Puerto Rico, para el cual hay disponible una filogenia, comparamos los resultados de un análisis Kruskal-Wallis, de un programa de simulación filogenética para generar una distribución de valores de dimorfismo, y de métodos de parsimonia para el análisis de caracteres discretos. Los resultados apoyan en términos generales un efecto significativo del habitáculo en el grado de dimorfismo sexual de *Anolis*.

Résumé.—Nous avons étudié les caractéristiques du dimorphisme sexuel adulte chez les lézards *Anolis* des Grandes Antilles (Cuba, Hispaniola, Jamaïque et Porto Rico). L'évolution des *Anolis* de ces îles converge vers l'un des six types d'habitats «écomorphes» (*sensu* Williams 1972). Chaque île abrite un ensemble représentatif d'écomorphes dont l'évolution semble résulter de radiations indépendantes sur chaque île. Cette réplication des écomorphes entre les îles conforte le pouvoir statistique de notre analyse. En nous basant sur la longueur museau-cloaque des adultes comme mesure corporelle, nous avons cherché à retracer l'évolution corrélée du dimorphisme sexuel et de l'écomorphe (habitat). Il nous manquait une phylogenèse pour compléter notre ensemble de données; la non indépendance statistique des espèces a donc été minimisée en choisissant au hasard un seul membre dans chaque groupe d'espèces étroitement apparentées. Les résultats des analyses de Kruskal-Wallis et ANOVA ont été comparés. Pour l'ensemble de données se rapportant à Porto Rico et à la Jamaïque (pour lesquelles nous disposions d'une phylogénie), nous avons comparé les résultats d'un test de Kruskal-Wallis (ne tenant pas compte de la phylogénie) avec ceux d'un test de randomisation basé sur un programme de simulation généalogique avec des données simulées (Martins et Garland 1991) pour répartir les valeurs de dimorphisme et ceux de méthodes de parcimonie pour analyser les caractères discrets. Les résultats confirment en règle générale l'effet significatif de l'habitat sur le degré de dimorphisme sexuel.

•

Social Control of Male Reproductive Success in *Ameiva exsul*

Allen R. Lewis,[1] Grissell Tirado,[1] Antonio Aponte,[1] Javier Blanco,[1] and José Sepulveda[1]

Abstract.—Like most large species of teiid lizards,

[1] Department of Biology, University of Puerto Rico, Mayagüez, Puerto Rico 00681–5000, USA.

Ameiva exsul is a wide-ranging intensive forager. We find that males and females of various sizes of this species coexist at high densities (up to $0.04/m^2$) on undefended, overlapping home ranges. Den holes often are closely aggregated. Male home ranges are larger on average than those of females, and for both sexes the largest home ranges we have measured are about 600 m^2. Males are larger than females, and larger individuals are usually socially dominant over smaller animals of either sex.

Copulation occurs in the open. A foraging female in breeding condition may be followed closely by a male who seems to use both visual and chemical cues to keep on the trail. Such following may terminate in copulation. The copulation may include the apparent cooperation of the female, or the male may run the female down and appear to overcome considerable resistance.

A potential problem of the teiid lifestyle from the perspective of large males is that, despite their social dominance, they may be unable to control access by small males to females on the undefended home ranges over which many animals forage. Thus we pose the general question: how is reproductive competition played out in wide-ranging teiids? And we ask the more specific question: are the ever-present subordinate males kept from reproduction, either by interference from dominant males or by female choice, or do the subordinates, far from being excluded, enjoy reproductive success comparable to that of dominant males?

In order to answer this question, we are relating actual reproductive success of males, as revealed by DNA typing of progeny, to size, social dominance, and expressed sexual behaviors.

Our study uses captive populations in order to limit the number of potential parents. We do this for two reasons: (1) unlike many fingerprinting studies, we will not know the identity of the mother *a priori*, the eggs are laid underground and apparently abandoned; thus having a small number of parents is essential in order to simplify the interpretation of bands, as well as to assure blood samples from all potential parents; and (2) a smaller number of individuals increases the chance of behavioral observations on all individuals.

We have four enclosures. Each measures 5 x 20 m, contains natural vegetation, and is stocked with three males and three females, resulting in a density of $0.06/m^2$. Enclosure one has consistently failed, leaving us with three functioning enclosures. From year one, 1993, we obtained limited behavioral data because of hawk predation, escapes, and the inability of observers to identify all participants in interactions. We did obtain clutches and have recently completed preliminary determinations of paternity using RAPDs and assuming bands are inherited as dominant markers. From year two, 1994, we have much more complete behavioral data as well as additional clutches, but have not yet completed analyses of paternity.

In year one, at least one of the three clutches from corral 2 was fathered by a subordinate male. In corral 3, both of the clutches appear to have been fathered by a subordinate. In corral 4, all three clutches were fathered by the smaller male.

In year two, in corral 2, the dominant male was never supplanted in any interaction and broke the tail of the second ranking male in fights. The third ranking male was never observed. The dominant male followed each female on several days and was seen to copulate once with the largest female. The second ranking male followed females on two days only and copulated once with the largest female. In corral 3, the dominant male chased and supplanted the second ranking male. The dominant male followed each female on several days and copulated with the largest female on two occasions. The second ranking male followed females on fewer days and was never seen to copulate. The third ranking male was never seen. In corral 4, the dominant male followed each of the females on several days and copulated once with the largest female. The second ranking male followed a small female on only one day and achieved copulation. Three, one, and three clutches were obtained from corrals 2, 3, and 4 in year two. Determination of paternity is not yet completed.

The paternity analysis and relatively incomplete behavioral observations from year one are complemented by the behavioral observations from year two. In the face of harrassment by dominant males, subordinate males obtain copulations with relatively little preliminary sexual following, and achieve paternity. Dominant males can supplant and injure subordinates that they encounter with receptive females, but, perhaps because of the active, dispersed mode of foraging, the dominant males cannot prevent access by subordinates to females nor, of course, can they control female choice.

Resumen.—Como la mayoría de las especies grandes de lagartos teiidos, *Ameiva exsul* es un forrajero intensivo de amplio radio de acción. Encontramos que machos y hembras de varios tamaños de esta especie coexisten a altas densidades (hasta 0.04 m^2) en áreas habituales solapadas sin defensas. Las madrigueras están a menudo estrechamente agregadas. Las áreas habituales de los machos son en promedio más grandes que aquellas de las hembras, y para ambos machos y hembras el área habitual mayor que hemos medido es cerca de 600 m^2. Los machos alcanzan tamaño corporal mayor que las hembras, y los individuos más grandes son, usualmente, socialmente dominantes sobre animales más pequeños de cualquier sexo.

Las copulaciones ocurren al descubierto. Una hembra forrajeadora en condición reproductiva puede ser seguida estrechamente por un macho que parece usar tanto señales visuales como químicas para mantenerse en el rastro. Tal seguimiento puede terminar en copulación. La copulación puede incluir la aparente cooperación de la hembra, o el macho puede perseguir y alcanzar la hembra y aparentemente supera considerable resistencia.

Un problema potencial del estilo de vida de los teiidos desde la perspectiva de los machos grandes es que, a pesar de su dominancia social, ellos podrían ser incapaces de controlar el acceso a las hembras por machos pequeños en el área habitual sin defensa en el cual muchos animales forrajean. De esta manera presentamos la pregunta general: cómo es que la competencia reproductiva se lleva a cabo en teiidos de amplio radio de acción? Y hacemos la pregunta más específica: son los siempre presentes machos subordinados impedidos de reproducirse, ya por interferencia de machos dominantes o por la preferencia de las hembras, o los machos subordi-nados, lejos de ser excluídos, gozan de éxito repro-ductivo comparable al de los machos dominantes?

Para contestar esta pregunta estamos relacionando el actual éxito reproductivo de los machos, según revelado por tipificación de ADN de la progenie, al tamaño, dominancia social y los comportamientos sexuales expresados.

Nuestro estudio usa poblaciones cautivas para limitar el número de padres potenciales. Hacemos esto por dos razones: (1) a diferencia de muchos estudios de tipificación de ADN, nosotros no sabemos la identidad de la madre a priori, los huevos son puestos bajo tierra y aparentemente abandonados; de esta manera tener un número pequeño de padres es esencial para simplificar la interpretación de bandas, así como asegurar muestras de sangre de todos los padres potenciales; y (2) un número más pequeño de individuos aumenta la oportunidad de observaciones de comportamiento de todos los individuos.

Tenemos cuatro cercados. Cada uno mide 5 x 20 m, contiene vegetación natural, y está poblado con tres machos y tres hembras, resultando en una densidad de 0.06/m^2. El cercado uno ha fallado consistentemente, dejandonos con tres cercados funcionales. Del año uno, 1993, obtuvimos datos de comportamiento limitados debido a depredación por halcones, escapatorias, y la inhabilidad de los observadores para identificar todos los participantes de las interacciones. Sí obtuvimos nidadas y recientemente completamos determinaciones de paternidad preliminares usando RAPDs, y asumiendo que las bandas son heredadas como marcadores dominantes. Del año dos, 1994, tenemos datos mucho más completos así como nidadas adicionales, pero no hemos completado aún los análisis de paternidad.

En el año uno, al menos una de las tres nidadas del corral 2 fue procreada por un macho subordinado. En el corral 3, ambas nidadas aparentan haber sido procreadas por un subordinado. En el corral 4, las tres nidadas fueron procreadas por el macho más pequeño.

En el año dos, en el corral 2, el macho dominante nunca fue suplantado en ninguna interacción y rompió la cola del macho segundo en rango durante peleas. El macho tercero en rango nunca fue observado. El macho dominante siguió cada hembra en varios días y fue visto copular una vez con la hembra más grande. El macho segundo en rango siguió hembras en dos días solamente y copuló una vez con la hembra más grande. En el corral 3, el macho dominante persiguió y suplantó al macho segundo en rango. El macho dominante siguió cada hembra en varios días y copuló con la hembra más grande en dos ocasiones. El macho segundo en rango siguió hembras en menos días y nunca se le vió copular. El macho tercero en rango nunca fue visto. En el corral 4, el macho dominante siguió cada una de las hembras en varios días y copuló una vez con la hembra más grande. El macho segundo en rango siguió a una hembra pequeña en un sólo día y logró copular. Se obtuvieron tres, una y tres nidadas de los corrales 2, 3, y 4 en el año dos. La determinación de paternidad no ha sido completada.

El análisis de paternidad y las observaciones de comportamiento relativamente incompletas para el año uno son complementadas por las observaciones de comportamiento del año dos. En vista del acoso por machos dominantes, los machos subordinados obtienen copulaciones con relativamente poco seguimiento sexual preliminar, y logran paternidad. Los machos dominantes pueden suplantar y herir subordinados que encuentran con hembras receptivas, pero, quizá debido al activo, dispersado modo de forrajeo, los machos dominantes no pueden prevenir el acceso por subordinados a las hembras ni, por supuesto, pueden ellos controlar la preferencia de las hembras.

Résumé.—À l'instar des grandes espèces de Téiidés, *Ameiva exsul* est un lézard très répandu. Les mâles et les femelles de différentes tailles coexistent à des densités élevées (jusqu'à 0,04 m^2), dans des secteurs non défendus souvent juxtaposés. Les terriers sont souvent très proches les uns des autres. Les territoires des mâles sont en moyenne plus grands que ceux des femelles et pour les mâles comme pour les femelles, le territoire le plus vaste que nous ayons mesuré couvre une superficie de 600 m^2. Les mâles sont plus gros que les femelles et les plus gros spécimens sont généralement ceux qui dominent les plus petits, quel que soit le sexe.

Les animaux copulent à l'air libre. Le mâle suit la femelle en chaleur et semble utiliser des indices visuels et chimiques pour ne pas la perdre de vue. Cette poursuite peut aboutir à la copulation. La copulation peut se faire avec la coopération apparente de la femelle ou au prix d'une résistance farouche de sa part.

Le problème potentiel du mode de vie des Téiidés, du point de vue des gros mâles, est qu'en dépit de la suprématie sociale qu'ils exercent, ils semblent incapables d'interdire l'accès aux femelles dans les territoires non défendus sur lesquels circulent de nombreux animaux. Aussi convient-il de s'interroger sur le jeu de la concurrence entre les Téiidés en matière de reproduction et de se demander notamment si les mâles dominés ne peuvent pas copuler parce que les mâles dominants les en empêchent ou parce que les femelles ne font pas porter leur choix sur eux ou si les mâles dominés, loin d'être exclus, parviennent à copuler avec autant de succès que les mâles dominants.

Pour répondre à cette question, nous avons établi un lien entre le succès effectif des mâles à se reproduire par analyse d'ADN de leur progéniture et la taille, la domination sociale et les comportements sexuels exhibés.

Notre étude repose sur des populations captives pour limiter le nombre de parents potentiels. Deux raisons motivent ce choix: (1) contrairement aux nombreuses études d'empreintes, nous ne connaîtrons pas l'identité de la mère a priori, les oeufs étant pondus sous terre et apparemment abandonnés; par conséquent, il est important d'avoir un nombre limité de parents pour simplifier l'interprétation des bandes et pour se procurer des échantillons sanguins de tous les parents potentiels; et (2) moins il y a d'individus plus les chances d'observer le comportement de tous les individus sont grandes.

Nous avons quatre enclos. Chacun mesure 5 × 20 m et abrite une végétation naturelle. Dans chaque enclos figurent trois mâles et trois femelles pour un densité de 0,06/m^2. L'enclos 1 n'a donné aucun résultat ce qui nous laisse trois enclos actifs. La première année (1993), nous avons recueilli des données comportementales limitées à cause des faucons prédateurs, des évasions et de l'incapacité des observateurs à identifier tous les participants et leurs interactions. Nous avons néanmoins obtenu des couvées et venons de terminer les déterminations préliminaires de paternité par RAPD en prenant les bandes héritées comme marqueurs dominants. Pour l'année 2 (1994), nous disposons de données comportementales plus complètes ainsi que de couvées supplémentaires mais nous n'avons pas encore terminé les analyses de paternité.

Pendant la première année, au moins une des trois couvées de l'enclos 2 a été conçue par un mâle dominé. Dans l'enclos 3, les deux couvées semblent avoir été conçues par des mâles dominés. Dans l'enclos 4, deux des trois couvées ont été conçues par des mâles dominés.

La deuxième année dans l'enclos 2, le mâle dominant n'a jamais été supplanté et a cassé la queue du deuxième mâle en importance à l'issue d'une bataille. Le troisième mâle en importance n'a jamais été observé. Le mâle dominant a suivi chaque femelle pendant plusieurs jours et on l'a vu copuler une fois avec la plus grosse femelle. Le deuxième mâle en importance a suivi les femelles pendant deux jours et a copulé une fois avec la plus grosse femelle. Dans l'enclos 3, le mâle dominant a pourchassé et vaincu le deuxième mâle en importance. Le mâle dominant a suivi chaque femelle pendant plusieurs jours et copulé avec la plus grosse femelle à deux reprises. Le deuxième mâle en importance a suivi les femelles pendant plusieurs jours mais on n'a pas observé de copulation. Le troisième mâle en importance n'a jamais été observé. Dans l'enclos 4, le mâle dominant a suivi chaque femelle pendant plusieurs jours et a copulé une fois avec la plus grosse femelle. Le deuxième mâle en importance a suivi une petite femelle pour une journée et a copulé. Trois, une et trois couvées ont été obtenues respectivement dans les enclos 2, 3 et 4 pendant la deuxième année. Les analyses de paternité ne sont pas encore achevées.

Les analyses de paternité et les observations comportementales relativement incomplètes de la première année ont été compensées par des observations comportementales la deuxième année. Compte tenu du harcèlement dont ils font l'objet de la part des mâles dominants, les mâles dominés réussissent à copuler moyennant des préliminaires relativement courts et parviennent à se reproduire. Les mâles dominants peuvent vaincre et blesser les mâles dominés qu'ils rencontrent auprès de femelles réceptives. Toutefois, compte tenu du mode actif et très dispersé de recherche de nourriture, les mâles dominants ne peuvent empêcher les mâles dominés d'avoir accès aux femelles, ni contrôler le choix de ces dernières.

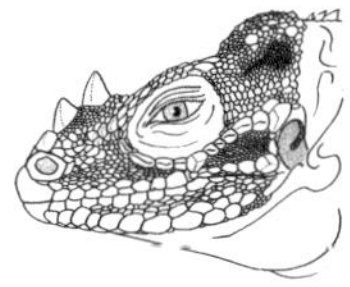

INDEX TO SCIENTIFIC NAMES

Names of amphibians and reptiles are in **boldface** type. Note that no distinction is made between genera and subgenera (however, species are allocated solely to genera) or between species, species groups, or species series. Names which appear in the plate contents (opposite page 94) are marked as such ("contents"). Species illustrated in plates are designated with plate numbers in **bold type**. Names that appear only in plate captions are indicated by plate number, but are not in bold type.

Acacia 307, 317, 321–323, 325–326, 407, 411, 418
  macrocantha 319
Acantholis loysiana 78
Acarina 222
Acrochordidae 341
Adenomera hylaedactyla 208
Aechmea 260, 263
  paniculigera 259, 261
Ahaetulla frenata 88
Alethinophidia 340–341
Alligatoridae 92, 99, 102
Aloponotus ricordii 73
Alsophis 21–22, 44, 99, 110, 218–219, 273, 327–338, 341–342, 344–346, 348–349, 365–366, contents, plates 7–8
  anomalus 86, 328–329, 340
  antiguae 86, 327–330
  antillensis 86, 327, 330, 332, **plate 7D**
  ater 31, 33, 86, 327–329, 340
  bruesi 88
  cantherigerus 86, 111, 328–335, 340, 342
  funereus 87
  leucomelas antiguae 86
  melanichnus 86, 327–329, 340
  portoricensis 86, 220–222, 327–330, 332, 334, 340, **plate 7E**
  portoricensis richardi 330
  rijersmai 86, 286, 327–329, 331
  rufiventris 86, 327, 329–333
  sanctaecrucis 86, 327–329
  vudii 86, 329–334, 340, 344, 349, 363
Ameiva 22, 24, 82–83, 86–88, 90, 99, 108, 114–117, 219, 279–281, 283, 285–286, 318–319, 323, 330–332, 334, 362, 366–367, contents, plate 8
  ameiva 31, 82, 108, 114, 118, 209, 212
  auberi 23, 26, 82, 331, 333, 363–364
  chrysolaema 22, 24, 82–83, 108, 317, 319–323, 326
  chrysolaema leberi 83
  cineracea 82
  corax 82
  corvina 82
  dorsalis 82
  erythrocephala 82, 331
  exsul 82, 108, 285, 330, 407, 445–448
  fuscata 82, 329
  griswoldi 83
  leberi 83
  lineolata 22, 83, 319–323, **plate 3D**
  major 83
  maynardi 83, 331
  plei 83, 277–285, 287–289, **plate 3C**
  plei var. ***exsul*** 82
  pluvianotata 83
  polops 83
  taeniura 22, 83
  vanzoi 83
  wetmorei 83, 330
Amphiacusta caraibea 438–440
Amphibia 21–23, 25, 52, 54, 57–62, 79, 100, 218–219, 221, contents, plate 1
Amphisbaena 64, 99, 103, 114, 116–117, 213, 218–219, contents, plate 2, plate 7
  alba 103, 209
  bakeri 64
  blanoides 64, **plate 2E**
  caeca 64, **plate 2D**
  caudalis 64
  cubana 64
  fenestrata 64
  fuliginosa fuliginosa 209
  gonavensis 64
  innocens 64
  innocens gonavensis 64
  manni 64
  palirostrata 64
  punctata 64
  schmidti 64, 103
  xera 64
Amphisbaenia 30, 64, contents, 99, 103
Amphisbaenidae 21, 64, contents, 99, 103
Amphiuma means 27
Amyris 418
Aneides aeneus 27
  hardyi 27
Anguidae 22–26, 64–66, 99, 103, contents, plate 3
Anguis Jamaicensis 91
  lumbricalis 91
  ventralis 66
Anolis 9–10, 17, 22, 24–25, 31, 38–39, 41, 43–45, 55, 57, 61–62, 74–81, 96, 99, 104–105, 107, 114, 116–117, 120, 177, 195, 197, 202–203, 210, 213, 217–219, 221–222, 224, 226–227, 279, 282, 285, 291–298, 319, 327, 329–338, 342, 346, 348–350, 365, 412–413, 419, 421, 442–445, contents, plates 5–8
  acutus 73, 220
  aeneus 43, 74, 207–210, 213, 216
  aeneus x ***trinitatis*** 210
  agrillaceus [sic] ***centralis*** 75
  ahli 74
  alayoni 74
  alfaroi 74
  aliniger 74, 300
  allisoni 74
  allogus 74
  altavelensis 74
  alumina 74
  alutaceus 74, 78
  alutaceus clivicolus 23
  anfiloquioi 74
  angusticeps 10, 23, 74, 79, 291–301, 329, **plate 5C**
  angusticeps paternus 80
  argenteolus 74
  argillaceus 74, 80
  armouri 74
  augusticeps [sic] 74
  bahorucoensis 74, 300, 442–443, **plate 5E**
  bahorucoensis bahorucoensis 20
  baleatus 24, 74
  baracoae 75
  barahonae 25, 75, 300, 442–443, **plate 5D**
  barbatus 75
  barbouri 17, 20, 75, **plate 5B**
  bartschi 75
  bimaculatus 75, 105
  birama 75
  bremeri 75
  brevirostris 75, 79, 82, 300, 303–305, 311, 313, 319–323
  brunneus 75, 331
  carolinensis 105
  caudalis 75, 303–305, 307–315
  centralis 75
  chamaeleonides 75, **plate 5A**
  chlorocyanus 75, 300
  chloro-cyanus aliniger 74
  christophei 75, 77, 300
  chrysolepis planiceps 209, 213
  clivicola 75
  clivicolus 75
  coelestinus 23, 76, 300, 442–443
  concolor 76
  confusus 76
  conspersus 76, 199, 329, **plate 5G**
  cooki 76, 203, 300
  cristatellus 76, 203, 300, 330, 407, 413
  cristatellus cooki 76
  cristatellus cristatellus 203
  cristatellus wileyi 202–203
  cupeyalensis 76
  cuvieri 76, 218–220, 300
  cyanopleurus 76–78
  cybotes 25, 76, 79, 300, 308, 442–443
  darlingtoni 77, 297
  delafuentei 76
  desechensis 76
  distichus 23, 76, 295, 300, 313, 319, 362–363, 365, 442–443
  distichus ravitergum 311
  dolichocephalus 76
  dominicensis caudalis 75
  equestris 9, 21, 22–23, 75, 76, 81, 273
  equestris baracoae 75
  equestris smallwoodi 81
  ernestwilliamsi 76, 203
  etheridgei 77
  eugenegrahami 77
  evermanni 77, 220–223, 300, 330
  extremus 77, 207–210, 213, 216
  fairchildi 77
  ferreus 77
  fowleri 20, 77, 297
  fugitivus 77
  garmani 77, 197, 300
  gingivinus 77, 331
  grahami 77, 195, 197–199, 205, 300
  griseus 77
  guafe 77
  guamuhaya 77
  guazuma 77
  gundlachi 77, 203, 220–221, 223, 300
  haetianus 78
  hendersoni 25, 76, 78
  hendersoni dolichocephalus 76
  homolechis 23, 78
  imias 78
  inexpectatus 78
  insolitus 78, 296–297, 300
  isolepis 78
  juangundlachi 78
  jubar 78
  koopmani 78
  krugi 78, 220, 300
  cf. ***lemurinus*** 209, 213

Anolis (continued)
- ***lineatopus*** 42, 78, 195, 197–199, 202, 300, 441–442
- ***lineatopus ahenobarbus*** 202
- ***lineatopus lineatopus*** 198, 202–203
- ***lineatopus lineatopus*** "***coxi***" 203
- ***lineatopus merope*** 202
- ***lineatopus neckeri*** 202, 204
- ***lividus*** 78, 329
- ***longiceps*** 78
- ***longitibialis*** 78, 300
- ***loysianus*** 78
- ***luciae*** 78
- ***lucius*** 78
- ***luteogularis*** 79
- ***macilentus*** 79
- ***marcanoi*** 79, 300
- ***marmoratus*** 79, 329, 332, **plate 5F**
- ***marmoratus caryae*** 203
- ***marmoratus chrysops*** 203
- ***marmoratus terraealtae*** 203
- ***marron*** 79, 304
- ***maynardi*** 79
- ***mestrei*** 79
- ***mimus*** 79
- ***monensis*** 79
- ***montanus*** 79
- ***monticola*** 22, 77, 79–80
- ***noblei*** 79
- ***nubilis*** 79
- ***occultus*** 42, 79, 220, 300
- ***oculatus*** 79, 329
- ***oculatus montanus*** 79
- ***olssoni*** 25, 79, 300, 319–323
- ***opalinus*** 79, 197–199, 202–203, 300
- ***ophiolepis*** 79, 329
- ***paternus*** 79
- ***pigmaequestris*** 80
- ***pinchoti*** 80
- ***placidus*** 80, 297
- ***poncensis*** 80, 300
- ***porcatus*** 80, 300, 329
- ***porcus*** 80
- ***principalis brunneus*** 75
- ***pulchellus*** 80, 220, 300, 329
- ***pumilus*** 80
- ***punctatissimus*** 199
- ***punctatus*** 77
- ***quadriocellifer*** 80, 329
- ***reconditus*** 80, 196–197, 202–203
- ***richardi*** 80, 207–210, 212, 216
- ***ricordii*** 24, 80
- ***ricordii barahonae*** 75
- ***rimarum*** 80
- ***roosevelti*** 80
- ***roquet*** 80, 105
- ***rubribarbus*** 80
- ***rupinae*** 80
- ***sabanus*** 81, 331
- ***sagrei*** 33–34, 81, 195, 274, 329–331, 350, 362, 365
- ***scriptus*** 81, 331
- ***semilineatus*** 81, 300
- ***sheplani*** 20, 81, 297
- ***shrevei*** 81
- ***singularis*** 81, 300
- ***smallwoodi*** 81
- ***smaragdinus*** 81, 362, 365
- ***spectrum*** 23, 81
- ***strahmi*** 81, 300
- ***stratulus*** 81, 220–224, 300, 330, 350
- ***striatulus*** [sic] 81
- ***trinitatis*** 81, 207–210, 213, 216
- ***trinitatis* x *aeneus*** 210
- ***valencienni*** 81, 292, 296–298, 300
- ***vanidicus*** 81
- ***vermiculatus*** 81
- ***vescus*** 81
- ***wattsi*** 81, 331
- ***websteri*** 81, 303–305, 307–315
- ***whitemani*** 25, 82, 319–323

Anolis (Acantholis) argillaceus 74
Anolis (Anolis) cybotes 76
Anolis (Ctenocercus)
- ***coelestinus*** 76

Anolis (Dracontura) alutaceus 74
- ***cyanopleurus*** 76
- ***ophiolepis*** 79

Anolis (Gastrotropis)
- ***argenteolus*** 74
- ***concolor*** 76

Antillophis 99, 110, 331–332, 341–342, 347, contents
- ***andreae*** 86, 329–330, 340
- ***parvifrons*** 26, 86, 335, 340, **plate 7F**

Anura 23–25, 30, 52–61, 100, 129, plate 1
Aporophis juliae 88
Apostolepis 349
Araneida 222
Aristelligella
- ***barbouri*** 66

Aristelliger 24, 65–66, 99, 106, 114, 116, contents, plate 4
- ***barbouri*** 66, 106
- ***cochranae*** 66, 106
- ***cochranae expectatus*** 66
- ***expectatus*** 66
- ***georgeensis*** 66, 106, 114
- ***hechti*** 66
- ***lar*** 66, **plate 4A**
- ***praesignis*** 66

Arrhyton 22, 25, 86–87, 99, 110, 218–219, 328, 341–342, 344–348, contents
- ***ainictum*** 86, 340
- ***callilaemum*** 86, 340
- ***dolichura*** 86, 340
- ***dolichurum*** 86
- ***exiguum*** 86, 330, 340
- ***funereum*** 87, 340
- ***landoi*** 87, 340
- ***polylepis*** 87, 340
- ***procerum*** 87, 340
- ***redimitum*** 87
- ***supernum*** 87, 340
- ***taeniatum*** 87, 340, 349, **plate 7G**
- ***tanyplectum*** 87, 340
- ***vittatum*** 87, 340
- ***vittatum landoi*** 87

Atractaspididae 341
Atractus trilineatus 209
- ***univittatus*** 209

Audantia armouri 74
- ***shrevei*** 81

Avicularia 221, 223
- *laeta* 438–440

B

Bachia 99, contents, plate 2
- ***flavescens*** 209
- ***heteropus*** 72, 104, 114, **plate 2G**
- ***heteropus alleni*** 209, 212, 214
- ***heteropus trinitatis*** 209

Bambusa 418
Barbus 61, 90
Batis maritima 426
Batrachia 63, 84, 89, 91
Blatteria 282, 285
Blattodea 222, 281–282
Boa 99, contents
- ***constrictor*** 30–31, 85, 109, 115, 212, 343, **plate 6D**
- ***constrictor constrictor*** 209
- ***dumerili*** 343
- ***Hortulana*** 85
- ***inornata*** 85
- ***madagascariensis*** 343
- ***mandrita*** 343
- ***melanura*** 90
- ***pardalis*** 90

Boidae 24, 85, 99, 108, 339–341, 417, contents, plate 6
Boiga dendrophila 335
- ***irregularis*** 335

Bolyeriidae 341
Bothrops 99, 112, 115–117, 213, 334, contents, plate 7
- ***asper*** 112, 209
- ***atrox*** 112
- ***caribbaea*** 92, **plate 7C**
- ***lanceolata*** 92

Brachycera 222
Brachymeles Leuckartii 64
Buchenavia capitata 218
Bucida buceras 411
Bufo 15, 52, 63, 97, 100, 218–220, contents, plate 1
- ***americanus*** 265
- ***beebei*** 208
- ***cataulaciceps*** 21, 52
- ***empusus*** 52
- ***fluviaticus*** 52
- ***fractus*** 52
- ***fustiger*** 52
- ***granulosus*** 100
- ***güntheri*** 52
- ***güntheri fractus*** 52
- ***guentheri*** 52, 271
- ***gundlachi*** 52
- ***lemur*** 52, 377
- ***longinasus*** 52
- ***marinus*** 31, 52, 100, 208, 211, 213, **plate 1B**
- ***peltacephalus fustiger*** 52
- ***peltocephalus*** 52, 100, 114, 116–117, **plate 1A**
- ***taladai*** 53

Bufonidae 23, 52, 97, 100, contents, plate 1
Bursera simaruba 411, 418
Buteo jamaicensis 219, 221
Byrsonima lucida 411

C

Cadea 21, 64, 103
- ***blanoides*** 21, 64, plate 2
- ***palirostrata*** 64

Caiman 99
- ***crocodilus*** 31, 92, 102, 211
- ***crocodilus crocodilus*** 209

Calamita [sic] ***cinereus*** 53
Callisaurus 16
Calyptahyla 97, 101, contents
- ***crucialis*** 53, 157, 163, **plate 1D**

Canella winterana 285
Capoëta 61, 90
Capparaceae 198
Capparis 198
- *cynophallophora* 411

Capromys 61, 90
Carassius 61,90
Caretta 383, 385
- ***caretta*** 208, 384, 388

Casearia 319
Castella erecta 278, 285
Cataphracta 63
Celestus 65–66, 99, 103, 114, 116, 329, 331, contents, plate 3
- ***agasepsoides*** 20, 64, **plate 3A**
- ***anelpistus*** 65
- ***barbouri*** 65
- ***carraui*** 65
- ***costatus*** 26, 65
- ***crusculus*** 65
- ***curtissi*** 65
- ***darlingtoni*** 65
- ***duquesneyi*** 65
- ***fowleri*** 65
- ***haetianus*** 65, 103
- ***hewardii*** 65
- ***macrotus*** 65
- ***marcanoi*** 65
- ***microblepharis*** 65
- ***occiduus*** 65
- ***sepsoides*** 65
- ***stenurus*** 65
- ***warreni*** 66

Centropyx 65–66
- ***Borckiana*** 83

Cephalocereus royenii 411
Cercopithecus mona 420–421
Chalcides heteropus 72
Chamaeleolis 17, 20, 22, 75, 77, 80, 104
- ***barbatus*** 75
- ***chamaeleonides*** plate 5
- ***guamuhaya*** 77
- ***porcus*** 80

Chamaeleonidae 275
Chamaelinorops 17, 25, 75, 104
- ***barbouri*** 17, 20, 75, plate 5

Channa 61, 90
Chelonia 383
- ***mydas*** 209, 382, 384, 386, 388, 440–441

Cheloniidae 29
Chelus fimbriatus 208
Chilabothrus gracilis 85
Chironius 99, 109, contents, plate 7
- ***carinatus*** 209
- ***exoletus*** 109
- ***grandisquamis*, plate 7H**
- ***vincenti*** 87, 109, 115–117, plate 7

Chironomidae 264
Clelia 99, 110, 212, contents, plate 7
- ***clelia*** 87, 110, 115, 212, 214, **plate 7I**
- ***clelia clelia*** 209
- ***equatoriana*** 110
- ***errabunda*** 87, 110, 115–116, 212, plate 7
- ***scytalina*** 110

Cnemidophorus 16, 99, contents, plate 3
- ***lemniscatus*** 31, 83, 108, 115, 209, **plate 3E**
- ***vanzoi*** 83, 108, 115–116, plate 3

Coccoloba 409
- *uvifera* 285, 407, 411, 413

Cocos sp. plate 4
Coleodactylus 179
Coleoptera 222, 281–282, 285
Colostethus 53, 97, 100–101, contents
- ***chalcopis*** 53, 100–101, 114, 116, **plate 1C**

Coluber cantherigerus 86
clelia 87
constrictor 335
cursor 88
guttatus 87
lanceolatus 92
Melanotus 88
Colubridae 22–26, 82–83, 86–88, 99, 109, 327, 340–341, contents, plates 7–8
Colubrinae 109
Coniophanes 87, 99, 110, contents, plate 8
andresensis 87, 110, 115–117, plate 8
fissidens 110
fissidens andresensis 87
imperialis plate 8A
Conocarpus erectus 197, 407, 411, 413, 426
Corallus 99, 343, 418, contents, plate 6
enydris 85, 417
hortulanus 31, 44, 85, 109, 115, 118, 209, 212–214, 417–421, 423, plate 6E
hortulanus cookii 213, 417, 420
hortulanus hortulanus 213
Cordia rickseckerii 413
Cornufer 142
unicolor 142
Craugaster 141, 147–148, 150–152, 154, 231–234
Cricosaura 98–99, 108, 116, 120, contents
typica 45, 85, 95, 108, 115–116, 121, 127–128, plate 5H
Crocodilus acutus 92
intermedius 92
rhombifer 92
Crocodylia 92, 99, 102, contents, plate 2
Crocodylidae 92, 99, 102, contents, plate 2
Crocodylus 99, contents, plate 2
acutus 38, 44, 92, 102–103, 114, 209, 425, 429–430, 432–433
intermedius 92, 103, 114, 209, 211
palustris 103
rhombifer 92, 103, 114, 116–117, plate 2C
Crotalus enyo 112
lepidus klauberi 27
Crotaphytidae 105
Crotaphytus 105
collaris 106
Croton 407
Cryptodacus vittatus 87
Ctenosaura 99, contents
similis 73, 105, 114, plate 6A
Culicidae 264
Cyclorrapha 222
Cyclura 11, 24, 29, 38, 46, 73, 99, 105, 114, 116–117, 219, 331–332, 361, 364, 395–396, contents, plate 6
carinata 43, 73, 398
collei 46, 73, 196–197, 395, 405–406
cornuta 73, 105, 426
cornuta stejnegeri 398
cychlura 73, 362
nubila 73, 105, 329, plate 6B
pinguis 43, 73
ricordii 73, 426
rileyi 73, 362
Cydnidae 282
Cylindropuntia 319
Cyprinodon 426
Cystignathus albilabris 62

Dacryodes excelsa 218
Darlingtonia 22, 26, 99, 110, 341, 346, contents, plate 8
haetiana 23, 26, 87, 340, plate 8B
Dasyprocta plate 6
Deirochelys reticularia 27
Deiroptyx 75
bartschi 75
Dendrobates 101, 267
Dendrobatidae 53, 97, 100, contents, plate 1
Dendrophis catesbyi 88
Dermochelys 383
coriacea 209, 382, 385
Desmognathus 15
Diadophis 99
punctatus 87, 109
Didelphis marsupialis 420–421
Diemictylus (=Notophthalmus)
viridescens 27
viridescens piaropicola 16
Diphalus fenestratus 64
Diploglossinae 24
Diploglossus 11, 23–26, 64–66, 99, 103, 114, 116, 218–220, 331, contents, plate 3
agasepsoides 64
anelpistus 65
carraui 65
costatus 22, 23
crusculus 65
delasagra 66, 103, 329
fowleri 65
marcanoi 65
microblepharis 65
montisserrati 66, 103
pleii 66, 103, plate 3B
stenurus 65
warreni 26, 65–66
Dipodomys 16
Dipsadidae 110
Dipsas variegata trinitatus 209
Diptera 222, 281–283
Dromicus 21–22, 61, 87, 90
clavatus 87
exiguus 86
ornatus 88
parvifrons 86
polylepis 87
rufiventris 86
Drymarchon corais corais 209
Dytiscidae 264

Elachistocleis ovalis 208
surinamensis 208
Elaphe 99
guttata 87, 109
quadrivirgata 351
Elapidae 89, 99, 111, 340, contents, plate 6
Elapomorphus 349
Elaps nigrocinctus 89
Eleutherodactylus 11, 21–25, 36, 38–39, 44–45, 54–61, 95–98, 101–102, 105, 107, 113–114, 116, 120–121, 127–142, 144–145, 147–148, 150–155, 177, 202, 213, 217–221, 226, 229–237, 239, 251, 253, 255, 258, 327, 329–334, 335–338, 365, 371–373, 375–377, 380, contents, plate 1, plates 7–8
abbotti 24, 54, 232
acerus 154
achatinus 151, 154
acmonis 54
actites 154
acuminatus 154
acutirostris 154
affinis 231
albipes 54
alcoae 24, 54
alfredi 151
altae 154
alticola 54, 231–232, 234
altamazonicus 154
amadeus 54, 231, 236, 238
amplinympha 54, 129–136, 139–140
andi 154
andrewsi 54
angelicus 154, 230–231, 233, 236
anolirex 154
anomalus 150, 152, 154, 232
antillensis 54, 130–131, 221, 233, 330, 377
appendiculatus 151, 154
apostates 24, 54
armstrongi 25, 54, plate 1H
atkinsi 54, 232
atratus 154
audanti 25, 54, 59
augusti 154, 232–233
auriculatoides 25, 55, 232
auriculatus 21–23, 55, 58, 136, 142, 144, 150, 153
azueroensis 154
babax 154
bacchus 154
bakeri 25, 55, 231, 235–236, 238
bakeri heminota 57
barlagnei 55, 129–136, 139–140
bartonsmithi 55
baryecuus 154
beebei 142–143
bellona 154
berkenbuschii 154
bicolor 154
bicumulus 154
biporcatus 152, 154
boconoensis 154
bocourti 154
bogotensis 154
boulengeri 154
bransfordii 154
bresslerae 55
brevirostris 25, 55
brittoni 55, 377
brocchi 154
bromeliaceus 154
buckleyi 154
bufoniformis 150, 154
cacao 154
cadenai 154
cajamarcensis 154
calcarulatus 154
caprifer 154
caribe 55
carmelitae 154
caryophyllaceus 154, 231, 235
cavernicola 55, 202–203
celator 154
cerasinus 154
cerastes 151, 154
chalceus 154
cheiroplethus 150, 154
chloronotus 154, 232
chlorophenax 55, 141–144, 146–150, 152, 155
cochranae 55, 377
colodactylus 154
condor 154
conspicillatus 154
cooki 55, 231, 235–236, 251–258, 372, 377, plate 1I
coqui 27, 43, 55, 130–131, 220–224, 229–231, 235–236, 241–256, 258, 330, 372–375, 377–378, 437–439
cornutus 154
corona 55
counouspeus 25, 55
crassidigitus 154, 233
cremnobates 154
crenunguis 154
cristinae 154
croceoinguinis 154
crucifer 154
cruentus 151, 154, 231
cruralis 154
cryophilius 154
cryptomelas 154
cubanus 56
cundalli 56, 202, 231, 233
cundalli glaucoreius 57
cuneatus 56
curtipes 152, 154, 231, 236
cystignathoides 232
danae 154
darlingtoni 25, 56, 58
daryi 154
decoratus 154, 231
delicatus 154
devillei 151, 154
diadematus 154
diaphonus 154
diastema 154, 232–233
dimidiatus 56, 151, 232
discoidalis 154
dolomedes 56
dolops 154
duellmani 154
eileenae 56
elassodiscus 154
elegans 154
emcelae 154
emiliae 56
eneidae 56, 371–372, 375–377, 379–380
erythropleura 154
escoces 154
etheridgei 56
eunaster 25, 56
euphronides 56, 129–136, 139–140, 211
fenestratus 154
fitzingeri 130, 134, 152, 154, 231
flavescens 25, 56
fleischmanni 154
fowleri 25, 56, 233
furcyensis 25, 56
fuscus 56
gaigei 154, 256
galdi 152, 154
ginesi 154
glandulifer 25, 56, 151, 231, 238

Eleutherodactylus (continued)
glanduliferoides 26, 57
glandulosus 154
glaphycompus 26, 57
glaucoreius 57, 231, 233
gollmeri 152, 154
gossei 57, 232, 234
grabhami 57, 202
gracilis 154
grahami 57
greggi 152, 154
greyi 57
griphus 57
gryllus 57, 373–377
guanahacabibes 57
guantanamera 57, 231, 234, 236
guentheri 231
gundlachi 57
haitianus 26, 57
hectus 154
hedricki 57, 220–221, 231, 235–236, 377
heminota 26, 57
hypostenor 26, 58, 141–150, 152, 155
imitatrix 154
incanus 154
incomptus 154
ingeri 154
inoptatus 58, 60, 141–153, 155, 231, 234
insignitus 154
intermedius 57–58, 232, 236
ionthus 58
jaimei 154
jamaicensis 58, 233
jasperi 58, 229–230, 236, 239, 371–373, 375, 377, 379–380
johnstonei 58, 129–136, 139–140, 207–208, 213, 216, 231–233, 235, 329, 331, 333, plate 1
jugans 58
junori 58
karlschmidti 58, 220, 255, 371–373, 375–377, 379–380
klinikowskii 58
labiosus 154
lacrimosus 154
lamprotes 58
lancinii 154
lanthanites 154
latens 154
laticeps 150, 154
leberi 58
lentus 58, 153
leoncei 58
leoni 154
leptolophus 154
leucopus 154
limbatus 33, 58
lineatus 154
locustus 58, 375–377
longirostris 154
loustes 154
lucioi 59
luteolateralis 154
luteolus 59
lymani 150, 154, 231
lynchi 154
malkini 154
mantipus 154
mariposa 59
marnocki 231
martiae 154
martinicensis 59, 129–136, 139–140, 329, **plate 1K**
matudai 154
maussi 154, 231
megalops 154
melacara 59, 233
melanostictus 154
merendonensis 152, 154
mexicanus 152, 154
milesi 154
minutus 59
modipeplus 154
monensis 59, 377
montanus 59, 232
myersi 154
nasutus 233
necerus 154
neodreptus 59
nervicus 154
new sp. A 232
new sp. P 232
new sp. TE 232, 236
nicefori 154, 232
nigrovittatus 151, 154
noblei 151, 154
nortoni 59, 141–142, 144–150, 152, 155
nubicola 59, 231
nyctophylax 154
obmutescens 154
occidentalis 154
ockendeni 154
ocreatus 154
orcesi 154
orcutti 59
orestes 154
ornatissimus 154
orpacobates 154
oxyrhynchus 59
palmeri 154
pantoni 59, 231
pantoni pentasyringos 60
parabates 59
parapelates 59, 141–142, 144–150, 152, 155
pardalis 154
parvus 56, 233
patriciae 60
paulsoni 60
pentasyringos 60, 231
peraticus 154
percultus 154, 232, 235
permixtus 154
pezopetrus 60
philipi 154
phoxocephalus 154
pictissimus 60
pictissimus probolaeus 60
pinarensis 21, 58, 60
pinchoni 60, 129–136, 139–140
pituinus 60
planirostris 24, 60, 231, 234, 329, 331, 363
platydactylus 155
plicatus 57
podiciferus 154
poolei 60
portoricensis 60, 220–221, 229, 231, 236, 372–377, 437–439
probolaeus 60
prolatus 155
prolixodiscus 155, 232
proserpens 155
psephosypharus 232
pseudoacuminatus 155
pugnax 155
punctariolus 154
pygmaeus 154
quaquaversus 155
quinquagesimus 155
racemus 155
raniformis 154
repens 155
rhabdolaemus 155
rhodesi 60
rhodopis 57, 154
richmondi 60, 153, 221, 233, 375–377
ricordii 21–22, 24, 57, 60–62, 150, 153
ridens 155
riveti 155
ronaldi 60
roseus 155
rostralis 152, 154
cf. ***rozei*** 208
rubicundus 155
rufifemoralis 60
rugulosus 154, 232–233
ruidus 155
ruizi 155
ruthae 22, 58, 61, 141–153, 155, 233
ruthae aporostegus 146
ruthae ruthae 146
ruthveni 155
sanctaemartae 155
sartori 154
satagius 155
schmidti 23, 61, 143
schwartzi 61, **plate 1J**
sciagraphus 61
scolodiscus 151
semipalmatus 61
sernai 155
shrevei 61, 129–136, 139–140, 211
sierramaestrae 61
simoterus 155, 232
sisyphodemus 61, 202
spatulatus 154, 233
spilogaster 155
spinosus 155
stejnegerianus 154
stenodiscus 155
sulcatus 152, 155
supernatis 155
surdus 152, 155
symingtoni 61
taeniatus 155
talamancae 154
tamsitti 155
taurus 154
tayrona 155, 231
tenebrionis 155
terraebolivaris 130, 155, 208, 231
cf. ***terraebolivaris*** 129–137, 139–140
thectopternus 155
thomasi 61
thorectes 61
thymalopsoides 155
thymelensis 155
toa 61
toftae 155
trachyblepharis 155
trepidotus 155
turquinensis 61
unicolor 61, 232, 236, 373–375, 377
unistrigatus 153, 155, 232, 251
uno 154
uranobates 155
urichi 129–137, 139–140, 208, 211, 233
urichi euphronides 56
urichi shrevei 61
variabilis 155
varians 61
varians ionthus 58
varleyi 62, 232
ventrilineatus 62, 231, 234, 238
cf. ***ventrilineatus*** 232, 238
ventrimarmoratus 155
versicolor 155
vertebralis 155
vicarius 155, 231
vidua 155
vocalis 154
walkeri 155
warreni 62
weinlandi 62
wetmorei 23–24, 62
wightmanae 62, 220–221, 231, 375–377, 437–439
w-nigrum 155
xucanebi 154
zeus 62
zugi 62
zygodactylus 154, 232
Embryopus Habichii 64
Emydidae 63, 99, 112, 169, contents, plate 2
Emys decussata 63
Engystoma carolinense 62
Epicrates 24, 44, 85, 99, 109, 115–117, 213, 218–220, 334, 339, 341–351, 356–357, contents, plate 6
angulifer 85, 339–340, 342, 344, 347, 412
cenchria 109, 339, 342–344, 346–347, 356–357
cenchria maurus 209
chrysogaster 85, 340, 342–345
exsul 85, 340, 342–346, 351
fordii 26, 85, 340, 342–344, 346, 350
gracilis 26, 85, 340, 342–344, 346, 350, **plate 6F**
inornatus 85, 219–220, 340, 342, 344, 346
monensis 85, 340, 342–344, 346, 350, 407–412, 414–416, 421
monensis granti 407, 409
monensis monensis 407
striatus 26, 30, 85, 340, 342, 344–346, 363, 365, 412
subflavus 85, 196, 340, 342, 344, 396
Epilobocera situatifrons 438–440
Eretomochelys 331–332, 381, 383, 390
imbricta 329, 333, 382, 384, 386, 388
imbricata imbricata 209
Erythrolamprus 349
aesculapii aesculapii 209
ocellatus 209
Euhyas 98, 101–102, 120–121, 141, 144, 148, 150–151, 153–155, 231–235, 237
Eunectes 343
deschauenseei 343
murinus 209, 343
notaeus 343
Euphorbia petiolaris 411

Euphorbiaceae 198
Euprepes 61, 90
Eupristis baleatus 74
Exiliboa 348

Falco sparvarius 286
Felis catus 362, 408, 410
Ficus 407, 418
Flectonotus fitzgeraldi 208

Gallus gallus 330
Gastrophryne 97
carolinensis 62, 102
Gecko mabouia 67
Geckonidae 65, 105
Gekko 99, 106
gecko 106
rapicauda 72
Gekkonidae 22–26, 66–72, 99, 275, contents, plate 4
Geocapromys brownii 397
Geochelone 99, 113, 115–116, contents, plate 2
carbonaria 29, 63, 113, 115, 209, 211, 213, **plate 2A**
denticulata 209, 211
Gonatodes 99, 175, 178–187, 193–194, contents, plate 4
albogularis 67, 106, 114, 209, **plate 4B**
ceciliae 209
humeralis 209
ocellatus 209–210
vittatus 193, 209
Gryllidae 281–282
Guaiacum officinale 411
Guarea trichilioides 218
Gymnodactylus albogularis 67
geckoides 209
Gymnophthalmidae 72, 99, 104, contents, plate 2
Gymnophthalmus 73, 99, contents, plate 2
cryptus 212
lineatus 104
pleii 72, 104, 114, 116
speciosus 209, 212
underwoodi 73, 104, 114, 118–119, 209, 212, 214, **plate 2F**

Helicops 349
angulatus 209
Helminthophis sp. 209
Hemidactylus 99, 106, 120, 219, contents, plate 4
brookii 67, 106
brookii haitianus 67, 106
garnotii 67
haitianus 67, 106, 114, 116, 118, **plate 4C**
mabouia 67, 106, 114, 118, 209, 211, 330
palaichthus 106, 114, 209, 211, 214
praesignis 66
turcicus 31, 67, 106
Hemiptera 282
Herangula 330
Herpestes 327
auropunctatus 32, 327, 395–396, 405–406, 408, 410, 420–421
Herpetodryas
carinatus var. ***vincenti*** 87
Heteropus Bifasciatus 73
Hippomane mancinella 278, 285, 411
Hispaniolus pratensis 84
Hohenbergia 259–261, 263–265, plate 1
Holcosus 82–83, 86–88
Holotropis herminieri 84
microlophus 23
Homalochilus chrysogaster 85
striatus 85
Homoptera 222, 282
Hyalinobatrachium orientalis 137
orientalis tobagoensis 208
Hydrodynastes 349
Hydrops 349
triangularis neglectus 209
Hyla 97, 101, 137, 163, 167, contents
albomarginata 101
boans 208
brunnea 53
crepitans 208
crucialis 53
crucifer 54
cinerea 53, 101
geographica 208
heilprini 53, 101, 113–114, 116, 157, 163, **plate 1E**
marianae 53, 157–158, 162–165, 167–168
marmorata 54
microcephala misera 208
minuscula 208
minuta 208
pulchrilineata 53, 157, 163–164
punctata 208
rubra 54
septentrionalis 27, 53
squirella 53, 101
vasta 53, 157, 163
wilderi 53, 157–158, 160, 162–168
Hylactophryne 229
Hylidae 53, 97, 101, 259, 269, 271, 275, contents, plate 1
Hylodes 61, 90
antillensis 54
auriculatus 55
cuneatus 56
dimidiatus 56
lentus 58
martinicensis 59, 77–78
monensis 59
oxyrhynchus 59
planirostris 60
plicatus 57
ricordii 60
varians 61
Hymenoptera 281–282
Hypsiboas dominicensis 53
Hypsirhynchus 23, 99, 110, 341, 346, contents, plate 8
ferox 37, 88, 340, 346, **plate 8C**

Ialtris 24, 88, 99, 110, 341, 346, contents, plate 8
agyrtes 88, 340
dorsalis 88, 340, **plate 8D**
parishi 88, 340
Idiodactylus georgeensis 66
Iguana 29, 99, 105, 331–332, contents, plate 6
cychlura 73
delicatissima 73, 105, 114, 116–117, 436–437, **plate 6C**
iguana 31, 73, 105, 114, 209, 212–213, 330, 436–437, plate 6
Iguana (Ctenosaura) similis 73
Iguana (Cyclura) nubila 73
Iguana (= Cyclura) collei 196–197
Iguanidae 22–26, 51, 73–82, 84, 99, 104, 275, contents, plate 6
Imantodes cenchoa cenchoa 209
Inga fagifolia 218

Kentropyx 99, contents, plate 3
borckiana 83, 108, 115, 118, **plate 3F**
striatus 209
Kinosternidae 63, 99, 112
Kinosternon 99
bauri palmarum 27
scorpioides 63, 112, 115
scorpioides scorpioides 209

Lacerta 73
ameiva 82
bimaculata 72, 75
cornuta 73
crocodilus 92
iguana 73
lemniscata 83
Occidua 65
roquet 80
sputator 72
turcica 67
Lacertilia 22–25, 30, 65, 70, 72, 74–78, 80, 83–84, contents, plates 2–5
Lachesis muta muta 209
Ladailadne 229
Lagunculariaceae 197
Lantana 319
Leiocephalinae 105
Leiocephalus 22–25, 84, 99, 105, 114, 116–117, 318–319, 329, 331–332, 362, 363, 366–367, contents, plate 3, plate 8
barahonensis 83
carinatus 9, 21, 83, 363–364, **plate 3G**
carinatus granti plate 3
carinatus punctatus 84
cubensis 21, 83, 85, 329
endomychus 84
eremitus 84
greenwayi 84
herminieri 84
inaguae 84
loxogrammus 84
lunatus 84
macropus 22, 84
melanochlorus 22, 84
onaneyi 84
personatus 22–23, 25–26, 84, **plate 3H**
personatus lunatus 84
pratensis 23, 84
psammodromus 84
punctatus 84, 331
raviceps 21, 23, 84
rhutidira 84
schreibersii 23, 84, 106
semilineatus 23, 85, 319–323
stictigaster 21, 23, 85
vinculum 85
vinculum endomychus 84
Leionotus maculatus 90
Lemairocereus 319
Lepidoblepharis 107, 175, 178–187, 193–194
xanthostigma 193
Lepidochelys 383
olivacea 209, 385
Lepidophyma 108
Lepidoptera 222, 281–282
Leposternon microcephalum 64
Leptocephalus 62, 90
Leptodactylidae 21–25, 54–62, 97, 101, 129, 141, contents, plate 1
Leptodactylus 62, 97, 100, 102, 137, 218–220, contents, plate 1
albilabris 62, 102, 114, 116–117, 121, 377, **plate 1L**
bolivianus 208
darlingtoni 58
dominicensis 62, 102, 121
fallax 62, 102, 114, 116–117
fuscus 102, 208
inoptatus 58, 142
insularum 62, 102, 114
knudseni 208
labarosus 102
macrosternum 208
melanotus 102
nesiotus 208
ocellatus 102
pentadactylus 102
stenodema 102
validus 62, 102, 114, 208, 211–212
ventrilineatus 62
wagneri 62, 102, 211
Leptodeira annulata ashmeadi 209
Leptophis 348
ahaetulla 348
ahaetulla coeruleodorsus 209
riveti 209
Leptotyphlopidae 89, 99, 111, contents, plate 7
Leptotyphlops 89, 99, 111, 213, contents, plate 7
albifrons 111, 115, 209, 212
asbolepis 89
bilineata 89, 111, 113, 115–116
calypso 89
columbi 89, 111, 115 116, **plate 7A**
goudotii 89, 111, 115
leptipilepta 89
pyrites 89
tenella 89, 111, 212
Leucaena 411
glauca 411
leucocephala 418
Liocephalus [sic] ***arenarius*** 84
eremitus 84
macropus 84
melanochlorus 84
personatus 84
raviceps 84
Liophis 99, 110, 349, contents, plate 8
andreae 86
cobella cobella 209
cursor 88, 110, 115–117
juliae 88, **plate 8E**
melanotus 88, 110, 115, 212
melanotus nesos 209
ornatus 88
perfuscus 88
reginae ssp. 209
reginae zweifeli 209
Lithodytes lineatus 208
Litoria luteola 59
Lonchocarpus 418

Mabuya 82, 99, 219, contents, plate 2
bistriata 82, 107, 114, 209, 212, 397
guaporicola 107
lineolata 82, 107, 113–114, 116, **plate 2I**
mabouya mabouya 212
Macrostomata 340–341
Macrotus 61, 90
Malaclemys terrapin 27
Malpighia emarginata 285
Malvaceae 197
Mangifera indica 421
Mannophryne olmonae 208
trinitatis 208
Margarops fuscatus 221, 286
Mastigodryas 99, 109, contents
bruesi 88, 109, 115–117, **plate 8F**
boddaerti boddaerti 209
boddaerti dunni 209
melanolomus 109
Meliococcus bijugatus 285, 411
Metopium toxiferum 411
Microcycas 98
Microhylidae 62, 97, 102
Micrurus 99, contents, plate 6
lemniscatus diutius 209
nigrocinctus 89, 111, 115, **plate 6H**
psyches circinalis 209
Mimosa 319
Mollossus 61, 90
Mormops 61, 90
Mus 332, 419, 421, plate 7
musculus 329, 331
Myctophum 61, 90
Mygale 69

Nasutitermes 407
Natricinae 109
Natrix atra 86
callilaema 86
capistrata 31
natrix 285
taxispilota 27
Neotarentola 107
Nerodia 99, 109, 341, 349
clarkii 88, 109, 115, 340, 349
Nesophontes 98
Neuroptera 282
Ninia atrata 209
Notechis ater 350
Notophthalmus viridescens piaropicola 16
Nyctanassa violacea 286

Olios antiguensis 438–440
Opheodrys aestivus 421
Ophiodes 103
Ophisaurus 99
ventralis 66, 103
Ornithodoros talaje 255
Orozomys 419–421
Orthoptera 222
Osteocephalus 157, 166, 168
taurinus 101, 158, 160, 162, 164–165, 167
Osteopilus 53, 97, 101, 114, 116–117, 157–159, 162–168, 267, 327, 331–334, 336, contents, plate 1
brunneus 53, 157-168, 259–269, **plate 1F**
dominicensis 53, 157–160, 163–168, 267, 333
septentrionalis 53, 101, 157–161, 163–168, 267, 271–276, 329, 331, 333, 363, 365, **plate 1G**
Otocryptis 61, 90
Otus nudipes 219, 438–440
Oxybelis 348
aeneus 209
fulgidus 348
Oxyrhopus petola petola 209

Paguristes 279, 285, plate 3
Panolopus 65–66
costatus 65–66
Pelecanus occidentalis 330
Pelomedusidae 63, 99, 112
Pelophilus fordii 85
Pelorius 100, 141–142, 144–155, 231, 233–235
Peltaphryne
empusa 52
lemur 52
Peltophryne 52, 100
guentheri 100
guentheri fracta 52
peltocephalus 100, plate 1
Pelusios 99
subniger 63, 112
Philodryas dorsalis 88
viridissimus 111
Phoenicopterus ruber 382
Phoridae 222
Phrynohyas venulosa 208
Phrynops gibbus 208
Phrynosomatidae 106
Phrynus longipes 222–223, 438–440
Phyllobates 101
limbatus 58
Phyllodactylus 25, 67, 99, 219, contents
pulcher 67, 106–107, 114, 116
wirshingi 67, 107, 113–114, 116, 121, **plate 4D**
Phyllodytes auratus 208
Phyllomedusa trinitatis 208
Phyllonycteris poeyi 329
Phymatidae 331
Physalaemus pustulosus 208
Physalopteroidea 271
Picrodendron 198
baccatum 198
Pictetia aculeata 411
Pipa pipa 208
Pisonia fragrans 418
subcordata 411, 413
Pithecellobium 418
Platydactylus 61, 90
americanus 72
Platymantis 142
Plethodon 15
jordani clemsonae 27
Pliocercus 110
Podocnemis expansa 209
unifilis 209
Poecilia 61, 90
Polychrotidae 73, 320, contents, plate 5
Polychrus marmoratus 209
Porrocaecum 272
Prestoea montana 218
Pristinotus schreibersii 84
Proctoporus shrevei 209
Psammophis antillensis 86
Pseudacris 97
brachyphona 27
crucifer 54, 101
nigrita 27
Pseudechis porphyriacus 350
Pseudemys decorata 63
stejnegeri 63
Pseudis paradoxa caribensis 208
Pseudoboa 99
coronata 110
neuwiedii 88, 110, 115, 209, 212
Pseudobranchus striatus 27
Pseudoeryx 349
Pseudogonatodes 179
Pseustes poecilonotus polylepis 209
sulphureus sulphureus 209
Psocoptera 282
Pythonidae 341

Rana 63, 97
capito 27
catesbeiana 63, 102
grylio 63, 102
halecina sphenocephala 63
marina 52
palmipes 208
sphenocephala 63, 102
sylvatica 266
Randia 418
Ranidae 63, 97, 102
Rattus 332, 419, 421, plates 6–7
rattus 330, 362, 408, 438–440
Regina Clarkii 88
Reptilia 22–26, 62–65, 67–68, 70, 72–73, 75–76, 79, 82, 84, 86–87, 89, 91, 103, 112, 218–219, 221, contents, plates 2–8
Rhadinea 110
Rhinoclemmys punctularia punctularia 209

Salientia 57
Saurenchelys 61, 90
Sauresia 65, 103
agasepsoides 20, plate 3
sepsoides 65
Sauria 22–26, 64–72, 74, 76–77, 79–83, 99, 103
Saurothera vieilloti 221
Scaphiopus 15
Sceloporus 105
Scinax 97
rubra 54, 101, 208, 211
Scincidae 65, 82, 99, 107, contents, plate 2
Scincus bistriatus 82
Scolecophidia 340
Scolopendra alternans 223
Scytale neuwiedii 88
Sebastes 61, 90
Seincus [sic] **(*Diploglossus*)** ***de la sagra*** 66
Serpentes 22–25, 85–87, 89, 91, 99, 109, 417, contents, plates 6–8
Sibon nebulata nebulata 209
Sillago 61, 90
Siphlophis cervinus 209
Skrjabinoptera 271, 275–276
scelopori 271–276
Sminthillus 33, 229
Solenodon 98
Solenopsis geminata 441–442
Solpyugidae 282
Sphaenorhynchus lacteus 208
Sphaeriodactylus [sic] ***elegans*** 69
notatus 70
Sphaerodactylus 11, 22–26, 33, 36, 65–72, 96, 99, 106–107, 114, 116–117, 175, 177–178, 184–185, 187, 189–191, 193–194, 213, 218–221, 328–332, 365–366, contents
altavelensis 67, 178–187, 189, 193
argivus 67, 189, 329
argus 67, 175–177, 185–187, 189–191, 194, 197–199
armasi 67, 176, 189
armstrongi 26, 68, 178–187, 189–190, 193
asterulus 68, 189
beattyi 68, 189
becki 68, 189–190
bromeliarum 68, 176, 189
caicosensis 68, 176, 189
callocricus 68, 189
celicara 68, 176, 189
cinereus 25, 68, 175–176, 178–191, 193–194
clenchi 25, 68, 189
cochranae 25, 68, 189
copei 22, 24–25, 68, 176, 178–191, 193
corticola 68, 176, 189
corticolus 68
cricoderus 68, 175, 177–187, 189, 193–194
cryphius 68, 178–187, 189, 193
darlingtoni 25, 69, 178–187, 189, 193
decoratus 21–23
difficilis 25, 69, 189
docimus 69, 176, 189
elasmorhynchus 69, 177, 189
elegans 69, 175–176, 178–187, 189, 191, 193–194
elegantulus 69, 178–190, 193
epiurus 69, 189
fantasticus 69, 175–190, 193–194
gaigeae 69, 176, 178–187, 189–190, 193
gilvitorques 24, 177, 189
goniorhynchus 69, 176–187, 189–191, 193
inaguae 69, 189
intermedius 69, 176, 178–189, 193
kirbyi 69, 189
klauberi 69, 176, 178–187, 189–190, 193
ladae 69, 189
lazelli 25, 69, 189
leucaster 69, 178–187, 189, 193
levinsi 70, 189
lineolatus 107, 189
macrolepis 70, 176, 178–187, 189, 193, 330, **plate 4F**
mariguanae 70, 176, 189
microlepis 22, 70, 189
micropithecus 70, 189
molei 107, 209
monensis 70, 189, 330
nicholsi 70, 178–186, 189–190, 193
nigropunctatus 24, 67–68, 70, 175–190, 193–194, 363
notatus 22–23, 68–71, 176–187, 189, 193, 331, 363
notatus randi 71

notatus savagei 71
nycteropus 70, 189
ocoae 70, 189
oliveri 70, 176, 178–189, 193
omoglaux 70, 189
oxyrhinus 70, 189
parkeri 70, 176, 189
parthenopion 70, 189
perissodactylius 71, 189
plummeri 71, 189
ramsdeni 71, 175, 177–187, 189–191, 193–194
randi 71, 189
rhabdotus 71, 188–189
richardi 71, 178–189, 193
richardsoni 71, 175–176, 178–190, 193–194, **plate 4E**
roosevelti 71, 189
rosaurae 189–190
ruibali 71, 176, 189
sabanus 71, 188–189
samanensis 71, 189
savagei 25, 71, 189
scaber 21, 71, 175–177, 186–190, 194
schwartzi 71, 175, 177–187, 189, 193–194
semasiops 71, 178–186, 189, 191, 193
shrevei 24–25, 68, 71, 175–176, 186–187, 189, 194
sommeri 71, 189
sputator 72, 107, 175–176, 178–190, 193–194
stejnegeri 25
storeyae 188–189
streptophorus 26, 72, 189
thompsoni 72, 175–176, 178–187, 189, 193–194
torrei 72, 176, 178–189, 193
townsendi 72, 189, 330
underwoodi 72, 189
vincenti 22, 72, 189
williamsi 72, 189
zygaena 72, 189
Spilotes pullatus pullatus 209
Squamata 64, 67, 69, 71, 99, 103, contents, plates 2–8
Stasina 221, 223
portoricensis 438–440
Steironotus arenarius 84
Stenostoma Goudotii 89
Suriana 411
Syrrhophus 101–102, 141, 148, 153, 229, 231–232, 234

Tabebuia pallida 418
Tamarindus indica 285, 411
Tantilla coronata 27
melanocephala 209
Tarentola 99, 107, 114, 116, 118, 120, 365–366, contents, plate 4
americana 23, 72, 107, 118, 363, **plate 4G**
Teiidae 22–24, 26, 73, 82–83, 99, 108, 275, 320, contents, plate 3
Testudines 30, 63, 99, 112, contents, plate 2
Testudinidae 63, 99, 113, contents, plate 2
Testudo carbonaria 63
scorpioides 63
scripta 63
subniger 63
terrapen 63
Thamnodynastes sp. 209
Thamnophis 109
Thecadactylus 99, 212, 214, contents
rapicauda 72, 107, 114, 209, 212, 214, **plate 4G**
Thespesia populnea 197, 411
Tilapia 429
mossambica 426
Tournefortia 411
Trachemys 44, 99, 112, 115–117, 169–174, 219, contents, plate 2
decorata 63, 169–170, **plate 2B**
decussata 44, 63, 112, 170–172
decussata angusta 169–171
decussata decussata 169–170
inflata 170
scripta 63, 112, 169–170
scripta elegans 171
stejnegeri 63, 121, 169–172
stejnegeri malonei 169–171
stejnegeri stejnegeri 169–171
stejnegeri vicina 169–170
terrapen 63, 169–173
terrapen* x *stejnegeri 171
Trachyboa 348
Tomodactylus 229
Trachycephalus marmoratus 54
Tretanorhinus 99, 110, 341, 349, contents, plate 8
variabilis 88, 110, 115–117, 340, 349, **plate 8G**
Tretioscincus 99, contents
bifasciatus 73, 104, 114, 118, plate 2
oriximinensis **plate 2H**
Trichechus manatus 381
Trigonocephalus caribbaea 92
Trimorphodon biscutatus 109
Trionyx 27
Tripanurgos compressus 209
Tropidonotus 61, 90
cubanus 87
Tropidophiidae 89, 99, 111, 340–341, contents, plate 6
Tropidophis 21, 24, 89–90, 99, 111, 115–117, 202, 330, 339, 341–342, 345–348, 357, contents, plate 6
canus 89, 340, 348
caymanensis 89
feicki 89, 340, **plate 6G**
fuscus 89, 340
greenwayi 22, 89, 340, 348
haetianus 24, 90, 111, 340, 344, 348
haitianus jamaicensis 202
haitianus stejnegeri 202
haitianus stullae 202–203
maculatus 21, 90, 340
maculatus pilsbryi 90
melanurus 90, 340, 348
melanurus caymanensis 89
nigriventris 90, 340
pardalis 21, 90, 340
pardalis greenwayi 89
paucisquamis 111
pilsbryi 90, 340
semicinctus 90, 340
wrighti 90, 340
Tropiduridae 83, 105, 320, contents, plate 3
Tropidurinae 105
Tropidurus plica 209
***Tropidurus* (*Leiolaemus*)**
cubensis 83
Tupinambis teguixin 209
Typha domingensis 426
Typhlonectes sp. 208
Typhlopidae 90–91, 99, 111, contents, plate 7
Typhlops 90–91, 99, 111–112, 115–117, 213, 218–219, 331, 342, 363, 365, contents, plate 7
bilineatus 89
biminiensis 90, 112
biminiensis epactia 90
brongersmianus 209
caecatus 112
capitulatus 90
catapontus 90
caymanensis 90
dominicana 90
dominicanus 90
epactia 90
gonavensis 90
granti 91
hectus 91
hypomethes 91
jamaicensis 91
lumbricalis 31, 91, 365, **plate 7B**
luzonensis 112
monastus 91
monensis 91
platycephalus 91, 112, 330
pusillus 91
richardi 91
richardi catapontus 90
rostellatus 91
schwartzi 91
sulcatus 91
syntherus 91
tasymicris 91
tetrathyreus 92
titanops 92
trinitatus 209

Ungalia 61, 90
cana 89
***Ungalia* (*Lionotus*)**
maculata var. ***semicincta*** 90
Ungaliophis 348
Ungualia [sic] ***haetiana*** 90
Urania 69
Uromacer 26, 99, 110, 341, 346, contents, plate 8
catesbyi 23, 26, 88, 333, 340, plate 8
dorsalis 25
frenatus 24–26, 88, 340, 346, **plate 8H**
oxyrhynchus 26, 88, 340, 345, 348

Vertebrata 62
Vesperus 61, 90
Viperidae 92, 99, 112, 340, contents, plate 7

Wetmorena 22, 65, 103
haetiana 65

Xantusiidae 85, 99, 108, contents, plate 5
Xenodon 61, 90
severus 110
Xenodontinae 110
Xiphocercus
darlingtoni 77
valencienni 81
Xiphosurus
ferreus 77
homolechis 78
oculatus 79

Zamenis anomalus 86
Zenaida aurita 331

PUBLICATIONS OF THE
SOCIETY FOR THE STUDY OF AMPHIBIANS AND REPTILES

SOCIETY PUBLICATIONS may be purchased from:
Dr. Robert D. Aldridge, Publications Secretary,
Department of Biology, St. Louis University,
St. Louis, Missouri 63103, USA.

Telephone: area code 314, 977–3910 or 977–3916.
Fax: area code 314, 977–3658.
E-mail: ssar@sluvca.slu.edu.

Prices are effective through December 1996. Make checks payable to "SSAR." Overseas customers must make payment in USA funds using a draft drawn on American banks (include an additional amount to cover bank conversion charges) or by International Money Order. All persons may charge to MasterCard or VISA (provide account number and expiration date); items marked "out-of-print" are no longer available.

Shipping and Handling Costs

Shipments inside the USA: Shipping costs are in addition to the price of publications. Add an amount for shipping of the first item ($3.00 for a book costing $10.00 or more or $2.00 if the item costs less than $10.00) plus an amount for any additional items ($1.00 each for books costing over $10.00 and $0.50 for each item costing less than $10.00).

Shipments outside the USA: Determine the cost for shipments inside USA (above) and then add 5% of the total cost of the order.

Large prints (marked *): For shipments inside the USA, add $3.00 for any quantity; outside the USA, instead add $7.00 for any quantity.

CONTRIBUTIONS TO HERPETOLOGY

Book-length monographs, comprising taxonomic revisions, results of symposia, and other major works. Prepublication discount to Society members.

Vol. 1. *Reproductive Biology and Diseases of Captive Reptiles,* by James B. Murphy and Joseph T. Collins (eds.). 1980. Results of a Society-sponsored symposium, including papers by 37 leading specialists. 287 p., illus. Paperbound. Out-of-print.

Vol. 2. *The Turtles of Venezuela,* by Peter C. H. Pritchard and Pedro Trebbau. 1984. An exhaustive natural history covering half of the turtle species of South America. 414 p., 48 color plates (25 watercolor portraits by Giorgio Voltolina and 165 photographs of turtles and habitats) measuring $8\frac{1}{2} \times 11$ inches, keys, 16 maps. Regular edition, clothbound $45.00; patron's edition, two leatherbound volumes in cloth-covered box, signed and numbered by authors and artist $300.00. (*Also*: set of 25 color prints of turtle portraits on heavy paper stock, in protective wrapper $30.00.)

Vol. 3. *Introduction to the Herpetofauna of Costa Rica / Introducción a la Herpetofauna de Costa Rica,* by Jay M. Savage and Jaime Villa R. 1986. Bilingual edition in English and Spanish, with distribution checklist, bibliographies, and extensive illustrated keys. 220 p., map. Clothbound $30.00.

Vol. 4. *Studies on Chinese Salamanders*, by Ermi Zhao, Qixiong Hu, Yaoming Jiang, and Yuhua Yang. 1988. Evolutionary review of all Chinese species with keys, diagnostic figures, and distribution maps. 80 p., 7 plates (including 10 color photographs of salamanders and habitats). Clothbound $12.00.

Vol. 5. *Contributions to the History of Herpetology*, by Kraig Adler, John S. Applegarth, and Ronald Altig. 1989. Biographies of 152 prominent herpetologists (with portraits and signatures), index to 2500 authors in taxonomic herpetology, and academic lineages of 1450 herpetologists. International coverage. 202 p., 148 photographs, 1 color plate. Clothbound $20.00.

Vol. 6. *Snakes of the* Agkistrodon *Complex: A Monographic Review*, by Howard K. Gloyd and Roger Conant. 1990. Comprehensive treatment of 33 taxa of pitvipers included in four genera: *Agkistrodon*, *Calloselasma*, *Deinagkistrodon*, and *Hypnale*. Also includes nine supplementary chapters by leading specialists. 620 p., 33 color plates (247 photographs of snakes and habitats), 20 uncolored plates, 60 text figures, checklist and keys, 6 charts, 28 maps. Clothbound $75.00. (*Also*: separate set of the 247 color photographs of snakes and habitats [on 32 plates], in protective wrapper. $30.00; limited-edition print of the book's frontispiece illustrating snakes of all four genera, from watercolor by David M. Dennis. Signed individually by Roger Conant and the artist $25.00.)

Vol. 7. *The Snakes of Iran*, by Mahmoud Latifi. 1991. Review of the 60 species of Iranian snakes, covering general biology, venoms, and snake bite. Appendix and supplemental bibliography by Alan E. Leviton and George R. Zug. 167 p., 22 color plates of snakes (66 figures), 2 color relief maps, 44 species range maps. Clothbound $22.00.

Vol. 8. *Handbook to Middle East Amphibians and Reptiles*, by Alan E. Leviton, Steven C. Anderson, Kraig Adler, and Sherman A. Minton. 1992. Annotated checklist, illustrated key, and identification manual covering 148 species and subspecies found in region from Turkish border south through the Arabian Peninsula (including Bahrain, Qatar, and United Arab Emirates) and the Arabian (Persian) Gulf. Chapters on venomous snakes and snakebite treatment plus extensive bibliography. 264 p., 32 color plates (220 photographs), maps, text figures. Clothbound $30.00.

Vol. 9. *Herpetology: Current Research on the Biology of Amphibians and Reptiles.* 1992. Proceedings of the First World Congress of Herpetology (1989), with a foreword by H.R.H. Prince Philip, Duke of Edinburgh. Includes the plenary lectures, a summary of the congress, and a list of delegates with their current addresses. 225 p., 28 photographs. Clothbound $28.00.

Vol. 10. *Herpetology of China*, by Ermi Zhao and Kraig Adler. 1993. Comprehensive review of Chinese amphibians and reptiles, including Hong Kong and Taiwan. 522 p., 48 color plates (371 photographs illustrating all 164 genera and half of the 661 species), portraits, text figures, maps. Clothbound $60.00.

Vol. 11. *Captive Management and Conservation of Amphibians and Reptiles*, by James B. Murphy, Kraig Adler, and Joseph T. Collins (eds.). 1994. Results of a Society-sponsored symposium, including chapters by 70 leading specialists. Foreword by Gerald Durrell. 408 p., 35 photographs, 1 color plate. Clothbound $58.00.

Vol. 12. *Contributions to West Indian Herpetology*, by Robert Powell and Robert W. Henderson (eds.). 1996. Results of a Society-sponsored symposium, including research chapters by 59 authors and a checklist of species with complete citations. Foreword by Thomas W. Schoener. 457 p., 28 photographs, 70 color photographs, index. Clothbound $60.00.

FACSIMILE REPRINTS IN HERPETOLOGY

Exact reprints of classic and important books and papers. Most titles have extensive new introductions by leading authorities. Prepublication discount to Society members.

ANDERSON, J. 1896. *Contribution to the Herpetology of Arabia*. Introduction and new checklist of Arabian amphibians and reptiles by Alan E. Leviton and Michele L. Aldrich. 160 p., illus. (one plate in color), map. Clothbound $25.00.

BELL, T. 1842–1843. *Herpetology of the "Beagle."* Part 5 of Charles Darwin's classic, "Zoology of the Voyage of H.M.S. Beagle," containing descriptions of amphibians and reptiles collected on the expedition. Introduction by Roberto Donoso-Barros. 100 p., 20 plates (measuring 8 1/2 × 11 inches), map. Paperbound $13.00.

BOGERT, C. M., and R. MARTÍN DEL CAMPO. 1956. *The Gila Monster and Its Allies*. The standard work on lizards of the family Helodermatidae. New preface by Charles M. Bogert and retrospective essay by Daniel D. Beck. 262 p., color plate, 62 photographs, 35 text figures, index. $38.00. (*Also*: limited-edition print of the book's frontispiece illustrating a Gila monster, from watercolor by David M. Dennis. Signed individually by Charles Bogert and the artist $25.00.)

BOJANUS, L. H. 1819–1821. *Anatome Testudinis Europaeae*. The standard atlas of turtle anatomy. Introduction by Alfred Sherwood Romer. 200 p., 40 foldout plates. Out-of-print.

BOULENGER, G. A. 1877–1920. *Contributions to American Herpetology*. A collection of papers (from various journals) covering North, Central, and South American species, with an introduction by James C. Battersby. Complete in 18 parts totalling 880 p., numerous illustrations, index. Paperbound. Complete set: 18 parts plus index and two tables of contents (for binding in two volumes), in parts as issued $55.00.

BULLETIN OF THE ANTIVENIN INSTITUTE OF AMERICA. Volumes 1–5, 1927–1932. Introduction by Sherman A. Minton. 575 p., 163 photographs, maps, index. Out-of-print.

CAMP, C. L. 1923. *Classification of the Lizards*. The foundation of modern lizard systematics. New preface by Charles L. Camp and an introduction by Garth Underwood. 220 p., 112 figures, index. Out-of-print.

CHANG, M. L. Y. 1936. *Amphibiens Urodèles de la Chine*. The classic work on Chinese salamanders, with a new checklist by Arden H. Brame. 168 p., illus., 5 plates. Out-of-print.

COPE, E. D. 1864. *Papers on the Higher Classification of Frogs*. Reprinted from Proceedings of the Academy of Natural Sciences of Philadelphia and Natural History Review. 32 p. Paperbound $3.00.

COPE, E. D. 1871. *Catalogue of Batrachia and Reptilia Obtained by McNiel in Nicaragua; Catalogue of Reptilia and Batrachia Obtained by Maynard in Florida*. 8 p. Paperbound $1.00.

COPE, E. D. 1892. *The Osteology of the Lacertilia*. An important contribution to lizard anatomy, reprinted from Proceedings of the American Philosophical Society. 44 p., 6 plates. Paperbound $4.00.

COWLES, R. B., and C. M. BOGERT. 1944. *A Preliminary Study of the Thermal Requirements of Desert Reptiles*. The foundation of thermoregulation biology, with extensive review of recent studies by F. Harvey Pough. Reprinted from Bulletin of American Museum of Natural History. 52 p., 11 plates. Paperbound $5.00.

DUNN, E. R. 1926. *Salamanders of the Family Plethodontidae*. The classic treatment of plethodontid salamanders. Introductions by David B. Wake and Arden H. Brame. 480 p., illus., 3 plates, 86 maps, index. Out-of-print.

ESCHSCHOLTZ, F. 1829–1833. *Zoologischer Atlas* (herpetological sections). Descriptions of new reptiles and amphibians from California and the Pacific. Introduction by Kraig Adler. 32 p., 4 plates (measuring 8 1/2 × 11 inches). Paperbound $3.00.

ESPADA, M. JIMÉNEZ DE LA. 1875. *Vertebrados del Viaje al Pacifico: Batracios*. Major taxonomic work on South American frogs. Introduction by Jay M. Savage. 208 p., 6 plates, maps. Clothbound $20.00.

FAUVEL, A.-A. 1879. *Alligators in China*. Original description of *Alligator sinensis*, including classical and natural history. 42 p., 3 plates. Paperbound $5.00.

FITZINGER, L. J. 1843. *Systema Reptilium*. An important nomenclatural landmark for herpetology, including Amphibia as well as reptiles; world-wide in scope. Introduction by Robert Mertens. 128 p., index. Paperbound $15.00.

GLOYD, H. K. 1940. *The Rattlesnakes, Genera* Sistrurus *and* Crotalus. Introduction and new checklist by Hobart M. Smith and Herbert M. Harris. 300 p., plus 31 plates of photographs, index. Out-of-print.

GRAY, J. E. 1825. *A Synopsis of the Genera of Reptiles and Amphibia*. Reprinted from Annals of Philosophy. 32 p. Paperbound $3.00.

GRAY, J. E. 1831–1844. *Zoological Miscellany*. A privately printed journal, devoted mostly to descriptions of amphibians, reptiles, and birds from throughout the world. Introduction by Arnold G. Kluge. 86 p., 4 plates. Paperbound $6.00.

GRAY, J. E., and A. GÜNTHER. 1845–1875. *Lizards of Australia and New Zealand*. The reptile section from "Voyage of H.M.S. Erebus and Terror," together with Gray's 1867 related book on Australian lizards. Introduction by Glenn M. Shea. 82 p., 20 plates (measuring 8 1/2 × 11 inches). Clothbound $20.00. (*Also*: set of the 20 plates in protective wrapper $12.00.)

GÜNTHER, A. 1885–1902. *Biologia Centrali-Americana. Reptilia and Batrachia*. The standard work on Middle American herpetology with 76 full-page plates measuring 8 1/2 × 11 inches (12 in color). Introductions by Hobart M. Smith, A. E. Gunther, and Kraig Adler. 575 p., photographs, maps. Clothbound $50.00. (*Also*: separate set of the 12 color plates, in protective wrapper $18.00.)

HOLBROOK, J. E. 1842. *North American Herpetology*. Five volumes bound in one. The classic work by the father of North American herpetology. Exact facsimile of the definitive second edition, including all 147 plates, measuring 8 1/2 × 11 inches (20 reproduced in full color). Introduction and checklists by Richard and Patricia Worthington and by Kraig Adler. 1032 p. Leatherbound patron's edition, out-of-print; regular edition, clothbound $60.00.

JUNIOR SOCIETY OF NATURAL SCIENCES (CINCINNATI, OHIO). 1930–1932. Herpetological papers from the society's Proceedings, with articles by Weller, Walker, Dury, and others. 56 p. Paperbound $3.00.

KIRTLAND, J. P. 1838. *Zoology of Ohio* (herpetological portion). 8 p. Paperbound $1.00.

LECONTE, J. E. 1824–1828. *Three Papers on Amphibians*, from the Annals of the Lyceum of Natural History, New York. 16 p. Paperbound $2.00.

LINNAEUS, C. 1766–1771. *Systema Naturae* (ed. 12) and *Mantissa Plantarum* (herpetological portions from both). 56 p. Introduction by Kraig Adler. Out-of-print.

LOVERIDGE, A. 1946. *Reptiles [and Amphibians] of the Pacific*. The standard review of the herpetofauna of the Pacific region including Australia and extending from Indonesia to

Hawaii and the Galápagos Islands. 271 p., 7 plates, 1 double-page map, index. Out-of-print.

McILHENNY, E. A. 1935. *The Alligator's Life History*. The most complete natural history of the American alligator. Introduction by Archie Carr and a review of recent literature by Jeffrey W. Lang. 125 p., 18 photographs and a portrait. Clothbound $20.00.

McLAIN, R. B. 1899. *Contributions to North American Herpetology*. Three privately printed papers containing important distributional records and the description of a new form. 28 p., index. Paperbound $2.00.

ORBIGNY, A. d' [and G. BIBRON]. 1847. *Voyage dans l'Amérique Méridionale*. This extract comprises the complete section on reptiles and amphibians from this voyage to South America. 14 p., 9 plates measuring 8½ × 11 inches. $3.00.

PETERS, W. 1838–1883. *The Herpetological Contributions of Wilhelm C. H. Peters (1815–1883)*. A collection of 174 titles, world-wide in scope, and including the herpetological volume in Peters' series, "Reise nach Mossambique." Biography, annotated bibliography, and synopsis of species by Aaron M. Bauer, Rainer Günther, and Meghan Klipfel. 714 pages, 114 plates, 9 photographs, maps, index. Clothbound $75.00.

RAFINESQUE, C. S. 1820. *Annals of Nature* (herpetological and ichthyological sections), 4 p. Paperbound $1.00.

RAFINESQUE, C. S. 1822. *On Two New Salamanders of Kentucky*. 2 p. Paperbound $1.00.

RAFINESQUE, C. S. 1832–1833. *Five Herpetological Papers from the Atlantic Journal*. 4 p. Paperbound $1.00.

SOWERBY, J. DeC., E. LEAR, and J. E. GRAY. 1872. *Tortoises, Terrapins, and Turtles Drawn From Life*. The finest atlas of turtle illustrations ever produced. Introduction by Ernest E. Williams. 26 p., 61 full-page plates (measuring 8½ × 11 inches). Clothbound $25.00.

SPIX, J. B. von, and J. G. WAGLER. 1824–1825. *Herpetology of Brazil*. The most comprehensive and important early survey of Brazilian herpetology. Introduction by P. E. Vanzolini. 400 p., 98 plates, one in color (each measuring 8½ × 11 inches), map. Clothbound $36.00.

TROSCHEL, F. H. 1850 [1852]. *Cophosaurus texanus, neue Eidechsengattung aus Texas*. 8 p. Paperbound $1.00.

TSCHUDI, J. J. von. 1838. *Classification der Batrachier*. A major work in systematic herpetology, with introduction by Robert Mertens. 118 p., 6 plates. Paperbound $18.00.

TSCHUDI, J. J. von. 1845. *Reptilium Conspectus*. Descriptions of new reptiles and amphibians from Peru. 24 p. Paperbound $2.00.

VanDENBURGH, J. 1895–1896. *Herpetology of Lower California*. Herpetology of Baja California, Mexico (collected papers). 101 p., 11 plates, index. Paperbound $8.00.

WAITE, E. R. 1929. *The Reptiles and Amphibians of South Australia*. Introduction by Michael J. Tyler and Mark Hutchinson. 282 p., color plate, portrait, 192 text figures including numerous photographs. Clothbound $35.00.

WIEGMANN, A. F. A. 1834. *Herpetologia Mexicana*. Introduction by Edward H. Taylor. 66 p., 10 plates (folio, measuring 10 × 14 inches). Out-of-print.

WILCOX, E. V. 1891. *Notes on Ohio Batrachians*. 3 p. Paperbound $1.00.

WILLISTON, S. W. 1925. *Osteology of the Reptiles*. Covers living and extinct forms, with introduction by Claude W. Hibbard. 304 p., 191 text figures, index. Out-of-print.

WRIGHT, A. H., and A. A. WRIGHT. 1962. *Handbook of Snakes of the United States and Canada, Volume 3, Bibliography*. Out-of-print since about 1969, this cross-indexed bibliography is a necessary companion to Volumes 1 and 2. 187 p. Clothbound $18.00.

HERPETOLOGICAL REVIEW AND H.I.S.S. PUBLICATIONS

The Society's official newsletter, international in coverage. In addition to news notes and feature articles, regular departments include regional societies, techniques, husbandry, life history, geographic distribution, and book reviews. Issued quarterly as part of Society membership or separately by subscription. All numbers are paperbound as issued and measure 8½ × 11 inches. In 1973, publications of the Herpetological Information Search Systems (*News-Journal* and *Titles and Reviews*) were substituted for *Herpetological Review*; content and format are the same.

Volume 1 (1967–1969), numbers 1–9, $5.00 per number.

Volumes 2–26 (1970–1995), four numbers in each volume (except volumes 3–4, with 6 numbers each), $5.00 per number.

The following numbers are out-of-print and *no longer available*: Volume 1(number 7), 3(2, 6), 4(1), 5(1, 2), 6(1, 2, 4), 7(3, 4), and 10(2).

Cumulative Index for Volumes 1–7 (1967–1976), 60 pages, $5.00.

Cumulative Index for Volumes 1–17 (1967–1986), 90 pages, $8.00.

H.I.S.S. Publications: News-Journal, volume 1, numbers 1–6, and Titles and Reviews, volume 1, numbers 1–2 (all of 1973–1974), complete set, $10.00.

Index to Geographic Distribution Records for Volumes 1–17 (1967–1986), including H.I.S.S. publications, 44 pages, $6.00.

CATALOGUE OF AMERICAN AMPHIBIANS AND REPTILES

Loose-leaf accounts of taxa (measuring 8½ × 11 inches) prepared by specialists, including synonymy, definition, description, distribution map, and comprehensive list of literature for each taxon. Covers amphibians and reptiles of the entire Western Hemisphere. Issued by subscription. Individual accounts are not sold separately.

CATALOGUE ACCOUNTS:

Complete set: Numbers 1–620, $320.00.
Partial sets: Numbers 1–190, $60.00.
Numbers 191–410, $70.00.
Numbers 411–620, $170.00.

INDEX TO ACCOUNTS 1–400: Cross-referenced, 64 pages, $6.00; accounts 401–600: Cross-referenced, 32 pages, $6.00.

IMPRINTED POST BINDER: $35.00. (*Note*: one binder holds about 200 accounts.)

SYSTEMATIC TABS: Ten printed tabs for binder, such as "Class Amphibia," "Order Caudata," etc., $6.00 per set.

JOURNAL OF HERPETOLOGY

The Society's official scientific journal, international in scope. Issued quarterly as part of Society membership. All numbers are paperbound as issued, measuring 7 × 10 inches.

Volume 1 (1968), numbers 1–4 combined.

Volumes 2–5 (1968–1971), numbers 1–2 and 3–4 combined, $8.00 per double number.

Volume 6 (1972), numbers 1, 2, and double number 3–4.

Volumes 7–29 (1973–1995), four numbers in each volume, $8.00 per single number.

The following volumes and numbers are out-of-print and are *no longer available*: Volume 1, 2, 3, 4, 5(3, 4), 6, 7(1), 8(1), 9(1, 2, 4), 10(1, 4), 11(4), and 12(1, 2).

Cumulative Index for Volumes 1–10 (1968–1976), 72 pages, $8.00.

HERPETOLOGICAL CIRCULARS

Miscellaneous publications of general interest to the herpetological community. All numbers are paperbound, as issued. Prepublication discount to Society members.

No. 1. *A Guide to Preservation Techniques for Amphibians and Reptiles* by George R. Pisani. 1973. 22 p., illus. $3.00.

No. 2. *Guía de Técnicas de Preservación de Anfibios y Reptiles* por George R. Pisani y Jaime Villa. 1974. 28 p., illus. $3.00.

No. 3. *Collections of Preserved Amphibians and Reptiles in the United States* compiled by David B. Wake (chair) and the Committee on Resources in Herpetology. 1975. 22 p. Out-of-print.

No. 4. *A Brief Outline of Suggested Treatments for Diseases of Captive Reptiles* by James Murphy. 1975. 13 p. Out-of-print.

No. 5. *Endangered and Threatened Amphibians and Reptiles in the United States* compiled by Ray E. Ashton, Jr. (chair) and 1973–74 SSAR Regional Herpetological Societies Liaison Committee. 1976. 65 p. Out-of-print.

No. 6. *Longevity of Reptiles and Amphibians in North American Collections* by J. Kevin Bowler. 1977. 32 p. $3.00. (See also number 21.)

No. 7. *Standard Common and Current Scientific Names for North American Amphibians and Reptiles* (1st ed.) by Joseph T. Collins, James E. Huheey, James L. Knight, and Hobart M. Smith. 1978. 36 p. $3.00. (See also numbers 12 and 19.)

No. 8. *A Brief History of Herpetology in North America Before 1900* by Kraig Adler. 1979. 40 p., 24 photographs, 1 map. $3.00.

No. 9. *A Review of Marking Techniques for Amphibians and Reptiles* by John W. Ferner. 1979. 42 p., illus. $3.00.

No. 10. *Vernacular Names of South American Turtles* by Russell A. Mittermeier, Federico Medem and Anders G. J. Rhodin. 1980. 44 p. $3.00.

No. 11. *Recent Instances of Albinism in North American Amphibians and Reptiles* by Stanley Dyrkacz. 1981. 36 p. $3.00.

No. 12. *Standard Common and Current Scientific Names for North American Amphibians and Reptiles* (2nd ed.) by Joseph T. Collins, Roger Conant, James E. Huheey, James L. Knight, Eric M. Rundquist, and Hobart M. Smith. 1982. 32 p. $3.00. (See also numbers 7 and 19.)

No. 13. *Silver Anniversary Membership Directory*, including addresses of all SSAR members, addresses and publications of the herpetological societies of the world, and a brief history of the Society. 1983. 56 p., 4 photographs. $3.00.

No. 14. *Checklist of the Turtles of the World with English Common Names* by John Iverson. 1985. 14 p. $3.00.

No. 15. *Cannibalism in Reptiles: A World-Wide Review* by Joseph C. Mitchell. 1986. 37 p. $4.00.

No. 16. *Herpetological Collecting and Collections Management* by John E. Simmons. 1987. 72 p., 6 photographs. $6.00.

No. 17. *An Annotated List and Guide to the Amphibians and Reptiles of Monteverde, Costa Rica* by Marc P. Hayes, J. Alan Pounds, and Walter W. Timmerman. 1989. 70 p., 32 figures. $5.00.

No. 18. *Type Catalogues of Herpetological Collections: An Annotated List of Lists* by Charles R. Crumly. 1990. 50 p. $5.00.

No. 19. *Standard Common and Current Scientific Names for North American Amphibians and Reptiles* (3rd ed.) compiled by Joseph T. Collins (coordinator for SSAR Common and Scientific Names List). 1990. 45 p. $5.00. (See also numbers 7 and 12.)

No. 20. *Age Determination in Turtles* by George R. Zug. 1991. 32 p., 6 figures. $5.00.

No. 21. *Longevity of Reptiles and Amphibians in North American Collections* (2nd ed.) by Andrew T. Snider and J. Kevin Bowler. 1992. 44 p. $5.00. (See also number 6.)

No. 22. *Biology, Status, and Management of the Timber Rattlesnake* (Crotalus horridus): *A Guide for Conservation* by William S. Brown. 1993. 84 p., 16 color photographs. $10.00.

No. 23. *Scientific and Common Names for the Amphibians and Reptiles of Mexico in English and Spanish / Nombres Científicos y Comunes en Ingles y Español de los Anfibios y los Reptiles de México* by Ernest A. Liner. Spanish translation by José L. Camarillo R. 1994. 118 p. $12.00.

No. 24. *Citations for the Original Descriptions of North American Amphibians and Reptiles*, by Ellin Beltz. 1995. 48 p. $7.00.

PUBLICATIONS OF THE OHIO HERPETOLOGICAL SOCIETY

OHS was the predecessor to the Society for the Study of Amphibians and Reptiles. All publications international in scope. Paperbound as issued.

Volume 1 numbers 1–4, plus Special Publications 1–2 (all 1958), facsimile reprint, out-of-print.

Volume 2 (1959–1960), four numbers, $2.00 per number; numbers 3 and 4 out-of-print.

Volume 3 (1961–1962), four numbers, $1.00 per number; numbers 1 and 3 out-of-print.

Volume 4 (1963–1964), four numbers; double number 1–2, $4.00, numbers 3 and 4 $2.00 each.

Volume 5 (1965–1966), four numbers, $2.00 per number.

Special Publications 3–4 (1961–1962), $2.00 per number; number 3 out-of-print.

OTHER MATERIALS AVAILABLE FROM THE SOCIETY

The following color prints and brochures may be purchased from the Society. (*Extra postage required; see "Shipping and Handling Costs.")

*SILVER ANNIVERSARY COMMEMORATIVE PRINT. Full-color print ($11\frac{1}{2} \times 15\frac{1}{4}$ inches) of a Gila Monster (*Heloderma suspectum*) on natural background, from a watercolor by David M. Dennis. Issued as part of Society's 25th Anniversary in 1982. Edition limited to 1000. $6.00 each or $5.00 in quantities of 10 or more.

*WORLD CONGRESS COMMEMORATIVE PRINT. Full-color print ($11\frac{1}{2} \times 15$ inches) of an Eastern Box Turtle (*Terrapene carolina*) in a natural setting, from a watercolor by David M. Dennis. Issued as part of SSAR's salute to the First World Congress of Herpetology, held at Canterbury, United Kingdom, in 1989. Edition limited to 1500. $6.00 each or $5.00 in quantities of 10 or more.

GUIDELINES FOR THE USE OF LIVE AMPHIBIANS AND REPTILES IN FIELD RESEARCH, by George R. Pisani, Stephen D. Busack, Herbert C. Dessauer, and Victor H. Hutchison, representing a joint committee of ASIH, HL, and SSAR. 1987. Brochure covers animal care, regulations, collecting, restraint and handling, marking, housing and maintenance in field, and final disposition of specimens. 16 p. $4.00 ($3.00 each in quantities of five or more copies).

•

Production Specifications

WORD PROCESSING AND FORMATTING OF TEXT: Avila College, Kansas City, Missouri, USA (Robert Powell) and Cornell University, Ithaca, New York, USA (Timothy D. Perry). Type was set in Times face, using the PageMaker 5.0 program on a Macintosh IIsi computer, and was provided to the printer on an 88 MB Syquest disk.

PRINTING OF TEXT AND BINDING: Thomson-Shore, Inc., Dexter, Michigan, USA (Ned Thomson, Lana Paton, Merry Sumner). The text is printed on 60-pound Joy White Offset paper, a recycled stock. The boards are covered in Roxite C cloth (vellum finish) with Multicolor Antique endpapers.

COLOR PLATES: Four Colour Imports, Ltd., Louisville, Kentucky, USA (Petrice Dahl, Amy Martin), agents for Everbest Printing Co., Ltd., Hong Kong. The color plates were laser-scanned from color transparencies by All Systems Color of Miamisburg, Ohio, USA (Thomas Kiser, Mindy Stelzer) and are printed on 128-gsm Korean Chon Ju Matte art paper.

PAPER: All paper in this book is acid- and groundwood-free. It meets the guidelines for permanence and durability of the Committee on Publication Guidelines for Book Longevity of the Council on Library Resources.

DATE OF PUBLICATION: 15 March 1996.

PLACE OF PUBLICATION: Ithaca, New York, USA.

NUMBER OF COPIES: 1500.